Ordinary Differential Equations and Mechanical Systems

Jan Awrejcewicz

Ordinary Differential Equations and Mechanical Systems

Springer

Jan Awrejcewicz
Department of Automation,
Biomechanics and Mechatronics
Łódź University of Technology
Łódź, Poland

ISBN 978-3-319-07658-4 ISBN 978-3-319-07659-1 (eBook)
DOI 10.1007/978-3-319-07659-1
Springer Cham Heidelberg New York Dordrecht London

Library of Congress Control Number: 2014947361

Mathematics Subject Classification (2010): 34-xx, 70-xx, 37-xx

Printed on acid-free paper

Springer is part of Springer Science+Business Media (www.springer.com)

Preface

Chapter 1 describes basic concepts of the theory of ordinary differential equations. Namely, solutions to ODEs, a notion of extended phase space, orbit (phase trajectory), motion, cascades and flows, trajectory, wandering point, $\alpha(\omega)$-limiting point, attractors and repellers are defined, among others. Picard's and Peano's theorems are introduced. Furthermore, global existence and uniqueness of solutions are defined and proved, and examples illustrating the introduced definitions are added.

Chapter 2 deals with ODEs of the first order. First, a general introduction to nonlinear ODEs is given, and a few illustrative examples are provided. Definitions and examples regarding separable homogenous and linear, as well as the exact and implicit differential equations are given.

Chapter 3 is concerned with second-order ODEs. Firstly, we begin with linear homogenous ODEs with time-dependent coefficients, and four illustrative examples are added. Then the hyper-geometric (Gauss) equation, Legendre equation and Bessel equation supplemented by examples are studied. Particular attention is paid to ODEs with periodic coefficients including the Hill equation, Meissner equation, Ince–Strutt and Kotowski diagrams. In addition, modelling of a generalized parametric oscillator using the nonlinear Milne–Pinney equation is carried out. Then ODEs with constant coefficients are briefly revisited.

Variational Hamiltonian principle, exhibiting physical aspects of a studied dynamical system, is applied to derive the Lagrange equations. Next, in Sect. 3.4, a reduction of the second-order ODEs to that of the first order is presented. Then, the canonical (Jacobi) form after application of a dual (Legendre) transformation is derived and discussed. Both Lagrangian and Hamiltonian functions, and their mutual relationships are presented including the illustrative geometric interpretation. Canonical transformations, Poisson brackets, as well as generating functions are described in Sect. 3.5, and two illustrative examples are provided (Sect. 3.6). Section 3.7 deals with normal forms of Hamiltonian systems. Furthermore, it is shown how to solve the Hamiltonian equations with damping terms. Problems related to Riemannian formulation of dynamics and chaos exhibited by Hamiltonian systems using the example of a swinging pendulum are described in Sect. 3.8. The Jacobi–Levi–Civita (JLC) equation is derived governing the evolution of geodesic separation. The original 2-DOF system oscillations have been reduced to consideration of one second-order homogenous ODE with time-dependent coefficient. Numerical results including the quasi-periodic and chaotic dynamics are reported. Next, the so far applied geometric analysis has been extended to study double pendulum dynamics. In particular, a connection between the JLC equation and the tangent dynamics equation is illustrated allowing us to define the Lyapunov exponents via the Riemannian geometry approach. This qualitatively different explanation and interpretation of chaotic dynamics as a parametric instability of geodesics has been illustratively supported by numerical results. Section 3.10 is devoted to a study of the linear second-order ODEs with constant coefficients presented in a matrix form. Both conservative and non-conservative non-autonomous systems are analysed and important question regarding their decoupling is rigorously considered. Then, a few problems of modal analysis and identification are addressed.

Chapter 4 is devoted to linear ODEs. Firstly, we show how a single nth order ODE is reduced to first-order ODEs. Then, normal and symmetric forms of ODEs are defined, and five illustrative examples are added. In Sect. 4.3 local solutions behaviour regarding the existence, extension and straightness are described. Classical theorems are formulated including a few well-known proofs. First-order linear ODEs with variable coefficients are studied in Sect. 4.4. It contains some classical theorems with their proofs, fundamental matrix of solutions, homogenous and non-homogenous differential equations and numerous examples. Particular attention is paid to homogenous and non-homogenous ODEs with periodic coefficients, the Floquet theory, characteristic multipliers and exponents structure of solutions, etc.

Chapter 5 focuses on higher order ODEs of a polynomial form. First, the Peano and Cauchy–Picard theorems are revisited. Second, a linear homogenous nth order equation is studied. Third, it is shown how an nth order differential equation is reduced to the nth order algebraic equation. Distinct and multiple roots of characteristic equation are discussed. This chapter includes also numerous examples.

Chapter 6 describes systems. It includes the system definition, as well as the asymptotic relations between Newton's and Einstein's theory, classical mechanics and quantum mechanics, and others.

Theory and criteria for similarity are introduced and discussed in Chap. 7. Geometrical, kinematic and dynamic similarities are illustrated. Three different approaches of obtaining similarity criteria are outlined. Several examples serving as a guide for intuition and physical interpretations are given.

Chapter 8 is devoted to a model definition and modelling. After an introductory Sect. 8.1, the mathematical modelling characterized by interdisciplinary and universal features is described (Sect. 8.2). Section 8.3 focuses on the modelling in mechanics, whereas the next section deals with general characteristics of mathematical modelling. Modelling approaches applied in a control are described in Sect. 8.5. Ordinary, adaptive and distributed control systems are discussed, and associated block diagrams are given. The mechanical engineering oriented results of control systems and their block diagrams are reported and discussed.

Chapter 9 deals with a phase plane and phase space based on the first-order ODEs. First, a general introduction of the phase plane concept is given in Sect. 9.1. Singular points are studied in Sect. 9.2, where also a classification of phase portraits is given. Section 9.3 yields the classification of singular points with a use of "Mathematica". An analysis of singular points governed by the three first-order ODEs is carried out in Sect. 9.4, and analytical and numerical solutions of phase space trajectories in the neighbourhood of singular points are derived.

Problems of stability are studied in Chap. 10. Introductory Sect. 10.1 gives a few classical stability definitions of Lyapunov, Lagrange, exponential, conditional and technical stability.

In addition, the limiting sets, attractors and repellers are also defined. Finally, a concept of Zhukovskiy's stability is illustrated and its impact on the stability of quasi-periodic and chaotic orbits is discussed. Lyapunov functions and second Lyapunov methods are discussed in Sect. 10.2, where also the so-called first and second Lyapunov theorems of stability/instability are formulated. Then a few other theorems are introduced including these of Chetayev and Barbashin–Krasovski, and two illustrative examples are added as well. Section 10.3 deals with the classical theories of stability and their impact on chaotic dynamics.

Chapter 11 is focused on modelling via perturbation methods. First section describes advantages and disadvantages of asymptotic methods. Section 11.2 presents some perturbation techniques including the Krylov method and Krylov–Bogolubov–Mitropolskiy method applied to autonomous and non-autonomous oscillators. In the latter case resonance and non-resonance oscillations are studied.

Continualization and discretization approaches are described in Chap. 12. Introduction is followed by a study of the 1D chain of coupled oscillators, which is then converted to a PDE governing propagation of waves (Sect. 12.2). The approach is extended to a model of planar hexagonal net of coupled oscillators (Sect. 12.3). The discretization approach is briefly commented on the basis of vibrations of a nonlinear shell with imperfections (Sect. 12.4). Finally, the modelling of 2D-structures governed by the amplitude equation is presented, and the latter equation is studied using a perturbation method.

Bifurcation phenomena are the subject of Chap. 13. Introductory Sect. 13.1 presents a relation between bifurcations and ODEs, stable and unstable manifolds, global and local bifurcation diagrams, as well as the classification of isolated solutions. Singular points of 1D and 2D vector fields are illustrated and analysed in Sect. 13.2. Examples taken from mechanics are given. Local bifurcations of hyperbolic and non-hyperbolic fixed points are studied, and next the double Hopf bifurcation is addressed. Section 13.3 deals with fixed points of maps and the associated bifurcations. Continuation (path following) approach using either the Galerkin approximation or shooting method is described in Sect. 13.4. An illustrative example from biomechanics is added. The next section aims at illustrating basic features of global bifurcations. Section 13.6 copes with bifurcations exhibited by piece-wise smooth dynamical systems. First, their importance is described and then stability of discontinuous systems useful for numerical applications is given. In the next section orbits exhibiting a degenerated contact with discontinuity surfaces, bifurcations in Filippov's systems, bifurcations of stationary points and periodic orbits are analysed.

The optimization of systems is a subject of Chap. 14. Firstly, historical roots are briefly revisited and variational principles are mentioned. Secondly, simple examples of optimization are reported (Sect. 14.2). Thirdly, conditional extremes are defined (Sect. 14.3), and then static optimization problems are revisited (Sect. 14.4), including local function approximation and definition of stationary points and quadratic forms. In the fourth place, Sect. 14.5 addresses problems associated with convexity of the sets of functions with a few definitions and theorems. Next, optimization without constraints and with conditions of local optimality is given in Sect. 14.6. Subsequently, optimality conditions regarding quadratic forms are presented in Sect. 14.7, whereas Sect. 14.8 deals with the equivalence constraints. Afterward, both Lagrange function and Lagrange multipliers, including their geometrical interpretations, are discussed in Sect. 14.9. Then, constraints of inequivalent types are described in Sect. 14.10. Finally, the shock work optimization is discussed in Sect. 14.11.

Chapter 15 describes phenomena of chaos and synchronization. Introductory Sect. 15.1 deals with a historical background and intuitive understanding of chaos and synchronization occurring in Nature as well as in pure and applied sciences. Modelling and identification of chaos is presented in Sect. 15.2, whereas Sect. 15.3 describes the role of Lyapunov exponents and their geometrical interpretation in quantifying chaotic orbits. Then, a frequency spectrum and autocorrelation function are described in Sects. 15.4 and 15.5, respectively. Modelling of nonlinear discrete systems with an emphasis on the chaotic dynamics is addressed in Sect. 15.6. It consists of an introduction, Bernoulli map, logistic map, map of a circle into circle, devil's stairs, Farey tree, Fibonacci numbers, Henon map and Ikeda map. Section 15.7 focuses on modelling chaotic ODEs and it includes a study of non-autonomous oscillator with different potentials, Melnikov's function approach, externally driven van der Pol's oscillator, Lorenz ODEs and their derivation. The synchronization phenomena of a mechanical system consisting of coupled triple pendulums with time-periodic mass distributions are analysed in Sect. 15.8, where many different kinds of synchrony as well as rich nonlinear dynamical effects including synchronization between chaotic as well as chaotic and regular dynamics have been numerically reported.

Finally, in Sect. 15.9 chaotic vibrations of flexible spherical rectangular shells loaded harmonically via the boundary conditions are investigated. In the first part one-layer shell made from an isotropic and homogeneous material is studied. The second part addresses nonlinear dynamics of multi-layer shells, taking into account gaps between the layers (design nonlinearity). Phase portraits, Fourier power spectra and wavelet spectra are constructed and investigated. Analysis of the shell curvature and shell design nonlinearity on the synchronization phenomena is carried out using also the phase difference as a new characteristic for monitoring and quantifying nonlinear vibrations.

The author wishes to express his thanks to T. Andrysiak, M. Kaźmierczak, G. Kudra and J. Mrozowski for their help in the book preparation.

Finally, I acknowledge the financial support of the National Science Centre of Poland under the grant MAESTRO 2, No. 2012/04/A/ST8/00738, for years 2013–2016.

Łódź, Poland Jan Awrejcewicz

Contents

1 Introduction 1
1.1 Existence of a Solution 5

2 First-Order ODEs 13
2.1 General Introduction 14
2.2 Separable Equation 16
2.3 Homogenous Equations 22
2.4 Linear Equations 27
2.5 Exact Differential Equations 39
2.6 Implicit Differential Equations Not Solved with Respect to a Derivative 45

3 Second-Order ODEs 51
3.1 Introduction 51
3.2 Linear ODEs 51
3.2.1 General Approaches 51
3.2.2 Hypergeometric (Gauss) Equation 60
3.2.3 The Legendre Equation and Legendre Polynomials 63
3.2.4 The Bessel Equation 69
3.2.5 ODEs with Periodic Coefficients 76
3.2.6 Modelling of Generalized Parametric Oscillator 91
3.2.7 ODEs with Constant Coefficient 100
3.3 Lagrange Equations and Variational Principle 107
3.4 Reduction to First-Order System of Equations 112
3.4.1 Mathematical Background 113
3.4.2 The Lagrangian Function 116
3.5 Canonical Transformations 118
3.6 Examples 120

3.7 Normal Forms of Hamiltonian Systems 124
3.7.1 Parametric Form of Canonical Transformations 125
3.7.2 Integration of Hamiltonian Equations Perturbated by Damping 126
3.8 Geometrical Approach to the Swinging Pendulum................ 130
3.8.1 The Analysed System and Geometrization 133
3.9 Geometric Analysis of a Double Pendulum Dynamics 138
3.9.1 The Pendulum and Geometrization 143
3.9.2 Numerical Simulations 146
3.10 A Set of Linear Second-Order ODEs with Constant Coefficients.. 148
3.10.1 Conservative Systems................................... 149
3.10.2 Non-conservative Systems 152
3.10.3 Modal Analysis and Identification 154

4 Linear ODEs 167
4.1 Introduction 167
4.2 Normal and Symmetric Forms 168
4.3 Local Solutions (Existence, Extensions and Straightness)........ 176
4.4 First-Order Linear Differential Equations with Variable Coefficients 187
4.4.1 Introduction 187
4.4.2 Fundamental Matrix of Solutions 193
4.4.3 Homogeneous Differential Equations................... 195
4.4.4 Examples of Homogeneous Linear Differential Equations 196
4.4.5 Non-homogeneous Differential Equations 206
4.4.6 Examples of Linear Non-homogenous Differential Equations 207
4.4.7 Homogeneous Differential Equations with Periodic Coefficients..................................... 209
4.4.8 The Floquet Theory 210
4.4.9 Reduction of Non-homogeneous Linear Differential Equations with Periodic Coefficients...... 212
4.4.10 Characteristic Multipliers................................ 213
4.4.11 Characteristic Exponents 214
4.4.12 Structure of Solutions for Simple Characteristic Multipliers............................... 214
4.4.13 Solutions Structure for Case of Multiple Multipliers .. 215
4.4.14 Stability 217
4.4.15 Periodic Solutions to Homogenous Differential Equations with Periodic Coefficients...... 217
4.4.16 Periodic Solutions to Non-homogenous Differential Equations with Periodic Coefficients...... 218

5 Higher-Order ODEs Polynomial Form 221
5.1 Introduction 221
5.2 Linear Homogeneous Differential Equations 223
5.3 Differential Equations with Constant Coefficients 226
5.4 Linear Non-homogeneous Differential Equations with Constant Coefficients 234
5.5 Differential Equations with Variable Coefficients 240

6 Systems 245

7 Theory and Criteria of Similarity 253

8 Model and Modelling 263
8.1 Introduction 263
8.2 Mathematical Modelling 266
8.3 Modelling in Mechanics 267
8.4 General Characteristics of Mathematical Modelling of Systems 273
8.5 Modelling Control Theory 273
8.5.1 Ordinary Automatic Control Systems 275
8.5.2 Distributed Automatic Control Systems 279
8.5.3 Classification of Control Systems with Respect to Another Criteria 280
8.5.4 Examples of Control Systems and Their Block Diagrams 282

9 Phase Plane and Phase Space 295
9.1 Introduction 295
9.2 Phase Plain and Singular Points 297
9.3 Analysis of Singular Points 304
9.3.1 Unstable Node 304
9.3.2 Stable Node 306
9.3.3 Critical Node 308
9.3.4 Degenerate Node 308
9.3.5 Saddle 310
9.3.6 Unstable Focus 313
9.3.7 Stable Focus 313
9.3.8 Centre 314
9.4 Analysis of Singular Points Governed by Three Differential Equations of First Order 317
9.4.1 Theory Concerning the Solving a System of Differential Equations and Method for Determining Roots of a Polynomial of Third Degree 317
9.4.2 Analysis of Singular Points Described by Three First-Order Differential Equations 318

10 Stability 329
10.1 Introduction 329
10.2 Lyapunov's Functions and Second Lyapunov's Method 348
10.3 Classical Theories of Stability and Chaotic Dynamics 358

11 Modelling via Perturbation Methods 363
11.1 Introduction 363
11.2 Selected Classical Perturbative Methods 364
11.2.1 Autonomous Systems 364
11.2.2 Nonautonomous Systems 379

12 Continualization and Discretization 395
12.1 Introduction 395
12.2 One-Dimensional Chain of Coupled Oscillators 395
12.3 Planar Hexagonal Net of Coupled Oscillators 400
12.4 Discretization 406
12.5 Modelling of Two-Dimensional Anisotropic Structures 410

13 Bifurcations 417
13.1 Introduction 417
13.2 Singular Points in 1D and 2D Vector Fields 421
13.2.1 1D Vector Fields 421
13.2.2 Two-Dimensional Vector Fields 429
13.2.3 Local Bifurcation of Hyperbolic Fixed Points 436
13.2.4 Bifurcation of a Non-hyperbolic Fixed Point (Hopf Bifurcation) 438
13.2.5 Double Hopf Bifurcation 451
13.3 Fixed Points of Maps 458
13.4 Continuation Technique 461
13.5 Global Bifurcations 467
13.6 Piece-Wise Smooth Dynamical System 469
13.6.1 Introduction 469
13.6.2 Stability 471
13.6.3 Orbits Exhibiting Degenerated Contact with Discontinuity Surfaces 477
13.6.4 Bifurcations in Filippov's Systems 480
13.6.5 Bifurcations of Stationary Points 480
13.6.6 Bifurcations of Periodic Orbits 483

14 Optimization of Systems 487
14.1 Introduction 487
14.2 Simple Examples of Optimization in Approximative Problems 488
14.3 Conditional Extrema 490
14.4 Static Optimization 494
14.4.1 Local Approximation of a Function 497
14.4.2 Stationary Points and Quadratic Forms 498

14.5 Convexity of Sets of Functions 499
14.6 Problems of Optimization Without Constraints 502
14.7 Optimality Conditions of a Quadratic Form 504
14.8 Equivalence Constraints 505
14.9 The Lagrange Function and Multipliers 507
14.10 Inequivalence Constraints 510
14.11 Shock Absorber Work Optimization 513
14.11.1 Introduction 513
14.11.2 Optimization Example 513

15 Chaos and Synchronization 527
15.1 Introduction 527
15.2 Modelling and Identification of Chaos 531
15.3 Lyapunov Exponents 539
15.4 Frequency Spectrum 540
15.5 Function of Autocorrelation 541
15.6 Modelling of Nonlinear Discrete Systems 542
15.6.1 Introduction 542
15.6.2 Bernoulli's Map 545
15.6.3 Logistic Map 548
15.6.4 Map of a Circle into a Circle 553
15.6.5 Devil's Stairs, Farey Tree and Fibonacci Numbers 557
15.6.6 Hénon Map 560
15.6.7 Ikeda Map 566
15.7 Modelling of Nonlinear Ordinary Differential Equations 568
15.7.1 Introduction 568
15.7.2 Non-autonomous Oscillator with Different Potentials 569
15.7.3 Melnikov Function and Chaos 571
15.7.4 Lorenz Attractor 575
15.8 Synchronization Phenomena of Coupled Triple Pendulums 578
15.8.1 Mathematical Model 578
15.8.2 Numerical Simulations 584
15.9 Chaos and Synchronization Phenomena Exhibited by Plates and Shells 586
15.9.1 Introduction 586
15.9.2 One Layer Shell 588
15.9.3 Two-Layer Shell 596

References 605

Chapter 1
Introduction

In this book we consider only processes, which future configurations are completely defined by their initial states, i.e. the so-called evolutionary deterministic processes. Since this chapter deals with basic concepts of theory of ordinary differential equations, the reader is encouraged to be familiar with the monographs [10, 11, 54, 63, 70, 84, 103, 105, 111, 114, 122, 193, 216, 217, 236, 257].

A solution $x = \varphi(t)$ to the system of ordinary differential equations

$$\dot{x} = F(t, x), \tag{1.1}$$

where $x = (x_1, \dots, x_n)$ and $F = (F_1, \dots, F_n)$, is the function $\varphi(t)$, which substituted into (1.1) satisfies it identically. We have assumed that $F_1, \dots, F_n$ are $C^r (r \geq 1)$ smooth functions.

The representation of $\varphi(t)$ in the space $\mathbb{R}^{n+1}$ of the variables (t, x) is called the *integral curve* of (1.1).

The solutions of (1.1) have the following properties:

(i) If $x = \varphi(t)$ is a solution to (1.1) then also $x = \varphi(t + t_0)$ is a solution to (1.1). Both of them correspond to the same initial point x_0 but for different time instant.
(ii) Let the solution satisfy the initial condition $x_0 = \varphi(t_0)$. Then it can be written in the form $x = \varphi(t - t_0, x_0)$, where $\varphi(0, x_0) = x_0$, $t \geq t_0$.
(iii) The following group property is satisfied:

$$\varphi(t_2, \varphi(t_1, x_0)) = \varphi(t_1 + t_2, x_0). \tag{1.2}$$

There exist two geometrical representations of a solution to (1.1). Let $\tilde{D} \subset \mathbb{R}^{n+1}$ denote the so-called extended phase space $\tilde{D} = D \times \mathbb{R}^1$. Changing the parameter t (time) we get points in phase space D for different values of the parameter t.

J. Awrejcewicz, *Ordinary Differential Equations and Mechanical Systems*,
DOI 10.1007/978-3-319-07659-1_1

When only smooth systems are considered, a velocity vector $F(t, x)$ is tangent to the phase trajectory at a given point x_0, and there is only one trajectory passing through such an arbitrary point (this question will be clarified in more detail later). The phase curves are also called *phase trajectories*.

As it has been mentioned earlier, the space $\tilde{D}$ contains *integral curves*. Their projection into the phase space D are phase trajectories or singular points (equilibrium states). If a non-singular trajectory corresponds to a solution $\varphi(t)$ of (1.1), and $\varphi(t_1) = \varphi(t_2)$ for $t_1 \neq t_2$, then $\varphi(t)$ is defined for all t and is periodic.

However, there are trajectories having no points of self-intersection, i.e. quasi-periodic or chaotic orbits. When any two distinct solutions corresponding to the same trajectory are identical up to a time shift $t \to t + t_0$, all solutions corresponding to the same periodic trajectory are periodic with the same period.

A solution also possesses a mechanical interpretation. Trajectory made by a *point* with an increase of time is called a *motion*. Sometimes it is useful to consider a 'time-reversed' system

$$\dot{x} = F(-t, x), \tag{1.3}$$

which can be obtained from (1.1), by reversing the direction of each tangent vector. Knowing the trajectories of one system we can easily find the corresponding trajectories of another system, simply by reversing the direction of the arrowheads.

Dynamical systems are the systems solution of which can be continued for time $t \in (-\infty, \infty)$, and the corresponding trajectories are called the *entire trajectories*.

Since in Eq. (1.1) the independent variable $x(t)$ as well as the function F are treated as vector functions, one can investigate the system state via its vector state. Hence, analysis of the function $x_t = \varphi_t(x_0)$ with respect to time t and the initial condition is referred as system's *dynamics* investigation.

Let Y be the metric space and $\varphi_t : Y \to Y$ be a family of transformations depending on the parameter t in a smooth way.

Definition 1.1. If $\forall z \in D$ and $t, t^* \in [0, \infty]$, family of transformations satisfies the identity

$$\varphi_t[\varphi_{t^*}(z)] = \varphi_{t+t^*}(z), \tag{1.4}$$

then the pair (Y, φ_t) is called dynamical system (with continuous time) or a flow. In a case of homeomorphism, when $t \in (-\infty, +\infty)$, we have $\varphi_t^{-1} = \varphi_{-t}$.

In a case, when time has discrete natural numbers, one obtains the definition of a cascade.

Definition 1.2. The pair (Y, φ), for natural values of the parameter t, where Y is the metric space and $\varphi : Y \to Y$, is called a cascade (or a dynamical system with a discrete time).

Let $\varphi^m = \underbrace{\varphi \circ \varphi \ldots \circ \varphi}_{m\text{-times}}$. (Since $\forall z \in Y$ and $n, m = 0, 1, 2, \ldots$, we have $\varphi^n[\varphi^m(z)] = \varphi^{n+m}(z)$). If we have homeomorphism, i.e. one-to-one continuous mappings with continuous inverse, then the above identity is true for all integer numbers n and m. The definition of cascades is used during analysis of all problems related to numerical solutions of the ordinary differential equations. Now we give some basic notations and definitions related to either *cascades* or *flows*.

The sequence $\{x_n\}_{n=-\infty}^{+\infty}$ is called a *trajectory* of the *point* x_0, where $x_{n+1} = \varphi(x_n)$. There are following distinct trajectories:

1. When $\psi(x_0) = x_0$, then x_0 is called a fixed point (or a periodic point with period 1).
2. When $x_i = \psi^i(x_0)$, $i = 0, 1, \ldots, k-1$ and $x_0 = \psi^k(x_0)$, ($x_i \neq x_j$ for $i \neq j$), then each x_i is called a periodic point of period k.
3. When for $k \to \pm\infty$, $x_i \neq x_j$ for $i \neq j$, then a sequence $\{x_k\}_{-\infty}^{+\infty}$ called a bi-infinite (or unclosed) trajectory.

Recall that a set B is called *invariant* if $B = \varphi(t, B)$ for any t. If $x \in A$, then the trajectory $\varphi(t, x) \in A$.

A point x_0 is called a *wondering point* if there is an open neighbourhood $U(x_0)$ of x_0 and $T > 0$, such that

$$U(x_0) \cap \varphi(t, U(x_0)) = \emptyset \quad \text{for } t > T. \tag{1.5}$$

A set of wondering points $\mathcal{W}$ is open and invariant. In contrary, a set of non-wondering points $\mathcal{M} = D/\mathcal{W}$ is closed and invariant (equilibrium states and periodic trajectories). The points on bi-asymptotic trajectories tending to equilibrium states and periodic trajectories as $t \to \pm\infty$ are also non-wondering.

Definition 1.3. A point x_0 is said to be positive Poisson-stable if for a given any neighbourhood $U(x_0)$ and any $T > 0$, there is $t > T$ such that $\varphi(t, x_0) \subset U(x_0)$. If for any $T > 0$ there exists t such that $t < -T$, then the point x_0 is called a negative Poisson-stable point. A point is said Poisson-stable (p) if it is positive (p^+) and negative (p^-) Poisson-stable. Note that p^+, p^- and p trajectories consist of non-wandering points.

A very important result has been obtained by Birkhoff.

Theorem 1.1. *If a $p(p^-, p^+)$-trajectory is unclosed, then its closure $\bar{p}(\bar{p}^-, \bar{p}^+)$ contains a continuum of unclosed p-trajectories.*

Considering ε-neighbourhood $U_\varepsilon(x_0)$ of a point x_0 and denoting by series $\{t_n(\varepsilon)\}_{-\infty}^{+\infty}$ the successive intersection of $U_\varepsilon(x_0)$, then values $\tau_n(\varepsilon) = t_{n+1}(\varepsilon) - t_n(\varepsilon)$ are called *the Poincare return times*. When the series $\{\tau_n(\varepsilon)\}$ is bounded for a finite ε, then the p-trajectory is said to be *recurrent*. A closure $\bar{p}$, corresponding to this trajectory, is non-empty, invariant and closed and it creates a so-called minimal set. The return time in this case is not constrained.

When the series $\{\tau_n(\varepsilon)\}$ is unbounded, then the closure $\bar{p}$ of D-trajectory is called *a quasi-minimal set*. In this case the closure may contain other objects, like periodic or quasi-periodic trajectories, which can be approached by a flow arbitrarily close.

Definition 1.4. A point x^* is called an ω-limit (α-limit) point of the trajectory $\{\varphi^n(x)\}$ if

$$\lim_{t_k \to \pm\infty} \varphi^{t_k}(x) = x^*. \tag{1.6}$$

A set of all ω-limit (α-limit) points of the trajectory L are denoted by $\Omega_L(A_L)$.

Observe that all of the points of the periodic trajectory L are both α and ω-limit points and $L = \Omega_L = A_L$. For an unclosed Poisson-stable trajectory $\bar{L} = \Omega_L = A_L$, where $\bar{L}$ is the closure of L. As it has been stated earlier, $\bar{L}$ is either minimal or quasi-minimal set.

Owing to the earlier investigations of Poincaré and Bendixon, only three topological cases can be found in two-dimensional (planar) systems: (a) equilibrium; (b) periodic orbit; (c) cycles. In the latter case one deals with either ω-limit homoclinic (one equilibrium state) or heteroclinic (two or more equilibrium states) cycles. A similar observation holds for negative semi-trajectories. The planar systems will be considered in more detail later.

Another important feature of the trajectory studies is a topological equivalence. Two objects in the phase space are topologically equivalent if there is a homeomorphism mapping the trajectories of one to another object. If the space D is compact one (for instance a closed and bounded subset of $\mathbb{R}^n$), then for $\forall x$ the set $\Omega(x)$ is non-empty and closed.

Definition 1.5. A subset $I \subset D$ is called an invariant set of the cascade (D, φ) if $\varphi(I) = I$.

Let us introduce a set of homeomorphism $H = \{h_i\}$ and the metric $\mathrm{dist}(h_1, h_2) = \sup_{x \in G} ||h_1 x - h_2 x||$.

Definition 1.6. If the condition $h_i L = L$ holds for all homomorphism h_i, satisfying $\mathrm{dist}(h_i, I) < \varepsilon$, where I is the identity homomorphism, then $L \in G$ is called the special trajectory.

Closed trajectories are equilibrium states and periodic orbits.

In the case of cascades, the fixed points, periodic points and ω-limit sets are invariant. If the mapping φ is invertible, then an *entire* trajectory $\{\varphi^k\}$, $k = 0, \pm 1, \pm 2, \ldots$ is also invariant. It can be shown that a sum or a product of invariant sets is also an invariant set.

Finally, let us introduce a definition of an *attractor (repiler)*.

Definition 1.7. A closed, bounded and invariant set $A \subset D$ is called attractor of the dynamical system (D, φ), when it has a neighbourhood $U(A)$ such that for any $x \in U(A)$ the trajectory $\{\varphi^n(x)\}$ remains in A and tends to A for $n \to \infty$, i.e.

$$\lim_{t \to +\infty} \rho((\varphi(t, x), A), A) = 0, \tag{1.7}$$

where

$$\rho(x, A) = \inf_{x_0 \in A} ||x - x_0||. \tag{1.8}$$

A set of all x for which $\{\varphi^n(x)\}$ tends to A is called *the basin of attraction of* A. A similar like definition can be formulated in a case of the repiler.

Definition 1.8. A closed, bounded and invariant set $R \subset D$ is called repiler of the dynamical system (D, φ), if there exists a neighbourhood $U(R)$ of R such that if $x \notin R$ and $x \in U(R)$, then there is n when $\varphi^k(x) \notin U(R)$ for $k > n$.

The equilibrium states, periodic and quasi-periodic orbits can be either attractors or repilers depending on their stability. The question how to find regular attractors or repilers will be addressed later.

1.1 Existence of a Solution

It is well known that Eq. (1.1) may have a unique solution on a given interval, may have no solution at all, may have infinitely many solutions or may have a few distinct solutions. However, finding an analytical solution explicitly to specific initial problems governed by (1.1) does not belong to easy tasks. It is important in many cases that even if we do not know a solution, we are still interested in getting the answer to the following problem: how we may know that a given ODE possesses a solution, and if it so, then we would like to know if it is a unique one.

Theorem 1.2 (Picard's Theorem). *Let a function $F(t, x)$ of (1.1) is continuous on the rectangular $\Pi = \{(t, x) : |t - t_0| \le a, |x - x_0| \le b, a > 0, b > 0\}$ and satisfies the Lipschitz conditions uniquely regarding x, i.e.*

$$|F(t, x_1) - F(t, x_2)| \le L|x_1 - x_2|$$

for all t, where $|t - t_0| \le a$, $|x_1 - x_0| \le b$, $|x_2 - x_0| \le b$. Let $M = \max_{(t,x)\in\Pi} |F(t, x)|$, $t^* = \min\left(a, \frac{b}{M}\right)$. Then the Cauchy problem associated with (1.1) has a unique solution in the interval

$$|t - t^*| \le \alpha, \ \alpha < \min\left(a, \tfrac{b}{M}, \tfrac{1}{L}\right)$$

A proof of Theorem 1.2 is given in [191] and it is omitted here. Picard's theorem allows not only to estimate a solution existence of (1.1), but it also guarantees its uniqueness.

In what follows we apply this theorem to find the solution, when we cannot find it using elementary approaches. If assumptions of Theorem 1.2 are satisfied, then a solution to (1.1) can be found as a limit of the uniformly convergent series $\lim_{n\to\infty}\{x_n(t)\}$ defined by the following recurrent rule

$$x_{n+1}(t) = x_0 + \int_{t_0}^{t} F(y, x_n(y))dy, \quad x_0(t) = x_0, \quad n = 1, 2, \ldots \tag{1.9}$$

Furthermore, one may also estimate an error introduced by the N approximation $\{x_n(t)\}$ via the following inequality

$$|x(t) - x_n(t)| \leq \frac{ML^{n-1}}{n!}(t^*)^n. \tag{1.10}$$

Example 1.1. Consider the one-dimensional initial value problem

$$\tfrac{dx}{dt} = t + x,\ x(0) = 1.$$

Equation (1.9) takes the form

$$x_{n+1}(t) = 1 + \int_0^t (y + x_n(y))dy,$$
$$n = 0, 1, 2, \ldots, x_0(t) = 1.$$

We substitute successively $n = 0, 1, 2, \ldots$ to equation in the above to get

$$x_0(t) = 1;$$

$$x_1(t) = 1 + \int_0^t (y + 1)dy = 1 + t + \frac{t^2}{2};$$

$$x_2(t) = 1 + \int_0^t \left(y + 1 + y + \frac{y^2}{2}\right)dy = 1 + t + t^2 + \frac{t^3}{3!};$$

$$x_3(t) = 1 + \int_0^t \left(y + 1 + y + y^2 + \frac{y^3}{3!}\right)dy = 1 + t + t^2 + \frac{2t^3}{3!} + \frac{t^4}{4!};$$

$$x_n(t) = 1 + \int_0^t \left(y + 1 + y + y^2 + \frac{y^3}{3} + \cdots + \frac{2y^{n-1}}{(n-1)!} + \frac{y^n}{n!}\right)dy$$
$$= 1 + t + t^2 + \frac{2t^3}{3!} + \cdots + \frac{2t^n}{n!} + \frac{t^{n+1}}{(n+1)!}$$

Final result can be presented in its equivalent form

$$x_n(t) = 2\sum_{k=0}^{n} \frac{t^k}{k!} + \frac{t^{k+1}}{(k+1)!} - t - 1$$

The solution to our problem follows

$$x(t) = \lim_{n\to\infty} x_n(t) = 2\sum_{k=0}^{\infty} \frac{t^k}{k!} - t - 1 = 2e^t - t - 1.$$

□

Example 1.2. Apply the method of successive approximations to the following Cauchy problem: $\frac{dx}{dt} = t - x^2$, $x(0) = 0$ defined on the rectangular $|t| \le 1$, $|x| \le 1$. Estimate an interval of the successive approximations convergence guaranteed by the Picard's theorem as well as an error between the exact solution and its second-order approximation.

Observe that the function $F(t,x) = t - x^2$ is continuously differentiable regarding x, and $\frac{dF}{dx} = -2x$. Function F satisfies the Lipschitz condition with $L = \max\left|\frac{\partial F}{\partial x}\right| = 2$. Since

$$M = \max_{|t|\le 1;|x|\le 1} |F(t,x)| = \max_{|t|\le 1;|x|\le 1} |t - x^2| = 2,$$

$$t^* = \min\left(a, \frac{b}{M}\right) = \min\left(1, \frac{1}{2}\right) = \frac{1}{2}.$$

Therefore, the Picard's approximation is convergent in the interval $[-\frac{1}{2}, \frac{1}{2}]$. Successive approximations obey the following rule

$$x_{n+1}(t) : \int_0^t (y - x_n(y))dy,\ n = 0, 1, 2, \dots,$$

and hence

$$n = 0: \qquad x_1(t) = \int_0^t (y - 0)dy = \tfrac{t^2}{2};$$

$$n = 1: x_2(t) = \int_0^t \left(y - \left(\tfrac{y^2}{2}\right)^2\right) = \tfrac{t^2}{2} - \tfrac{t^5}{20}.$$

Equation (1.10) takes the form

$$|x(t) - x_2(t)| = \frac{ML^1}{2!}(t^*)^2 = \frac{2\cdot 2}{2}\left(\frac{1}{2}\right)^2 = \frac{1}{2}.$$

□

Example 1.3. Show that the so-called Riccati equation

$$\frac{dx}{dt} = a(t)x^2 + b(t)x + c(t) \equiv F(t,x),$$

where $a(t)$, $b(t)$ and $c(t)$ are continuous functions, cannot have a singular solution.

Function $F(t,x)$ is continuously differentiable with respect to x, and $\frac{\partial F}{\partial x} : 2ax + b$ is bounded in an arbitrary rectangular

$$\Pi = \{(t,x) \in R^2; |t - t_0| \le a, |x - x_0| \le b\}.$$

Function $F(t,x)$ satisfies the Lipschitz condition in the rectangular Π with respect to x, and hence it satisfies the Picard's assumptions. The studied equation does not have singular solutions. □

In the given below Peano theorem one may guarantee existence of a solution but its uniqueness is not defined.

Theorem 1.3 (Peano's Theorem). *Let function $F(t,x)$ of (1.1) is continuous on the rectangular*

$$\Pi = \{(t,x) : t \in [t_0, t_0 + a], |x - x_0| \le b\},$$

where $\sup_{(t,x)\in\Pi} |F(t,x)| = M$. *Then the Couchy problem regarding (1.1) has a solution in the interval* $[t_0, t_0 + \alpha]$, *where* $\alpha = \min\left(a, \frac{b}{M}\right)$.

Example 1.4. Find singular solutions to the equation $\frac{dx}{dt} = 1 + \frac{3}{2}(x-t)^{\frac{1}{3}}$.

Let us introduce the new variable $x - t = y$, and the problem boils down to investigation of the following equation $\frac{dy}{dt} = \frac{3}{2}y^{\frac{1}{3}}$. For $y = 0$ the last equation does not satisfy the Lipschitz conditions, i.e. assumption of the Picard's theorem are not satisfied, although the Peano theorem assumptions are satisfied. The studied equation can be solved by the following steps.

$$\int y^{-\frac{1}{3}}dy = \frac{3}{2}\int dt;\ y^{\frac{2}{3}} = (t - C);\ y = (\sqrt{t - C})^3.$$

Initial condition $y(t_0) = 0$ is satisfied by $y = (\sqrt{t - t_0})^3$ and by $y = 0$. Therefore, $y = 0$ is a singular solution to equation $\frac{dy}{dt} = \frac{3}{2}y^{\frac{1}{3}}$, and the function $x = t$ is a singular solution to the initial equation. The remaining solutions are defined by the formula $x = t + (t - C)^{\frac{3}{2}}$. □

In general, if the function $F(t,x)$ satisfies the Picard's theorem on the closed rectangle Π, then its any solution $x = x(t)$, $x(t_0) = x_0$, $(t_0, x_0) \in \Pi$ can be extended outside the rectangle. Furthermore, if the function $F(t,x)$ in a slab $\alpha_1 \le t \le \alpha_2$, $|x| < \infty$ $(\alpha_1 \ge -\infty, \alpha_2 \le +\infty)$ is continuous and satisfies the inequality $|F(t,x)| \le a(t)|x| + b(t)$, where $a(t)$, $b(t)$ are continuous functions, then any solution of Eq. (1.1) can be extended in the interval $\alpha_1 < t < \alpha_2$.

Theorem 1.4 (Global Existence of Solutions). *Let F be a vector-valued function of $n+1$ real variables, and let I be an open interval containing $t = t_0$. If $F(t,x)$ is continuous and satisfies the Lipschitz condition for all t in I and for all $x_1, x_2 \in \mathbb{R}^n$, then the initial value problem $x(t_0) = x_0$ has a solution in the entire interval I.*

Proof ([84]). In what follows we demonstrate that the series $\{x_n(t)\}_0^\infty$ of successive approximation

$$x(t_0) = x_0,\ x_{n+1} = x_0 + \int_0^t F(x_n(s), s)ds$$

converges to a solution $x(t)$ of $\frac{dx}{dt} = F(t,x),\ x(t_0) = x_0$. □

We take $t_0 = 0$ and consider $t \geq 0$. We show that if $[0, t^*]$ is a closed and bounded interval of I, then $\{x_n(t)\}$ converges uniformly on $[0, t^*]$ to a limit $x(t)$. In other words, given $\varepsilon > 0$, there is an integer N such that

$$|x_n(t) - x(t)| < \varepsilon,$$

for all $n \geq N$ and all $t \in [0, t^*]$. Let $M = \max |F(x_0, t)|$ for $t \in [0, t^*]$, then

$$|x_1(t) - x_0(t)| = \left| \int_0^t F(x_0(s), s)ds \right| \leq \int_0^t |F(x_0(s), s)|ds \leq Mt,$$

and similarly

$$|x_2(t) - x_1(t)| = \left| \int_0^t [F(x_1(s), s) - F(x_0(s), s)]ds \right| \leq m \int_0^t |x_1(s) - x_0(s)|ds$$

and therefore

$$|x_2(t) - x_1(t)| \leq m \int_0^t Msds = \frac{1}{2} mMt^2.$$

Assuming that

$$|x_n(t) - x_{n-1}(t)| \leq \frac{M}{m} \frac{(kt)^n}{n!},$$

it follows (using induction)

$$|x_{n+1}(t) - x_n(t)| = \left| \int_0^t [F(x_n(s), s) - F(x_{n-1}(s), s)]ds \right| \leq m \int_0^t |x_n(s) - x_{n-1}(s)|ds,$$

and finally

$$|x_{n+1}(t) - x_n(t)| \leq m \int_0^t \frac{M}{m} \frac{(ms)^n}{n!} ds = \frac{M}{m} \frac{(mt)^{n+1}}{(n+1)!} = \frac{M}{m}(e^{mt^*} - 1).$$

It means that the terms of the series

$$x_0(t) + \sum_{n=1}^{\infty} [x_n(t) - x_{n-1}(t)]$$

are dominated in the convergent series with positive constants. Since the sequence of this uniformly convergent series on $[0, t^*]$ is the original sequence $\{x_n(t)\}_0^\infty$, then its uniform convergence has been proved. Standard theorems of advanced calculus yield the following conclusions:

(i) The limit function $x(t)$ is continuous on $[0, t^*]$.
(ii) The Lipschitz continuity of F gives the estimation

$$|F(x_n(t), t) - F(x(t), t)| \leq m|x_n(t) - x(t)| < m\varepsilon$$

for $t \in [0, t^*]$ and $n \geq N$; it means that the sequence $\{F(x_n(t), t)\}_0^\infty$ converges uniformly to $F(x(t), t)$ on $[0, t^*]$.
(iii) It follows that

$$x(t) = \lim_{n\to\infty} x_{n+1}(t) = x_0 + \lim_{n\to\infty} \int_0^t F(x_n(s), s) ds = x_0 + \int_0^t \lim F(x_n(s), s) ds$$

$$= x_0 + \int_0^t F(x(s), s) ds.$$

(iv) Since $x(t)$ is continuous on $[0, t^*]$, then the latter results imply that $dx/dt = F(x(t), t)$ on $[0, t^*]$. Because this is true on every closed subinterval of I, then it is true on the entire I.

Theorem 1.5 (Global Existence of a Solution of Linear Systems). *Let the $n \times n$ matrix-valued function $A(t)$ and the vector-valued function $f(t)$ are continuous on the open interval I containing $t = t_0$. Then the Cauchy problem*

$$\frac{dx}{dt} = A(t)x + f(t), \quad x(t_0) = x_0$$

has a solution on the entire interval I.

Proof. We need to show that for each closed and bounded subinterval of I there is a Lipschitz constant L such that

$$||[A(t)x_1 + f(t)] - [A(t)x_2 + f(t)]|| \leq L|x_1 - x_2|.$$

It means that

$$|A(t)x| \leq L|x| = ||A|| \cdot |x|,$$

where $||A|| = \sqrt{\sum_{i,j=1}^{n} (a_{ij})^2}$. Further, it means that $L = ||A||$, and because $A(t)$ is continuous on the closed and bounded subinterval I, then its norm $||A||$ is bounded on the considered subinterval. The global existence theorem for linear system has been proved. □

In the case of nonlinear ordinary differential equations a solution may exist only on a small neighbourhood of $t = t_0$, and the length of existence interval can depend on a nonlinear differential equation and on the initial condition $x(t_0) = x_0$.

If $F(t,x)$ of (1.1) is continuously differentiable in vicinity of the point (x_0, t_0) in $(n+1)$-dimensional space, then it can be concluded that $F(x,t)$ satisfies the Lipschitz condition on a rectangle Π centered at (x_0, t_0) of the form $|t - t_0| < \alpha$, $|x_i - x_{0i}| < \beta_i$, $i = 1, \ldots, n$. If one applies the Lipschitz condition

$$x_{n+1}(t) = x_0 + \int_{t_0}^{t} F(x_n(s)s)ds,$$

then the point $(x_n(t), t)$ lies in the rectangle Π only for a suitable choice of t.

Theorem 1.6 (Local Solutions Existence). *If the first-order partial derivatives of F in (1.1) all exist and are continuous in a neighbourhood of the point (x_0, t_0), then the Cauchy problem (1.1) has a solution on some open interval containing $t = t_0$.*

However, if the Lipschitz condition is satisfied, then an investigated solution is in addition *unique*.

Theorem 1.7 (Uniqueness). *Let on some region Q in $(n+1)$-space the function $F(x,t)$ in (1.1) is continuous and satisfies the Lipschitz condition*

$$|F(x_1,t) - F(x_2,t)| \leq L(x_1 - x_2).$$

If $x_1(t)$ and $x_2(t)$ are solutions to the Cauchy problem (1.1) on some open interval I, where $t = t_0 \in I$ such that the solution curves $(x_1(t), t)$ and $(x_2(t), t)$ lie in Q for all t in I, then $x_1(t) = x_2(t)$ for all t in I.

Proof. We consider only 1D case, where x is real and we follow the steps given in [191].

Consider the function

$$\phi(t) = [x_1(t) - x_2(t)]^2,$$

where $x_1(t_0) = x_2(t_0) = x_0$, i.e. $\phi(t_0) = 0$.
Differentiating equation in the above one gets

$$|\dot{\phi}(t)| = |2(x_1 - x_2)\cdot(\dot{x}_1 - \dot{x}_2)| = |2(x_1 - x_2)\cdot(F(x_1, t) - F(x_2, d))| \le 2L|x_1 - x_2|^2 = 2L\phi(t).$$

On the other hand a solution to the differential equation

$$\dot{\varphi}(t) = 2L\varphi(t), \quad \varphi(t_0) = \varphi_0$$

is as follows

$$\varphi(t) = \varphi_0 e^{2L(t-t_0)}.$$

For $\phi(t_0) = \varphi(t_0)$ it yields

$$\phi(t) \le \varphi(t) \text{ for } t \ge t_0.$$

Therefore

$$0 \le (x_1(t) - x_2(t))^2 \le (x_1(t_0) - x_2(t_0))^2 e^{2L(t-t_0)},$$

and taking into account square roots we finally obtain

$$0 \le |x_1(t) - x_2(t)| \le |x_1(t_0) - x_2(t_0)| e^{L(t-t_0)}.$$

Because $x_1(t_0) - x_2(t_0) = 0$, then $x_1(t) \equiv x_2(t)$. □

The carried out so far proof allows to illustrate how solutions of (1.1) depend continuously on the initial value $x(t_0)$. Namely, if we take $|x_1(t_0) - x_2(t_0)| \le \delta$, then the last inequality implies that

$$|x_1(t) - x_2(t)| \le \delta e^{L(t^* - t_0)} = \varepsilon$$

for all $t_0 \le t \le t^*$. The Cauchy problems are said to be *well posed* as mathematical model for real-world processes if the considered differential equation has unique solutions that are continuous with respect to initial values.

As we will see further, through a point (t_0, x_0) may pass only one integral curve of Eq. (1.1) satisfying a given initial condition, which in many cases corresponds to a proper modelling of real-world processes. However, more archetypical questions are valid before starting to solve a given differential equation. The so far discussed theorems allow to verify if a solution actually *exists*, and if it is *unique*.

As it will be shown further, one may deal with (a) failure of existence; (b) failure of uniqueness; (c) one, a few or infinitely many solutions.

Chapter 2
First-Order ODEs

Modelling of various problems in engineering, physics, chemistry, biology and economics allows formulating of differential equations, where a being searched function is expressed via its time changes (velocities). One of the simplest example is that given by a first-order ODE of the form

$$\frac{dy}{dt} = F(y), \tag{2.1}$$

where $F(t)$ is a known function, and we are looking for $y(t)$. Here by t we denote time. In general, any given differential equation has infinitely many solutions. In order to choose from infinite solutions those corresponding to a studied real process, one should attach initial conditions of the form $y(t_0) = y_0$.

In general, there is no direct rule/recipe for construction of an ODE. Let $y = y(t)$ be a dependence between t and y of the investigated process. We are going to monitor the difference $y(t + \Delta t) - y(t)$ caused by the disturbance Δt. Then, if we take

$$\dot{y} \equiv \frac{dy}{dt} = \lim_{\Delta t \to 0} \frac{y(t + \Delta t) - y(t)}{\Delta t},$$

we obtain a differential equation, i.e. dependence of the process velocity in the point t governed by the function $F(y)$.

There are also cases where a function $y(t)$ appears under an integral and the obtained equation is called *the integral equation*, which in simple cases can be transformed to a differential equation.

J. Awrejcewicz, *Ordinary Differential Equations and Mechanical Systems*,
DOI 10.1007/978-3-319-07659-1_2

2.1 General Introduction

A differential equation of the form

$$f\left(t, y, \frac{dy}{dt}\right) = 0 \tag{2.2}$$

is called the first-order ordinary differential equation, where t is the independent variable (here referred to time, but in general it can be taken as a space variable x), and $y(t)$ is the unknown function to be determined. Observe that Eq. (2.2) is not solved with respect to its derivative dy/dt. In many cases, however, one deals with the following differential equation

$$\frac{dy}{dt} = f(t, y), \tag{2.3}$$

which is called the first-order ODE solved with respect to the derivative. Alternatively, one may deal often with the following form of first-order ODE

$$P(t, y)dt + Q(t, y)dy = 0, \tag{2.4}$$

where P, Q are given functions.

We say that $y = \phi(t)$ is a solution to either (2.2) or (2.3) in an interval J, if

$$f\left(t, \phi(t), \frac{d\phi(t)}{dt}\right) \equiv 0, \tag{2.5}$$

or

$$\frac{d\phi(t)}{dt} = f(t, \phi(t)), \tag{2.6}$$

for all $t \in J$.

One may also find a solution to Eq. (2.2) in the *implicit form* $\varphi(t, \phi(t))$, where $\phi(t) = y$ is a solution to Eq. (2.2). Solution in the form of $\varphi(t, \phi(t))$ is also referred to as *the integral of Eq. (2.2)*.

A graph of solution $y = \phi(t)$ of Eq. (2.2) is called *the integral curve* of the studied differential equation. Projection of the solution graph onto the plane (t, y) is called *the phase curve* (or *trajectory*) of the investigated first-order ODE.

A problem related to finding a solution $y = \phi(t)$ satisfying the initial condition $y(t_0) = y_0$ is called the Cauchy problem.

If we take a point (t, y) for $t \in J$, then a tangent line passing through this point creates with the axis t an angle α, then $\tan\alpha = f(t, y)$. A family of all tangent lines defines a *direction field* for the studied differential equation. If we draw a short line segment possessing the slope $f(t, y)$ through each of representative collection of points (t, y), then all line segments constitute *a slope field* for the investigated ODE.

A curve constituting of points with the same slope field is called the *isocline*. In other words all integral curves passing through an isocline intersect the axis t with the same angle.

Example 2.1. Prove that the function $y = \phi(t)$ given in the parametric form $t = xe^x$, $y = e^{-x}$ satisfies the following differential equation

$$(1 + ty)\frac{dy}{dt} + y^2 = 0.$$

We have

$$\begin{aligned}(1 + ty)\frac{dy}{dt} + y^2 &= (1 + xe^x e^{-x})\frac{dy}{dx}\frac{dx}{dt} + e^{-2x} \\ &= -(1 + x)e^{-x} \cdot \frac{1}{(1 + x)e^x} + e^{-2x} = 0,\end{aligned}$$

which proves that $\phi(t)$ satisfies the studied equation. □

Example 2.2. Construct a differential equation of a family of ellipses of the following canonical form

$$\frac{x^2}{a^2} + \frac{y^2}{b^2} = 1,$$

where $0 < b < a$.

Acting by d/dx on both sides of this algebraic equation yields

$$\frac{x}{a^2} + \frac{y\frac{dy}{dx}}{b^2} = 0.$$

Solving both equations we get

$$\sqrt{a^2 - x^2}\frac{dy}{dx} + \frac{b}{a}x = 0.$$

□

Example 2.3. Construct a differential equation of the force lines of a dipole constituted by two electric charges $(+q, -q)$ located on the distance $2a$, where the force lines satisfy the Coulomb algebraic equation of the form

$$\frac{x + a}{r_1} - \frac{x - a}{r_2} = C,$$

where: $r_1^2 = (x + a)^2 + y^2, \quad r_2^2 = (x - a)^2 + y^2.$

A differentiation of the algebraic equation yields

$$\frac{r_1 - (x+a)\frac{dr_1}{dx}}{r_1^2} - \frac{r_2 - (x-a)\frac{dr_2}{dx}}{r_2^2} = 0,$$

and also

$$\frac{dr_1}{dx} = \frac{x + a + y\frac{dy}{dx}}{r_1}, \qquad \frac{dr_2}{dx} = \frac{x - a + y\frac{dy}{dx}}{r_2}.$$

Finally, after a few of transformations we get

$$\left(\frac{x-a}{r_2^3} - \frac{x-a}{r_1^3}\right)\frac{dy}{dx} + \left(\frac{1}{r_2^3} - \frac{1}{r_1^3}\right)y = 0.$$

□

Example 2.4. How many solutions of the equation $(x-1)\frac{dy}{dx} + y = 0$ defines the relation

$$y(x-1) = C,$$

for each fixed $C \in \mathbb{R}$. Find the solutions associated with the initial conditions $y(0) = 0$, $y(0) = -1$, $y(2) = 1$. Define intervals of solution existence as well as the corresponding integral and phase curves.

First we verify that $\varphi(x) = \frac{C}{x-1}$ satisfies the given differential equation. We have $\varphi_1(x) = \frac{C}{x-1}$ with $x \in (C, +\infty)$ and $\varphi_2(x) = \frac{C}{x-1}$ with $x \in (1, +\infty)$.

The initial condition $y(0) = 0$ is satisfied by the solution $y = 0$. Its integral curve corresponds to the axis of abscissa, whereas its phase corresponds to a projection of the integral curve into the axis of ordinates, i.e. the point $y = 0$.

In the case of $y(0) = -1$ we find that $C = 1$. It means that the integral curve of this solution corresponds the hyperbola branch $y(x-1) = 1$ for $x \in (-\infty, 1)$. The phase curve of this solution is the ray $y < 0$.

Finally, in the case $y(2) = 1$ we obtain $C = 1$. Integral curve of the solution $y = \frac{1}{1-x}$ is the hyperbola $y(x-1) = 1$ branch, where $x \in (1, +\infty)$ phase curve is the ray $y > 0$. □

2.2 Separable Equation

The first-order differential equation of the form

$$\frac{dy}{dx} = f(x)g(y) \tag{2.7}$$

is called a *separable differential equation*.

If $g(C_0) = 0$ in the point $y = C_0$, then the function $y = C_0$ is the solution to Eq. (2.7). If $g(y) \neq 0$, then the following relation is obtained

$$\int \frac{dy}{g(y)} - \int f(x)dx = C. \tag{2.8}$$

Theorem 2.1. *Let the function $f(x)$ and $g(x)$ are continuously differentiable in the vicinity of points $x = x_0$, $y = y_0$ respectively, where $g(y_0) \neq 0$. Therefore, there is a unique solution $y = \phi(x)$ of Eq. (2.7) with the attached initial condition $\phi(x_0) = y_0$ in the vicinity of the point $x = x_0$, satisfying the relationship*

$$\int_{y_0}^{\phi(x)} \frac{dy}{g(y)} = \int_{x_0}^{x} f(x)dx.$$

If we have the equation

$$\frac{dy}{dx} = f(ax + by + c), \tag{2.9}$$

then introducing a new variable

$$z = ax + by + c, \tag{2.10}$$

we get

$$\frac{dz}{dx} = bf(z) + a, \tag{2.11}$$

i.e. the problem is reduced to Eq. (2.7).

One may use the following physical interpretation of the differential equation

$$\frac{dy}{dx} = f(y). \tag{2.12}$$

Let us attach to each point y a vector of the length $|f(y)|$, which direction is defined by the axis $0y$ providing that $f(y) > 0$. Therefore, a set of all vectors defines a vector field. The points $f(y) = 0$ are called *singular points* of the vector field (or *its equilibrium positions* in the case when we deal with time). Having drawn the vector field of the given Eq. (2.12) one may draw schematically the integral curves.

Example 2.5. Find a solution of the following differential equation

$$x(1 + y^2) + y(1 + x^2)\frac{dy}{dx} = 0.$$

We transform the studied equation to the form

$$\int \frac{xdx}{1+x^2} + \int \frac{ydy}{1+y^2} = 2\ln C$$

and hence after integration we get

$$\ln(1+x^2) + \ln(1+y^2) = \ln C,$$

which means that

$$(1+x^2)(1+y^2) = C.$$

□

Example 2.6. Solve the following ODE

$$\frac{dy}{dx} + y = 2x + 1.$$

In order to transform the given ODE into that of separable variables we introduce the following new variable

$$y - 2x - 1 = z,$$

and hence

$$\frac{dz}{dx} + z + 2 = 0.$$

Separating variables and integrating we get

$$\int \frac{dz}{z+2} + \int dx = 0,$$

which means that

$$\ln|z+2| + x = \ln C_0, \quad |z+2| = C_0 e^{-x}, \quad C_0 > 0.$$

Observe that $z = -2$ satisfies the studied equation directly, and therefore, all its solutions are given by the following formula

$$z = -2 + Ce^{-x}, \quad C \in \mathbb{R},$$

and finally we get

$$y = 2x - 1 + Ce^{-x}.$$

□

In what follows we proceed with a few examples of real-world applications.

Example 2.7. A particle of mass m is subjected to action of a constant force, and it moves with the constant acceleration a. The viscous damping of the surrounding medium is c. Find the particle velocity providing that $v(0) = 0$.

The second Newton law gives

$$\frac{dv(t)}{dt} = \frac{ma - cv(t)}{m},$$

or equivalently

$$\frac{dv}{dt} = -\frac{c}{m}v + a.$$

The trivial (time independent solution) is

$$v(t) = \frac{m}{c}a,$$

and hence all solutions are given by the formula

$$v(t) = \frac{m}{c}a + Ce^{-\frac{c}{m}t}$$

The initial condition allows to find $C = -\frac{m}{c}a$, and finally

$$v(t) = \frac{m}{c}a\left(1 - e^{-\frac{c}{m}t}\right),$$

which means also that

$$\lim_{t\to\infty} v(t) = \frac{m}{c}a.$$

□

Example 2.8. A meteorite of mass M starts to move from its rest position into the Earth centre linearly from the height h (Fig. 2.1). Determine the meteorite velocity, when it touches the Earth surface assuming the Earth radius R.

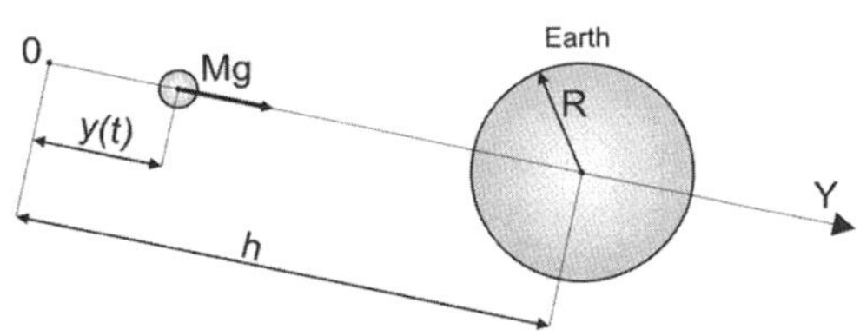

Fig. 2.1 Meteorite movement towards Earth centre

We denote by $y = y(t)$ the meteorite distance from its movement beginning point $y(0) = 0$, and by $h - y(t)$ we denote the meteorite distance from the Earth centre in time instant t. The meteorite is subjected to action of two forces: Ma and Mg. Owing to the Newton principle we have

$$\frac{Ma}{R^2} = \frac{Mg}{(h-y)^2},$$

and hence

$$a = \frac{gR^2}{(h-y)^2}.$$

Therefore,

$$a = \frac{dv}{dt} = \frac{dv}{dy} \cdot \frac{dy}{dt} = \frac{dv}{dy}v,$$

and the following governing ODE is obtained

$$v\frac{dv}{dy} = \frac{gR^2}{(h-y)^2},$$

or equivalently

$$\frac{1}{2}\frac{d(v)^2}{dy} = \frac{gR^2}{(h-y)^2}.$$

Integration of the obtained equation yields

$$v^2 = \frac{2gR^2}{h-y} + C.$$

Taking into account $y(0) = 0$, we get $C = -\frac{2gR^2}{h}$, and finally

$$v^2 = \frac{2gR^2 y}{h(h-y)}.$$

On the Earth surface $y = h - R$, and we get

$$v = \sqrt{2gR\left(1 - \frac{R}{h}\right)}$$

Taking into account that $h \to \infty$, the last formula yields

$$v = \sqrt{2gR}.$$

□

Example 2.9. Two substances A and B undergo a chemical reaction yielding a substance C. We assume amount of the C substance by $y(t)$ in the time instant t after the reaction, and we denote by α and β the amount of substance A and B, in the beginning of reaction, respectively. Find $\frac{dy}{dt}$ assuming that the reaction velocity is proportional to the product of reacting masses.

The governing equation is

$$\frac{dy}{dt} = p(\alpha - y)(\beta - y),\ p > 0,$$

and p is the proportionality coefficient. Separation of the variables yields

$$\frac{dy}{y-\alpha} - \frac{dy}{y-\beta} = -p(\beta - \alpha)dt.$$

After integration one gets

$$\frac{y-\alpha}{y-\beta} = Ce^{-p(\beta-\alpha)t}.$$

Taking into account the initial condition $y(0) = 0$ we obtain the constant $C = \alpha/\beta$, i.e.

$$\frac{y-\alpha}{y-\beta} = \frac{\alpha}{\beta}e^{-p(\beta-\alpha)t},$$

or equivalently

$$y(t) = \alpha\beta\frac{1 - e^{-p(\beta-\alpha)t}}{\beta - \alpha e^{-p(\beta-\alpha)t}}.$$

Observe that for $\beta > \alpha$ we have

$$\lim_{t\to\infty} y(t) = \alpha,$$

whereas for $\beta < \alpha$ we obtain

$$\lim_{t\to\infty} y(t) = \lim \alpha\beta\frac{e^{p(\beta-\alpha)t} - 1}{\beta e^{p(\beta-\alpha)t} - \alpha} = \beta.$$

In the case when $\alpha = \beta$ the governing equation is

$$\frac{dy}{dt} = p(\alpha - y)^2.$$

Separation of the variables of this equation and the integration allows to find the following dependence

$$\frac{1}{\alpha - y} = pt + C.$$

Since $y(0) = 0$, therefore $C = 1/\alpha$. In this case the reaction B governed by the equation

$$y(t) = \alpha \left(1 - \frac{1}{1 + \alpha pt}\right),$$

which for $t \to \infty$ yield

$$\lim_{t \to \infty} y(t) = \alpha.$$

2.3 Homogenous Equations

A function $F(x, y)$ is called homogenous of order k, if for all $\sigma > 0$ the following property holds [208]

$$F(\sigma x, \sigma y) = \sigma^k F(x, y) \tag{2.13}$$

For instance the functions

$$\frac{x + y}{x - y}, \quad \frac{x^2 + xy}{y - x}, \quad x^2 + y^2 + 2xy \tag{2.14}$$

are homogenous of order $k = 0, 1, 2,$, respectively.

A differential equation

$$\frac{dy}{dx} = F(x, y) \tag{2.15}$$

is called *homogenous*, if the function $F(x, y)$ is of order zero.

Equation

$$F_1(x, y)dx + F_2(x, y)dy = 0 \tag{2.16}$$

is called homogeneous, if the function F_1, F_2 are homogeneous of the same order.

In the case of a homogeneous equation the introduction of a new variable $y = zx$ allows to get en equation with separable variables. One may use also polar coordinates (ϱ, φ) and by substitution $x = \varrho \cos \varphi$, $y = \varrho \sin \varphi$ again an equation with separable variables is obtained.

It should be mentioned that the equation

$$\frac{dy}{dx} = F\left(\frac{a_1x + b_1y + c_1}{a_2x + b_2y + c_2}\right) \tag{2.17}$$

can also be transformed to a homogeneous equation through the following linear transformation

$$x = x_0 + X,\ y = y_0 + Y, \tag{2.18}$$

where (x_0, y_0) is the point of intersection of straight lines $a_1x + b_1y + c_1 = 0$ and $a_2x + b_2y + c_2 = 0$. If the lines do not intersect then $a_1/b_1 = a_2/b_2$, and in this case Eq. (2.17) is transformed to that with separable variables using

$$a_1x + b_2y + c_1 = X. \tag{2.19}$$

The function $G(x, y)$ is called quasi-homogenous of order k, if for certain α and β the following relation holds

$$G(\sigma^\alpha x, \sigma^\beta y) = \sigma^k G(x, y), \tag{2.20}$$

for all $k > 0$.

Exponents α, β are called *weights*. We say that $x(y)$ has weight $\alpha(\beta)$, and for instance $7x^2y^5$ has the weight $2\alpha + 5\beta$.

Differential equation (2.15) is called quasi-homogeneous if the associated function $F(x, y)$ is quasi-homogeneous with weights α and β of order $\beta - \alpha$, i.e. $F(\sigma^\alpha x, \sigma^\beta y) = \sigma^{\beta-\alpha} F(x, y)$.

A quasi-homogeneous differential equation can be reduced to a homogeneous one. However, in many practical cases one may use the direct variables change $y = zx^{\frac{\beta}{\alpha}}$ allowing to get an equation with separable variables.

Example 2.10. Find a solution of the following ODE

$$\frac{dy}{dx} = \frac{xy + y^2e^{-\frac{x}{y}}}{x^2}.$$

We introduce the new variable $y = zx$, and obtain

$$x\frac{dz}{dx} + z = z + z^2e^{-\frac{1}{z}},$$

or equivalently

$$\frac{e^{\frac{1}{z}}}{z^2}dz = \frac{dx}{x}.$$

Integration of the last equation yields

$$-e^{\frac{1}{z}} = \ln|x| - C,$$

or equivalently

$$e^{\frac{x}{y}} + \ln|x| = C.$$

□

Example 2.11. Solve the following equation

$$\frac{dy}{dx} = 2\left(\frac{y+1}{x+y-2}\right)^2.$$

We introduce the following variables

$$y+1 = Y,\ x-3 = X,$$

and we get

$$\frac{dY}{dX} = 2\frac{Y^2}{(X+Y)^2}.$$

Now we introduce the following new variable

$$Y = uX,$$

and the following ODE is obtained

$$X\frac{du}{dX} + u = \frac{2u^2}{(1+u)^2},$$

or equivalently

$$\ln|u| + 2\arctan u + \ln|X| = \ln C,$$

which means that

$$uX = C\exp(-2\arctan u).$$

In the original variable the solution is

$$(y+1)\exp\left(2\arctan\frac{y+1}{x-3}\right) = C.$$

□

Example 2.12. Prove that integral curves of the equation

$$[2x(x^2-axy+y^2)-y^2\sqrt{x^2+y^2}]dx+y[2(x^2-axy+y^2)+x\sqrt{x^2+y^2}]dy=0$$

are closed curves surrounding the coordinates origin for $|a|<2$.

Since the studied equation is homogenous, then we introduce polar coordinates to get

$$\varrho^3[2(1-a\sin\varphi\cos\varphi)\cos\varphi-\sin^2\varphi](\cos\varphi d\varrho-\varrho\sin\varphi d\varphi)$$
$$+\varrho^3[2\sin\varphi(1-a\sin\varphi\cos\varphi)+\cos^2\varphi](\sin\varphi d\varrho+\varrho\cos\varphi d\varphi)=0$$

or equivalently

$$2(1-a\sin\varphi\cos\varphi)d\varrho+\varrho\sin\varphi d\varphi=0.$$

Separating the variables we obtain

$$\frac{d\varrho}{\varrho}+\frac{\sin\varphi}{2-a\sin 2\varphi}d\varphi=0,$$

and after integration we get

$$\ln\varrho+\int_0^{\varphi}\frac{\sin u du}{2-a\sin 2u}=\ln\varrho_0,\quad \varrho_0=\varrho(0),$$

or equivalently

$$\varrho=\varrho_0\exp\left(\int_0^{\varphi}\frac{\sin u}{2-a\sin 2u}du\right).$$

If we prove that the function $\int_0^{\varphi}\frac{\sin u du}{2-a\sin 2u}$ is periodic regarding φ with the period 2π, then $\varrho=\varrho(\varphi)$ for arbitrary $\varrho_0>0$ is the 2π periodic function and its integral curve is closed. We have

$$\int_0^{\varphi+2\pi}\frac{\sin u du}{2-a\sin 2u}=\int_0^{2\pi}\frac{\sin u du}{2-a\sin 2u}+\int_{2\pi}^{\varphi+2\pi}\frac{\sin u du}{2-a\sin 2u}$$
$$=\int_0^{\pi}\frac{\sin u du}{2-a\sin 2u}-\int_{\pi}^{2\pi}\frac{\sin u du}{2-a\sin 2u}+\int_0^{\varphi}\frac{\sin(2\pi+u)du}{2-a\sin 2(2\pi+u)}$$

$$= \int_0^{\pi} \frac{\sin du}{2 - a\sin 2u} - \int_0^{\pi} \frac{\sin u du}{2 - a\sin 2u} + \int_0^{\varphi} \frac{\sin u du}{2 - a\sin 2u} = \int_0^{\varphi} \frac{\sin u du}{2 - a\sin 2u}$$

which proves that $\varrho(\varphi) = \varrho(\varphi + 2\pi)$. □

Example 2.13. Solve the following differential equation

$$\frac{dy}{dx} = \frac{4x^6 - y^4}{2x^4 y}.$$

Let $x(y)$ assign the weight $\alpha(\beta)$. Then

$$F(x, y) = \frac{4\sigma^{6\alpha} x^6 - \sigma^{4\beta} y^4}{2\sigma^{4\alpha+\beta} x^4 y} = \sigma^{\beta-\alpha} \frac{4x^6 - y^4}{2x^4 y}.$$

This equation is satisfied when

$$6\alpha - 4\alpha - \beta = 4\beta - 4\alpha - \beta = \beta - \alpha$$

which means that $\beta/\alpha = 3/2$. Therefore, it has been proved that the studied equation is homogeneous. In order to separate its values, the following variable is introduced $y = zx^{\frac{3}{2}}$. We obtain

$$\frac{dz}{dx} x^{\frac{3}{2}} + \frac{3}{2} x^{\frac{1}{2}} z = \frac{4x^6 - z^4 x^6}{2x^4 z x^{\frac{3}{2}}},$$

and multiplying both sides by $x^{-\frac{1}{2}}$ we get

$$x\frac{dz}{dx} + \frac{3}{2} z = \frac{4 - z^4}{2z}$$

or equivalently

$$\frac{2z dz}{(z^2+4)(z^2-1)} + \frac{dx}{x} = 0, \quad z \neq \pm 1.$$

Direct integration yields

$$\ln \frac{|z^2 - 1|}{z^2 + 4} + 5\ln|x| = \ln C. \tag{$*$}$$

In order to verify the obtained result we use the following differentiation formulas:

$$(\ln|y|)' = \frac{y'}{y}, \quad y = \frac{u}{v}, \quad u = z^2 - 1, \quad v = z^2 + 4.$$

Since

$$y' = \left(\frac{u}{v}\right)' = \frac{vu' - uv'}{v} = \frac{2z(z^2+4) - 2z(z^2-1)}{(z^2+4)} = \frac{10z}{(z^2+4)^2},$$

hence

$$\frac{y'}{y} = \frac{10z}{(z^2+4)(z^2-1)}.$$

Full differentiation of $(*)$ yields

$$\frac{10z\,dz}{(z^2+4)(z^2-1)} + \frac{5}{x}dx = 0.$$

Formula $(*)$ yields

$$\frac{z^2-1}{z^2+4}x^5 = C.$$

Since $z^2 = y^2/x^3$, therefore

$$\frac{y^2-x^3}{y^2+4x^3}x^5 = C.$$

2.4 Linear Equations

Linear first-order ODE has the following form

$$\frac{dy}{dx} + a(x)y = f(x). \tag{2.21}$$

There exist three different methods yielding a solution of Eq. (2.21)

(i) *The Lagrange method.* This method is based on a constant variation. We consider first a homogeneous equation associated with (2.21) of the form

$$\frac{dy}{dx} + a(x)y = 0, \tag{2.22}$$

and its solution is

$$y = C\exp\left[-\int a(x)dx\right]. \tag{2.23}$$

We are looking for solution to Eq. (2.21) by variation of the constant $C = C(x)$, namely

$$y = C(x)\exp\left[-\int a(x)dx\right]. \tag{2.24}$$

Substituting (2.24) to (2.21) we obtain

$$\frac{dC(x)}{dx} = f(x)\exp\left[\int a(x)dx\right], \tag{2.25}$$

and hence

$$C(x) = C + \int\left[f(x)\exp\left[\int a(x)dx\right]\right]dx, \tag{2.26}$$

where C is the arbitrary constant.

Finally, substitution of (2.26) into (2.24) yields

$$y = \exp\left[-\int a(x)dx\right]\left\{C + \int\left[f(x)\exp\left(\int a(x)dx\right)\right]\right\}. \tag{2.27}$$

Any solution passing through the point (x_0, y_0) can be written in the following form

$$y = \exp\left[-\int_{x_0}^{x} a(z)dz\right]\left\{y_0 + \int_{x_0}^{x} f(u)\left[\exp\left(\int_{x_0}^{u} a(x)dx\right)du\right]\right\}. \tag{2.28}$$

(ii) *The Bernoulli method.* We are looking for a solution of (2.21) in the following form

$$y = u(x)v(x). \tag{2.29}$$

Substitution of (2.29) to (2.21) gives

$$\frac{du}{dx}v + u\frac{dv}{dx} + a(x)uv = f(x). \tag{2.30}$$

If we take $u(x)$ as the solution of equation

$$\frac{du}{dx} + a(x)u = 0, \tag{2.31}$$

then

$$u(x) = \exp\left[-\int a(x)dx\right]. \tag{2.32}$$

Substituting (2.32) into (2.30) gives

$$\exp\left[-\int a(x)dx\right]\frac{dv}{dx} = f(x), \tag{2.33}$$

and therefore

$$v(x) = C + \int f(x)\exp\left[\int a(x)dx\right]dx, \tag{2.34}$$

where C is a constant.

(iii) *The method of an integrating multiplier.* We multiply both parts of Eq. (2.21) by $\exp\left(\int a(x)dx\right)$, and we get

$$\frac{d}{dx}\left[y\exp\left(\int a(x)dx\right)\right] = f(x)\exp\left(\int a(x)dx\right) \tag{2.35}$$

or equivalently

$$y = \exp\left(-\int a(x)dx\right)\left[C + \int f(x)\exp\left(\int a(x)dx\right)dx\right] \tag{2.36}$$

Equation of the form

$$A(y) + [B(y)x - C(y)]\frac{dy}{dx} = 0 \tag{2.37}$$

can be transformed to the form (2.21). We multiply both sides by $\frac{1}{A}\frac{dx}{dy}$ and we get

$$\frac{dx}{dy} + \alpha(y)x = \beta(y) \tag{2.38}$$

where

$$\alpha(y) = \frac{B}{A}, \quad \beta(y) = \frac{C}{A}. \tag{2.39}$$

It should be emphasized that equations of the form

$$F'(y)\frac{dy}{dx} + F(y)a(x) = b(x), \tag{2.40}$$

where $'$ denotes $\frac{d}{dy}$ can be transformed to the linear equation by introduction of the relation $u = f(y)$.

A particular role in theory of first-order differential equations play *the Bernoulli and Riccati equations*. The equation

$$\frac{dy}{dx} + a(x)y = b(x)y^n, \; n \neq 0, 1 \tag{2.41}$$

is called *the Bernoulli equation*. It is transformed to the following form

$$y^{-n}\frac{dy}{dx} + a(x)y^{1-n} = b(x), \; y \neq 0, \tag{2.42}$$

and it is reduced to a linear equation via the variable change $u = y^{1-n}$. This approach will be illustrated through examples. One may also apply here *the Bernoulli method*.

The equation

$$\frac{dy}{dx} + a(x)y + b(x)y^2 = C(x) \tag{2.43}$$

is called a *Riccati equation*. In general it cannot be solved in quadratures. However, if one of its particular solutions is known, say $y_1(x)$ then the transformation $y = y_1 + u$ allows reduction of the problem to that of finding solution to the Bernoulli equation.

Example 2.14. A current in the electrical network with the resistance R, induction L and excitation voltage $u(t) = u_0 \sin \omega t$ is governed by the following equation

$$L\frac{di}{dt} + Ri = u_0 \sin \omega t, \quad i(0) = 0.$$

Find $i = i(t)$.

We have

$$\frac{di}{dt} + \alpha i = \beta \sin \omega t,$$

where $\alpha = \frac{R}{L}$, $\beta = \frac{u_0}{L}$. We apply here *the Bernoulli method*, i.e. we assume

$$i(t) = u(t)v(t).$$

Substitution of $i(t)$ into the governing equation yields

$$\frac{du}{dt}v + u\frac{dv}{dt} + \alpha uv = \beta \sin \omega t. \tag{$*$}$$

We consider a solution of the homogeneous equation

$$\frac{du}{dt} + \alpha u = 0$$

of the form

$$u = \exp(-\alpha t).$$

We substitute it to $(*)$ and we obtain $\frac{dv}{dt} = \beta e^{\alpha t} \sin \omega t$, what means that

$$v(t) = \beta \left[\int e^{\alpha t} \sin \omega t dt + C \right].$$

We successively compute

$$\begin{aligned} V(t) = \int e^{\alpha t} \sin \omega t dt &= -e^{\alpha t} \frac{1}{\omega} \cos \omega t + \frac{\alpha}{\omega} \int e^{\alpha t} \cos \omega t dt \\ &= -\frac{1}{\omega} e^{\alpha t} \cos \omega t + \frac{\alpha}{\omega^2} e^{\alpha t} \sin \omega t - \frac{\alpha^2}{\omega^2} \int e^{\alpha t} \sin \omega t dt \\ &= -\frac{1}{\omega} e^{\alpha t} \cos \omega t + \frac{\alpha}{\omega^2} e^{\alpha t} \sin \omega t - \frac{\alpha^2}{\omega^2} V(t), \end{aligned}$$

and therefore

$$V(t) = \frac{e^{\alpha t}(-\omega \cos \omega t + \alpha \sin \omega t)}{\omega^2 + \alpha^2}.$$

Finally, we find

$$v(t) = \beta \left[\frac{e^{\alpha t}(-\omega \cos \omega t + \alpha \sin \omega t)}{\omega^2 + \alpha^2} + C \right],$$

and

$$i(t) = u(t)v(t) = \beta \left(\frac{-\omega \cos \omega t + \alpha \sin \omega t}{\omega^2 + \alpha^2} + Ce^{-\alpha t} \right).$$

Since $i(0) = 0$, $C = \frac{\omega}{\omega^2 + \alpha^2}$, and therefore

$$i(t) = \frac{\beta}{\omega^2 + \alpha^2}(-\omega \cos \omega t + \alpha \sin \omega t + \omega e^{-\alpha t}).$$

Observe that

$$\lim_{t \to \infty} i(t) = \frac{u_0}{L(\omega^2 + \alpha^2)}(-\omega \cos \omega t + \alpha \sin \omega t) = \frac{u_0}{\sqrt{(L\omega)^2 + R^2}} \sin(\omega t - \varphi),$$

where $\tan\varphi = \frac{\omega}{\alpha}$ denotes the initial current phase. □

Example 2.15. Show that equation

$$\frac{dy}{dx} + \alpha y = f(x), \quad \alpha > 0,$$

possesses only one bounded solution assuming that $f(x)$ is bounded for all $x \in \mathbb{R}$. Find this solution, and show that if $f(x + x_0) = f(x)$, then $y(x) = y(x + x_0)$, where x_0 is a period.

First we find a solution to the homogeneous equation

$$\frac{dy}{dx} + \alpha y = 0.$$

After variables separation we get

$$\frac{dy}{y} = -\alpha dx,$$

and hence

$$\ln |y| + \alpha x = \ln C,$$

which means that

$$y = Ce^{-\alpha x}$$

assuming that $y \neq 0$.

We apply here *the Lagrange's method*. Namely, we have

$$y(x) = C(x)e^{-\alpha x},$$

and substitution of $y(x)$ into the governing equation gives

$$\frac{dC}{dx} = e^{\alpha x} f(x).$$

It means that

$$C(x) = C(x_0) + \int_{x_0}^{x} e^{\alpha z} f(z) dz.$$

The sought solution has the following form

$$y(x) = C(x_0)e^{-\alpha x} + \int_{x_0}^{x} e^{-\alpha(x-z)} f(z) dz. \qquad (*)$$

Assuming $y(x_0) = y_0$ we obtain

$$y_0 = C(x_0)e^{-\alpha x_0},$$

or equivalently

$$C(x_0) = y_0 e^{\alpha x_0}.$$

Therefore, solution $(*)$ takes the following form

$$y(x) = e^{-\alpha(x-x_0)} y_0 + \int_{x_0}^{x} e^{-\alpha(x-z)} f(z)dz.$$

We multiply both sides of the last equation by $e^{\alpha(x-x_0)}$ to get

$$e^{\alpha(x-x_0)} y(x) = y_0 + \int_{x_0}^{x} e^{\alpha(z-x_0)} f(z)dz.$$

We consider the case $x \to -\infty$ (the case of $x \to +\infty$ can be studied in the similar way). We have

$$\lim_{x\to-\infty} e^{\alpha(x-x_0)} y(x) = y_0 + \int_{x_0}^{-\infty} e^{\alpha(z-x_0)} f(z)dz,$$

and hence

$$y_0 = \int_{-\infty}^{x_0} e^{\alpha(z-x_0)} f(z)dz,$$

because for a bounded solution $\lim_{x\to-\infty} e^{\alpha(x-x_0)} y(x) = 0$. It means that

$$\lim_{x\to-\infty} y(x) = Y(x) = \int_{-\infty}^{x} \exp(-\alpha(x-z)) f(z)dz \qquad (**)$$

is bounded, assuming that $f(z)$ is bounded.

In what follows we show that $Y(x)$ is the only bounded solution of the studied equation. Let us assume that there exists one more bounded solution denoted by $Y_*(x)$. It means that the difference

$$\Delta Y = Y(x) - Y_*(x)$$

is bounded. We also have

$$\frac{dY}{dx} + \alpha Y = f(x),$$
$$\frac{dY_*}{dx} + \alpha Y_* = f(x),$$

which means that

$$\frac{d(\Delta Y)}{\Delta Y} + \alpha(\Delta Y) = 0,$$

and hence

$$\Delta Y(x) = Ce^{\alpha x}.$$

Owing to our introduced assumption $\Delta Y(x)$ is bounded for all $x \in \mathbb{R}$, which means that $Ce^{\alpha x}$ must be bounded. This is true only if $C = 0$, which yields $Y(x) = Y_*(x)$.

Let us now show that if $f(x + x_0) = f(x)$ than $Y(x) = Y(x + x_0)$. It follows from $(**)$ that

$$Y(x + x_0) = \int_{-\infty}^{x+x_0} e^{-\alpha(x+x_0-z)} f(z)dz = \int_{-\infty}^{x} e^{-\alpha(x-\tau)} f(\tau + x_0)d\tau$$
$$= \int_{-\infty}^{x} e^{-\alpha(x-\tau)} f(\tau)d\tau = Y(x).$$

□

Example 2.16. Solve the following Bernoulli equation

$$x\frac{dy}{dx} + y = y^2 \ln x.$$

We use the Bernoulli method, and we look for a solution of the form

$$y = u(x)v(x).$$

Substitution of $y(x)$ into the studied equation yields

$$xu\frac{dv}{dx} + xv\frac{du}{dx} + uv = u^2v^2 \ln x \qquad (*)$$

We take

$$x\frac{du}{dx} + u = 0,$$

and hence

$$u = \frac{1}{x}.$$

Substituting this result into (∗) we get

$$x^2 \frac{dv}{dx} = v^2 \ln x,$$

and hence

$$\frac{1}{v^2} dv = \frac{1}{x^2} \ln x dx.$$

Integration of the last obtained equation gives

$$-\frac{1}{v} = \int \frac{\ln x}{x^2} dx = -\frac{\ln x}{x} - \frac{1}{x} - C,$$

which means that

$$v(x) = \frac{x}{1 + Cx + \ln x},$$

and finally

$$y(x) = u(x)v(x) = \frac{1}{1 + Cx + \ln x}.$$

□

Example 2.17. Solve the following Bernoulli equation

$$(1 + x^2)\frac{dy}{dx} - 2xy = 4\sqrt{y(1 + x^2)}\arctan x.$$

Assuming

$$y(x) = u(x)v(x)$$

we get

$$(1 + x^2)\left(\frac{du}{dx} v + u\frac{dv}{dx}\right) - 2xuv = 4\sqrt{uv(1 + x^2)}\arctan x$$

or equivalently

$$(1 + x^2)\frac{du}{dx} v + (1 + x^2)\left(\frac{dv}{dx} - \frac{2x}{1 + x^2} v\right) u = 4\sqrt{uv(1 + x^2)}\arctan x.$$

We take an arbitrary solution to the equation

$$\frac{dv}{dx} - \frac{2x}{1+x^2}v = 0,$$

i.e. for example the following one

$$v(x) = 1 + x^2.$$

Therefore, we get

$$(1+x^2)^2\frac{du}{dx} = 4(1+x^2)\sqrt{u}\arctan x.$$

One of the solution is $u = 0$, and the other solutions are found through the successive transformations

$$\frac{du}{dx} = \frac{4\arctan x}{1+x^2}\sqrt{u},$$

$$\frac{du}{2\sqrt{u}} = \frac{2\arctan x}{1+x^2}dx,$$

$$\sqrt{u} = \arctan^2 x + C.$$

Finally, the solutions are

$$y = 0,$$

$$y = (1+x^2)(\arctan^2 x + C)^2.$$

□

Example 2.18. Solve the following Riccati equation

$$\frac{dy}{dx} + y^2 = \frac{2}{x^2}.$$

Let us look for a particular solution of the form

$$y_1 = \frac{A}{x}.$$

Substituting y_1 into the studied equation yields

$$-\frac{A}{x^2} + \frac{A^2}{x^2} = \frac{2}{x^2}.$$

The second-order algebraic equation yields two roots $A_1 = -1$, $A_2 = 2$. Let us introduce a new variable z of the form

$$y = z - \frac{1}{x},$$

and therefore

$$\frac{dz}{dx} + \frac{1}{x^2} + z^2 - \frac{2z}{x} + \frac{1}{x^2} = \frac{2}{x^2},$$

or equivalently

$$\frac{dz}{dx} - \frac{2}{x}z = -z^2.$$

We multiply both sides of the obtained equation by x^2 to get

$$x\frac{d}{dx}(zx) = 3zx - (zx)^2.$$

We take

$$zx = u,$$

and integrate the following equation

$$x\frac{du}{dx} = u(3-u).$$

Separation of the variables yields

$$\frac{du}{u(3-u)} = \frac{dx}{x}.$$

Since

$$\frac{1}{u(3-u)} = \frac{1}{3u} + \frac{1}{3(3-u)},$$

therefore

$$\frac{1}{3}\int\frac{du}{u} + \frac{1}{3}\int\frac{du}{3-u} = \int\frac{dx}{x},$$

and consequently

$$\frac{1}{3}\left[\ln|u| - \ln|3-u|\right] = \ln|x| + \ln C_1, \qquad C_1 > 0.$$

Finally, we find

$$\ln\left|\frac{u}{3-u}\right| = 3\ln|C_1 x|,$$

or

$$\ln\left|\frac{u}{3-u}\right| = \ln|(C_1 x)^3|.$$

We consider two cases:

(i)

$$\frac{u}{3-u} \geq 0.$$

In this case we have

$$\frac{u}{3-u} = Cx^3,$$

which means that

$$\frac{zx}{3-zx} = Cx^3,$$

and hence

$$z = \frac{3Cx^2}{Cx^3+1}.$$

We finally get

$$y = z - \frac{1}{x} = \frac{2}{x}$$

and

$$y = \frac{3Cx^2}{Cx^3+1} - \frac{1}{x} = \frac{2Cx^3-1}{x(1+Cx^3)}.$$

(ii)

$$\frac{u}{3-u} < 0.$$

In this case we have

$$\frac{u}{u-3} = Cx^3,$$

which means that

$$zx = C(zx - 3)x^3,$$

and hence

$$z = \frac{3Cx^2}{Cx^3 - 1}.$$

We finally obtain

$$y = \frac{2}{x}$$

and

$$y = \frac{2Cx^3 + 1}{x(Cx^3 - 1)}$$

□

2.5 Exact Differential Equations

The differential equation

$$M(x, y)dx + N(x, y)dy = 0 \tag{2.44}$$

is called an *exact differential equation* if its left-hand side is the full differential of a certain function $V(x, y)$ such that

$$dV(x, y) \equiv \frac{\partial V}{\partial x}dx + \frac{\partial V}{\partial y}dy = M(x, y)dx + N(x, y)dy = 0. \tag{2.45}$$

A necessary condition that Eq. (2.44) is exact one follows

$$\frac{\partial M(x, y)}{\partial y} = \frac{\partial N(x, y)}{\partial x}. \tag{2.46}$$

If $V(x, y)$ is known than all solutions of (2.44) satisfy the condition

$$V(x, y) = C, \tag{2.47}$$

where C is an arbitrary constant.

We show how we can find the function $V(x, y)$. Since

$$\frac{\partial V}{\partial x} = M(x, y), \quad \frac{\partial V}{\partial y} = N(x, y), \tag{2.48}$$

then

$$V(x, y) = \int M(x, y)dx = \psi(x, y) + \psi(y). \tag{2.49}$$

We differentiate (2.49) to get

$$\frac{\partial \psi(x, y)}{\partial y} + \frac{\partial \psi(y)}{\partial y} = N(x, y). \tag{2.50}$$

In some cases the general form given by (2.44) can be transformed to an exact differential equation by introduction of a so-called integrating multiplier $m(x, y)$ [208]. In Eq. (2.46) we introduce $m(x, y)$, and we obtain the following exact differential equation

$$\frac{\partial}{\partial y}(mM) = \frac{\partial}{\partial x}(mN), \tag{2.51}$$

which means that m should satisfy the following equation

$$m\left(\frac{\partial M}{\partial y} - \frac{\partial N}{\partial x}\right) = N\frac{\partial m}{\partial x} - M\frac{\partial m}{\partial y}. \tag{2.52}$$

The obtained general form (2.52) can be simplified in the following cases

(i) If $m(x, y) = m(x)$ then

$$\frac{1}{m}\frac{dm}{dx} = \frac{\frac{\partial M}{\partial y} - \frac{\partial N}{\partial x}}{N}. \tag{2.53}$$

(ii) If $m(x, y) = m(y)$ then

$$-\frac{1}{m}\frac{dm}{dx} = \frac{\frac{\partial M}{\partial y} - \frac{\partial N}{\partial x}}{M}. \tag{2.54}$$

(iii) If $m(x, y) = m(r(x, y))$, where $r(x, y)$ is a known function then

$$\frac{1}{m}\frac{dm}{dr} = \frac{\frac{\partial M}{\partial y} - \frac{\partial N}{\partial x}}{N\frac{\partial r}{\partial x} - M\frac{\partial r}{\partial y}}. \tag{2.55}$$

Example 2.19. Solve the differential equation

$$(2xy + 3y^2)dx + (x^2 + 6xy - 3y^2)dy = 0.$$

We have

$$M(x, y) = 2xy + 3y^2,\ N(x, y) = x^2 + 6xy - 3y^2,$$

and hence

$$\frac{\partial M}{\partial y} = \frac{\partial N}{\partial x} = 2x + 6y.$$

It means that the left-hand side of the differential equation is a full differential of a certain function $V(x, y)$.

We have

$$\frac{\partial V}{\partial x} = 2xy + 3y^2, \quad \frac{\partial V}{\partial y} = x^2 + 6xy - 3y^2.$$

First equation of the above yields

$$V(x, y) = x^2 y + 3xy^2 + \psi(y).$$

We differentiate the last equation with respect to y and thus

$$\frac{\partial V}{\partial y} = x^2 + 6xy + \frac{\partial \psi(y)}{\partial y} = x^2 + 6xy - 3y^2.$$

It means that

$$\psi(y) = -y^3 + C.$$

Hence

$$V(x, y) = x^2 y + 3xy^2 - y^3 + C,$$

and a general solution to the studied ODE is defined implicitly by the equation

$$x^2 y + 3xy^2 - y^3 = C.$$

□

Example 2.20. Solve the differential equation

$$2x\left(1 + \sqrt{x^2 - y}\right)dx - \sqrt{x^2 - y}dy = 0.$$

Observe that

$$\frac{\partial}{\partial y}[2x(1 + \sqrt{x^2 - y})] = \frac{\partial}{\partial x}(-\sqrt{x^2 - y}) = -\frac{x}{\sqrt{x^2 - y}},$$

and hence we deal with the exact differential equation. We have

$$\frac{\partial V}{\partial x} = 2x(1 + \sqrt{x^2 - y}), \quad \frac{\partial V}{\partial y} = -\sqrt{x^2 - y}, \qquad (*)$$

and integration of the first equation yields

$$V(x, y) = \int (2x + 2x\sqrt{x^2 - y})dx = x^2 + \frac{2}{3}(x^2 - y)^{\frac{3}{2}} + \psi(y).$$

Substitution of $V(x, y)$ into the second equation of $(*)$ gives

$$\frac{\partial}{\partial y}[x^2 + \frac{2}{3}(x^2 - y)^{\frac{3}{2}} + \psi(y)] = -\sqrt{x^2 - y},$$

or equivalently

$$-\sqrt{x^2 - y} + \frac{d\psi}{dy} = -\sqrt{x^2 - y},$$

which means that

$$\psi(y) = C.$$

Finally, we have

$$V(x, y) = x^2 + \frac{2}{3}(x^2 - y)^{\frac{3}{2}},$$

and a general solution to the studied differential equation is

$$x^2 + \frac{2}{3}(x^2 - y)^{\frac{3}{2}} = C$$

or

$$y = x^2 - [\frac{3}{2}(C - x^2)]^{\frac{2}{3}}.$$

□

Example 2.21. A mirror reflects solar radiation in a way that a light ray coming from a source 0 after the reflection is parallel to a given direction $0X$, which is the rotation axis. Figure 2.2 shows a scheme of the light ray $0A$ coming from the light source 0, and the rectangular coordinates $0XY$. Derive the mirror shape analytically.

Since A belongs to the mirror surface, the marked angles φ before and after reflection are equal, and $n(t)$ denotes a normal (tangent) to the curve being intersection of the mirror and surface $0XY$.

Owing to the reflection principle (the angle of incidence is equal to the reflection angle) $0A = 0B$, and hence

$$\tan\varphi = \frac{AA'}{B0 + 0A'} = \frac{AA'}{\sqrt{(0A')^2 + (A'A)^2} + 0A'},$$

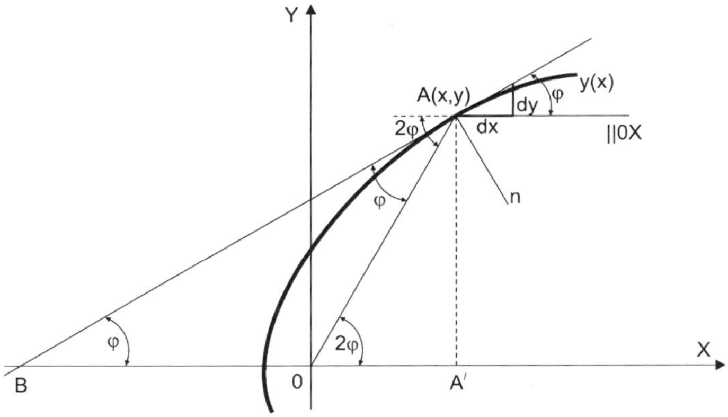

Fig. 2.2 The mirror shape $y(x)$ and light rays

or equivalently

$$\frac{dy}{dx} = \frac{y}{x + \sqrt{x^2 + y^2}}.$$

We may rewrite the latter equation in the following way

$$xdx + ydy = \sqrt{x^2 + y^2}dx,$$

because

$$dy = \frac{y(x - \sqrt{x^2 + y^2})}{(x + \sqrt{x^2 + y^2})(x - \sqrt{x^2 + y^2})}dx = \frac{x - \sqrt{x^2 + y^2}}{-y}dx.$$

Applying the integrating multiplier

$$m(x, y) = \frac{1}{\sqrt{x^2 + y^2}},$$

we get

$$\frac{xdx + ydy}{\sqrt{x^2 + y^2}} - dx = 0$$

or equivalently

$$\frac{d(x^2 + y^2)}{2\sqrt{x^2 + y^2}} - dx = 0.$$

It means that

$$\sqrt{x^2 + y^2} = x + C,$$

which allows to find the mirror surface as a paraboloid that intersects with the surface $0XY$ yielding a parabola governed by the equation

$$y^2 = 2Cx + C^2.$$

□

Example 2.22. Solve the differential equation

$$ydx - (x + x^2 + y^2)dy = 0$$

assuming the integrating multiplier $m = m(r(x, y))$, where $r(x, y) = x^2 + y^2$.

We apply formula (2.55) directly, and we get

$$\frac{1}{m}\frac{dm}{dr} = \frac{1 + 1 + 2x}{-2(x + x^2 + y^2)x - 2y^2} = \frac{2(1 + x)}{-2(1 + x)(x^2 + y^2)} = -\frac{1}{r}.$$

Therefore, the following differential equation is obtained

$$\frac{dm}{m} + \frac{dr}{r} = 0,$$

which yields

$$m(x, y) = \frac{1}{r(x, y)} = \frac{1}{x^2 + y^2}.$$

Now, we multiply by m the studied differential equation to get

$$\frac{ydx}{x^2 + y^2} - \left(\frac{x}{x^2 + y^2} + 1\right)dy = 0,$$

which is an exact differential equation, i.e.

$$\frac{\partial V}{\partial x} = \frac{y}{x^2 + y^2}, \quad \frac{\partial V}{\partial y} = -\left(\frac{x}{x^2 + y^2} + 1\right).$$

Integration of the first equation in the above gives

$$V(x, y) = \int \frac{y}{x^2 + y^2}dx = \arctan\frac{x}{y} + \psi(y),$$

and hence

$$\frac{\partial V}{\partial y} = \frac{\partial}{\partial y}\left(\arctan\frac{x}{y} + \psi(y)\right) = -\frac{x}{x^2+y^2} - 1.$$

It means that

$$-\frac{1}{1+\frac{x^2}{y^2}}\frac{x}{y^2} + \frac{d\psi}{dy} = -\frac{x}{x^2+y^2} - 1,$$

and finally

$$\frac{d\psi}{dy} = -1, \quad \psi(y) = -y + C_1,$$

and

$$V(x, y) = \arctan\frac{x}{y} - y + C_1.$$

We have the following solutions: one given explicitly $y = 0$, and other given implicitly

$$\arctan\frac{x}{y} - y = C.$$

2.6 Implicit Differential Equations Not Solved with Respect to a Derivative

We consider here the differential equation (2.2), which cannot be solved with respect to $\frac{dy}{dt}$, i.e. we cannot reduce the problem to that of Eq. (2.3). It may happen, however, that Eq. (2.2) can be solved with respect to either x or y. In what follows we describe briefly the method of the parameter introduction yielding a solution in the latter case. Let

$$y = f(x, y'), \quad y' \equiv \frac{dy}{dx} = p, \tag{2.56}$$

where p is *the introduced parameter*. The full differential of $y = f(x, y')$ follows

$$pdx = \frac{\partial f}{\partial x}dx + \frac{\partial(x,p)}{\partial p}dp. \tag{2.57}$$

It means that we have got the exact differential equation form (2.44), where

$$M(x, y) = \frac{\partial f}{\partial x}, \quad N(x, y) = \frac{\partial(x,p)}{\partial p}. \tag{2.58}$$

In the previous section supplemented by many examples we have described various methods yielding solutions to Eq. (2.56). Namely, we can take

$$x = \psi(p, c), \quad y = f(x, p), \tag{2.59}$$

where $x = \psi(p, c)$ is the implicit form of solution governed by Eq. (2.56).

Theorem 2.2. *Suppose that the function $f(x, y, y')$ in a neighbourhood of the point (x_0, y_0, y_0'), where y_0' is one of the roots of the equation $f(x_0, y_0, y_0') = 0$, is continuous regarding x and it is continuously differentiable with respect to y, y', and $\frac{\partial f}{\partial y'}(x_0, y_0, y_0') \neq 0$. Then there exists a unique solution $y = \psi'(x)$ of the Cauchy problem $f(x, y, y') = 0$, $y(x_0) = y_0$ defined in a satisfactorily close neighbourhood of the point x_0, where $\psi'(x_0) = y_0'$.*

Recall that the uniqueness of problem of Eq. (2.2) means that the point (x_0, y_0) is a point of the solution uniqueness, i.e. there are no other integral curves of (2.2) which pass through the point (x_0, y_0) and have the same slope in this point. Otherwise, the solution uniqueness is violated.

Theorem 2.2 yields sufficient conditions of a solution existence and uniqueness for Eq. (2.2).

Assuming that the function $f(x, y, y')$ is continuous with respect to x and continuously differentiable with respect to y and y', then a possible set of singular points is defined via the following system of algebraic equations

$$\begin{aligned} f(x, y, y') &= 0, \\ \frac{\partial f}{\partial y'}(x, y, y') &= 0. \end{aligned} \tag{2.60}$$

It is required, while solving Eq. (2.2) to find singular solution, i.e. we remove y' from Eq. (2.60) and we get a so-called discriminant-type curve. Each branch of this curve should be verified if it is a solution to Eq. (2.2). Assuming a positive reply, our next step consists of checking if its points correspond to the solution non-uniqueness.

The method of parameter introduction can be directly applied either to the so-called Claurait equation

$$y = xy' + \psi(y'), \tag{2.61}$$

or to the so-called Lagrange equation

$$y = x\varphi(y') + \psi(y'). \tag{2.62}$$

Example 2.23. Solve the following Claurait equation

$$\sqrt{(y')^2 + 1} + xy' - y = 0.$$

We introduce $p = y'$ to get

$$y = xp + \sqrt{1 + p^2}.$$

Differentiation of the last equation with respect to x yields

$$\frac{dy}{dx} = p + x\frac{dp}{dx} + \frac{p\frac{dp}{dx}}{\sqrt{1 + p^2}},$$

and hence

$$\left(x + \frac{p}{\sqrt{1 + p^2}}\right)\frac{dp}{dx} = 0.$$

It means that either

$$x = -\frac{p}{\sqrt{1 + p^2}}.$$

or

$$p = C.$$

A solution to the problem is as follows:

$$y = Cx + \sqrt{1 + C^2}$$

or equivalently

$$x = -\frac{p}{\sqrt{1 + p^2}},$$
$$y = px + \sqrt{1 + p^2}.$$

□

Example 2.24. Solve the following Lagrange equation

$$y' + y = x(y')^2.$$

It is easily solved with respect to y, i.e.

$$y = x(y')^2 - y'$$

or equivalently

$$y = xp^2 - p,$$

where $p = y'$. Differentiation of this algebraic equation yields

$$p \equiv \frac{dy}{dx} = p^2 + 2px\frac{dp}{dx} - \frac{dp}{dx},$$

or equivalently

$$p(p-1)\frac{dx}{dp} = (1 - 2px),$$

$$\frac{dx}{dp} + \frac{2x}{p-1} = \frac{1}{p(p-1)}.$$

In other words, the problem has been reduced to a linear differential equation with the following solution

$$x = \frac{p - \ln p + C}{(p-1)^2}.$$

□

Example 2.25. Derive an equation governing a family of equipotential curves of the electric field generated by a dipole. Recall that the equipotential curves are orthogonal to force curves of the electric field (see Example 2.3).

As it has been shown previously in Example 2.3, we have

$$\left(\frac{x-a}{r_2^3} - \frac{x+a}{r_1^3}\right)\frac{dy}{dx} - \left(\frac{1}{r_2^3} - \frac{1}{r_1^3}\right) = 0,$$

where

$$r_1^2 = (x+a)^2 + y^2, \quad r_2^2 = (x-a)^2 + y^2. \qquad (*)$$

We may generalize the studied case in Example 2.3. Namely, we began with the algebraic problem governed by the following equation

$$F(x, y, a) = 0,$$

where

$$F(x, y, a) = \frac{x+a}{\sqrt{(x+a)^2 + y^2}} - \frac{x-a}{\sqrt{(x-a)^2 + y^2}} - C.$$

For a given C, we have a family of one parameter curves. In what follows we define another family of the *isogonal curves*, which interset the first family curves with the same angle φ, for $\varphi = \pi/2$ we say that both trajectories (curves) are orthogonal.

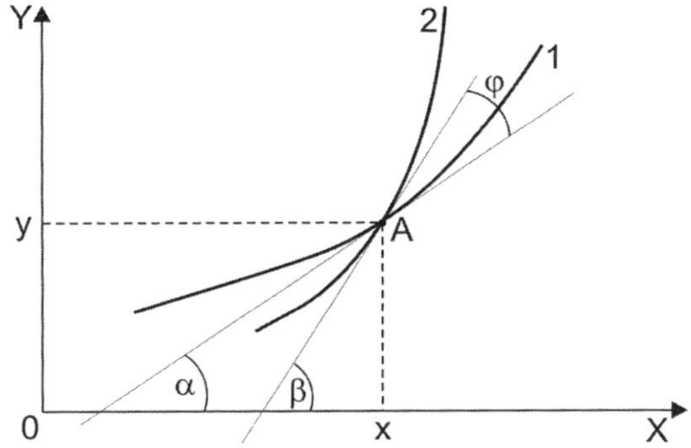

Fig. 2.3 Two curves 1 and 2 intersecting in point A

We differentiate the algebraic equation to get

$$\frac{dF}{dx} \equiv \frac{\partial F(x,y,a)}{\partial x} + \frac{\partial F(x,y,a)}{\partial y}\frac{dy}{dx} = 0.$$

We may also exclude the parameter a using the equation $F = 0$. In our case we have

$$\frac{\partial F}{\partial x} = \left(\frac{1}{r_2^3} - \frac{1}{r_1^3}\right) y, \quad \frac{\partial F}{\partial y} = \frac{x-a}{r_2^3} - \frac{x-a}{r_1^3}.$$

In Fig. 2.3 two curves belonging to both families are shown intersecting in the point $A = A(x, y)$.

The angle between two curves at point A is φ (known), which is given by the formula

$$\pi = \alpha + \varphi + \pi - \beta.$$

Therefore, we get

$$\tan\beta = \tan(\alpha + \varphi) = \frac{\tan\alpha + \tan\varphi}{1 - \tan\alpha\tan\varphi}.$$

We apply the following notation $\tan\alpha = y'$, $\tan\beta = y'_*$, $\tan\varphi = m$, and hence

$$y'_* = \frac{y' + m}{1 - my'}.$$

In a case of orthogonal trajectories we have $\varphi = \pi/2$, and therefore

$$\tan\beta = \frac{\tan\alpha + \tan\frac{\pi}{2}}{1 - \tan\alpha\tan\frac{\pi}{2}}$$

$$= \lim_{\varphi \to \frac{\pi}{2}} \frac{1 + \frac{\tan\alpha}{\tan\varphi}}{\frac{1}{\tan\varphi} - \tan\alpha} = -\frac{1}{\tan\alpha}$$

or equivalently

$$y'_* = -\frac{1}{y'}.$$

The so far consideration implies a simple recipe. In order to find a differential equations of the family of *isogonal trajectories* to the trajectories (curves) governed by equation $F(x, y, a) = 0$, we need to substitute the term $y' = \frac{dy}{dx}$ standing in equation $\partial F/\partial x + \partial F/\partial y y' = 0$, by the term y'_*. In a case for $\varphi = \frac{\pi}{2}$ (orthogonal trajectories) we substitute y' by $-\frac{1}{y'} = -\frac{dx}{dy}$.

In the studied case, using the so far described orthogonality property we obtain the following differential equation

$$\left(\frac{x-a}{r_2^3} - \frac{x+a}{r_1^3}\right)\left(-\frac{1}{\frac{dy}{dx}}\right) - \left(\frac{1}{r_2^3} - \frac{1}{r_1^3}\right) y = 0,$$

or equivalently

$$(x-a)r_1^3 - (x+a)r_2^3 + y(r_1^3 - r_2^3)\frac{dy}{dx} = 0.$$

From $(*)$ we get

$$r_1 dr_1 = (x+a)dx + ydy,$$
$$r_2 dr_2 = (x-a)dx + ydy,$$

therefore the problem is reduced to the following differential equation

$$r_1^3 r_2 dr_2 = r_2^3 r_1 dr_1,$$

which yields the following solution

$$\frac{1}{r_2} - \frac{1}{r_1} = C,$$

and hence

$$\frac{1}{\sqrt{(x-a)^2 + y^2}} - \frac{1}{\sqrt{(x+a)^2 + y^2}} = C.$$

□

Chapter 3
Second-Order ODEs

3.1 Introduction

One may wonder why we introduce this chapter, since second-order systems of differential equations are reducible to the earlier discussed first-order systems of differential equations. The reason has at least two main sources. First of all, they appear in a natural traditional way beginning with the works of D'Alembert, Fermat, Maupertuis, Jean Bernoulli, Hamilton and Lagrange in that period, when mathematics and mechanics have inspired each other very strongly. The second reason is that the second-order differential equations are obtained from Newton's second law or from Lagrange's equations and they have a direct physical meaning. In addition, there exist some direct methods to deal with the second-order differential equations without their reduction to a set of first-order equations [59, 130, 160, 242].

3.2 Linear ODEs

3.2.1 General Approaches

We show that the general form of a second-order linear homogeneous ODE

$$a(t)\frac{d^2y}{dt^2} + b(t)\frac{dy}{dt} + c(t)y = 0, \tag{3.1}$$

where $a(t)$, $b(t)$, $c(t)$ are continuous function on a certain internal $[t_0; t_1]$, can be transformed to so-called self-conjugated equation of the form

$$\frac{d}{dt}\left(p(t)\frac{dy}{dt}\right) + q(t)y = p(t)\frac{d^2y}{dt^2} + \frac{dp}{dt}\frac{dy}{dt} + q(t)y = 0. \tag{3.2}$$

J. Awrejcewicz, *Ordinary Differential Equations and Mechanical Systems*,
DOI 10.1007/978-3-319-07659-1_3

We multiply both sides of (3.1) by $a^{-1}(t)e^{\int b(t)a^{-1}(t)dt}$ to get

$$\frac{d^2y}{dt^2}\exp\left(\int\left(\frac{b}{a}\right)dt\right)+\frac{b}{a}\frac{dy}{dt}\exp\left(\int\left(\frac{b}{a}\right)dt\right)+\frac{c}{a}y\exp\left(\int\left(\frac{b}{a}\right)dt\right)=0. \tag{3.3}$$

Observe that (3.3) can be presented in the form of (3.2), where

$$p(t)=\exp\left(\int\frac{b(t)}{a(t)}dt\right),\quad q(t)=\frac{c(t)}{a(t)}\exp\left(\int\frac{b(t)}{a(t)}dt\right) \tag{3.4}$$

Example 3.1. Solve the equation $2t\frac{d^2y}{dt^2}+\frac{dy}{dt}-2y=0$.

We multiply both sides of this equation by $\frac{1}{2\sqrt{t}}$, $t>0$, to get

$$\sqrt{t}\frac{d^2y}{dt^2}+\frac{1}{2\sqrt{t}}\frac{dy}{dt}-\frac{1}{\sqrt{t}}y=\frac{d}{dt}\left(\sqrt{t}\frac{dy}{dt}\right)-\frac{1}{\sqrt{t}}y=0.$$

Introducing $\tau=2\sqrt{t}$ ($\sqrt{t}d\tau=dt$) we obtain

$$\frac{d^2y}{d\tau^2}-y=0,$$

and finally

$$y=C_1e^{\tau}+C_2e^{-\tau}.$$

□

In what follows we show that if functions $\phi(t)$ and $\psi(t)$ are solutions of the self-conjugated equation (3.2), then there exists a constant C such that this equation possesses the following first integral

$$\phi(t)\frac{d\psi(t)}{dt}-\frac{d\phi}{dt}\psi(t)=\frac{C}{p(t)}. \tag{3.5}$$

Note that the Wronskian

$$W(t)=\begin{vmatrix}\phi(t) & \psi(t)\\ \frac{d\phi}{dt} & \frac{d\psi}{dt}\end{vmatrix}=\phi(t)\frac{d\psi}{dt}-\psi\frac{d\phi}{dt}=\frac{C}{\exp\left(\int\frac{b(t)}{a(t)}dt\right)}$$
$$=C\exp\left(-\int\frac{\frac{dp}{dt}(t)}{p(t)}dt\right)=C\exp\left(-\ln p(t)\right)=\frac{C}{p(t)}. \tag{3.6}$$

On the other hand we have

$$W(t) = W(t_0)\exp\left(-\int_{t_0}^{t}\frac{\frac{dp}{ds}(s)}{p(s)}ds\right) = \frac{W(t_0)p(t_0)}{p(t)}. \tag{3.7}$$

Comparison of (3.6) and (3.7) yields

$$C = W(t_0)p(t_0). \tag{3.8}$$

Because $W(t_0)$ depends on $\phi(t)$ and $\psi(t)$, than the constant C depends also on $\phi(t)$ and $\psi(t)$.

We consider now a non-homogeneous equation corresponding to (3.2) of the form

$$\frac{d}{dt}\left(p(t)\frac{dy}{dt}\right) + q(t)y = F(t). \tag{3.9}$$

Assuming that $p(t) \neq 0$ we introduce the new variable

$$\tau = \tau(t) = \int_{t_0}^{t}\frac{dr}{p(r)}, \tag{3.10}$$

and let the function $t = \gamma(\tau)$, where $\gamma = \tau^{-1}$. Equation (3.9) takes the form

$$\frac{d}{dt}\left(p(t)\frac{dy}{dt}\right) = \frac{d}{p(t)d\tau}\left(p(t)\frac{dy}{p(t)d\tau}\right) = \frac{1}{p(t)}\frac{d^2y}{d\tau^2}, \tag{3.11}$$

where $p(t)d\tau = dt$. Therefore, Eq. (3.9), taking into account (3.11), can be cast to the following form

$$\frac{d^2y}{d\tau^2} + p(t)q(t)y = p(t)F(t) \tag{3.12}$$

or equivalently

$$\frac{d^2y}{d\tau^2} + Q(\tau)y = R(\tau), \tag{3.13}$$

where

$$\begin{aligned} Q(\tau) &= p(\gamma(\tau))q(\gamma(\tau)), \\ R(\tau) &= p(\gamma(\tau))F(\gamma(\tau)). \end{aligned} \tag{3.14}$$

Assuming $a(t) \neq 0$ and adding the non-homogeneous term, Eq. (3.1) can be presented in the following form

$$\frac{d^2y}{dt^2} + b(t)\frac{dy}{dt} + c(t)y = f(t). \tag{3.15}$$

Alternatively, in what follows we show that Eq. (3.15) can be reduced to Eq. (3.13). Namely, introduction of a new variable of the form

$$y = x \exp\left(-\frac{1}{2}\int_{t_0}^{t} b(\tau)d\tau\right) \tag{3.16}$$

to Eq. (3.15) yields

$$\frac{d^2x}{dt^2} + Q(t)x = R(t) \tag{3.17}$$

where we have

$$Q(t) = c(t) - \frac{1}{4}b^2(t) - \frac{1}{2}\frac{db(t)}{dt},$$
$$R(t) = f(t)\exp\left(\frac{1}{2}\int_{t_0}^{t} b(\tau)d\tau\right). \tag{3.18}$$

It should be emphasized that second-order ODEs with time-dependent coefficients only in rare cases can be solved in elementary functions. Let us find solution to the equation

$$\frac{d^2y}{dt^2} + b(t)\frac{dy}{dt} + c(t)y = 0, \tag{3.19}$$

where $b(t) = \sum_{i=0}^{\infty} b_i t^i$, $c(t) = \sum_{i=0}^{\infty} c_i t^i$.

Theorem 3.1. *If functions $b(t)$ and $c(t)$ are analytical for $|t - t_0| < T$ (here $t_0 = 0$), then any solution $y = y(t)$ of Eq. (3.19) can be presented by the series*

$$y(t) = \sum_{i=0}^{\infty} a_i t^i, \tag{3.20}$$

having a limit for $|t - t_0| < T$.

We substitute (3.20) into (3.19) to get

$$\sum_{i=2}^{\infty} i(i-1)a_i t^{i-2} + \sum_{i=0}^{\infty} b_i t^i \sum_{i=1}^{\infty} i a_i t^{i-1} + \sum_{i=0}^{\infty} c_i t^i \sum_{i=0}^{\infty} a_i t^i = 0. \tag{3.21}$$

We compare next terms standing by powers t^0, t^1, t^2, t^3, ..., and we obtain the following recurrent set of unknown coefficients a_0, a_1, a_2, ...

$$
\begin{aligned}
c_0 a_0 + b_0 a_1 + 1 \cdot 2 a_2 &= 0 \quad : t^0 \\
c_1 a_0 + (c_0 + b_1) a_1 + 2 b_0 a_2 + 2 \cdot 3 a_3 &= 0 \quad : t^1 \\
&\dots
\end{aligned}
\tag{3.22}
$$

Although a_0 and a_1 can be arbitrarily taken, but at least one of them should not be equal to zero. They serve as the initial conditions: $y(0) = a_0$, $\frac{dy}{dt}(0) = a_1$. Then we successively compute $a_2, a_3, \dots$

Example 3.2. Find a solution of the equation

$$\frac{d^2 y}{dt^2} + ty = 0.$$

We differentiate two times the solution

$$y = \sum_{i=0}^{\infty} a_i t^i$$

to get

$$\sum_{i=2}^{\infty} i(i-1) a_i t^{i-2} + t \sum_{i=0}^{\infty} a_i t^i = 0.$$

Comparison of terms standing by the same powers of t yields the following algebraic equations

$$
\begin{aligned}
2 \cdot 1 \cdot a_2 &= 0, \\
3 \cdot 2 \cdot a_3 + a_0 &= 0, \\
4 \cdot 3 \cdot a_4 + a_1 &= 0, \\
i(i-1) a_i + a_{i-3} &= 0, \\
&\dots\dots\dots\dots
\end{aligned}
$$

Linear algebraic equations have the following solutions

$$a_2 = 0, \quad a_{i+3} = -\frac{a_i}{(i+2)(i+3)}, \quad i = 0, 1, 2, \dots$$

If we take $a_0 = 1, a_1 = 0$, then we obtain

$$a_{3(m+1)} = -\frac{a_{3m}}{(3m+2)(3m+3)}, \qquad m = 0, 1, 2, \ldots$$

$$a_{3m} = \frac{(-1)^m}{2 \cdot 3 \cdot 5 \cdot 6 \ldots (3m-1) \cdot 3m}, \qquad m = 1, 2, \ldots$$

First solution has the form

$$y_1 = 1 + \sum_{m=1}^{\infty} \frac{(-1)^m t^{3m}}{2 \cdot 3 \cdot 5 \cdot 6 \ldots (3m-1)3m}.$$

Taking $a_0 = 0, a_1 = 1$ yields

$$a_{3m+4} = -\frac{a_{3m+1}}{(3m+3)(3m+4)}, \qquad m = 0, 1, 2, \ldots$$

It means that a second solution is

$$y_2(t) = t + \sum_{m=1}^{\infty} \frac{(-1)^m t^{3m+1}}{3 \cdot 4 \cdot 6 \cdot 7 \ldots 3m \cdot (3m+1)}.$$

All solutions of the investigated equation are governed by the following formula

$$y(t) = C_1 y_1(t) + C_2 y_2(t),$$

where C_1, C_2 are arbitrary constants. □

Example 3.3. Find all solutions to the equation

$$\frac{d^2y}{dt^2} + t\frac{dy}{dt} + y = 0.$$

We are looking for solutions in the following form

$$y = \sum_{i=0}^{\infty} a_i t^i,$$

and substitution of this series to the studied equation yields:

$$\sum_{i=2}^{\infty} a_i i(i-1)t^{i-2} + t\sum_{i=1}^{\infty} a_i i t^{i-1} + \sum_{i=0}^{\infty} a_i t^i = 0.$$

Comparison of terms standing by some powers of t gives

$$(i+1)(i+2)a_{i+2} + (i+1)a_i = 0, \quad i = 0, 1, 2, \ldots$$

We take $a_0 = 1$, $a_1 = 0$ and we obtain

$$a_{2m} = \frac{(-1)^m}{2 \cdot 4 \cdot 6 \ldots (2m)}, \quad m = 1, 2, \ldots, \quad a_{2m+1} = 0,$$

and hence

$$y_1(t) = 1 - \frac{t^2}{2} + \frac{t^4}{2 \cdot 4} - \frac{t^6}{2 \cdot 4 \cdot 6} + \cdots + \frac{(-1)^m t^{2m}}{2 \cdot 4 \cdot 6 \cdot \ldots \cdot (2m)} = e^{-\frac{t^2}{2}}$$

On the other hand, by taking $a_0 = 0$, $a_1 = 1$, we get

$$a_{2m} = 0, \quad a_{2m+1} = \frac{(-1)^m}{1 \cdot 3 \cdot 5 \ldots (2m+1)}, \quad m = 0, 1, 2, \ldots,$$

and therefore

$$y_2(t) = \sum_{m=0}^{\infty} \frac{(-1)^m t^{2m-1}}{1 \cdot 3 \cdot 5 \ldots (2m+1)}.$$

Finally, all solutions are given by

$$y(t) = C_1 y_1(t) + C_2 y_2(t).$$

□

One may deal with the equation

$$t^2 \frac{d^2 y}{dt^2} + t b(t) \frac{dy}{dt} + c(t) y = 0, \tag{3.23}$$

and for analytical $b(t)$, $c(t)$ in the interval $|t| < T$, then for $b(0) \neq 0$, $c(0) \neq 0$ the point $t = 0$ is called the *regular singular point*. If $b(t) = \frac{b_1(t)}{b_0(t)}$, and $c(t) = \frac{c_1(t)}{c_0(t)}$, then the point t, where either $b_0(t) = 0$ or $c_0(t) = 0$, is called a *singular point*. In what follows we show how to find a solution in the neighbourhood of singular point $t = t_0$ in the form of a power series. We take the following series

$$y = (t - t_0)^{\sigma} \sum_{i=0}^{\infty} a_i (t - t_0)^i, \tag{3.24}$$

where σ and a_0, a_1, a_2, a_3 are going to be determined. Substitution of (3.24) into (3.23) yields

$$\begin{aligned}[\sigma(\sigma-1)+b_0\sigma+c_0]a_0+\{[\sigma(\sigma+1)+b_0(\sigma+1)+c_0]a_1+(\sigma b_0+c_0)a_0\}t\\ +\cdots+\{[(\sigma+i)(\sigma+i-1)+\cdots+b_0(\sigma+i)+c_0]a_i+\cdots\\ +(\sigma b_0+c_0)a_0\}t^i+\cdots=0.\end{aligned} \tag{3.25}$$

Comparison of terms standing by the same powers of t allows to get the following recurrent set of nonlinear algebraic equations

$$\begin{gathered}F_0(\sigma)a_0=0,\\ F_0(\sigma+1)a_1+F_1(\sigma)a_0=0,\\ \cdots\\ F_0(\sigma+i)a_i+F_1(\sigma+i-1)a_{i-1}+F_2(\sigma+i-2)a_{i-2}+\cdots+F_i(\sigma)a_0=0,\end{gathered} \tag{3.26}$$

where

$$\begin{aligned}F_0(\sigma)&=\sigma(\sigma-1)+b_0\sigma+c_0,\\ F_m(\sigma)&=\sigma b_m+c_m,\quad m=1,2,\ldots\end{aligned} \tag{3.27}$$

Assuming $a_0\neq 0$ one gets the following *characteristic equation*

$$\sigma(\sigma-1)+b_0\sigma+c_0=0. \tag{3.28}$$

The following two possibilities exist depending on the difference of roots $\sigma_1-\sigma_2$:

(i) If $\sigma_1-\sigma_2\neq i$ (integer), then $f_0(\sigma_1+i)\neq 0$, $f_0(\sigma_2+i)\neq 0$. In this case two non-dependent solutions have the following forms

$$\begin{aligned}y_1(t)&=t^{\sigma_1}\sum_{i=0}^{\infty}a_{1i}t^i,\\ y_2(t)&=t^{\sigma_2}\sum_{i=0}^{\infty}a_{2i}t^i;\end{aligned} \tag{3.29}$$

(ii) If $\sigma_1-\sigma_2=i$ (integer), then only one solution $y_1(t)$ can be constructed. Second solution has the following form

$$y_2(t)=y_1(t)\int\frac{e^{-\int b(t)dt}}{y_1^2(t)}dt. \tag{3.30}$$

It means that the second solution is

$$y_2(t)=Cy_1(t)\ln(t)+t^{\sigma_1}\sum_{i=0}^{\infty}a_it^i, \tag{3.31}$$

where C is a constant, which can be equal to zero.

Example 3.4. Solve the following ODE

$$2t^2\frac{d^2y}{dt^2} + (3t - 2t^2)\frac{dy}{dt} - (t+1)y = 0.$$

Point $t = 0$ is a regular singular point. The corresponding characteristic equation is

$$2\sigma(\sigma - 1) + 3\sigma - 1 = 0,$$

having roots $\sigma = \frac{1}{2}$ and $\sigma_2 = -1$. Let us construct a solution associated with the root σ_1:

$$y_1 = t^{\frac{1}{2}} \sum_{i=0}^{\infty} a_i t^i,\ a_0 \neq 0,\ t > 0.$$

A successive differentiation gives

$$\frac{dy_1}{dt} = \sum_{i=0}^{\infty}\left(i + \frac{1}{2}\right) a_i t^{i-\frac{1}{2}},$$

$$\frac{d^2y_1}{dt^2} = \sum_{i=0}^{\infty}\left(i + \frac{1}{2}\right)\left(i - \frac{1}{2}\right) a_i t^{i-\frac{3}{2}}.$$

Substitution of y_1, $\frac{dy_1}{dt}$, $\frac{d^2y_1}{dt^2}$ to the studied equation yields

$$2t^2\sum_{i=0}^{\infty}\left(i^2 - \frac{1}{4}\right) a_i t^{i-\frac{3}{2}} + t(3-2t)\sum_{i=0}^{\infty}\left(i + \frac{1}{2}\right) a_i t^{i-\frac{1}{2}} - (t+1)\sum_{i=0}^{\infty} a_i t^{i+\frac{1}{2}} = 0,$$

or equivalently

$$\sum_{i=0}^{\infty} i(2i+3)a_i t^i - \sum_{i=0}^{\infty} 2(i+1)a_i t^{i+1} = 0.$$

Comparison of terms standing by the same powers of t gives the following recurrent set of linear algebraic equations

$$i(2i+3)a_i = 2ia_{i-1}, \quad i = 1, 2, \ldots$$

Taking $a_0 = 1$ we get

$$a_i = \frac{2^i}{5 \cdot 7 \cdot 9 \cdot \ldots \cdot (2i+3)}, \quad i = 1, 2, \ldots$$

and hence

$$y_1(t) = \sqrt{t}\left(1 + \sum_{i=1}^{\infty} \frac{(2t)^i}{5 \cdot 7 \cdot 9 \cdot \ldots \cdot (2i+3)}\right).$$

A solution corresponding to the root σ_2 has the form

$$y_2(t) = \frac{1}{t}\sum_{i=0}^{\infty} a_i t^i.$$

Substitution of y_2, $\frac{dy_2}{dt}$, $\frac{d^2y_2}{dt^2}$ to the studied ODE gives

$$2(i-1)(i-2)a_i + 3(i-1)a_i - 2(i-2)a_{i-1} - a_{i-1} - a_i = 0,$$

or equivalently

$$i(2i-3)a_i = (2i-3)a_{i-1}, \quad i = 1, 2, 3, \ldots$$

Taking $a_0 = 1$, we get

$$a_1 = 1, \quad a_2 = \frac{1}{2!}, \quad a_3 = \frac{1}{3!}, \quad \ldots, \quad a_i = \frac{1}{i!}, \quad \ldots,$$

and therefore

$$y_2(t) = \frac{1}{t}\left(1 + t + \frac{t^2}{2!} + \cdots + \frac{t^i}{i!} + \cdots\right) = \frac{e^t}{t}.$$

Finally, the being sought general solution has the following form

$$y = C_1 y_1(t) + C_2 y_2(t),$$

where C_1, C_2 are constants.

□

3.2.2 Hypergeometric (Gauss) Equation

A hypergeometric (or a Gauss) equation has the following form

$$t(t-1)\frac{d^2y}{dt^2} + [-\gamma + (\alpha + \beta + 1)t]\frac{dy}{dt} + \alpha\beta y = 0. \tag{3.32}$$

Observe that points $t = 0$ and $t = 1$ are singular. In the vicinity of $t = 0$ Eq. (3.32) can be presented in the following form

$$\frac{d^2y}{dt^2} + \frac{[\gamma - (\alpha + \beta + 1)t] \sum_{i=0}^{\infty} t^i}{t} \frac{dy}{dt} - \frac{\alpha\beta \sum_{i=0}^{\infty} t^{i+1}}{t^2} y = 0. \tag{3.33}$$

The characteristic equation associated with (3.33) has the form

$$\sigma(\sigma - 1) + \gamma\sigma = 0, \tag{3.34}$$

and its roots are: $\sigma_1 = 0$, $\sigma_2 = 1 - \gamma$. For $\gamma \in N$, $N > 0$ one may construct two linearly independent solutions of (3.33) in the series form being convergent for $|t| < 1$.

We show how to solve Eq. (3.32) when γ is not a non-positive integer number. A solution corresponding to $\sigma_1 = 0$ has the following form

$$y_1(t) = \sum_{i=0}^{\infty} a_i t^i. \tag{3.35}$$

Substituting y_1, $\frac{dy_1}{dt}$, $\frac{d^2y_1}{dt^2}$ to (3.32), and comparing the terms standing by the same powers of t we obtain

$$i(i-1)a_i - (i+1)ia_{i+1} - \gamma(i+1)a_{i+1} + (\alpha + \beta + 1)ia_i + \alpha\beta a_i = 0,$$

$$\begin{aligned} a_{i+1} &= \frac{i(i-1) + i(\alpha + \beta + 1) + \alpha\beta}{(i+1)(i+\gamma)} a_i \\ &= \frac{(\alpha + i)(\beta + i)}{(i+1)(\gamma + i)} a_i, \quad i = 0, 1, 2, \ldots \end{aligned} \tag{3.36}$$

Taking $a_0 = 1$ we get

$$a_1 = \frac{\alpha\beta}{\gamma}, \quad a_2 = \frac{\alpha(\alpha+1)\beta(\beta+1)}{2!\gamma(\gamma+1)}, \quad a_3 = \frac{\alpha(\alpha+1)(\alpha+2)\beta(\beta+1)(\beta+2)}{3!\gamma(\gamma+1)(\gamma+2)},$$
$$\ldots, \quad a_i = \frac{\alpha(\alpha+1)\ldots(\alpha+i-1)\beta(\beta+1)(\beta+2)\ldots(\beta+i-1)}{i!\gamma(\gamma+1)(\gamma+2)\ldots(\gamma+i-1)}. \tag{3.37}$$

Therefore

$$\begin{aligned} y_1(t) &= 1 + \sum_{i=1}^{\infty} \frac{\alpha(\alpha+1)(\alpha+2)\ldots(\alpha+i-1)\beta(\beta+1)(\beta+2)\ldots(\beta+i-1)}{i!\gamma(\gamma+1)(\gamma+2)\ldots(\gamma+i-1)} t^i \\ &= F(\alpha, \beta, \gamma, t). \end{aligned} \tag{3.38}$$

The function $F(\alpha, \beta, \gamma, t)$ is called the hyper-geometric function. The corresponding series on the right-hand side of (3.38) is called the hyper-geometric series and it is convergent for $|t| < 1$.

A second solution is sought in the form

$$y_2(t) = t^{1-\gamma} \sum_{i=0}^{\infty} a_i t^i . \tag{3.39}$$

We introduce a new variable

$$y = t^{1-\gamma} x, \tag{3.40}$$

and hence

$$\begin{aligned} \frac{dy}{dt} &= t^{1-\gamma} \frac{dx}{dt} + (1-\gamma) t^{-\gamma} x, \\ \frac{d^2 y}{dt^2} &= t^{1-\gamma} \frac{d^2 x}{dt^2} + 2(1-\gamma) t^{-\gamma} \frac{dx}{dt} - \gamma(1-\gamma) t^{-\gamma-1} x. \end{aligned} \tag{3.41}$$

Substituting (3.40), (3.41) to (3.32) one gets

$$\begin{aligned} t(t-1)\frac{d^2 x}{dt^2} + \{-(2-\gamma) + [1 + (\alpha+1-\gamma) + (\beta+1-\gamma)]t\} \frac{dx}{dt} \\ +(\alpha+1-\gamma)(\beta+1-\gamma)x = 0. \end{aligned} \tag{3.42}$$

Observe that Eq. (3.42) is also a hypergeometric equation with the parameters $\alpha + 1 - \gamma, \beta + 1 - \gamma, 2 - \gamma$.

Finally, the general solution of Eq. (3.32) has the following form

$$y(t) = C_1 F(\alpha, \beta, \gamma, t) + C_2 t^{1-\gamma} F(\alpha + 1 - \gamma, \beta + 1 - \gamma, 2 - \gamma, t). \tag{3.43}$$

We show also how to solve the hypergeometric equation in the neighbourhood of the singular point $t = 1$. We transform Eq. (3.32) to its equivalent form

$$\tau(\tau - 1)\frac{d^2 y}{d\tau^2} + [-(\alpha + \beta + 1 - \gamma) + (\alpha + \beta + 1)\tau]\frac{dy}{d\tau} + \alpha\beta y = 0 \tag{3.44}$$

by introduction of the independent variable $t = 1 - \tau$, which has the singularity in the point $\tau = 0$ (the case previously studied).

Two linearly independent solutions have the form

$$\begin{aligned} y_1(\tau) &= F(\alpha, \beta, \alpha + \beta + 1 - \gamma, \tau), \\ y_2(\tau) &= \tau^{\gamma-\alpha-\beta} F(\gamma - \beta, \gamma - \alpha, \gamma + 1 - \alpha - \beta, \tau). \end{aligned} \tag{3.45}$$

It should be emphasized that in many cases hypergeometric functions can be expressed by elementary functions for certain particular values of α, β and γ.

For instance, the following observation holds for $|t| < 1$:

(i) $F(1, \beta, \beta, t) = \frac{1}{1-t}$,
(ii) $\ln(1+t) = tF(1, 1, 2, -t)$,
(iii) $F(\alpha, \beta, \alpha, t) = (1-t)^{-\beta}$,
(iv) $\ln\frac{1+t}{1-t} = 2tF\left(\frac{1}{2}, 1, \frac{3}{2}, t^2\right)$.

3.2.3 The Legendre Equation and Legendre Polynomials

The Legendre equation has the following form

$$(1-t^2)\frac{d^2y}{dt^2} - 2t\frac{dy}{dt} + n(n+1)y = 0,\ n \in N, \tag{3.46}$$

and its solutions are expressed by the so-called Legendre polynomials of the form

$$P_n(t) = \frac{1}{2^n n!}\frac{d^n(t^2-1)^n}{dt^n}, \tag{3.47}$$

being known as the Rodrigues representation, and n stands for the equation order.

We show that Legendre polynomials $P_n(t)$, $P_m(t)$ are orthogonal in the interval $(-1, 1)$, i.e. the following relation is satisfied

$$\int_{-1}^{1} P_m(t)P_n(t)dt = 0,\ \text{for } m \neq n. \tag{3.48}$$

Substitution of the Legendre polynomials to the Legendre equation (3.46) yields

$$\begin{aligned}\frac{d}{dt}\left[(1-t^2)\frac{dP_m(t)}{dt}\right] + m(m+1)P_m(t) &= 0,\\ \frac{d}{dt}\left[(1-t^2)\frac{dP_n(t)}{dt}\right] + n(n+1)P_n(t) &= 0.\end{aligned} \tag{3.49}$$

We multiply first (second) equation by $P_n(t)$ ($P_m(t)$) in the interval $(-1, 1)$, and we obtain

$$[m(m+1) - n(n+1)]\int_{-1}^{1} P_m(t)P_n(t)dt = I_1 - I_2, \tag{3.50}$$

where

$$\begin{aligned}I_1 &= \int_{-1}^{1}\left\{P_m(t)\frac{d}{dt}\left[(1-t^2)\frac{dP_n(t)}{dt}\right]\right\}dt,\\ I_2 &= \int_{-1}^{1}\left\{P_n(t)\frac{d}{dt}\left[(1-t^2)\frac{dP_m(t)}{dt}\right]\right\}dt.\end{aligned} \tag{3.51}$$

Applying an integration by parts we get

$$I_1 = (1-t^2)P_m(t)\left.\frac{dP_n(t)}{dt}\right|_{t=-1}^{t=1} - \int_{-1}^{1}(1-t^2)\frac{dP_m(t)}{dt}\frac{dP_n(t)}{dt}dt = -\int_{-1}^{1}(1-t^2)\frac{dP_m(t)}{dt}\frac{dP_n(t)}{dt}dt,$$
$$I_2 = (1-t^2)P_n(t)\left.\frac{dP_m(t)}{dt}\right|_{t=-1}^{t=1} - \int_{-1}^{1}(1-t^2)\frac{dP_n(t)}{dt}\frac{dP_m(t)}{dt}dt = -\int_{-1}^{1}(1-t^2)\frac{dP_n(t)}{dt}\frac{dP_m(t)}{dt}dt. \tag{3.52}$$

Taking into account (3.52) in (3.50) we obtain

$$[m(m+1)-n(n+1)]\int_{-1}^{1} P_m(t)P_n(t)dt = 0, \tag{3.53}$$

and because $m \neq n$ we finally get

$$\int_{-1}^{1} P_m(t)P_n(t)dt = 0. \tag{3.54}$$

One may prove additionally that

$$P_n(-t) = (-1)^n P_n(t). \tag{3.55}$$

Now we are going to show that (3.47) satisfies (3.46). We introduce the following new variable

$$\tau = \frac{1-t}{2}, \tag{3.56}$$

and hence

$$\frac{dy}{dt} = -\frac{1}{2}\frac{dy}{d\tau}, \quad \frac{d^2y}{dt^2} = \frac{1}{4}\frac{d^2y}{d\tau^2}. \tag{3.57}$$

Substituting (3.56), (3.57) into (3.46) we get

$$\frac{1}{4}[1-(1-2\tau)^2]\frac{d^2y}{d\tau^2} + (1-2\tau)\frac{dy}{d\tau} + n(n+1)y = 0, \tag{3.58}$$

or equivalently

$$\tau(\tau-1)\frac{d^2y}{d\tau^2} + (-1+2\tau)\frac{dy}{d\tau} - n(n+1)y = 0. \tag{3.59}$$

Consider a polynomial

$$P_n(t) = a_0 + a_1t + a_2t^2 + \cdots + a_nt^n. \tag{3.60}$$

We say that $P_n(t)$ satisfies (3.46) whenever $P_n(t) = 1$, and then $P_n(t)$ is called a Legendre polynomial. If we make the following choice

$$a_n = \frac{(2n)!}{2^n(n!)^2} = \frac{1 \cdot 3 \cdot 5 \cdot \cdots \cdot (2n-1)}{n!}, \tag{3.61}$$

then $P_n(t) = 1$ is satisfied.

It can be shown (see Example 3.5) that

$$a_{k+2} = -\frac{(n-k)(n+k+1)}{(k+1)(k+2)}, \quad k = 0, 1, 2, \ldots. \tag{3.62}$$

Taking $k = n - 4$, and then $n \to n + 2$, we obtain

$$a_{n-2} = -\frac{(n-1)n}{2^2(2n-1)} a_n = -\frac{(2n-2)!}{2^n(n-1)!(n-2)!}, \tag{3.63}$$

which can be generalized for $n - 2m \geq 0$ to the following formula

$$a_{n-2m} = (-1)^m \frac{(2n-2m)!}{2^n m!(n-m)!(n-2m)!}, \tag{3.64}$$

and the series takes the form

$$\sum_{m=0}^{M} (-1)^m \frac{(2n-2m)!}{2^n m!(n-m)!(n-2m)!}, \tag{3.65}$$

where $M = \frac{n}{2}$ (n is even), $M = \frac{n-1}{2}$ (n is odd).

It can be proved that for n even (odd), any polynomial solution $y(t)$ of (3.46) which has only even (odd) powers of t is a multiple of $P_n(t)$.

In what follows we prove that (3.60) is the solution to the Legendre equations (3.46).

We take

$$R(t) = (t^2 - 1)^n, \tag{3.66}$$

and hence

$$\frac{d}{dt} R = 2nt(t^2 - 1)^{n-1}. \tag{3.67}$$

Multiply both sides of by $t^2 - 1$ we obtain

$$(t^2 - 1)\frac{dR}{dt} = 2nt(t^2 - 1) = 2ntR, \tag{3.68}$$

and hence

$$(t^2 - 1)\frac{dR}{dt} - 2ntR = 0. \tag{3.69}$$

We apply the Leibniz rule for differentiation

$$(f \cdot g)^n = \binom{n}{k} f^{(k)} g^{(n-k)}, \tag{3.70}$$

and we obtain

$$\begin{aligned}(t^2 - 1)\frac{d^{n+2}R}{dt^{n+2}} + 2(n+1)t\frac{d^{n+1}R}{dt^{n+1}} + \frac{2n(n+1)}{1 \cdot 2}\frac{d^n R}{dt^n} \\ + -2nt\frac{d^{n+1}R}{dt^{n+1}} - 2n(n+1)\frac{d^n R}{dt^n} = 0.\end{aligned} \tag{3.71}$$

Substituting $V = \frac{d^n R}{dt^n}$ to (3.71) yields

$$(1 - t^2)\frac{d^2 V}{dt^2} - 2t\frac{dV}{dt} + n(n+1)V = 0. \tag{3.72}$$

It means that $V = \frac{d^n R}{dt^n}$ is the solution to the Legendre equation.
We may take

$$P_n(t) = \alpha V(t) = \alpha \frac{d^n}{dt^n}(t^2 - 1)^n, \quad \alpha \in R. \tag{3.73}$$

Observe that

$$\frac{d^n}{dt^n}(t^2 - 1)^n = \frac{d^n}{dt^n}[(t-1)(t+1)]^n = n!(t+1) + \text{terms with} \quad (t-1). \tag{3.74}$$

We take, however, $t = 1$ and hence

$$\left.\frac{d^n}{dt^n}(t^2 - 1)^n\right|_{t=1} = 2^n n!. \tag{3.75}$$

Therefore, since $P_n(t) = 1$, we obtain

$$P_n(t) = \frac{1}{2^n n!}\frac{d^n}{dt^n}(t^2 - 1)^n. \tag{3.76}$$

Equation (3.59) is a hyper-geometric equation with parameters $\alpha = n + 1$, $\beta = -n$, $\gamma = 1$. Therefore, its solution is

$$y_1(t) = F\left(n + 1, -n, 1, \frac{1-t}{2}\right) = P_n(t). \tag{3.77}$$

For instance, for $n = 1, 2, 3, 4$ we obtain

$$
\begin{aligned}
P_1(t) &= F\left(2, -1, 1, \frac{1-t}{2}\right) = t, \\
P_2(t) &= F\left(3, -2, 1, \frac{1-t}{2}\right) = \frac{1}{2}(3t^2 - 1), \\
P_3(t) &= F\left(4, -3, 1, \frac{1-t}{2}\right) = \frac{1}{2}(5t^3 - 3t), \\
P_4(t) &= F\left(5, -4, 1, \frac{1-t}{2}\right) = \frac{1}{8}(35t^4 - 30t^2 + 3), \\
&\dots
\end{aligned}
\tag{3.78}
$$

Example 3.5. Find a general solution to the Legendre equation (3.46) using the following series

$$y = \sum_{k=0}^{\infty} a_k t^k.$$

We have

$$\frac{dy}{dt} = k a_k t^{k-1}, \qquad \frac{d^2 y}{dt^2} = k(k-1)a_k t^{k-2},$$

and substituting the formulas given in the above to Eq. (3.46) we get

$$\sum_{k=0}^{\infty} \left[(1-t^2)k(k-1)a_k t^{k-2} - 2tka_k t^{k-1} + n(n+1)a_k t^k\right] = 0,$$

$$\sum \left[k(k-1)a_k t^{k-2} - k(k-1)a_k t^k - 2ka_k t^k + n(n+1)a_k t^k\right] = 0,$$

$$\sum \left[(k+2)(k+1)a_{k+2} + [-k^2 + k - 2k + n(n+1)]a_k\right] t^k$$

$$= \sum_{k=0}^{\infty} [(k+2)(k+1)a_{k+2} + (n-k)(n+k+1)a_k]\, t^k.$$

Therefore, we find

$$a_{k+2} = -\frac{(n-k)(n+k+1)}{(k+1)(k+2)}, \qquad k = 0, 1, 2, \dots.$$

Observe that

$$a_2 = -\frac{n(n+1)}{2!}a_0, \qquad a_3 = -\frac{(n-1)(n+2)}{3!}a_1,$$
$$a_4 = -\frac{(n-2)(n+3)}{3\cdot 4}a_2 = (-1)^2\frac{n(n-2)(n+1)(n+3)}{4!}a_0,$$
$$a_5 = (-1)^2\frac{(n-1)(n-3)(n+2)(n+4)}{5!}a_1, \ldots.$$

By deduction we have

$$a_{2m} = (-1)^m\frac{n(n-2)\ldots(n-2m+2)(n+1)(n+3)\ldots(n+2m+1)}{(2m)!}a_0,$$

$$a_{2m+1} = (-1)^m\frac{(n-1)(n-3)\ldots(n-2m+1)(n+2)(n+4)\ldots(n+2m)}{(2m+1)!}a_1.$$

We may choose a_0 and a_1 in an arbitrary manner. Taking $a_0 = 1$, $a_1 = 0$ we obtain

$$y_1 = 1 - \frac{n(n+1)}{2!}t^2 + \cdots + (-1)^m\frac{(n-2m+2)\ldots(n+2m-1)}{(2m)!}t^{2m} + \ldots,$$

whereas taking $a_0 = 0$, $a_1 = 0$ we get

$$y_2 = t - \frac{(n-1)(n+2)}{3!}t^3 + \cdots + (-1)^m\frac{(n-2m+1)\ldots(n+2m)}{(2m+1)!}t^{2m+1} + \cdots.$$

The general solution is

$$y = C_1y_1 + C_2y_2,$$

where C_1, C_2 are arbitrary constants.

Example 3.6. Solve the Cauchy problem of the following Legendre equation

$$(1-t^2)\frac{d^2y}{dt^2} - 2t\frac{dy}{dt} + 6y = 0,$$

for $y(0) = 0$, $\frac{dy}{dt}(0) = -2$.

Owing to the earlier considerations we have $n = 2$, and therefore we have the following solution

$$y_1 = Cp_2(t) = \frac{C}{2}(3t^2 - 1),$$

but this solution does not satisfy the initial conditions. Taking into account the initial conditions and observing that $t = 0$ is a regular (non-singular) point we are looking for a solution in the following form

$$y(t) = -2t + \sum_{i=2}^{\infty} a_i t^i.$$

Substituting y, $\frac{dy}{dt}$ and $\frac{d^2y}{dt^2}$ into the investigated equation we get

$$(1-t^2)\sum_{i=2}^{\infty} i(i-1)a_i t^{i-2} - 2t\left(-2 + \sum_{i=2}^{\infty} i a_i t^{i-1}\right) + 6\left(-2t + \sum_{i=2}^{\infty} a_i t^i\right) = 0,$$

and this equation yields relations

$$a_1 = -2,\ a_2 = 0,\ (i+2)(i+1)a_{i+2} = (i+3)(i-2)a_i,\ i = 1, 2, \ldots$$

allowing to find

$$a_{2i+1} = \tfrac{1}{2}\left(\tfrac{3}{2i-1} - \tfrac{1}{2i+1}\right),\ i = 1, 2, \ldots$$

Finally, the solution is

$$\begin{aligned} y = \sum_{i=0}^{\infty} \frac{1}{2}\left(\frac{3}{2i-1} - \frac{1}{2i+1}\right) t^{2i+1} &= -\frac{3}{2}t + \frac{1}{2}(3t^2-1)\sum_{i=1}^{\infty}\frac{t^{2i-1}}{i} \\ &= \frac{1}{4}(3t^2-1)\ln\frac{1+t}{1-t} - \frac{3}{2}t. \end{aligned}$$

□

3.2.4 The Bessel Equation

The Bessel equation has the following form

$$t^2\frac{d^2y}{dt^2} + t\frac{dy}{dt} + (t^2 - m^2)y = 0. \tag{3.79}$$

One may observe that Eq. (3.79) is not changed during transformation $t \to -t$, therefore we consider only the case $t > 0$. This equation possesses one singular point $t = 0$ and its characteristic equation corresponding to $t = 0$ has the following form

$$\sigma(\sigma - 1) + \sigma - m^2 = 0, \tag{3.80}$$

or equivalently

$$\sigma^2 - m^2 = 0. \tag{3.81}$$

It means that $\sigma_{1,2} = \pm m$ assuming $m \neq 0$. A solution to (3.79) is sought in the following series form

$$y = \sum_{i=0}^{\infty} a_i t^{i+\sigma}, \; a_0 \neq 0. \tag{3.82}$$

We obtain

$$\frac{dy}{dt} = \sum_{i=0}^{\infty}(i+\sigma)a_i t^{i+\sigma-1}, \quad \frac{d^2y}{dt^2} = \sum_{i=0}^{\infty}(i+\sigma)(\sigma+i-1)a_i t^{i+\sigma-2}, \tag{3.83}$$

and substituting (3.82), (3.83) into (3.79) we get

$$t^{\sigma}\sum_{i=0}^{\infty}(i+\sigma)(\sigma+i-1)a_i t^i + t^{\sigma}\sum_{i=0}^{\infty}(i+\sigma)a_i t^i + t^{\sigma}\sum_{i=0}^{\infty}a_i t^{i+2} - t^{\sigma}\sum_{i=0}^{\infty}m^2 a_i t^i = 0, \tag{3.84}$$

or equivalently

$$\sum_{i=0}^{\infty}[(\sigma+i)^2 - m^2]a_i t^i + \sum_{i=0}^{\infty}a_i t^{i+2} = 0. \tag{3.85}$$

Comparing terms standing by the same powers of t we obtain

$$\begin{aligned} (\sigma^2 - m^2)a_0 &= 0, \qquad [(\sigma+1)^2 - m^2]a_1 = 0, \\ \left[(\sigma+i)^2 - m^2\right]a_i + a_{i-2} &= 0, \qquad i = 2, 3, \ldots \end{aligned} \tag{3.86}$$

Firstly, let us construct a solution corresponding to $\sigma = m$. We take $a_0 \neq 0$, $a_1 = 0$, and we get

$$a_i = -\frac{a_{i-2}}{i(2m+i)}, \quad i = 2, 3, \ldots \tag{3.87}$$

Therefore, we obtain

$$\begin{aligned} a_{2k+1} &= 0, \quad k = 0, 1, 2, \ldots, \\ a_2 &= -\frac{a_0}{2^2 \cdot 1 \cdot (m+1)}, \quad a_4 = \frac{a_0}{2^4 \cdot 2!(m+1)(m+2)}, \ldots, \\ a_{2i} &= (-1)^i \frac{a_0}{2^{2i} \cdot i!(m+1)(m+2)\ldots(m+i)}, \end{aligned} \tag{3.88}$$

and the solution has the form

$$y_1(t) = \sum_{i=0}^{\infty}(-1)^i \frac{a_0}{2^{2i} \cdot i!(m+1)(m+2)\ldots(m+i)} t^{2i+m}. \tag{3.89}$$

One may verify using, for instance, the D'Alembert's principle, that series (3.89) is uniformly convergent on a finite arbitrary taken interval $t \in (0, T)$.

In what follows we take

$$a_0 = \frac{1}{2^m \Gamma(m+1)}, \tag{3.90}$$

where $\Gamma(\alpha)$ is well-known Gamma-function

$$\Gamma(\alpha) = \int_0^{\infty} t^{\alpha-1} e^{-t} dt, \quad \alpha > 0. \tag{3.91}$$

Using the property

$$\Gamma(m+1) = m\Gamma(m), \tag{3.92}$$

we get

$$\begin{aligned} y_1(t) &= \sum_{i=0}^{\infty}(-1)^i \frac{2^m a_0}{i!(m+1)(m+2)\ldots(m+i)} \left(\frac{t}{2}\right)^{2i+m} \\ &= \sum_{i=0}^{\infty}(-1)^i \frac{2^m}{2^m i!\Gamma(m+1)(m+1)(m+2)\ldots(m+i)} \left(\frac{t}{2}\right)^{2i+m} \\ &= \sum_{i=0}^{\infty}(-1)^i \frac{1}{i!\Gamma(m+i+1)} \left(\frac{t}{2}\right)^{2i+m} \equiv J_m(t), \end{aligned} \tag{3.93}$$

where $J_m(t)$ is called *the mth order Bessel function.*

A second solution $y_2(t)$ being linearly independent on $y_1(t)$ is assumed in the following form

$$y_2(t) = \sum_{i=0}^{\infty} a_i t^{i-m}. \tag{3.94}$$

The following algebraic equation serve for determination of a_i coefficients for $\sigma = -m$:

$$(m^2 - m^2)a_0 = 0, \; (1-2m)a_1 = 0, \; \ldots, \; [(-m+i)^2 - m^2]a_i + a_{i-2} = 0. \tag{3.95}$$

Taking $a_0 \neq 0, a_1 = 0$ we get

$$a_{2i+1} = 0, \quad a_{2i} = -\frac{a_{2i-2}}{2^2 i(-m+i)}. \tag{3.96}$$

For $m \neq C$, where C is integer, we obtain

$$a_{2i} = (-1)^i \frac{a_0}{2^{2i} i!(-m+1)(-m+2)\ldots(-m+i)}, \tag{3.97}$$

and hence

$$y_2(t) = \sum_{i=0}^{\infty} (-1)^i \frac{a_0}{2^{2i} i!(-m+1)(-m+2)\ldots(-m+i)} t^{2i-m}. \tag{3.98}$$

Assuming

$$a_0 = \frac{1}{2^{-m}\Gamma(-m+1)}, \tag{3.99}$$

we get

$$y_2(t) = \sum_{i=0}^{\infty} (-1)^i \frac{1}{i!\Gamma(-m+i+1)} \left(\frac{t}{2}\right)^{2i-m} \equiv J_{-m}(t), \tag{3.100}$$

where $J_{-m}(t)$ is called *the mth order Bessel function with the negative index.*

Finally, a general solution of the Bessel equation (3.79) has the following form

$$y(t) = C_1 J_m(t) + C_2 J_{-m}(t). \tag{3.101}$$

Although the majority of the studied through special functions cases refer to real-world applications modelling of mechanical continuous objects and governed by partial differential equations, there are also direct problems yielding the Bessel equation.

Example 3.7 ([108]). Consider a vertical beam of mass m and length L which is deflected due to its own weight.

It is known from the theory of materials strength that, assuming the beam stiffness EI, where E is the Young modulus and I is the moment of inertia of the beam cross section regarding its neutral axis, the beam curvature is governed by the following differential equation

$$\frac{\ddot{y}}{(1+\dot{y}^2)^{\frac{3}{2}}} = \pm\frac{M(x)}{EI},$$

Fig. 3.1 The beam deflection caused by its own weight

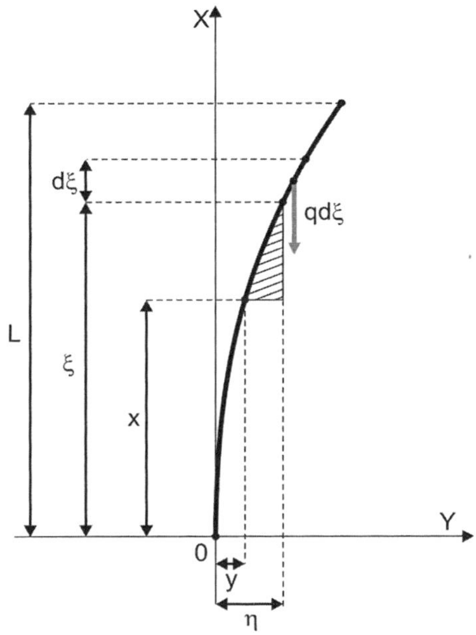

where dot denotes $\frac{dy}{dx}$, and $M(x)$ is the bending moment. Assuming small beam deflection the second-order differential equation is simplified to the form

$$\frac{d^2y}{dx^2} = \pm\frac{M(x)}{EI}, \qquad (*)$$

and this equation is called the differential equation of the *deflected beam axis*. If we take the beam element $d\eta$ located in the distance η from the introduced rectangular coordinates origin (see Fig. 3.1), then

$$dM(x) = -(qd\xi)(\eta - y)$$

or equivalently

$$M(x) = -\int_x^L q(\eta - y)d\xi,$$

where $q = \frac{mg}{L}$. For a small deflection we have $\eta = \xi\frac{dy}{dx}$. We substitute $M(x)$ to Eq. (∗), and then we differentiate both sides of the obtained equation with respect to $\frac{d}{dx}$ to get

$$EI\frac{d^3y}{dx^3} = -q\left|\xi\frac{dy}{dx} - y\right|_x^L = -q\left[L\frac{dy}{dx} - y - x\frac{dy}{dx} + y\right] = -q(L-x)\frac{dy}{dx}. \qquad (**)$$

We introduce the new variable

$$z = \frac{2}{3}\sqrt{\frac{q}{EI}}(L-x)^{\frac{3}{2}}, \qquad \frac{dz}{dx} = -\sqrt{\frac{q}{EI}}(L-x)^{\frac{1}{2}}.$$

We successively proceed with the differentiation process of the form

$$\frac{dy}{dx} = \frac{dy}{dz}\frac{dz}{dx} = -\sqrt{\frac{q}{EI}}(L-x)^{\frac{1}{2}}\dot{y} = -\dot{y}z^{\frac{1}{3}}\sqrt[3]{\frac{3}{2}\frac{q}{EI}},$$

$$\frac{d^2y}{dx^2} = \frac{d^2y}{dz^2}\left(\frac{dz}{dx}\right) + \frac{dy}{dz}\frac{d^2z}{dx^2} = \left(\frac{3}{2}\frac{q}{EI}\right)^{\frac{2}{3}}\left(\ddot{y}z^{\frac{2}{3}} + \frac{1}{3}\dot{y}z^{-\frac{1}{3}}\right),$$

$$\frac{d^3y}{dx^3} = \frac{d^3y}{dz^3}\left(\frac{dy}{dx}\right)^3 + 3\frac{d^2y}{dz^2}\frac{dz}{dx}\frac{d^2z}{dx^2} + \frac{dy}{dz}\frac{d^3z}{dx^3} = \frac{3}{2}\frac{q}{EI}\left(-\dddot{y}z - \ddot{y} + \frac{1}{9}\dot{y}z^{-1}\right).$$

Substituting the derived formulas to (∗∗) we get

$$\dddot{y} + \frac{1}{z} + \left(1 - \frac{1}{9z^2}\right)\dot{y} = 0,$$

and by introducing $\dot{y} = u$ we finally obtain

$$\frac{d^2u}{dz^2} + \frac{1}{z}\frac{du}{dz} + \left(1 - \frac{1}{9z^2}\right)u = 0.$$

We have got the Bessel equation of $m = \frac{1}{3}$ order. Its solution has the following form

$$u(z) = C_1 J_{1/3}(z) + C_2 J_{-1/3}(z).$$

□

The so far studied examples show that the problem of finding solutions to the Bessel equation is reduced to that of finding Bessel functions. In what follows we show that if $m \in C$ then

$$J_{-n}(t) = (-1)^n J_n(t). \tag{3.102}$$

Note that $\Gamma(-n+i+2) = \infty$ for $i = 0, 1, \ldots, n-1$, and hence

$$J_{-n}(t) = \sum_{i=0}^{\infty} \frac{(-1)^n \left(\frac{t}{2}\right)^{-n+2i}}{i!\Gamma(-n+i+1)} = \sum_{i=n}^{\infty} \frac{(-1)^i \left(\frac{t}{2}\right)^{-n+2i}}{\Gamma(i+1)\Gamma(-n+i+1)}.$$

Introducing $i = n + m$ we obtain

$$J_{-n}(t) = \sum_{k=0}^{\infty} \frac{(-1)^{k+n}}{\Gamma(k+1)\Gamma(k+n+1)} \left(\frac{t}{2}\right)^{2k+n} = (-1^n)J_n(t). \tag{3.103}$$

One may prove also the following relations

$$\begin{aligned} \frac{d}{dt}[t^m J_m(t)] &= t^m J_{m-1}(t), \\ \frac{d}{dt}[t^{-m} J_m(t)] &= -t^{-m} J_{m+1}(t), \end{aligned} \tag{3.104}$$

for arbitrary m.

There is also the following formula allowing to find the $m+1$ order Bessel function assuming that the Bessel functions of order m and $m-1$ are known:

$$J_{m-1}(t) + J_{m+1}(t) = \frac{2m}{t} J_m(t). \tag{3.105}$$

Example 3.8. Find the following Bessel functions: $J_{1/2}(t), \quad J_{-1/2}(t), \quad J_{3/2}(t)$.

$$\begin{aligned} J_{1/2}(t) &= \frac{\sqrt{t}}{\sqrt{2}\Gamma\left(\frac{3}{2}\right)} \left[1 - \frac{t^2}{2\cdot 3} + \frac{t^4}{2\cdot 4\cdot 3\cdot 5} - \frac{t^6}{2\cdot 4\cdot 6\cdot 3\cdot 5\cdot 7} + \cdots\right] \\ &= \frac{1}{\sqrt{2t}\Gamma\left(\frac{3}{2}\right)} \left[t - \frac{t^3}{3!} + \frac{t^5}{5!} - \frac{t^7}{7!} + \cdots\right] = \frac{1}{\sqrt{2t}\Gamma\left(\frac{3}{2}\right)} \sin(t). \end{aligned}$$

Since

$$\Gamma\left(\frac{3}{2}\right) = \frac{1}{2}\Gamma\left(\frac{1}{2}\right) = \frac{1}{2}\int_0^{\infty} e^{-t}\frac{dt}{\sqrt{t}} = \int_0^{\infty} e^{-s^2} ds = \frac{\sqrt{\pi}}{2},$$

therefore

$$\begin{aligned} J_{1/2}(t) &= \sqrt{\frac{2}{\pi t}} \sin(t), \\ J_{-1/2}(t) &= \frac{\sqrt{2}}{\sqrt{t}\Gamma\left(\frac{1}{2}\right)} \left(1 - \frac{t^2}{2!} + \frac{t^4}{4!} - \frac{t^6}{6!} + \cdots\right) = \sqrt{\frac{2}{\pi t}} \cos(t). \end{aligned}$$

Owing to formula (3.105) for $m = \frac{1}{2}$ we obtain

$$J_{3/2}(t) = \frac{2 \cdot \frac{1}{2}}{t} J_{1/2}(t) - J_{-1/2}(t)$$
$$= \frac{1}{t}\sqrt{\frac{2}{\pi t}} \sin(t) - \sqrt{\frac{2}{\pi t}} \cos(t) = \sqrt{\frac{2}{\pi t}} \left(\frac{\sin(t)}{t} - \cos(t) \right).$$

□

3.2.5 ODEs with Periodic Coefficients

In order to familiarize with the problems of systems governed by linear ODE with periodic coefficients we consider one degree-of-freedom mechanical system small vibrations (see Fig. 3.2). There is a body of mass m at distance l from the axis of rotation. The body is put in a massless box by means of two springs (each one of stiffness k).

During the motion a body of mass m moves along the slideways and its moment of inertia B with respect to the point 0 varies. In order to derive the equations of motion we make use of a theorem, which states that derivative of the angular momentum with respect to time equals the sum of all torques acting on the system

$$\frac{d(B\dot{\varphi})}{dt} = \sum_i M_i. \tag{3.106}$$

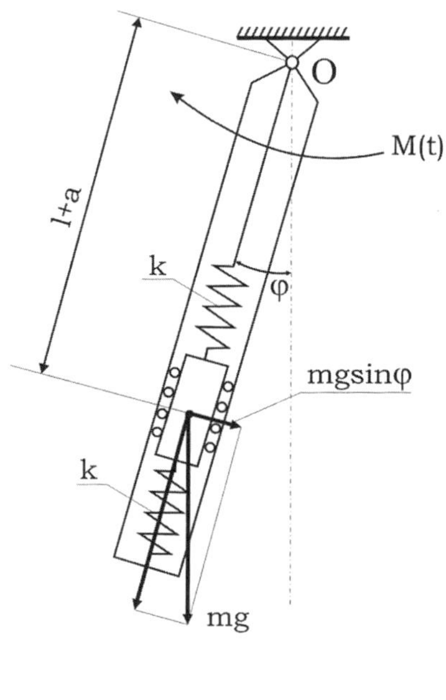

Fig. 3.2 Parametric vibrations under excitation $M(t)$

Let the motion of the mass in the sideways be described by a function $a \sin \omega t$, then a moment of inertia reads

$$B = m(l + a \sin \omega t)^2. \tag{3.107}$$

Left-hand side of Eq. (3.107) is described by the formula

$$\frac{d}{dt}(B\dot{\varphi}) = B\ddot{\varphi} + \frac{dB}{dt}\dot{\varphi} = m(l + a \sin \omega t)^2 \ddot{\varphi} + 2m(l + a \sin \omega t) a\omega \dot{\varphi} \cos \omega t, \tag{3.108}$$

where: $\omega^2 = 2km^{-1}$. Since, the torque comes from earthpull and from external torque $M(t)$, we have

$$\sum M_i = -mg(l + a \sin \omega t) \sin \varphi + M(t). \tag{3.109}$$

Assuming $\sin \varphi \approx \varphi$ by Eq. (3.106) we find

$$\ddot{\varphi} + P(t)\dot{\varphi} + Q(t)\varphi = R(t), \tag{3.110}$$

where

$$P(t) = \frac{2a\omega \cos \omega t}{l + a \sin \omega t}, \quad Q(t) = \frac{g}{l + a \sin \omega t}, \quad R(t) = \frac{M(t)}{m(l + a \sin \omega t)^2}. \tag{3.111}$$

Next, we will consider Eq. (3.110). Let us introduce a new variable $y(t)$ defined as follows

$$\varphi(t) = y(t) \exp\left[-\frac{1}{2}\int_{t_0}^{t} P(\tau)d\tau\right]. \tag{3.112}$$

By Eq. (3.112) we obtain

$$\dot{\varphi}(t) = \left(\dot{y} - \frac{1}{2}Py\right) \exp\left[-\frac{1}{2}\int_{t_0}^{t} P(\tau)d\tau\right],$$

$$\ddot{\varphi} = \left(\ddot{y} - P\dot{y} - \frac{1}{2}\dot{P}y + \frac{1}{4}P^2 y\right) \exp\left[-\frac{1}{2}\int_{t_0}^{t} P(\tau)d\tau\right]. \tag{3.113}$$

Substituting (3.113) into Eq. (3.110), the term with the first-order derivative $\dot{\varphi}$ is cancelled. We obtain

$$\ddot{y} + p(t)y = r(t), \tag{3.114}$$

where

$$p(t) = Q + \frac{1}{4}P^2 - \frac{1}{2}\dot{P}, \quad r(t) = R\exp\left[\frac{1}{2}\int\limits_{t_0}^{t} P(\tau)d\tau\right]. \tag{3.115}$$

If $p(t) = p(t+T)$, where T is a period of changes of the coefficient p, then Eq. (3.114) will be called *Hill's equation* [162]. Consider the case $r(t) = 0$. Let $y_1(t)$ and $y_2(t)$ be solutions of Eq. (3.114). We choose the solution so that it satisfies the following initial conditions

$$\begin{aligned} y_1(0) &= 1, \quad \dot{y}_1(0) = 0, \\ y_2(0) &= 0, \quad \dot{y}_2(0) = 1. \end{aligned} \tag{3.116}$$

If $y_1(t)$ and $y_2(t)$ are the independent solutions, then $y_1(t+T)$ and $y_2(t+T)$ are the independent solutions as well. The latter can be expressed as linear combinations

$$\begin{aligned} y_1(t+T) &= \varphi_{11}y_1(t) + \varphi_{12}y_2(t), \\ y_2(t+T) &= \varphi_{21}y_1(t) + \varphi_{22}y_2(t). \end{aligned} \tag{3.117}$$

Then, taking into account (3.116) we get

$$\begin{aligned} y_1(T) &= \varphi_{11}, \quad \dot{y}_1(T) = \varphi_{12}, \\ y_2(T) &= \varphi_{21}, \quad \dot{y}_2(T) = \varphi_{22}. \end{aligned} \tag{3.118}$$

By Eq. (3.114) we get

$$\ddot{y}_1 + p(t)y_1 = 0, \; \ddot{y}_2 + p(t)y_2 = 0. \tag{3.119}$$

Next multiplying the first equation (3.119) by $(-y_2)$, the second one by y_1 and adding them, we find

$$\ddot{y}_2 y_1 - \ddot{y}_1 y_2 = 0. \tag{3.120}$$

Integration of the above equation yields

$$\dot{y}_2 y_1 - \dot{y}_1 y_2 = C, \tag{3.121}$$

where C is a constant.

Substituting 0 and T into (3.121) for periodic solutions we obtain

$$\dot{y}_2(T)\,y_1(T) - \dot{y}_1(T)\,y_2(T) = \dot{y}_2(0)\,y_1(0) - \dot{y}_1(0)\,y_2(0), \tag{3.122}$$

and taking into account (3.116) and (3.118) we get

$$\varphi_{22}\varphi_{11} - \varphi_{12}\varphi_{21} = 1. \tag{3.123}$$

Let us introduce the following multiplier μ:

$$y_1(t+T) = \mu y_1(t),\ y_2(t+T) = \mu y_2(t). \tag{3.124}$$

After taking into account the above equations in (3.117) we get a characteristic equation

$$\begin{vmatrix} \varphi_{11} - \mu & \varphi_{12} \\ \varphi_{21} & \varphi_{22} - \mu \end{vmatrix} = 0 \tag{3.125}$$

or in the expanded form

$$\mu^2 - \mu(\varphi_{22} + \varphi_{11}) + 1 = 0, \tag{3.126}$$

because $\varphi_{22}\varphi_{11} - \varphi_{12}\varphi_{21} = 1$.

Roots of Eq. (3.126) are called multipliers and take the values

$$\mu_{1,2} = \alpha \pm \sqrt{\alpha^2 - 1}, \tag{3.127}$$

where

$$\alpha = \frac{1}{2}\left(\varphi_{11} + \varphi_{22}\right). \tag{3.128}$$

Both multipliers are either real or complex. Their dependence on the parameter α is presented in Fig. 3.3

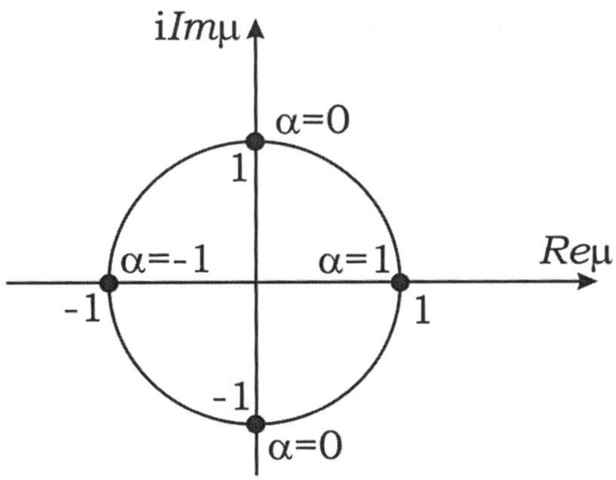

Fig. 3.3 Graph of dependence of the multipliers on the parameter α

It follows from Eq. (3.127) that for any value of α we have

$$\mu_1\mu_2 = 1. \tag{3.129}$$

In the interval $|\alpha| < 1$ multiplier μ is complex of a unit absolute value and we will write it in the form

$$\mu_{1,2} = \exp(\pm i\lambda_I T), \tag{3.130}$$

where

$$e^{\pm i\lambda_I T} = \cos\lambda_I T \pm i \sin\lambda_I T = \alpha \pm i\sqrt{1-\alpha^2},$$

$$0 < |\lambda_I T| < T,$$

hence $0 < |\lambda_I| < 1$. According to Vieta's formulas (3.129) and $\mu_1 + \mu_2 = 2\alpha$, the sign of α decides whether both real roots are negative or positive.

In the interval $|\alpha| > 1$ multipliers μ are real and can be presented as follows

$$\mu_{1,2} = \pm\exp(\pm\lambda_R T), \tag{3.131}$$

where λ_R takes on values $0 < \lambda_R < \infty$. For $\alpha = 1$ we have $\mu_1 = \mu_2 = 1$, and for $\alpha = -1$ we have $\mu_1 = \mu_2 = -1$. For $\mu = 1$ or $\mu = -1$ the solution is periodic and in the first case has the form

$$y(t+T) = y(t) \tag{3.132}$$

while in the second case

$$y(t+2T) = -y(t+T) = y(t), \tag{3.133}$$

and this means a solution of period T and $2T$, respectively. If $\mu > 1$ ($|\alpha| > 1$), then the solution $y(t)$ grows unbounded as time increases and it is unstable. A stable solution occurs for $\mu \le 1$ ($\mu < 1$ occurs for a system with damping). On the boundary between stable and unstable solution a periodic one appears (for $|\mu| = 1$, $|\alpha| = 1$).

Introducing a new function $\Phi(t)$ such that

$$y(t) = \Phi(t)\exp(\lambda t) = \Phi(t)\exp(\lambda_R + i\lambda_I)t \tag{3.134}$$

we obtain

$$y(t+T) = \Phi(t+T)\exp(\lambda(t+T)) = y(t)\exp(\lambda T) = \Phi(t)\exp(\lambda t)\exp(\lambda T). \tag{3.135}$$

Equating the second and last part of the equality (3.135) we obtain $\Phi(t+T) = \Phi(t)$, so the function $\Phi(t)$ is periodic of period T.

Since the function $y(-t)$ is also a solution, the complete solutions of Hill's equation has the form

$$y(t) = C_1 \exp(\lambda t)\,\Phi(t) + C_2 \exp(-\lambda t)\,\Phi(-t) \tag{3.136}$$

where C_1 and C_2 are constant determined by initial conditions. As one can see, the solution is always unstable unless λ is imaginary.

For $\mu_{1,2} = 1$ ($\mu_{1,2} = e^{\lambda T}$, $1 = e^{\lambda T}$, hence $\lambda = 0$) a solution of Hill's equation has the form

$$y(t) = C_1\Phi_1(t) + C_2 t\Phi_2(t), \tag{3.137}$$

while for $\mu_{1,2} = -1$ ($-1 + i0 = e^{\lambda T}$, $\cos T \pm i \sin T = e^{\pm iT}$, hence $\lambda = \pm i$) a solution is

$$y(t) = C_1 \exp(it)\Phi_1(t) + C_2 t \exp(it)\Phi_2(t), \tag{3.138}$$

provided that $\Phi_1 = \Phi_1(t+2T)$ and $\Phi_2 = \Phi_2(t+2T)$. Both solutions described by the formulas (3.137) and (3.138) are unstable because of the occurrence of t in the second terms of both formulas.

To sum up, one needs to emphasize that a periodic solution appears on the boundary of stability loss. Finding of periodic solutions of periods T and $2T$ determine the boundary of stability loss (points $\mu = \alpha = 1$ and $\mu = \alpha = -1$ from Fig. 3.3 determine the boundary of stability).

Now, we consider two particular cases of Hill's equation. First, we discuss a so-called Meissner's equation (see [167]):

$$\ddot{y} + p(t)\,y = 0, \tag{3.139}$$

where a graph of $p(t)$ is depicted in Fig. 3.4.

The equation of motion has the form

$$\begin{aligned} \ddot{y}_a + a^2 y_a &= 0 \text{ for } 0 < t - mT \le T_1 \\ \ddot{y}_b + b^2 y_b &= 0 \text{ for } T_1 < t - mT \le T \end{aligned} \tag{3.140}$$

where $m = 0, 1, 2, \ldots$. A general solution of the first equation of (3.140) can be written as

$$y_a = C_a^{(1)} \cos at + C_a^{(2)} \sin at. \tag{3.141}$$

Hence, we have

$$\{X_a\} = \begin{pmatrix} y_a \\ \dot{y}_a \end{pmatrix} = \begin{bmatrix} \cos at & \sin at \\ -a \sin at & a \cos at \end{bmatrix} \begin{pmatrix} C_a^{(1)} \\ C_a^{(2)} \end{pmatrix} \tag{3.142}$$

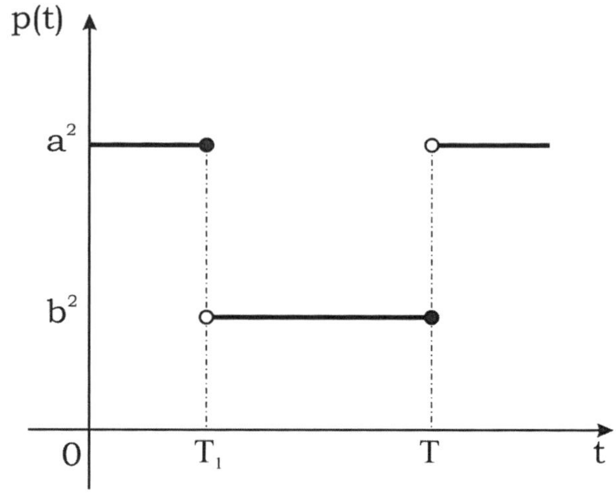

Fig. 3.4 Periodic changes of the coefficient $p(t)$ (Meissner's equation)

Similarly, we get for the second equation of (3.141)

$$\{X_b\} = \begin{pmatrix} y_b \\ \dot{y}_b \end{pmatrix} = \begin{bmatrix} \cos bt & \sin bt \\ -b\sin bt & b\cos bt \end{bmatrix} \begin{pmatrix} C_b^{(1)} \\ C_b^{(2)} \end{pmatrix}. \tag{3.143}$$

For $t = T_1$ we have the condition

$$\{X_a(T_1)\} = \{X_b(T_1)\}. \tag{3.144}$$

Let $\{\varphi_1(t)\}$ designate the first general solution of the initial conditions $\{\varphi_1(0)\} = [1 \quad 0]^T$. By Eq. (3.142) we get

$$\begin{pmatrix} 1 \\ 0 \end{pmatrix} = \begin{bmatrix} 1 & 0 \\ 0 & a \end{bmatrix} \begin{pmatrix} C_a^{(1)} \\ C_a^{(2)} \end{pmatrix}, \tag{3.145}$$

and hence

$$\begin{pmatrix} C_a^{(1)} \\ C_a^{(2)} \end{pmatrix} = \begin{pmatrix} 1 \\ 0 \end{pmatrix}. \tag{3.146}$$

Substituting (3.146) into (3.142) we obtain

$$\{\varphi_{1a}(t)\} = \begin{pmatrix} \cos at \\ -a\sin at \end{pmatrix}. \tag{3.147}$$

Bearing in mind that we have

$$\mathbf{A} = [A_{jk}] = \begin{bmatrix} b\cos bT_1 & b\sin bT_1 \\ -\sin bT_1 & \cos bT_1 \end{bmatrix}, \quad [A_{jk}]^T = \begin{bmatrix} b\cos bT_1 & -\sin bT_1 \\ b\sin bT_1 & \cos bT_1 \end{bmatrix},$$

and hence $\mathbf{A}^{-1} = \left[\frac{A_{jk}}{\det \mathbf{A}}\right]^T$.

The condition (3.144) has the form

$$\{\varphi_{1a}(T_1)\} = \{\varphi_{1b}(T_1)\}. \tag{3.148}$$

Thus, Eqs. (3.147) and (3.143) yield

$$\begin{pmatrix} \cos aT_1 \\ -a\sin aT_1 \end{pmatrix} = \begin{bmatrix} \cos bT_1 & \sin bT_1 \\ -b\sin bT_1 & b\cos bT_1 \end{bmatrix} \begin{pmatrix} C_b^{(1)} \\ C_b^{(2)} \end{pmatrix}. \tag{3.149}$$

Solving Eq. (3.149) we get

$$\begin{pmatrix} C_b^{(1)} \\ C_b^{(2)} \end{pmatrix} = \frac{1}{b} \begin{bmatrix} b\cos bT_1 & -\sin bT_1 \\ b\sin bT_1 & \cos bT_1 \end{bmatrix} \begin{pmatrix} \cos aT_1 \\ -a\sin aT_1 \end{pmatrix}. \tag{3.150}$$

Taking into account the Eqs. (3.143) and (3.150) we find

$$\{\varphi_{1b}(t)\} = \begin{pmatrix} \cos aT_1 \cos b(t-T_1) - \frac{a}{b}\sin aT_1 \sin b(t-T_1) \\ -b\cos aT_1 \sin b(t-T_1) - a\sin aT_1 \cos b(t-T_1) \end{pmatrix}. \tag{3.151}$$

Let us verify the formula (3.151). We have

$$C_b^{(1)} = \frac{1}{b}[b\cos bT_1 \cos aT_1 + a\sin bT_1 \sin aT_1],$$

$$C_b^{(2)} = \frac{1}{b}[b\sin bT_1 \cos aT_1 - a\cos bT_1 \sin aT_1],$$

and

$$\{\varphi_{1b}(t)\} = \begin{bmatrix} \cos bt & \sin bt \\ -b\sin bt & b\cos bt \end{bmatrix} \begin{pmatrix} C_b^{(1)} \\ C_b^{(2)} \end{pmatrix}.$$

After transformation we obtain

$$\begin{aligned}\{\varphi_{1b}^{(1)}(t)\} &= \frac{1}{b}[b\cos bT_1 \cos aT_1 + a\sin bT_1 \sin aT_1]\cos bt \\ &\quad + \frac{1}{b}[b\sin bT_1 \cos aT_1 - a\sin aT_1 \cos bT_1]\sin bt,\end{aligned}$$

and next

$$\begin{aligned}\{\varphi_{1b}^{(1)}(t)\} &= \frac{1}{b}\cos aT_1[b\cos bT_1 \cos bt + b\sin bT_1 \sin bt] \\ &\quad + \frac{1}{b}\sin aT_1[a\sin bT_1 \cos bt - a\cos bT_1 \sin bt] \\ &= \cos aT_1 \cos b(t-T_1) - \frac{a}{b}\sin aT_1 \sin b(t-T_1).\end{aligned}$$

Similarly, we can perform the procedure for the second independent solution $\{\varphi_2(t)\}$ with the initial condition $\{\varphi_2(0)\} = [0 \quad 1]^T$. We find

$$\begin{aligned}\{\varphi_{2a}(t)\} &= \begin{pmatrix} \frac{1}{a}\sin at \\ \cos at \end{pmatrix},\\ \{\varphi_{2b}(t)\} &= \begin{pmatrix} \frac{1}{a}\sin aT_1\cos b(t-T_1) + \frac{1}{b}\cos aT_1 \sin b(t-T_1) \\ \frac{-b}{a}\sin aT_1 \sin b(t-T_1) + \cos aT_1\cos b(t-T_1)\end{pmatrix}.\end{aligned} \tag{3.152}$$

The solution in the end of period T can be expressed by means of a matrix

$$[\Phi_*] = [\{\varphi_{1b}(T)\}, \{\varphi_{2b}(T)\}]. \tag{3.153}$$

By Eqs. (3.151), (3.152) and (3.128) we obtain

$$\alpha = \frac{1}{2}\left[2\cos aT_1 \cos b(T-T_1) - \left(\frac{a}{b} + \frac{b}{a}\right)\sin aT_1 \sin b(T-T_1)\right]. \tag{3.154}$$

As we know by the earlier considerations, on the boundary of stability we have $\alpha = \pm 1$, so

$$2\cos aT_1 \cos b(T-T_1) - \left(\frac{a}{b} + \frac{b}{a}\right)\sin aT_1 \sin b(T-T_1) = \pm 2. \tag{3.155}$$

The above equation contains the following control parameters: a, b, T_1, T. Equation (3.155) can be also treated as an equation of surface in three-dimensional parameter space, which separates the parameter spaces corresponding to stable and unstable solutions.

For the sake of simplicity we assume that $b = 0$. By Eq. (3.155) we get

$$2\cos aT_1 - a(T-T_1)\sin aT_1 = \pm 2 \tag{3.156}$$

since

$$\lim_{b\to 0} \frac{\sin b(T-T_1)}{b} = T - T_1. \tag{3.157}$$

By (3.156) we get a curve, which is a boundary of stability loss in the plane aT_1, aT of the form

$$aT = aT_1 - \frac{2(\cos aT_1 \mp 1)}{\sin aT_1}. \tag{3.158}$$

Equation (3.156) is satisfied for

$$aT_1 = n\pi,\ n = 1, 2, 3, \ldots \tag{3.159}$$

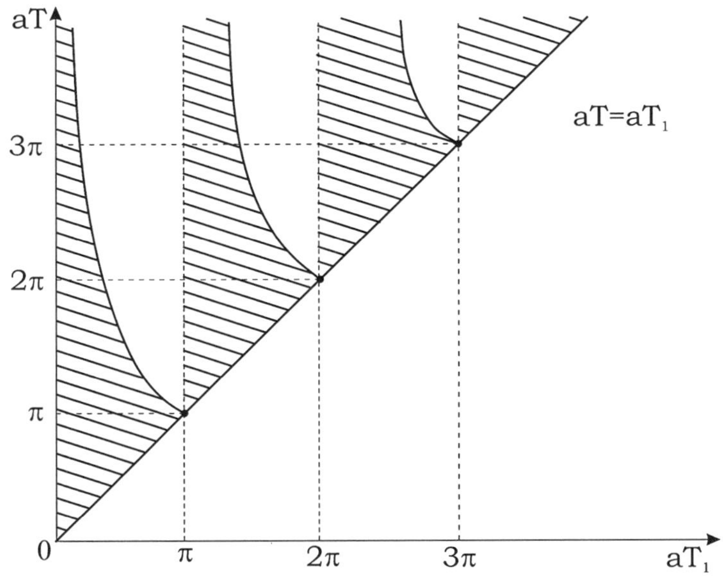

Fig. 3.5 Stability boundaries for Meissner's equation in the case $b = 0$ (stable regions are cross-hatched)

and Eq. (3.159) defines boundaries of stability loss. The results of the above analysis along with the marked regions of stability are depicted in Fig. 3.5.

In order to determine which of the regions in Fig. 3.5 are stable, we choose any point from a given region and evaluate the value of α.

Let us finally emphasize that recently a particular aspect of the Meissner's equation [167] stability regarding the invariance of the trace of a product of matrices has been addressed in [4].

The second particular case of Hill's equation is Mathieu's equation of the form

$$\ddot{y} + \left(\lambda^* + \gamma^* \cos \Omega t\right) y = 0, \tag{3.160}$$

or introducing $\tau = \Omega t$ we have

$$\ddot{y} + (\lambda + \gamma \cos \tau)\, y = 0, \tag{3.161}$$

where

$$\lambda = \frac{\lambda^*}{\Omega^2}, \quad \gamma = \frac{\gamma^*}{\Omega^2}.$$

Regions of instability in the plane λ and γ were determined for Mathieu equation for the first time by Ince and Strutt (Fig. 3.6).

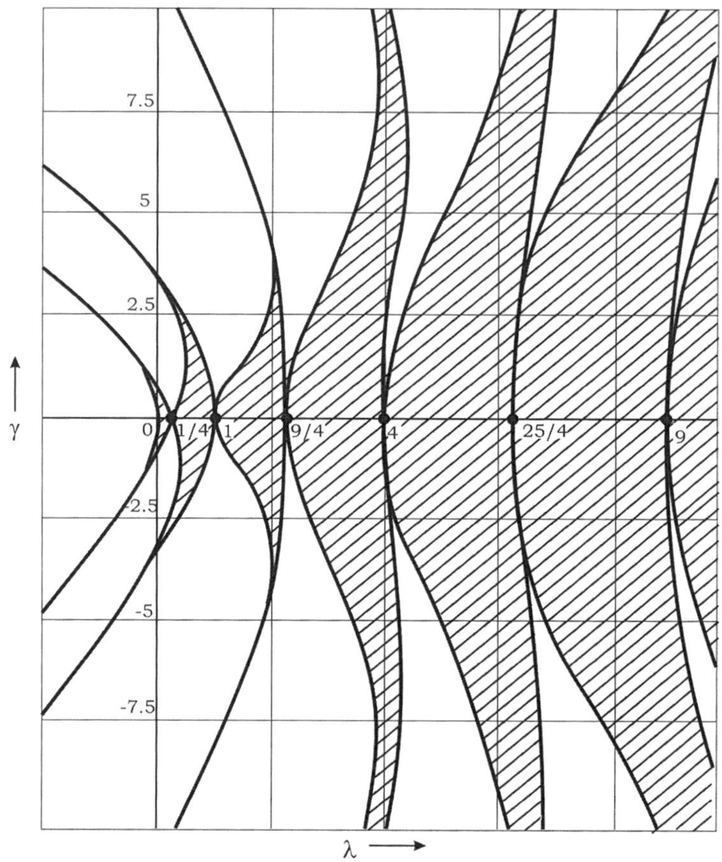

Fig. 3.6 Ince–Strutt diagram—stable regions are cross-hatched

The boundary curves nearby the axis λ can be determined with the help of perturbative method. A particular case of the Mathieu equation (for $\gamma = 0$) is an equation

$$\ddot{y} + \lambda y = 0. \tag{3.162}$$

It is easy to see that for $\lambda \leq 0$ solutions of Eq. (3.162) are unstable. It is interesting that occurrence of γ (parametric excitation) makes a region of stability appear. One can conclude that one can stabilize the motion by means of parametric excitation. In technics, the case $\lambda > 0$ plays a very important role. The most dangerous case is the one with $\lambda = \frac{1}{4}$.

Now, consider an influence of linear damping on regions of stability loss. On this purpose, let us analyse the equation

$$\ddot{w} + P_0 \dot{w} + Q(t) w = 0. \tag{3.163}$$

This equation can be obtained from Eq. (3.110) by assuming $w = \varphi$, $P(t) = P_0 = \text{const}$, and $R(t) = 0$. Next, we can make a substitution analogously to (3.112) and cancel a term with the first derivative. However, we proceed in a different way, namely we will write Eq. (3.163) in the form

$$\begin{pmatrix} w \\ \dot{w} \end{pmatrix}^{\cdot} = [A(t)] \begin{pmatrix} w \\ \dot{w} \end{pmatrix}, \tag{3.164}$$

where

$$[A(t)] = \begin{bmatrix} 0 & 1 \\ -Q(t) & -P_0 \end{bmatrix}. \tag{3.165}$$

Trace of the matrix $Sp[A(t)] = -P_0 = \text{const}$ and after Liouville we get

$$\det[\Phi(t, t_0)] = \det[\Phi(t_0, t_0)] \exp\left[\int_{t_0}^{t} Sp[A(\tau)]d\tau\right], \tag{3.166}$$

where $\Phi(t)$ was introduced in the formula (3.134).

If we assume

$$\det[\Phi(t_0 + T, t_0)] = \det[\Phi_*], \tag{3.167}$$

then according to (3.166) we get

$$\det[\Phi_*] = \det\begin{bmatrix} \varphi_{11} & \varphi_{12} \\ \varphi_{21} & \varphi_{22} \end{bmatrix} = e^{-P_0 T}, \tag{3.168}$$

hence

$$\varphi_{11}\varphi_{22} - \varphi_{12}\varphi_{21} = e^{-P_0 T}. \tag{3.169}$$

The characteristic equation has the form

$$\det([\Phi_*] - \mu[I]) = 0, \tag{3.170}$$

by which we obtain

$$\mu_{1,2} = \alpha \pm \sqrt{\alpha^2 - (\varphi_{11}\varphi_{22} - \varphi_{12}\varphi_{21})}, \tag{3.171}$$

where

$$\alpha = \frac{1}{2}(\varphi_{11} + \varphi_{22}). \tag{3.172}$$

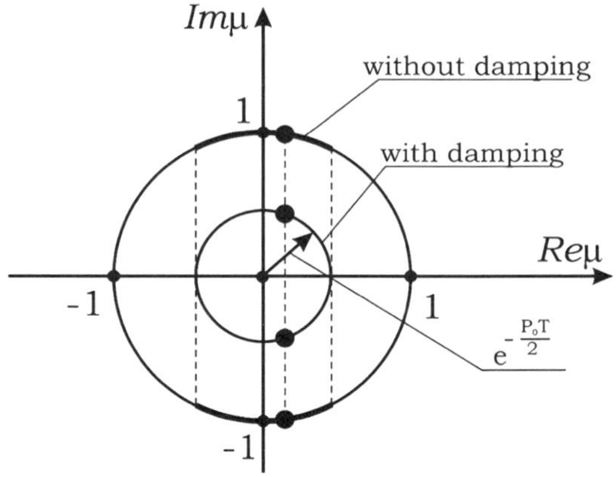

Fig. 3.7 Position of characteristic multipliers in the parametric system with and without damping

Taking into account Eq. (3.169) in (3.171) we find

$$\mu_{1,2} = \alpha \pm \sqrt{\alpha^2 - \exp(-P_0T)}. \tag{3.173}$$

It follows from Eq. (3.173) that if $|\alpha| \le \exp\left(\frac{-P_0T}{2}\right)$, then roots μ_1 and μ_2 are complex conjugate. Furthermore, (3.173) implies

$$|\mu_1| = |\mu_2| = \exp\left(\frac{-P_0T}{2}\right) < 1. \tag{3.174}$$

It follows from the above considerations that characteristic multipliers lie within a unit circle of the complex plane Imμ, Reμ. Figure 3.7 illustrates this situation.

The solutions corresponding to μ_1 and μ_2 are asymptotically stable. The second case to consider is α belonging to the interval

$$e^{\frac{-P_0T}{2}} < |\alpha| < \frac{1}{2}[1 + e^{-P_0T}]. \tag{3.175}$$

We have (assuming $\mu = 1$ in the formula (3.173))

$$(1 - \alpha)^2 = \alpha^2 - e^{-P_0T},$$

hence

$$\alpha = \frac{1}{2}(1 + e^{-P_0T}).$$

In this case, characteristic multipliers are real and lie within a unit circle. Finally, in the last case, i.e. for

$$|\alpha| > \frac{1}{2}[1 + e^{-P_0T}] \tag{3.176}$$

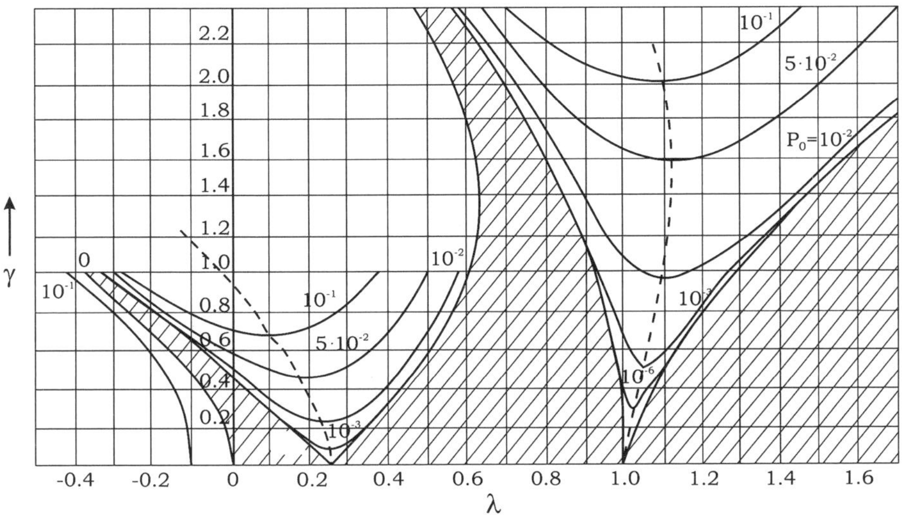

Fig. 3.8 Boundaries of stability loss for Mathieu equation determined by Kotowski for different values of damping P_0

an absolute value of one of the multipliers exceeds 1, i.e. the corresponding solution grows in time. It is noteworthy that in this case the system is unstable despite the positive damping in this system. This result cannot be obtained, when in Eq. (3.163) $Q(t) = Q_0 = \text{const.}$

Now, consider a generalized Mathieu equation containing a linear damping term of the form

$$y'' + P_0 y' + (\lambda + \gamma \cos \tau) y = 0, \tag{3.177}$$

where: $(') = \frac{d}{d\tau}$, $\tau = \Omega t$, $\lambda = \lambda^* \Omega^{-2}$, $\gamma = \gamma^* \Omega^{-2}$.

This problem has been completely solved by Kotowski [137], and the results of his calculations are given in Fig. 3.8.

It follows from the figure that regions of stability expand as damping grows. As λ grows, another regions of stability loss appear. The regions gets more narrow and hence these are less dangerous.

Example 3.9. A rotor of rectangular cross-section $b \times h$ and length l rotates at angular velocity ω. In the middle of its length, there is a mass m which can move only along the y-axis (see Fig. 3.9). During the motion in this direction, the mass is exerted to the spring force and damping coming from two springs and the dampers. Derive the governing equation of the rotor dynamics.

Let us take a cartesian coordinate systems x_1, y_1 (moving with the rotor) and absolute x, y (Fig. 3.9).

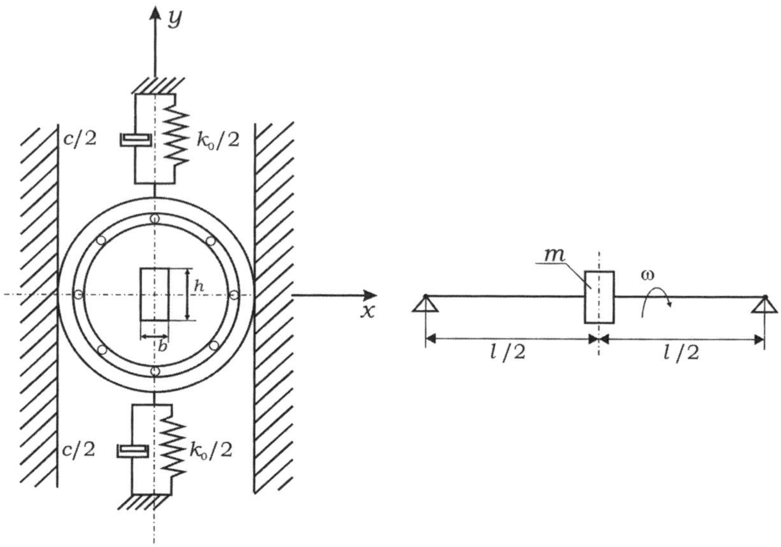

Fig. 3.9 Scheme of the rotor vibrations (the rotor has a rectangular cross section)

The relationship between the coordinates is expressed by

$$x = x_1 \cos \omega t - y_1 \sin \omega t, \quad y = y_1 \cos \omega t + x_1 \sin \omega t.$$

The moment of inertia of cross-section with respect to the x-axis equals

$$I_x = \int_S y^2 dS,$$

and hence

$$I_x = \cos^2 \omega t \int_S y_1^2 dS + \sin^2 \omega t \int_S x_1^2 dS + \sin 2\omega t \int_S x_1 y_1 dS.$$

Denoting

$$I_{x_1} = \int_S y_1^2 dS, \quad I_{y_1} = \int_S x_1^2 dS \quad \text{and} \quad I_{x_1 y_1} = \int_S x_1 y_1 dS = 0,$$

we get

$$I_x = \frac{I_{x_1} + I_{y_1}}{2} + \frac{I_{x_1} - I_{y_1}}{2} \cos 2\omega t.$$

For the rotor position determined by the angle α, the stiffness is

$$k_{x_1} = \frac{48EI_{x_1}}{l^3} \quad (\alpha = 0), \quad k_{y_1} = \frac{48EI_{y_1}}{l^3} \quad (\alpha = \frac{\pi}{2}),$$

The equation of motion of mass m body takes the form

$$m\ddot{y} + c\dot{y} + \left(k_0 + \frac{k_{x_1} + k_{y_1}}{2} + \frac{k_{x_1} - k_{y_1}}{2}\cos 2\omega t\right) y = 0.$$

In order to transform the above equation to (3.177) we make the following substitution

$$\tau = 2\omega t$$

and we get

$$P_0 = \frac{c}{2\omega m}, \quad \lambda = \frac{k_0}{4m\omega^2} + \frac{k_{x_1} + k_{y_1}}{8m\omega^2}, \quad \gamma = \frac{k_{x_1} - k_{y_1}}{8m\omega^2}.$$

□

3.2.6 Modelling of Generalized Parametric Oscillator

In what follows we consider a general form of the equation of parametric oscillator

$$\ddot{x} + \omega^2(t)x = 0, \tag{3.178}$$

where: $\omega(t) > 0$ and its graph is presented in Fig. 3.10 (note that, for $\omega_+ = \omega_-$ and $\omega(t) = \omega(t+T)$ the problem of modelling reduces to Meissner's equation).

The model equation (3.178) plays an important role in both classical and quantum mechanics. In classical mechanics, the presented variation of $\omega(t)$ is related to the variation of an oscillator stiffness, whilst in quantum mechanics Eq. (3.178) governs a reduced time-independent one-dimensional Schrödinger equation.

It follows from Fig. 3.10 that $\omega(t)$ possesses the following limiting form

$$\lim_{t \to \pm\infty} \omega(t) = \omega_\pm. \tag{3.179}$$

In other words, $\omega(t)$ presents an asymmetrical smooth and rectangular pulse. In further considerations, we will base ourselves on [229,230] and its generalization. We will assume the following analytic functions describing an asymmetric smooth pulse

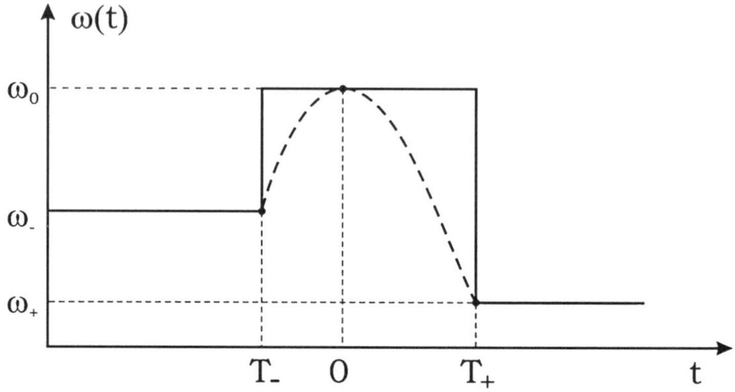

Fig. 3.10 Piecewise smooth (*solid line*) and smooth (*dashed curve*) function $\omega(t)$

$$\begin{aligned}
\omega(t) &= \omega_-, && -\infty < t \le T_- \\
\omega(t) &= \sqrt{\omega_-^2 + (\omega_0^2 - \omega_-^2)\cos^2\left(\frac{\pi t}{2T_-}\right)}, && T_- < t < 0, \\
\omega(t) &= \sqrt{\omega_+^2 + (\omega_0^2 - \omega_+^2)\cos^2\left(\frac{\pi t}{2T_+}\right)}, && 0 < t < T_+, \\
\omega(t) &= \omega_+, && T_+ < t \le +\infty
\end{aligned} \tag{3.180}$$

Our goal is to find a solution of Eq. (3.178), when all the parameters describing the pulse are known and the asymptotic properties are taken into account (3.179). We seek a solution of (3.178) in the form

$$x(t) = \rho e^{\pm i\phi}, \tag{3.181}$$

where: $\rho = \rho(t)$, $\phi = \phi(t)$.

Differentiating $x(t)$ we obtain

$$\begin{aligned}
\dot{x} &= \dot{\rho}e^{\pm i\phi} \pm i\rho\dot{\phi}e^{\pm i\phi}, \\
\ddot{x} &= \ddot{\rho}e^{\pm i\phi} \pm 2i\dot{\rho}\dot{\phi}e^{\pm i\phi} + i\rho\ddot{\phi}e^{\pm i\phi} - \rho\dot{\phi}^2e^{\pm i\phi}.
\end{aligned} \tag{3.182}$$

Substituting (3.182) into (3.178) we obtain

$$\ddot{\rho} + 2i\dot{\rho}\dot{\phi} + i\rho\ddot{\phi} - \rho\dot{\phi}^2 + \omega^2(t)\rho = 0 \tag{3.183}$$

and excluding real and imaginary parts, we have

$$\ddot{\rho} + \omega^2(t)\rho = \rho\dot{\phi}^2, \tag{3.184}$$

$$2\dot{\rho}\dot{\phi} + \rho\ddot{\phi} = 0. \tag{3.185}$$

Note that if (3.181) satisfies (3.178), then it satisfies (3.184) and (3.185) as well. Consider a particular solution of the form

$$\dot{\phi} = \frac{1}{\rho^2(t)} \tag{3.186}$$

Substituting

$$\ddot{\phi}(t) = \frac{d}{dt}\left(\frac{1}{\rho^2(t)}\right) = -2\dot{\rho}\rho^{-3}$$

into Eq. (3.185) we get the identity

$$2\dot{\rho}\frac{1}{\rho^2} + \rho\left(-2\dot{\rho}\rho^{-3}\right) \equiv 0,$$

whereas by Eq. (3.184) we obtain the Milne–Pinney [172, 200] equation of the form

$$\ddot{\rho} + \omega^2(t)\rho = \frac{1}{\rho^3(t)}. \tag{3.187}$$

Instead of the linear equation (3.178) we have obtained the nonlinear equation (3.187). However, it turns out that in the latter case any two solutions of Eq. (3.187), i.e. $\rho(t)$ and $u(t)$ are connected by the relationship

$$\rho(t) = \pm u(t)\sqrt{M \pm \sqrt{M^2 - 1}\cos\left(2\int^t u^{-2}(\tau)\right)d\tau}, \tag{3.188}$$

where M is invariant. On the other hand (see [24, 25, 28, 29]) it is known that a general solution of Eq. (3.187) can be expressed through two particular solutions of Eq. (3.188) according to the following nonlinear superposition principle

$$\rho(t) = \sqrt{Ax_1^2(t) + 2Bx_1(t)x_2(t) + Cx_2^2(t)}. \tag{3.189}$$

Since the solutions $x_1(t)$, $x_2(t)$ are linearly independent, the Wronski relationship is satisfied

$$x_1\dot{x}_2 - \dot{x}_1x_2 = 1,$$

and moreover, the coefficients satisfy the equation

$$AC - B^2 = 1.$$

Note that according to (3.181), we can assume the following form of the sought solutions

$$x_1(t) = u(t)\cos\Theta(t),$$
$$x_2(t) = u(t)\sin\Theta(t), \tag{3.190}$$

while according to (3.186) we have

$$\dot{\Theta} = u^{-2}(t), \tag{3.191}$$

and $u(t)$ satisfies Milne–Pinney equation (3.187). Substituting (3.190) into (3.189) we get

$$\rho(t) = \pm u(t)\sqrt{A\cos^2\Theta + 2\sqrt{AC-1}\sin\Theta\cos\Theta + C\sin^2\Theta}.$$

The above expression can be transformed into the following form

$$\rho(t) = \pm u(t)\sqrt{\frac{1}{2}(A+C) + \frac{1}{2}\sqrt{(A+C+2)(A+C-2)}\cos 2(\Theta(t)-\gamma)},$$

where

$$\tan 2\gamma = 2\frac{\sqrt{AC-1}}{A-C}.$$

Note that, according to (3.191) we have

$$\Theta(t) - \gamma = \int^t u^{-2}(\tau)\,d\tau + C - \gamma = \int^t u^{-2}(\tau)\,d\tau,$$

for $C = \gamma$. Assuming $A + C = 2M$ we get

$$\rho(t) = \pm u(t)\sqrt{M + \frac{1}{2}\sqrt{(2M+2)(2M-2)\cos(2\int^t u^{-2}(\tau))d\tau}}$$
$$= \pm u(t)\sqrt{M + \sqrt{(M^2-1)\cos(2\int^t u^{-2}(\tau))d\tau}},$$

which is in agreement with (3.188). According to (3.187), any solution $u(t)$ generates an associated solution $\rho(t)$. However, in our case we must define two quantities: an invariant M and a lower limit of integration. A sign ($\pm$) means that any two solutions $u(t)$, $\rho(t)$ possess also two solutions, which are a mirror reflection of them.

Any two solutions $u(t)$, $\rho(t)$ of Milne–Pinney equation (3.187) defines the following invariant

$$M = \frac{1}{2}\left[(\dot{\rho}u - \rho\dot{u})^2 + \left(\frac{\rho}{u}\right)^2 + \left(\frac{u}{\rho}\right)^2\right]. \tag{3.192}$$

In order to determine the value of M, it is sufficient to take an arbitrary instant of time. Note that any solution of Eq. (3.187) defines a pair of independent solutions of Eq. (3.178) according to (3.181).

Our task is to analyse the following three solutions:

(i) a pre-pulse solution corresponding to the constant ω_-;
(ii) a post-pulse solution corresponding to the constant ω_+;
(iii) a pulse solution $\rho_0(t)$ corresponding to the constant $\omega(0) = \omega_0$.

Let us begin from the case (iii). If $\rho_0(t)$ is approximately constant value, then $\dot{\rho}_0(0) = \ddot{\rho}_0(0) = 0$. By (3.187) we get

$$\omega^2(0) = \frac{1}{\rho_0^4(0)},$$

or

$$\rho_0(0) = \frac{1}{\sqrt{\omega(0)}}, \qquad \dot{\rho}_0(0) = 0. \tag{3.193}$$

Our goal is to estimate asymptotically $\rho_0(t)$ as $t \to \pm\infty$, provided that we know asymptotic values of $u(t)$ connected with (i) and (ii), and the associated solution $\rho_0(t)$ defined by Eq. (3.192). Constant values (asymptotic) of the solution $u(t)$ are determined by the values $\omega_\pm$ and Eq. (3.187). We have

$$\lim_{t\to\pm\infty} \omega_\pm^2 u(t) = \lim_{t\to\pm\infty} \frac{1}{u^3(t)},$$

and hence

$$\lim_{t\to\pm\infty} u_\pm(t) = \frac{1}{\sqrt{\omega_+}}, \qquad \lim_{t\to\pm\infty} \dot{u}_\pm(t) = 0. \tag{3.194}$$

Two different solutions $u_+(t)$ and $u_-(t)$ remain constant as $t \to +\infty$ and $t \to -\infty$. Two pairs of the solutions $\{\rho_0(t), u_+(t)\}$ and $\{\rho_0(t), u_-(t)\}$ are connected via the nonlinear superposition (3.188). On the other hand, pairs of the conditions (3.193) and (3.194) serve for determining the lower limit of integration and the invariants $M_\pm$.

Now, let us estimate the asymptotic values of limits $\lim_{t\to\pm\infty} \rho_0(t)$ provided that we know the asymptotic values (3.194). Let us denote $t_\pm$ as lower limits of the integral. We obtain

$$\lim_{t\to\pm\infty}\int_{t_\pm}^{t}\frac{d\tau}{u_\pm^2(\tau)}=\int_{t_\pm}^{t}(\sqrt{\omega_\pm})^2=\omega_\pm(t-t_\pm).$$

By Eqs. (3.188) we get

$$\lim_{t\to\pm\infty}\rho_0(t)=\frac{1}{\sqrt{\omega_\pm}}\sqrt{M_\pm-\sqrt{M_\pm^2-1}\cos(2\omega_\pm(t-t_\pm))}\ . \tag{3.195}$$

In the next step, we will determine the constants $t_\pm$ and $M_\pm$. First, we determine $M_\pm$ by (3.195) after taking into account the following equality

$$\rho_0\left(-T_-\right)=\rho\left(T_+\right)=\frac{1}{\sqrt{\omega_0}}\ . \tag{3.196}$$

Since as $t=\pm\infty$ the solution $u\left(t\right)=\left(\omega_\pm\right)^{-\frac{1}{2}}$, and for $-\infty\leq t\leq T_-$, $T_+<t<+\infty$ the frequency of the oscillator equals ω_- and ω_+, respectively, then we have

$$u\left(-T_-\right)=\frac{1}{\sqrt{\omega_-}},\quad u\left(T_+\right)=\frac{1}{\sqrt{\omega_+}}\ . \tag{3.197}$$

By (3.192), taking into account (3.196), (3.197) and assuming $t=T_-$ and $t=T_+$ we get

$$M_\pm=\frac{1}{2}\left[\left(\frac{1}{\sqrt{\omega_0}}\right)^2\cdot\left(\frac{1}{\sqrt{\omega_+}}\right)^{-2}+\left(\frac{1}{\sqrt{\omega_+}}\right)^2\cdot\left(\frac{1}{\sqrt{\omega_0}}\right)^{-2}\right],$$

hence

$$M_\pm=\frac{1}{2}\left(\frac{\omega_\pm}{\omega_0}+\frac{\omega_0}{\omega_\pm}\right). \tag{3.198}$$

All the solutions serving for determining the invariant value M are constant solutions, which does not depend on $T_\pm$. By Eq. (3.195) we get the value of ρ_0 for $t=T_+$ (or $t=-T_-$). Further considerations will be carried out for $t=T_+$. In this case, the Milne invariant is (see (3.192))

$$M_+=\frac{1}{2}\left[\left(\dot{\rho}_0\left(T_+\right)\cdot\frac{1}{\sqrt{\omega_+}}\right)^2+\left(\sqrt{\omega_+}\rho_0\right)^2+\left(\frac{1}{\sqrt{\omega_+}\rho_0}\right)^2\right],$$

and after transformation

$$M_+=\frac{1}{2\omega_+}\left[\dot{\rho}_0(T_+)+\rho_0^2(T_+)\omega_+^2+\frac{1}{\rho_0^2}(T_+)\right]. \tag{3.199}$$

By Eq. (3.195) we get

$$\dot{\rho}_0^+ (T_+) = \frac{d\rho_0^+}{dt} (T_+) = \frac{1}{\sqrt{\omega_+}} \frac{\sqrt{M_+^2 - 1} \sin 2\omega_+ (T_+ - t_+) 2\omega_+}{2\sqrt{M_+ - \sqrt{M_+^2 - 1} \cos (2\omega_+ (T_+ - t_+))}},$$

and hence

$$\rho_0^+ \cdot \dot{\rho}_0^+ = \sqrt{M_+^2 - 1} \sin 2\omega_+ (T_+ - t_+), \tag{3.200}$$

where ρ_0^+ is defined by (3.195) for $t = T_+$. By the above equation we find

$$\arcsin \frac{\rho_0^+ \dot{\rho}_0^+}{\sqrt{M_+^2 - 1}} = 2\omega_+ T_+ - 2\omega_+ t_+,$$

hence

$$t_+ = T_+ - \frac{1}{2\omega_+} \arcsin \frac{\rho_0^+ \dot{\rho}_0^+}{\sqrt{M_+^2 - 1}}. \tag{3.201}$$

Similarly, one can show that

$$t_- = T_- + \frac{1}{2\omega_-} \arcsin \frac{\rho_0^- \dot{\rho}_0^-}{\sqrt{M_-^2 - 1}} \tag{3.202}$$

In the considered case $\dot{\rho}_0 (T_+) = \dot{\rho}_0 (T_-) = 0$, and thus we have $t_+ = T_+$. Now, we go back to a solution of Eq. (3.178) described by the formula (3.181). Let us remind that asymptotics of the function $\rho (t)$ are determined by the limit (3.195). A general solution of Eq. (3.178) is a linear combination of $\cos (\omega_+ t)$ and $\sin(\omega_+ t)$, hence a pair of real fundamental solutions reads

$$\begin{aligned} A (t) &= \rho_0 (t) \cos \varphi (t), \\ B (t) &= \rho_0 (t) \sin \varphi (t), \end{aligned} \tag{3.203}$$

where

$$\phi (t) = \int_0^t \frac{d\tau}{\rho_0^2 (\tau)}. \tag{3.204}$$

Note that a phase-shift between the exact solution and the corresponding asymptotic solution $\rho_{0,+\infty} (t)$ reads

$$\alpha^+ = \int_0^t \left(\frac{1}{\rho_0^2 (\tau)} - \frac{1}{\rho_{0,\infty}^2 (\tau)} \right) d\tau,$$

or

$$\int_0^t \frac{d\tau}{\rho_0^2(\tau)} = \int_0^t \frac{1}{\rho_{0,\infty}^2} d\tau + \alpha^+ \quad \text{as} \quad t \to +\infty. \tag{3.205}$$

By Eq. (3.203) as $t \to +\infty$ we get

$$\lim_{t\to+\infty} A(t) = A^+(t) = A^+(t) = \rho_{0,\infty}(t)\cos\left(\int_0^t \rho_{0,\infty}^{-2}(\tau)d\tau + \alpha^+\right),$$

$$\lim_{t\to+\infty} B(t) = A^+(t) = B^+(t) = \rho_{0,\infty}(t)\cos\left(\int_0^t \rho_{0,\infty}^{-2}(\tau)d\tau + \alpha^+\right). \tag{3.206}$$

According to (3.195) we have

$$\int_0^t \frac{d\tau}{\rho_{0,\infty}^2} = \int_0^t \frac{\omega_+ d\tau}{M_+ - \sqrt{M_+^2 - 1}\cos\left(2\omega_+\left(\tau - t_+\right)\right)} \tag{3.207}$$

In the handbook [15] we find that

$$\int \frac{dx}{b + c\cos ax} = \frac{2}{a\sqrt{b^2 - c^2}}\arctan\frac{(b-c)\tan\frac{1}{2}ax}{\sqrt{b^2 - c^2}}, \quad b^2 > c^2,$$

and

$$\int \frac{dx}{b + c\cos ax} = \frac{1}{a\sqrt{c^2 - b^2}}\ln\left|\frac{(c-b)\tan\frac{1}{2}ax + \sqrt{c^2 + b^2}}{(c-b)\tan\frac{1}{2}ax + \sqrt{c^2 - b^2}}\right|, \quad b^2 < c^2.$$

In our case (see (3.208)) we have

$$b^2 - c^2 = M_+^2 - \left(M_+^2 - 1\right) = 1,$$

and furthermore

$$x = \tau - t_+,\ dx = d\tau,\ a = 2\omega_+.$$

By Eq. (3.207) we obtain

$$\begin{aligned}
\int_0^t \frac{d\tau}{\rho_{0,\infty}^2} &= \omega_+ \int_0^t \frac{d\tau}{b + c\cos a(\tau - t_+)} = \int_{-t_+}^{t-t_+} \frac{dx}{b + c\cos ax} \\
&= \omega_+ \left\{\frac{1}{\omega_+}\arctan\left[\left(M_+ - \sqrt{M_+^2 - 1}\right)\tan\omega_+ x\right]\right\}_{-t_+}^{t-t_+} \\
&= \arctan\left[\left(M_+ - \sqrt{M_+^2 - 1}\right)\tan\omega_+(t - t_+)\right]
\end{aligned}$$

$$+ - \arctan\left[\left(M_{+} - \sqrt{M_{+}^{2} - 1}\right)\tan\omega_{+}(-t_{+})\right]$$
$$= \phi_{+}(t) + \Delta_{+}, \tag{3.208}$$

where

$$\phi_{+}(t) = \arctan[(M_{+} - \sqrt{M_{+}^{2} - 1})\tan\omega_{+}(t - t_{+})],$$
$$\Delta_{+} = \arctan[(M_{+} - \sqrt{M_{+}^{2} - 1})\tan\omega_{+}t_{+}]. \tag{3.209}$$

By (3.206), (3.208) and (3.209) we get

$$A^{+}(t) = \rho_{0,\infty}(t)\cos[\phi_{+}(t) + \Delta_{+} + \alpha_{+}],$$
$$B^{+}(t) = \rho_{0,\infty}(t)\sin[\phi_{+}(t) + \Delta_{+} + \alpha_{+}]. \tag{3.210}$$

Assuming that the phase-shift is constant $\Delta_{+} + \alpha_{+} = 0$, by Eq. (3.210) we get

$$A^{+}(t) = \rho_{0,\infty}(t)\cos\varphi_{+}(t),$$
$$B^{+}(t) = \rho_{0,\infty}(t)\sin\varphi_{+}(t), \tag{3.211}$$

where

$$\phi_{+}(t) = \arctan[(M_{+} - \sqrt{M_{+}^{2} - 1})\tan\omega_{+}(t - t_{+})],$$
$$\rho_{0,\infty} = \frac{1}{\sqrt{\omega_{+}}}\sqrt{M_{+} - \sqrt{M_{+}^{2} - 1}\cos[2\omega_{+}(t - t_{+})]}. \tag{3.212}$$

Equation (3.211) could be transformed into the following form

$$A_{+}(t) = \sqrt{\frac{M_{+} - \sqrt{M_{+}^{2} - 1}}{\omega_{+}}}\cos[\omega_{+}(t - t_{+})],$$
$$B_{+}(t) = \sqrt{\frac{M_{+} - \sqrt{M_{+}^{2} - 1}}{\omega_{+}}}\sin[\omega_{+}(t - t_{+})]. \tag{3.213}$$

Moreover, the considerations carried out in this chapter have been discussed in more detail in this reference.

3.2.7 ODEs with Constant Coefficient

Theory of vibrations is one of the developed branches of mechanics, applied mathematics and physics. There is a large amount of literature devoted to the theory of vibrations of lumped (modeled by ordinary differential equations) and continuous systems (modeled by partial differential equations). In this work, we will give some fundamentals concerning vibrations of lumped systems from mechanics viewpoint (see [12, 13]).

Assume that particles of a lumped material system have been connected by massless elastic-damping elements. The connections generate forces and torques, which depend on displacements and velocities of these points. Imposing initial conditions, for example initial displacement and velocity of particles at particular values of the system parameters (mass, stiffness, damping, geometry of the system), can result in vibrations of a considered mechanical system. We will confine ourselves only to small vibrations in neighbourhood of a particular static configuration of the system (equilibrium position). Let us remind that a system (in nonlinear systems) can possess a few different equilibrium positions.

In the case of small vibrations around the analysed equilibrium position, one can perform the linearization procedure which relies on Taylor (Maclaurin) expanding of some functions up to linear terms (though the linearization is not always possible). Consequently, the problem reduces to the analysis of linear differential equations with constant or variable coefficients.

Let us consider the case of driven vibrations of one degree-of-freedom system, depicted in Fig. 3.11.

The equation of motion takes the form

$$\ddot{x} + 2h\dot{x} + \alpha^2 x = q \cos \omega t, \tag{3.214}$$

where

$$\frac{c}{m} = 2h, \quad \alpha^2 = \frac{k}{m}, \quad q = \frac{F_0}{m}.$$

First, let us consider free vibrations ($F_0 = 0$). We seek solutions of the form

$$x(t) = e^{rt}. \tag{3.215}$$

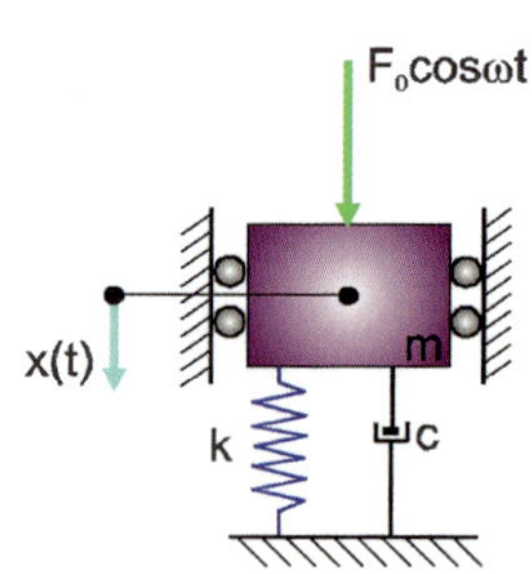

Fig. 3.11 Vibrations of one degree-of-freedom system with damping under harmonic excitation

Substituting (3.215) into (3.214) we obtain the following characteristic equation

$$r^2 + 2hr + \alpha^2 = 0. \tag{3.216}$$

We get the following roots of the above equation

$$r_{1,2} = -h \pm \sqrt{h^2 - \alpha^2}. \tag{3.217}$$

A general solution has the form

$$x(t) = A_1 e^{r_1 t} + A_2 e^{r_2 t}. \tag{3.218}$$

If $h > \alpha$, then the solution has the form

$$x(t) = e^{-ht}\left(A_1 e^{\sqrt{h^2-\alpha^2}t} + A_2 e^{-\sqrt{h^2-\alpha^2}t}\right), \tag{3.219}$$

and as one can easily see $\lim_{t\to\infty} x(t) = 0$. The function $x(t)$ approaches zero without oscillations.

If $\alpha > h$, then Eq. (3.217) takes the form

$$r_{1,2} = -h \pm i\lambda, \quad i^2 = -1, \tag{3.220}$$

where $\lambda = \sqrt{\alpha^2 - h^2}$ and the solution has the form

$$x(t) = \frac{1}{2}e^{-ht}\mathrm{Re}(\bar{A}e^{i\lambda t} + Ae^{-i\lambda t}), \tag{3.221}$$

where $\bar{A}$ and A are complex conjugate.

Euler's formula enables us to find a real form of the solution

$$e^{i\lambda t} = \cos\lambda t + i\sin\lambda t. \tag{3.222}$$

We have

$$\begin{gathered} x(t) = \frac{1}{2}e^{-ht}\mathrm{Re}[(A_R - A_I i)(\cos\lambda t + i\sin\lambda t) \\ +(A_R + A_I i)(\cos\lambda t - i\sin\lambda t)] = e^{-ht}(A_1\cos\lambda t + A_2\sin\lambda t), \end{gathered} \tag{3.223}$$

where we assumed: $A = A_R + A_I i$, $A_R = A_1$, $A_I = A_2$.

We have the third case left, namely $h = \alpha$. This corresponds to the critical damping coefficient c_{cr} determined by the formula below

$$c_{cr} = 2m\sqrt{\frac{k}{m}} = 2\sqrt{km}. \tag{3.224}$$

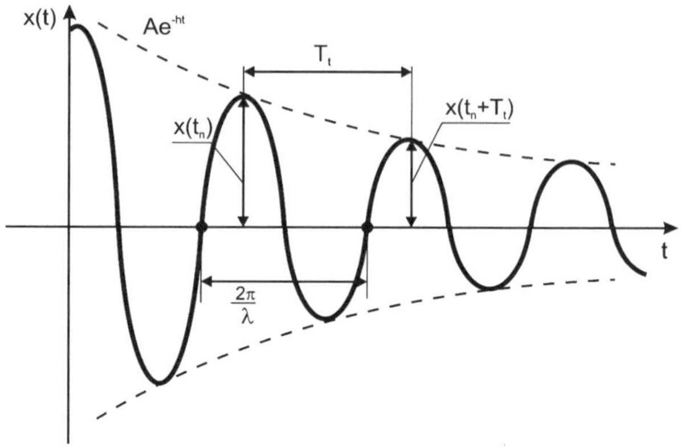

Fig. 3.12 Free damped vibrations of one degree-of-freedom system

In this case we have a double root of the characteristic equation and a solution has the form

$$x(t) = (A_1 + A_2 t)e^{-ht}. \tag{3.225}$$

There are no oscillations in the case of critical damping. The solution (3.223) describes damped harmonic oscillations and its graph is depicted in Fig. 3.12.

A characteristic feature of damped vibrations is to attain maxima (minima), which are distant from each other by the so-called period of damped vibrations $T_t = \frac{2\pi}{\lambda}$. Thus, one can introduce a notion of logarithmic decrement

$$\delta = \ln \frac{x(t)}{x(t+T_t)} = \ln \frac{(A_1 \cos \lambda t_n + A_2 \sin \lambda t_n)e^{-ht_n}}{\left[A_1 \cos \lambda \left(t_n + \frac{2\pi}{\lambda}\right) + A_2 \sin \lambda \left(t_n + \frac{2\pi}{\lambda}\right)\right] e^{-h\left(t_n + \frac{2\pi}{\lambda}\right)}}$$
$$= \ln \frac{1}{e^{-h\frac{2\pi}{\lambda}}} = \frac{2\pi}{\lambda} h. \tag{3.226}$$

Logarithmic decrement can serve for determination of a viscous damping coefficient of vibrations. When δ is known (a natural logarithm of a ratio of two sequential maximal amplitudes and the corresponding time of their occurrence) then the coefficient of damping can be found as

$$c = 2m\frac{\delta}{T_t}. \tag{3.227}$$

At the end, we consider the case of driven oscillations, namely $F_0 \neq 0$. In order to get a solution we make use of complex numbers. Equation (3.214) takes the form

$$\ddot{x} + 2h\dot{x} + \alpha^2 x = q(\cos \omega t + i \sin \omega t) = qe^{i\omega t}. \tag{3.228}$$

The above second-order differential equation is non-homogeneous. Its solution is a superposition of a general solution of the homogeneous differential equation (Eq. (3.228) as $q = 0$) and a particular solution of the non-homogeneous equation (3.228). The latter is sought in the form

$$x = \bar{A}e^{i\omega t}. \tag{3.229}$$

Substituting (3.229) into (3.228) we get

$$(-\omega^2 + 2h\omega i + \alpha^2)\bar{A} = q, \tag{3.230}$$

where $\bar{A}$ is complex conjugate of $A = A_R + A_I i$.

By the above equation

$$\bar{A} = A_R - iA_I = \frac{q}{(\alpha^2 - \omega^2) + 2h\omega i}. \tag{3.231}$$

In order to find A_R and A_I let us multiply the nominator and the denominator of the right-hand side (3.231) by $[(\alpha^2 - \omega^2) - 2h\omega i]$. We obtain

$$A_R - iA_I = \frac{q(\alpha^2 - \omega^2)}{(\alpha^2 - \omega^2)^2 + 4h^2\omega^2} - \frac{2h\omega q}{(\alpha^2 - \omega^2)^2 + 4h^2\omega^2}i, \tag{3.232}$$

and hence

$$A_R = \frac{q(\alpha^2 - \omega^2)}{(\alpha^2 - \omega^2)^2 + 4h^2\omega^2}, \quad A_I = \frac{2h\omega q}{(\alpha^2 - \omega^2)^2 + 4h^2\omega^2}. \tag{3.233}$$

The solution (3.229) possesses a real interpretation of the form

$$\begin{aligned} \operatorname{Re} x(t) &= \operatorname{Re}[(A_R - A_I i)(\cos\omega t + i\sin\omega t)] \\ &= A_R\cos\omega t + A_I\sin\omega t = a\cos(\omega t - \beta). \end{aligned} \tag{3.234}$$

Since

$$A_R\cos\omega t + A_I\sin\omega t = a\cos\beta\cos\omega t + a\sin\beta\sin\omega t, \tag{3.235}$$

then

$$A_R = a\cos\beta,\ A_I = a\sin\beta, \tag{3.236}$$

and hence

$$a = \sqrt{A_R^2 + A_I^2}, \quad \tan\beta = \frac{A_I}{A_R}. \tag{3.237}$$

Since a general solution of the homogeneous equation (3.223) decays as $t \to \infty$, only a particular solution (3.234) of the inhomogeneous equation is left. As one can see, the response of the system is harmonic and shifted with the phase angle β with respect to the excitation. By Eq. (3.237) we get

$$a = \frac{q}{\sqrt{(\alpha^2-\omega^2)^2+4h^2\omega^2}} = \frac{1}{\alpha^2}\frac{1}{\sqrt{\left[1-\left(\frac{\omega}{\alpha}\right)^2\right]^2+4\left(\frac{h}{\alpha}\right)^2\left(\frac{\omega}{\alpha}\right)^2}},$$

$$\beta = \text{arc tan}\frac{2h\omega}{\alpha^2-\omega^2} = \text{arc tan}\frac{2\frac{h}{\alpha}\frac{\omega}{\alpha}}{1-\left(\frac{\omega}{\alpha}\right)^2}. \tag{3.238}$$

The coefficient $\frac{q}{\alpha^2} \equiv \frac{F_0}{k} = x_{\text{st}}$ describes static displacement. The first equation of (3.238) is transformed into the form

$$\nu \equiv \nu\left(\frac{\omega}{\alpha}\right) = \frac{a}{x_s} = \frac{1}{\sqrt{\left[1-\left(\frac{\omega}{\alpha}\right)^2\right]^2+4\left(\frac{h}{\alpha}\right)^2\left(\frac{\omega}{\alpha}\right)^2}}. \tag{3.239}$$

The above function $\nu = \nu\left(\frac{\omega}{\alpha}\right)$ describes *amplitude–frequency* characteristics (see Fig. 3.13).

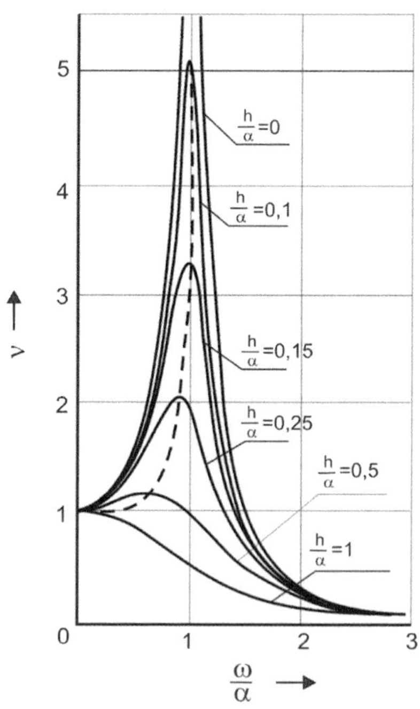

Fig. 3.13 The amplitude–frequency characteristics of a harmonically driven oscillator with damping for selected values of the parameter $\frac{h}{\alpha}$

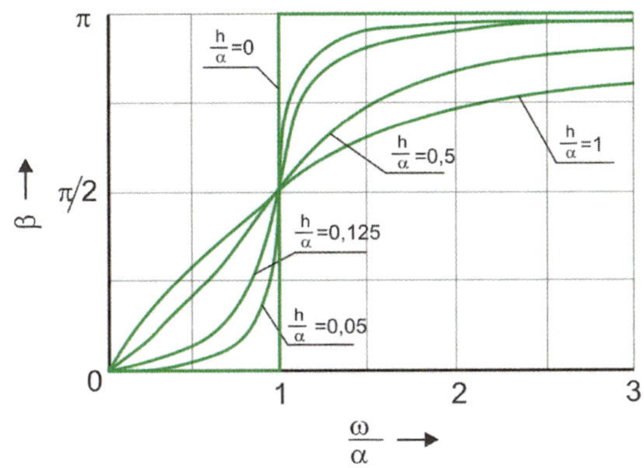

Fig. 3.14 The phase-frequency characteristics of a harmonically driven oscillator with damping for selected values of $\frac{h}{\alpha}$

The second equation of the system (3.238) describes phase-frequency characteristics (see Fig. 3.14).

As $h = 0$, we get the case of harmonically driven oscillator without damping. According to (3.234) we have respectively

$$x(t) = a\cos\omega t, \tag{3.240}$$

and

$$x(t) = -a\cos\omega t, \tag{3.241}$$

as $\omega > \alpha$. This means that before resonance the vibrations are in-phase with the external excitation, whereas after resonance the vibrations are anti-phase with the excitation.

Resonance appears when $\omega = \alpha$ and then the amplitude attains infinity. The damping makes the amplitudes of resonant vibrations decrease. The case $h = 0$ has been also presented in Figs. 3.13 and 3.14. If in Fig. 3.11 we take any time-dependent excitation $F(t)$ instead harmonic one, then the equation of motion of this system takes the form

$$\ddot{x} + 2h\dot{x} + \alpha^2 x = q(t), \tag{3.242}$$

where $q(t) = \frac{1}{m}F(t)$.

In order to solve Eq. (3.242) we apply Laplace transform, i.e. a function $x(t)$ is associated with a function $X(s)$ of a complex variable via the following integral transformation

$$X(s) = \int_{-\infty}^{+\infty} x(t)e^{-st}\,dt, \tag{3.243}$$

where $s = c + i\omega$. The initial conditions for Eq. (3.242) will be determined for $t_0 = 0$. Taking algebraic transformations related to Eq. (3.242) and the variable s in complex domain, we can make an inverse transformation

$$x(t) = L^{-1}[X(s)] = \frac{1}{2\pi i} \int\limits_{c-i\omega}^{c+i\omega} X(s)e^{-st}dt. \tag{3.244}$$

In practice, such integrals are not calculated. We usually make use of tables of originals and transforms. Since, we are interested only in a particular solution, we put the following initial conditions $x(0) = \dot{x}(0) = 0$. Applying Laplace transform to Eq. (3.242), we obtain

$$(s^2 + 2hs + \alpha^2)X(s) = Q(s). \tag{3.245}$$

Consequently, we obtain

$$X(s) = \frac{Q(s)}{\alpha^2} \frac{1}{\frac{1}{\alpha^2}s^2 + 2\frac{h}{\alpha}\frac{1}{\alpha}s + 1}. \tag{3.246}$$

The right-hand side of Eq. (3.246) is a quotient of two Laplace transforms. According to *Borel's theorem on convolution*

$$f_1(t) * f_2(t) = \int\limits_0^t f_1(\tau)f_2(t-\tau)d\tau = \int\limits_0^t f_1(t-\tau)f_2(\tau)d\tau,$$

$$L[f_1(t) * f_2(t)] = F_1(s)F_2(s),$$

$$L^{-1}[F_1(s)F_2(s)] = f_1(t) * f_2(t). \tag{3.247}$$

In the latter case the original of such product is a convolution of time-dependent functions, which are the originals of the mentioned transforms. On the other hand

$$L^{-1}\left[\frac{1}{\alpha^2}Q(s)\right] = \frac{1}{\alpha^2}L^{-1}[Q(s)] = \frac{1}{\alpha^2}q(t)$$

$$L^{-1}\left[\frac{1}{T^2s^2 + 2\xi Ts + 1}\right] = \frac{1}{T\sqrt{1-\xi^2}}e^{\frac{\xi t}{T}} \sin\left(\sqrt{1-\xi^2}\frac{t}{T}\right), \tag{3.248}$$

where: $\xi = \frac{h}{\alpha}$, $T = \frac{1}{\alpha}$.

According to Eq. (3.247) we get

$$x(t) = \int\limits_0^t \frac{q(\tau)}{\lambda} e^{-h(t-\tau)} \sin\lambda(t-\tau)d\tau, \tag{3.249}$$

where: $\lambda = \sqrt{\alpha^2 - h^2}$.

3.3 Lagrange Equations and Variational Principle

We are going to show that a trajectory of any dynamical system is a result of minimization of the action

$$S = \int_{t_0}^{t_1} L(\mathbf{r}, \dot{\mathbf{r}}, t)\, dt \tag{3.250}$$

in the time interval $t \in [t_0, t_1]$, where $\mathbf{r}$ is position vector of a particle with mass m.

Assume that the motion of the particle is governed by the dynamical system and by the space properties in which the particle moves. Consider different and possible trajectories of the particle starting from the point $A\,(q(t_0), \dot{q}(t_0), t_0)$ and achieving point $B\,(q(t_1), \dot{q}(t_1), t_1)$, as it has been shown in Fig. 3.15.

The position vector $\mathbf{r}$ as well as its time derivative $\dot{\mathbf{r}}$ are represented by

$$\mathbf{r} = \sum_{i=1}^{n} q^i \mathbf{a_i}, \tag{3.251}$$

$$\dot{\mathbf{r}} = \sum_{i=1}^{n} \frac{dq^i}{dt}\mathbf{a_i} = \sum_{i=1}^{n} \dot{q}^i \mathbf{a_i}, \quad i = 1, 2, \ldots, n, \tag{3.252}$$

where $\mathbf{a_i}$ are the bases vectors of the taken coordinate system.

It appears that although there are infinitely many paths from A to B (in general), but it is known that the physical dynamical system realizes only by one of them. Note that Eq. (3.250) n generalized coordinates q^i and n generalized velocities $\dot{q}^i$ appear, which are independent, and their variations are also independent.

Let us introduce now a small parameter ε in the way that for a true trajectory we have $\varepsilon = 0$, whereas for other ones we have $\varepsilon \neq 0$. Observe that for all trajectories which meet in points A and B, we have also $\varepsilon = 0$. From (3.250) one gets

$$S(\mathbf{r}(\varepsilon), \dot{\mathbf{r}}(\varepsilon), \varepsilon; t_0, t_1) = \int_{t_0}^{t_1} L(\mathbf{r}(\varepsilon), \dot{\mathbf{r}}(\varepsilon), \varepsilon, t)\, dt, \tag{3.253}$$

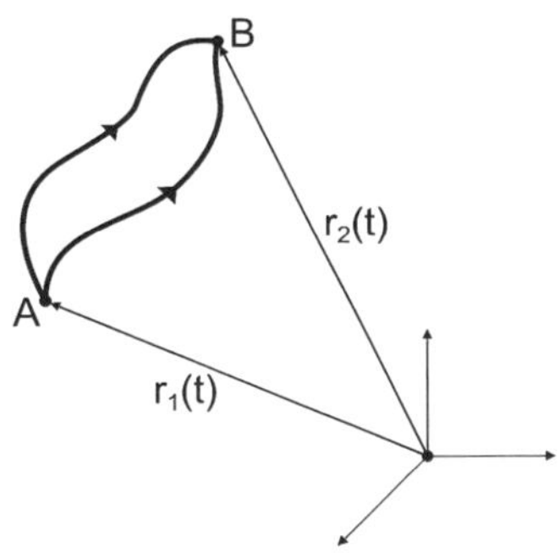

Fig. 3.15 Examples of two different paths realized by the particle m while moving between points A and B

and equivalently

$$
\begin{aligned}
&S\left(\mathbf{r},\dot{\mathbf{r}},0;t_0,t_1\right)+\varepsilon\left.\frac{dS}{d\varepsilon}\right|_{\varepsilon=0}+\frac{1}{2}\varepsilon^2\left.\frac{d^2S}{d\varepsilon^2}\right|_{\varepsilon=0}+\dots\\
&=\int_{t_0}^{t_1}L\left(\mathbf{r},\dot{\mathbf{r}},0,t\right)dt+\varepsilon\int_{t_0}^{t_1}\left.\frac{dL}{d\varepsilon}\right|_{\varepsilon=0}dt+\frac{\varepsilon^2}{2}\int_{t_0}^{t_1}\left.\frac{d^2L}{d\varepsilon^2}\right|_{\varepsilon=0}dt\\
&+\varepsilon\int_{t_0}^{t_1}\left(\sum_{i=1}^{n}\frac{\partial L}{\partial q^i}\left.\frac{dq^i}{d\varepsilon}\right|_{\varepsilon=0}+\sum_{i=1}^{n}\frac{\partial L}{\partial \dot{q}^i}\left.\frac{d\dot{q}^i}{d\varepsilon}\right|_{\varepsilon=0}\right)dt\\
&+\frac{\varepsilon^2}{2}\int_{t_0}^{t_1}\left(\sum_{i=1}^{n}\frac{\partial^2 L}{\partial\left(q^i\right)^2}\left.\frac{dq^i}{d\varepsilon}\frac{dq^i}{d\varepsilon}\right|_{\varepsilon=0}+\sum_{i=1}^{n}\frac{\partial L}{\partial q^i}\left.\frac{d^2q^i}{d\varepsilon^2}\right|_{\varepsilon=0}\right.\\
&+2\sum_{i=1}^{n}\frac{\partial^2 L}{\partial q^i\partial\dot{q}_i}\left.\frac{d\dot{q}^i}{d\varepsilon}\right|_{\varepsilon=0}\left.\frac{dq^i}{d\varepsilon}\right|_{\varepsilon=0}+\sum_{i=1}^{n}\frac{\partial^2 L}{\partial \dot{q}^i\partial\dot{q}_i}\left.\frac{d\dot{q}^i}{d\varepsilon}\right|_{\varepsilon=0}\left.\frac{dq^i}{d\varepsilon}\right|_{\varepsilon=0}\\
&\left.+\sum_{i=1}^{n}\frac{\partial L}{\partial \dot{q}^i}\left.\frac{d^2\dot{q}^i}{d\varepsilon^2}\right|_{\varepsilon=0}\right)+\cdots
\end{aligned}
\tag{3.254}
$$

Now we apply the Hamiltonian principle. Since we consider the different functionals for different values of ε and due to Hamiltonian principle for a physical (true) trajectory, the functional S achieves its extremum (it will be shown further that for almost all physical systems it is minimum). This principle is represented by the following formula

$$
\left.\frac{dS}{d\varepsilon}\right|_{\varepsilon=0}=\left[\frac{d}{d\varepsilon}\int_{t_0}^{t_1}L\left(\mathbf{r},\dot{\mathbf{r}},t\right)dt\right]_{\varepsilon=0}. \tag{3.255}
$$

Remark 3.1. Observe that now we might introduce $\left[\frac{d}{d\varepsilon}\right]_{\varepsilon=0}=\delta$, and this operator is well known in mechanics. This is infinite small variation of any quantity, on which it acts.

Remark 3.2. In most books of mechanics formula (3.255) is used as the fundamental principle, and it serves for derivation of Lagrange equation in different manner.

Remark 3.3. The use of perturbation approach allows for generalization and for clear and continuous statements and conclusions.

Remark 3.4. Since the perturbation techniques belong to extremely developed, it gives good starting points for many other generalizations.

Comparing the terms standing by the parameter ε in (3.254) we get

$$\left.\frac{dS}{d\varepsilon}\right|_{\varepsilon=0} = \int_{t_0}^{t_1} \left.\frac{dL}{d\varepsilon}\right|_{\varepsilon=0} dt + \int_{t_0}^{t_1} \left(\sum_{i=1}^{n} \frac{\partial L}{\partial q^i} \left.\frac{dq^i}{d\varepsilon}\right|_{\varepsilon=0} + \underline{\sum_{i=1}^{n} \frac{\partial L}{\partial \dot{q}^i} \left.\frac{d\dot{q}^i}{d\varepsilon}\right|_{\varepsilon=0}} \right) dt. \tag{3.256}$$

Consider now the underlined term in (3.256), which can be transformed in the following way

$$\frac{\partial L}{\partial \dot{q}^i} \left.\frac{d\dot{q}^i}{d\varepsilon}\right|_{\varepsilon=0} = \frac{\partial L}{\partial \dot{q}^i} \frac{d}{dt} \left(\left.\frac{dq^i}{d\varepsilon}\right|_{\varepsilon=0} \right) = \frac{d}{dt} \left(\frac{\partial L}{\partial \dot{q}^i} \left.\frac{dq^i}{d\varepsilon}\right|_{\varepsilon=0} \right) - \left(\left.\frac{d}{dt} \frac{\partial L}{\partial \dot{q}^i}\right|_{\varepsilon=0} \right) \left.\frac{dq^i}{d\varepsilon}\right|_{\varepsilon=0}, \tag{3.257}$$

because

$$\frac{d}{dt}\left(\frac{\partial L}{\partial \dot{q}^i}\right)\frac{dq^i}{d\varepsilon} + \frac{\partial L}{\partial \dot{q}^i}\frac{d^2 q^i}{dt d\varepsilon} - \left.\frac{d}{dt}\frac{\partial L}{\partial \dot{q}^i}\frac{dq^i}{d\varepsilon}\right|_{\varepsilon=0} = \frac{\partial L}{\partial \dot{q}^i}\frac{d}{dt}\left(\left.\frac{dq^i}{d\varepsilon}\right|_{\varepsilon=0}\right).$$

Taking into account (3.257) in (3.256) we get

$$\left.\frac{d}{d\varepsilon}\right|_{\varepsilon=0} \int_{t_0}^{t} L dt = \delta \int_{t_0}^{t} L dt = 0.$$

Assuming that $\left.\frac{dS}{d\varepsilon}\right|_{\varepsilon=0} = \int_{t_0}^{t_1} \left.\frac{dL}{d\varepsilon}\right|_{\varepsilon=0} dt = 0$ and taking into account (3.257), from (3.256) one gets

$$\int_{t_0}^{t_1} \left[\sum_{i=1}^{n} \frac{\partial L}{\partial q^i} \left.\frac{dq^i}{d\varepsilon}\right|_{\varepsilon=0} + \sum_{i=1}^{n} \frac{d}{dt} \left(\frac{\partial L}{\partial \dot{q}^i} \left.\frac{dq^i}{d\varepsilon}\right|_{\varepsilon=0} \right) - \sum_{i=1}^{n} \left(\left.\frac{d}{dt} \frac{\partial L}{\partial \dot{q}^i}\right|_{\varepsilon=0} \right) \left.\frac{dq^i}{d\varepsilon}\right|_{\varepsilon=0} \right] dt = 0,$$

or in the equivalent form

$$\int_{t_0}^{t_1} \sum_{i=1}^{n} \left(\left[\frac{\partial L}{\partial q^i} - \left(\frac{d}{dt} \frac{\partial L}{\partial \dot{q}^i} \right) \right] \left.\frac{dq^i}{d\varepsilon}\right|_{\varepsilon=0} \right) dt + \underline{\int_{t_0}^{t_1} \sum_{i=1}^{n} \frac{d}{dt} \left(\frac{\partial L}{\partial \dot{q}^i} \left.\frac{dq^i}{d\varepsilon}\right|_{\varepsilon=0} \right) dt} = 0. \tag{3.258}$$

Observe that for any $t \in (t_0, t_1)$ we have

$$q^i(t) = q^i_\varepsilon(t) = q^i(t)\big|_{\varepsilon=0} + \varepsilon \left.\frac{\partial q^i(t)}{\partial \varepsilon}\right|_{\varepsilon=0} + \frac{1}{2}\varepsilon^2 \left.\frac{\partial q^i(t)}{\partial \varepsilon^2}\right|_{\varepsilon=0} + \ldots \tag{3.259}$$

Consider now the underlined term in (3.258). This integral has the value

$$\begin{aligned}
&\frac{\partial L}{\partial \dot{q}^i}\bigg|_{\varepsilon=0,t_1} \cdot \frac{dq^i}{d\varepsilon}\bigg|_{\varepsilon=0,t_1} - \frac{\partial L}{\partial \dot{q}^i}\bigg|_{\varepsilon=0,t_0} \cdot \frac{dq^i}{d\varepsilon}\bigg|_{\varepsilon=0,t_0} \\
&= \frac{\partial L}{\partial \dot{q}^i}\bigg|_{\varepsilon=0,t_1} \cdot \delta q^i\big|_{\varepsilon=0,t_1} - \frac{\partial L}{\partial \dot{q}^i}\bigg|_{\varepsilon=0,t_0} \cdot \delta q^i\big|_{t_0} = 0,
\end{aligned} \tag{3.260}$$

because we are in the time instants t_0 and t_1 on the investigated (true) trajectory and in these points the infinite small variations $\delta q^i\big|_{t_0} = \delta q^i\big|_{t_1} = 0, i = 1, \ldots, n$.

Therefore, the investigated first-order asymptotic approximation takes the form

$$\int_{t_0}^{t_1} \sum_{i=1}^{n} \left(\left[\frac{\partial L}{\partial q^i} - \left(\frac{d}{dt} \frac{\partial L}{\partial \dot{q}^i} \right) \right] \frac{dq^i}{d\varepsilon}\bigg|_{\varepsilon=0} \right) dt = 0. \tag{3.261}$$

Recall the Gelfand and Fomin theorem proved in 1963 [98]. Let f_i be a set of n integrable functions of a real variable t on the interval $[t_0, t_1]$. Suppose that

$$\int_{t_0}^{t_1} f_i h_i dt = 0 \equiv (f_i, h_i = 0) \tag{3.262}$$

holds for every arbitrary set of integrable functions h_i on the same interval and all of them vanish at the points t_0 and t_1. The theorem states that $f_i = 0$ for all $i = 1, \ldots, n$. Note that in our case we have $h_i \equiv \frac{dq^i}{d\varepsilon}\big|_{\varepsilon=0}$. From (3.261) we obtain a set of Lagrange equations

$$\frac{dL}{dq^i} - \frac{d}{dt}\left(\frac{\partial L}{\partial \dot{q}^i} \right) = 0, \quad i = 1, \ldots, n. \tag{3.263}$$

Consider now the second-order approximations:

$$\begin{aligned}
\frac{d^2 S}{d\varepsilon^2}\bigg|_{\varepsilon=0} = \int_{t_0}^{t_1} \frac{d^2 L}{d\varepsilon^2}\bigg|_{\varepsilon=0} dt + \int_{t_0}^{t_1} \sum_{i=1}^{n} \Bigg(&\frac{\partial^2 L}{\partial (q^i)^2} \frac{dq^i}{d\varepsilon}\bigg|_{\varepsilon=0} \frac{dq^i}{d\varepsilon}\bigg|_{\varepsilon=0} \\
&+ \frac{\partial L}{\partial q^i} \frac{d^2 q^i}{d\varepsilon^2}\bigg|_{\varepsilon=0} + 2\frac{\partial^2 L}{\partial q^i \partial \dot{q}_i} \frac{d\dot{q}^i}{d\varepsilon}\bigg|_{\varepsilon=0} \frac{dq^i}{d\varepsilon}\bigg|_{\varepsilon=0} \\
&+ \frac{\partial^2 L}{\partial \dot{q}_i \partial \dot{q}_i} \frac{d\dot{q}^i}{d\varepsilon}\bigg|_{\varepsilon=0} \frac{d\dot{q}^i}{d\varepsilon}\bigg|_{\varepsilon=0} + \frac{\partial L}{\partial \dot{q}^i} \frac{d^2 \dot{q}^i}{d\varepsilon^2}\bigg|_{\varepsilon=0} \Bigg) dt.
\end{aligned} \tag{3.264}$$

Now we use the properties of the earlier introduced operator $\delta = \frac{d}{d\varepsilon}\big|_{\varepsilon=0}$, and from (3.264) we obtain

$$\delta^2 S = \delta\left(\delta\int_{t_0}^{t^1} L dt\right) + \delta^2 \int_{t_0}^{t_1}\sum_{i=1}^{n}\left(\frac{\partial^2 L}{\partial q^i \partial q^i} q^i q^i + \underline{\frac{\partial L}{\partial q^i} q^i}\right.$$

$$\left. + 2\frac{\partial^2 L}{\partial q^i \partial \dot{q}^i}\dot{q}^i q^i + \frac{\partial^2 L}{\partial \dot{q}^i \partial \dot{q}^i}\dot{q}^i \dot{q}^i + \underline{\underline{\frac{\partial^2 L}{\partial \dot{q}^i}\dot{q}^i}}\right) dt \tag{3.265}$$

The infinite small variation of both underlined terms in (3.265) corresponds to the considered already situation. In words, an action of δ on the underlined terms gives Eq. (3.258), and then (3.261). Assuming that the system realizes the motion governed by (3.258), the action of δ on (3.258) gives zero, and the underlined terms in (3.265) vanish.

However, consider a possible physical interpretation of the omitted terms. Let us begin with

$$\left(\frac{\partial L}{\partial \dot{q}^i}\frac{d\dot{q}^i}{d\varepsilon}\right)\frac{d\dot{q}^i}{\partial \varepsilon}\bigg|_{\varepsilon=0} = \frac{d\dot{q}^i}{\partial \varepsilon}\bigg|_{\varepsilon=0}\left(\left[\frac{\partial L}{\partial q^i} - \frac{d}{dt}\left(\frac{\partial L}{\partial \dot{q}^i}\right)\right]\frac{dq^i}{d\varepsilon}\right).$$

Here we give an interpretation of $\delta^2 S \equiv \frac{d^2 S}{d\varepsilon^2}\Big|_{\varepsilon=0}$. It can be treated as $\delta(\delta S)$, and one can understand this as the perturbation of $\delta S = 0$, or in other words of a trivial value. Therefore, analysis of perturbation of $\delta S = 0$ allows to estimate stability of $\delta S = 0$, i.e. stability of (3.258), and consequently, stability of the Lagrange equations. On the other hand, since the functional can be considered in a similar way to a function, therefore a second derivative $\frac{d^2 S}{d\varepsilon^2}\Big|_{\varepsilon=0}$ decides about stability. We say that the Lagrangian function (and then Lagrange equations) is stable when $\delta^2 S > 0$, and taking into account (3.265) one gets

$$\int_{t_0}^{t_1}\sum_{i=1}^{n}\left(\frac{\partial^2 L}{\partial q^i \partial q^i} q^i q^i + 2\frac{\partial^2 L}{\partial q^i \partial \dot{q}^i}\dot{q}^i q^i + \frac{\partial^2 L}{\partial \dot{q}^i \partial \dot{q}^i}\dot{q}^i \dot{q}^i\right) dt > 0. \tag{3.266}$$

To sum up, the obtained set of n differential equations (3.263), after introduction of Lagrangian L and after carrying out the differentiation gives the set of second-order differential equations with respect to time.

Recall that the generalized force Q_i corresponds to a general coordinate q^i. The work done by generalized force on the virtual (generalized) displacement is equal to the work done by the active forces along the Cartesian coordinates. The Lagrange equations of a second kind have the form

$$\frac{d}{dt}\left(\frac{\partial T}{\partial \dot{q}^i}\right) - \frac{\partial T}{\partial q^i} = Q_i, \quad i = 1, \ldots, s, \tag{3.267}$$

$$Q_i = \sum_{j=1}^{j*} F_j \frac{\partial u_j}{\partial q^i}, \tag{3.268}$$

and $j*$ denotes number of active forces. If we deal only with potential forces then

$$Q_i = -\frac{\partial V}{\partial q^i} \tag{3.269}$$

and (3.267) has the following form

$$\frac{d}{dt}\left(\frac{\partial T}{\partial \dot{q}^i}\right) - \frac{\partial T}{\partial q^i} + \frac{\partial V}{\partial q^i} = 0, \quad i = 1, \ldots, s. \tag{3.270}$$

3.4 Reduction to First-Order System of Equations

In mechanical and civil engineering [13] many problems can be reduced to analysis of second-order linear differential equations with variable coefficients of the form

$$M(t)\ddot{y} + C(t)\dot{y} + K(t)y = F(t), \tag{3.271}$$

where $M(t)$, $C(t)$ and $K(t)$ are square real matrices of order n. The Lagrangian coordinate y and the generalized force F are n-dimensional vectors. The matrices are called mass, damping and stiffness matrices, respectively. The system of Eq. (3.271) is reduced to the following set of first-order differential equations

$$\dot{x} = A(t)x + f(t), \tag{3.272}$$

where:

$$A(t) = \begin{bmatrix} 0 & I \\ -M^{-1}k & -M^{-1}C \end{bmatrix},$$

$$f(t) = \begin{bmatrix} 0 \\ -M^{-1}F(t) \end{bmatrix},$$

$$x(t) = \begin{bmatrix} y \\ \dot{y} \end{bmatrix}.$$

If the reduced Eq. (3.272) is complex-valued, then one writes

$$x = x_R + ix_I, \quad A = A_R + iA_I, \quad f = f_R + if_I. \tag{3.273}$$

Substituting (3.273) into (3.272) we get

$$\begin{bmatrix} \dot{x}_R \\ \dot{x}_I \end{bmatrix} = \begin{bmatrix} A_R & -A_I \\ A_I & A_R \end{bmatrix} \begin{bmatrix} x_R \\ x_I \end{bmatrix} + \begin{bmatrix} f_R \\ f_I \end{bmatrix}. \tag{3.274}$$

Hence, instead of the complex-valued differential equations (3.272) we have got real-valued system of differential equations (3.274) but with dimension two times larger. It means that in both cases a real-valued first-order system of differential equations can be taken into further considerations.

In [160] the authors discuss some advantages of a direct use of second-order system of differential equations instead of a reduction to first-order system of differential equations. Namely, they show that in this case the generalized modal analysis very widely used in mechanical engineering, direct consideration of the second-order differential equations can offer substantial reduction in computational effort and ample clear physical insight.

The earlier introduced reduction from a second-order system of differential equations to a first-order system of differential equations is a standard one. In a mechanical system, where the external forces are conservative, i.e. the forces arise from potential functions, the equations of motion corresponding to n-degree-of-freedom system, can be written as a Hamiltonian system in R^{2n}. The Hamiltonian system has some special mathematical structure, and hence it can be further exploited to gain more insight to the Hamiltonian system in order to simplify the equations of motion, where certain resultant coordinates are constant.

Although the Hamiltonian approach is nowadays very much saturated by geometric theory of the paths or curves, theory of systems of particles, and Lie'algebra, we focus our considerations on more basic approaches related to differential equations.

3.4.1 Mathematical Background

The Lagrangian equations have two remarkable properties. First, they are derived on a basis that a definite integral is stationary. Second, the Lagrangian function is quadratic regarding the velocities.

Hamilton's transformation reduces all Lagrangian problems to the 'canonical' form (after Jacobi). The n original Lagrangian differential equations are transformed to the $2n$ first-order differential equations having special simple and symmetric structure. Now we follow the considerations given in [148, 171, 237].

Consider a function

$$F = F(u_1, \ldots, u_n). \tag{3.275}$$

Let us introduce the new variables $v_1, \ldots, v_n$ in the following way

$$v_i = \frac{\partial F}{\partial u_i}(u_1, \ldots, u_n). \tag{3.276}$$

We assume that the determinant formed by the second-order partial derivatives of (3.276) is not equal to zero. This implies that the n variables v_i are independent. Using this assumption one can solve (3.276) to get

$$u_1 = \hat{u}_1(v_1, v_2, \ldots, v_n), \ldots, u_n = \hat{u}_n(v_1, \ldots, v_n). \tag{3.277}$$

Let us define a new function

$$G = \sum_{i=1}^{n} u_i v_i - F. \tag{3.278}$$

From (3.277) and (3.278) the function G can be expressed in terms of the new values only

$$G = G(v_1, \ldots, v_n). \tag{3.279}$$

Now we apply the principle of *least action* which, according to Hamilton, allows to transform the problems of mechanics to those of the calculus of variations. Consider the infinitesimal variation δG produced by arbitrary and infinitesimal variations δv_i:

$$\begin{aligned}\delta G &= \sum_{i=1}^{n} \frac{\partial G}{\partial v_i}\delta v_i = \sum_{i=1}^{n} \delta(u_i v_i - F) = \sum_{i=1}^{n} (u_i \delta v_i + v_i \delta u_i) - \delta F \\ &= \sum_{i=1}^{n} \left[u_i \delta v_i + \left(v_i - \frac{\partial F}{\partial u_i}\right) \partial u_i\right] = \sum_{i=1}^{n} u_i \delta v_i.\end{aligned} \tag{3.280}$$

Recall that $G = G(v_i, \ldots, v_n)$, and therefore we need to express all u_i as the functions of v_i. However, since (3.276) holds we have got (3.280) and hence

$$\delta G \equiv \sum_{i=1}^{n} \frac{\partial G}{\partial v_i}\delta v_i = \sum_{i=1}^{n} u_i \delta v_i, \tag{3.281}$$

which means that

$$u_i = \frac{\partial G}{\partial v_i}. \tag{3.282}$$

The described transformation was introduced by the French mathematician Legendre and is sometimes called dual transformation.

$$\begin{array}{cc} \multicolumn{2}{c}{\underline{\underline{\text{Duality}}}} \\ \underline{\text{Old System}} & \underline{\text{New System}} \\ u_1, \ldots, u_n & v_1, \ldots, v_n \\ F = F(u_1, \ldots, u_n) & G = G(v_1, \ldots, v_n) \\ u_i = \frac{\partial G}{\partial v_i} & v_i = \frac{\partial F}{\partial u_i} \\ F = \sum_{i=1}^{n} u_i v_i - G & G = \sum_{i=1}^{n} u_i v_i - F \end{array} \tag{3.283}$$

New variables are the partial derivatives of the old function with respect to the old variables and vice versa. 'Old' and 'new' systems are entirely equivalent for the transformation (3.283).

We introduce now two sets of 'old' variables $u_1, \ldots, u_n$ and $w_1, \ldots, w_m$. The additional variables are considered as parameters and they do not participate in the transformation

$$F = F(w_1, \ldots, w_m; u_1, \ldots, u_n). \tag{3.284}$$

The u_i are active, whereas w_i are passive variables of the transformation. We have

$$\begin{aligned} G &= G(v_1, \ldots, v_n; w_1, \ldots, w_m), \\ v_i &= \frac{\partial F}{\partial u_i}, \\ G &= \sum_{i=1}^{n} u_i v_i - F, \end{aligned} \tag{3.285}$$

$$\begin{aligned} \delta G &= \sum_{i=1}^{n} \frac{\partial G}{\partial v_i} \delta v_i + \sum_{j=1}^{m} \frac{\partial G}{\partial w_j} \delta w_j = \sum_{i=1}^{n} \delta(u_i v_i - F) \\ &= \sum_{i=1}^{n} (u_i \delta v_i + v_i \delta u_i) - \delta F \\ &= \sum_{i=1}^{n} (u_i \delta v_i + v_i \delta u_i) - \sum_{i=1}^{n} \frac{\partial F}{\partial u_i} \delta u_i - \sum_{j=1}^{m} \frac{\partial F}{\partial w_j} \delta w_j \\ &= \sum_{i=1}^{n} \left[u_i \delta v_i + \left(v_i - \frac{\partial F}{\partial u_i} \right) \delta u_i \right] - \sum_{j=1}^{m} \frac{\partial F}{\partial w_j} \delta w_j \\ &= \sum_{i=1}^{n} u_i \delta v_i - \sum_{j=1}^{m} \frac{\partial F}{\partial w_j} \delta w_j, \end{aligned} \tag{3.286}$$

which means that in addition to the earlier obtained results we have also

$$\frac{\partial G}{\partial w_j} = -\frac{\partial F}{\partial w_j}. \tag{3.287}$$

The Legendre's transformation is entirely symmetrical in both systems, i.e. it leads from the old to the new system and vice versa. It changes a given function of a given set of variables into a new function of a new set of variables.

3.4.2 The Lagrangian Function

Consider the Lagrangian function

$$L = L(q_1, \ldots, q_n, \dot{q}_1, \ldots, \dot{q}_n, t), \tag{3.288}$$

where q_i are the generalized coordinates and $\dot{q}_i$ are the corresponding velocities, and $i = 1, \ldots, n$. In words, we have mechanical system with n-degrees-of-freedom. We consider $\dot{q}_1, \dot{q}_2, \ldots, \dot{q}_n$ as the active variables, whereas $q_1, q_2, \ldots, q_n$ and t as the passive variables. Coming back to our earlier consideration one convinced that now $m = n + 1$. Let us now repeat the introduced earlier Legendre's steps. We denote the 'new variables' by p_i (the so-called momenta), and L corresponds to 'old' function. Hence

$$p_i = \frac{\partial L}{\partial \dot{q}_i}. \tag{3.289}$$

We introduce the 'new' function (corresponding to G), which is now called 'total energy' of the form

$$H = \sum_{i=1}^{n} p_i \dot{q}_i - L. \tag{3.290}$$

The variables $\dot{q}_i$ in (3.290) are expressed in terms of the new variables p_i by solving Eq. (3.289), and we get

$$H = H(q_1, \ldots, q_n; p_1, \ldots, p_n; t), \tag{3.291}$$

and H is called the Hamiltonian function. The dual nature of transformation follows:

$$\begin{array}{cc} \underline{\text{Old System}} & \underline{\text{New System}} \\ \dot{q}_i = \frac{\partial H}{\partial p_i} & p_i = \frac{\partial L}{\partial \dot{q}_i} \\ L = \sum\limits_{i=1}^{n} p_i \dot{q}_i - H & H = \sum_{i=1}^{n} p_i \dot{q}_i - L \\ L = L(q_1, \ldots, q_n; \dot{q}_1, \ldots, \dot{q}_n; t) & H = H(q_1, \ldots, q_n; p_1, \ldots, p_n; t) \end{array} \tag{3.292}$$

Observe that using the Lagrangian function L one can construct the Hamiltonian function H, and vice versa. Equation (3.287) has the form

$$\frac{\partial L}{\partial q^i} = -\frac{\partial H}{\partial q^i}, \quad \frac{\partial L}{\partial t} = -\frac{\partial H}{\partial t}. \tag{3.293}$$

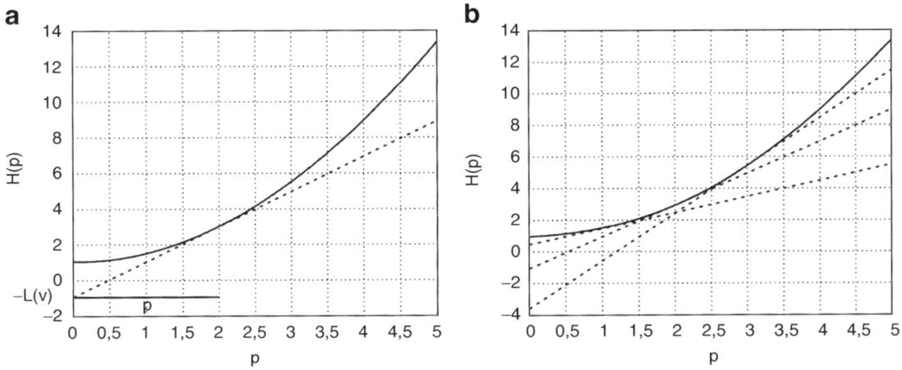

Fig. 3.16 Geometrical interpretation of $H(p)$

It happens also, especially in many problems in physics, that for a given Hamiltonian one needs to find Lagrangian. In this case the Hessian determinant Γ should be not equal to zero. Note that in deriving the Hamiltonian from the Lagrangian, the Hessian determinant, $J \neq 0$, where $\Gamma = J^{-1}$:

There is an interesting geometric interpretation of the relationship between the Lagrangian and the Hamiltonian. For simplicity in presentation, let us reduce the dimension of the system to 1 and hold q fixed. If $H(p)$ is convex, i.e. $\frac{\partial^2 H}{\partial p^2} > 0$, and at a value p, we draw a tangent line of slope v, then the y-intercept of the tangent line has a value of $-L(v)$. Since the slope of a tangent is the ratio of the vertical height to the horizontal, we have $v = \frac{H(p)+L(v)}{p}$. Rearranging, we get $H(p) = pr - L(v)$. By replacing v with $\dot{q}$, we get the Legendre transformation.

It is seen from Fig. 3.16 that the supremum of the set of the tangent lines gives $H(p)$, which is a conjugate function of $L(v) : H(p) = \sup_v(pv - L(v))$. This is another reason that H and L are called dual functions of each other.

The transformation theory that operates on Hamiltonian systems gives rights to mix the coordinates q and p, and for instance p is not necessarily needed to be the momentum of the system, whereas q need not be the configuration of the system. However, when the Hamiltonian does have a physical interpretation of total energy; i.e. $H = T + U, \quad L = T - U$ and when p is the momentum, then the total energy is constant $\left(\frac{dE}{dt} = 0\right)$ in a time-invariant conservative system. In this case also H is always conserved when it is time-invariant and the states of the system evolve along the contours of constant H. In other words $H(q, p, t) = \text{const}$, which gives a first integral. Another interpretation of H is from convex analysis.

We will illustrate it geometrically with Fig. 3.16b. For every value p, we draw a tangent line of slope v, so by varying the values of p, we get a set of tangent lines which are shown as dashed lines above. The supremum of the set of the tangent lines gives us $H(p)$. Mathematically, this means that $H(p)$ is a conjugate function of $L(v)$, i.e. $H(p) = \sup_v(pv - L(v))$.

3.5 Canonical Transformations

Now we introduce the basics of transformation theory for Hamiltonian systems.

Definition 3.1 (Poisson Bracket). Let f and g be functions of $(\mathbf{q}, \mathbf{p}, t)$. The Poisson bracket of f and g is defined as:

$$[f, g] = \frac{\partial f}{\partial \mathbf{p}} \frac{\partial g}{\partial \mathbf{q}} - \frac{\partial f}{\partial \mathbf{q}} \frac{\partial g}{\partial \mathbf{p}}. \tag{3.294}$$

In particular, note that

$$\dot{q}_h = [H, q_h], \quad \dot{p}_h = [H, p_h].$$

In order to preserve the Hamilton's equations structure, we would be looking for coordinate transformations that preserve the Poisson brackets. Suppose that new coordinates system is $(\mathbf{q}'', \mathbf{p}'', t'')$, we want $[f, g] = [f, g]''$ for every f and g, where $[f, g]''$ is defined as:

$$[f, g]'' = \frac{\partial f}{\partial \mathbf{p}''} \frac{\partial g}{\partial \mathbf{q}''} - \frac{\partial f}{\partial \mathbf{q}''} \frac{\partial g}{\partial \mathbf{p}''}. \tag{3.295}$$

The main idea of this type of transformation is as follows: First, we introduce an intermediate coordinate system $(\mathbf{q}', \mathbf{p}', t')$, with the following relationship:

$$\mathbf{q}' = \mathbf{q}, \quad \mathbf{p}' = \mathbf{p}'', \quad t' = t = t''. \tag{3.296}$$

This means that the first transformation from $(\mathbf{q}, \mathbf{p}, t)$ to $(\mathbf{q}', \mathbf{p}', t')$ involves keeping $\mathbf{q} = \mathbf{q}'$ and $t = t'$ invariant and transforming $\mathbf{p}$ to $\mathbf{p}'$. The second transformation from $(\mathbf{q}', \mathbf{p}', t')$ to $(\mathbf{q}'', \mathbf{p}'', t'')$ involves keeping $\mathbf{p}' = \mathbf{p}''$ and $t' = t''$ invariant and transforming $\mathbf{q}'$ to $\mathbf{q}''$. How we go about achieving this is through a generating function F, as stated below.

Theorem 3.2 (Canonical Transformation). *If the transformation from the coordinate system $(\mathbf{q}, \mathbf{p}, t)$ to new coordinate system $(\mathbf{q}'', \mathbf{p}'', t'')$, satisfies $[f, g] = [f, g]''$ for every f and g, then the transformation is canonical. In addition, the transformation is canonical if only if there exists a function $F : (\mathbf{q}', \mathbf{p}', t') \to \mathbb{R}$ such that*

$$\mathbf{p} = \frac{\partial F}{\partial \mathbf{q}'} \quad \textit{and} \quad \mathbf{q}'' = \frac{\partial F}{\partial \mathbf{p}'}. \tag{3.297}$$

F is called a generating function for the transformation.

The proof of this theorem is given in [254]. In practice, knowing which F that generates the coordinate transformation is not obvious. However, the method given below makes finding F more straightforward [254].

1. Pick a family of functions $S(q_1,\ldots,q_n,k_1,\ldots,k_n,t)$, parametrized by $k_1,\ldots,k_n$.
2. Construct the functions $p'_1,\ldots,p'_n$ by solving the system of equations $\mathbf{p} = \frac{\partial S}{\partial \mathbf{q}}\Big|_{\mathbf{k}=\mathbf{p}'}$ for $\mathbf{p}'$ as functions of $(\mathbf{q},\mathbf{p},t)$.
3. Define the intermediate coordinate system $(\mathbf{q}',\mathbf{p}',t')$ by noting that $\mathbf{q}' = \mathbf{q}$. Substitute $\mathbf{q}' = \mathbf{q}$ and $\mathbf{p}' = \mathbf{k}$ into S, which gives us $F(q'_1,\ldots,q'_n,p'_1,\ldots,p'_n)$.
4. Construct the final coordinate system $(\mathbf{q}'',\mathbf{p}'',t'')$ as

$$\mathbf{p}'' = \mathbf{p}', \quad \mathbf{q}'' = \frac{\partial F}{\partial \mathbf{p}'}, \quad t'' = t' = t. \tag{3.298}$$

With the canonical transformations introduced, we shall now examine how Hamilton's equations behave under canonical transformation.

Theorem 3.3. *Let $G(\mathbf{p},\mathbf{q},t)$ be any function such that $G = \frac{\partial F}{\partial t'}$, where F is the generating function. Then for any function $f(\mathbf{p},\mathbf{q},t)$ we have*

$$\frac{\partial f}{\partial t} = \frac{\partial f}{\partial t''} + [G, f]. \tag{3.299}$$

For any function $f(\mathbf{p},\mathbf{q},t)$, its total derivative in time is

$$\begin{aligned}\frac{df}{dt} &= \frac{\partial f}{\partial \mathbf{q}}\dot{\mathbf{q}} + \frac{\partial f}{\partial \mathbf{p}}\dot{\mathbf{p}} + \frac{\partial f}{\partial t}\\ &= \frac{\partial f}{\partial \mathbf{q}}\frac{\partial H}{\partial \mathbf{p}} - \frac{\partial f}{\partial \mathbf{p}}\frac{\partial H}{\partial \mathbf{q}} + \frac{\partial f}{\partial t}\\ &= [H, f] + \frac{\partial f}{\partial t}.\end{aligned} \tag{3.300}$$

Applying the proposition above (3.299), we have

$$\begin{aligned}\frac{df}{dt} &= [H, f] + \frac{\partial f}{\partial t}\\ &= [H, f] + [G, f] + \frac{\partial f}{\partial t''}\\ &= [H + G, f] + \frac{\partial f}{\partial t''}.\end{aligned}$$

As such we obtain a new Hamiltonian $H'' = H + G$ with Hamilton's equations in the new coordinates:

$$\frac{d}{dt}p''_h = -\frac{\partial H''}{\partial q''_h} \quad \text{and} \quad \frac{d}{dt}q''_h = \frac{\partial H''}{\partial p''_h}. \tag{3.301}$$

Of special interest is to find a transformation resulting in $H'' = 0$, then the trajectories of the equations of motion will be given by $\mathbf{p}'' =$ constant and $\mathbf{q}'' =$ constant. The transformation that results in $H'' = 0$ satisfies the Hamilton–Jacobi equation, which is a first-order, nonlinear partial differential equation. The main idea in developing the Hamilton–Jacobi equation is briefly described here.

Using the method described earlier to construct the family of functions $S(\mathbf{q}, \mathbf{k}, t)$, the set of points $\mathbf{p} = \left.\frac{\partial S}{\partial \mathbf{q}}\right|_{\mathbf{k}=\mathbf{p}'}$, form a $n+1$ dimensional surface. The original Hamiltonian on this surface is $H = H(\mathbf{q}, \frac{\partial S}{\partial \mathbf{q}}, t)$. Since $G = \frac{\partial F}{\partial t'}$, it is also equivalent to $G = \frac{\partial S}{\partial t}$ in the original coordinate system. We want $H'' = 0$, so it means that $H + G = 0$, i.e.

$$H\left(\mathbf{q}, \frac{\partial S}{\partial \mathbf{q}}, t\right) + \frac{\partial S}{\partial t} = 0. \tag{3.302}$$

This is the Hamiltonian–Jacobi equation that will result in $H'' = 0$.

3.6 Examples

We will now return to the examples.

Example 3.10 ([254]). A simple harmonic oscillator.

If the potential function is $U = \frac{1}{2}\alpha q^2$, then the Hamiltonian is now

$$H(p, q) = \frac{p^2}{2m} + \frac{1}{2}\alpha q^2.$$

The Hamiltonian–Jacobi equation is (by substituting $p = \frac{\partial S}{\partial q}$ into (3.302))

$$\frac{\partial S}{\partial t} + \frac{1}{2m}\left(\frac{\partial S}{\partial q}\right)^2 + \frac{1}{2}\alpha q^2 = 0.$$

The solution to this partial differential equation is parameterized by k (which can be checked by substituting the solution back in the Hamilton–Jacobi equation above).

$$S = -kt + g(q), \quad g(q) = \int \left(2km - \alpha m q^2\right)^{\frac{1}{2}} dq.$$

Following the steps of transformation, we have

$$\begin{aligned} q &= q' \quad \text{and} \quad t = t', \\ p &= \left.\frac{\partial S}{\partial q}\right|_{k=p'} = -p't' + \int \left(2p'm - \alpha m q^2\right)^{\frac{1}{2}}, \\ p' &= \frac{p^2}{2m} + \frac{1}{2}\alpha q^2. \end{aligned}$$

The generating function is

$$F = S|_{k=p'} = S = -p't' + \int \left(2p'm - \alpha m(q')^2\right)^{\frac{1}{2}} dq'.$$

The Hamiltonian's equations in the new coordinates are

$$\begin{aligned}
p'' &= p' = \frac{p^2}{2m} + \frac{1}{2}\alpha q^2, \\
q'' &= \frac{\partial F}{\partial p'} = -t' + \int \frac{m}{(2p'm - \alpha m(q')^2)^{\frac{1}{2}}} dq' \\
&= -t' + \sqrt{\frac{m}{\alpha}} \int \frac{1}{(\frac{2}{\alpha}p' - (q')^2)^{\frac{1}{2}}} dq' \\
&= -t' + \sqrt{\frac{m}{\alpha}} \sin^{-1}\left(\frac{q'}{\sqrt{2p'/\alpha}}\right) \\
&= -t + \sqrt{\frac{m}{\alpha}} \sin^{-1}\left(\frac{q}{\sqrt{2p''/\alpha}}\right).
\end{aligned}$$

Since $H'' = 0$, we have $p'' = E =$ constant and $q'' = \varepsilon =$ constant. If we non-dimensionalize the equations by introducing $\omega^2 = \frac{\alpha}{m}$ and $A = \sqrt{\frac{2E}{\alpha}}$, we can rewrite the equation for p'' and q'' as

$$\begin{aligned}
A^2 &= \frac{p^2}{m\alpha} + q^2, \\
\varepsilon &= -t + \frac{1}{\omega} \sin^{-1}\left(\frac{q}{A}\right).
\end{aligned}$$

We can solve the equations of motion in the original p and q coordinates by rearranging equation for ε

$$\begin{aligned}
q &= A \sin\left(\omega(t + \varepsilon)\right), \\
p &= m\omega A \cos\left(\omega(t + \varepsilon)\right).
\end{aligned}$$

These equations can be interpreted as follows:

1. For a given initial condition $p(t_0)$ and $q(t_0)$, we have constant E (and equivalently A) which determines the constant energy level or the contour that the particle is orbiting.
2. The constant ε determines the initial 'phase' or angle between p and q for a given initial conditions.

3. The equations of motion describe the motion of an elliptical orbit of constant E, which is consistent with the motion that this system is conservative. This system is stable, but not asymptotically stable.

□

Example 3.11. Central force motion.

Consider the motion of the particle moving on a plane, subjected to a force of magnitude $\frac{\lambda m}{r^2}$ directed towards a fixed point, O. Using polar coordinates, the Lagrangian is

$$L = T - U = \frac{1}{2}m\left(\dot{r}^2 + r^2\dot{\theta}^2\right) - \left(-\frac{\lambda m}{r}\right).$$

Here, the generalized coordinates are $q = [r, \theta]^T$. The generalized momenta are:

$$p_1 = \frac{\partial L}{\partial \dot{q}_1} = \frac{\partial L}{\partial \dot{r}} = m\dot{r},$$

$$p_2 = \frac{\partial L}{\partial \dot{q}_2} = \frac{\partial L}{\partial \dot{\theta}} = mr^2\dot{\theta}.$$

The Hamiltonian is computed as follows:

$$\begin{aligned} H(q, p) &= p_1\dot{q}_1 + p_2\dot{q}_2 - \left(\frac{1}{2}m\left(\dot{r}^2 + r^2\dot{\theta}^2\right) - \left(-\frac{\lambda m}{r}\right)\right) \\ &= m\dot{r}^2 + mr^2\dot{\theta}^2 - \frac{1}{2}m\dot{r}^2 - \frac{1}{2}mr^2\dot{\theta}^2 - \frac{\lambda m}{r} \\ &= \frac{1}{2}m\dot{r}^2 + \frac{1}{2}mr^2\dot{\theta}^2 - \frac{\lambda m}{r} \\ &= \frac{1}{2m}\left(p_1^2 + \frac{p_2^2}{q_1^2}\right) - \frac{\lambda m}{r}, \end{aligned}$$

and hence we obtain

$$\dot{p}_1 = -\frac{\partial H}{\partial q_1} = \frac{p_2^2}{mq_1^3} - \frac{\lambda m}{q_1^2},$$

$$\dot{p}_2 = -\frac{\partial H}{\partial q_2} = 0,$$

$$\dot{q}_1 = \frac{\partial H}{\partial p_1} = \frac{p_1}{m},$$

$$\dot{q}_2 = \frac{\partial H}{\partial p_2} = \frac{p_2}{mq_1^2}.$$

The Hamilton–Jacobi equation is obtained from (3.302):

$$\frac{\partial S}{\partial t} + \frac{1}{2m}\left(\frac{\partial S}{\partial q_1}\right)^2 + \frac{1}{2mq_1^2}\left(\frac{\partial S}{\partial q_2}\right)^2 - \frac{\lambda m}{q_1} = 0.$$

The solution of the Hamilton–Jacobi equation can be found by separating the variables. Taking S is of the form $S = f(t) + g_1(q_1) + g_2(q_2)$, we get

$$-\frac{df}{dt} = \frac{1}{2m}\left(\frac{dg_1}{dq_1}\right)^2 + \frac{1}{2mq_1^2}\left(\frac{dg_2}{dq_2}\right)^2 - \frac{\lambda m}{q_1}.$$

Since the LHS of the above equation is a function of t only and the RHS is a function of q_1 and q_2 only, therefore both sides must be equal to a constant. As a result, both sides can be integrated separately, with the constant of integration being equal. Thus, $f = (-k_1 t + \text{constant})$ and RHS is

$$\frac{dg_2}{dq_2} = q_1\left[2mk_1 - \left(\frac{dg_1}{dq_1}\right)^2 + \frac{2\lambda m^2}{q_1}\right]^{\frac{1}{2}}.$$

Again, both sides must be equal to a constant, so we have

$$g_2 = (k_2 q_2 + \text{const}), \quad \frac{dg_1}{dq_1} = Q(q_1), \quad Q(q_1) = \left[\frac{2\lambda m^2}{q_1} - \frac{k_2^2}{q_1^2} + 2k_1 m\right]^{\frac{1}{2}}.$$

Hence, S has the form

$$S(q_1, q_2, k_1, k_2, t) = -k_1 t + k_2 q_2 + \int Q(q_1) dq_1 + \text{constant}.$$

Following the steps of transformation, we have

$$q_1 = q_1', \quad q_2 = q_2' \quad \text{and} \quad t = t'$$

$$p_1 = \left.\frac{\partial S}{\partial q_1}\right|_{k_1 = p_1',\, k_2 = p_2'} = Q|_{k_1 = p_1',\, k_2 = p_2'},$$

$$p_1 = \left[\frac{2\lambda m^2}{q_1} - \frac{p_2'^2}{q_1^2} + 2p_1' m\right]^{\frac{1}{2}},$$

$$p_1'' = p_1' = \frac{p_1^2}{2m} - \frac{\lambda m}{q_1} + \frac{(p_2')^2}{2mq_1^2} = \text{const},$$

$$p_2 = \left.\frac{\partial S}{\partial q_2}\right|_{k_1 = p_1',\, k_2 = p_2'} = k_2 = p_2',$$

$$p_2'' = p_2' = p_2 = \text{const}.$$

Using the fact that the transformed Hamiltonian H'' is zero, which means that $q_1'' = \text{const}$ and $q_2'' = \text{const}$, we have

$$q_1'' = \frac{\partial F}{\partial p_1'} = \frac{\partial S}{\partial k_1} = -t + \int \frac{m dq_1}{Q(q_1)} = \text{const},$$
$$q_2'' = \frac{\partial F}{\partial p_2'} = \frac{\partial S}{\partial k_2} = q_2 - \int \frac{k_2 dq_1}{q_1^2 Q(q_1)} = \text{const}\,.$$

The following observations can be made about the Hamilton's equations in the transformed coordinates (q'', p'', t''):

1. We have $p_1'' = H = \text{constant}$, but since $H = E$, the total energy is conserved, which is the case for this conservative Hamiltonian system.
2. We have also $p_2'' = p_2 = mr^2\dot{\theta} = \text{constant}$, which means that the angular momentum of the system about the axis perpendicular to the plane is conserved.
3. Differentiating the equation for q_1'' with respect to time, we have the differential equation below, which is a function of only one variable $q_1 = r$.

$$\frac{dq_1}{dt} = \frac{Q(q_1)}{m}, \quad \left(q_1 = r \Rightarrow \frac{dr}{dt} = \frac{Q(r)}{m}\right).$$

4. Differentiating the equation for q_2'' with respect to q_2, we have a differential equation of $q_2 (= \theta)$ that explicitly depends on $q_1 = r$. Note that $k_2 = p_2'' =$ constant.

$$\frac{dq_2}{dq_1} = \frac{k_2}{q_1^2 Q(q_1)}, \quad \left(\frac{d\theta}{dr} = \frac{k_2}{r^2 Q(r)}\right).$$

We can see that by using canonical transformation systematically, this problem can be solved analytically.

□

3.7 Normal Forms of Hamiltonian Systems

In this section a method to analyse the dynamics of Hamiltonian systems with a periodically modulated Hamiltonian is presented [32, 255]. The method is based on a special parametric form of the canonical transformation

$$\mathbf{q}, \mathbf{p} \to \mathbf{Q}, \mathbf{P}, \quad H(t, \mathbf{q}, \mathbf{p}) \to \bar{H}(t, \mathbf{Q}, \mathbf{P})$$
$$\begin{cases} \mathbf{q} = \mathbf{x} - \frac{1}{2}\Psi_{\mathbf{x}} \\ \mathbf{p} = \mathbf{y} + \frac{1}{2}\Psi_{\mathbf{y}} \end{cases}, \quad \begin{cases} \mathbf{P} = \mathbf{x} + \frac{1}{2}\Psi_{\mathbf{x}} \\ \mathbf{Q} = \mathbf{y} - \frac{1}{2}\Psi_{\mathbf{y}} \end{cases} \tag{3.303}$$
$$\bar{H}(t, \mathbf{Q}, \mathbf{P}) = \Psi_t(t, \mathbf{x}, \mathbf{y}) + H(t, \mathbf{Q}, \mathbf{P})$$

using Poincaré generating function $\Psi(t, \mathbf{x}, \mathbf{y})$. As a result, stability problem of a periodic solution is reduced to finding a minimum of the Poincaré function.

The proposed method can be used to find normal forms of Hamiltonians. It should be emphasized that we apply the modified concept of Zhuravlev [32, 255] to define an invariant normal form, which does not require any partition on either autonomous–non-autonomous, or resonance–non-resonance cases, but it is treated in the frame of one approach. In order to find the corresponding normal form asymptotics, a system of equations is derived analogues to Zhuravlev's chain of equation. Instead of the generator method and guiding Hamiltonian, a parameterized guiding function is used. It enables a direct (without the transformation to an autonomous system as in Zhuravlev's method) computation of the chain of equations for non-autonomous Hamiltonians. For autonomous systems, the methods of computation of normal forms coincide in the firs and second approximations.

3.7.1 Parametric Form of Canonical Transformations

A general result of the Hamiltonian systems parameterization by canonical transformation is formulated in the frame of the following theorem [195–197].

Theorem 3.4. *Let the transformation of variables* $\mathbf{q}, \mathbf{p} \to \mathbf{Q}, \mathbf{P}$ *be given in the following parametric form*

$$\begin{array}{ll} \mathbf{q} = \mathbf{x} - \frac{1}{2}\Psi_{\mathbf{x}}, & \mathbf{P} = \mathbf{x} + \frac{1}{2}\Psi_{\mathbf{x}} \\ \mathbf{p} = \mathbf{y} + \frac{1}{2}\Psi_{\mathbf{y}}, & \mathbf{Q} = \mathbf{y} - \frac{1}{2}\Psi_{\mathbf{y}} \end{array}. \tag{3.304}$$

Then, for any arbitrary function $\Psi(t, \mathbf{x}, \mathbf{y})$ *the following property holds: Jacobians of two transformations* $\mathbf{q} = \mathbf{q}(t, \mathbf{x}, \mathbf{y})$, $\mathbf{p} = \mathbf{p}(t, \mathbf{x}, \mathbf{y})$ *and* $\mathbf{Q} = \mathbf{Q}(t, \mathbf{x}, \mathbf{y})$, $\mathbf{P} = \mathbf{P}(t, \mathbf{x}, \mathbf{y})$ *are identity ones:*

$$\frac{\partial(\mathbf{q}, \mathbf{p})}{\partial(\mathbf{x}, \mathbf{y})} = \frac{\partial(\mathbf{Q}, \mathbf{P})}{\partial(\mathbf{x}, \mathbf{y})} = J(t, \mathbf{x}, \mathbf{y}). \tag{3.305}$$

In the space $J > 0$ *relation (3.304) with respect to variables* $\mathbf{q}, \mathbf{p} \to \mathbf{Q}, \mathbf{P}$ *transforms the Hamiltonian system* $H = H(t, \mathbf{q}, \mathbf{p})$ *by the following rule*

$$\Psi_t(t, \mathbf{x}, \mathbf{y}) + H(t, \mathbf{q}, \mathbf{p}) = \tilde{H}(t, \mathbf{Q}, \mathbf{P}), \tag{3.306}$$

where arguments $\mathbf{q}, \mathbf{p}$ *and* $\mathbf{Q}, \mathbf{P}$ *in the Hamiltonians* H *and* $\tilde{H}$ *are expressed via the parameters* $\mathbf{x}, \mathbf{y}$ *in formulas* (3.304).

In the next step we investigate for which canonical transformations the parameterization exists.

3.7.2 Integration of Hamiltonian Equations Perturbated by Damping

Below, we consider dynamics of a unit material mass driven by a potential periodic force $-\frac{\epsilon \partial F(t,q)}{\partial q}$ embedded in the medium with damping $-\delta \dot{q}$ governed by the equation

$$\ddot{q} = -\frac{\epsilon^2 \partial F(t,q)}{\partial q} - \delta \dot{q}, \tag{3.307}$$

where ϵ and δ are the small parameters.

Equation (3.307) is transformed to the following form

$$\dot{q} = \frac{\partial H}{\partial p}, \quad \dot{p} + \delta p = -\frac{\partial H}{\partial q}, \quad H = \frac{\epsilon p^2}{2} + \epsilon F(t,q). \tag{3.308}$$

Note that (3.308) is not Hamiltonian, since damping force is not potential. However, applying the following noncanonical transformation

$$q = \tilde{q}, \quad p = \tilde{p}\exp(-\delta t) \tag{3.309}$$

the system assumes the following Hamiltonian form

$$\dot{\tilde{q}} = \frac{\partial \tilde{H}}{\partial \tilde{p}}, \quad \dot{\tilde{p}} = -\frac{\partial \tilde{H}}{\partial \tilde{q}}, \quad \tilde{H} = \epsilon \left[\frac{\exp(-\delta t)\tilde{p}^2}{2} + \exp(\delta t)\tilde{F}(t,\tilde{q}) \right]. \tag{3.310}$$

Our target is to analyse an asymptotic solution of Hamiltonian (3.310) on the period $0 \leq t \leq 2\pi$. We remain only the terms ϵ^3 and $\epsilon\delta$, in the Hamiltonian $\tilde{H}$, i.e.

$$\tilde{H}(t,x,y,\epsilon) = \epsilon H_1(t,x,y) + \epsilon^2 H_2(t,x,y) + \epsilon^3 H_3(t,x,y) + \epsilon\delta H_4(t,x,y) + \cdots, \tag{3.311}$$

The system of Eq. (3.310) has the standard form and we apply the Poincaré transformation with respect to the period in the following parametric form

$$\begin{cases} \tilde{q}_{n-1} = x - \frac{1}{2}\Psi_y \\ \tilde{p}_{n-1} = y + \frac{1}{2}\Psi_x \end{cases}, \quad \begin{cases} \tilde{q}_n = x + \frac{1}{2}\Psi_y \\ \tilde{p}_n = y - \frac{1}{2}\Psi_x \end{cases}, \tag{3.312}$$

where function $\Psi(\tau, x, y)$ is found solving a Jacobian type equation.

Points $\tilde{q}_n$, $\tilde{p}_n$ lie on the trajectory $\tilde{q}(t)$, $\tilde{p}(t)$, defined by the solution of Hamiltonian (3.310) and they are called Poincaré points. A distance in time between them is equal to period 2π: $\tilde{q}_n = \tilde{q}(2\pi n)$, $\tilde{p}_n = \tilde{p}(2\pi n)$. In what follows recurrent relations (3.308) may be obtained for periodic and Poincaré points. For this purpose the function Ψ is sought in the following form

$$\Psi = \epsilon\Psi_1 + \epsilon^2\Psi_2 + \epsilon^3\Psi_3 + \epsilon\delta\Psi_4 + \cdots, \tag{3.313}$$

Substituting series (3.313) into Jacobi equation, the following series coefficients are found [196]

$$
\begin{aligned}
\Psi_1(t,x,y) &= \int_{t_0}^{t} H_1 dt, \\
\Psi_2(t,x,y) &= \int_{t_0}^{t} \left[H_2 - \tfrac{1}{2}\{H_1, \Psi_1\}\right] dt, \\
\Psi_3(t,x,y) &= \int_{t_0}^{t} \left[H_3 - \tfrac{1}{2}(\{H_2, \Psi_1\} + \{H_1, \Psi_2\}) + \right. \\
&\quad \left. + \tfrac{1}{8}(H_{1xx}\Psi_{1y}\Psi_{1y} - 2H_{1xy}\Psi_{1x}\Psi_{1y} + H_{1yy}\Psi_{1x}\Psi_{1x})\right] dt, \\
\Psi_4(t,x,y) &= \int_{t_0}^{t} H_4 dt, \quad \text{etc.}
\end{aligned}
\tag{3.314}
$$

where $\{f, g\} = f_y g_x - f_x g_y$ denotes the Poisson bracket.

The obtained formulas allow to carry out the full system analysis, which will be illustrated by the following example.

Example 3.12. Investigate dynamics of a material point driven by the force $-\epsilon^2 \cos t \cos q$ and damped by $-\delta\dot{q}$.

In this case for the coefficients of the series (3.311) one gets

$$
H_1 = \frac{y^2}{2} + \cos t \sin x, \quad H_2 = 0, \quad H_3 = 0, \quad H_4 = t\left(-\frac{y^2}{2} + \cos\tau \sin x\right).
$$

Substituting the above expressions to (3.314), one gets the following coefficients for the time instant $t = 2\pi$

$$
\begin{aligned}
\Psi &= \epsilon\Psi_1 + \epsilon^2\Psi_2 + \epsilon^3\Psi_3 + \epsilon\delta\Psi_4, \\
\Psi_1 &= 2\pi\frac{y^2}{2}, \quad \Psi_2 = 0, \quad \Psi_3 = 2\pi\left(\frac{1}{8}\cos 2x - y^2 \sin x\right), \quad \Psi_4 = -\pi^2 y^2.
\end{aligned}
$$

Removing from (3.312) parameters x and y and using (3.309) the following recurrent relations are obtained for Poincaré points q_n, p_n, $n = 0, 1, 2, \ldots$ of Eq. (3.308)

$$
\begin{aligned}
q_n &= q_{n-1} + 2\pi\epsilon p_{n-1} + O(\epsilon^3), \\
p_n &= p_{n-1}(1 - 2\pi\delta) + \frac{1}{4}\epsilon^3 \sin 2q_n + O(\epsilon^4),
\end{aligned}
$$

where ϵ is the order of impulse $p_n = O(\epsilon)$ (invariant curves are shown in Fig. 3.17).

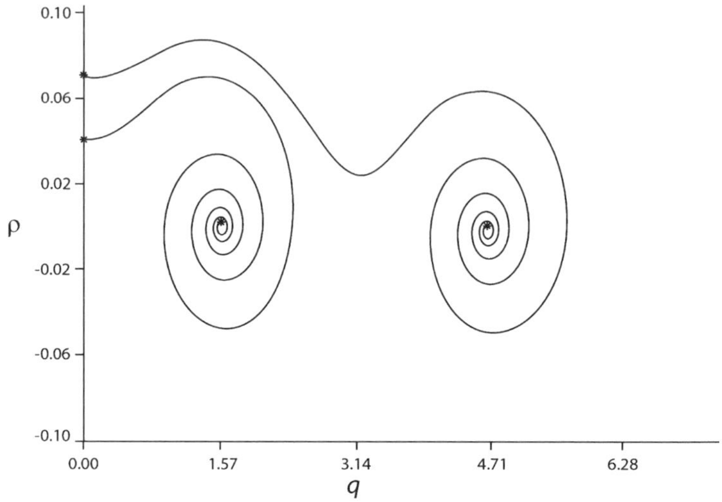

Fig. 3.17 Invariant curve with damping

For $\delta = 0$ we deal with a Hamiltonian system. It follows from Theorem 5 in [196] that Poincaré points lie on closed invariant curve $\Psi(q_n, p_n) = \text{const} = 2\pi\epsilon^3 C$ with the accuracy of ϵ^5 in the form

$$\frac{p_n^2}{2\epsilon^2} + \frac{1}{8}\cos 2q_n = C. \qquad (*)$$

A family of invariant curves is shown in Fig. 3.18.

For $\delta \neq 0$ our system is non-Hamiltonian and Poincaré points with an increase of n start to lie on curves defined by the preceding equation with a constant $C = c_n$ depended on n. For c_n one may obtain the recurrent relations in the following way. Let a pair $\tilde{q}_{n-1} = q_{n-1}$, $\tilde{p}_{n-1} = p_{n-1}$ and $\tilde{q}_n$, $\tilde{p}_n$ of Hamiltonian (3.310) lie on the following invariant curve

$$\Psi(\tilde{q}_{n-1}, \tilde{p}_{n-1}) = \Psi(\tilde{q}_n, \tilde{p}_n) = 2\pi\epsilon^3 c_{n-1}.$$

The next pair ($q_n = \tilde{q}_n$, $p_n = \tilde{p}_n(1 - 2\pi\delta)$) and ($\tilde{q}_{n+1} = q_{n+1}$, $\tilde{p}_{n+1}$) lie on the invariant curve with modified constant $2\pi\epsilon^3 c_n$ of the form

$$\Psi(\tilde{q}_n, \tilde{p}_n(1 - 2\pi\delta)) = 2\pi\epsilon^3 c_n.$$

Consequently, taking into account two last formulas and applying the Lagrange theorem one gets

$$-2\pi\delta p_n \frac{\partial\Psi}{\partial p_n} = 2\pi\epsilon^3(c_n - c_{n-1}).$$

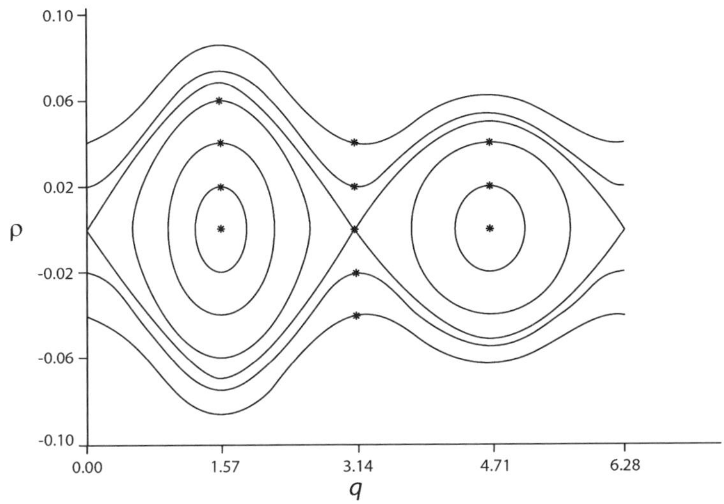

Fig. 3.18 Invariant curve without damping

Substituting the above function Ψ the following recurrent solution is found

$$c_n = c_{n-1} - \frac{2\pi\delta p_n^2}{\epsilon^2}. \qquad (**)$$

For $n = 0$ and having the initial values of q_0, p_0, one finds c_0

$$c_0 = \frac{p_0^2}{2\epsilon^2} + \frac{1}{8}\cos 2q_0,$$

and then constants q_1, p_1, c_1 are found from previous relations $(*)$ and preceding one.

It follows from $(**)$ that constant c_n decreases monotonously and tends to minimum of the function.

$$f(x, y) = \frac{y^2}{2\epsilon^2} + \frac{1}{8}\cos 2x.$$

Minimum $x = \frac{\pi}{2} + \pi n y = 0$ corresponds to stable periodic solution of (3.307). Poincaré points approach a stable point in a spiral way. Therefore, the found minimum is stable focus (see Fig. 3.17).

Note that a similar equation is studied in monographs [96, 182]. However, three approximations of averaging KBM method [51] are required in order to get the same result.

3.8 Geometrical Approach to the Swinging Pendulum

Riemannian formulation of dynamics makes use of possibility of studying the instability of system motions through the instability properties of geodesics in a suitable manifold. It is believed that geometrical approach may provide a good explanation of the onset of chaos in Hamiltonian systems as an alternative point of view. This technique has been recently applied to study chaos in Hamiltonian systems [56–58, 61], however other mechanical systems can be also analysed in this manner, e.g. systems with velocity-dependent potentials that are described by Finslerian geometry. The most important tool in this approach is the Jacobi–Levi–Civita (JLC) equation which describes how nearby geodesics locally scatter. The instability properties are completely determined by curvature properties of the underlying manifold due to the occurrence of the Riemann tensor in the JLC equation. Analysing the JLC equation we observe that manifolds with negative curvature induce chaos, however a chaotic behaviour may occur in systems with positive curvature manifolds due to curvature fluctuations along geodesics [56,57,179,198]. Since a generic Riemannian space consists of the ambient space and a metric tensor, so in order to apply this technique we need to define these quantities. We have a few choices at our disposal. We can use configuration space endowed with a Jacobi metric, enlarged space–time manifold with an Eisenhart metric or tangent bundle of a configuration manifold with a Sasaki metric. However, no matter the ambient space we choose for geometrization, the geodesics projected onto the configuration space of a system have to correspond the trajectories of an investigated system. In other words, the geodesics equations should give us the equations of motion. It should be emphasized that this technique does not provide a new method of solving differential equations. It provides a qualitative description of the behaviour of systems using formalism of differential geometry (Riemannian geometry in this case). This method is still under development and other types of spaces are sought for a purpose of geometrization. So far, Hamiltonian systems with many degrees-of-freedom have been investigated with the use of this approach [56–58,198]. However, Hamiltonian systems with a few degrees-of-freedom are not investigated so often as the previous ones in this manner except in paper [61] that deals with a two degrees-of-freedom system (Hénon–Heiles model). Note that Hénon–Heiles model dynamics is described by a Hamiltonian with a coordinate-independent kinetic energy term. Here, we analyse a two degrees-of-freedom system (the swinging pendulum) of the coordinate-dependent kinetic energy term.

In what follows we introduce basic tools for geometrization. Let us consider a conservative system with n degrees-of-freedom, which is described by the following Lagrangian:

$$L = T - V = \tfrac{1}{2}a_{\mu\nu}(x)\dot{x}^{\mu}\dot{x}^{\nu} - V(x), \tag{3.315}$$

where $\dot{x}^{\mu} \equiv \frac{dx^{\mu}}{dt}$.

For this conservative system Hamilton's principle can be cast in Maupertuis' principle:

$$\delta \mathscr{S} = \delta \int \frac{\partial L}{\partial \dot{x}^\mu} \dot{x}^\mu dt = 0. \tag{3.316}$$

It is well known that equations of motion can be obtained using the rules of variational calculus for the functional $\mathscr{S}$. We have the analogous situation in Riemannian geometry, where geodesics equations are obtained from:

$$\delta \int ds = 0, \tag{3.317}$$

where ds is arc length. Hence, we can identify these two situations in the following way:

$$\delta \int ds = \delta \mathscr{S} = 0. \tag{3.318}$$

The kinetic energy T is a homogeneous function of degree two in the velocities, hence:

$$\frac{\partial L}{\partial \dot{x}^\mu} \dot{x}^\mu = 2T. \tag{3.319}$$

Using the above result and substituting it to (3.316), we get Maupertuis' principle:

$$\delta \mathscr{S} = \delta \int 2T dt = 0 \tag{3.320}$$

Making use of (3.318) and the fact that the system is conservative, one can easily verify that:

$$2T dt = \sqrt{2(E - V(x)) a_{\mu\nu}(x) dx^\mu dx^\nu} = \sqrt{g_{\mu\nu}(x) dx^\mu dx^\nu} = ds, \tag{3.321}$$

where E is the total energy. Hence, we find the metric tensor which is referred as the Jacobi metric:

$$g_{\mu\nu}(x) = 2(E - V(x)) a_{\mu\nu}(x). \tag{3.322}$$

Note that the substitution $E - V = T$ made here is essential. As we said earlier, the geometry used in this approach is a Riemannian one hence a metric tensor should not be velocity-dependent. However, the kinetic energy is velocity- dependent by definition and that is why we must put $E - V(x)$ instead of T in (3.322).

Now, we present a brief derivation of the main tool of this approach, namely the JLC equation [56–58, 61]. Let us define a congruence of geodesics as a family of geodesics $\{x_\mu(s) = x(s,u) : u \in \mathscr{R}\}$. The geodesics are parameterized by a parameter u whereas s is arc length. Let $J(s) = \frac{d}{du}$ denote a tangent vector at u (s is fixed). $J(s)$ is called a Jacobi vector that can locally measure the distance between two nearby geodesics. Since $\frac{d}{ds}$ is tangent to a geodesic, we get:

$$\nabla_s \frac{d}{ds} = 0, \tag{3.323}$$

where $\nabla_s \equiv \nabla_{\frac{d}{ds}}$.

It is easy to verify that

$$\nabla_s \frac{d}{du} = \nabla_u \frac{d}{ds}. \tag{3.324}$$

Hence, we obtain

$$\nabla_s^2 \frac{d}{du} = \nabla_s \nabla_u \frac{d}{ds}. \tag{3.325}$$

Let us introduce the Riemann curvature tensor [39]:

$$R\left(\frac{d}{ds}, \frac{d}{du}\right)\frac{d}{ds} = \nabla_s \nabla_u \frac{d}{ds} - \nabla_u \nabla_s \frac{d}{ds} - \nabla_{\left[\frac{d}{ds}, \frac{d}{du}\right]} \frac{d}{ds}. \tag{3.326}$$

Making use of (3.323) and the fact that $\left[\frac{d}{ds}, \frac{d}{du}\right] = 0$ we get:

$$R\left(\frac{d}{ds}, \frac{d}{du}\right)\frac{d}{ds} = \nabla_s \nabla_u \frac{d}{ds}. \tag{3.327}$$

Substituting the obtained result to (3.325), we find:

$$\nabla_s^2 \frac{d}{du} + R\left(\frac{d}{du}, \frac{d}{ds}\right)\frac{d}{ds} = 0, \tag{3.328}$$

where we use the antisymmetry of the Riemann curvature tensor with respect to its first two arguments. The obtained equation is called JLC equation and it describes the evolution of geodesics separation along geodesic. In order to make any further calculation Eq. (3.328) will be expressed in local coordinate system later on.

3.8.1 The Analysed System and Geometrization

We study again, the two degrees-of-freedom, conservative mechanical system which consists of elastic swinging pendulum with nonlinear stiffness (see Fig. 3.19) [35]. The nonlinearity of the spring has the form:

$$k(y) = k_1(y + y_{st}) + k_2(y + y_{st})^3, \tag{3.329}$$

where y_{st} is a position of the mass at rest. The Lagrangian describing the dynamics of our system has the form:

$$\mathscr{L} = \frac{1}{2}m(\dot{r}^2 + r^2\dot{\varphi}^2) - \frac{k_1}{2}(r - l_0)^2 - \frac{k_2}{4}(r - l_0)^4 + mgr\cos\varphi. \tag{3.330}$$

From Euler–Lagrange equations one can easily obtain:

$$\begin{aligned}\ddot{r} &= r\dot{\varphi}^2 - \frac{k_1}{m}(r - l_0) - \frac{k_2}{m}(r - l_0)^3 + g\cos\varphi,\\ \ddot{\varphi} &= -\frac{2}{r}\dot{r}\dot{\varphi} - \frac{g}{r}\sin\varphi.\end{aligned} \tag{3.331}$$

In order to obtain dimensionless equations we apply the following substitutions:

$$z = \frac{r - l_0}{l} - z_0, \quad z_0 = \frac{y_{st}}{l}, \quad \alpha = \frac{k_2 l^3}{mg}, \quad \beta^2 = \frac{k_1 l}{mg}, \quad t \to t\sqrt{\frac{g}{l}}. \tag{3.332}$$

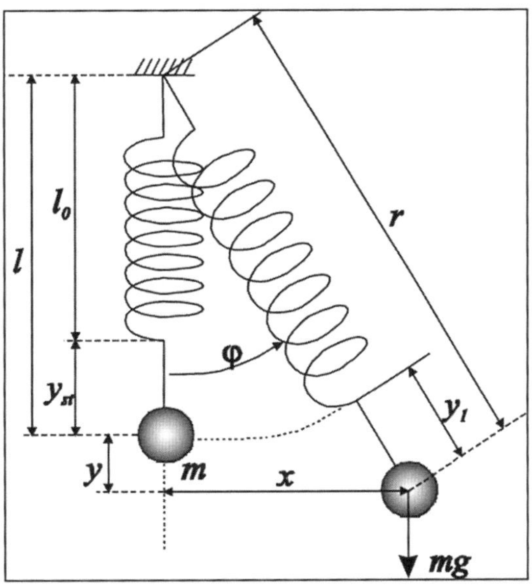

Fig. 3.19 The investigated system

Hence, we get

$$\ddot{z} = (1+z)\dot{\varphi}^2 - \beta^2(z+z_0) - \alpha(z+z_0)^3 + \cos\varphi,$$
$$\ddot{\varphi} = -\frac{2}{1+z}\dot{z}\dot{\varphi} - \frac{1}{1+z}\sin\varphi, \tag{3.333}$$

where the corresponding dimensionless Lagrangian reads

$$\tilde{\mathscr{L}} = \dot{z}^2 + (1+z)^2\dot{\varphi}^2 - \beta^2(z+z_0)^2 - \frac{\alpha}{2}(z+z_0)^4 + 2(1+z)\cos\varphi. \tag{3.334}$$

As the Riemannian space manifold we choose a configuration manifold endowed with a Jacobi metric g. In our case the Jacobi metric has the following form:

$$g = 4(\mathscr{E} - \mathscr{V})\begin{pmatrix} 1 & 0 \\ 0 & (1+z)^2 \end{pmatrix}, \tag{3.335}$$

where $\mathscr{V}$ is dimensionless potential:

$$\mathscr{V} = \beta^2(z+z_0)^2 + \frac{\alpha}{2}(z+z_0)^4 - 2(1+z)\cos\varphi, \tag{3.336}$$

and $\mathscr{E}$ is total energy of the system. In order to find JLC equation, first we have to find coefficients of the Riemann curvature tensor. It is easy to verify that connection coefficients have the following form:

$$\Gamma^1_{11} = -\frac{1}{2\mathscr{W}}\frac{\partial\mathscr{V}}{\partial z}, \quad \Gamma^1_{12} = -\frac{1}{2\mathscr{W}}\frac{\partial\mathscr{V}}{\partial\varphi},$$
$$\Gamma^1_{22} = -\frac{1}{2\mathscr{W}}\left(2(1+z)\mathscr{W} - (1+z)^2\frac{\partial\mathscr{V}}{\partial z}\right), \quad \Gamma^2_{11} = \frac{1}{2\mathscr{W}(1+z)^2}\frac{\partial\mathscr{V}}{\partial\varphi}$$
$$\Gamma^2_{12} = \frac{1}{2\mathscr{W}(1+z)^2}\left(2(1+z)\mathscr{W} - (1+z)^2\frac{\partial\mathscr{V}}{\partial z}\right), \quad \Gamma^2_{22} = \frac{1}{2\mathscr{W}}\frac{\partial\mathscr{V}}{\partial\varphi},$$

where $\mathscr{W} = \mathscr{E} - \mathscr{V}$. Since the Riemannian space manifold is two-dimensional, there is only one nonzero coefficient of the Riemann curvature tensor. It has the following form:

$$R_{2121} = 2\frac{\partial^2\mathscr{V}}{\partial\varphi^2} + \frac{2}{\mathscr{W}}\left(\frac{\partial\mathscr{V}}{\partial\varphi}\right)^2 + 2(1+z)^2\frac{\partial^2\mathscr{V}}{\partial z^2} + 2(1+z)\frac{\partial\mathscr{V}}{\partial z} + 2\frac{(1+z)^2}{\mathscr{W}}\left(\frac{\partial\mathscr{V}}{\partial z}\right)^2. \tag{3.337}$$

Let us choose special base vectors $\{E_1, E_2\}$, such that $E_1 = \frac{d}{ds}$ and $g(E_1, E_2) = 0$, $g(E_2, E_2) = 1$.

In other words the base is formed by the orthonormal set of vectors. In local coordinate system these base vectors have the form (Einstein's summation conversion is used in this section)

$$E_i = Y_i^j e_j, \tag{3.338}$$

where $i = 1, 2$ and

$$\begin{pmatrix} Y_1^1 & Y_1^2 \\ Y_2^1 & Y_2^2 \end{pmatrix} = \begin{pmatrix} \frac{dz}{ds} & \frac{d\varphi}{ds} \\ (1+z)\frac{d\varphi}{ds} & -\frac{1}{1+z}\frac{dz}{ds} \end{pmatrix}, \; e_1 = \frac{\partial}{\partial z}, \; e_2 = \frac{\partial}{\partial \varphi}. \tag{3.339}$$

The Jacobi vector in this base reads $J = J^n E_n$ (J^n should not be confused with n-power of J). Substituting this into the JLC Eq. (3.328) we get

$$\nabla_S^2(J^n E_n) + R(J^n E_n, E_1)E_1 = 0. \tag{3.340}$$

Since E_1 is tangent to the geodesics, one gets

$$\frac{d^2 J^n}{ds^2} E_n + J^i R(E_i, E_1)E_1 = 0. \tag{3.341}$$

Since $R(E_n, E_1)E_1$ is a vector, so it can be decomposed in the orthonormal base as follows

$$R(E_i, E_1)E_1 = \sum_{n=1}^{2} g(R(E_i, E_1)E_1, E_n)E_n. \tag{3.342}$$

We put sum explicitly here because i indices are on the same level so we cannot apply Einstein's summation convention. Substituting (3.342) to (3.341) one gets

$$\frac{d^2 J^n}{ds^2} + J^i g(R(E_i, E_1)E_1, E_n) = 0, \tag{3.343}$$

where $n = 1, 2$.

Due to the antisymmetry of the Riemann tensor the following equation is obtained

$$\frac{d^2 J^n}{ds^2} + J^2 g(R(E_2, E_1)E_1, E_n) = 0. \tag{3.344}$$

Making use of (3.338) one finds

$$g(R(E_2, E_1)E_1, E_n) = R_{klij} Y_n^k Y_1^l Y_2^i Y_1^j. \tag{3.345}$$

Taking into account (3.339) two uncoupled equations JLC equations are obtained:

$$\frac{d^2 J^1}{ds^2} = 0,$$
$$\frac{d^2 J^2}{ds^2} + \frac{R_{2121}}{\det g} J^2 = 0. \tag{3.346}$$

Actually, we are only interested in the evolution of J^2 along geodesics because this component of Jacobi vector is a coefficient standing by the vector that is orthogonal to the tangent direction of geodesics. Hence, from now on we consider only the following equation

$$\frac{d^2 J^2}{ds^2} + \frac{R_{2121}}{\det g} J^2 = 0. \tag{3.347}$$

Due to the dimension of the Riemannian space manifold we find

$$\frac{R_{2121}}{\det g} = \frac{1}{2} R, \tag{3.348}$$

where R is a scalar curvature. In our case, the scalar curvature R takes the form

$$R = \frac{2R_{2121}}{\det g} = \frac{1}{8\mathscr{W}^2(1+z)^2}\left(\frac{\partial^2 \mathscr{V}}{\partial \varphi^2} + \frac{1}{\mathscr{W}}\left(\frac{\partial \mathscr{V}}{\partial \varphi}\right)^2 + (1+z)^2 \frac{\partial^2 \mathscr{V}}{\partial z^2} + (1+z)\frac{\partial \mathscr{V}}{\partial z}\right)$$
$$+ \frac{1}{8\mathscr{W}^3}\left(\frac{\partial \mathscr{V}}{\partial z}\right)^2. \tag{3.349}$$

Let us go back to the JLC equation

$$\frac{d^2 J^2}{ds^2} + \frac{1}{2} J^2 R(s) = 0. \tag{3.350}$$

The above equation is a differential equation with respect to the arc length s. In order to make further calculation we have to transform this equation into a differential equation with respect to time t. The metric used in this section is the Jacobi metric, so we have

$$ds = 2\mathscr{W}\, dt. \tag{3.351}$$

Making use of the above identity we get the desired differential equation with respect to time

$$\frac{d^2 J^2}{dt^2} - \frac{\dot{\mathscr{W}}}{\mathscr{W}} \frac{dJ^2}{dt} + 2\mathscr{W}^2 R J^2 = 0. \tag{3.352}$$

To obtain an oscillator equation we apply the substitution [33–35]:

$$J^2 = J \exp\left(\frac{1}{2}\int \frac{\dot{\mathscr{W}}}{\mathscr{W}} dt\right) = J\sqrt{\mathscr{W}}. \tag{3.353}$$

The quantity $\mathscr{W}$ is never negative since its values equal the kinetic energy by definition, so $\sqrt{\mathscr{W}}$ always exists. Hence we find the oscillator equation

$$\ddot{J} + \frac{1}{2}J\left(\frac{\ddot{\mathscr{W}}}{\mathscr{W}} + 4\mathscr{W}^2 R - \frac{3}{2}\left(\frac{\dot{\mathscr{W}}}{\mathscr{W}}\right)^2\right) = 0. \tag{3.354}$$

Observe that $\sqrt{\mathscr{W}} \le \mathscr{E}$ so we can examine only J instead of J^2. Eq. (3.354) is an oscillator equation with time-dependent frequency

$$\Omega = \frac{1}{2}\left(\frac{\ddot{\mathscr{W}}}{\mathscr{W}} + 4\mathscr{W}^2 R - \frac{3}{2}\left(\frac{\dot{\mathscr{W}}}{\mathscr{W}}\right)^2\right). \tag{3.355}$$

Thus, the analysis of behaviour of the system is transformed to the analysis of solutions of the following equation

$$\ddot{J} + \Omega J = 0. \tag{3.356}$$

As we mentioned earlier, the above equation is an oscillator equation with time-dependent frequency Ω. Note that the frequency is not explicitly time-dependent. However, it consists of functions of positions and velocities of the spring mass that are time-dependent. Hence, in order to solve (3.356) we must know the solutions of dynamics equations (3.333). In what follows the JLC equation is numerically solved, some numerical results are illustrated and discussed.

In order to solve numerically Eq. (3.356) we put the following values of the parameters: ($\alpha = 2, \beta = 1$), ($\alpha = 0.1, \beta = 0.2$) and ($\alpha = 1, \beta = 2$). The results are displayed in four figures for each case (except the last case, where six figures are presented). The first figure shows the projection of the phase trajectories onto $(z, \dot{z})$ subspace. The second one presents a corresponding Poincaré section or another projection onto $(\varphi, \dot{\varphi})$ subspace. Next figures present the evolution of J, $\ln|J|$.

Our first case is a quasiperiodic one (see Fig. 3.20). It is easy to see that J evolves rapidly and fluctuates along the trajectory. The rate of the growth is presented in next figure, where $\ln|J|$ is put on the vertical axis. The second case is a weak-chaotic one (see Fig. 3.21). Observe that behaviour of J is similar to the previous one (see Fig. 3.20), i.e. J evolves rapidly as well. Hence, it is rather hard to distinguish between these two cases. In the last case we have a qualitatively different situation

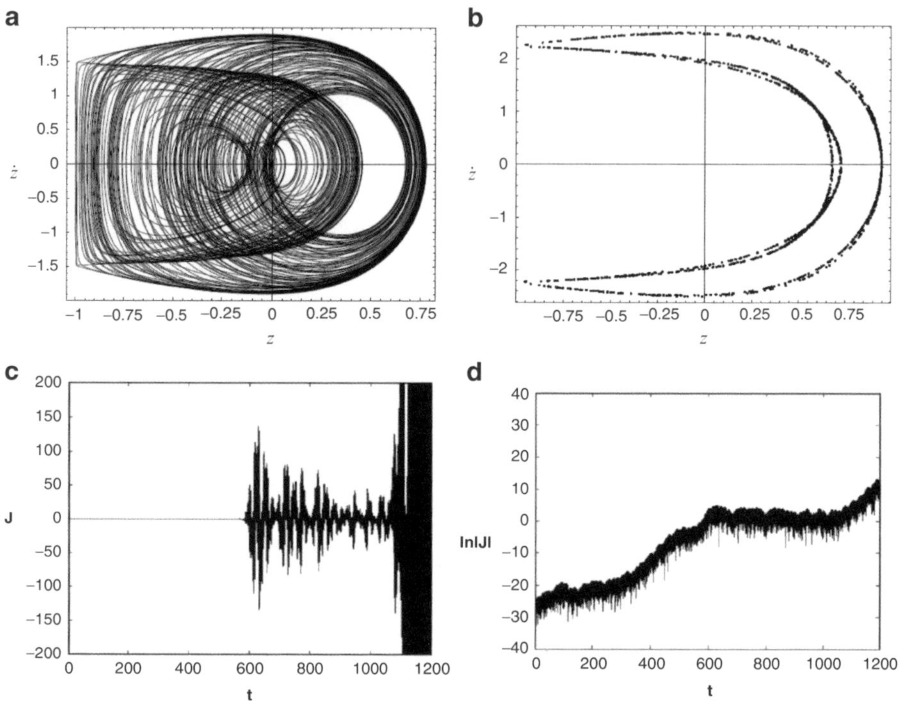

Fig. 3.20 Quasiperiodic case: $\alpha = 2$, $\beta = 1$, $z(0) = 0.2$, $\dot{z}(0) = 0.4$, $\varphi(0) = 1.7$, $\dot{\varphi}(0) = 0.8$. (**a**) Projection $(z, \dot{z})$. (**b**) Poincaré map $(z, \dot{z})$. (**c**) Evolution of J. (**d**) Evolution of $\ln|J|$

(see Fig. 3.22). One can observe that the evolution of J is bounded and resembles the vibrations with amplitude modulation. Such evolution of J should indicates that a behaviour of the system is regular.

However, we can observe (see Fig. 3.22a–d) in phase portraits and Poincaré sections that we have rather quasiperiodic behaviour than periodic one. In order to explain this fact we must take a look at the $(z, \dot{z})$ and $(\varphi, \dot{\varphi})$ Poincaré sections and phase portraits. One can observe that the phase trajectories do not penetrate the whole energy surface. The phase trajectories are bounded to a small region (see the scale in Fig. 3.22a–d). Hence, there cannot be any rapid growth of J (see Fig. 3.22e).

3.9 Geometric Analysis of a Double Pendulum Dynamics

The classical approach to analysis of Hamiltonian systems has been widely applied, providing a classical explanation of the onset of chaos in these systems. In addition to the classical techniques for analysing Hamiltonian systems, the geometric approach plays an important role. The geometric approach is based on the relation

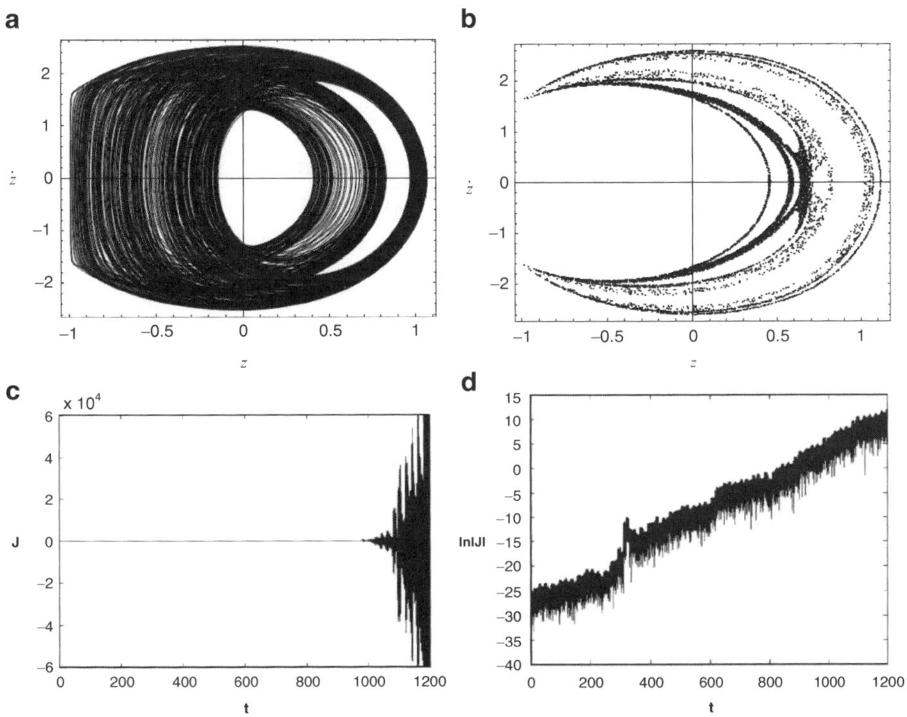

Fig. 3.21 Weak-chaotic case: $\alpha = 1$, $\beta = 2$, $z(0) = 0.2$, $\dot{z}(0) = 2.4$, $\varphi(0) = 0.7$, $\dot{\varphi}(0) = 0.4$. (**a**) Projection $(z, \dot{z})$. (**b**) Poincaré map $(z, \dot{z})$. (**c**) Evolution of J. (**d**) Evolution of $\ln|J|$

between Riemannian geometry and Hamiltonian dynamics, but is distinct from the incorporated footnote geometric formulation of Hamiltonian mechanics in terms of symplectic geometry. This technique has into text been successfully applied [42, 57, 60, 61], especially to systems with many degrees of freedom. It has also been widely applied in general relativity [226, 227] and to low dynamical systems with a nondiagonal metric tensor [35]. It is believed that the geometric approach can provide an alternative to slight revision for the classical explanation for the onset of chaos in Hamiltonian systems, which involves the homoclinic ease of reading intersections [155].

As it has been already mentioned in the geometric approach to Hamiltonian dynamics, the analysis of dynamical trajectories and behaviour of a system is cast into the analysis of a geodesic flow in a corresponding Riemannian space. The main tool of this approach is the so-called JLC equation [42, 55]. In general, the JLC equation is a system of second-order differential equations with respect to a geodesic length, and it describes the evolution of a tangent vector (so-called Jacobi vector) along the geodesic. Although there are many dynamical systems that can be described in this manner, there are some that cannot, namely systems with velocity-dependent potentials. However, this kind of dynamical system can be analysed by means of the Finslerian geometry [42]. Here, we study conservative Hamiltonian systems, which can be geometrized within the Riemannian geometry approach [33].

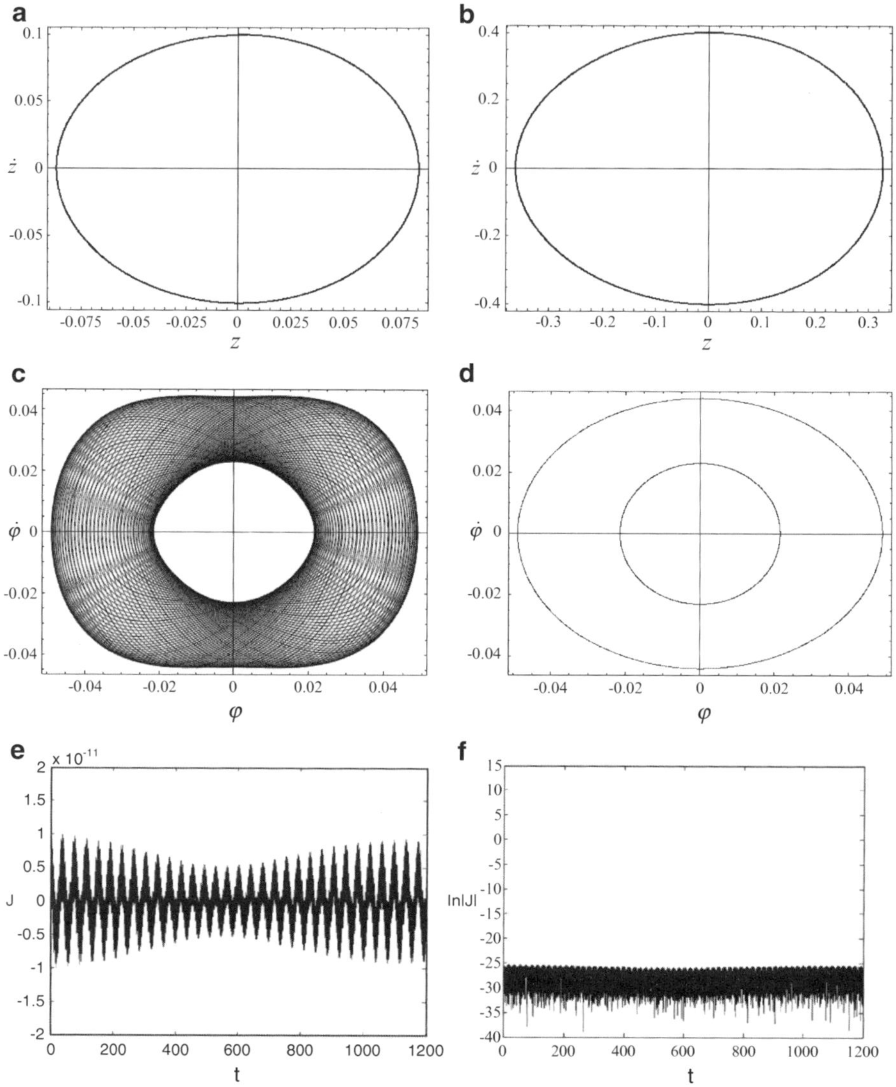

Fig. 3.22 Regular system case: $\alpha = 0.1$, $\beta = 0.2$, $z(0) = 0$, $\dot{z}(0) = 0.1$, $\varphi(0) = 0$, $\dot{\varphi}(0) = 0.03$. (**a**) Projection $(z, \dot{z})$. (**b**) Poincaré map $(z, \dot{z})$. (**c**) Projection $(\varphi, \dot{\varphi})$. (**d**) Poincaré map $(\varphi, \dot{\varphi})$. (**e**) Evolution of J. (**f**) Evolution of $\ln |J|$

The point is that motion of a Hamiltonian system can be viewed as the motion of a single virtual particle along a geodesic in a suitable Riemannian space $\mathcal{Q}$. From the above condition one can obtain the geodesics equation, which has the following form in local coordinates [55]

$$\frac{d^2q^i}{ds^2} + \Gamma^i_{jk}\frac{dq^j}{ds}\frac{dq^k}{ds} = 0. \tag{3.357}$$

The Riemannian space is endowed with a metric tensor, which is obtained from the dynamics of the analysed system. Since we choose a configuration space of an analysed system for a Riemannian manifold, the metric tensor is the Jacobi metric $\mathfrak{g}$, which is connected to the dynamics by the following relationship [57]

$$\mathfrak{g}_{ij} = 2W\mathfrak{a}_{ij}(q), \quad W \equiv E - V, \tag{3.358}$$

where E is a total energy and V is potential energy. The matrix 'a' is a kinetic energy matrix (we use the brought footnote into Einstein summation convention):

$$L = \frac{1}{2}\dot{q}^i\mathfrak{a}_{ij}\dot{q}^j - V. \tag{3.359}$$

This relationship follows from the Maupertuis principle, which gives

$$ds = 2Wdt. \tag{3.360}$$

The main tool of the geometric approach, namely the JLC equation in a local coordinate system, has the following form [55, 57]

$$\frac{\delta^2 J^n}{\delta s^2} + J^i\frac{dq^j}{ds}\frac{dq^k}{ds}R^n_{kij} = 0, \; n = 1, 2, \ldots, \dim\mathcal{Q}, \tag{3.361}$$

where q^j satisfy the geodesics equation (3.357), J^n are components of the Jacobi vector, R^n_{kij} are components of the Riemann curvature tensor, and

$$\frac{\delta J^n}{\delta s} = \frac{dJ^n}{ds} + \Gamma^n_{kl}J^n\frac{dq^l}{ds} \tag{3.362}$$

are so-called absolute derivatives. The above equation has a similar form to the tangent dynamics equation, which is used to evaluate Lapunov's exponents. In fact, Eq. (3.361) takes exactly the same form in the case of the Eisenhart metric [56]. This means that there is a connection between the JLC equation and the tangent dynamics equation. Moreover, it is possible to find Lapunov's exponents using the Riemannian geometry approach. This has been done only for systems with many degrees of freedom and diagonal metric tensor [57].

Because we are interested in systems of only two degrees of freedom, the Riemannian space $\mathcal{Q}$ is two dimensional. This implies that we have only one nonzero component R_{2121} of the Riemann curvature tensor [55, 179]. In this case, the JLC equation (3.361) takes the form

$$\frac{d^2\Psi}{ds^2} + \frac{R_{2121}}{\det\mathfrak{g}}\Psi = 0, \tag{3.363}$$

where Ψ is a normal component of the Jacobi vector relative to the geodesic. The tangent component of the Jacobi vector evolves only linearly in a geodesic length, so it does not contribute to the character of the solution [42]. Next, making use of the fact that

$$\mathscr{R} = \frac{2R_{2121}}{\det \mathfrak{q}}, \tag{3.364}$$

we obtain a single differential equation which carries information about the system behaviour

$$\frac{d^2\Psi}{ds^2} + \frac{1}{2}\mathscr{R}\Psi = 0, \tag{3.365}$$

where $\mathscr{R}$ is the scalar curvature which, in general, is not periodic in τ. At this point, we can see where a possible explanation of the onset of chaos in Hamiltonian system lies. The component Ψ of the Jacobi vector represents a distance between two nearby geodesics, which in turn represent trajectories of the analysed system. The solutions of Eq. (3.365) can exhibit exponential growth due to parametric excitations in the scalar curvature. Hence, this formulation and description of Hamiltonian dynamics gives us a qualitatively different explanation of the onset of chaos as a parametric instability of geodesics [61].

In order to solve Eq. (3.365), we need to transform it into a differential equation with respect to the real time, t. Taking into account Eq. (3.359) we find

$$\ddot{\Psi} - \frac{\dot{W}}{W}\dot{\Psi} + 2\mathscr{R}W^2\Psi = 0. \tag{3.366}$$

The above equation can be easily transformed into another form by means of the following substitution [61]

$$\Psi = J\sqrt{W}, \tag{3.367}$$

which gives

$$\ddot{J} + \Omega(\tau)J = 0, \tag{3.368}$$

where

$$\Omega(\tau) \equiv \frac{1}{2}\left(\frac{\ddot{W}}{W} - \frac{1}{2}\left(\frac{\dot{W}}{W}\right)^2 + 4\mathscr{R}W^2\right). \tag{3.369}$$

It should be emphasized here that Ω is not, in general, periodic in τ-time. Although Ω is written as a function of τ, it does not depend on τ explicitly. In fact, it depends on a particular trajectory of the system.

3.9.1 The Pendulum and Geometrization

In this section we analyse a mechanical system with constraints, namely a double physical pendulum. The dynamics of the pendulum are described by the following Lagrangian L

$$L = \frac{1}{2}(m_1c_1^2+J_1+m_2l_1^2)\dot{\varphi}_1^2+\frac{1}{2}(m_2c_2^2+J_2)\dot{\varphi}_2^2+m_2c_2l_1\dot{\varphi}_1\varphi_2\cos(\varphi_1-\varphi_2)-V(\varphi_1,\varphi_2), \tag{3.370}$$

where

$$V(\varphi_1,\varphi_2) = g(m_2l_1+m_2c_2+m_1c_1)-g(m_1c_1+m_2l_1)\cos\varphi_1-m_2gc_2\cos\varphi_2, \tag{3.371}$$

m_1 and m_2 are masses, J_1 and J_2 are moments of inertia, and c_1 and c_2 are the positions of centers of masses of the first and second link, respectively (see Fig. 3.23). In order to cast the above Lagrangian into a non-dimensional form, we introduce the following scaling

$$\begin{aligned} \tau &\equiv t\sqrt{\frac{m_1gc_1+m_2gl_1}{J_1+m_1c_1^2+m_2l_1^2}}, \quad &\beta \equiv \frac{J_2+m_2c_2^2}{J_1+m_1c_1^2+m_2l_1^2}, \\ \kappa &\equiv \frac{m_2c_2l_1}{J_1+m_1c_1^2+m_2l_1^2}, \quad &\mu \equiv \frac{m_2c_2}{m_1c_1+m_2l_1}. \end{aligned} \tag{3.372}$$

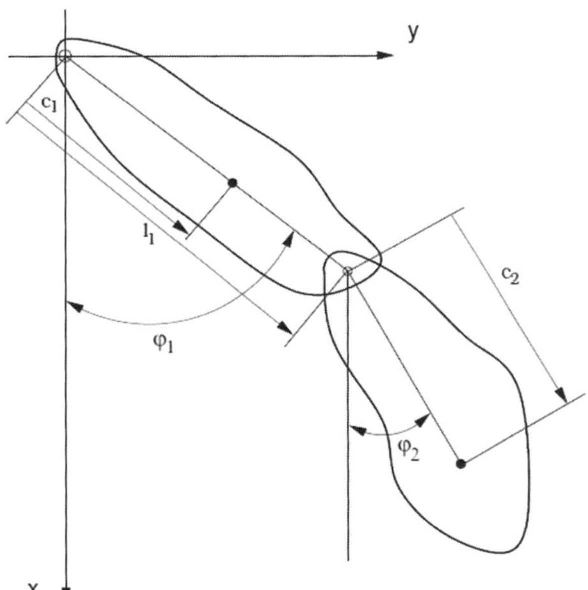

Fig. 3.23 Double physical pendulum

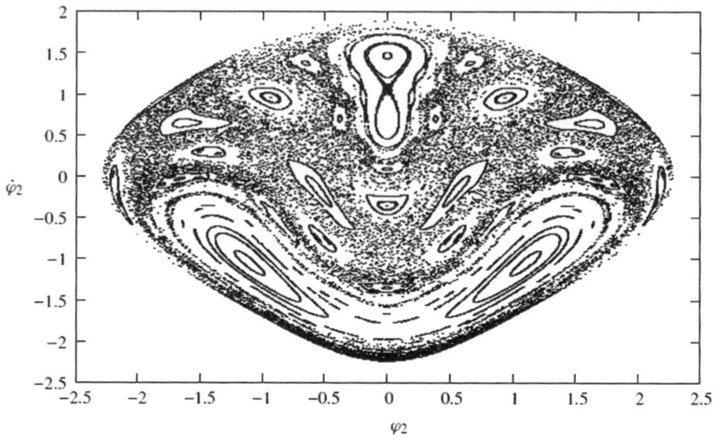

Fig. 3.24 Poincaré section for $\mathscr{E} = 1.1$

Hence, the Lagrangian takes the non-dimensional form

$$
\begin{aligned}
L &= \frac{1}{2}\dot{\varphi}_1^2 + \frac{\beta}{2}\dot{\varphi}_2^2 + \kappa\dot{\varphi}_1\dot{\varphi}_2\cos\phi - 1 - \mu + \cos\varphi_1 + \mu\cos\varphi_2 \\
&= \frac{1}{2}(\dot{\varphi}_1\dot{\varphi}_2)\mathfrak{a}\begin{pmatrix}\dot{\varphi}_1\\ \dot{\varphi}_2\end{pmatrix} - 1 - \mu + \cos\varphi_1 + \mu\cos\varphi_2, \\
\phi &\equiv \varphi_1 - \varphi_2
\end{aligned}
\tag{3.373}
$$

where

$$
\mathfrak{a} = \begin{pmatrix} 1 & \kappa\cos\phi \\ \kappa\cos\phi & \beta \end{pmatrix}, \tag{3.374}
$$

The dot over φ denotes τ derivative. Using the Euler–Lagrange equations we obtain the equations of motion

$$
\begin{aligned}
\ddot{\varphi}_1 &= \frac{-\kappa\sin\phi(\kappa\cos\phi\dot{\varphi}_1^2 + \beta\dot{\varphi}_2^2) - \beta\sin\varphi_1 + \kappa\mu\sin\varphi_2\cos\phi}{\beta - \kappa^2\cos^2\phi}, \\
\ddot{\varphi}_2 &= \frac{\kappa\sin\phi(\kappa\cos\phi\dot{\varphi}_2^2 + \dot{\varphi}_1^2) - \mu\sin\varphi_2 + \kappa\sin\varphi_1\cos\phi}{\beta - \kappa^2\cos^2\phi}.
\end{aligned}
\tag{3.375}
$$

Let us consider the Jacobi metric $\mathfrak{g}$ of the physical pendulum

$$
\mathfrak{g} = 2\mathscr{W}\mathfrak{a} = 2\mathscr{W}\begin{pmatrix} 1 & \kappa\cos\phi \\ \kappa\cos\phi & \beta \end{pmatrix} \quad \mathscr{W} \equiv \mathscr{E} - 1 - \mu + \cos\varphi_1 + \mu\cos\varphi_2. \tag{3.376}
$$

Next, we find the connection coefficients Γ^i_{jk}:

$$
\begin{aligned}
\Gamma^1_{11} &= \frac{1}{2\mathscr{W}\det\mathfrak{a}}(2\kappa^2\sin\varphi_1\cos^2\phi + \mathscr{W}\kappa^2\sin(2\phi) - \mu\kappa\sin\varphi_2\cos\phi - \beta\sin\varphi_1),\\
\Gamma^2_{22} &= \frac{1}{2\mathscr{W}\det\mathfrak{a}}(2\mu\kappa^2\sin\varphi_2\cos^2\phi - \mathscr{W}\kappa^2\sin(2\phi) - \beta\kappa\sin\varphi_1\cos\phi - \beta\mu\sin\varphi_2),\\
\Gamma^2_{11} &= \frac{1}{2\mathscr{W}\det\mathfrak{a}}(\mu\sin\varphi_2 - \kappa\sin\varphi_1\cos\phi - 2\mathscr{W}\kappa\sin\phi),\\
\Gamma^1_{22} &= \frac{\beta}{2\mathscr{W}\det\mathfrak{a}}(\beta\sin\varphi_1 - \mu\kappa\sin\varphi_2\cos\phi + 2\mathscr{W}\kappa\sin\phi),\\
\Gamma^2_{12} &= \frac{1}{2\mathscr{W}\det\mathfrak{a}}(\mu\kappa\sin\varphi_2\cos\phi - \beta\sin\varphi_1),\\
\Gamma^1_{12} &= \frac{\beta}{2\mathscr{W}\det\mathfrak{a}}(\kappa\sin\varphi_1\cos\phi - \mu\sin\varphi_2).
\end{aligned}
\tag{3.377}
$$

In a two-dimensional space there is only one nonzero component of the Riemann curvature tensor, namely,

$$
\begin{aligned}
R_{2121} &= \mu\cos\varphi_2 + 2\mathscr{W}\kappa\cos\phi + \beta\cos\varphi_1\\
&\quad + \frac{1}{\mathscr{W}}(\kappa\sin\varphi_1\cos\phi - \mu\sin\varphi_2)^2 + \frac{\sin^2\varphi_1\det\mathfrak{a}}{\mathscr{W}}\\
&\quad + -\frac{\kappa\sin\phi}{\det\mathfrak{a}}(\beta\kappa\sin\varphi_1\cos\phi - \mu\kappa\sin\varphi_2\cos\phi - \beta\mu\sin\varphi_2 + \beta\sin\varphi_1)\\
&\quad + -\frac{2\mathscr{W}\kappa^3\sin^2\phi\cos\phi}{\det\mathfrak{a}}.
\end{aligned}
\tag{3.378}
$$

Making use of Eq. (3.364) we find the scalar curvature:

$$
\begin{aligned}
\mathscr{R} &= \frac{\kappa\cos\phi}{\mathscr{W}\det\mathfrak{a}} - \frac{\kappa^3\sin^2\phi\cos\phi}{\mathscr{W}\det^2\mathfrak{a}} + \frac{\mu\cos\varphi_2 + \beta\cos\varphi_1}{2\mathscr{W}^2\det\mathfrak{a}}\\
&\quad + -\frac{\kappa\sin\phi(\beta\kappa\sin\varphi_1\cos\phi - \mu\kappa\sin\varphi_2\cos\phi - \beta\mu\sin\varphi_2 + \beta\sin\varphi_1)}{2\mathscr{W}^2\det^2\mathfrak{a}}\\
&\quad + \frac{\sin^2\varphi_1}{2\mathscr{W}^3} + \frac{(\kappa\sin\varphi_1\cos\phi - \mu\sin\varphi_2)^2}{2\mathscr{W}^3\det\mathfrak{a}}.
\end{aligned}
\tag{3.379}
$$

Finally, inserting the obtained scalar curvature into Eq. (3.368) we get the JLC equation for the physical pendulum.

3.9.2 Numerical Simulations

The equations of motion have been numerically solved by means of the symplectic algorithm of Stromer–Verlet [110] and the obtained JLC equation (3.368) by the Dormand–Prince 5(4) algorithm with algorithm with variable time-step and the energy correction. Numerical simulation parameters were given the following values: $\beta = 0.6$, $\kappa = 0.4$, $\mu = 0.66667$. The simulation was performed for the total energy $\mathscr{E} = 1.1$. Below, we present the Poincaré section (Fig. 3.24), in which one can observe chaotic regions as well as islands of regular behaviour. Thus, we can analyse the system's behaviour on the same energy level. The numerical results include three cases, namely two of them (Figs. 3.25 and 3.26) from regions of regular behaviour and the last (Fig. 3.27) one from the chaotic region. The initial conditions of the regular behaviour cases have been taken from the interior of the regular islands, so that trajectories stay in regular regions regardless of numerical errors.

The presented figures include two projections of the phase trajectories (only in the case of regular behaviour), the corresponding Poincaré section of a particular trajectory, and the graph, which presents the evolution rate of a solution of the JLC equation. One can easily observe that in the case of regular trajectories (Figs. 3.25 and 3.26) the evolution of the Jacobi vector along the geodesic is bounded. However, in Fig. 3.27 we can observe the unbounded evolution of the Jacobi vector, which means that two nearby geodesics originating from the neighbourhood of the initial

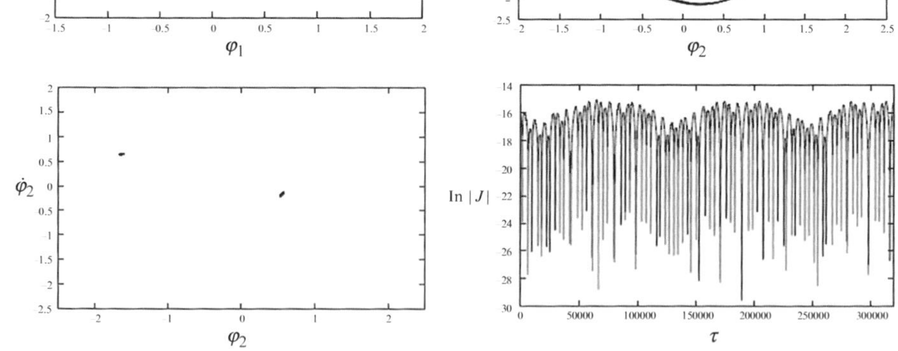

Fig. 3.25 Initial conditions: $\varphi_2 = -1.63$, $\dot{\varphi}_2 = 0.63$

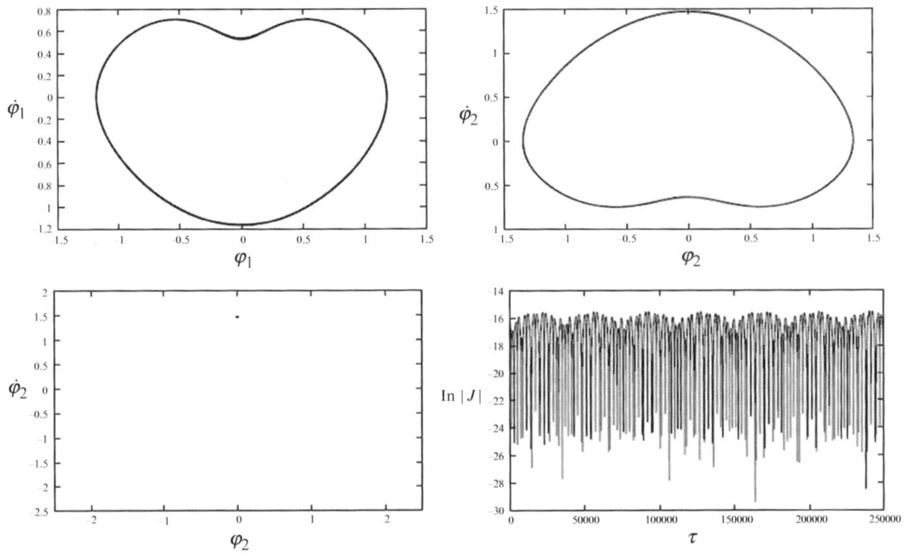

Fig. 3.26 Initial conditions: $\varphi_2 = 0$, $\dot{\varphi}_2 = 1.46$

condition move away from each other and hence the distance between them grows exponentially. This is caused by the parametric resonance occurring in the JLC equation (3.365).

The obtained results can be briefly summarized as follows. We have applied the Riemannian approach to a low dimensional system with constraints and have shown that the geometric approach gives results that are in qualitative agreement with those obtained from the classical approach. The existence of constraints is manifest in the metric tensor, which has a nondiagonal form in this case. Although the obtained results show that there is an agreement between classical and geometric approaches, a more thorough analysis is needed. The aim of this approach is to make use of the Riemannian geometry tools to gain information about a system's behaviour without referring to the geodesic evolution. The geometric approach has already been applied to systems that have no constraints and many degrees of freedom [42, 57]. However, systems with approach few degrees of freedom and constraints are more difficult to analyse in this manner.

Lyapunov exponents vs. geometric instability exponents of a double physical pendulum has been further studied in [34]. Within the Riemannian geometry formulation, the geometrical instability exponents are defined as geometric Lyapunov exponents, since the tangent dynamics equation has the same form as the JLC equation in the Eisenhart metric.

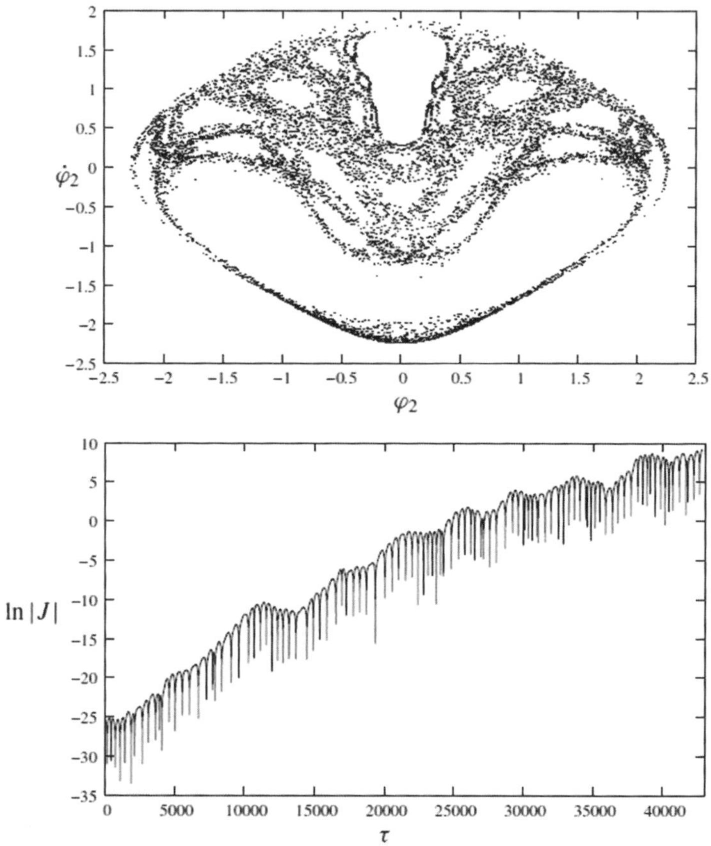

Fig. 3.27 Initial conditions: $\varphi_2 = -1.93$, $\dot{\varphi}_2 = -0.23$

3.10 A Set of Linear Second-Order ODEs with Constant Coefficients

Consider the systems governed by the following set of second-order differential equations

$$A\ddot{q} + B\dot{q} + Cq = f(t), \tag{3.380}$$

where the matrices A, B and C do not depend on time t and they are arbitrary square matrices of nth order with real coefficients. Here our considerations will follow reference [72] (see references therein and [97]). According to the authors sometimes a direct application of the second-order form (3.380) may lead to essential reduction of a computational time preserving a clear physical insight. In other words, sometimes a reduction to the first-order differential equations is not required and may cause more problems in analysis and understanding a particular with respect to application oriented researches and engineers. The classical modal analysis, which has long tradition in mechanics, can be directly applied to a lumped dynamical system (3.380) instead of using the state-space Hamiltonian approach. Although the Hamiltonian approach possesses its beauty and attracts theoretically oriented researchers, engineers rely on traditional terms like Lagrangian coordinates (here q), generalized excitation (here $f(t)$), modes and natural frequencies. Since we consider here non-conservative linear systems (note the occurrence of matrix B), hence also two other important questions appear (see also [87, 126]).

(i) When does a linear non-conservative system possess classical normal modes?
(ii) What are free frequencies of non-conservative systems? Note that model analysis of a symmetric and positive definite system is a process of diagonalizing two matrices at the same time. Then if the system (3.380) is decoupled, it may be treated as n independent single-degree-of-freedom systems.

These two questions will be explained further. Traditionally, here are three modal analysis approaches devoted to system (3.380):

(iii) Simultaneous reduction of A, B and C to symmetric matrices by a common *similarity* transformation;
(iv) Simultaneous reduction of the coefficient matrices to diagonal forms;
(v) Simultaneous reduction of matrices A, B and C to upper triangular matrices by a common similarity transformation [72].

However, the restrictions required for application of the mentioned approaches show that only a small subclass of linear non-conservative systems may be analysed.

Following reference [72], instead of A, B and C we take matrices M, D and K. The matrix M is called *mass inertial matrix* and is positive definite, whereas D (damping matrix) and K (stiffness matrix) are positive semidefinite.

3.10.1 Conservative Systems

We consider the system

$$A\ddot{q} + Cq = f(t), \tag{3.381}$$

which is obtained from (3.380) for $B = 0$. In a mechanical language Eq. (3.381) govern dynamics of an undamped non-gyroscopic conservative and non-autonomous system. The system (3.381) can be decoupled if and only if there are two non-singular matrices U and V such that VAU and VCU are diagonal.

Recall that two square matrices P and Q are said to be *equivalently transformed* if $P = VQU$. The *equivalent transformation* preserves the rank of the matrices.

If $V = U^{-1}$ then we have a similarity transformation. If $V = U^T$ then we have a *congruence transformation*. Note that the classical model analysis is based on the *congruence transformation*.

If U is an orthogonal matrix then a congruence transformation is also a similarity transformation, i.e. $U^{-1} = U^T$.

Consider an autonomous case ($f = 0$) and assume a solution in the form of a column vector

$$q = ue^{\alpha t}. \tag{3.382}$$

Substituting (3.382) to (3.381) we get

$$\alpha^2 Au + Cu = 0 \tag{3.383}$$

and hence

$$Cu = \lambda Au, \tag{3.384}$$

where $\lambda = -\alpha^2$. The adjoint eigenvalue problem has the form

$$C^T v = \lambda A^T v, \tag{3.385}$$

and (3.384) and (3.385) have the same determinant and hence they have the same eigenvalues. If one assumes that A is non-singular there exists a full set (n) of eigenvalues. An eigenvalue problem is said to be *defective* if it does not have a full set of independent eigenvalues. A sufficient (but not necessary) condition under which (3.385) is not defective is that the eigenvalues are distinct. On the other hand, (3.384) is not defective if and only if (3.385) is not defective. Every λ_j possesses the corresponding eigenvectors u_j and v_j:

$$Cu_i = \lambda_i Au_i, \tag{3.386}$$

$$C^T v_j = \lambda_j A^T v_j. \tag{3.387}$$

Each column vector $u_i(v_i)$ is undetermined to the extent of an arbitrary multiplicative constant. Recall that $(AB)^T = B^T A^T$, and hence $(C^T v_j)^T = v_j^T C$. Applying transpose to (3.387) we obtain

$$v_j^T C = \lambda_j v_j^T A. \tag{3.388}$$

Multiplying both sides of (3.388) right-handly by u_i we get

$$v_j^T C u_i = \lambda_j v_j^T A u_i. \tag{3.389}$$

Multiplying both sides of (3.386) left-handly by v_j^T we get

$$v_j^T C u_i = \lambda_i v_j^T A u_i. \tag{3.390}$$

Comparing (3.389) and (3.390) we obtain

$$(\lambda_i - \lambda_j) v_j^T A u_i = 0. \tag{3.391}$$

Now, assuming $\lambda_i \neq \lambda_j$ the eigenvectors u_i and v_j are biorthogonal (because $v_j^T \cdot (A u_i) = 0$) with respect to matrix A. For $\lambda_i = \lambda_j$ a so-called multiple eigenvalue problem appears. Let λ_k be a repeated eigenvalue of multiplicity m. Then, there exist m independent eigenvectors u_k and m independent vectors v_k associated with λ_k. The u_k eigenvectors are biorthogonal to other $n-m$ eigenvectors, but (in general) they are not biorthogonal to v_k. However, both u_k and v_k may be orthogonalized [72].

To sum up, a biorthogonality relation between u_j and v_j holds always. Note that a positive definiteness of the mass matrix M ensures that a being considered system is not inertially degenerate. A symmetry of M and K ensures that the system is not defective.

Normalization of the eigenvectors u_i and v_j gives

$$v_j^T A u_i = \delta_{ij}, \quad i, j = 1, 2, \ldots, n. \tag{3.392}$$

Observe that u_i is adjoint to v_j and both of them have an arbitrary multiplicative constant.

Taking into account (3.389) and (3.392) we get

$$v_j^T C u_i = \lambda_j \delta_{ij}, \quad i, j = 1, 2, \ldots, n. \tag{3.393}$$

Define square matrices

$$U = [u_1, u_2, \ldots, u_n], \tag{3.394}$$

$$V = [v_1, v_2, \ldots, v_n]^T, \tag{3.395}$$

$$\Lambda = \mathrm{diag}[\lambda_1, \lambda_2, \ldots, \lambda_n]. \tag{3.396}$$

From (3.392) we get

$$VAU = I \tag{3.397}$$

and from (3.393) we obtain

$$VCU = \Lambda. \tag{3.398}$$

Letting $q = Up$ from (3.381) we get

$$AU\ddot{p} + CUp = f(t). \tag{3.399}$$

Multiplying (3.399) left-handly by V we get

$$VAU\ddot{p} + VCUp = Vf(t), \tag{3.400}$$

or equivalently (see (3.397) and (3.398))

$$\ddot{p} + \Lambda p = Vf(t). \tag{3.401}$$

To conclude, an undamped non-gyroscopic (not degenerated and not defectived) system can always be decoupled by equivalence transformation. The equivalence transformation is defined by (3.384) and (3.385). For a symmetric and definite system Eq. (3.384) takes the form

$$Ku = \lambda Mu. \tag{3.402}$$

Since K and M are symmetric and definite, and Eqs. (3.384) and (3.385) are identical, a solution of one of them can be taken. The modes and adjoint modes are $V = U^T$. Therefore, the decoupling eigenvalue transformation reduces to the classical modal transformation if the matrices are symmetric and definite.

The homogeneous system of (3.401) has the following solutions

$$p_i = A_i \cos(\omega_i t - \Phi_i), \quad i = 1, 2, \ldots, n \tag{3.403}$$

where: A_i—amplitude, Φ_i—phase angle, p_i is a component of the vector p and $\lambda_i = \omega_i^2$. Recalling $q = Up$ we obtain

$$q = \sum_{i=1}^{n} u_i p_i = \sum_{i=1}^{n} u_i A_i \cos(\omega_i t - \Phi_i), \tag{3.404}$$

which means that a general response of an autonomous conservative system is a superposition of n harmonic oscillations.

3.10.2 Non-conservative Systems

In general, when $D \neq 0$ the system

$$M\ddot{q} + D\dot{q} + Kq = f(t), \tag{3.405}$$

may be diagonalized if and only if $M^{-1}D$ and $M^{-1}K$ commute in multiplication, i.e.

$$DM^{-1}K = KM^{-1}D. \tag{3.406}$$

The condition (3.406) is necessary and sufficient one for the modal transformation to decouple a damped symmetric and definite system. However, this condition in practice is rarely satisfied.

Applying the equivalence transformation to Eq. (3.380) we use two matrices U and V to diagonalize A and C. Following steps from previous section we get

$$AU\ddot{p} + BU\dot{p} + CUp = f(t) \tag{3.407}$$

and next

$$VAU\ddot{p} + VBU\dot{p} + VCUp = Vf(t). \tag{3.408}$$

Taking into account (3.397) and (3.398) we obtain

$$\ddot{p} + VBU\dot{p} + \Lambda p = Vf(t). \tag{3.409}$$

It is clear that the system (3.409) is decoupled if and only if VBU is diagonal, which in general is not true. On the other hand, one can approximate VBU by its diagonal part when other non-diagonal elements are small enough.

Theorem 3.5. *The linear non-conservative system* (3.380) *can be decoupled by an equivalence transformation if and only if the matrices $A^{-1}B$ and $A^{-1}C$ commute in multiplication*

$$BA^{-1}C = CA^{-1}B. \tag{3.410}$$

Proof. From (3.397) we get $AU = V^{-1}$, and hence $A^{-1} = UV$.

From (3.410) we obtain

$$BUVC = CUVB. \tag{3.411}$$

Premultiply Eq. (3.411) by V and postmultiply by U to get

$$VBU\ VCU = VCU\ VBU. \tag{3.412}$$

Taking into account (3.412) and (3.398) we obtain

$$S\Lambda = \Lambda S. \tag{3.413}$$

Note that (3.413) is satisfied when S is diagonal, i.e. VBU is diagonal. On the other hand it follows from (3.413)

$$VBA^{-1}CU = VCA^{-1}BU, \tag{3.414}$$

which shows that (3.410) is satisfied. □

Let us use an explicit expression of right- and left-hand sides of (3.413)

$$\begin{bmatrix} s_{11} & s_{12} & \dots & s_{1n} \\ s_{21} & s_{22} & \dots & s_{2n} \\ \vdots & \vdots & & \vdots \\ s_{n1} & s_{n2} & \dots & s_{nn} \end{bmatrix} \cdot \begin{bmatrix} \lambda_1 & 0 & 0 & \dots & 0 \\ 0 & \lambda_2 & 0 & \dots & 0 \\ \vdots & \vdots & \vdots & & \vdots \\ 0 & 0 & 0 & \dots & \lambda_n \end{bmatrix} = \begin{bmatrix} s_{11}\lambda_1 & s_{12}\lambda_2 & \dots & s_{1n}\lambda_n \\ s_{21}\lambda_1 & s_{22}\lambda_2 & \dots & s_{2n}\lambda_n \\ \vdots & \vdots & & \vdots \\ s_{n1}\lambda_1 & s_{n2}\lambda_2 & \dots & s_{nn}\lambda_n \end{bmatrix} \tag{3.415}$$

$$\begin{bmatrix} \lambda_1 & 0 & 0 & \dots & 0 \\ 0 & \lambda_2 & 0 & \dots & 0 \\ \vdots & \vdots & \vdots & & \vdots \\ 0 & 0 & 0 & \dots & \lambda_n \end{bmatrix} \cdot \begin{bmatrix} s_{11} & s_{12} & \dots & s_{1n} \\ s_{21} & s_{22} & \dots & s_{2n} \\ \vdots & \vdots & & \vdots \\ s_{n1} & s_{n2} & \dots & s_{nn} \end{bmatrix} = \begin{bmatrix} s_{11}\lambda_1 & s_{12}\lambda_1 & \dots & s_{1n}\lambda_1 \\ s_{21}\lambda_2 & s_{22}\lambda_2 & \dots & s_{2n}\lambda_2 \\ \vdots & \vdots & & \vdots \\ s_{n1}\lambda_n & s_{n2}\lambda_n & \dots & s_{nn}\lambda_n \end{bmatrix} \tag{3.416}$$

Comparing the corresponding elements of (3.415) and (3.416) we obtain

$$\lambda_j s_{ij} = \lambda_i s_{ij}, \quad i, j = 1, 2, \dots, n. \tag{3.417}$$

If $\lambda_i \neq \lambda_j$, then $s_{ij} = 0$ and hence S must be diagonal. If we have k distinct eigenvalues, $k < n$, then S has a block diagonal form

$$S = \mathrm{diag}[S_1, S_2, \dots, S_k]. \tag{3.418}$$

3.10.3 Modal Analysis and Identification

Recall the considerations given in previous sections and consider the mechanical multibody system with the corresponding kinetic T and potential V energies defined by the formulas

$$T = \frac{1}{2}\sum_{r,s=1}^{n} a_{rs}\dot{q}_s\dot{q}_r, \tag{3.419}$$

$$V = \frac{1}{2}\sum_{r,s=1}^{n} c_{rs}q_s q_r, \tag{3.420}$$

where the coefficients a_{rs} describe inertial properties, and c_{rs} are stiffness coefficients. We consider also the simplest case where the following symmetry holds: $a_{rs} = a_{sr}$, $c_{rs} = c_{sr}$. According to the previous results (Sect. 3.10.1) one can find the non-singular $n \times n$ matrix U that

$$q = Up \tag{3.421}$$

and the kinetic and potential energies will get the so-called canonical forms

$$T = \frac{1}{2}\sum_{s=1}^{n} \mu_s \dot{p}_s^2, \tag{3.422}$$

$$V = \frac{1}{2}\sum_{s=1}^{n} \gamma_s p_s^2. \tag{3.423}$$

Our conservative autonomous system (see Sect. 3.10.1) gets the form

$$\mu_s \ddot{p}_s + \gamma_s p_s = 0, \quad s = 1, 2, \ldots, n. \tag{3.424}$$

Hence, in the case of our linear system the transformation (3.421) decouples the system (3.381) and each of n Eq. (3.424) governs a separated form (mode) of a harmonic oscillation with the frequency $\omega_s = \sqrt{\gamma_s/\mu_s}$, $s = 1, \ldots, n$.

Note that all components of the transformation (3.421) are real, and hence each of the modes is characterized by a movement of all points of the system, i.e. all of the points reach simultaneously the extremal configurations of the system as well as the equilibrium positions. They are called *normal oscillations*.

In the case when damping as well as the gyroscopic and circulatory forces appear [166, 176], the following system of second-order differential equations governs the dynamics of a lumped system with n-degrees-of-freedom (see (3.380))

$$\sum_{s=1}^{n} (a_{rs}\ddot{q}_s + b_{rs}\dot{q}_s + c_{rs}q_s) = 0, \tag{3.425}$$

where $b_{rs} = b_{sr}$ and $c_{rs} \neq c_{sr}$. Although this system of equations can be diagonalized using a certain complex matrix [62], but the described earlier normal modes cannot be realized. In general, the modal analysis can be applied only to the systems governed by a self-adjoint operator [136].

In the case of identification of the non-conservative system (3.405) we apply the so-called vector of testing inputs $f(t) = \mathrm{col}\{f_1(t), \ldots, f_n(t)\}$. Since we have three square matrices $M = [m_{ik}]$, $D = [d_{ik}]$ and $K = [k_{ik}]$ with the orders $n \times n$, we would like to define $3n^2$ unknown elements of the matrices. During identification process we use inputs $f_k(t)$ to get outputs $q_i(t)$, $k, i = 1, 2, \ldots, n$. Knowing $q_i(t)$, we can differentiate them to get $\dot{q}_i(t)$, and $\ddot{q}_i(t)$ in the time instants t_l, $l = 1, 2, \ldots, 3s$, where $s \gg n$. Knowing accelerations, velocities and displacements for a given time instants t_l one obtains the following set of algebraic non-homogenous linear set of equations

$$\sum_{k=1}^{n} (m_{ik}\ddot{q}_k(t_l) + d_{ik}\dot{q}_k(t_l) + k_{ik}q(t_l)) = f_i(t_l), \quad i = 1, 2, \ldots, n, \tag{3.426}$$

which serves to find the unknown elements m_{ik}, d_{ik} and k_{ik}. Since we measure experimentally accelerations, velocities and displacements then some errors are introduced denoted here by

$$\varepsilon_i(t_l) = \sum_{k=1}^{n} (m_{ik}\ddot{q}_k(t_l) + d_{ik}\dot{q}_k(t_l) + k_{ik}q(t_l) - f_i(t_l)). \tag{3.427}$$

The function

$$\phi(m_{ik}, d_{ik}, k_{ik}) = \sum_{l=1}^{3s} \sum_{i=1}^{n} \varepsilon_i^2(t_l) \tag{3.428}$$

may be used to find the elements m_{ik}, d_{ik}, k_{ik}. To achieve this we differentiate (3.428) successively by m^*_{jk}, d^*_{jk} and k^*_{jk} and we obtain $3n^2$ linear algebraic equations with $3n^2$ unknowns. Since we differentiate in time the experimentally obtained processes, one can get even relatively high errors. To achieve high order reliability of the obtained results, applications of other tools from mechanics are highly recommended (integral characteristics, correlation functions, fast Fourier transform and the amplitude–frequency characteristics). A classical approach uses the Fourier transformation, and from (3.426) we obtain (we take t instead of t_l)

$$\sum_{k=1}^{n} (-\omega^2 m_{jk} + i\omega d_{jk} + k_{jk}) Q_k(i\omega) = F_j(i\omega), \tag{3.429}$$

where

$$Q_k(i\omega) = \frac{1}{2\pi} \int_{-\infty}^{\infty} q_k(t) e^{-i\omega t} dt, \tag{3.430}$$

$$F_j(i\omega) = \frac{1}{2\pi} \int_{-\infty}^{\infty} f_j(t) e^{-i\omega t} dt, \quad i = \sqrt{-1}. \tag{3.431}$$

Observe that we have got algebraic equations with complex coefficients in a frequency domain. We separate real and imaginary parts to obtain

$$\sum_{k=1}^{n}[-\omega^2 m_{jk}\mathrm{Re}Q_k(i\omega) - \omega d_{jk}\mathrm{Im}Q_k(i\omega) + k_{jk}\mathrm{Re}Q_k(i\omega)] = \mathrm{Re}F_j(i\omega),$$

$$\sum_{k=1}^{n}[-\omega^2 m_{jk}\mathrm{Im}Q_k(i\omega) + \omega d_{jk}\mathrm{Re}Q_k(i\omega) + k_{jk}\mathrm{Im}Q_k(i\omega)] = \mathrm{Im}F_j(i\omega),$$

$$j = 1, 2, \ldots, n. \tag{3.432}$$

In order to obtain $3n^2$ unknown elements in the frequency domain one needs to have the input–output characteristics for at least $1.5n$ different values of the excitation frequency ω_l ($l \gg 1.5n$). Introducing the matrix S and two vectors Y and X one obtains the following set of algebraic equations

$$SY = X, \tag{3.433}$$

where:

$$\begin{aligned} Y_{2ln} &= Y(m_{jk}, d_{jk}, k_{jk}), \\ S_{2ln \times 2ln} &= S(\mathrm{Im}Q(i\omega_l), \mathrm{Re}Q(i\omega_l), \omega_l), \quad 3n = 2l. \\ X_{2ln} &= X(\mathrm{Im}F(i\omega_l), \mathrm{Re}F(i\omega_l)). \end{aligned} \tag{3.434}$$

Multiplying both sides of (3.433) by S^T we get

$$S^T SY = S^T X. \tag{3.435}$$

It is clear that in order to get reliable results one needs to carry out properly the measurement in a frequency domain. The measurement are 'enough good' if $S^T S \approx I$, where I is the identity matrix. It is achieved if all resonance peaks are well represented. Note that number of peaks on the amplitude–frequency characteristics corresponds to a number of degrees-of-freedom of the system.

The dynamical lumped system (3.426) can also be interpreted in the following way. Apply an excitation only to the kth point and measure a reaction in point j. Hence from (3.426) we obtain

$$\begin{bmatrix} a_{11} & a_{12} & \ldots & a_{1j} & \ldots & a_{1n} \\ a_{21} & a_{22} & \ldots & a_{2j} & \ldots & a_{2n} \\ \vdots & \vdots & & \vdots & & \vdots \\ a_{k1} & a_{k2} & \ldots & a_{kj} & \ldots & a_{kn} \\ \vdots & \vdots & & \vdots & & \vdots \\ a_{n1} & a_{n2} & \ldots & a_{nj} & \ldots & a_{nn} \end{bmatrix} \begin{bmatrix} q_1 \\ q_2 \\ \vdots \\ q_j \\ \vdots \\ q_n \end{bmatrix} = \begin{bmatrix} 0 \\ 0 \\ \vdots \\ f_k \\ \vdots \\ 0 \end{bmatrix}, \tag{3.436}$$

where:

$$a_{jk} = m_{jk}s^2 + d_{jk}s + c_{jk}. \tag{3.437}$$

Therefore, from (3.436) one obtains

$$q_j = \frac{W_{kj}}{W} = \frac{(-1)^{k+j} W_{kj}}{W} f_k, \tag{3.438}$$

where: $W = \det[A]$, and W_{kj} is obtained from W by cancelling the kth row and jth column of W.

Applying the Fourier transform to (3.438) we obtain

$$Q_j(i\omega) = H_{kj}(i\omega)F_k(i\omega), \tag{3.439}$$

and

$$H_{kj}(i\omega) = \frac{W_{kj}(i\omega)}{W(i\omega)}, \quad j,k = 1,2,\ldots,n \tag{3.440}$$

is the frequency characteristic between kth input and jth output. The complex matrix $H(i\omega)$ is known in mechanics as the matrix of dynamical elasticity (or receptances), and it can be separated into two parts

$$H(i\omega) = \phi(\omega) + i\psi(\omega). \tag{3.441}$$

Taking into account (3.429) and (3.441) we get two algebraic equations:

$$\begin{aligned} \text{Re} &: (-\omega^2 M + K)\phi(\omega) - \omega D\psi(\omega) = 1, \\ \text{Im} &: (-\omega^2 M + K)\psi(\omega) + \omega D\phi(\omega) = 0, \end{aligned} \tag{3.442}$$

and they yield

$$\begin{aligned} \phi(\omega) &= -(\omega D)^{-1}(-\omega^2 M + K)\psi(\omega), \\ \psi(\omega) &= -\{(-\omega^2 M + K)[(\omega D)^{-1}(-\omega^2 M + K)] + \omega D\}^{-1}. \end{aligned} \tag{3.443}$$

Real and imaginary parts of $H(i\omega)$ are expressed by the matrices M, D, K and the frequency ω. If one uses a harmonic excitation (input) of the form

$$f_k(t) = a\sin(\omega t), \tag{3.444}$$

then the jth output is defined by

$$x_j(t) = b(\omega)\sin(\omega t - \Theta(\omega)), \tag{3.445}$$

where:

$$|H_{kj}(i\omega)| = \frac{b(\omega)}{a}, \quad \arg H_{kj}(i\omega) = \Theta(\omega). \tag{3.446}$$

Hence, knowing input (3.444) and output (3.445), one can define $b(\omega)$, a, ω and $\Theta(\omega)$, i.e. one can define the complex matrix $H_{kj}(i\omega)$.

Let us apply now the state variables approach to the system (3.426) to get

$$\begin{bmatrix} 0 & M \\ M & D \end{bmatrix} \begin{bmatrix} \dot{v}_1 \\ \vdots \\ \dot{v}_n \\ \dot{q}_1 \\ \vdots \\ \dot{q}_n \end{bmatrix} + \begin{bmatrix} -M & 0 \\ 0 & K \end{bmatrix} \begin{bmatrix} v_1 \\ \vdots \\ v_n \\ q_1 \\ \vdots \\ q_n \end{bmatrix} = \begin{bmatrix} 0 \\ \vdots \\ 0 \\ f_1 \\ \vdots \\ f_n \end{bmatrix} \tag{3.447}$$

or equivalently

$$P\dot{x} + Rx = S(t), \tag{3.448}$$

where:

$$v_j = \dot{q}_j, \quad j = 1, 2, \ldots, n. \tag{3.449}$$

It is clear that the measured electric signals $y(t)$ are proportional to the displacements and velocities of the considered system, and hence

$$\begin{bmatrix} y_1(t) \\ y_2(t) \\ \vdots \\ y_m(t) \end{bmatrix} = \begin{bmatrix} g_{11} & g_{12} & \cdots & g_{1n} \\ g_{21} & g_{22} & \cdots & g_{2n} \\ \vdots & \vdots & \vdots & \vdots \\ g_{m1} & g_{m2} & \cdots & g_{mn} \end{bmatrix} \begin{bmatrix} v_1 \\ \vdots \\ v_n \\ q_1 \\ \vdots \\ q_n \end{bmatrix}, \quad (\text{or } y = Gx). \tag{3.450}$$

Note that we measure outputs in m points and the matrix G is rectangular one.

Let

$$x(t) = A\xi(t), \tag{3.451}$$

where

$$\xi = \{\xi_1, \ldots, \xi_{2n}\}^T. \tag{3.452}$$

From (3.448) and (3.451) we obtain

$$PA\dot{\xi} + RA\xi = S(t). \tag{3.453}$$

From (3.448) and (3.450) we obtain

$$y(t) = G(e^{-P^{-1}R(t-t_0)}x_0 + \int_{t_0}^{t} e^{-P^{-1}R(t-\tau)}P^{-1}S(\tau)d\tau), \tag{3.454}$$

where

$$x_0 = x(t_0). \tag{3.455}$$

On the other hand taking into account the linear transformation (3.451) and the differential equation (3.453) we obtain

$$\tilde{y}(t) = G\left[Ae^{-A^{-1}P^{-1}RA(t-t_0)}\xi_0 + \int_{t_0}^{t} e^{-A^{-1}P^{-1}RA(t-\tau)}A^{-1}P^{-1}S(\tau)d\tau\right]. \tag{3.456}$$

Recall that from matrices theory one gets

$$f(A^{-1}BA) = A^{-1}f(B)A \tag{3.457}$$

and hence

$$\begin{aligned} & e^{-A^{-1}P^{-1}RA(t-t_0)} \\ & = A^{-1}e^{-P^{-1}R(t-t_0)}A, \end{aligned} \tag{3.458}$$

$$\int_{t_0}^{t} e^{-A^{-1}P^{-1}RA(t-\tau)}A^{-1}P^{-1}S(\tau)d\tau = \int_{t_0}^{t} A^{-1}e^{-P^{-1}R(t-\tau)}\underbrace{AA^{-1}}P^{-1}S(\tau)d\tau. \tag{3.459}$$

From (3.456), taking into account (3.458) and (3.459) we get

$$\begin{aligned} \tilde{y}(t) &= G\underbrace{AA^{-1}}e^{-P^{-1}R(t-t_0)}A\xi_0 + G\underbrace{AA^{-1}}\int_{t_0}^{t} e^{-P^{-1}R(t-\tau)}P^{-1}S(\tau)d\tau \\ &= G(e^{-P^{-1}R(t-t_0)}x_0 + \int_{t_0}^{t} e^{-P^{-1}R(t-\tau)}P^{-1}S(\tau)d\tau). \end{aligned} \tag{3.460}$$

Comparing (3.454) and (3.460) we observe that

$$y(t) = \tilde{y}(t), \tag{3.461}$$

which means that linear differential equations (3.448) and (3.453) govern the same dynamical systems from a point of view of experiment (they are not distinguishable experimentally).

We consider again the non-autonomous system governed by the set of differential equations (3.426), where the excitation is composed of harmonic forces variously distributed with regard to amplitudes and phases, i.e. we consider the following mechanical system

$$M\ddot{q} + D\dot{q} + Kq = Fe^{i\Theta}e^{i\nu t}. \tag{3.462}$$

The matrices M, D and K are square, symmetric and positive definite of the nth order (we, as earlier, consider damped oscillations of n degree-of-freedom mechanical system), q is n-dimensional vector of displacements; ν is excitation frequency and $Fe^{i\Theta} = [F_1 e^{i\Theta_1}, \ldots, F_n e^{i\Theta_n}]^T$ is the vector describing complex amplitudes of excitations with different phases Θ_j, $j = 1, \ldots, n$. Note that in general, the monophase oscillations realized in the system (3.462) do not overlap with the corresponding normal oscillations exhibited by conservative systems. However, in some cases the monophase and normal oscillations are identical. It means that we can use the system (3.462) to identify the normal oscillations.

Let

$$Fe^{i\Theta} = F^{(c)} + iF^{(s)} \tag{3.463}$$

and assume that

$$\dot{q} = (U + iV)e^{i\nu t}, \quad q = \frac{1}{i\nu}(U + iV)e^{i\nu t}, \quad \ddot{q} = i\nu(U + iV)e^{i\nu t}. \tag{3.464}$$

Above we have used the notation

$$\dot{X} = U + iV \tag{3.465}$$

and

$$U = [U_1, \ldots, U_n]^T, \quad V = [V_1, \ldots, V_n]^T, \tag{3.466}$$

and we have assumed that a velocity is measured experimentally and then it is differentiated (to get $\ddot{q}$) and integrated (to get q).

Even if we realize the monophase excitations ($\Theta_j = \Theta_0$, $j = 1, \ldots, n$) a response of the system may be not monophase. The oscillations of the system (3.462) will be monophased, when the vectors V and U are collinear, i.e.

$$V = \lambda U, \tag{3.467}$$

and λ is a real number. Substituting (3.463) and (3.464) into (3.462) and separating the real imaginary parts one gets

$$(K - M\nu^2)U - \nu DV = -\nu F^s, \tag{3.468}$$

$$\nu DU + (K - M\nu^2)V = \nu F^c. \tag{3.469}$$

Recall that introducing the concept of receptance matrix $H(i\omega)$ (see (3.441)) its real and imaginary parts are defined by (3.443). Now, taking into account the collinearity requirement (3.467) from (3.468) and (3.469) we obtain the real and imaginary parts of exciting forces

$$F^c = \frac{1}{\nu}[\nu D + (K - M\nu^2)\lambda]U, \tag{3.470}$$

$$F^s = \frac{1}{\nu}[\nu\lambda D - (K - M\nu^2)]U, \tag{3.471}$$

needed to realize monophase oscillations using non-monophase excitations. Howe-ver, we can even realize a monophase oscillations applying a monophase excitations. In the later case we take $F^s = 0$, and from (3.471) we obtain

$$[\nu\lambda D - (K - M\nu^2)]U = 0. \tag{3.472}$$

The characteristic equation

$$\det[\nu\lambda D - (K - M\nu^2)] = 0, \tag{3.473}$$

yields λ_j and

$$\lambda_j = \tan\Theta_j, \quad j = 1, \ldots, n. \tag{3.474}$$

Note that in this case to every λ_j corresponds Θ_j, i.e. one can realize n different monophase excited oscillations. In addition, the roots of characteristic equations (3.473) are real because M, D and K are symmetric and positive definite (we assume that the roots λ_j are distinct). The U_j corresponding to every λ_j may be found from the homogeneous algebraic equation (3.472)

$$[\lambda_j \nu D - (K - M\nu^2)]U_j = 0. \tag{3.475}$$

Since from (3.475) one gets

$$\lambda_j = \frac{U_j^T(K - M\nu^2)U_j}{\nu U_j^T D U_j}, \tag{3.476}$$

then from (3.465) we obtain

$$\dot{X}_j = (1 + i\lambda_j)U_j, \tag{3.477}$$

and hence

$$|\dot{X}_j| = \sqrt{1 + \lambda_j^2}U_j. \tag{3.478}$$

The amplitude of the monophases excitation is defined by (3.470)

$$F_j^c = \frac{1}{\nu}[\nu D + (K - M\nu^2)\lambda_j]U_j. \tag{3.479}$$

Note that an orthogonality condition of the vectors U_j $(j = 1, 2, \ldots, n)$ holds, because we have assumed that the matrices M, D and K are symmetric. Also the vectors corresponding to different λ_j are mutually orthogonal and therefore we get

$$\nu\lambda_j U_j^T DU_k = U_j^T(K - M\nu^2)U_k = 0, \quad for\ j \neq k, \tag{3.480}$$

$$U_j^T DU_j = \beta_j, \quad U_j^T(K - M\nu^2)U_j = \gamma_j - \mu_j\nu^2, \quad for \quad j = k, \tag{3.481}$$

and β_j, γ_j, μ_j are positive real numbers. Observe that all relations depend on the excitation frequency ν.

Now we briefly discuss the following important question: when the monophases oscillations overlap with normal oscillations? There are two possibilities to solve this problem.

(i) *Phase resonance*. The phase resonance is defined when $V = 0$. In accordance with (3.467) and for $U \neq 0$ we get $\lambda = 0$. Hence, from (3.475) and (3.479) we obtain

$$(K - M\nu^2)U_j = 0, \tag{3.482}$$

$$F_j^c = DU_j. \tag{3.483}$$

Note than when we treat ν as an unknown the obtained Eq. (3.480) is the same as that in a case of conservative oscillations.

A solution to (3.482) yields $\nu_j = \omega_j$, $j = 1, 2, \ldots, n$, where ω_j are eigenfrequencies of the associated conservative system.

It means that a phase resonance is realized when the excitation frequency is equal to one of the eigenfrequencies. Equation (3.483) possesses also clear physical interpretation. The external forces compensate the damping forces, which may be of an arbitrary value.

(ii) *Special case*. Assume that the following relation holds

$$D = s_1 M + s_2 K. \tag{3.484}$$

In words, damping matrix is a linear combination of both inertial and stiffness matrices (s_1 and s_2 are scalars).

From (3.475) and (3.479) by taking into account (3.484) we obtain

$$[\lambda_j \nu(s_1 M + s_2 K) - (K - M\nu^2)]U_j = 0, \tag{3.485}$$

or equivalently

$$(K - \Omega_j^2 M)U_j = 0, \tag{3.486}$$

where:

$$\Omega_j^2 = \frac{\nu(\lambda_j s_1 + \nu)}{1 - \lambda_j \nu s_2}. \tag{3.487}$$

Similarly to the previous case, it is easy to observe that Eq. (3.487) has the solution $\Omega_j = \omega_j$.
From (3.476) and (3.487) we get

$$\begin{aligned}\lambda_j = \tan\Theta_j &= \frac{U_j^T(K - M\nu^2)U_j}{\nu U_j^T(s_1 M + s_2 K)U_j}\\ &= \frac{U_j^T(\Omega_j^2 - \nu^2)MU_j}{\nu U_j^T(s_1 + s_2\Omega_j^2)MU_j} = \frac{\omega_j^2 - \nu^2}{\nu(s_1 + s_2\omega_j^2)}.\end{aligned} \tag{3.488}$$

In our next step we calculate the required forcing amplitude by substituting (3.483) into (3.479):

$$\begin{aligned}F_j^c &= \frac{1}{\nu}[\nu(s_1 M + s_2 K) + (K - M\nu^2)\lambda_j]U_j)\\ &= [(s_1 - \nu\lambda_j) + \omega_j^2(s_2 + \frac{\lambda_j}{\nu})]MU_j.\end{aligned} \tag{3.489}$$

Note that in this case F_j^c can realized for any frequency ν, when the formula (3.489) is satisfied.

Finally we use non-monophase excitations in order to realize monophase oscillations (outputs) which overlap with the jth normal mode. First we take monophase forces F^c which realize the jth phase resonance. Second, we take the forces F^s shifted in phase $\frac{\pi}{2}$. The forces F^s are chosen in such a way that the normal modes excited by F^c remain unchanged. The first step corresponds to the satisfaction of the following analytical requirements (see (3.482) and (3.483))

$$\begin{aligned}(K - M\nu^2)U_j &= 0,\\ F_j^c &= DU_j,\\ V_j &= 0.\end{aligned} \tag{3.490}$$

In words, the inertial forces are balanced by elasticity forces, whereas damping forces are balanced by external forces. Substituting into (3.468) and (3.469) $V = 0$, $F^s = F_j^s$, and $F^c = F_j^c$ we get (the second step)

$$\begin{aligned}(K - M\nu^2)U &= -\nu F_j^s,\\ DU &= F_j^c,\\ V &= 0.\end{aligned} \tag{3.491}$$

The following physical interpretation is associated with Eq. (3.491). The difference between inertial and elasticity forces is balanced by the forces F_j^s, shifted by $\frac{\pi}{2}$ in comparison to F_j^c. The damping forces are balanced by the forces F_j^c. The vector F_j^s depends on frequency ν. The realized state of the system is called a *fictional phase resonance*.

Chapter 4
Linear ODEs

4.1 Introduction

First we show that a single nth-order ordinary differential equation can be reduced to a first-order system of differential equations.

Consider an nth-order linear differential equation with variable coefficients

$$L_n(y) = y^{(n)} + p_{n-1}(t)y^{(n-1)} + \cdots + p_1(t)\dot{y} + p_0(t)y = f(t). \tag{4.1}$$

The Cauchy problem (or initial value problem) requires the initial values of the function y and their derivatives up to $(n-1)$th-order to be known

$$\begin{aligned} y(t_0) &= y_0^0, \\ \dot{y}(t_0) &= y_1^0, \\ &\vdots \\ y^{(n-1)}(t_0) &= y_{n-1}^0. \end{aligned} \tag{4.2}$$

Introducing a change of variables

$$y_k = y^{(k)}, \quad k = 0, 1, \ldots, n-1, \tag{4.3}$$

Equation (4.1) is transformed to the system of first-order differential equations

$$\begin{aligned} \dot{y}_0 &= y_1, \\ \dot{y}_1 &= y_2, \\ &\vdots \end{aligned} \tag{4.4}$$

J. Awrejcewicz, *Ordinary Differential Equations and Mechanical Systems*,
DOI 10.1007/978-3-319-07659-1_4

$$\dot{y}_{n-2} = y_{n-1},$$
$$\dot{y}_{n-1} = -p_{n-1}(t)y_{n-1} - \cdots - p_0(t)y_0 + f(t),$$

and the initial conditions (4.2) have the form

$$y_k(t_0) = y_k^0, \quad k = 0, 1, \ldots, n-1. \tag{4.5}$$

In the case of homogeneous differential equation $L_n(y) = 0$ with constant coefficients, it can be transformed to nth-order system with constant coefficients

$$\dot{Y} = PY, \tag{4.6}$$

where

$$Y = \begin{Bmatrix} y \\ \dot{y} \\ \vdots \\ \vdots \\ y^{(n-1)} \end{Bmatrix}, \qquad P = \begin{bmatrix} 0 & 1 & 0 & \cdots & 0 \\ 0 & 0 & 1 & \cdots & 0 \\ \vdots & \vdots & \vdots & \ddots & \vdots \\ \vdots & \vdots & \vdots & & 1 \\ -p_0 & -p_1 & -p_2 & \cdots & -p_{n-1} \end{bmatrix}. \tag{4.7}$$

The characteristic polynomial of the matrix P is

$$W(r) = \det(rI - P) = r^n + \sum_{i=0}^{n-1} p_i r^i, \tag{4.8}$$

which is exactly the same as that corresponding to $L_n(y) = 0$. The matrix approach is widely used in engineering since there are many commercial programs and subroutines to find eigenvalues of matrices. However, one should observe that this approach is somehow more difficult, because it requires a knowledge of eigenvectors corresponding to each of the eigenvalues of the matrix P. A problem of eigenvectors do not appears in the case of a characteristic polynomial.

4.2 Normal and Symmetric Forms

System of ODEs:

$$\frac{dy_i}{dt} = F_i(t, y_1, \ldots, y_n), \quad i = 1, \ldots, n \tag{4.9}$$

is said to be the normal form system. In other words it is solved with respect to derivatives of unknown functions $y_i = y_i(t)$.

A solution to the system (4.9) on the interval $J \subset \mathbb{R}$ is the set of continuously differentiated functions on J, which satisfy (4.9), i.e.

$$\frac{d\varphi_i}{dt} = F_i(t, y_1, \ldots, y_n), \quad i = 1, \ldots, n \tag{4.10}$$

for all $t \in J$. A function $\phi(t, y_1, \ldots, y_n)$ being continuously differentiated on J is called the *first integral of (4.9)*, if the following formula holds

$$\frac{d\phi}{dt} = \frac{\partial\phi}{\partial t} + \sum_{i=1}^{n} \frac{\partial\phi}{\partial y_i} F_i(t, y_1, y_2, \ldots, y_n) = 0. \tag{4.11}$$

If we know n independent first integrals of (4.9), i.e. $\phi_1, \phi_2, \ldots, \phi_n$, then the system

$$\phi_i(t, y_1, \ldots, y_n) = C_i,\ i = 1, \ldots, n \tag{4.12}$$

defines the general integral solution of ODEs (4.9) (where C_i are constants). Observe that knowing (4.12) all solutions $y_1(t), y_2(t), \ldots, y_n(t)$ are known and can be found either analytically or numerically.

Furthermore, if we know one of the first integrals of (4.9) of the form

$$\phi(t, y_1, \ldots, y_n) = C, \tag{4.13}$$

then we may derive from (4.13) for instance y_n:

$$y_n = \phi^*(t, y_1, y_2, \ldots, y_{n-1}, C). \tag{4.14}$$

Substituting (4.14) into first $(n-1)$ equations of (4.9) one obtains a system of ODEs with $(n-1)$ independent functions $y_1, y_2, \ldots, y_{n-1}$. It means that dimensions of the original system (4.9) has been reduced due to the known first integral (4.13).

In general, there are two methods of solutions to a system of ODEs in the normal form.

(i) *Reduction* of (4.9) into either one differential equation of order n or to a few differential equations of an order less than n.

The first equations of (4.9) is differentiated $(n-1)$-times, and after each of the differentiation process we substitute $\frac{dy_i}{dt}$ by their values from remaining equations, and finally the following set of equations are obtained

$$\begin{aligned}
\frac{dy_1}{dt} &= f_1(t, y_1, \ldots, y_n),\\
\frac{d^2y_1}{dt^2} &= f_2(t, y_1, \ldots, y_n),\\
&\vdots\\
\frac{d^{n-1}y_1}{dt^{n-1}} &= f_{n-1}(t, y_1, \ldots, y_n),\\
\frac{d^n y_1}{dt^n} &= f_n(t, y_1, \ldots, y_n).
\end{aligned} \tag{4.15}$$

Now one may find $y_2, y_3, \ldots y_n$ from sequence of $n-1$ first equations of (4.15), and substituting obtained formulas to the last equation of (4.15), the following nth order one differential equation is obtained

$$\frac{d^n y_1}{dt^n} = f\left(t, y_1, \frac{dy_1}{dt}, \ldots, \frac{d^{n-1}y_1}{dt^{n-1}}\right). \tag{4.16}$$

A solution to (4.16) allows to find all of the solutions of system (4.9).

(ii) Method of *integral combination*. This method relies on application of arithmetic combinations in order to find the so-called integral combination yielding easily integrable equations regarding a new unknown function $u = u(t, y_1, \ldots, y_n)$. Sometimes we have ODEs presented in the so-called symmetric form

$$\frac{dy_1}{Y_1(y_1, \ldots, y_n)} = \frac{dy_2}{Y_2(y_1, \ldots, y_n)} = \ldots = \frac{dy_n}{Y_n(y_1, \ldots, y_n)}. \tag{4.17}$$

In order to solve Eq. (4.17) the following rule of equal fractions can be applied. Namely, assuming that

$$\frac{A_1}{B_1} = \frac{A_2}{B_2} = \ldots = \frac{A_n}{B_n}, \tag{4.18}$$

and having arbitrary numbers $m_1, m_2, \ldots, m_n$, the following formula holds

$$\frac{A_1}{B_1} = \frac{A_2}{B_2} = \ldots = \frac{A_n}{B_n} = \frac{m_1A_1 + m_2A_2 + \cdots + m_nA_n}{m_1B_1 + m_2B_2 + \cdots + m_nB_n}. \tag{4.19}$$

Example 4.1. Show that functions $\phi_1 = tx$, $\phi_2 = ty + x^2$ are first independent integrals of the following ODEs

$$\frac{dx}{dt} = -\frac{x}{t},$$

$$\frac{dy}{dt} = \frac{2x^2 - ty}{t^2}.$$

Differentiation of ϕ_1 and ϕ_2 yields

$$\frac{d\phi_1}{dt} = t\frac{dx}{dt} + x = -x + x = 0,$$

$$\frac{d\phi_2}{dt} = t\frac{dy}{dt} + y + 2x\frac{dx}{dt} = \frac{2x^2}{t} - y + y - \frac{2x^2}{t} = 0.$$

Since

$$\operatorname{rank}\left[\frac{D(\phi_1, \phi_2)}{D(t, x, y)}\right] = \operatorname{rank}\begin{bmatrix} \frac{\partial\phi_1}{\partial t} & \frac{\partial\phi_1}{\partial x} & \frac{\partial\phi_1}{\partial y} \\ \frac{\partial\phi_2}{\partial t} & \frac{\partial\phi_2}{\partial x} & \frac{\partial\phi_2}{\partial y} \end{bmatrix} = 2,$$

therefore integrals ϕ_1 and ϕ_2 are independent. □

Example 4.2. Show that the function

$$\phi(x, y) = x^2 + y^2 - 2\ln|xy - 1|$$

is constant along an arbitrary solution to the given ODEs

$$\frac{dx}{dt} = x + y - xy^2,$$

$$\frac{dy}{dt} = -x - y + x^2 y.$$

We compute

$$\begin{aligned}
\frac{d}{dt}\phi(x(t), y(t)) &= 2x\frac{dx}{dt} + 2y\frac{dy}{dt} - \frac{2(xy-1)\left(x\frac{dy}{dt} + y\frac{dx}{dt}\right)}{(xy-1)^2} \\
&= 2\left(x - \frac{y}{xy-1}\right)\frac{dx}{dt} + 2\left(y - \frac{x}{xy-1}\right)\frac{dy}{dt} \\
&= \frac{2}{xy-1}\left[(-x - y + x^2 y)\frac{dx}{dt} + (x - y + xy^2)\frac{dy}{dt}\right] \\
&= \frac{2}{xy-1}\left[(-x-y+x^2 y)(x + y - xy^2)\right. \\
&\quad \left. +(-x - y + xy^2)(-x - y + x^2 y)\right] = 0.
\end{aligned}$$

It means that for all $x(t)$, $y(t)$ we have $\frac{d}{dt}\phi(x(t), y(t)) \equiv 0$, which means that $\phi(x(t), y(t)) = \text{const}$. □

Example 4.3. Determine one of the first integrals of the system of ODEs of the form

$$(x^2 + y^2 - t^2)\frac{dx}{dt} = tx,$$
$$(x^2 + y^2 - t^2)\frac{dy}{dt} = ty.$$

Both equations are multiplied by $2x$ and $2y$, respectively:

$$(x^2 + y^2 - t^2)\frac{dx^2}{dt} = 2tx^2,$$
$$(x^2 + y^2 - t^2)\frac{dy^2}{dt} = 2ty^2.$$

Introducing $z = x^2 + y^2$ and making a sum of both equations one obtains

$$(z - t^2)\frac{dz}{dt} - 2tz = 0,$$

and therefore

$$z\frac{dz}{dt} - \left(t^2\frac{zdz}{dt} + 2tz\right) = \frac{d}{dt}\left(\frac{z^2}{2} - t^2 z\right) = 0.$$

It means that $z^2 - 2t^2 z = C$ or equivalently

$$(x^2 + y^2)(x^2 + y^2 - 2t^2) = C.$$

□

Example 4.4. Solve the following ODEs

$$\frac{dx}{dt} = y + (1 - x^2 - y^2)x,$$
$$\frac{dy}{dt} = -x + (1 - x^2 - y^2)y$$

and show a direction of motion along trajectories.

A critical solution is $x = y = 0$.
Introducing polar coordinates $x = \varrho\cos\varphi$, $y = \varrho\sin\varphi$ one gets ($\varrho > 0$):

$$\begin{aligned}\frac{d\varrho}{dt}\cos\varphi - \varrho\sin\varphi\frac{d\varphi}{dt} &= \varrho\sin\varphi + (1 - \varrho^2)\varrho\cos\varphi,\\ \frac{d\varrho}{dt}\sin\varphi - \varrho\cos\varphi\frac{d\varphi}{dt} &= -\varrho\cos\varphi + (1 - \varrho^2)\varrho\sin\varphi.\end{aligned} \qquad (*)$$

Multiplying Eq. $(*)$ by $\cos\varphi$ and $-\sin\varphi$, respectively, one gets

$$\frac{d\varrho}{dt} = \varrho(1-\varrho^2).$$

Multiplying Eq. $(*)$ by $\sin\varphi$ and $\cos\varphi$, respectively, and after a simple computation one obtains

$$\frac{d\varphi}{dt} = -1.$$

Therefore, instead of initial ODEs we deal with the following ones

$$\frac{d\varrho}{dt} = \varrho(1-\varrho^2), \quad \frac{d\varphi}{dt} = -1. \tag{$**$}$$

First of two equations in the above is

$$\frac{d\varrho^2}{dt} = 2\varrho^2(1-\varrho^2),$$

what means that $\varrho = 1$ is a solution. After variables separation we get

$$\left(\frac{1}{\varrho^2-1} - \frac{1}{\varrho^2}\right) d\varrho^2 = -2dt.$$

After an integration one gets

$$\ln|\varrho^2 - 1| - \ln\varrho^2 = -2t + \ln C_1,$$

and hence

$$\left|1 - \frac{1}{\varrho^2}\right| = C_1 e^{-2t}, \quad C_1 > 0.$$

Therefore, in the circle $0 < \varrho \leqslant 1$ the solution of $(**)$ has the form

$$\varrho = \frac{1}{\sqrt{1 + C_1 e^{-2t}}}, \quad \varphi = C_2 - t, \quad C_1 \geqslant 0,$$

whereas outside the circle ($\varrho > 1$) the solution of $(**)$ is

$$\varrho = \frac{1}{\sqrt{1 - C_1 e^{-2t}}}, \quad \varphi = C_2 - t, \quad 0 < C_1 < 1.$$

Coming back to original variables we have the following solutions:

(i) If $x_0^2 + y_0^2 \leqslant 1$ then $x = \frac{\cos(t-C_2)}{\sqrt{1+C_1e^{-2t}}}$, $y = \frac{\sin(t-C_2)}{\sqrt{1+C_1e^{-2t}}}$, $C_1 \geqslant 1$;

(ii) If $x_0^2 + y_0^2 \geqslant 1$ then $x = \frac{\cos(t-C_2)}{\sqrt{1-C_1e^{-2t}}}$, $y = \frac{\sin(t-C_2)}{\sqrt{1-C_1e^{-2t}}}$, $0 \leqslant C_1 < 1$.

In addition, the origin $(0,0)$ corresponds to solution trajectory $x = y = 0$. Trajectory of solution lying on the circle $x^2 + y^2 = 1$ ($x = \cos(t - C_2)$, $y = \sin(t - C_2)$) is invariant. In fact, it is a stable limit cycle. In polar coordinates trajectories of solutions are governed by the following equations:

(i) For $0 < \varrho_0 < 1$ we have

$$\varrho = \frac{1}{\sqrt{1 + C_1 e^{2(\varphi - C_2)}}}, \qquad C_1 > 0;$$

(ii) For $\varrho_0 > 1$ we have

$$\varrho = \frac{1}{\sqrt{1 - C_1 e^{2(\varphi - C_2)}}}, \qquad 0 < C_1 < 1.$$

Observe that for $t \to +\infty$, $\varphi = C_2 - t$ tends to $-\infty$. This means that a phase point moving on the spiral approaches the circle $x^2 + y^2 = 1$ earlier referred as the stable limit cycle shown in Fig. 4.1 □

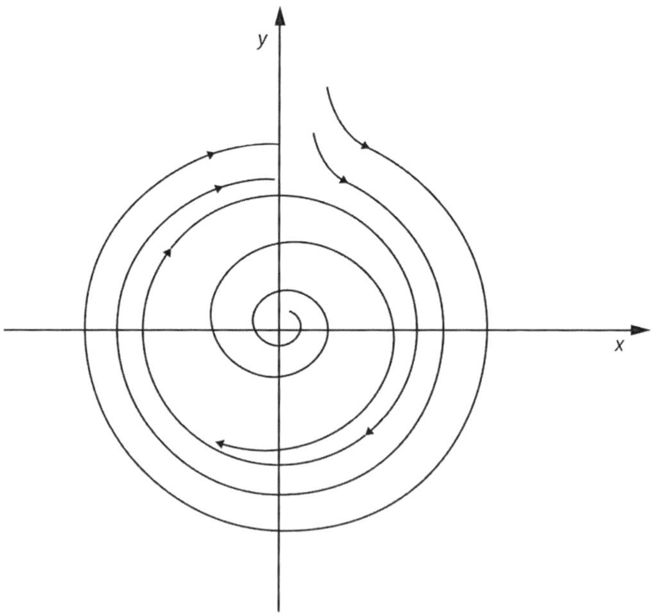

Fig. 4.1 Trajectories and the limit cycle

Example 4.5. Find a solution to the following equation

$$\frac{dx}{x+y-xy^2} = \frac{dy}{x^2y-x-y} = \frac{dz}{y^2-x^2}.$$

In what follows we apply here the rule of equal fractions. We construct the following integrable combination

$$\frac{2xdx+2ydy}{2x(x+y-xy^2)+2y(x^2y-x-y)} = \frac{dz}{y^2-x^2}$$

or equivalently

$$\frac{d(x^2+y^2)}{d(x^2-y^2)} = -\frac{dz}{x^2-y^2},$$

which finally yields

$$d(x^2+y^2) = -2dz.$$

The first integral has the form

$$x^2+y^2-2z = C_1.$$

The second integrable combination follows:

$$\frac{ydx+xdy}{xy+y^2-xy^3+x^3y-x^2-xy} = \frac{dz}{y^2-x^2},$$

and hence

$$\frac{d(xy)}{(y^2-x^2)(1-xy)} = \frac{dz}{y^2-x^2},$$

which finally yields

$$\frac{d(xy)}{1-xy} = dz.$$

It allows to find the second integral of the form

$$\ln|1-xy|+z = C_2.$$

□

4.3 Local Solutions (Existence, Extensions and Straightness)

We follow here mainly approach described in the book [191]. Consider a very general case of a system of nonlinear ordinary differential equations

$$\dot{y} = f(t, y), \tag{4.20}$$

where $y(t) \in \mathbb{R}^n$ and $f : \mathbb{R}^{n+1} \to \mathbb{R}^n$.

Theorem 4.1 (Picard). *Given the function $f(t, y) : \mathbb{R}^{n+1} \to \mathbb{R}^n$ which is continuous within the set $S = \{(t, y) : |t - t_0| \le a, \ |y - y_0| \le b\}$, where $\sup_{(t,y)\in S} |f(t, y)| = N$ and which satisfies the Lipschitz condition because of t in S, i.e.*

$$|f(t, y_1) - f(t, y_2)| \le M |y_1 - y_2|, \tag{4.21}$$

for a certain number M. The Cauchy problem

$$\begin{aligned} \dot{y} &= f(t, y), \\ y(t_0) &= y_0 \end{aligned} \tag{4.22}$$

possesses a unique solution in the interval $|t - t_0| \le \alpha$, $\alpha < \min\left(a, \frac{b}{N}, \frac{1}{M}\right)$.

Proof. Consider a subset of the metric space of continuous functions

$$E = \{y(t) : \ y(t_0) = y_0, \ |y(t) - y_0| \le b, \ |t - t_0| \le \alpha\}. \tag{4.23}$$

Observe that E constitutes the closed subset of continuous functions space and forms as complete space. Let us consider the following transformation in E

$$F(y)(t) = y_0 + \int_{t_0}^{t} f(s, y(s))ds. \tag{4.24}$$

Notice that if there is a fixed point of the transformation

$$y(t) = y_0 + \int_{t_0}^{t} f(s, y(s))ds, \tag{4.25}$$

then it satisfies (4.22). Continuity of the function f and integral property yields that the function $y(t)$ defined by (4.25) is the differentiable function and its derivative is continuous. Differentiating (4.25) one gets (4.22). We need only to show that the transformation F has a fixed point in the space E. We are going to check the inequality

$$|F(y)(t) - y_0| \le b. \tag{4.26}$$

Observe that

$$\begin{aligned} \sup_{|t-t_0|\le\alpha} |F(y)(t) - y_0| &= \sup_{|t-t_0|\le\alpha} \left| \int_{t_0}^{t} f(s, y(s))ds \right| \\ &\le \sup_{|t-t_0|\le\alpha} \int_{t_0}^{t} \sup_{s\in[t,t_0]} |f(s, y(s))|\, ds \le M\alpha \le b, \end{aligned} \tag{4.27}$$

which means that F maps the space E into E. Now, we show that this transformation is contracting. Observe that

$$\begin{aligned} &\sup_{|t-t_0|\leqslant\alpha} |F(y_1)(t) - F(y_2)(t)| \\ &= \sup_{|t-t_0|\leqslant\alpha} \left| \int_{t_0}^{t} [f(s, y_1(s)) - f(s, y_2(s))]ds \right| \\ &\leqslant \sup_{|t-t_0|\leqslant\alpha} \int_{t_0}^{t} |f(s, y_1(s)) - f(s, y_2(s))|\, ds \\ &\leqslant \sup_{|t-t_0|\leqslant\alpha} \int_{t_0}^{t} M\, |y_1(s) - y_2(s)|\, ds \\ &\leqslant \sup_{|t-t_0|\leqslant\alpha} M \int_{t_0}^{t} \sup_{|s-t_0|\leqslant\alpha} |y_1(s) - y_2(s)|\, ds \\ &\leqslant M \sup_{|t-t_0|\leqslant\alpha} |y_1(t) - y_2(t)| \sup_{|t-t_0|\leqslant\alpha} \int_{t_0}^{t} ds \\ &\leqslant M\alpha \sup_{|t-t_0|\leqslant\alpha} |y_1(t) - y_2(t)|\, . \end{aligned} \tag{4.28}$$

The obtained inequality estimations show that the map F is contracting if $\alpha < \frac{1}{M}$. The Banach theorem on fixed point says that F has a fixed point being a limit of the series $y^{n+1}(t) = F(y^n)(t)$, where $y^0(t) = y_0$ and that this is the only one fixed point of this map in E. This proves an existence of solution to the problem (4.22), as well as a uniqueness of this solution. □

If in Theorem 4.1 the Lipschitz inequality is omitted one gets the so-called Peano theorem. This theorem certainly extends the class of functions satisfying the nonlinear differential equation (4.20), but a very important property of uniqueness is lost. However, this situation very often appears in engineering.

Theorem 4.2 (Peano). *Let the function* $f(t, y) : \mathbb{R}^{n+1} \to \mathbb{R}^n$ *be continuous in* $S = \{(t, y) : t \in [t_0, t_0 + a],\ |y - y_0| \leq b\}$ *and* $\sup_{(t,y)\in S} |f(t, y)| = N$. *Then the Cauchy problem (4.22) has a solution in the interval* $[t_0, t_0 + \alpha]$, *where* $\alpha = \min\left(a, \frac{b}{N}\right)$.

Proof. In order to prove the theorem we use the Euler scheme. Divide the interval $[t_0, t_0 + a]$ into n_1 subintervals with the ends $t_i^{(1)}$

$$t_0 = t_0^{(1)} < t_1^{(1)} < \ldots < t_{n_1}^{(1)} = t_0 + \alpha, \tag{4.29}$$

and construct a piecewise linear function approximating a solution to (4.22) in accordance with the Euler scheme, i.e.

$$\begin{aligned} &\varphi_1(t_0) = y_0, \\ &\varphi_1(t) = \varphi_1(t_i^{(1)}) + f\left(t_i^{(1)}, \varphi_1(t_i^{(1)})\right)\left(t - t_i^{(1)}\right), \quad t \in (t_i^{(1)}, t_{i+1}^{(1)}]. \end{aligned} \tag{4.30}$$

The piecewise function is obtained by linking points obtained using Euler's approximation. In a similar way one obtains the function $\varphi_k(t)$ being kth-order approximation. The latter is obtained dividing the interval $[t_0, t_0 + \alpha]$ into n_k elements $t_0 = t_0^{(k)} < t_1^{(k)} < \cdots < t_{n_k}^{(k)} = t_0 + \alpha$. We are going to consider the limit case $n_k \to \infty$ for which $\sup_i \left|t_{i+1}^{(k)} - t_i^{(k)}\right| \to 0$.

First, we observe that the functions $\varphi_k(t)$ have the following properties:

(i) Since $\varphi_k(t)$ are piecewise linear, hence they are continuous in $t \in [t_0, t_0 + \alpha]$ and differentiable everywhere except for the points $t_i^{(k)}$;

(ii) The Euler scheme yields the estimation

$$|\varphi_k(t)| \leq |y_0| + N\alpha; \tag{4.31}$$

(iii) In addition, the following inequality holds

$$|\varphi_k(t_2) - \varphi_k(t_1)| \leq N(t_2 - t_1), \tag{4.32}$$

which does not depend on k and hence the functions $\varphi_k(t)$ are similarly uniformly continuous.

The mentioned properties (i)–(iii) and the known Arzeli–Ascoli theorem led to conclusion that there is a subseries $\varphi_{k_j}(t)$ uniformly convergent to $\varphi(t)$ in the interval $[t_0, t_0 + \alpha]$.

Now we show that $\varphi(t)$ is a solution of the Cauchy problem (4.22). Since all of the functions $\varphi_k(t)$ satisfy the condition $\varphi_k(t_0) = y_0$, hence also $\varphi(t_0) = y_0$. We need to prove that

$$\lim_{h\to 0} \frac{\varphi(t + h) - \varphi(t)}{h} = f(t, \varphi(t)), \tag{4.33}$$

for $t \in [t_0, t_0 + \alpha]$. This is equivalent to the inequality

$$\left| \frac{\varphi(t+h) - \varphi(t)}{h} - f(t, \varphi(t)) \right| < \varepsilon, \tag{4.34}$$

for every $\varepsilon > 0$, for every $0 < h < h_0$ and for $t \in [t_0, t_0 + \alpha]$. If k_j is sufficiently large, then

$$\begin{aligned} &\sup_t \left(\varphi_{k_j}(t) - \varphi(t)\right) < \frac{\varepsilon h}{4}, \\ &\left| f\left(t + \theta h, \varphi_{k_j}(t + \theta h)\right) - f\left(t, \varphi_{k_j}(t)\right) \right| \le \frac{\varepsilon}{4} \end{aligned} \tag{4.35}$$

for $\theta \in [0, 1]$. The above inequality is true, because if t_1, t_2 and y_1, y_2 are close enough to each other, then

$$|f(t_1, y_1) - f(t_2, y_2)| < \frac{\varepsilon}{4}, \tag{4.36}$$

and $f(t, y)$ and $\varphi_{k_j}(t)$ are uniformly continuous. Observe that

$$\begin{aligned} &\left| \tfrac{\varphi(t+h)-\varphi(t)}{h} - f\left(t, \varphi(t)\right) \right| \\ &= \left| f\left(t, \varphi_{k_j}(t)\right) - f\left(t, \varphi(t)\right) + \tfrac{\varphi(t+h)-\varphi_{k_j}(t+h)}{h} + \right. \\ &\left. -\tfrac{\varphi(t)-\varphi_{k_j}(t)}{h} + \tfrac{\varphi_{k_j}(t+h)-\varphi_{k_j}(t)}{h} - f\left(t, \varphi_{k_j}(t)\right) \right| \\ &\le \tfrac{3}{4}\varepsilon + \left| \tfrac{\varphi_{k_j}(t+h)-\varphi_{k_j}(t)}{h} - f\left(t, \varphi_{k_j}(t)\right) \right|. \end{aligned} \tag{4.37}$$

On the other hand, for $t \in (t_i^{(k_j)}, t_{i+1}^{(k_j)})$ and $t + h \in \left[t_i^{(k_j)}, t_{i+1}^{(k_j)}\right]$, we have

$$\begin{aligned} &\left| \varphi_{k_j}(t+h) - \varphi_{k_j}(t) - hf\left(t, \varphi_{k_j}(t)\right) \right| \\ &= \left| f\left(t_i^{(k_j)}, \varphi_{k_j}(t_i^{(k_j)})\right) \left(t_{i+1}^{(k_j)} - t\right) + \cdots + \right. \\ &\left. + f\left(t_n^{(k_j)}, \varphi_{k_j}(t_n^{(k_j)})\right) \left(t + h - t_n^{(k_j)}\right) - hf\left(t, \varphi_{k_j}(t)\right) \right| = \\ &= \left| f\left(t_i^{(k_j)}, \varphi_{k_j}(t_i^{(k_j)})\right) - f\left(t, \varphi_{k_j}(t)\right) \left(t_{i+1}^{(k_j)} - t\right) + \ldots \right. \\ &\left. + \left(f\left(t_n^{(k_j)}, \varphi_{k_j}(t_n^{(k_j)})\right) - f\left(t, \varphi_{k_j}(t)\right)\right) \left(t + h - t_n^{(k_j)}\right) \right| \leqslant \frac{\varepsilon h}{4}. \end{aligned} \tag{4.38}$$

So, we have proved that

$$\left| \frac{\varphi_{k_j}(t+h) - \varphi_{k_j}(t)}{h} - f(t, \varphi_{k_j}(t)) \right| \le \frac{\varepsilon}{4}, \tag{4.39}$$

and consequently

$$\left| \frac{\varphi(t+h) - \varphi(t)}{h} - f(t, \varphi(t)) \right| \le \frac{3}{4}\varepsilon + \frac{\varepsilon}{4} = \varepsilon. \tag{4.40}$$

□

Remark 4.1. The differential equations having (only) continuous right-hand side (the Lipschitz condition is not required) may possess many solutions corresponding to the same initial conditions.

Recall that in both Picard's and Peano's theorems we have considered a bounded interval $[t_0, t_0 + \alpha]$ in which $\varphi(t)$ is defined. A natural question arises, how long the solution can be extended into left and right starting from the limiting values of the previous interval into intervals $[t_1, t_1 + \alpha]$ and $[t_0 - \alpha, t_0]$, respectively.

Definition 4.1. A solution $\varphi(t)$ defined in the interval $J \subset \mathbb{R}$ is called saturated solution, if its extension into interval J_1, where $J \subset J_1$ (or J is the proper subset of J_1) does not exist. The interval J is called the maximal interval of existence of solution $\varphi(t)$.

In order to investigate a behaviour of a saturated solution on the boundaries of its existence we prove the following lemma.

Lemma 4.1. *Given the continuous and bounded function $f(t, y)$ defined on an open set $E \subset \mathbb{R}^{n+1}$. Let $\varphi(t)$ be the solution of (4.22) in the interval (α_-, α_+). Then, there are limits $\varphi(\alpha_- + 0)$ and $\varphi(\alpha_+ - 0)$. If the function $f(t, y)$ is continuous in the point $(\alpha_-, \varphi(\alpha_- + 0))$ or the point $(\alpha_+, \varphi(\alpha_+ - 0))$, then the solution $\varphi(t)$ can be extended into the interval $[\alpha_-, \alpha_+)$ or into the interval $(\alpha_-, \alpha_+]$, respectively. This remains true if $f(t, y)$ is not defined in the point $(\alpha_-, \varphi(\alpha_- + 0))$ or in the point $(\alpha_+, \varphi(\alpha_+ - 0))$, but it can be extended into this point continuously.*

Proof. From the assumption, $\varphi(t)$ satisfies the differential equation (4.22), and hence

$$\varphi(t) = \varphi(t_0) + \int_{t_0}^{t} f(s, \varphi(s))ds, \tag{4.41}$$

for $\alpha_- < t_0 \le t < \alpha_+$. Because $f(t, y)$ is bounded on the set E, hence for $\alpha_- < t_1 \le t_2 < \alpha_+$ we have the estimation

$$|\varphi(t_2) - \varphi(t_1)| \le \left| \int_{t_1}^{t_2} f(s, \varphi(s))ds \right| \le N(t_2 - t_1), \tag{4.42}$$

where $N = \sup_{(t,y)\in E} |f(t,y)|$. Hence, $\varphi(t_2) - \varphi(t_1) \to 0$ for $t_1, t_2 \to \alpha_- + 0$ and $t_1, t_2 \to \alpha_+ - 0$. This implies existence of the limits $\varphi(\alpha_- + 0)$ and $\varphi(\alpha_+ - 0)$.

The first integral equation of our proof guarantees that if $f(t,y)$ is continuous up to the point $(\alpha_+, \varphi(\alpha_+ - 0))$, then

$$\varphi(\alpha_+) = \varphi(t_0) + \int_{t_0}^{\alpha_+} f(s, \varphi(s))ds. \tag{4.43}$$

An extension of solution into the interval $[\alpha_-, \alpha_+)$ can be shown in a similar way.

□

It is very important to predict a solution behaviour on a boundary of the existence limit. This is done by the following theorem.

Theorem 4.3. *Given a continuous function $f(t,y)$ in the open set $E \subset \mathbb{R}^{n+1}$ and given a solution $\varphi(t)$ of the differential equation (4.22) in the interval $[t_0, t_0 + \alpha]$. Then the function $\varphi(t)$ can be extended to a saturated solution with the maximal interval of existance (β_-, β_+). If the series $\{t_n\}$ is convergent to one of the ends of (β_-, β_+), then the series $\{(t_n, \varphi(t_n))\}$ is convergent to the edge of the set E. If the set E is unbounded, then the series $\{(t_n, \varphi(t_n))\}$ can be unbounded for $t_n \to \beta_-$ or $t_n \to \beta_+$.*

Proof. Let $U \subset E$ be a compact set. Let $U \subset V$, where V is open and bounded set, and $\bar{V} \subset E$. If $(t_0, y_0) \in U$, then a solution $\varphi(t)$ starting from the point (t_0, y_0) can be extended into interval $[t_0, t_1]$, and $(t_1, \varphi(t_1)) \notin U$.

To show this, one needs to consider a differential equation (4.22) in V and apply the Peano theorem for $a = b = dist(\bar{V}, \partial E)$ and $N = \sup_{(t,y)\in V} |f(t,y)|$. The Peano theorem says that $\varphi(t)$ exists in the interval $[t_0, t_0 + \alpha]$, where α depends only on a, b and N, i.e. it depends on the set V. If $(t_0 + \alpha, \varphi(t_0 + \alpha)) \in U$, then taking this point as the new initial value one extends the solution into the interval $[t_0, t_0 + 2\alpha]$, and so on. Since U is compact, then after a finite number on extensions we get the interval $[t_0, t_1]$, where $(t_1, \varphi(t_1)) \notin U$.

Let us cover the set E by an ascending series of sets $E = \bigcup_{n=1}^{\infty} E_n$, where E_n are open, bounded and $\bar{E}_n \subset E_{n+1}$. Taking into account the earlier part of our proof one concludes that there exists the series $\{t_i\}$ and the series of indices $\{n_i\}$, such that $(t_i, \varphi(t_i)) \in E_{n_i}$ and $(t_i, \varphi(t_i)) \notin E_{n_{i-1}}$. Since the series $\{t_i\}$ is monotonous, it has a limit. Assume that $\{t_i\}$ is bounded from a top. Then there is a finite limit $\beta_+ = \lim_{i\to\infty} t_i$. If the series $(t_i, \varphi(t_i))$ is unbounded, then theorem is proved.

If the series $(t_i, \varphi(t_i))$ is bounded, then Lemma 4.1 states that the function $\varphi(t)$ has the boundary $\varphi(\beta_+ - 0)$. The point $(\beta_+, \varphi(\beta_+ - 0))$ belongs to the edge of E. If it is an internal point, then due to Lemma 4.1 $(\beta_+, \varphi(\beta_+))$ must be an element of a certain E_k, $\bar{E}_k \subset E$. In the latter case $\varphi(t)$ can be extended to interval larger than $[t_0, \beta_+)$, which is in contradiction to the maximal value of β_+.

One can follow similar steps in a space of extension into the left side up to the point β_-. □

Very often in real engineering systems we have parameters, which are positive (like a mass or moment of inertia), or they can be both positive and negative (like damping or stiffness). Therefore, the next important question appears, how a solution of (4.22) depends on the initial conditions and on the right-hand side of the following Cauchy problem

$$\begin{aligned}\dot{y} &= f(t, y, \lambda),\\ y(t_0) &= y_0,\end{aligned} \tag{4.44}$$

where λ is a parameter. Now a solution of (4.22) will be treated as a function of all quantities, i.e. $y(t) = \varphi(t, t_0, y_0, \lambda)$.

We show that a solution dependence on the initial conditions is equivalent to the dependence of the right-hand side of Eq. (4.33) on the parameter. And vice versa, a dependence of a solution on a parameter is equivalent (and can be transformed) to the dependence on an initial condition.

In the first case, by the variables change

$$t = t_n - t_0, \quad y = y_n - y_0, \tag{4.45}$$

the initial value problem of (4.44) is reduced to the following

$$\begin{aligned}\dot{y} &= f(t - t_0, y - y_0, \lambda),\\ y(0) &= 0,\end{aligned} \tag{4.46}$$

and in the above we have taken $t = t_n$, $y = y_n$.

In the second case, we take $\lambda = \lambda(t)$ and now instead of (4.44) we get

$$\begin{aligned}\dot{y} &= f(t, y, \lambda),\\ \dot{\lambda} &= 0,\\ y(t_0) &= y_0,\\ \lambda(t_0) &= \lambda_0.\end{aligned} \tag{4.47}$$

Now, taking

$$y^* = \begin{Bmatrix} y \\ \lambda \end{Bmatrix}, \quad f^* = \begin{Bmatrix} f \\ 0 \end{Bmatrix}, \tag{4.48}$$

we obtain the following initial problem

$$\begin{aligned}\dot{y}^* &= f^*\left(t, y^*\right),\\ y^*(t_0) &= y_0^*.\end{aligned} \tag{4.49}$$

Theorem 4.4. *Given $\{f_n(t, y)\}$, $n \in \mathbb{N}$, a series of the functions defined in the open set $S \subset \mathbb{R}^{n+1}$ and being continuous in this set. Let $\lim_{n\to\infty} f_n(t, y) = f(t, y)$, where a convergence is uniform in each of a compact set in S. Consider the series $\{t_n^0, y_n^0\} \in S$ convergent to $(t_0, y_0) \in S$. Let $y_n(t)$ be a solution to the initial value problem*

$$\begin{aligned}\dot{y}_n &= f_n(t, y_n),\\ y_n(t_n^0) &= y_n^0.\end{aligned} \tag{4.50}$$

Then the Cauchy problem

$$\begin{aligned}\dot{y} &= f(t, y),\\ y(t_0) &= y_0,\end{aligned} \tag{4.51}$$

has a solution in a certain interval $[\delta, \gamma]$. In addition, there exists a subseries (n_k) of natural numbers that $y_{n_k}(t)$ are defined in the interval $[\delta, \gamma]$, $y_{n_k}(t) \underset{k\to\infty}{\rightarrow} y(t)$, and a convergence is uniform in the interval $[\delta, \gamma]$. If $y(t)$ is a unique solution of (4.51) in the interval $[\delta, \gamma]$, then the series $y_n(t)$ is defined for $t \in [\delta, \gamma]$, and is uniformly convergent in this interval to $y(t)$.

Proof. Since the series f_n is uniformly convergent on the compact sets, then $f(t, y)$ is continuous on a certain compact set $U \subset S$, which possesses the point (t_0, y_0). Then, there is such a number N that $\sup_{(t,y)\in U} |f(t, y)| < N$. For large enough n ($n > n_k$) we have also $\sup_{(t,y)\in U} |f_n(t, y)| < N$. The Peano theorem guarantees that there are solutions $y_n(t)$ of the problem (4.50) defined in the interval $[t_n^0, t_n^0 + \alpha]$. Let $[\delta, \gamma] = \bigcap_{n>n_k} [t_n^0, t_n^0 + \alpha]$. Since for $n > n_k$ the series (t_n^0, y_n^0) is convergent to (t_0, y_0), hence the interval $[\delta, \gamma]$ possesses non-empty inside. Observe that in the interval $[\delta, \gamma]$ the functions $y_n(t)$ are commonly bounded and similarly continuous (see the proof of the Peano theorem). The Arzeli–Ascoli theorem yields conclusion that there is the subseries $y_{n_k}(t)$ convergent uniformly in the interval $[\delta, \gamma]$ to the function $\varphi(t)$. After integration of (4.50) and (4.51) we obtain

$$y_n(t) = y_n^0 + \int_{t_n^0}^{t} f_n(s, y_n(s))ds,$$

$$y(t) = y_0 + \int_{t_0}^{t} f(s, y(s))ds. \tag{4.52}$$

The proved convergence leads to conclusion that $\varphi(t)$ is a solution to (4.51) in the interval $[\delta, \gamma]$. If (4.51) has a unique solution $y(t)$ in the interval $[\delta, \gamma]$ and since each convergent subseries of the series $y_n(t)$ tends to a certain solution of (4.51), then all such subseries tend to a common limit. It means that $y_n(t)$ converges to $y(t)$. □

Remark 4.2. If the function $f(t, y, \lambda)$ is bounded and continuous in a certain open set E, and each point $(t_0, y_0, \lambda_0) \in E$ is associated with exactly one integral curve $\varphi(t, t_0, y_0, \lambda_0)$ of (4.51), then φ depends continuously on the point (t_0, y_0, λ_0).

Lemma 4.2 (Gronwall). *Given a real positive function $u(t)$ in the interval $J \subset \mathbb{R}$. If $u(t)$ satisfies the integral inequality*

$$u(t) \leq a + b \int_{t_0}^{t} u(\tau) d\tau \tag{4.53}$$

for $t_0, t \in J$, $t > t_0$, where $a \geq 0$, $b > 0$, then for $t \in J$ the following estimation holds

$$u(t) \leq a e^{b(t-t_0)}. \tag{4.54}$$

Proof. Multiplying both sides of (4.53) by b we obtain

$$bu(t) \leq b \left(a + b \int_{t_0}^{t} u(\tau) d\tau \right), \tag{4.55}$$

and hence

$$\frac{bu(t)}{a + b \int_{t_0}^{t} u(\tau) d\tau} \leq b. \tag{4.56}$$

After integration of the above inequality we get

$$\ln \left(a + b \int_{t_0}^{t} u(\tau) d\tau \right) - \ln a \leq b(t - t_0), \tag{4.57}$$

or

$$a + b \int_{t_0}^{t} u(\tau) d\tau \leq a e^{b(t-t_0)}. \tag{4.58}$$

□

Theorem 4.5. *Given the function $f(t, y, \lambda)$ of the class C^1 with respect to its arguments $(t, y) \in S \subset \mathbb{R}^{n+1}$ and $\lambda \in G \subset \mathbb{R}^k$, where the sets S and G are open. Then the solution $y = y(t, t_0, y_0, \lambda)$ of the initial value problem (4.44) is of the class C^1 with respect to variables t, t_0, y_0 and λ in an open set, on which it is defined. If the Jacobi matrix $x(t) = \frac{\partial y(t,t_0,y_0,\lambda)}{\partial \lambda}$, then it satisfies the matrix equation*

$$\frac{dx}{dt} = \frac{\partial f(t, y, \lambda)}{\partial y} x + \frac{\partial f(t, y, \lambda)}{\partial \lambda} \tag{4.59}$$

with the initial condition

$$x(t_0) = \frac{\partial y(t_0, t_0, y_0, \lambda)}{\partial \lambda} = 0. \tag{4.60}$$

The Jacobi matrix $z(t) = \frac{\partial y(t, t_0, y_0, \lambda)}{\partial y_0}$ *satisfies the equation*

$$\frac{dz}{dt} = \frac{\partial f(t, y, \lambda)}{\partial y} z \tag{4.61}$$

with the initial condition

$$z(t_0) = \frac{\partial y(t_0, t_0, y_0, \lambda)}{\partial y_0} = I, \tag{4.62}$$

where I *is the identity matrix of the order* $n \times n$.

The proof is omitted here.

Remark 4.3. If the function (Theorem 4.5) $f(t, y, \lambda)$ is of the C^r class, where $r \geq 1$, then a solution $y(t, t_0, y_0, \lambda)$ is also of the C^r class.

Theorem 4.5 yields another important result stating that a nonlinear equation (4.22) can be locally described by a linear equation. In fact, this observation is very often used during a so-called continuation or path following method, where behaviour of either periodic or quasiperiodic orbits of any nonlinear systems are tracked with a change of some control parameters. This theorem is also called the theorem on straightening, because of its geometrical interpretation.

Theorem 4.6. *Given in open set* $S \subset \mathbb{R}^{n+1}$ *the non-autonomous equation (4.20) with* C^r*-smooth function* f, $r \geq 1$, *and given a certain initial point* $(t_0, y_0) \in S$. *Then there exists a surroundings* V *of the point* (t_0, y_0), $V \subset S$, *and diffeomorphism* $g : V \to W$ *of the* C^r*-smoothness, where* $W \subset \mathbb{R}^{n+1}$ *is such that if* $(s, u_1, u_2, \ldots, u_n)$ *is a local coordinate system in* W, *then the diffeomorphism* g *transforms (4.20) to the equation*

$$\frac{du}{ds} = 0. \tag{4.63}$$

Proof. (We follow the steps from [191].) Let $\varphi(t, t_0, p)$ be a solution of (4.20) with the initial condition $y(t_0) = y_0 = p$. Since the integral curves of Eq. (4.63) are represented by straight lines then our aim is to find a transformation from the integral curve $\varphi(t, t_0, p)$ to a straight line $u(s) = p$ (see Fig. 4.2).

It follows from Remark 4.3 that the mapping g has C^r-smoothness, and in addition, since the Jacobian of an inverse transformation

$$g^{-1}(t, p) = (t, \varphi(t, t_0, p)) \tag{4.64}$$

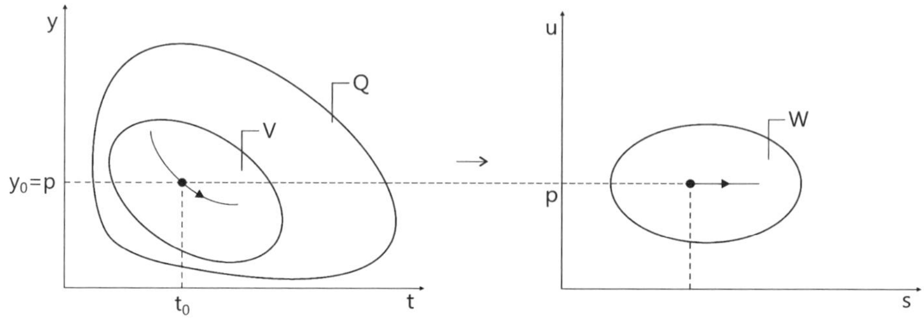

Fig. 4.2 Geometrical interpretation of straightening

is different from zero, then g is a diffeomorphism. The vector field $[1, f(t, y)]$ is transformed to the vector field $[1, 0]$. This is shown in Fig. 4.2, where the tangent vectors $[1, f(t, y)]$ are transformed to the horizontal tangent vectors $[1, 0]$. In other words, the diffeomorphism g transforms the non-autonomous equation (4.20) with the vector field $[1, f(t, y)]$ into the equation $\frac{du}{ds} = 0$ with the constant vector field $[1, 0]$. □

A similar theorem can be formulated in a case of an autonomous system.

Theorem 4.7. *Given in an open set $S \subset \mathbb{R}^{n+1}$ the autonomous differential equation $\dot{y} = f(y)$, where f is C^r-smooth ($r \geq 1$) function and given a certain non-singular point $y_0 \in Q$ of the vector field $f(y)$. Then there exists a surroundings V of the point y_0, $V \subset Q$, and the diffeomorphism $g : V \to W$ of the C^r-smoothness, where $W \subset \mathbb{R}^n$, which transforms $\dot{y} = f(y)$ into the equation*

$$\dot{u} = e_1, \tag{4.65}$$

where $(u_1, \ldots, u_m)$ is the local coordinate system in W, and e_1 is the versor of u_1. The equation $\dot{y} = f(y)$ can be presented in the form

$$\begin{aligned} \dot{u}_1 &= 1, \\ \dot{u}_i &= 0, \quad i = 2, \ldots, n. \end{aligned} \tag{4.66}$$

Proof is omitted here.

The last two theorems describe a local behaviour of the integral curves of both non-autonomous and autonomous nonlinear ordinary differential equations. In the first case, one can find an appropriate diffeomorphism transforming the integral curves into the straight parallel lines to the axis t. In the second case (autonomous), a family of integral curves can be transformed into a family of straight lines parallel to u_1.

4.4 First-Order Linear Differential Equations with Variable Coefficients

4.4.1 Introduction

Consider a non-homogeneous system of differential equations

$$\dot{y} = P(t)y + F(t) \tag{4.67}$$

with the attached initial condition

$$y(t_0) = y_0. \tag{4.68}$$

The matrix $P(t)$ has dimension $n \times n$ and the elements $p_{ki}(t)$, where k denotes row number and i denotes column number. Functions $y(t)$ and $F(t)$ are n-dimensional vector functions and functions $p_{ki}(t)$ are continuous in a given interval $a \leq t \leq b$.

Theorem 4.8. *If vector functions $P(t)$ and $F(t)$ are continuous for $t \in [a, b]$, then each point of the set $[a, b] \times \mathbb{R}^n$ belongs to only one integral curve of the vector equation (4.67).*

Proof. We consider here the one-dimensional case, since this is more illustrative. In the nth-order case one follows steps given below but instead of modulus of a real number $|\cdot|$ one can use a norm $\|\cdot\|$ of a vector or a matrix.

An existence and uniqueness of a local solution results from the Picard's theorem, because $f(t) - p(t)y$, where $F_1 = f$, and $P_1 = p$ satisfies locally the Lipschitz condition. Now we show that it can be extended into the whole interval $[a, b]$. In other words, we need to prove that a local solution $y(t)$ can be extended into the whole interval $[a, b]$. This is equivalent to show that it is bounded in each of the interval points of $[a, b]$. Let $y(t)$ contains the point (t_0, y_0). The one-dimensional equation (4.67) yields

$$y(t) = y_0 - \int_{t_0}^{t} p(s)y(s)ds + \int_{t_0}^{t} f(s)ds, \tag{4.69}$$

and hence

$$|y(t_1)| \leq \left| y_0 + \int_{t_0}^{t_1} f(s)ds \right| + K \int_{t_0}^{t_1} |y(s)|\, ds, \tag{4.70}$$

where $K = \sup_{t \in [t_0, t_1]} |p(t)|$ and $t_1 \in [a, b]$. Since the interval $[t_0, t_1]$ is compact, hence the function $f(t)$ is bounded for $t \in [t_0, t_1]$ and

$$\left| y_0 + \int_{t_0}^{t_1} f(s)ds \right| = c < +\infty. \tag{4.71}$$

So we obtain the following estimation

$$|y(t_1)| \leq c + K \int_{t_0}^{t_1} |y(s)|\, ds, \tag{4.72}$$

which means that $y(t)$ is bounded in each of $t_1 \in [a, b]$. □

Observe that Eq. (4.67) can be written in the form

$$L(y) = \dot{y} - P(t)y, \tag{4.73}$$

where L is a first-order linear vector differential operator.

Theorem 4.9. *If the vectors $C_1, C_2, \dots, C_n$ are linearly independent, then the corresponding solutions $y_1, y_2, \dots, y_n$ to the Cauchy homogeneous problem*

$$\begin{aligned} &L(y_i) = \dot{y}_i - P(t)y_i = 0, \quad t \in [a, b], \\ &y_i(t_0) = C_i, \quad i = 1, \dots, n \end{aligned} \tag{4.74}$$

are linearly independent for each $t \in [a, b]$.

Proof (We follow the steps given in [191]). A necessary condition for a linear independence of the solution is that Wronskian determinant

$$W(t) = \begin{vmatrix} y_1^1 & y_1^2 & \cdots & y_1^n \\ y_2^1 & y_2^2 & \cdots & y_2^n \\ \vdots & \vdots & \ddots & \vdots \\ y_n^1 & y_n^2 & \cdots & y_n^n \end{vmatrix} \neq 0. \tag{4.75}$$

Recall that if $W(t_0) \neq 0$, $a \leq t_0 \leq b$ and y_i^k, $(k, i = 1, \dots, n)$ are solutions to (4.74), then the solutions are linearly independent. In addition, a necessary and sufficient condition for linear independence of the solutions $y_i^k(t)$ is that for any arbitrary point t_0 we have $W(t_0) \neq 0$, which means that also $W(t) \neq 0$.

Observe that in accordance with (4.74) we have

$$\dot{y}_i^j = \sum_{k=1}^{n} p_{ik} y_k^j. \tag{4.76}$$

Let us differentiate the Wronskian determinant

$$\dot{W}(t) = \sum_{i=1}^{n} \begin{vmatrix} y_1^1 & y_1^2 & \dots & y_1^n \\ \vdots & \vdots & & \vdots \\ (y_i^1)' & (y_i^2)' & \cdots & (y_i^n)' \\ \vdots & \vdots & \ddots & \vdots \\ y_n^1 & y_n^2 & \cdots & y_n^n \end{vmatrix}$$

$$= \sum_{i=1}^{n} \begin{vmatrix} y_1^1 & y_1^2 & \dots & y_1^n \\ \vdots & \vdots & \dots & \vdots \\ \sum_{j=1}^{n} p_{ij}(t) y_j^1 & \sum_{j=1}^{n} p_{ij}(t) y_j^2 & \cdots & \sum_{j=1}^{n} p_{ij}(t) y_j^n \\ \vdots & \vdots & \ddots & \vdots \\ y_n^1 & y_n^2 & \cdots & y_n^n \end{vmatrix}$$

$$= \sum_{i=1}^{n} \sum_{j=1}^{n} p_{ij}(t) \begin{vmatrix} y_1^1 & y_1^2 & \dots & y_1^n \\ \vdots & \vdots & & \vdots \\ y_j^1 & y_j^2 & \cdots & y_j^n \\ \vdots & \vdots & \ddots & \vdots \\ y_n^1 & y_n^2 & \cdots & y_n^n \end{vmatrix}. \tag{4.77}$$

One can observe that for $i \neq j$ the rows ith and jth are the same and for $i = j$ we get $W(t)$. Hence from the latter result we get

$$\dot{W}(t) = \sum_{i=1}^{n} \sum_{j=1}^{n} p_{ij}(t) \begin{vmatrix} y_1^1 & y_1^2 & \dots & y_1^n \\ \vdots & \vdots & & \vdots \\ y_j^1 & y_j^2 & \cdots & y_j^n \\ \vdots & \vdots & \ddots & \vdots \\ y_n^1 & y_n^2 & \cdots & y_n^n \end{vmatrix} = \sum_{k=1}^{n} p_{kk}(t) W(t) \tag{4.78}$$

or

$$\frac{\dot{W}(t)}{W(t)} = \sum_{k=1}^{n} p_{kk}(t). \tag{4.79}$$

After integration of (4.79) we get

$$W(t) = W(t_0) e^{\int_{t_0}^{t} \mathrm{tr}(P(s)) ds}, \tag{4.80}$$

where $\mathrm{tr}(P) = \sum_{k=1}^{n} p_{kk}(t)$ denotes trace of the matrix P.

Since $W(t_0) \neq 0$ then, as it can be seen in (4.80), $W(t) \neq 0$ for $t \in [a,b]$. It means also that the set of solutions $\{y_1(t), y_2(t), \ldots, y_n(t)\}$ is linearly independent. □

Remark 4.4. If $y(t)$ is a solution to (4.74) and $y(t_0) = 0$ for a certain $t_0 \in [a,b]$, then $y(t) \equiv 0$.

Remark 4.5. If $W(t_0) \neq 0$ for a certain $t_0 \in [a,b]$, then $W(t) \neq 0$ for each $t \in [a,b]$.

Remark 4.6. The formula (4.80) is known as the Liouville's formula (sometimes referred as the Ostrogradskiy–Liouville formula).

Example 4.6. Derive the Liouville's formula for $n = 2$ and a homogeneous differential equation.

For $n = 2$ we have

$$\begin{aligned}\dot{W}_2(t) &= \frac{d}{dt}\begin{vmatrix} y_1^1 & y_1^2 \\ y_2^1 & y_2^2 \end{vmatrix} = \begin{vmatrix} \dot{y}_1^1 & \dot{y}_1^2 \\ y_2^1 & y_2^2 \end{vmatrix} + \begin{vmatrix} y_1^1 & y_1^2 \\ \dot{y}_2^1 & \dot{y}_2^2 \end{vmatrix} \\ &= \begin{vmatrix} p_{11}y_1^1 + p_{12}y_2^1 & p_{11}y_1^2 + p_{12}y_2^2 \\ y_2^1 & y_2^2 \end{vmatrix} + \begin{vmatrix} y_1^1 & y_1^2 \\ p_{21}y_1^1 + p_{22}y_2^1 & p_{21}y_1^2 + p_{22}y_2^2 \end{vmatrix} \\ &= p_{11}\begin{vmatrix} y_1^1 & y_1^2 \\ y_2^1 & y_2^2 \end{vmatrix} + p_{22}\begin{vmatrix} y_1^1 & y_1^2 \\ y_2^1 & y_2^2 \end{vmatrix} = (p_{11} + p_{22})W_2(t),\end{aligned}$$

where

$$W_2(t) = \begin{vmatrix} y_1^1 & y_1^2 \\ y_2^1 & y_2^2 \end{vmatrix}.$$

After separation of the variables and integration we get

$$W(t) = W(t_0)e^{\int_{t_0}^t (p_{11}(\tau)+p_{22}(\tau))d\tau}.$$

□

Definition 4.2. The square $(n \times n)$ matrix $Y(t)$ satisfying the differential equation

$$\dot{Y} = P(t)Y, \tag{4.81}$$

composed of n functions $y_1(t), y_2(t), \ldots, y_n(t)$ being the columns of Y, for which $W(t) \neq 0$, is called a fundamental matrix of $L(y_i) = 0$, $i = 1, \ldots, n$. The vectors $y_1(t), y_2(t), \ldots, y_n(t)$ are called the fundamental solutions of $L(y_i) = 0$, $i = 1, \ldots, n$. The determinant $W(t) = \det(Y(t))$ is called the Wronskian determinant of the functions $y_1(t), y_2(t), \ldots, y_n(t)$.

Theorem 4.10. *Any linear homogeneous equation*

$$\dot{y} = P(t)y \tag{4.82}$$

has an associated fundamental system of solutions.

Proof. Take n linearly independent vectors in $\mathbb{R}^n$, $y_0^1, y_0^2, \ldots, y_0^n$. Then we solve (4.82) with the successive initial conditions

$$y(t_0) = y_0^i, \quad i = 1, \ldots, n, \; t_0 \in [a, b]. \tag{4.83}$$

The obtained solutions $y^i(t)$ form the fundamental system of equations, since

$$W(t_0) \neq 0 \tag{4.84}$$

and hence $W(t) \neq 0$ for each $t \in [a, b]$ (see Theorem 4.9). □

Consider now the non-homogeneous system of Eq. (4.67), which can be written in the form

$$\dot{y}_k = \sum_{i=1}^{n} p_{ki}(t) y_i + f_k(t) = L_k(y) + f_k(t), \quad k = 1, \ldots, n. \tag{4.85}$$

We assume that the function $p_{ki}(t)$ and $f_k(t)$ are continuous in the interval $a \leq t \leq b$. Hence, the Picard's theorem guarantees an existence of a solution to the Cauchy problem for arbitrary but finite values of the initial conditions $y_k(t_0) = y_k^0$ (we follow here the considerations from monograph [191]).

Theorem 4.11. *Let $\bar{y} = (\bar{y}_1, \ldots, \bar{y}_n)$ be a particular solution to (4.85), i.e.*

$$\dot{\bar{y}}_k = L_k(\bar{y}) + f_k(t), \quad k = 1, \ldots, n. \tag{4.86}$$

A general solution to (4.85) has the form

$$y_k = z_k(t) + \bar{y}_k, \quad k = 1, \ldots, n, \tag{4.87}$$

where $z = (z_1(t), \ldots, z_n(t))$ is a general solution to the homogeneous equation

$$\dot{y}_k = L_k(y). \tag{4.88}$$

Proof. Substituting (4.87) into (4.85) we get

$$\dot{z}_k(t) + \dot{\bar{y}}_k(t) = L_k(z) + L_k(\bar{y}) + f_k(t), \quad k = 1, \ldots, n \tag{4.89}$$

and hence

$$\dot{z}_k(t) = L_k(z). \tag{4.90}$$

Assume that the fundamental solutions $z_{l1}, \ldots, z_{ln}$, $l = 1, \ldots, n$ are known. Hence, a general solution to homogeneous equation (4.90) is

$$z_k = \sum_{v=1}^{n} c_v z_{vk}, \quad k = 1, \ldots, n. \tag{4.91}$$

Differentiating (4.91), and noticing that $\dot{z}_k = L_k(z)$, we obtain

$$\sum_{v=1}^{n} c_v \dot{z}_{vk} = L_k(z), \quad k = 1, \ldots, n. \tag{4.92}$$

Applying the Lagrange (or variable coefficients) method we are looking for the following solutions

$$z_k(t) = \sum_{v=1}^{n} c_v(t) z_{vk}, \quad k = 1, \ldots, n. \tag{4.93}$$

Differentiation of (4.93) yields

$$\dot{z}_k(t) = \sum_{v=1}^{n} \dot{c}_v(t) z_{vk} + \sum_{v=1}^{n} c_v(t) \dot{z}_{vk}, \quad k = 1, \ldots, n. \tag{4.94}$$

Substituting (4.94) to (4.85) we obtain

$$\sum_{v=1}^{n} \dot{c}_v(t) z_{vk} + \sum_{v=1}^{n} c_v(t) \dot{z}_{vk} = L_k(z) + f_k(t), \quad k = 1, \ldots, n. \tag{4.95}$$

Since

$$L_k(z) = \sum_{v=1}^{n} c_v(t) \dot{z}_{vk}, \quad k = 1, \ldots, n, \tag{4.96}$$

then

$$\sum_{v=1}^{n} \dot{c}_v(t) z_{vk} = f_k(t), \quad k = 1, \ldots, n. \tag{4.97}$$

Since the Wronskian determinant is not equal to zero (z_{vk} are the components of fundamental solutions), one can solve (4.97) to get $c_1(t), \ldots, c_n(t)$, and hence $z_k(t)$ in accordance with formula (4.93). □

After the general introduction and some preliminary theorems and definitions, now we can proceed with much more advanced results.

Let us consider a set of K differential equations of the first order

$$\dot{y}(t) = P(t)y, \tag{4.98}$$

where $y \in \mathbb{R}^K$. We assume that time-dependent matrix is integrable and is piecewise continuous. Then the Cauchy problem $y_0 = y(t_0)$ has a unique solution $y(t, t_0)$.

Recall that any functions is said to be piecewise continuous if it has at each finite time interval (t_1, t_2) a finite number of discontinuities.

Any function piecewise continuous is called integrable, if in each continuity point t^*, where $t_1 \leq t^* \leq t_2$, there exist limits

$$\begin{aligned} \lim_{\varepsilon \to 0-} \int_{t_1}^{t^*-\varepsilon} |f(t)|\,dt &= \int_{t_1}^{t^*} |f(t)|\,dt, \\ \lim_{\varepsilon \to 0+} \int_{t^*+\varepsilon}^{t_2} |f(t)|\,dt &= \int_{t^*}^{t_2} |f(t)|\,dt, \end{aligned} \tag{4.99}$$

where (t_1, t_2) does not include any discontinuities.

4.4.2 Fundamental Matrix of Solutions

Assume that K solutions of the form $\Gamma_k(t)$ to the system (4.98) are known. Then, by assumption

$$\dot{\Gamma}_k(t) = A(t)\Gamma_k, \quad k = 1, \ldots, K, \tag{4.100}$$

any linear combination

$$x(t) = \sum_{k=1}^{K} c_k \Gamma_k(t) \tag{4.101}$$

is also a solution to (4.98), where $c_1, \ldots, c_k$ are constants, which must satisfy

$$\begin{aligned} &|c_1| + \cdots + |c_k| > 0, \\ &\sum_{k=1}^{K} c_k \Gamma_k \neq 0, \end{aligned} \tag{4.102}$$

i.e. they are linearly independent.

Observe that any system of K linearly independent solutions is called *a fundamental system* (or a main system), and the corresponding solutions are called *fundamental ones.*

The *fundamental solutions* create a $K \times K$ matrix called the *fundamental matrix*

$$\dot{\Gamma}(t) = A(t)\Gamma(t), \tag{4.103}$$

and $\Gamma \equiv \Gamma(t, t_0)$.

Observe that any matrix

$$\tilde{\Gamma} = \Gamma C, \tag{4.104}$$

is also a fundamental matrix, where C is a regular and non-singular matrix. To convince yourself with this observation put (4.104) into (4.98) to get

$$\dot{\tilde{\Gamma}} = \dot{\Gamma} C = A \Gamma C = A \tilde{\Gamma}. \tag{4.105}$$

We are going to consider a particular fundamental matrix $\Phi(t, t_0)$ such that

$$\dot{\Phi}(t, t_0) = A(t)\Phi(t, t_0), \tag{4.106}$$

$$\Phi(t_0, t_0) = I. \tag{4.107}$$

Any fundamental solution satisfies

$$\begin{aligned} \dot{\varphi}_k(t, t_0) &= A(t)\varphi_k(t, t_0), \\ \varphi_k(t_0, t_0) &= i_k, \quad k = 1, 2, \ldots, K, \end{aligned} \tag{4.108}$$

and i_k denotes kth unit vector. The matrix $\Phi(t, t_0)$ is called *matricant* or *the transition* matrix.

Each of the fundamental matrices can be found using a transition matrix

$$\Gamma(t, t_0) = \Phi(t, t_0)\Gamma(t_0, t_0). \tag{4.109}$$

The Wronskian determinant has the form

$$W(t, t_0) = \det \Gamma(t, t_0) = W(t_0, t_0) \exp\left(\int_{t_0}^{t} tr\,(A(\tau))\, d\tau \right). \tag{4.110}$$

The associated Wronskian with matricant Φ has the form

$$W(t_0, t_0) = \det \Phi(t_0, t_0) = \det I = 1, \tag{4.111}$$

and

$$\det \Phi(t, t_0) = \exp\left(\int_{t_0}^{t} tr\,(A(\tau))\, d\tau \right) \neq 0, \tag{4.112}$$

which means that the matricant cannot be singular.

4.4.3 Homogeneous Differential Equations

From (4.103) one concludes that

$$x(t) = \Phi(t, t_0)x_0, \tag{4.113}$$

where $x_0 = x(t_0)$.

Matricant is a regular matrix (see (4.112)) and moreover

$$\dot{\Phi}(t, t_0) = A(t)\Phi(t, t_0), \tag{4.114}$$

$$\Phi(t_0, t_0) = I. \tag{4.115}$$

The solution (4.113) has the following geometrical interpretation (Fig. 4.3).

It is seen that the matricant transforms the vector $x(t_0)$ into a new vector $x(t_1)$ in time interval $[t_0, t_1]$. Equation (4.113) yields

$$x(t_1) = \Phi(t_1, t_0)x(t_0). \tag{4.116}$$

On the other hand

$$x(t_2) = \Phi(t_2, t_1)x(t_1) = \Phi(t_2, t_1)\Phi(t_1, t_0)x(t_0), \tag{4.117}$$

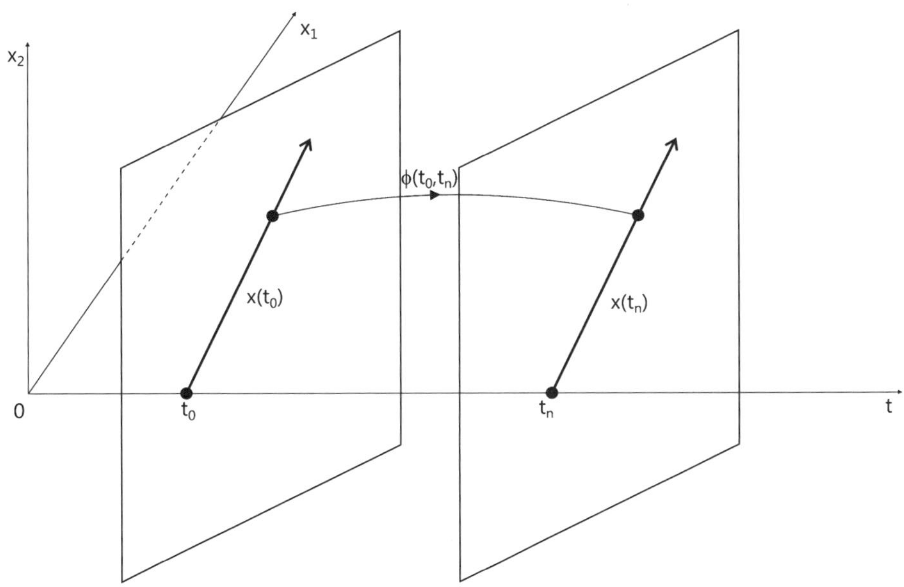

Fig. 4.3 A role of matricant (geometrical interpretation)

and

$$x(t_2) = \Phi(t_2, t_0)x(t_0). \tag{4.118}$$

Comparing (4.117) and (4.118) we get

$$\Phi(t_2, t_0) = \Phi(t_2, t_1)\Phi(t_1, t_0), \tag{4.119}$$

which agrees with our group properties mentioned in Chap. 1. Observe that a transition matrix can be expressed as a product of many matrices from the considered time interval. The group property allows also to find an inverse transformation

$$x(t_0) = \Phi(t_0, t_1)x(t_1) \tag{4.120}$$

and taking into account (4.116) one gets

$$\Phi(t_0, t_1) = \Phi^{-1}(t_1, t_0). \tag{4.121}$$

The obtained result is important, since it means that any inverse matrix can be obtained easily by changing its arguments.

4.4.4 Examples of Homogeneous Linear Differential Equations

As it has been mentioned, a linear system of homogeneous differential equations is defined as follows

$$\frac{dy}{dt} = P(t)y, \ y \in \mathbb{R}^n, \ t \in J, \tag{4.122}$$

where $P(t)$ is a square matrix of $n \times n$ dimension with its elements $p_{ij}(t)$, $i, j = 1, \ldots, n$ being continuous on the interval J. A set of all solutions of (4.122) defined on J creates a linear space. Any solution of (4.122) defined on J can be extended on entire J. In addition, the linear space of all solutions of (4.122) is isomorphic to a phase space $\mathbb{R}^n$ of system (4.122).

Fundamental system of solutions of the homogeneous equations (4.122) creates the linear space, i.e. there are n linearly independent solutions of (4.122). A matrix $\phi(t)$ consisting of solutions situated on its rows is called the *fundamental matrix*. If $\phi(t_0) = I$, where I is a unit matrix, then $\phi(t)$ is called the *matricant*.

Any solution to (4.122) can be presented as a linear combination of the fundamental matrix solutions. It means that if one finds $n + 1$ solutions, then they are linearly dependent.

If one takes vector-function $\varphi_1(t), \ldots, \varphi_n(t)$, then the Wronskian is defined by

$$W(t) = \begin{vmatrix} \varphi_{11}(t), & \ldots, & \varphi_{n1}(t) \\ \vdots & & \\ \varphi_{1n}(t), & \ldots, & \varphi_{nn}(t) \end{vmatrix} \tag{4.123}$$

and any

$$\varphi_k(t) = \varphi_1 E_1 + \cdots + \varphi_n E_n,$$

where unit vectors $E_1, \ldots, E_n$ create the linear basis. If for arbitrary point t we get $W(t) \neq 0$, then the system of solutions $\varphi_1(t), \ldots, \varphi_n(t)$ is the fundamental system of solutions. On the other hand, if $W(t) = 0$ for a point t, then it is identically equal to zero for all t. As it has been already mentioned

$$W(t) = W(t_0) \exp\left(\int_{t_0}^{t} trP(s)ds\right) = W(t_0) \exp\left(\int_{t_0}^{t} \sum_{i=1}^{n} p_{ii}(s)ds\right) \tag{4.124}$$

Example 4.7. Verify that $x_1 = -\sin t$, $y_1 = \cos t$ are solutions of the following two equations

$$\frac{dx}{dt} = x\cos^2 t - (1 - \sin t \cos t)y,$$
$$\frac{dy}{dt} = (1 + \sin t \cos t)y + y\sin^2 t.$$

Find the fundamental matrix of solutions.

Assume that $x = \varphi(t)$, $y = \psi(t)$ are solutions and $\varphi(0) = 1$, $\psi(0) = 1$. The corresponding Wronskian is

$$\begin{vmatrix} \varphi(t) & -\sin t \\ \psi(t) & \cos t \end{vmatrix} = \begin{vmatrix} 1 & 0 \\ 0 & 1 \end{vmatrix} \exp\left(\int_0^1 (\sin^2 \tau + \cos^2 \tau)d\tau\right) = e^t,$$

and hence

$$\varphi(t)\cos t + \psi(t)\sin t = e^t.$$

Substituting $x = \varphi$ and $y = \psi$ to two first-order differential equations one gets

$$\frac{d\varphi}{dt} = -\psi + \cos t(\varphi\cos t + \psi\sin t) = -\psi + e^t\cos t,$$
$$\frac{d\psi}{dt} = \varphi + \sin t(\varphi\cos t + \psi\sin t) = \varphi + e^t\sin t.$$

Observe that

$$\frac{d}{dt}(\varphi - e^t \cos t) = \frac{d\varphi}{dt} + e^t(\sin t - \cos t) = -(\psi - e^t \sin t),$$
$$\frac{d}{dt}(\psi - e^t \sin t) = \frac{d\psi}{dt} - e^t(\cos t + \sin t) = \varphi - e^t \cos t,$$

and hence we have found two additional solutions: $\varphi(t) = e^t \cos t$, $\psi(t) = e^t \sin t$. The fundamental matrix of solutions is

$$\phi(t) = \begin{bmatrix} e^t \cos t & -\sin t \\ e^t \sin t & \cos t \end{bmatrix}.$$

□

Example 4.8. Determine a general solution of the following equations

$$\frac{dx}{dt} = t(1-t)x + (t^3 - t^2 + t + 1)y$$
$$\frac{dy}{dt} = (1-t)x + (t^2 - t + 1)y$$

assuming that $x \equiv x_1 = a + bt + ct^2$, $y \equiv y_1 = d + ft$ satisfy the equations.

Substitution of (x_1, y_1) into the set of two first-order ODEs yields

$$\begin{array}{ccc} -c + f = 0, & c - b + d - f = 0, & b - a - d + f = 0, \\ a + f + d - 2c = 0, & b = d, & a + d - f = 0, \end{array}$$

and hence $b = d = 0$, $a = f = c$. Taking $a = f = c = 1$ we get $x_1 = 1 + t^2$, $y_1 = t$. We are looking for remaining solutions $x = \varphi(t)$, $y = \psi(t)$, where $\varphi(0) = 0$, $\psi(0) = 1$. The Wronskian

$$\begin{vmatrix} 1+t^2 & \varphi(t) \\ t & \psi(t) \end{vmatrix} = \begin{vmatrix} 1 & 0 \\ 0 & 1 \end{vmatrix} \exp\left[\int_0^t [s(1-s) + s^2 - s + 1]ds\right] = \exp(t),$$

or equivalently

$$(1+t^2)\psi(t) - t\varphi(t) = e^t.$$

The second considered ODE yields

$$\frac{d\psi}{dt} = (1-t)\varphi + (1+t^2)\psi - t\psi = \varphi - t\psi + e^t,$$

and hence

$$t\frac{d\psi}{dt} = t\varphi - t^2\psi + te^t = \psi + (t-1)e^t.$$

The obtained equation satisfied by $\psi(t) = e^t$, $\psi(0) = 1$.

The Wronskian formula yields

$$t\varphi = (1+t^2)\psi - e^t = (1+t^2)e^t - e^t = t^2e^t,$$

i.e. we have found two remaining solutions $\psi(t) = e^t$, $\varphi(t) = te^t$.

Therefore, the sought general solution has the following form

$$x = C_1(1+t^2) + C_2te^t,$$
$$y = C_1t + C_2e^t.$$

□

In the case when in (4.122) $P(t) = P = \text{const}$, the problem of finding solutions can be solved using directly the *Euler method*. The following form of solution is being looked for

$$y = ae^{\sigma t}, \tag{4.125}$$

where σ is an eigenvalue of the matrix P, and the vector $a[a_1, \ldots, a_n]$ corresponds to the eigenvalue σ. Now, if we deal with $\sigma_1, \ldots, \sigma_n$ and each of them being associated with the vector $a_1, \ldots, a_n$, then a general solution to Eq. (4.122) with $P(t) = P$ has the following form

$$y = C_1a_1e^{\sigma_1 t} + C_2a_2e^{\sigma_2 t} + \cdots + C_na_ne^{\sigma_n t}, \tag{4.126}$$

where $C_1, C_2, \ldots, C_n$ are arbitrary numbers. If there is a multiple root σ^k of the k-multiplicity with k-linearly independent vectors $a_1, a_2, \ldots, a_k$, then this root corresponds to k-linearly independent solutions $e^{\sigma t}a_1, e^{\sigma t}a_2, \ldots, e^{\sigma t}a_k$.

If for a root σ^k with multiplicity k there are only m $(m < k)$ linearly independent eigenvectors, then the corresponding solution has the following form

$$y = (a_0 + a_1t + \cdots + a_{k-m}t^{k-m})e^{\sigma t}. \tag{4.127}$$

If among eigenvalues are complex numbers, then each complex number corresponds to a complex eigenvector. Since matrix P is real, then $\operatorname{Re}\sigma$, and $\operatorname{Im}\sigma$ corresponds to real eigenvectors.

There is also a direct approach of finding the fundamental matrix of solutions to Eq. (4.122) with $P(t) = P$. Namely, we introduce a notion of the exponent matrix e^{At} of the form

$$e^{At} = I + \frac{1}{1!}A + \frac{1}{2!}A^2 + \cdots = \sum_{k=0}^{\infty}\frac{1}{k!}A^k \tag{4.128}$$

The following properties hold:

(i) If $AB = BA$, then $e^{AB} = e^A \cdot e^B = e^B \cdot e^A$;
(ii) If $A = T^{-1}JT$, then $e^A = T^{-1}e^J T$;
(iii) Matrix $Y(t) = e^{Pt}$ is a solution to the Cauchy problem (4.122) with $P(t) = P$, i.e.

$$\tfrac{dY}{dt} = PY,\ Y(0) = I.$$

Third property means that $y(t) = e^{Pt}y_0$, where $y_0 = y(0)$. Therefore, the problem reduces to that of finding the matrix e^{Pt}. In order to compute e^{Pt} one may apply the property (ii), where J is the so-called Jordan cell

$$J = \begin{bmatrix} \sigma\ 1\ 0 \dots 0 \\ 0\ \sigma\ 1 \dots 0 \\ \vdots \\ 0\ 0\ 0 \dots \sigma \end{bmatrix}, \tag{4.129}$$

and $P = T^{-1}J_P T$, where J_P is the Jordan cell corresponding to the matrix P. Introducing

$$E = \begin{bmatrix} 0\ 1\ 0 \dots 0 \\ 0\ 0\ 1 \dots 0 \\ \vdots \\ 0\ 0\ 0 \dots 0 \end{bmatrix}, \tag{4.130}$$

one gets

$$J = \sigma I + E, \tag{4.131}$$

and hence

$$e^{Jt} = e^{\sigma t}e^{Et}, \tag{4.132}$$

where e^{Et} can be found using series (4.128)

Example 4.9. Solve the following system of ODEs

$$\frac{dx}{dt} = 5x + 2y,$$
$$\frac{dy}{dt} = -4x - y.$$

Assuming $x = Ae^{\sigma t}$, $y = Be^{\sigma t}$, governing equations yield

$$(5-\sigma)A + 2B = 0,$$
$$-4A - (1+\sigma)B = 0,$$

and the determinant

$$\begin{vmatrix} 5-\sigma & 2 \\ -4 & -1-\sigma \end{vmatrix} = 0$$

gives $\sigma_1 = 1, \sigma_2 = 3$. For $\sigma = 1$ we have $4A + 2B = 0$. We take $A = 1$, $B = -2$, and the solution is $x_1 = e^t$, $y_1 = -2e^t$. For $\sigma_2 = 3$ one gets $A + B = 0$; we take $A = 1$, $B = 1$, and the solution is $x_2 = e^{3t}$, $y_2 = -e^{3t}$. One may check that the obtained solutions are linearly independent because

$$\begin{vmatrix} e^t & e^{3t} \\ -2e^t & -e^{-3t} \end{vmatrix} \neq 0.$$

All possible solutions to the studied equation are governed by formulas

$$x = C_1e^t + C_2e^{3t},$$
$$y = -2C_1e^t - C_2e^{3t},$$

where C_1, C_2 are arbitrary constants. □

Example 4.10. Find e^{Pt}, where

(i) $P = \begin{bmatrix} 0 & 1 \\ 1 & 0 \end{bmatrix}$; (ii) $P = \begin{bmatrix} 0 & 1 \\ -1 & 0 \end{bmatrix}$; (iii) $P = \begin{bmatrix} 3 & -2 \\ 4 & -3 \end{bmatrix}$; (iv) $P = \begin{bmatrix} 3 & 1 \\ -1 & 1 \end{bmatrix}$.

We use definition (4.128), where

$$e^{Pt} = I + Pt + \frac{P^2t^2}{2} + \frac{P^3t^3}{6} + \cdots$$

(i) We have

$$P^2 = \begin{bmatrix} 0 & 1 \\ 1 & 0 \end{bmatrix}\begin{bmatrix} 0 & 1 \\ 1 & 0 \end{bmatrix} = \begin{bmatrix} 1 & 0 \\ 0 & 1 \end{bmatrix} = I,$$
$$P^3 = PP^2 = IP = P,\ P^4 = P^2P^2 = II = I.$$

Therefore $P^k = P$, if $k = 2p + 1$, and $P^k = I$, if $k = 2p$.

Finally, we obtain

$$
\begin{aligned}
e^{Pt} &= I + Pt + I\frac{t^2}{2!} + P\frac{t^3}{3!} + I\frac{t^4}{4!} + \cdots \\
&= \begin{bmatrix} 1 & 0 \\ 0 & 1 \end{bmatrix} + \begin{bmatrix} 0 & 1 \\ 1 & 0 \end{bmatrix} t + \begin{bmatrix} 1 & 0 \\ 0 & 1 \end{bmatrix} \frac{t^2}{2} + \begin{bmatrix} 0 & 1 \\ 1 & 0 \end{bmatrix} \frac{t^3}{6} + \cdots \\
&= \begin{bmatrix} 1 + \frac{t^2}{2!} + \frac{t^4}{4!} + \cdots & t + \frac{t^3}{3!} + \frac{t^5}{5!} + \cdots \\ t + \frac{t^3}{3!} + \frac{t^5}{5!} + \cdots & 1 + \frac{t^2}{2!} + \frac{t^4}{4!} + \cdots \end{bmatrix} = \begin{bmatrix} \cosh t & \sinh t \\ \sinh t & \cosh t \end{bmatrix}.
\end{aligned}
$$

(ii) We have

$$
P^2 = \begin{bmatrix} 0 & 1 \\ -1 & 0 \end{bmatrix} \begin{bmatrix} 0 & 1 \\ -1 & 0 \end{bmatrix} = \begin{bmatrix} -1 & 0 \\ 0 & -1 \end{bmatrix},
$$

$$
P^3 = P^2 P = \begin{bmatrix} -1 & 0 \\ 0 & -1 \end{bmatrix} \begin{bmatrix} 0 & 1 \\ -1 & 0 \end{bmatrix} = \begin{bmatrix} 0 & -1 \\ 1 & 0 \end{bmatrix},
$$

$$
P^4 = P^3 P = \begin{bmatrix} 0 & -1 \\ 1 & 0 \end{bmatrix} \begin{bmatrix} 0 & 1 \\ -1 & 0 \end{bmatrix} = \begin{bmatrix} 1 & 0 \\ 0 & 1 \end{bmatrix}.
$$

Therefore, we have

$$
P^{2k} = (-1)^k I, \quad P^{2k+1} = (-1)^k P, \quad k = 0, 1, 2, \ldots
$$

Finally, one obtains

$$
\begin{aligned}
e^{Pt} &= I + Pt - I\frac{t^2}{2!} - P\frac{t^3}{3!} + I\frac{t^4}{4!} + \cdots \\
&= \begin{bmatrix} 1 & 0 \\ 0 & 1 \end{bmatrix} + \begin{bmatrix} 0 & 1 \\ -1 & 0 \end{bmatrix} t - \begin{bmatrix} 1 & 0 \\ 0 & 1 \end{bmatrix} \frac{t^2}{2!} - \begin{bmatrix} 0 & 1 \\ -1 & 0 \end{bmatrix} \frac{t^3}{3!} + \begin{bmatrix} 1 & 0 \\ 0 & 1 \end{bmatrix} \frac{t^4}{4!} + \cdots \\
&= \begin{bmatrix} 1 - \frac{t^2}{2!} + \frac{t^4}{4!} + \cdots & t - \frac{t^3}{3!} + \frac{t^5}{5!} + \cdots \\ -t + \frac{t^3}{3!} - \frac{t^5}{5!} + \cdots & 1 - \frac{t^2}{2!} + \frac{t^4}{4!} + \cdots \end{bmatrix} = \begin{bmatrix} \cos t & \sin t \\ -\sin t & \cos t \end{bmatrix}.
\end{aligned}
$$

(iii) In this case one computes $\sigma_1 = 1$, $\sigma_2 = -1$, and $J = \begin{bmatrix} 1 & 0 \\ 0 & -1 \end{bmatrix}$. We are going to find a non-singular matrix $T = \begin{bmatrix} a & b \\ c & d \end{bmatrix}$ satisfying the condition

$$
P = T^{-1} J T,
$$

or equivalently

$$\begin{bmatrix} a & b \\ c & d \end{bmatrix}\begin{bmatrix} 3 & -2 \\ 4 & -3 \end{bmatrix} = \begin{bmatrix} 1 & 0 \\ 0 & -1 \end{bmatrix}\begin{bmatrix} a & b \\ c & d \end{bmatrix},$$

which can be presented in the form

$$\begin{bmatrix} 3a+4b & -2a-3b \\ 3c+4d & -2c-3d \end{bmatrix} = \begin{bmatrix} a & b \\ -c & -d \end{bmatrix}.$$

The problem boils down to solving of algebraic equations

$$a + 2b = 0, \; c + d = 0.$$

Taking $a = 2, b = 1, c = -1, d = 1$ we get

$$T = \begin{bmatrix} 2 & -1 \\ -1 & 1 \end{bmatrix}, \; T^{-1} = \begin{bmatrix} 1 & 1 \\ 1 & 2 \end{bmatrix}.$$

Since $e^{Pt} = T^{-1}e^{Jt}T$, therefore

$$e^{Pt} = \begin{bmatrix} 1 & 1 \\ 1 & 2 \end{bmatrix}\begin{bmatrix} e^t & 0 \\ 0 & e^{-t} \end{bmatrix}\begin{bmatrix} 2 & -1 \\ -1 & 1 \end{bmatrix} = \begin{bmatrix} 2e^t - e^{-t} & -e^t + e^{-t} \\ 2e^t - 2e^{-t} & -e^t + 2e^{-t} \end{bmatrix}.$$

(iv) The characteristic equation is

$$\begin{vmatrix} 3-\sigma & 1 \\ -1 & 1-\sigma \end{vmatrix} = \sigma^2 - 4\sigma + 4 = 0,$$

and therefore $\sigma_1 = \sigma_2 = 2$. Because

$$\operatorname{rank}(P - \sigma I) = \operatorname{rank}\begin{bmatrix} 1 & 1 \\ -1 & -1 \end{bmatrix} = 1,$$

then $J = \begin{bmatrix} 2 & 1 \\ 0 & 2 \end{bmatrix}$. We are going to find $T = \begin{bmatrix} a & b \\ c & d \end{bmatrix}$ such that $P = T^{-1}JT$, or equivalently

$$\begin{bmatrix} a & b \\ c & d \end{bmatrix}\begin{bmatrix} 3 & 1 \\ -1 & 1 \end{bmatrix} = \begin{bmatrix} 2 & 1 \\ 0 & 2 \end{bmatrix}\begin{bmatrix} a & b \\ c & d \end{bmatrix}.$$

The following linear equations

$$\begin{aligned} 3a - b &= 2a + c, \; a + b = 2b + d, \\ 3c - d &= 2c, \qquad c + d = 2d \end{aligned}$$

being equivalent to two equations

$$a - b - c = 0, \; c - d = 0$$

yield the solution $a = 3, b = 2, c = 1, d = 1$. Therefore, we define

$$T = \begin{bmatrix} 3 & 2 \\ 1 & 1 \end{bmatrix}, \; T^{-1} = \begin{bmatrix} 1 & -2 \\ -1 & 3 \end{bmatrix}.$$

One may verify that

$$P = T^{-1} J T$$

or

$$\begin{bmatrix} 3 & 1 \\ -1 & 1 \end{bmatrix} = \begin{bmatrix} 1 & -2 \\ -1 & 3 \end{bmatrix} \begin{bmatrix} 2 & 1 \\ 0 & 2 \end{bmatrix} \begin{bmatrix} 3 & 2 \\ 1 & 1 \end{bmatrix}.$$

Taking into account that

$$e^{Jt} = e^{2t} \begin{bmatrix} 1 & t \\ 0 & 1 \end{bmatrix},$$

we have

$$e^{Pt} = \begin{bmatrix} 1 & -2 \\ -1 & 3 \end{bmatrix} \begin{bmatrix} e^{2t} & te^{2t} \\ 0 & e^{2t} \end{bmatrix} \begin{bmatrix} 3 & 2 \\ 1 & 1 \end{bmatrix} = \begin{bmatrix} (t+1)e^{2t} & te^{2t} \\ -te^{2t} & (1-t)e^{2t} \end{bmatrix}$$

□

Example 4.11. Find a solution to the following ODEs

$$\begin{bmatrix} \dot{x}_1 \\ \dot{x}_2 \\ \dot{x}_3 \end{bmatrix} = \begin{bmatrix} 1 & 1 & -1 \\ -1 & 2 & -1 \\ 2 & -1 & 4 \end{bmatrix} \begin{bmatrix} x_1 \\ x_2 \\ x_3 \end{bmatrix},$$

where $[x_1(0), x_2(0), x_3(0)]^T = [0, 1, 0]^T$.

Eigenvalues of the matrix P follow

$$\begin{vmatrix} 1-\sigma & 1 & -1 \\ -1 & 2-\sigma & -1 \\ 2 & -1 & 4-\sigma \end{vmatrix} = (\sigma - 2)^2(\sigma - 3) = 0,$$

and hence $\sigma_1 = \sigma_2 = 2, \sigma_3 = 3$. Observe that

$$\operatorname{rank}(P - \sigma_1 I) = \operatorname{rank}\begin{bmatrix} -1 & 1 & -1 \\ -1 & 0 & -1 \\ 2 & -1 & 2 \end{bmatrix} = 2.$$

Therefore the Jordan form corresponding to P is

$$J = \begin{bmatrix} 2 & 1 & 0 \\ 0 & 2 & 0 \\ 0 & 0 & 3 \end{bmatrix}.$$

We are going to find a matrix T defined by the formula $P = T^{-1}JT$:

$$\begin{bmatrix} t_{11} & t_{12} & t_{13} \\ t_{21} & t_{22} & t_{23} \\ t_{31} & t_{32} & t_{33} \end{bmatrix}\begin{bmatrix} 1 & 1 & -1 \\ -1 & 2 & -1 \\ 2 & -1 & 4 \end{bmatrix} = \begin{bmatrix} 2 & 1 & 0 \\ 0 & 2 & 0 \\ 0 & 0 & 3 \end{bmatrix}\begin{bmatrix} t_{11} & t_{12} & t_{13} \\ t_{21} & t_{22} & t_{23} \\ t_{31} & t_{32} & t_{33} \end{bmatrix}.$$

Resulting linear algebraic equations yield: $t_{11} = 3, t_{12} = t_{32} = 2, t_{21} = t_{22} = t_{23} = t_{31} = t_{33} = 1$, and hence

$$T = \begin{bmatrix} 3 & 0 & 2 \\ 1 & 1 & 1 \\ 1 & 0 & 1 \end{bmatrix}, \; T^{-1} = \begin{bmatrix} 1 & 0 & -2 \\ 0 & 1 & -1 \\ -1 & 0 & 3 \end{bmatrix}.$$

Because

$$e^{Jt} = \begin{bmatrix} e^{2t} & te^{2t} & 0 \\ 0 & e^{2t} & 0 \\ 0 & 0 & e^{3t} \end{bmatrix},$$

therefore

$$e^{Pt} = \begin{bmatrix} 1 & 0 & -2 \\ 0 & 1 & -1 \\ -1 & 0 & 3 \end{bmatrix}\begin{bmatrix} e^{2t} & te^{2t} & 0 \\ 0 & e^{2t} & 0 \\ 0 & 0 & e^{3t} \end{bmatrix}\begin{bmatrix} 3 & 0 & 2 \\ 1 & 1 & 1 \\ 1 & 0 & 1 \end{bmatrix}$$

$$= \begin{bmatrix} e^{2t}(3+t) - 2e^{3t} & te^{2t} & e^{2t}(2+t) - 2e^{3t} \\ e^{2t} - e^{3t} & e^{2t} & e^{2t} - e^{3t} \\ -(3+t)e^{2t} + 3e^{3t} & -te^{2t} & -e^{2t}(2+t) + 3e^{3t} \end{bmatrix}.$$

Finally, the sought solution has the following form

$$x(t) = e^{Pt}[0, 1, 0]^T = [te^{2t}, e^2, -te^{2t}]^T. \qquad \square$$

4.4.5 Non-homogeneous Differential Equations

As it has been shown earlier, any solution to Eq. (4.67) consists of two parts: a general solution to the homogeneous equation and a particular solution to the non-homogeneous equation, i.e.

$$y(t) = y_g(t) + y_p(t). \tag{4.133}$$

Assume that

$$y_p(t_0) = 0, \quad y_g(t_0) = y(t_0). \tag{4.134}$$

Let us find $u(t)$ which satisfies the equation

$$y_p(t) = \Phi(t, t_0)u(t). \tag{4.135}$$

Differentiating the above equation gives

$$\dot{y}_p = \dot{\Phi}u + \Phi(t, t_0)\dot{u}, \tag{4.136}$$

and from (4.67) one obtains

$$\dot{\Phi}u + \Phi\dot{u} = A\Phi u + F. \tag{4.137}$$

Note that $\dot{\Phi} = A\Phi$ and hence

$$\dot{u}(t) = \Phi^{-1}(t, t_0)F(t). \tag{4.138}$$

An integration of (4.138) gives

$$u(t) = u(t_0) + \int_{t_0}^{t} \Phi^{-1}(\tau, t_0)F(\tau)d\tau. \tag{4.139}$$

From (4.115), (4.134), (4.135) and (4.139) one obtains $u(t_0) = 0$. Taking into account (4.135) and (4.139) one gets

$$\begin{aligned} y_p(t) &= \Phi(t, t_0) \int_{t_0}^{t} \Phi^{-1}(\tau, t_0)F(\tau)d\tau \\ &= \int_{t_0}^{t} \Phi(t, t_0)\Phi(t_0, \tau)F(\tau)d\tau = \int_{t_0}^{t} \Phi(t, \tau)F(\tau)d\tau. \end{aligned} \tag{4.140}$$

Finally, keeping in mind (4.133), we obtain

$$y(t) = \Phi(t, t_0)y_0 + \int_{t_0}^{t} \Phi(t, \tau)F(\tau)d\tau. \tag{4.141}$$

Observe that the obtained solution can be expressed analytically only in particular cases. In general, it is found numerically.

4.4.6 Examples of Linear Non-homogenous Differential Equations

The system of Eq. (4.67) can be presented in the following first-order form

$$\frac{dy_i}{dt} = p_{i1}(t)y_1 + p_{i2}(t)y_2 + \cdots + p_{in}(t)y_n + F_i(t), \quad i = 1, \ldots, n. \tag{4.142}$$

As it has been already mentioned there are two main approaches aimed on solving ODEs (4.142) in getting an equation of a higher order: (i) method of integrable combinations and (ii) method of variation of arbitrary constants.

The Lagrange method (ii) can be applied if the associated homogeneous problem can be solved. In other words if

$$y = Y(t)c, \tag{4.143}$$

where $Y(t)$ is the fundamental matrix of solutions, we are looking to solve the original non-homogenous problem in the form

$$y(t) = Y(t)c(t), \tag{4.144}$$

where now the vector c is time dependent. Elements of the vector $c(t)$ are defined by solving the following system of equation

$$Y(t)\frac{dc}{dt} = F(t). \tag{4.145}$$

A general solution to the original non-homogeneous problem consists of a sum of a general solution to the associated homogenous problem and an arbitrary one solution of the non-homogeneous problem.

Example 4.12. Find a solution to the following ODEs

$$\frac{dx}{dt} = y - 5\cos t,$$
$$\frac{dy}{dy} = 2x + y.$$

We find first a general solution to the associated homogeneous equation (see Sect. 4.4.4)

$$x_g = C_1e^{-t} + C_2e^{2t},$$
$$y_g = -C_1e^{-t} + 2C_2e^{2t}.$$

A particular solution of the non-homogeneous problem is sought in the form

$$x = A_1 \sin t + B_1 \cos t,$$
$$y = A_2 \sin t + B_2 \cos t. \qquad (*)$$

Substituting $(*)$ to the initial non-homogeneous set of equations we get

$$A_1 - B_2 = -5, \qquad B_1 + A_2 = 0,$$
$$A_2 - 2B_1 - B_2 = 0,\ 2A_1 + B_2 + A_2 = 0.$$

Obtained linear algebraic equations are satisfied for $A_1 = -2$, $B_1 = -1$, $A_2 = 1$, $B_2 = 3$. Therefore, the general solution of the problem is

$$x = C_1 e^{-t} + C_2 e^{2t} - 2\sin t - \cos t,$$
$$y = -C_1 e^{-t} + 2C_2 e^{2t} + \sin t + 3\cos t.$$

□

Example 4.13. Find a solution to the following ODEs

$$\frac{dx}{dt} = 2x + y + 2e^t,$$
$$\frac{dy}{dt} = x + 2y - 3e^{4t}.$$

The characteristic equation of the associated homogeneous problem

$$\begin{vmatrix} 2-\sigma & 1 \\ 1 & 2\sigma \end{vmatrix} = \sigma^2 - 4\sigma + 3 = 0$$

yields the eigenvalues $\sigma_1 = 1$, $\sigma_2 = 3$. Therefore, a general solution of the homogeneous problem is

$$x_g = C_1 e^t + C_2 e^{3t},$$
$$y_g = -C_1 e^t + C_2 e^{3t}.$$

Since $\sigma = 1$ corresponds to the excitation e^t, therefore we are looking for a particular solution of the non-homogeneous problem of the following form

$$x_p = (A_1 + A_2 t)e^t + A_3 e^{4t}, \quad y_p = (B_1 + B_2 t)e^t + B_3 e^{4t}.$$

Substituting x_p, y_p into initial ODEs the following algebraic equations are obtained:

$$-A_1 + A_2 - B_1 = 2, \quad A_1 + B_1 - B_2 = 0, \quad A_2 + B_2 = 0,$$
$$2A_3 - B_3 = 0, \quad A_3 - 2B_3 = 3.$$

The algebraic equations yield the solution $A_1 = 0$, $A_2 = 1$, $A_3 = -1$, $B_1 = -1$, $B_2 = -1$, $B_3 = -2$. Finally, the general solution to the non-homogeneous system is

$$x = C_1 e^t + C_2 e^{3t} + te^t - e^{4t},$$
$$y = -C_1 e^t + C_2 e^{3t} - (t+1)e^t - 2e^{4t}. \qquad \square$$

4.4.7 Homogeneous Differential Equations with Periodic Coefficients

The homogeneous differential equations with periodic coefficients play an important role in many applications. Consider a system of Eq. (4.98) of the form

$$\dot{y}(t) = P(t)y, \tag{4.146}$$

where $P(t) = P(t+T)$. According to our earlier consideration we get

$$\begin{aligned} x(t) &= \Phi(t, t_0)x_0, \\ \dot{\Phi}(t, t_0) &= P(t)\Phi(t, t_0). \end{aligned} \tag{4.147}$$

Since $P(t+T) = P(t)$, hence we get

$$\dot{\Phi}(t+T, t_0) = P(t)\Phi(t+T, t_0). \tag{4.148}$$

It means that $\Phi(t+T, t_0)$ is a fundamental matrix, and therefore

$$\Phi(t+T, t_0) = \Phi(t, t_0)\Phi_*, \tag{4.149}$$

where Φ_* is constant and non-singular matrix. For $t = t_0$ one gets

$$\Phi(t_0+T, t_0) = \Phi(t_0, t_0)\Phi_* = I\Phi_*. \tag{4.150}$$

The matrix Φ_* is called the monodromy matrix. Observe that

$$\begin{aligned} &\Phi(t+T, t_0) = \Phi(t, t_0)\Phi(t_0+T, t_0) = \Phi(t, t_0)\Phi_*, \\ &\Phi(t+2T, t_0) = \Phi((t+T)+T, t_0) = \Phi(t+T, t_0)\Phi_* = \Phi(t, t_0)\Phi_*^2, \\ &\vdots \\ &\Phi(t+nT, t_0) = \Phi(t, t_0)\Phi_*^n, \end{aligned} \tag{4.151}$$

and hence

$$x(t+nT) = \Phi(t+nT, t_0)x(t_0) = \Phi(t, t_0)\Phi_*^n x(t_0). \tag{4.152}$$

The obtained result suggests that a knowledge of the matricant in the time interval $t_0 \le t \le t_0 + T$ is sufficient to find a solution in the time interval $t_0 \le t < +\infty$.

4.4.8 The Floquet Theory

Assume that a $K \times K$-dimensional matrix $\bar{F}$ is defined in the following way

$$\Phi_* = \Phi(t_0 + T, t_0) = e^{T\bar{F}}. \tag{4.153}$$

Observe that in general $\bar{F}$ is complex. Let us define now the following matrix

$$\Psi(t) = \Phi(t, t_0)e^{-(t-t_0)\bar{F}}. \tag{4.154}$$

It can be proved that

$$e^{(t_1+t_2)\bar{F}} = e^{t_1\bar{F}}e^{t_2\bar{F}} = e^{t_2\bar{F}}e^{t_1\bar{F}}. \tag{4.155}$$

Now we are going to show that $\Psi(t) = \Psi(t+T)$. From (4.154) one gets

$$\Psi(t+T) = \Phi(t+T, t_0)e^{-(t-t_0+T)\bar{F}}. \tag{4.156}$$

Hence, from $(4.151)_1$, (4.153), (4.154) and (4.156) we get

$$\Psi(t+T) = \Phi(t, t_0)e^{T\bar{F}}e^{-T\bar{F}}e^{-(t-t_0)\bar{F}} = \Phi(t, t_0)e^{-(t-t_0)\bar{F}} = \Psi(t). \tag{4.157}$$

Since $\det \Phi(t, t_0) \ne 0$ and $\det(e^{-(t-t_0)\bar{F}}) \ne 0$, hence

$$\det \Psi(t) = \det \Phi(t, t_0) \det e^{-(t-t_0)\bar{F}} \ne 0. \tag{4.158}$$

From (4.154) we also get

$$\Psi(t_0) = \Phi(t_0, t_0)I = I. \tag{4.159}$$

Finally, from (4.154) we obtain

$$\Phi(t, t_0) = \Psi(t)e^{(t-t_0)\bar{F}}. \tag{4.160}$$

It means that the matricant can be expressed in the form of (4.160), where

$$\Psi_{K\times K} = \Psi_{K\times K}(t+T) \tag{4.161}$$

is regular ($\Psi(t_0) = I$), and $\bar{F}_{K\times K}$ is a constant matrix.

According to (4.160), having any constant and regular matrix C one can construct the fundamental matrix of solutions

$$\Gamma(t,t_0) = \Phi(t,t_0)C, \tag{4.162}$$

and hence

$$\Gamma(t,t_0) = \Psi(t)e^{(t-t_0)\bar{F}}C. \tag{4.163}$$

Recall that for any square matrices $\bar{F}$ and C we have

$$e^{C^{-1}\bar{F}C} = C^{-1}e^{\bar{F}}C \tag{4.164}$$

and hence (4.163) can be transformed to the form

$$\Gamma(t,t_0) = \Psi(t)CC^{-1}e^{(t-t_0)\bar{F}}C = \Psi(t)Ce^{(t-t_0)C^{-1}\bar{F}C}. \tag{4.165}$$

Now, introducing new matrices

$$\begin{aligned} \Psi_* &= \Psi(t)C, \\ F_* &= C^{-1}\bar{F}C, \end{aligned} \tag{4.166}$$

we get

$$\Gamma(t,t_0) = \Psi_*(t)e^{(t-t_0)F_*}. \tag{4.167}$$

It is seen that $\Gamma(t,t_0)$ is similar to $\Phi(t,t_0)$ (see (4.160)) and in addition

$$\Psi_*(t+T) = \Psi_*(t). \tag{4.168}$$

Since both matrices C and $\Psi(t)$ are regular, hence from (4.166) we get

$$\det\Psi_* = \det\Psi\det C \neq 0, \tag{4.169}$$

and one can conclude that also $\Psi_*(t)$ is regular.

4.4.9 Reduction of Non-homogeneous Linear Differential Equations with Periodic Coefficients

We are going to show that any non-homogeneous dynamical system

$$\begin{aligned} \dot{y} &= P(t)y + F(t), \\ P(t) &= P(t+T), \end{aligned} \tag{4.170}$$

can be reduced to a system with constant matrix coefficients. Hence, a stability problem of (4.170) can be reduced to that of the earlier discussed for singular points.

Let

$$y(t) = Z(t)x(t), \tag{4.171}$$

where $Z(t)$ is a regular matrix. Hence, from (4.170) and (4.171) one gets

$$\dot{Z}x + Z\dot{x} = PZx + F, \tag{4.172}$$

which yields

$$\dot{x} = Z^{-1}(PZ - \dot{Z})x + Z^{-1}F. \tag{4.173}$$

Now we prove that $Z^{-1}(PZ - \dot{Z})$ is the matrix which constant coefficients, where

$$Z(t) = \Phi(t, t_0)e^{-(t-t_0)\bar{F}}. \tag{4.174}$$

Observe that

$$\begin{aligned} \frac{d}{dt}\left(e^{-(t-t_0)\bar{F}}\right) &= \frac{d}{dt}\left(\sum_{n=0}^{\infty}\frac{(t_0-t)^n}{n!}\bar{F}^n\right) = -\sum_{n=0}^{\infty}\frac{n(t_0-t)^{n-1}}{n!}\bar{F}^{n-1}\bar{F} \\ &= -\left(\sum_{n=0}^{\infty}\frac{(t_0-t)^{n-1}}{(n-1)!}\bar{F}^{n-1}\right)\bar{F} = -\left(\sum_{k=0}^{\infty}\frac{(t_0-t)^k}{k!}\bar{F}^k\right)\bar{F} = -e^{-(t-t_0)\bar{F}}\bar{F}. \end{aligned} \tag{4.175}$$

It means that

$$\frac{d}{dt}\left(e^{-(t-t_0)\bar{F}}\right) = -e^{-(t-t_0)\bar{F}}\bar{F} = -\bar{F}e^{-(t-t_0)\bar{F}}. \tag{4.176}$$

Hence the investigated matrix (see (4.173)) has the form

$$\begin{aligned} Z^{-1}(PZ - \dot{Z}) = \Phi^{-1}(t,t_0)e^{(t-t_0)\bar{F}}\Big(&P\Phi(t,t_0)e^{-(t-t_0)\bar{F}} - \dot{\Phi}(t,t_0)e^{-(t-t_0)\bar{F}} + \\ &+ \Phi(t,t_0)\bar{F}e^{-(t-t_0)\bar{F}}\Big) = \Phi^{-1}\Phi\bar{F} = \bar{F}. \end{aligned} \tag{4.177}$$

Taking into account the obtained results Eq. (4.173) has the following form

$$\dot{x} = \bar{F}x + \Phi^{-1}(t, t_0)e^{(t-t_0)\bar{F}}\bar{F}. \tag{4.178}$$

To sum up, it has been shown that a change of variables (4.171), where $Z(t)$ has the form (4.174), leads to reduction of the non-homogeneous set of differential equations with periodic coefficients (4.173) to the non-homogeneous set of differential equations (4.178) with constant coefficients.

4.4.10 Characteristic Multipliers

The eigenvalues μ of a monodromy matrix Φ_* are called the characteristic multipliers, which are found from the equation

$$\Phi_* a = \mu a, \tag{4.179}$$

where a is the vector associated with an eigenvalue. The characteristic equation

$$R(\mu) = \det(\Phi_* - \mu I) = 0 \tag{4.180}$$

has K roots. Each of the eigenvalues μ_k is associated with an eigenvector a_k

$$\Phi_* a_k = \mu_k a_k. \tag{4.181}$$

Observe that the eigenvectors a_k corresponding to the distinct μ_k are linearly independent. In a case of multiple root μ_k, where μ_k denotes its multiplicity, a solution is more complicated. One has to check in addition a matrix $(\Phi_* - \mu_k I)$ defect d_k, where

$$d_k = k - r(\Phi_* - \mu_k I), \tag{4.182}$$

where

$$1 \le d_k \le \nu_k, \quad k = 1, 2, \ldots, k^*, \; k_* \le k, \tag{4.183}$$

and k^* is the number of distinct multiplicators.

If μ_k has a multiplicity ν_k, then there exist d_k linearly independent eigenvectors. If $d_k = \nu_k$, then μ_k is associated with ν_k independent vectors.

Let us take a particular solution $y_k(t)$ of a homogeneous equation with the initial conditions $y_k(t_0) = Ca_k$, where C is a certain constant. Then

$$y_k(t) = \Phi(t, t_0)y_k(t_0) = C\Phi(t, t_0)a_k, \tag{4.184}$$

and taking into account (4.152) one obtains

$$y_k(t+T) = C\Phi(t,t_0)\Phi_* a_k. \tag{4.185}$$

Taking into account (4.181) one gets from (4.185)

$$y_k(t+T) = \mu_k C\Phi(t,t_0)a_k = \mu_k y_k(t). \tag{4.186}$$

The latter result has the following interpretation. A solution $y_k(t+T)$ shifted in time of the period T can be obtained as a product of μ_k and $y_k(t)$. Therefore, the eigenvalues μ_k are called *the characteristic multipliers*.

4.4.11 Characteristic Exponents

The characteristic exponents λ_k are the eigenvalues of the matrix F:

$$\det(\bar{F} - \lambda I) = 0. \tag{4.187}$$

They are associated with the multipliers by the relations

$$\mu_k = \exp(\lambda_k T), \tag{4.188}$$

or

$$\lambda_k = \frac{1}{T}\ln \mu_k. \tag{4.189}$$

Note that any one value of μ_k corresponds to infinitely many characteristic exponents $\lambda_k = \lambda_{ko} + im\frac{2\pi}{T}$, where m is integer number, and $i^2 = -1$. It can be also proven that any multiple multiplier corresponds to a characteristic exponent with the same multiplicity.

4.4.12 Structure of Solutions for Simple Characteristic Multipliers

Let the matrix A_* contains the eigenvectors a_k

$$A_* = (a_1, \ldots, a_k). \tag{4.190}$$

Owing to (4.181) we get

$$\phi_* A_* = A_* \operatorname{diag}(\mu_i), \tag{4.191}$$

and hence

$$\phi_* = A_* \operatorname{diag}(\mu_i) A_*^{-1}. \tag{4.192}$$

From (4.152) one obtains

$$y(t + nT) = \psi(t, t_0) A_* \operatorname{diag}(\mu_i)^n A_*^{-1} y_0. \tag{4.193}$$

A similar like solution structure is obtained also for the case $d_k = \nu_k$, since in this case there are also k linearly independent eigenvectors.

4.4.13 Solutions Structure for Case of Multiple Multipliers

We consider here only the case, where μ_k is associated with a number of independent vectors.

Note that any characteristic multiplier μ_j of multiplicity Θ_j is associated with $d_j < \upsilon_j$ linearly independent eigenvectors $a_{j1}, \ldots, a_{jd_j}$. One needs to increase a number $\upsilon_j - d_j$ of independent vectors.

Recall that from the matrix theory one has

$$(\phi_* - \mu_j I) a_{j\nu}^{(l)} = a_{j\nu}^{(l-1)}, \tag{4.194}$$

where:

$$a_{j\nu}^{(l)} = a_{j\nu},\ l = 2, 3, \ldots, J_{\nu\nu},\ \sum_{\nu=1}^{d_j} (J_{j\nu} - 1) = \nu_j - d_j. \tag{4.195}$$

In the above $J_{j\nu} - 1$ denoted the number of main vectors $a_{j\nu}^{(l)}$ corresponding to each of the eigenvectors $a_{j\nu}$, whereas $J_{j\nu}$ denotes the largest number of nonlinearly independent vectors. Obtained vectors can be used to construct a regular matrix

$$\tilde{A} = (\tilde{A}_1, \tilde{A}_2, \ldots, \tilde{A}_{K*}), \tag{4.196}$$

where the block matrices dimensions $(k \times \nu_j)$ possess the following structure

$$\tilde{A}_j = (a_{j1}^{(1)}, a_{j1}^{(2)}, \ldots, a_{j1}^{(J_{j1})}, \ldots, a_{jd_j}^{(1)}, \ldots, a_{jd_j}^{(J_{id_j})}), \tag{4.197}$$

where: $j = 1, 2, \ldots, K^*$, and K^* is the number of different multipliers. The matrix ϕ_* in this case has the following structure

$$\phi_* = \tilde{A} J \tilde{A}^{-1}, \tag{4.198}$$

and the matrix J is a quasi-diagonal Jordan matrix

$$J = \mathrm{diag}(\ldots, J_{j1}, \ldots, J_{jd_j}, \ldots), \tag{4.199}$$

where the blocks have the form

$$J_{jv} = \begin{bmatrix} \mu_j & 1 & 0 \ldots & 0 \\ 0 & \mu_j & 1 \ldots & 0 \\ \vdots & \vdots & \vdots & \\ 0 & 0 & 0 \ldots & \mu_j \end{bmatrix}. \tag{4.200}$$

In the case of simple eigenvalue the matrix blocks are simplified to one element μ_j.

One gets form (4.198)

$$\phi_*^n = \tilde{A} J^n \tilde{A}^{-1}, \tag{4.201}$$

where:

$$J^n = \mathrm{diag}(\ldots, J_{j1}^n, \ldots, J_{jd_j}^n, \ldots) \tag{4.202}$$

Hence, similarly to (4.193), one gets

$$y(t + nT) = \phi(t, t_0)\tilde{A}\,\mathrm{diag}(\ldots, J_{j1}^n, \ldots, J_j^n d_j, \ldots)\tilde{A}^{-1} y_0. \tag{4.203}$$

Taking into account (4.200) we get

$$J_{jv} = \mu_j I + K_{jv}, \tag{4.204}$$

where:

$$K_{jv} = \begin{bmatrix} 0 & 1 & 0 \ldots & 0 \\ 0 & 0 & 1 \ldots & 0 \\ \vdots & \vdots & \vdots & \\ 0 & 0 & 0 \ldots & 1 \\ 0 & 0 & 0 \ldots & 0 \end{bmatrix}, \tag{4.205}$$

is the matrix of dimension $J_{jv} \times J_{jv}$.

It is worthless to note that the matrix K_{jv}^n is obtained by a shift of a line consisting of n steps towards right.

Since

$$K_{jv}^n = 0 \text{ for } n \geqslant J_{iv}, \tag{4.206}$$

then

$$J_{j\nu}^{n} = (\mu_j I + K_{j\nu})^n = \mu_j^n I + \sum_{\sigma=1}^{J_{j\nu}-1} \binom{n}{s} \mu_j^{n-\sigma} K_{j\nu}^{\sigma}. \tag{4.207}$$

Expanding the formula (4.207) we get

$$\begin{aligned} J_{j\nu}^{n} &= \mu_j^n I + n\mu_j^{-1} K_{j\nu} + \tfrac{n(n-1)}{2!}\mu_j^{-2} K_{j\nu}^2 + \cdots \\ &+ \tfrac{n(n-1)\ldots(n-J_{j\nu}+1)}{(J_{j\nu}-1)!}\mu_j^{1-J_{j\nu}} K_{j\nu}^{J_{j\nu}-1}. \end{aligned} \tag{4.208}$$

Observe that for $n \to 0$ we have $J_{j\nu}^{n} \to 0$, if $|\mu_j| < 1$. For $|\mu_j| \geqslant 1$ the values of $J_{j\nu}^{n} \to \infty$.

4.4.14 Stability

The following stability conditions can be formulated for the solutions to homogeneous ODEs with periodic coefficients.

(i) If each of the multipliers satisfies the inequality $|\mu_i| < 1$ (or, equivalently, $\operatorname{Re} x_j < 0$), then all solutions tend to zero and hence they are asymptotically stable.
(ii) If there exists at least one of the multipliers $|\mu_i| > 1$ (or, equivalently, the characteristic exponent with $\operatorname{Re} \lambda_i > 0$), then the considered dynamical system is unstable.
(iii) Assume that there are no multipliers $|\mu_j| > 1$. If there are $|\mu_j| = 1$, which are simple or multiple but with a full defect, then the considered system is stable, but not asymptotically stable. If there are multiple multipliers without a full defect, then the considered system is unstable.

4.4.15 Periodic Solutions to Homogenous Differential Equations with Periodic Coefficients

Recall that periodic solutions can appear in the case if only simple or multiple with a full defect multiplier satisfying $|\mu_j| = 1$ exists. All other multipliers satisfy the inequality $|\mu_j| > 1$.

According to (4.186) one value $\mu_k = 1$ corresponds to infinitely many T periodic solutions $y_k(t_0) = Ca_k$, where a_k is the eigenvector associated with μ_k and C is a certain arbitrary constant.

In the case $\mu_k = -1$ we have infinitely many 2π-periodic solutions, because

$$\begin{aligned} y(t+T) &= -y(t), \\ y(t+2\pi) = -y(t+2\pi) &= y(t), \end{aligned} \tag{4.209}$$

where $T = 2\pi$.

In the general case, each of two complex conjugate multipliers of the form

$$\mu = \cos\frac{2\pi}{\nu} \pm i\sin\frac{2\pi}{\nu} \tag{4.210}$$

is associated with periodic oscillations with the period νT, where $T = \frac{2\pi}{\nu}$ is an integer, according to (4.152) we get

$$y(t+\nu T) = \psi(t, t_0)\psi_*^{\nu} y(t_0), \tag{4.211}$$

and taking into account periodicity conditions one gets

$$(\psi_*^{\nu} - I)y(t_0) = 0, \tag{4.212}$$

which defines the initial conditions required for finding the periodic solution.

4.4.16 Periodic Solutions to Non-homogenous Differential Equations with Periodic Coefficients

Consider the system (4.67) where $P(t) = P(t+T)$, $F(t) = F(t+\frac{p}{q}T)$, and p, q are integers.

Let T be the smallest period of $P(t)$ and $F(t)$ is equal to pT. The fundamental question appears. Does the system (4.67) have also a solution with the period pT?

Assume that it is true. Hence,

$$y(t_0 + pT) = y(t_0). \tag{4.213}$$

According to (4.141) we havev

$$y(t) = \psi(t, t_0)y_0 + \int_{t_0}^{t} \psi(t, \tau)F(\tau)d\tau, \tag{4.214}$$

and hence

$$(I - \psi(t_0 + pT, t_0))y_0 = \int_{t_0}^{t_0+pT} \psi(t_0 + pT, \tau)d\tau. \tag{4.215}$$

On the other hand

$$\psi(t_0 + pT, t_0) = \psi_*^p. \tag{4.216}$$

Since we are going to find y_0 from (4.215), then the matrix $(I - \psi(t_0 + pT, t_0))$ must be regular one. It means that in this case the associated homogeneous differential equation

$$\dot{y} = P(t)y, \tag{4.217}$$

with $P(t) = P(t + pT)$ should not have the multiplier $\mu = 1$, i.e. it should not have a periodic solution with the period pT. If the later requirement is true, then

$$y_0 = (I - \psi_*^p)^{-1} \int_{t_0}^{t_0+pT} \psi(t_0 + pT, \tau)F(\tau)d\tau \tag{4.218}$$

Now, from (4.214) we find that

$$y(t) = \psi(t, t_0)(I - \psi_*^p)^{-1} \int_{t_0}^{t_0+pT} \psi(t_0 + pT, \tau)F(\tau)d\tau + \int_{t_0}^{t} \psi(\tau, \tau)F(\tau)d\tau. \tag{4.219}$$

Chapter 5
Higher-Order ODEs Polynomial Form

5.1 Introduction

If a function $f(t, x, \dot{x}, \ldots, x^{(n)})$ is defined and is continuous in a subset of $\mathbb{R}^{n+2} (n \geq 1)$, then the equation

$$f(t, x, \dot{x}, \ldots, x^{(n)}) = 0 \tag{5.1}$$

is said to be *ordinary differential equation of nth-order*.

Fortunately, it happens very often that (5.1) can be transformed to the form

$$x^{(n)} = f_1(t, x, \dot{x}, \ldots, x^{(n-1)}). \tag{5.2}$$

The Cauchy problem for Eq. (5.2) is that of finding a solution $x(t)$ satisfying the series of the following initial conditions

$$x(t_0) = x_0, \quad \dot{x}(t_0) = \dot{x}_0, \ldots, x^{(n-1)}(t_0) = x_0^{(n-1)}. \tag{5.3}$$

Theorem 5.1 (Peano). *If the function f_1 is continuous in certain open subset, then for an arbitrary point $(t_0, x_0, \dot{x}_0, \ldots, x_0^{(n-1)})$ belonging to this subset there is a solution to Eq. (5.2) defined in a neighbourhood of t_0, which satisfies (5.3).*

Theorem 5.2 (Cauchy–Picard). *If the function f_1 satisfies both the conditions of Theorem 5.1 and the Lipschitz conditions with regard to the variables $x, \dot{x}, \ldots, x^{(n-1)}$, then for arbitrary initial conditions (5.3) there is only one solution (uniqueness) to Eq. (5.2).*

J. Awrejcewicz, *Ordinary Differential Equations and Mechanical Systems*,
DOI 10.1007/978-3-319-07659-1_5

Recall that the Lipschitz constant L for the first-order case can be found from the inequality

$$\left|\frac{f(t,x_1)-f(t,x_2)}{x_1-x_2}\right| = \left|\frac{\partial f}{\partial x}(x)\right| \leq L. \tag{5.4}$$

In words the Lipschitz constant can be defined by an upper bound of $\frac{\partial f}{\partial x}$. In the nth-order case, when $f = f(t, x_1, \ldots, x_n)$, the Lipschitz constant L can be defined by the inequality

$$\text{Max}\left\{\left|\frac{\partial f}{\partial x_1}\right|, \ldots, \left|\frac{\partial f}{\partial x_n}\right|\right\} \leq L. \tag{5.5}$$

To show a uniqueness of a solution let us recall the example of the equation $\frac{\partial f}{\partial x} = 1+x^2$ given in the book [191]. In this case $f(t,x_1) = 1+x_1^2$ and $f(t,x_2) = 1+x_2^2$. Hence $|f(t,x_1) - f(t,x_2)| = \left|x_1^2 - x_2^2\right| = |x_1 - x_2|\,|x_1 + x_2| \leq 2\,|x_1 - x_2|$ in the rectangle $|t| < 1, |x| < 1$. Theorem 5.2 states that there exists only one solution passing through (0,0). In fact, this solution has the following analytical form: $x(t) = \text{tg}(t)$. Let D be a domain composed of points, where any point corresponds to only one solution of the Cauchy problem. The function

$$x = \phi(t, C_1, \ldots, C_n) \tag{5.6}$$

is said to be a *general solution* to Eq. (5.2), if the following assumptions are satisfied:

(i) the function ϕ has the nth-order derivative with respect to t;
(ii) for any point defined by the series (5.3) the following equations

$$\begin{aligned}
x_0 &= \phi(t, C_1, \ldots, C_n),\\
\dot{x}_0 &= \dot{\phi}(t, C_1, \ldots, C_n),\\
&\vdots\\
x_0^{(n-1)} &= \phi^{(n-1)}(t, C_1, \ldots, C_n),
\end{aligned} \tag{5.7}$$

have the unique solutions with regard to the constants $C_1^0, \ldots, C_n^0$;
(iii) the function $\phi(t, C_1^0, \ldots, C_n^0)$ is the solution to (5.2) for arbitrary constants $C_1^0, \ldots, C_n^0$, which are the solutions to (5.7).

If a general solution is given in the implicit way

$$\phi_0(t, x, C_1, \ldots C_n) = 0, \tag{5.8}$$

then Eq. (5.8) defines a general solution to Eq. (5.2).

Any arbitrary solution (5.6) for the specified values of constants $C_1, \ldots, C_n$ is said to be the *particular solution* of Eq. (5.2). The following steps are required in order to find a particular solution knowing a general one:

(i) the algebraic set of Eq. (5.7) should be derived from either (5.6) or (5.8), which yields the constants $C_1, \ldots, C_n$;
(ii) the found specified values $C_1^0, \ldots, C_n^0$ are substituted to (5.6) or (5.8), which are now solutions to a Cauchy problem.

Note that sometimes a general solution can be represented in a parametric form

$$\begin{aligned} t &= t(p, C_1, \ldots, C_n), \\ x &= x(p, C_1, \ldots, C_n), \end{aligned} \tag{5.9}$$

where p is a parameter. In the most general case governed by Eq. (5.1), the following theorem satisfies an existence and uniqueness of a Cauchy problem.

Theorem 5.3. *Assume that the function defined by (5.1) is continuous and posses-ses the continuous derivatives with regard to $x, \dot{x}, \ldots, x^{(n)}$. Hence, for an arbitrary point $(t_0, x_0, \dot{x}_0, \ldots, x_0^{(n)})$ such that*

$$f(t_0, x_0, \dot{x}_0, \ldots, x_0^{(n)}) = 0, \quad \frac{\partial f}{\partial x^{(n)}}(t_0, x_0, \dot{x}_0, \ldots, x_0^{(n)}) \neq 0, \tag{5.10}$$

there is exactly only one solution to Eq. (5.1) defined in the neighbourhood of t_0 and satisfying the initial conditions (5.3).

A general solution constitutes of a family of integral curves in the plane (t, x) with n parameters $C_1, \ldots, C_n$.

5.2 Linear Homogeneous Differential Equations

Consider the following homogeneous nth-order differential equation

$$L_n(y) \equiv y^{(n)} + p_1(t)y^{(n-1)} + \cdots + p_n(t)y = 0, \tag{5.11}$$

where L_n is called the nth-order linear differential operator (here $y^{(n)} = \frac{d^n y}{dt^n}$). Recall that in general, a linear differential operator (a function) has the following properties:

(i) $L(Cy) = CL(y)$,
(ii) $L(y_1 + y_2) = L(y_1) + L(y_2)$,
(iii) $L(C_1 y_1 + \cdots + C_m y_m) = C_1 L(y_1) + \cdots + C_m L(y_m)$, for any y_i and C_i.

Theorem 5.4. *If $y_1, \ldots, y_n$ are the solutions to homogeneous equation (5.11), then $y = C_1 y_1 + \cdots + C_n y_n$ is also a solution to $L_n(y) = 0$ for arbitrary constant numbers $C_1, \ldots, C_n$.*

Remark 5.1. The set of functions $\{y_1(t), \ldots, y_n(t)\}$ is said to be a *fundamental set of solutions* of $L_n(y) = 0$.

Remark 5.2. The function $y(t) = C_1 y_1(t) + \cdots + C_n y_n(t)$ is called the *complementary* function of $L_n(y) = 0$

Definition 5.1 (Linearly Independent Functions). Let $u_1(t), \ldots, u_m(t)$ are the functions defined in the interval $a \le t \le b$. If there are numbers $\alpha_1, \ldots, \alpha_m$, not all of them equal to zero, and the following equation is satisfied

$$z = \alpha_1 u_1 + \cdots + \alpha_m u_m \equiv 0, \qquad a \le t \le b, \tag{5.12}$$

then we say that the functions $u_1(t), \ldots, u_m(t)$ are linearly dependent in the considered interval.

The vector z is called a *linear combination* of the members of the set $\{u_1, \ldots, u_m\}$, whereas the set $\{\alpha_1, \ldots, \alpha_m\}$ is called a *set of coefficients* of the linear combination.

Recall that:

(i) the set of all linear combinations of members of $\{y_1, \ldots, y_m\}$ is called the *linear span* (or *simply span*) of $\{y_1, \ldots, y_m\}$;
(ii) a linear space U is a set that satisfies the following properties:
   (a) if $u, v \in U$ then $u + v \in U$;
   (b) if $u \in U$ and $c \in R$, then $Cu \in U$;
(iii) the span defined in (i) forms a linear space.

Assuming that Eq. (5.12) is satisfied if and only if $\alpha_1 = \cdots = \alpha_m \equiv 0$, then the functions $u_1, \ldots, u_m$ are called *linearly independent*. A linearly independent spanning set for a linear space $U = \{u_1, \ldots, u_m\}$ is called *a basis for* U. A dimension of U is defined by the number of vectors in a basis of vector space U. A necessary condition for linear dependence of the functions $u_1, \ldots, u_n$ can be obtained in the following way. Let the given functions $u_1, \ldots, u_n$ to be are linearly dependent. It means that they satisfy (5.12). Differentiating $n-1$ times Eq. (5.12) one gets

$$\alpha_1 u_1^{(k)} + \cdots + \alpha_n u_n^{(k)} = 0, \quad k = 1, \ldots, n-1. \tag{5.13}$$

Of course, we have assumed that the functions $\{u_1, \ldots, u_n\}$ have $n-1$ continuous derivatives on some interval (they are C^{n-1} smooth). It is known from algebra that the system of Eqs. (5.12) and (5.13) has non-trivial solution of $\alpha_1, \ldots, \alpha_n$, when the following determinant function (called Wronskian) is equal to zero:

$$W(u_1,\dots,u_n) = \begin{vmatrix} u_1 & \cdots & u_n \\ u_1' & \cdots & u_n' \\ \vdots & & \\ u_1^{(n-1)} & \cdots & u_n^{(n-1)} \end{vmatrix} = 0, \quad a \le t \le b. \tag{5.14}$$

Theorem 5.5. *If the functions $y_1,\dots,y_n$ are the solutions of $L_n(y) = 0$ on $a \le t \le b$, then $W(y_1(t),\dots,y_n(t))$ is either zero for every t from [a,b] or is never zero for any $t \in [a,b]$.*

Proof. We take $n = 2$ and follow the proof given in the book [191]. Here we omit a generalization for any n. Consider the following linear operator with variable coefficients

$$L_2(y) \equiv y'' + p_1(t)y' + p_2(t)y = 0.$$

The Wronskian

$$W(y_1,y_2) = \begin{vmatrix} y_1 & y_2 \\ y_1' & y_2' \end{vmatrix} = y_1y_2' - y_2y_1'.$$

On the other hand

$$W' = (y_1y_2' - y_2y_1')' = y_1y''_2 - y''_1y_2 = y_1(-p_1y_2' - p_2y_2) - y_2(-p_1y_1' - p_2y_1)$$
$$= -p_1(y_1y_2' - y_2y_1') = -p_1W,$$

and hence

$$\ln W = -\int p_1 dt + \ln C.$$

Finally, one gets

$$W(y_1,y_2) = Ce^{-\int p_1(t)dt},$$

which for $C = 0$ is identically zero, or for $C \neq 0$ is never zero. □

Theorem 5.6. *If $y_1,\dots,y_n$ are solutions of the equation $L_n(y) = 0$ and $W(y_1,\dots,y_n)|_{t=t_0} = 0$, then $W(y_1,\dots,y_n) \equiv 0$ for $a \le t \le b$, and $y_1,\dots,y_n$ are linearly dependent solutions in the interval $t \in [a,b]$.*

Theorem 5.7. *A necessary and sufficient condition for solutions $z_1,\dots,z_n$ of $L_n(z) = 0$ to be linearly independent in an arbitrary point is $W(z_1,\dots,z_n)|_{t=t_0} \neq 0$. If $W(t_0) \neq 0$, then also $W(t) \neq 0$ for $a \le t \le b$, and if $W(t_0) = 0$, then also $W(t) = 0$.*

Theorem 5.8. *If $y_1, \ldots, y_n$ are linearly independent solutions of the equation $L_n(y) = 0$, then $y = C_1 y_1 + \cdots + C_n y_n$ is a general solution for $a \leq t \leq b$, $-\infty < y_k < +\infty$, $k = 0, 1, \ldots, n-1$.*

In the case of a homogeneous linear nth-order equation $L_n(y) = 0$, there are n linearly independent solutions $y_1, \ldots, y_n$ forming the *fundamental set of solutions* (proof is omitted here). It is interesting to note that the Wronskian can be used to define a differential equation, when a fundamental set is known.

Recall that a set of all solutions of the linear equation $L(y) = 0$ is called a *null space* or *kernel* of the operator L.

Theorem 5.9. *A homogeneous linear differential equation $L_n(y) = 0$ has n linearly independent solutions that form a basis for the set of all solutions. The dimension of the kernel of an* nth-order *linear differential operator is n.*

Example 5.1. Define a linear differential equation possessing the fundamental solution: $y_1(t) = \sin t$, $y_2(t) = e^t$.

We define the Wronskian

$$
\begin{aligned}
W(\sin t, e^t, y) &= \begin{vmatrix} \sin t & e^t & y \\ \cos t & e^t & y' \\ -\sin t & e^t & y'' \end{vmatrix} e^t \begin{vmatrix} \sin t & 1 & y \\ \cos t & 1 & y' \\ -\sin t & 1\& y'' \end{vmatrix} \\
&= e^t \left(\sin t \begin{vmatrix} 1 & y' \\ 1 & y'' \end{vmatrix} - 1 \begin{vmatrix} \cos t & y' \\ -\sin t & y'' \end{vmatrix} + y \begin{vmatrix} \cos t & 1 \\ -\sin t & 1 \end{vmatrix} \right) \\
&= e^t[\sin t(y'' - y') - (y'' \cos t + y' \sin t) + y(\cos t + \sin t)] \\
&= e^t[(\sin t - \cos t)y'' - 2\sin t + y' + (\sin t + \cos t)y] = 0,
\end{aligned}
$$

and the being sought differential equation is defined immediately as

$$L_2(y) = (\sin t - \cos t)y'' - 2\sin t y' + (\sin t + \cos t)y = 0.$$

In order to verify the obtained results we check:

$$L_2(\sin t) = -(\sin t - \cos t)\sin t - 2\sin t \cos t + (\sin t + \cos t)\sin t = 0,$$

$$L_2(e^t) = e^t L_2(\sin t) = 0.$$

□

5.3 Differential Equations with Constant Coefficients

Since characteristic equation corresponding to an nth-order homogeneous differential equation cannot be solved using radicals already for $n \geq 5$, very often numerical or approximate analytical methods are used. The most important observation is that the nth-order differential equation is reduced to the nth-order polynomial (algebraic) equation. The solutions differ from each other qualitatively for distinct and repeated (multiple) roots of a characteristic equation, which we are going to discuss.

Theorem 5.10. *Given the nth-order linear differential equation* $L_n(y) = y^{(n)} + p_1 y^{(n-1)} + \cdots + p_{n-1}y' + p_n y = 0$. *Let the characteristic polynomial corresponding to* $L_n(y) = 0$ *possess:*

(i) n distinct real roots r_i, $i = 1, \ldots, n$. *Then the n functions* $y_i(t) = e^{r_i t}$ *form a basis for the kernel of* $L_n(y) = 0$. *A general solution has the form*

$$y(t) = \sum_{i=1}^{n} C_i e^{r_i t};$$

(ii) Single root repeated n times. Then the n functions $y_1(t) = e^{rt}, y_2(t) = te^{rt}, \ldots, y_n(t) = t^{n-1}e^{rt}$ *form a basis for the kernel of* $L_n(y) = 0$. *A general solution can be written as*

$$y(t) = (C_1 + c_2 t + \cdots + C_n t^{n-1})e^{rt};$$

(iii) Single root repeated m times and there are n − m distinct roots $r_{m+1}, r_{m+2}, \ldots, r_n$. *Then the n functions*

$$\begin{aligned} y_1(t) &= e^{rt}, \\ y_2(t) &= te^{rt}, \\ &\vdots \\ y_m(t) &= t^{m-1}e^{rt}, \\ y_{m+1}(t) &= e^{r_{m+1}t}, \\ &\vdots \\ y_n(t) &= e^{r_n t} \end{aligned}$$

form the kernel of $L_n(y) = 0$. *A general solution has the form*

$$y(t) = (C_1 + C_2 + \cdots + C_m t^{m-1})e^{rt} + C_{m+1}e^{r_{m+1}t} + \cdots + C_n e^{r_n t}.$$

Example 5.2. Consider the following third-order differential equation

$$L_3(y) = y''' - 8y'' + 19y' - 12y = 0.$$

Introducing $y = e^{rt}$ one obtains the characteristic polynomial $r^3 - 8r^2 + 19r - 12 = 0$, with three distinct roots: $r_1 = 1, r_2 = 3, r_3 = 4$. The linearly independent solutions are: $y_1(t) = e^t, y_2(t) = e^{3t}, y_3(t) = e^{4t}$, and every solution is written as $y(t) = C_1 e^t + C_2 e^{3t} + C_3 e^{4t}$. □

Example 5.3. Consider the fifth-order differential equation $L_5(y) = y^{(5)} - 12y^{(4)} + 56y''' - 126y'' + 135y' - 54 = 0$.

The characteristic equation corresponding to $L_5(y)$ has the following roots: $r_1 = 1$, $r_2 = 2$ and $r_3 = r_4 = r_5 = 3$ (multiplicity 3). The following set forms a basis $\{e^t,\ e^{2t},\ e^{3t},\ te^{3t},\ t^2e^{3t}\}$. Every solution in the kernel has the form

$$y(t) = C_1e^t + C_2e^{2t} + C_3e^{3t} + C_4te^{3t} + C_5t^2e^{3t}.$$

In order to investigate their linear dependence one calculates

$$W(e^t, e^{2t}, e^{3t}, te^{3t}, t^2e^{3t})$$

$$= \begin{vmatrix} e^t & e^{2t} & e^{3t} & te^{3t} & t^2e^{3t} \\ e^t & 2e^{2t} & 3e^{3t} & e^{3t} + 3te^{3t} & 2te^{3t} + 3t^2e^{3t} \\ e^t & 4e^{2t} & 9e^{3t} & 6e^{3t} + 9te^{3t} & 2e^{3t} + 12te^{3t} + 9t^2e^{3t} \\ e^t & 8e^{2t} & 27e^{3t} & 27e^{3t} + 27te^{3t} & 16e^{3t} + 54te^{3t} + 27t^2e^{3t} \\ e^t & 16e^{2t} & 81e^{3t} & 108e^{3t} + 81te^{3t} & 108e^{3t} + 216te^{3t} + 81t^2e^{3t} \end{vmatrix}$$

$$= e^te^{2t}e^{3t}e^{3t}e^{3t} \begin{vmatrix} 1 & 1 & 1 & t & t^2 \\ 1 & 2 & 3 & 1+3t & 2t+3t^2 \\ 1 & 4 & 9 & 6+9t & 2+12t+9t^2 \\ 1 & 8 & 27 & 27+27t & 16+54t+27t^2 \\ 1 & 16 & 81 & 108+81t & 108+216t+81t^2 \end{vmatrix} \neq 0.$$

Theorem 5.11. *Given the* nth-*order linear differential equation* $L_n(y) = y^{(n)} + p_1y^{(n-1)} + \cdots + p_{n-1}y' + p_ny = 0$. *Let the characteristic equation corresponding to* $L_n(y) = 0$ *to have:*

(i) A pair of complex roots $r = a \pm bi$, $i^2 = -1$. *Then the functions* $y_1(t) = e^{at}\cos bt$ *and* $y_2(t) = e^{at}\sin bt$ *are linearly independent solution of* $L_n(y) = 0$.
(ii) A pair of complex conjugate roots repeated m times. Then the 2m functions

$$y_1(t) = e^{at}\cos bt,\quad y_2(t) = te^{at}\cos bt\quad ,\ldots,\quad y_m(t) = t^{m-1}e^{at}\cos bt,$$

$$y_{m+1}(t)=e^{at}\sin bt,\quad y_{m+2}(t)=te^{at}\sin bt\quad ,\ldots,\quad y_{2m}(t) = t^{m-1}e^{at}\sin bt$$

are linearly independent solutions of $L_n(y) = 0$. *In addition, every function* $y(t) = (C_1 + C_2t + \cdots + C_mt^{m-1})e^{at}\cos bt + (C_{m+1} + C_{m+2}t + \cdots + C_{2m}t^{m-1})e^{at}\sin bt$ *is a solution to the studied equation.*

In order to get a better imagination of understanding of repeated roots of a characteristic equation we review some properties of polynomials.

(i) A polynomial $w(t)$ has a root r if $w(t) = (t - r)w_0(t)$.
(ii) A polynomial $w(t)$ has a double root if $w(t) = (t - r)^2 w_1(t)$. In this case $w(t)$ and $w'(t) = 2(t - r)w_1(t)$ have a common root r.
(iii) In general, a polynomial $w(t)$ has a root r of multiplicity k if $w(t) = (t-r)^k w_{k-1}(t)$. In this case, $w(t), w'(t), \dots, w^{k-1}(t)$ have a common root r.

Now we focus our attention on a second polynomial corresponding to differential operator $L_2(y) = y'' - 2ry' + r^2 y$. The corresponding quadratic equation is $w(t) = t^2 - 2rt + r^2 = (t - r)^2$. According to Theorem 5.10 we have the solutions: $y_1(t) = e^{rt}, y_2(t) = te^{rt}$. It is true, because $L_2(y_1) = e^{rt}(r^2 - 2r^2 + r^2) = 0$, and $L_2(te^{rt}) = (te^{rt})'' - 2r(te^{rt})' + r^2 te^{rt} = (e^{rt} + rte^{rt})' - 2r(e^{rt} + rte^{rt}) + r^2 te^{rt} = 2re^{rt} + r^2 te^{rt} - 2re^{rt} - 2r^2 te^{rt} + r^2 te^{rt} = 0$. The above example leads to more general observation. Namely, have got $L_2(e^{rt}) = e^{rt}w(r)$. Observe that $\frac{\partial}{\partial r}L_2(e^{rt}) = \frac{\partial}{\partial r}[e^{rt}w(r)] = te^{rt}w(r) + w'(r)e^{rt}$. But recall that r is the double root of the characteristic equation $w(r) = 0$ and hence $w(r) = w'(r) = 0$, i.e. $\frac{\partial}{\partial r}L_2(e^{rt}) = 0$. On the other hand $\frac{\partial}{\partial r}L_2(e^{rt}) = L_2(\frac{\partial}{\partial r}e^{rt}) = L_2(te^{rt}) = 0$, which shows that te^{rt} is a solution of the differential operator L_2. Finally, let us check a linear dependence of both solutions. The Wronskian determinant gives

$$w(e^{rt}, te^{rt}) = \begin{vmatrix} e^{rt} & te^{rt} \\ e^{rt} & e^{rt} + te^{rt} \end{vmatrix} = e^{2rt} \neq 0,$$

which means that they are independent. This observation can be generalized by the following theorem.

Theorem 5.12. *If r_0 is a root of m-multiplicity of the characteristic equation*

$$w(r) = r^n + a_1 r^{n-1} + a_2 r^{n-2} + \dots + a_{n-1}r + a_n = 0,$$

where $1 \le m \le n$, then the functions $e^{r_0 t}, te^{r_0 t}, \dots, r^{m-1}e^{r_0 t}$ are the linearly independent solutions of the characteristic equation $w(r)$.

Proof. If r_0 is the root of $w(r)$ of multiplicity m, then

$$w(r_0) = \frac{dw}{dr}(r_0) = \frac{d^2 w}{dr^2}(r_0) = \dots = \frac{d^{k-1}w(r_0)}{dr^{k-1}} = 0.$$

Consider now the differential operator

$$L(y) \equiv y^{(n)} + a_1 y^{(n-1)} + \dots + a_n y.$$

It is easy to check that $L(e^{rt}) = w(r)e^{rt}$. Consider now $L(t^i e^{rt}) = L\left(\frac{\partial^i e^{rt}}{\partial r^i}\right) = \frac{\partial^i}{\partial r^i}L(e^{rt}) = \frac{\partial^i}{\partial r^i}(w(r)e^{rt})$. We have

$$\begin{aligned} i = 0: &\quad w(r)e^{rt} = 0; \\ i = 1: &\quad (w' + tw)e^{rt} = 0; \\ i = 2: &\quad (w'' + 2tw' + w(1 + t^2))e^{rt} = 0; \\ &\vdots \end{aligned}$$

Since for $i = k - 1$ we obtain $w^{k-1} = \cdots = w'' = w' = w = 0$, hence we have also $L(t^i e^{r_0 t}) = 0$. It means that $t^i e^{r_0 t}$ are solutions to $L(y)$. In addition, they are linearly independent, because the polynomials of different orders are linearly independent. □

In the case of complex conjugate roots (Theorem 5.12) we recall some fundamental properties known from algebra. According to the Euler's formula we have

$$e^{i\beta} = \cos\beta + i\sin\beta,$$

where: $\mathrm{Re}(e^{i\beta}) = \cos\beta$, $\mathrm{Im}(e^{i\beta}) = \sin\beta$ and $e^{a+ib} = e^a(\cos b + i\sin b)$.

Now let the characteristic equation representing the differential equation $L(y) = 0$ to have conjugate complex roots $a \pm ib$, and we obtain the following solutions to the differential equation: $e^{(a+ib)t} = e^{at}(\cos bt + i\sin bt)$ and $e^{(a-ib)t} = e^{at}(\cos bt - i\sin bt)$. Observe that instead of taking this solution one can take $y_1 = e^{at}\cos bt$ and $y_2 = e^{at}\sin bt$. Both of them are real-valued independent solutions (check), and they span the same real-valued solutions as the complex-valued solutions $e^{(a\pm ib)t}$. This observation can be generalized by the following theorem.

Theorem 5.13. *If u and v are in the domain composed of real elements of a real-valued function L, and we take $y = u + iv$, then $L(y) = L(u) + iL(v)$. In addition, if $L(y) = 0$, then $L(u) = L(v) = 0$.*

Remark 5.3. Theorem 5.12 shows that the domain of L can be extended to include complex-valued component.

Our considerations can be briefly outlined in the following way [191]. The solutions to homogeneous linear nth-order ordinary differential equation (5.11) are sought in forms of exponential functions

$$y = e^{rt}, \tag{5.15}$$

where r is constant. We have

$$y^{(k)} = \frac{\partial^k y}{\partial t^k} = r^k e^{rt}, \quad k = 1, \ldots, n. \tag{5.16}$$

Substituting the assumed solution (5.15) to (5.16) we get

$$L(e^{rt}) = e^{rt} P_n(r) = 0, \tag{5.17}$$

where the polynomial

$$P_n(r) = r^n + p_1 r^{n-1} + \cdots + p_{n-1} r + p_n = 0 \tag{5.18}$$

is called the *characteristic equation* (or *the characteristic polynomial*). The roots of the characteristic equation (5.18) are called the *characteristic roots*. If the roots $r_1, \dots, r_n$ are real and different, then we have n independent solutions of the form

$$y_1 = e^{r_1 t}, \dots, y_n = e^{r_n t}. \tag{5.19}$$

The Wronsky determinant has the form

$$w(e^{r_1 t}, \dots, e^{r_n t}) = \begin{vmatrix} e^{r_1 t} & \dots & e^{r_n t} \\ r_1 e^{r_1 t} & \dots & r_n e^{r_n t} \\ \vdots & \vdots & \vdots \\ r_1^{n-1} e^{r_1 t} & \dots & r_n^{n-1} e^{r_n t} \end{vmatrix}$$

$$= e^{(r_1+r_2+\dots+r_{n-1})t} \begin{vmatrix} 1 & 1 & \dots & 1 \\ r_1 & r_2 & \dots & r_n \\ \vdots & \vdots & \vdots & \vdots \\ r_1^{n-1} & r_2^{n-1} & \dots & r_n^{n-1} \end{vmatrix}$$

$$= e^{(r_1+r_2+\dots+r_{n-1})t} (r_n - r_1)(r_n - r_2) \dots (r_n - r_{n-1})(r_{n-2} - r_1) \dots$$
$$\dots (r_{n-1} - r_{n-2}) \dots (r_2 - r_1) \neq 0. \tag{5.20}$$

Observe that $w(t) \to 0$, if $t \to +\infty$, when $r_k < 0, k = 1, \dots, n$. A general (or every) solution to (5.11) has the form

$$y = C_1 e^{r_1 t} + \dots + C_n e^{r_n t}. \tag{5.21}$$

If $r_k < 0, k = 1, \dots, n$, then every solution y governed by (5.21) $y \to 0$ for $t \to +\infty$.

Consider the case, when $r_1, \dots, r_n$ are different, but some of them are complex conjugate. Let $r_1 = a + ib$. Since the coefficients of the differential equation (5.11) are real, then also $r_2 = \bar{r}_1 = a - ib$ exists. A general solution has the form $y = u(t) + iv(t)$, where $u(t)$ and $v(t)$ are real-valued functions. However, if $y(t_0) = u(t_0) + iv(t_0) = y_0$, $y^{(k)}(t_0) = u^{(k)}(t_0) + iv^{(k)}(t_0) = y_0^{(k)}$, $k = 1, \dots, n-1$, where y_0 and $y_0^{(k)}$ are real-valued, then $v(t_0) = v'(t_0) = \dots = v^{(n-1)}(t_0) = 0$, and hence $v(t) \equiv 0$.

Given the initial conditions one finds the being sought constants $C_1, \dots, C_n$ from the algebraic equations

$$y_0 = C_1 e^{r_1 t_0} + \dots + C_n e^{r_1 t_0},$$
$$y_0^{(k)} = C_1 r_1^k e^{r_1 t_0} + \dots + C_n r_n^k e^{r_n t_0}, \quad k = 1, \dots, n-1. \tag{5.22}$$

It can be shown that the solution (5.21), after substituting the constant $C_i i = 1, \ldots, n$ from (5.22), is real-valued. Hence, since the solution is real-valued then instead of complex-valued form of the solution we can use the real-valued form of the solution. For our case $r_1 = a + ib$, $r_2 = a - ib$ we get

$$\begin{aligned} C_1 e^{r_1 t} + C_2 e^{r_2 t} &= C_1 e^{at}(\cos bt + i \sin bt) + C_2 e^{at}(\cos bt - i \sin bt) \\ &= (C_1 + C_2)e^{at} \cos bt + i(C_1 - C_2)e^{at} \sin bt = C_{10} e^{at} \cos bt + C_{20} \sin bt, \end{aligned}$$

where: $C_{10} = C_1 + C_2$, $C_{20} = i(C_1 - C_2)$. It is easy to check that $C_{1,2} = C_{10} \pm C_{20}/\mathrm{i}$, i.e. they are complex conjugate, but $C_{i0}, i = 1, 2$ are real-valued. In other words

$$y = C_{10} e^{at} \cos bt + C_{20} e^{at} \sin bt + C_3 e^{r_3 t} + \cdots + C_n e^{r_n t}. \tag{5.23}$$

For real-value initial conditions $y_0, \ y_0', \ldots, \ y^{(n-2)}$ the solution (5.23) is real-valued and also C_{10} and C_{20} should be real-valued. Since $e^{r_1 t} = e^{at}(\cos bt + i \sin bt)$ is the solution, then also $y_1 = e^{at} \cos bt$ and $y_2 = e^{at} \sin bt$ are the solutions generated by the roots $r_1 = a + ib$, $r_2 = a - bt$. Recall now the discussed case of the root $r_1 = \cdots = r_m$ with mth-multiplicity. The kth-order derivative of (5.11) yields

$$\frac{d^k L(e^{rt})}{dr^k} = L\left[\frac{d^k e^{rt}}{dr^k}\right] = L(t^k e^{rt}), \tag{5.24}$$

and

$$L(t^k e^{rt}) = (t^k e^{rt})^{(n)} + p_1 (t^k e^{rt})^{n-1} + \cdots + p_{n-1}(t^k e^{rt})' + p_n (t^k e^{rt}). \tag{5.25}$$

Owing to the Leibniz formula applied to dot product of two functions we have

$$\begin{aligned} (uv)^{(n)} &= uv^{(n)} + \binom{n}{1} u' v^{(n-1)} + \binom{n}{2} u'' v^{(n-2)} + \ldots \\ &\ldots + \binom{n}{n-1} u^{(n-1)} v' + u^{(n)} v = \sum_{k=0}^{n} u^{(k)} v^{(n-k)}, \end{aligned} \tag{5.26}$$

where: $u^{(0)} = u$, $v^{(0)} = v$. Applying Leibniz formula (5.26) to (5.25) we obtain

$$\begin{aligned} L(t^k e^{rt}) &= t^k e^{rt} P_n(r) + C_1^{(k)} t^{k-1} e^{rt} P'_n(r) \\ &+ C_2^{(k)} t^{k-2} e^{rt} P''_n(r) + \cdots + e^{rt} P_n^{(k)}(r), \end{aligned} \tag{5.27}$$

where the numbers $C_l^{(k)}$ depend on k and l. Since $r_1 = \cdots = r_m$, hence the characteristic equation has the following property

$$P_n(r_1) = P_n'(r_1) = \cdots = P_n^{m-1}(r_1) = 0. \tag{5.28}$$

It means that $L(t^k e^{rt}) = 0$ for $k = 1, 2, \ldots, m-1$ and $r = r_1$, where r_1 is the root of mth multiplicity. In other words, m times repeated root generates m solutions of the form $e^{r_1 t}, te^{r_1 t}, \ldots, t^{m-1}e^{r_1 t}$. If $r_1, \ldots, r_l$ are different roots of the corresponding multiplicity $m_1, \ldots, m_l$ and there are no other roots (i.e. $m_1 + \cdots + m_l = n$), then every solution to homogeneous linear differential equation (5.11) has the form

$$y = P_{m_1-1}(t)e^{r_1 t} + P_{m_2-1}(t)e^{r_2 t} + \cdots + P_{m_l-1}(t)e^{r_l t}, \tag{5.29}$$

where: $P_{m_1-1}(t), P_{m_2-1}(t), \ldots, P_{m_l-1}(t)$ are the polynomials of t of the corresponding orders: $m_1 - 1, m_2 - 1, \ldots, m_l - 1$. It can be checked that a set $\{e^{r_1 t}, re^{r_1 t}, \ldots, t^{m_1-1}e^{r_1 t}, \ldots, e^{r_l t}, te^{r_l t}, \ldots, t^{m_l-1}e^{r_l t}\}$ has members which are linearly independent.

Recall now our considerations related to repeated complex roots. If among the roots $r_1, \ldots, r_l$ a complex one appears, say $r_1 = a + ib$, then also exists $r_2 = a - ib$ with the same multiplicity. The used real initial conditions $y_0, y_0', \ldots, y_0^{(n-1)}$ allow to find a set of constants to define a real solution.

Assume that $r = a + ib$ and $\bar{r} = a - ib$ are m times repeated roots. Every solution includes the functions

$$\begin{aligned} &e^{rt}, te^{rt}, \ldots, t^{m-1}e^{rt}, \\ &e^{\bar{r}t}, te^{\bar{r}t}, \ldots, t^{m-1}e^{\bar{r}t}. \end{aligned} \tag{5.30}$$

Since $t^k e^{rt} = t^k e^{at} e^{ibt} = t^k e^{at}(\cos bt + i \sin bt)$ is the solution, then its real and imaginary part are also solutions. The same consideration holds for $\bar{r}$. In the case when a root is purely imaginary with mth multiplicity ($a = 0$), then every solution is composed of the functions $t^k \cos bt, t^k \sin bt, k = 0, 1, \ldots, m$.

Let us sum up our brief conclusions related to a repeated root. Any real-valued root r with mth multiplicity generates m solutions, whereas a complex-valued root r with mth multiplicity generates $2m$ solutions. Every solution corresponding to m-times repeated root is represented by

$$y = \sum_{k=0}^{m-1} C_k t^k e^{rt}. \tag{5.31}$$

Every solution corresponding to complex root with mth multiplicity has the form

$$y = \sum_{k=0}^{m-1} \left(A_k t^k e^{at} \cos bt + B_k t^k e^{at} \sin bt\right). \tag{5.32}$$

Finally, every solution corresponding to repeated imaginary roots $r = ib, \bar{r} = -ib$ has the form

$$y = \sum_{k=0}^{m-1} \left(A_k t^k \cos bt + B_k t^k \sin bt\right). \tag{5.33}$$

In both latter cases, A_k and B_k are defined by the initial conditions.

Example 5.4. Using formula (5.25) and (5.26) for $k = 2$ find a corresponding number C_l^k when

$$L(y) = y''' + p_1 y'' + p_2 y' + p_3 y.$$

Formula (5.26) gives

$$(uv)''' = uv''' + \binom{3}{1} u'v'' + \binom{3}{2} u''v' + u'''v,$$

$$(uv)'' = uv'' + \binom{2}{1} u'v' + u''v',$$

$$(uv)' = u'v + v'y,$$

whereas from (5.25) we get

$$\begin{aligned} L(t^k e^{rt}) &= (t^k e^{rt})'' + p_1(t^k e^{rt})'' + p_2(t^k e^{rt})' + p_3(t^k e^{rt}) \\ &= (t^2r^3+6tr^2+6r)e^{rt}+p_1(t^2r^2+4tr+2)e^{rt}+p_2(t^2r+2t)e^{rt}+p_3t^2e^{rt} \\ &= e^{rt}[t^2(r^3 + p_1r^2 + p_2r + p_3) + t(6r^2 + 4rp_1 + 2p_2) + 6r + 2p_1]. \end{aligned}$$

Since

$$\begin{aligned} P_3(r) &= r^3 + p_1r^2 + p_2r + p_3, \\ P'_3(r) &= 3r^2 + 2p_1r + p_2, \\ P''_3(r) &= 6r + 2p_1, \end{aligned}$$

therefore

$$L(t^2e^{rt}) = t^2e^{rt}P_3(r) + C_1^2 P_3'(r)e^{rt} + e^{rt}P_3''(r),$$

and we have $C_1^2 = 2$.

5.4 Linear Non-homogeneous Differential Equations with Constant Coefficients

In the previous section we have considered linear homogeneous operator defined by Eq. (5.11). Here we are going to consider non-homogeneous problems.

Theorem 5.14. *If the fundamental solutions $y_1(t), \ldots, y_n(t)$ to Eq. (5.11) are known, then a general solution to the following non-homogeneous equation*

$$L_n(y) = y^{(n)} + p_1(t)y^{(n-1)} + \cdots + p_n(t)y = f(t) \tag{5.34}$$

has the form

$$y = y_p + C_1 y_1 + C_2 y_2 + \cdots + C_n y_n, \tag{5.35}$$

where $C_1, C_2, \ldots, C_n$ are arbitrary numbers, and y_p is an arbitrary particular solution of (5.34).

There are two general methods to solve a non-homogeneous problem, i.e. the method of *undetermined coefficients* and the method of *variations of parameters*. The method of undetermined coefficients relies on observation that a being sought solution is somehow similar to the right-hand side function $f(t)$.

Example 5.5. Find a general solution to the non-homogeneous differential equation

$$L(y) = y'' - 3y' + 2y = 3t^2 - 2t + 4.$$

First we consider the corresponding homogeneous equation $L(y) = y'' - 3y' + 2y = 0$. The corresponding characteristic equation is $r^2 - 3r + 2 = 0$ with the roots $r_1 = 1$ and $r_2 = 2$, and hence the general solution of the homogeneous equation $y = C_1e^t + C_2e^{2t}$. Let us look for a particular solution of the form $y_p = at^2 + bt + c$. The simple calculus yields

$$\begin{aligned} L(y_p) &= 2at - 3(2at + b) + 2(at^2 + bt + c) \\ &= 2at^2 + t(2b - 6a) + 2c + 2a - 3b. \end{aligned}$$

Equating the coefficients standing by the same powers of t we obtain:

$$\begin{aligned} t^0 &: 2c + 2a - 3b = 4, \\ t^1 &: \quad 2b - 6a = -2, \\ t^2 &: \qquad 2a = 3. \end{aligned}$$

The above linear algebraic equations give: $a = 1.5$, $b = 3.5$, $c = 5.75$. The general solution of the non-homogeneous problem is $y = C_1e^t + C_2e^{2t} + 1.5t^2 + 3.5t + 5.75$. □

Example 5.6. Find a general solution of the non-homogeneous equation

$$L(y) = y'' + 2hy' + \alpha_0^2 y = q \cos \omega t.$$

This equation governs oscillation of a one-degree-of-freedom mechanical system with mass m and with a viscous positive damping, a linear stiffness and a harmonic excitation. The parameters have the following physical meaning: c—damping ($2h = c/m$); k—stiffness ($\alpha_0^2 = k/m$); P_0—amplitude of exciting force ($q = p_0/m$) and ω is frequency of excitation. The characteristic equation is

$$r^2 + 2hr + \alpha_0^2 = 0,$$

which gives the roots $r_{1,2} = -h \pm \sqrt{h^2 - \alpha_0^2}$. If $h > \alpha_0$, then we have two real roots and oscillations do not appear. If $h = \alpha_0$ we have so-called critical damping $c_{cr} = 2\sqrt{km}$ and the root is double. Here we consider the case $\alpha_0^2 > h^2$ which corresponds to oscillations and $r_{1,2} = -h \pm i\lambda$, where $\lambda = \sqrt{\alpha_0^2 - h^2}$. Since $e^{(-h+i\lambda)t} = e^{-ht}(\cos\lambda t + i\sin\lambda t)$ and according to our earlier discussion we can take the following real-valued general solution corresponding to homogeneous equation: $y = C_1 e^{-ht}\cos\lambda t + C_2 e^{-ht}\sin\lambda t$. As it has been mentioned, the form of the particular solution to the non-homogeneous equation is similar to the right-hand side excitation and is assumed to be: $y_p = A\cos\omega t + B\sin\omega t$. Since

$$\begin{aligned} y' &= -A\omega\sin\omega t + B\omega\cos\omega t, \\ y'' &= -A\omega^2\cos\omega t - B\omega^2\sin\omega t, \end{aligned}$$

hence

$$\begin{aligned} L(y_p) = {} & -A\omega^2\cos\omega t - B\omega^2\sin\omega t - 2hA\omega\sin\omega t + 2hB\omega\cos\omega t + \alpha_0^2 A\cos\omega t \\ & + \alpha_0^2 B\sin\omega t = q\cos\omega t. \end{aligned}$$

Equating the terms standing by $\sin\omega t$ and $\cos\omega t$ yields

$$\begin{aligned} A(\alpha_0^2 - \omega^2) + 2h\omega B &= q, \\ -2h\omega A + (\alpha_0^2 - \omega^2)B &= 0, \end{aligned}$$

and therefore

$$A = \frac{W_A}{W} = \frac{\begin{vmatrix} q & 2h\omega \\ 0 & \alpha_0^2 - \omega^2 \end{vmatrix}}{\begin{vmatrix} \alpha_0^2 - \omega^2 & 2h\omega \\ -2h\omega & \alpha_0^2 - \omega^2 \end{vmatrix}} = \frac{q(\alpha_0^2 - \omega^2)}{(\alpha_0^2 - \omega^2)^2 + 4h^2\omega^2},$$

$$B = \frac{W_B}{W} = \frac{\begin{vmatrix} \alpha_0^2 - \omega^2 & q \\ -2h\omega & 0 \end{vmatrix}}{\begin{vmatrix} \alpha_0^2 - \omega^2 & 2h\omega \\ -2h\omega & \alpha_0^2 - \omega^2 \end{vmatrix}} = \frac{2h\omega q}{(\alpha_0^2 - \omega^2)^2 + 4h^2\omega^2}.$$

Finally, the general solution to homogeneous equation has the form

$$y = C_1 e^{-ht} \cos \lambda t + C_2 e^{-ht} \sin \lambda t + A \cos \omega t + B \sin \omega t.$$

□

The obtained solution possesses clear physical interpretation. It is composed of oscillation of autonomous damped system (this part vanishes, when $t \to +\infty$) and a particular solution generated by the harmonic excitation. A steady state oscillation is represented by the solution

$$\lim_{t\to\infty} y(t) = A \cos \omega t + B \sin \omega t = a \cos(\omega t + \varphi).$$

Observe that

$$a \cos(\omega t + \varphi) = a \cos \varphi \cos \omega t - a \sin \varphi \sin \omega t = A \cos \omega t + B \sin \omega t.$$

Equating terms standing by $\cos \omega t$ and $\sin \omega t$ we obtain

$$A = a \cos \varphi, \quad B = -a \sin \varphi,$$

and hence

$$a = \sqrt{A^2 + B^2} = \frac{q}{\sqrt{(\alpha_0^2 - \omega^2)^2 + 4h^2\omega^2}},$$

$$\varphi = -\arctan \frac{B}{A} = -\arctan \frac{2h\omega}{\alpha_0^2 - \omega^2}.$$

This result indicates that the oscillations are harmonic with the amplitude a and are delayed by the phase shift ϕ in comparison to excitation. The latter example allows to solve more general problem, which possesses many applications in engineering, and particularly in mechanics.

Consider now the following non-homogeneous problem

$$L_n(y) = e^{at}[P(t) \cos bt + Q(t) \sin bt], \tag{5.36}$$

where $a, b \in R$ and $P(t)$ and $Q(t)$ are polynomials of orders p and q, respectively. Let the characteristic equation corresponding to the homogeneous equation has the form

$$r^n + p_1 r^{n-1} + \cdots + p_n = 0. \tag{5.37}$$

By s we denote the largest values among p and q (for $p = q$ we take $s = p = q$). If $a + ib$ is not a solution to the characteristic equation (5.38), then also $a - ib$ is not a solution. In this case the being sought particular solution has the form

$$y_p = e^{at}[R(t) \cos bt + S(t) \sin bt],$$

where $R(t)$ and $S(t)$ are polynomials of an order not higher than s. If $a + ib$ is repeated root of multiplicity k, then

$$y_p = t^k e^{at}[R(t)\cos bt + S(t)\sin bt], \tag{5.38}$$

where $R(t)$ and $S(t)$ are polynomials of an order not higher than s.

The steps of the method of undetermined coefficients can be even more detailed presented by introducing the following theorem (see [191]).

Theorem 5.15. *Apply the following steps to solve $L_n(y) = f(t)$ using the method of undetermined coefficients:*

(i) Factor the characteristic polynomial $w_{L_n}(r)$ of L_n into linear and irreducible quadratic factors and form a basis B_L for the kernel of L.
(ii) Separate f(t) into groups of terms that are annihilated by a single annihilator. Form $w_A(r)$, the characteristic polynomial for an annihilator of $f(t)$ factored into linear and irreducible quadratic factors. Recall that for a given function f, an annihilator of f is a linear operator A such that $A(f) = 0$. For instance, $y'' + 16y$ is an annihilator of $\sin\phi t$ and $\cos\phi t$, because the characteristic equation is $(r^2 + 16) = (r + 4i)(r - 4i)$.
(iii) The characteristic polynomial of A_L is defined by $w_A(r)w_L(r)$. Form the basis B_{AL}, and find B consisting of the functions in B_{AL} that are not in B_L.
(iv) Form $y_p(t)$ as a linear combination of the functions in B and equate coefficients of $L(y_p) = f(t)$.
(v) Determine values for the coefficients and define a particular solution.
(vi) Find $y(t) = y_g(t) + y_p(t)$.

Example 5.7. Solve the differential equation $L(y) = y''' - 2y'' + 9y' - 18y = e^t + 3\sin 2t$ using the method of undetermined coefficients.

(i) The characteristic polynomial $w_L(r) = r^3 - 2r^2 + 9r - 18 = (r-2)(r^2+9)$, and hence $\{B_L = e^{2t}, \sin 3t, \cos 3t\}$.
(ii) The function $f(t) = e^t + 3\sin 2t$ is annihilated by the characteristic polynomial $w_A = (r-1)(r^2+4)$, and the basis for kernel of A is $B_A = \{e^t, \sin 2t, \cos 2t\}$.
(iii) The characteristic polynomial of A_L is $w_A w_L = (r-2)(r^2+9)(r-1)(r^2+4)$, and the kernel of A_L is $B_{AL} = \{e^t, e^{2t}, \sin 2t, \cos 2t, \sin 3t, \cos 3t\}$. We have found also that $B = B_A = \{e^t, \sin 2t, \cos 2t\}$.
(iv) The particular solution $y_p = ae^t + b\sin 2t + c\cos 2t$. The successive differentiation gives

$$y'_p = ae^t + 2b\cos 2t - 2c\sin 2t,$$

$$y''_p = ae^t - 4b\sin 2t - 4c\cos 2t,$$

$$y'''_p = ae^t - 8b\sin 2t + 8c\cos 2t,$$

and

$$\begin{aligned}L(y_p) &= ae^t - 8b\sin 2t + 8c\cos 2t - 2(ae^t - 4b\sin 2t + 4c\cos 2t)\\ &\quad +9(ae^t + 2b\cos 2t - 2c\sin 2t) - 18(ae^t + b\sin 2t + c\cos 2t)\\ &= -10ae^t - 18(b+c)\sin 2t + 2(9b-c)\cos 2t.\end{aligned}$$

Equating the appropriate coefficients with those of $f(t)$ gives

$$\begin{aligned}e^t &: \quad -10a = 1,\\ \sin 2t &: -18(b+c) = 3,\\ \cos 2t &: \quad 9b - c = 0.\end{aligned}$$

(v) The solution of the latter linear algebraic equation is: $a = -10, b = -\frac{1}{60}$, $c = -\frac{3}{20}$, and hence the particular solution is

$$y_p(t) = -10e^t - \frac{1}{60}\sin 2t - \frac{3}{20}\cos 2t.$$

(vi) Finally, the general solution is

$$y = C_1e^{2t} + C_2\sin 3t + C_3\cos 3t - 10e^t - \frac{1}{60}\sin 2t - \frac{3}{20}\cos 2t.$$

□

Now we briefly describe the method of *variation of parameters*. Recall that even in the method of undetermined coefficients we have found the kernel of a homogeneous equation $L_n(y) = 0$ first.

Theorem 5.16. *A particular solution $y_p(t)$ of non-homogeneous equation $L_n(y) = f(t)$ can be found by quadratures if it will be sought in the form*

$$y_p(t) = C_1(t)y_1(t) + C_2(t)y_2(t) + \cdots + C_n(t)y_n(t), \tag{5.39}$$

where $\{y_1(t), y_2(t), \ldots, y_n(t)\}$ is the basis of the kernel of $L_n = 0$.

Theorem 5.17. *If the functions $C_1(t), \ldots, C_n(t)$ satisfy the system of equations*

$$\begin{aligned}&C'_1y_1 + C'_2y_2 + \cdots + C'_ny_n\\ &C'_1y'_1 + C'_2y'_2 + \cdots + C'_ny'_n\\ &\qquad\vdots\\ &C'_1y_1^{(n-1)} + C'_2y_2^{(n-1)} + \cdots + C'_ny_n^{(n-1)},\end{aligned} \tag{5.40}$$

then the function y_p defined by (5.39) satisfies the non-homogeneous equation $L_n(y) = f(t)$.

Remark 5.4. The function $y_p(t)$ possesses the form of a general integral of $L_n(y) = 0$, where now the constant are the functions depending on t.

Remark 5.5. Observe that $C_1', C_2', \ldots, C_n'$ are uniquely defined, since the determinant of the system (5.41) is the Wronskian determinant, and by definition it never equals zero.

In order to get $C_1(t), \ldots, C_n(t)$ we integrate the functions $C_1'(t), \ldots, C_n'(t)$ without the constant of integrations, because they appear already in the part of n general solutions of the homogeneous equation.

Example 5.8. Find a general solution of the equation

$$L_2(y) = y'' + y = t^2.$$

The kernel of $L_2(y) = 0$ is $\{\sin t, \cos t\}$. The general solution of $L_2(y) = t^2$ is $y = y_p + C_1 \cos t + C_2 \sin t$.

The functions $y_p(t)$ is sought in the form

$$y_p(t) = C_1(t) \cos t + C_2(t) \sin t.$$

From (5.41) we obtain

$$C'_1 \cos t + C'_2 \sin t = 0,$$
$$-C'_1 \cos t + C'_2 \sin t = t^2,$$

and hence

$$C_1' = -t^2 \sin t, \quad C_2' = t^2 \cos t.$$

Integration of C_1', C_2' yields

$$C_1(t) = (t^2 - 2) \cos t - 2t \sin t, \quad C_2(t) = (t^2 - 2) \sin t + 2t \cos t$$

and

$$y = t^2 - 2 + C_1 \cos t + C_2 \sin t.$$

5.5 Differential Equations with Variable Coefficients

We begin with a special class of higher order homogeneous differential equations, which can be reduced to the differential equations with constant coefficients. A solution to the differential equation

$$L_n(y) = p_n t^n \frac{d^n y}{dt^n} + p_{n-1} t^{n-1} \frac{d^{n-1} y}{dt^{n-1}} + \cdots + p_1 t \frac{dy}{dt} + p_0 y, \tag{5.41}$$

known as the Cauchy–Euler differential equation, can be sought in the form

$$y = t^r. \tag{5.42}$$

Substituting (5.42) to (5.41) gives

$$L_n(t^r) = w(r)t^r, \tag{5.43}$$

and hence $L_n = 0$, when $w(r) = 0$. The analysis is similar to previous one, but now instead of the function e^{rt}, we have t^r.

(i) Assume that there are $r_1, r_2, \dots, r_m$ distinct real roots ($m \leq n$), then $t^{r_1}, t^{r_2}, \dots, t^{r_m}$ are real independent solutions, and

$$y(t) = C_1 t^{r_1} + C_2^{r_2} + \cdots + C_m t^{r_m} \tag{5.44}$$

is a general solution (it defines an m-dimensional subspace of the kernel of (5.41)).

(ii) Assume that r_0 is a root with kth multiplicity, then its subkernel follows

$$\{t^{r_0}, t^{r_0} \ln t, \dots, t^{r_0} (\ln t)^{k-1}\},$$

and a general solution

$$y(t) = C_1 t^{r_0} + C_2 t^{r_0} \ln t + \cdots + C_k t^{r_0} (\ln t)^{k-1} \tag{5.45}$$

defines a k-dimensional subspace.

(iii) Assume that $r = a \pm bi$, then $t^a \cos(b \ln t)$ and $t^a \sin(b \ln t)$ are real linearly independent solutions. In addition, if $a \pm bi$ has multiplicity k, then

$$\begin{aligned} &t^a \cos(b \ln t), t^a \ln t \cos(b \ln t), \dots, t^a (\ln t)^{k-1} \cos(b \ln t), \\ &t^a \sin(b \ln t), t^a \ln t \sin(b \ln t), \dots, t^a (\ln t)^{k-1} \sin(b \ln t), \end{aligned} \tag{5.46}$$

are $2k$ real linearly independent solutions in the kernel of (5.41). Their linear combinations span a $2k$-dimensional subspace.

Now we show how to reduce the homogeneous and non-homogeneous Cauchy–Euler equation to an equation with constant coefficients.

Theorem 5.18. *Introducing a variable* $x = \ln t$ *in the non-homogeneous Cauchy–Euler equation*

$$L_n(y) = p_n t^n \frac{d^n y}{dt^n} + p_{n-1} t^{n-1} \frac{d^{n-1} y}{dt^{n-1}} + \cdots + p_1 t \frac{dy}{dt} + p_0 y = f(t), \tag{5.47}$$

leads to conversion to the differential equation with constant coefficients

$$L[y(e^x)] = g(x), \tag{5.48}$$

where $g(x) = f(e^x)$.

Example 5.9. Transform the Cauchy–Euler differential equation

$$t^3 y''' - 4t^2 y'' + 3ty' + y = 0$$

to a differential equation with constant coefficients.

Let $t + e^x$, and use the chain rule to get

$$\frac{dy}{dt} = \frac{dy}{dx}\frac{dx}{dt} = \frac{dy}{dx}\frac{1}{t}, \quad \text{i.e.} \quad t\frac{dy}{dt} = \frac{dy}{dx};$$

$$\frac{d^2y}{dt^2} = \frac{d}{dt}\left(\frac{dy}{dt}\right) = \frac{d}{dt}\left(\frac{dy}{dx}\frac{1}{t}\right) = \frac{d}{dt}\left(\frac{dy}{dx}\right)\frac{1}{t} + \frac{dy}{dx}\frac{d}{dt}\left(\frac{1}{t}\right)$$

$$= \frac{1}{t^2}\left(\frac{d^2y}{dx^2} - \frac{dy}{dx}\right) \quad \text{i.e.} \quad t^2\frac{d^2y}{dt^2} = \frac{d^2y}{dx^2} - \frac{dy}{dx};$$

$$\frac{d^3y}{dt^3} = \frac{d}{dt}\left(\frac{d^2y}{dt^2}\right) = \frac{d}{dt}\left[\frac{1}{t^2}\left(\frac{d^2y}{dx^2} - \frac{dy}{dx}\right)\right]$$

$$= \frac{-2}{t^3}\left(\frac{d^2y}{dx^2} - \frac{dy}{dx}\right) + \frac{1}{t^2}\left[\frac{d}{dt}\left(\frac{d^2y}{dx^2}\right) - \frac{d}{dt}\left(\frac{dy}{dx}\right)\right]$$

$$= \frac{-2}{t^3}\left(\frac{d^2y}{dx^2} - \frac{dy}{dx}\right) + \frac{1}{t^3}\left(\frac{d^3y}{dx^3} - \frac{d^2y}{dx}\right)$$

$$= \frac{1}{t^3}\left(\frac{d^3y}{dx^3} - 3\frac{d^2y}{dx^2} - 2\frac{dy}{dx}\right), \quad \text{i.e.} \quad \frac{d^3y}{dt^3} = \frac{d^3y}{dx^3} - 3\frac{d^2y}{dx^2} - 2\frac{dy}{dx}.$$

Our investigated equation becomes

$$\frac{d^3y}{dx^3} - 7\frac{d^2y}{dx^2} + 5\frac{dy}{dx} + y = 0.$$

The characteristic polynomial is

$$r^3 - 7r^2 + 5r + 1 = (r-1)(r^2 - 6r - 1)$$

and possesses the characteristic roots

$$r_2 = 1, \quad r_1 = 3 - \sqrt{10}, \quad r_3 = 3 + \sqrt{10}.$$

Hence the solution is

$$y = C_1 e^{r_1 x} + C_2 e^{r_2 x} + C_3 e^{r_3 x} = C_1 e^{r_1 \ln t} + C_2 e^{r_2 \ln t} + C_3 e^{r_3 \ln t}.$$

□

A key role in mechanics plays second-order differential equation, since a second derivative corresponds to acceleration of a moving rigid body with one degree of freedom. It can happen that one of its solutions is known (for example, from theoretical considerations or from an experiment), and a question appears, how to find a second linearly independent solution?

Consider $L_2(y) = p_2(t)y'' + p_1(t)y' + p_0(t)y = 0$, and assume $y(t) = v(t)y_1(t)$, where $y_1(t)$ satisfies $L_2(y_1) = 0$. The assumed form of solution refers to the so-called Bernoulli method, which is often applied to solve first-order non-homogeneous differential equations. The successive differentiation gives

$$\begin{aligned} y' &= v'y_1 + vy'_1, \\ y'' &= v''y_1 + 2v'y'_1 + vy''_1, \end{aligned} \tag{5.49}$$

and hence

$$\begin{aligned} L_2(vy_1) &= p_2(t)(v''y_1 + 2v'y'_1 + vy''_1) + p_1(t)(v'y_1 + vy'_1) + p_0(t)vy_1 \\ &= v''(p_2y_1) + v'(2p_2y'_1 + p_1y_1) + v(p_2y''_1 + p_1y'_1 + p_0y_1) \\ &= p_2y_1v'' + (2p_2y'_1 + p_1y_1)v' = 0. \end{aligned} \tag{5.50}$$

Denoting $v' = w$ one can solve first-order linear differential equation with regard to w' to give

$$w(t) = y_1^{-2} e^{-\int p_1(t)/p_2(t)} dt. \tag{5.51}$$

Hence

$$v(t) = \int y_1^{-2} e^{-\int p_1(t)/p_2(t)dt} dt, \tag{5.52}$$

and the second solution is

$$y_2(t) = v(t)y_1(t) = y_1(t) \int y_1^{-2}(t) e^{-\int p_1(t)/p_2(t)dt} dt. \tag{5.53}$$

Chapter 6
Systems

In the free encyclopedia "Wikipedia", one can find that the entry "system" is ambiguous and old. The word "system" comes from Greek and means a "compound object" (σύτημα), and this explains why it occurs in many different branches of science such as: *social science* (social system, law system), *astronomy* (solar system, system of planets, heliocentric system), *philosophy* (philosophical system), *geology* (geological system), *anatomy* (nervous system), *mathematics* (number system, decimal system, binary system), *computer science* (operating system), *logic* (deductive system) and many others. It is difficult to give a single definition of a system suitable for any branch of science. That is why we give only a few definitions and choose the most suitable one for our purposes (see [45, 48, 107, 132, 150, 168, 235]). There is a tendency to use a concept of system as a primary one, i.e. indefinable. However, in this case it is a composed object which contains interacting elements. We give our own definition within this background.

(i) *Definition (generalizing, asymptotical).* System is a collection of material elements or marks, which undergo slow evolutionary processes in space–time. The interchange of interactions among the systems, ones extracted from the environment (one being a paradigm determined by a particular branch of science), is of order ε, where $\varepsilon \ll 1$. Below, we give some features of such defined system.

  (a) elements of the system can vanish or appear in time of order $(1/\varepsilon)^n$, $n \in N$;
  (b) in the system one can distinguish subsystems, which interact in a region of space and time. The way they interact obeys the rules related to a particular branch of science. The role of a subsystem can be understood as "system residence" in a particular region of space and time of order $(1/\varepsilon)^n$ as *opened* or *closed* subsystem;
  (c) the function and purpose of the system is determined by a generating entity (human, laws of the nature);

J. Awrejcewicz, *Ordinary Differential Equations and Mechanical Systems*,
DOI 10.1007/978-3-319-07659-1_6

(d) there exists a border between the system extracted from the environment and the rest of the environment. Matter and energy (the order of magnitude ε) or information (depending on a given paradigm represented by a particular branch of science) are interchanged through the border. Note that sometimes in particular cases the interchange may lead to qualitative changes of the system (or another one, quantitative changes can transfer into qualitative ones);
(e) we will assume that the mentioned borders undergo slow changes in parallel with the mentioned changes of the system state;
(f) one can use the definitions from Wikipedia, i.e. a system can (but in a defined time interval of order $(1/\varepsilon)^n$) characterize with morphostasis (keeping form and shape the same) or morphogenesis (tendency to changes);
(g) equilibrium state (dynamical or static state, understood as a particular case of the dynamical one between the system and the rest of the environment is homeostatic). Note that from Greek *homoios* means similar (new), and *stasis* means residing in time interval $(1/\varepsilon)^n$;
(h) one admits the possibility of various results (after exceeding time $(1/\varepsilon)^n$) or preservation of the system with the same causes (perturbations of order ε), i.e. so-called equifinality holds (it is just a result of slow evolution of the considered system);
(i) similarly, one can explain the *equipotentiality* principle. The causes (perturbations) originating from the same source can cause various effects (responses or behaviour) to the system. It depends on places in the border region where the causes arrive to the analysed system.

Now, we give three definitions of a system used in mathematics, cybernetics and natural sciences.

(ii) Definition (mathematical)
System (S) is a subset of N-elements relation, which is the Cartesian product ($\times$) of properties set (objects of a system or a process), namely

$$S \subset (U_1 \times U_2 \times \cdots \times U_N), \tag{6.1}$$

where U_i denotes ith set of properties.

In many branches of science, an object is regarded as a "black box", i.e. we do not know a mathematical model of an examined object or a process but we know completely a set of the inputs X and the outputs Y (responses) of the system. There also exists a concept of a "white box", when the knowledge of our object is complete. Such defined system has the form

$$S \subset (X \times Y) \tag{6.2}$$

The aim of every examination is to determine the black box. The examination is equivalent to identification of the system. The identification can be performed basing itself on the knowledge, investigations, development of science, leading to a kind of mathematical model, e.g. based on algebraic, differential, differential-integral or integral equations. Some parameters of the system are determined by the *identification*, i.e. based on the measured or known inputs X and outputs Y (this problem will be considered in more detail). It can happen that behaviour of the black box is so complicated that it is impossible to assume a priori a model.

Then, it is useful to apply the concept of neural networks. One needs to "teach" the network, usually basing oneself on a large amount of input signals, so that the response of the system is generated by the taught network. In other words, when we teach our network we make it change its dye from black to white, i.e. a white box.

Note that this method is often used in the behaviour analysis in psychology, medicine, chemistry and even in mechanics (though mechanics is highly saturated by mathematics). This way is often more economic with regard to the time duration of solving the problem or it is not possible to perform the identification of the whole object due to the costs (e.g. bridge support are immersed deep in water or some of them are not accessible for examining due to detrimental conditions such as chemical or nuclear). There is a concept of relation used in the given definition. Below, we give the definition of relation

Definition of relation (set theory).

Relation ρ between elements of the sets X and Y is any subset of the Cartesian product of X and Y, i.e. $\rho \subset X \times Y$. The relation defines any set of the ordered pairs *(x,y)*, where $x \in X$ and $y \in Y$. Notation $(x, y) \in \rho$ is equivalent to $x\rho y$.

As an example, consider relation in a set of integer numbers C such that $y = kx$. Such defined relation is called the *relation of divisibility*. For example $(2, -6)$ is the element of the relation of divisibility ρ, whereas $(-5, 3) \notin \rho$.

(iii) *Definition (natural).* System is a collection of interacting material elements, whose mutual interactions make a common goal (function), which cannot be reduced to a function of a single or a few selected elements.

It is easy to give an example of a natural object, namely a plant or an animal. For example, the brain consists of cells that are connected by cells of the nervous system and it receives stimulus's. The brain would not be able to function without cells of the nervous system.

(iv) *Definition (cybernetic).* System is a functional entirety that is extracted from environment, which influences the system by means of stimulus's (actions, signals, input quantities), and the system influences the environment through "feedback loop" by means of the outputs (reactions).

The notion of cybernetics comes from Greek (kybernetes—steersman, manager; kyberan—to control) and is a science of control systems, transmitting and processing information.

Cybernetics is one of the mainstreams of the so-called systematic research and is an interdisciplinary science. It is assumed that it was Andre Marie Ampere (1775–1836), who first used this concept in "Essay on philosophy of science".

The Polish philosopher Bronislaw F. Trentowski used this concept for the first time in 1843, in "The relation of philosophy to cybernetics".

The American mathematician Norbert Wiener is regarded as a creator of cybernetics as a sole branch of science. Now, one can distinguish *theoretical cybernetics* (emerged from biology and mathematics) and *applied cybernetics* (technical, economical or biocybernetics).

To sum up, it seems that the definition (i) has the broadest range and can be applied in many different branches of science. Many of the derived properties (a)–(i) will be explained further with the help of many examples.

At the end of this chapter, we refer to the genesis of the earlier given definition (i). The genesis is related to asymptotological understanding of the nature laws and to the conception of quantitative changes and their transformations into qualitative changes. In classical textbooks, one usually contrasts traditional Aristotle's mechanics with Newtonian mechanics. However, both conceptions can be seen by asymptotology. One can show that rough Aristotle's approach could have been a stimulus that made modern mechanics arise after asymptotological pattern. The way the science develops is not like pulling down old theories and building new ones.

It is not unusual for an old theory to be a source of new ideas and concepts. Although there are many various opinions on development of science today's, one can see asymptotological relations between them. Aristotle thought that force is a cause of motion and a state of rest is when no force is applied. According to Aristotle, a dropped ball from a horizontally moving object falls downward. Despite the beliefs outwardly distant from present conceptions, it turns out that one can point some asymptotological relations between Aristotle's and Newtonian theory. In order to make a thorough study of such relations, we consider the motion of a body experiencing the force F and the viscous resistance with the coefficient c. Aristotle's idea can be expressed through the relationship

$$F = c\dot{x} = cv. \tag{6.3}$$

The relationship confirms the earlier described Aristotle's ideas. When there is no force ($F = 0$), there is no motion (since $\dot{x} = 0$). If the force is constant, then it makes the motion velocity grow. To Aristotle, the resistance force was an attribute of motion and was not external. Newton introduced inertial force and resistance force which are external. The equation of motion of a massive body reads

$$m\frac{dv}{dt} = F - cv, \tag{6.4}$$

and after integrating it leads to the relationship

$$v = \frac{F}{c}\left(1 - e^{-\frac{c}{m}t}\right). \tag{6.5}$$

One can see as $t \to \infty$ we get Aristotle's law. Aristotle did not know the law of inertia and it was Galileo, who made this breakthrough discovery. Thus, Aristotle's approximation is a kind of asymptotics of motion for sufficiently long time intervals. The larger resistance and lower mass, the greater difference is between Aristotle's and Galileo's principle. For small values $\frac{c}{m}t$ we obtain

$$v \cong \frac{F}{c}\left(1 - 1 + \frac{c}{m}t\right) \cong \frac{F}{m}t. \tag{6.6}$$

The above equation describes additional asymptotics. This equation holds for sufficiently small time t intervals. Moreover, the lower resisting coefficient c is, the longer time will pass before Aristotle's asymptotics comes into play. This example shows the described asymptotological transition between two theories, namely Aristotle's and Galileo–Newton's. Aristotle's theory "generated" the asymptotics, and after differentiating we get

$$m\frac{dv}{dt} = F, \tag{6.7}$$

that is definitely Newton's second law. Did Aristotle discover Newton's law without noticing it?

The way through the asymptotics led us to a completely different world, alien to Aristotle, to the world of Hamiltonian mechanics. Theory of conservative systems is one of the asymptotics of Newtonian mechanics. In this new world, completely new notions emerged such as vibrations, periodic and quasiperiodic dynamics, chaotic orbits. This new, idealized world is a reflection of the real one and approximates it.

Is Aristotle's theory already dead? Does not it have a reflection in the reality? On the contrary, it does. Movements of polymer molecules in solutions and relaxation movements can be well approximated by Aristotle's principle.

It was Einstein, who made the next asymptotological transformation arise. This led to the Special Relativity. Let us discuss the asymptotic transition between Newton's and Einstein's theory.

Consider a particle of mass m, moving at the velocity v and experiencing the force F. According to the Special Relativity, the velocity is described by the formula

$$v = \frac{v_0}{\sqrt{1 + \left(\frac{v_0}{c}\right)^2}}, \tag{6.8}$$

where $v_0 = \frac{Ft}{m_0}$ and c is the speed of light. For small values of the velocity we stay in the world of Newton's theory and then $v \cong v_0$.

The equation allows to discover additional asymptotics. It is easy to see that

$$\lim_{t\to\infty} v = \lim \frac{1}{\sqrt{\varepsilon + \left(\frac{1}{c}\right)^2}} = c, \tag{6.9}$$

where $\varepsilon = v_0^{-2}$.

We seek the solution in the form

$$v(\varepsilon) = v(0) + \frac{dv}{d\varepsilon}(0)\varepsilon + 0(\varepsilon)^2. \tag{6.10}$$

Since

$$\frac{dv}{d\varepsilon}(\varepsilon) = -\frac{1}{2}\frac{1}{\left(\sqrt{\varepsilon + \left(\frac{1}{c}\right)^2}\right)^3}, \tag{6.11}$$

then

$$\frac{dv}{d\varepsilon}(0) = -\frac{1}{2}c^3. \tag{6.12}$$

By (6.10) we get

$$v = c\left(1 - \frac{1}{2}\frac{c^2}{v_0^2}\right). \tag{6.13}$$

As one can see, as $t \to \infty$ we get completely new effects. With the help of the asymptotics, one succeeded to either discover or explain a few new physical phenomena, namely deviation of light rays in gravitational fields, retardation of electromagnetic signals propagating in gravitational fields and many others.

Another example of asymptotical relations is classical mechanics and quantum mechanics. First, let us investigate analogous relations between wave optics and geometrical one. It turns out that there exists a justified, asymptotical transition between these theories. The transition from wave optics to geometrical optics is related to neglecting of the wave of the length $\varepsilon(\varepsilon \approx 10^{-7}m)$ in comparison with the size of the object. Let a light wave be described by the following formula:

$$u = A(x, y, z, \varepsilon)e^{\frac{i\varphi(x,y,z)}{\varepsilon}}, \tag{6.14}$$

where A is the amplitude and it can be put into the form of series

$$A = A_0 + A_1\varepsilon + \cdots \tag{6.15}$$

and $i^2 = -1$ and $\varphi(x, y, z)$ is a phase-shift at the point (x, y, z). One puts the sought solution into the wave equation. Next, one equates the terms occurring by the same powers of ε. Thus we obtain a nonlinear differential equation to determine $\varphi(x, y, z)$. This is the equation which corresponds to the approximation introduced by theory of geometrical optics. It was E. Schrödinger, one of the first, who transferred asymptotical similarities between the mentioned theories to asymptotical relation between quantum mechanics and classical mechanics. Similarly, one can find asymptotological relations between both mechanics. The point is that the initially given probability distribution of a particle position varies according to the classical mechanics laws.

At the end, it is worth mentioning that the methodology of asymptotical methods enables to comprehend the real world deeply. There are no eternally accurate rules and laws—they are less or more approximated as Einstein noticed. Such fundamental laws as follows: Ohm's law, Hooke's law, Coulomb's friction law $T = \mu N$ are the approximated laws and the given relations are obtained through linearization of nonlinear laws.

Even one of the fundamental models of fluid mechanics, namely the Navier–Stokes model, is the asymptotics of gas flow in the framework of Boltzmann's theory as $\varepsilon \to 0$, where ε is the mean free path of molecules. Such asymptotical transition allows many completely new concepts to come into being; the concepts which could not appear in the framework of old theories.

Chapter 7
Theory and Criteria of Similarity

For many years, theory of similarity has been used either in aware or unaware manner in many branches of science. It turns out that it is impossible to realize the examining real phenomena and processes, or objects due to their complexity and costs. For example, the examining water flow in rivers or behaviour of manned and unmanned flying objects during the flight. In such cases, one usually builds a smaller object (or larger) in such a way that one could carry out measurements in labs by means of modern apparatus. The obtained results should be reliable, i.e. there should exist the possibility of either direct or indirect transition to the real object. Theory of similarity gives an answer to the essential question: what range and under what conditions does a model represent a real object? In practice, a researcher fixes his attention on the object he examines, i.e. on the *similar* object. Sometimes, it is difficult for an individual to examine the proper object or the object is out of reach.

It may happen that engineer's intuition allows building a device that could not have been built on the basis of mathematical models due to the lack of sufficient development in a particular scientific discipline. The primary object (real or imagined) is called *original* and the object similar to the original is called *model*.

Let us notice that theory of similarity is based on dimensional analysis. The theory determines relationships between physical parameters, which influence the phenomenon under consideration. This approach is often used in nonlinear issues, especially in fluid mechanics and aerodynamics, or hydrodynamics.

It should be emphasized that *theory of similarity* is mainly supported by *measurement analysis* and defines dependencies between physical parameters having influence on a being investigated phenomenon. Such approach has been used often in nonlinear problems especially matched with fluid mechanics, aeromechanics or hydromechanics. Obtained results due to theory of similarity can be transformed from laboratory devices to industry and allow for minimization of measurement numbers and costs.

J. Awrejcewicz, *Ordinary Differential Equations and Mechanical Systems*,
DOI 10.1007/978-3-319-07659-1_7

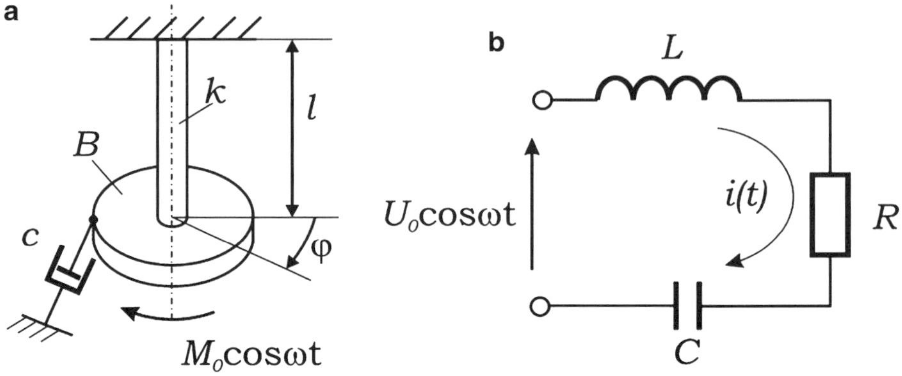

Fig. 7.1 The original (mechanical system) and the model (electrical system)

Traditionally the following partition of *similarity* has been introduced.

(i) *Geometrical*—shapes and dimensions of two objects are similar, e.g. a ratio of two characteristic dimensions;
(ii) *Kinematical*—distributions of physical fields lines are similar, e.g. velocity, pressure or acceleration;
(iii) *Dynamical*—scale of similarity of different characteristic quantities is the same. In the case of incompressible fluids, Reynolds number *Re* is such number and in the case of compressible fluids one uses Mach, Strouhal or Prandtl numbers. This description requires introducing concepts of force and torque, or tension.

Let us remind that Reynolds number $Re = \frac{\rho l v}{\mu}$ where l is characteristic dimension (e.g. diameter of a pipe), v is characteristic velocity of fluid and μ is kinematical viscosity, and ρ is density. For $Re < 2{,}300$ we have a laminar flow, for $2{,}300 < Re < 10{,}000$ transition flow and for $Re > 10{,}000$ turbulent flow appears.

Similarity numbers allow for qualitative evaluation of the examined phenomena. They are often used in applications. They follow: Abbe, Archimedes, Arrhenius, Biot, Euler, Fourier, Rayleigh or Weber numbers. Note that relationships original-model can concern both material systems and processes (e.g. flow), i.e. phenomena in general.

As an example, we consider a vibration phenomenon described by a second-order differential equation in the mechanical and electrical systems (Fig. 7.1).

The mechanical system equations of motion, based on Newton's second law, can be written in the form

$$B\ddot{\varphi} + c\dot{\varphi} + k\varphi = M_0 \cos \omega t, \tag{7.1}$$

where $(\dot{\cdot}) \equiv \frac{d}{dt}$, B is moment of inertia of a shield suspended on a weightless rod of torsion rigidity $k = (GI_0)l^{-1}$, G is shear modulus, I_0 is moment of inertia of cross-section and l is length of the rod. M_0 and ω are amplitude and frequency of harmonic excitation respectively, φ is generalized coordinate (angle measured in radian).

The electrical system equation of motion is based on Kirchhoff's law, i.e. we equate the sum of voltage drops on all elements of the circuit to the voltage on terminals. We obtain

$$L\frac{di}{dt} + Ri + \frac{1}{C}\int i dt = U_0 \cos \omega t, \tag{7.2}$$

where L is self-induction coefficient, R—resistance, C—capacitance, i—current intensity, U_0 and ω are amplitude and frequency of external voltage (driving), respectively.

Since $i = \frac{dq}{dt}$, where q is electric charge, then by Eq. (7.2) one gets

$$L\ddot{q} + R\dot{q} + \frac{1}{C}q = U_0 \cos \omega t. \tag{7.3}$$

It is easy to see that Eqs. (7.1) and (7.3) are similar. We have the following similar quantities: the rotation angle φ and the charge q; the angular velocity $\dot{\varphi}$ and the current intensity i; the moment of inertia B and the self-induction coefficient L; the viscous damping coefficient c and the resistance R; torsion rigidity k and reciprocal of the capacitance C^{-1}; the amplitude of the torque M_0 and the voltage U_0; frequency ω of torque and voltage.

We aim to determine the "period of damped vibrations in both systems". On this purpose, we transform the homogeneous equation, obtained from (7.1) and (7.3), into the form

$$\ddot{\varphi} + 2h\dot{\varphi} + \alpha^2\varphi = 0, \tag{7.4}$$

$$q'' + 2h'q' + \alpha'^2 q = 0, \tag{7.5}$$

where $\alpha^2 = \frac{k}{B}$, $2h = \frac{c}{B}$; $\alpha'^2 = (LC)^{-1}$, $2h' = \frac{R}{L}$, $' = \frac{d}{dt'}$, $^{\cdot} = \frac{d}{dt}$.

In order to determine relationships between the original and the model, we make use of *dimensional analysis*. Let us introduce the following values of the scales

$$k_\varphi = \frac{\varphi}{q}, \quad k_t = \frac{t}{t'}, \quad k_h = \frac{h}{h'}, \quad k_\alpha = \frac{\alpha}{\alpha'}, \tag{7.6}$$

where ($'$) refers to the model. Making use of (7.6) in (7.4) we get

$$\frac{k_\varphi}{k_t^2}q'' + \frac{k_\varphi k_h}{k_t}2h'q' + k_\varphi k_\alpha^2 \alpha'^2 q = 0. \tag{7.7}$$

The equations of the original and the model are identical when there are the following relationships between the scales

$$\frac{k_\varphi}{k_t^2} = \frac{k_\varphi k_h}{k_t} = k_\varphi k_\alpha^2. \tag{7.8}$$

The above relationships imply

$$k_t k_h = 1, \quad k_\alpha k_t = 1,$$
$$k_\alpha = k_h \tag{7.9}$$

which means

$$ht = h't' \quad \text{and} \quad \alpha t = \alpha' t',$$

or

$$\frac{h}{\alpha} = \frac{h'}{\alpha'}. \tag{7.10}$$

We have obtained the criterion of similarity (7.10). This means that some dimensionless combinations of the signal parameters must equal to the analogous combinations of the model parameters. By Eq. (7.10) we get as well

$$\frac{c}{\sqrt{kB}} = R\sqrt{\frac{C}{L}} \tag{7.11}$$

The above equation implies that two of the three parameters R, C, L can be chosen freely but the third one is obtained from (7.11).

When we have a model (the electrical system) we can measure the period of damping vibrations T', and then determine the period of vibrations in the mechanical system (the original) in the following way

$$T = k_t T' \tag{7.12}$$

where

$$k_t = \frac{1}{k_\alpha} = \frac{\alpha'}{\alpha} = \frac{\sqrt{\frac{1}{LC}}}{\sqrt{\frac{k}{B}}} = \sqrt{\frac{B}{kLC}}. \tag{7.13}$$

There is another often applied method of obtaining similarity criteria, namely the transforming equations of both the original and the model into the same dimensionless form, and the imposing of identity conditions on the form of the equations and their coefficients. The vibration phenomenon will be described by the only one dimensionless differential equation of second order. We introduce to the original and to the model, the following dimensionless values

$$\varphi_1 = \frac{\varphi}{\varphi_0}, \quad t_1 = \frac{t}{T}, \quad q_1 = \frac{q}{q_0}, \quad t_1' = \frac{t'}{T'}, \tag{7.14}$$

where φ_0, q_0, T' and T are values of the angle and the charge, and the time (period).

Inserting (7.14) into Eqs. (7.1) and (7.3) we get

$$\frac{B\ddot{\varphi}_1}{kT^2} + \frac{c}{kT}\dot{\varphi}_1 + \varphi_1 = \frac{M_0}{k\varphi_0} \cdot \cos 2\pi t_1, \tag{7.15}$$

$$\frac{LC\ddot{q}_1}{T'^2} + \frac{RC}{T'}\dot{q}_1 + q_1 = \frac{U_0 C}{q_0} \cdot \cos 2\pi t_1, \tag{7.16}$$

since the cosine input values are assumed to be the same, namely $\omega T = 2\pi$, $\omega' T' = 2\pi, \omega' t' = \omega t$.

One can assume that $\frac{\omega}{\omega'} = \frac{t'}{t} = \frac{T'}{T} = 1$, which implies $t_1 = t_1{}'$.

Equations (7.15) and (7.16) possess the same independent variable t_1. The quantities φ_1 and q_1 are dimensionless, so the original and the model can be represented by the second-order differential equation

$$(T_1^2 p^2 + T_0 p + 1)x = K \cos 2\pi t_1 \tag{7.17}$$

where

$$\begin{aligned} p &= \frac{d}{dt_1}, \\ x &= \varphi_1 = q_1, \\ T_1 &= \frac{B}{kT^2} = \frac{LC}{T'^2}, \\ T_0 &= \frac{c}{kT} = \frac{RC}{T'}, \\ K &= \frac{M_0}{k\varphi_0} = \frac{U_0 C}{q_0}. \end{aligned} \tag{7.18}$$

One can say that the obtained relationships are equivalent to (7.10), (7.11).

Now, let us come to the third way of determining similarity criteria between the original and the model, using Eqs. (7.4) and (7.5). Each of the equation terms (7.4) or (7.5) possesses the same dimension. The terms can be written in the following way

$$\left[\frac{\varphi}{t^2}\right] = \left[\frac{h\varphi}{t}\right] = [\alpha^2 \varphi], \tag{7.19}$$

dividing the above equality by $\left[\frac{\varphi}{t^2}\right]$ we get

$$[1] = [ht] = [\alpha^2 t^2]. \tag{7.20}$$

This implies the products ht and αt are dimensionless combinations of dimensional quantities for the mechanical system. Similarly, with regard to Eq. (7.5) we get that the products $h't'$ and $\alpha' t'$ are dimensionless combinations of dimensional quantities for the electric system.

Newton formulated the following law of similarity of two phenomena.

> We say that two phenomena are similar to each other when the quantities occurring in the equations which describe the phenomena form dimensionless combinations are of equal values.

In regard to the analysed example, the similarity criteria are the following

$$ht = h't' \quad \text{and} \quad \alpha t = \alpha' t' \tag{7.21}$$

which lead to (7.10).

The example was based on the knowledge of equations describing the phenomena in the original and the model.

Now, we will try to determine similarity criteria only through the analysis of physical quantities characterizing the phenomenon (the equations describing the phenomena are unknown), basing ourselves on dimensional analysis.

Every dimensional quantity can be presented in the following way

$$Q = q x_1^{a_1} x_2^{a_2} \ldots x_m^{a_m}, \tag{7.22}$$

where $x_1, \ldots, x_m$ are dimensional units, q is dimensionless, and $a_1, \ldots, a_m$ are real numbers. As an example, let us put $1\,W = 1\,\mathrm{N\,ms^{-1}}$ or $g = 9.81\,\mathrm{ms^{-2}}$.

Let n dimensional quantities $Q_1, \ldots, Q_n$ be given in the description of the examined phenomenon, however only m measurement units are required to complete the description, $n > m$. It turns out that the choice of these m measuring units is free provided that the condition of dimension independence holds.

We say that the quantities $Q_1, \ldots, Q_m$ are dimension-independent if the equality

$$Q_1^{r_1} Q_2^{r_2} \ldots Q_n^{r_n} = \gamma,\ \gamma > 0,\ \gamma, r_1 \ldots r_m \in R \tag{7.23}$$

implies $r_1 = r_2 = \ldots = r_m = 0$ and $\gamma = 1$.

Example 7.1. Show that the displacement s, the acceleration p and the force F are dimension-independent.

According to (7.23) we have

$$s^{r_1} p^{r_2} F^{r_3} = \gamma,$$

and after taking into account the dimensions

$$m^{r_1} \left(ms^{-2}\right)^{r_2} \left(kgms^{-2}\right)^{r_3} = m^o kg^o s^o,$$

or equivalently

$$m^{r_1+r_2+r_3} kg^{r_3} s^{-2r_2-2r_3} = m^o kg^o s^o.$$

Equating the exponents of the same bases we get

$$\begin{aligned} r_1 + r_2 + r_3 &= 0, \\ r_3 &= 0, \\ -2\,(r_2 + r_3) &= 0, \end{aligned}$$

hence

$$r_1 = r_2 = r_3 = 0$$

which proves that s, p and F are dimension-independent.

One can do it in another way. On this purpose, one must determine the rank of the exponents matrix of each dimensional quantity. For the considered case from the example 7.1 we have

	m	kg	s
s	1	0	0
p	1	0	−2
F	1	1	−2

Three quantities s, p and F are linearly independent because the determinant of third order is nonzero, i.e.

$$\begin{vmatrix} 1 & 0 & 0 \\ 1 & 0 & -2 \\ 1 & 1 & -2 \end{vmatrix} = \begin{vmatrix} 0 & -2 \\ 1 & -2 \end{vmatrix} = 2$$

If there are m of the n dimension-independent quantities, then it is easy to determine $n - m$ non-dimensional combinations of these quantities referred to the form

$$\begin{aligned} Q_{m+1} Q_1^{r_{1.1}} Q_2^{r_{2.1}} \dots Q_m^{r_{m.1}} &= \gamma_1, \\ Q_{m+2} Q_1^{r_{1.2}} Q_2^{r_{2.2}} \dots Q_m^{r_{m.2}} &= \gamma_2, \\ &\vdots \\ Q_n Q_1^{r_{1,n-m}} Q_2^{r_{2,n-m}} \dots Q_m^{r_{m,n-m}} &= \gamma_{n-m}, \end{aligned} \tag{7.24}$$

or of analogous form referred to the model (prime mark):

$$
\begin{aligned}
Q'_{m+1} Q_1'^{r_{1.1}} Q_2'^{r_{2.1}} \ldots Q_m'^{r_{m.1}} &= \gamma'_1, \\
Q'_{m+2} Q_1'^{r_{1.2}} Q_2'^{r_{2.2}} \ldots Q_m'^{r_{m.2}} &= \gamma'_2, \\
&\vdots \\
Q'_n Q_1'^{r_{1.n-m}} Q_2'^{r_{2.n-m}} \ldots Q_m'^{r_{m.n-m}} &= \gamma'_{n-m}.
\end{aligned}
\tag{7.25}
$$

The similarity criteria in number $n - m$ are obtained from the equations

$$
\gamma_1 = \gamma'_1, \ldots, \gamma_{n-m} = \gamma'_{n-m}. \tag{7.26}
$$

Let us introduce the following scales of independent quantities

$$
k_i = \tfrac{Q_i}{Q'_i}, \; i = 1, \ldots, m \tag{7.27}
$$

and dependent quantities

$$
k_{m+1} = \frac{Q_{m+1}}{Q'_{m+1}}, \tag{7.28}
$$

After dividing both sides of Eqs. (7.24) and (7.25), and making use of (7.26)–(7.28) we obtain

$$
k_{m+1} k_1^{r_{1.1}} k_2^{r_{2.1}} \ldots k_m^{r_{m.1}} = 1, \tag{7.29}
$$

hence

$$
k_{m+1} = k_1^{-r_{1.1}} k_2^{-r_{2.1}} \ldots k_m^{-r_{m.1}}. \tag{7.30}
$$

Product of type (7.24) takes the form

$$
\alpha t^a = \gamma_1, \quad h t^b = \gamma_2, \tag{7.31}
$$

where α and h are dimension-dependent.

From the first equation (7.31) we have

$$
s^{-1} s^a = s^0, \tag{7.32}
$$

hence $a = 1$, and the first invariant takes the form

$$
\alpha t = \gamma_1. \tag{7.33}
$$

The second equation (7.31) leads to determining the second invariant of the form

$$ht = \gamma_2. \tag{7.34}$$

Analogous considerations, carried out for the electrical system, lead to determining invariants of the form

$$\alpha' t' = \gamma'_1, \quad \text{and} \quad h' t' = \gamma'_2. \tag{7.35}$$

Since we get the following similarity criteria

$$\alpha t = \alpha' t', \quad \text{and} \quad ht = h' t', \tag{7.36}$$

dividing both sides we get the invariant

$$\frac{\alpha}{\alpha'}\frac{t}{t'} = \frac{\gamma_1}{\gamma_1'} \equiv 1, \quad \frac{h}{h'}\frac{t}{t'} = \frac{\gamma_2}{\gamma_2'} \equiv 1, \tag{7.37}$$

thus $k_\alpha k_t = 1$, $k_h k_t = 1$. □

At the end, we consider the procedure of transforming Duffing's equation into dimensionless form.

Example 7.2. Let us analyse an oscillator described by the second-order differential equation

$$\frac{d^2\bar{u}}{d\bar{t}^2} + f(\bar{u}) = 0.$$

We Maclaurin-expand the nonlinear function

$$f(\bar{u}) = f(0) + \bar{u}\frac{df}{d\bar{u}}(0) + \frac{1}{2}\bar{u}^2\frac{d^2 f}{d\bar{u}^2}(0) + \frac{1}{3!}\bar{u}^3\frac{d^3 f}{d\bar{u}^3}(0) + \cdots .$$

Let us take

$$f(0) = \frac{d^2 f}{d\bar{u}^2}(0) = 0 \quad \text{and} \quad \frac{df}{d\bar{u}}(0) > 0,$$

hence the resulting equation takes the form

$$\frac{d^2\bar{u}}{d\bar{t}^2} + \bar{u}\frac{df}{d\bar{u}}(0) + \frac{1}{6}\bar{u}^3\frac{d^3 f}{d\bar{u}^3}(0) = 0.$$

After introduction of the following non-dimensional quantities $u = \bar{u}l^{-1}$, $t = \bar{t}T^{-1}$ one gets

$$\frac{l}{T^2}\frac{d^2u}{dt^2} + ul\frac{df}{d\bar{u}}(0) + \frac{1}{6}u^3l^3\frac{d^3f}{d\bar{u}^3}(0) = 0,$$

which leads to the equation

$$\frac{d^2u}{dt^2} + uT^2\frac{df}{d\bar{u}}(0) + \frac{1}{6}u^3T^2l^2\frac{d^3f}{d\bar{u}^3}(0) = 0. \qquad (*)$$

Assume that

$$\left[T^2\frac{df}{d\bar{u}}(0)\right] = 1,$$

what follows from the often applied combination $[FT^2/ML] = 1$, where L is the length, T the time, M the mass and F the force. Note that $[f] = Nkg^{-1}$, $[\bar{u}] = m$, $\left[T^2\right] = s^2$, and hence

$$\left[T^2\frac{df}{d\bar{u}}(0)\right] = s^2(kgms^{-2})kg^{-1}m^{-1} = 1.$$

Equation $(*)$ takes the following dimensionless Duffing-type form

$$\frac{d^2u}{dt^2} + au + bu^3 = 0,$$

where:

$$a = \frac{df^*}{du}(0), \quad b = \frac{1}{6}\frac{d^3f^*}{du^3}(0), \quad f^* = \frac{fT^2}{l}.$$

Transformation of the dimensional equation into dimensionless one has a significant meaning. It enables to generalize the results, i.e. transition from one scientific discipline to another one, e.g. from mechanics to electrics. Moreover, transforming equations into dimensionless form sometimes allow the number of parameters to be reduced significantly. It has non-trivial meaning in analysis, especially during numerical computations. □

Chapter 8
Model and Modelling

8.1 Introduction

The concept of model has deep historical roots and definitely it is not unambiguous. Especially it is used in both common life and science. In this book, we think of a model as a simplified representation of the reality. Depending on the goal of investigation, a human makes some simplifications intentionally leaving the real system without many details and many features, which are unnecessary from the awaited goal viewpoint.

The concept of model is very broad. In sociology, culture is treated as a collection of behaviour models and human activities. In linguistics, in the case of colloquial language, grammatical rules and writing are models.

The ideal scientific theory should include a collection of minimal axioms (principles and concepts taken without proof). One can obtain, basing oneself on the axioms, a solution of any problem with the use of formal logic, i.e. mathematically. It turns out that complexity of phenomena around us, as well as limitations originating from creative abilities of Man makes a scientific theory impossible to apply.

The examining processes proceeding in the reality is performed by means of suitable mathematical apparatus, based on the earlier chosen mathematical models. The structure of mathematical model is the key to describing phenomena and processes. It should be emphasized that any mathematical model is approximated and is not adequate to the process it describes.

When one constructs a mathematical model, one tends to catch the most characteristic features of the analysed process. On the other hand, the mathematical model should be quite simple and it should provide necessary information about the process under consideration. Consequently, some singularities of a process are completely taken into account, others at a certain degree and the rest is ignored. We say about the so-called idealization procedure and the success of the investigation strongly depends on this procedure.

J. Awrejcewicz, *Ordinary Differential Equations and Mechanical Systems*,
DOI 10.1007/978-3-319-07659-1_8

In order to make use of mathematical apparatus in practice, any problem should be simplified. Only experiments can verify the assumptions, the ones that should be verified not only physically but also mathematically.

One does not require experiments related to problems, which do not differ from each other too much, while qualitatively new problems require new experiments since the introduced simplifications can lead to physically unrealizable conclusions. During the idealization procedure it is necessary to introduce a kind of order of its elements by comparing them.

For example, if one of the system elements is 1 cm long then a natural question arises, namely: is this length small or large? Only initial formulation of the problem can give an answer to this question. It is obvious that when we examine the motion of an orbiting satellite, we can assume that 1 cm is negligible. On the other hand, if we consider distances between molecules then 1 cm is huge.

Let us take another example. It is well known that air is compressible. Does one always need to take into account the compressibility of air? Again, it depends on initially formulated problem. If an object moves in air at small velocity V, then we can neglect the compressibility. However, if the velocity is high, even close to the speed of sound or higher, then we must take into account the compressibility. In this case, it is convenient to introduce a dimensionless quantity $M = \frac{V}{a}$ called Mach number. It plays an important role in aerodynamics.

For example, as $M \ll 1$ one can use the idealized model of incompressible fluid, however at larger values of Mach number the compressibility of air should be taken into account. We have the similar situation during the building of a mathematical model and in other branches of science or technology, where other characteristic dimensionless numbers play an important role. These numbers are formed as combinations of three dimensionless quantities, such as: length L, time T and mass M. For the sake of convenience, one assumes that the dimension of the combination FT^2/ML equals 1 (where F is force). In other words, one of the quantities F, T, L and M can be chosen freely.

The concept of *model* can be defined in regard to a language L. In this case, the model consists of the universal set U containing all the objects under examination (symbols), which are significant from the goal of examinations viewpoint, and the map from L to U, called an interpretation function.

From the viewpoint of logic, a theory is defined as a consistent collection of sentences. The Austrian scientist Kurt Gödel stated the theorem of consistency, which says that a theory possesses a model if and only if it is consistent. This means that theory can lead to the contradiction. Gödel's theorem, proved in 1931, allows to verify correctness of a theory through the examining of its models. It follows from the Gödel's theorem that the theory T proves the sentence X if and only if each model of T fulfils X. Validity of a statement X is proved on the basis of statements belonging to the theory T, axioms and proving rules of classical logic.

Besides Gödel, the Polish mathematician and logician Alfred Tarski had an impact on the theory of models, called *logical semantics*. Tarski claimed that in a theory there must exist true statements, which are not proved by the theory. He emphasized the importance of meaning (*semantics*) of the examined statements and not only their *syntactical* relations. He pointed the attention that equivalence of syntactic and semantic consequences exhibits only internal equivalence of the theory but it does not allow to conclude about the validity of a sentence as a result of the analysis of syntactic relations.

In the 1960s of twentieth century, mathematicians and logicians obtained many significant results so that the theory of models isolated from logic as an individual science.

The reality around us is so rich that sciences which so far have not undergone absolute formalization and their results are based on deductive reasoning and intellectual evaluations dependent on subjective media. With regard to rich development of mathematics, a natural tendency of a researcher is to introduce a *mathematical model*, i.e. applying mathematical apparatus for description of the examined object or process. Most of the sciences undergo such an approach. One needs to mention such sciences as: mechanics, electrical engineering, physics, biophysics and biomechanics, biology, economy, sociology, medicine or even politics sciences.

Most of the examples given in this book and concerning the modelling will be related to applied sciences. Nowadays, it is difficult to imagine life without engineering achievements based on mathematical fundamentals. This concerns civil engineering, construction of airplanes and spaceships, transport of people and resources, production of machines and mechanisms, control of complex technological processes, novel solutions in bioengineering, or various solutions in electronics and high technology.

With regard to rapid changes and achievements in science, particular scientific disciplines underwent the transformations, which originate from the expectations imposed by the industry, as well as tendency to being interdisciplinary. The field of interests of mechanics ranges from atomic to astronomic phenomena. The tools of mechanics are used by physicists, biologists, chemists, ecologists, physicians, economists and even cryptologists which proves its interdisciplinary.

There exist indissoluble relations between mathematics and mechanics in historical course and one pointed out how the development of one of the mentioned scientific disciplines positively influenced the development of the second one. For example, problems appearing in a field of interests of mechanics stimulated the development of such branches of science as theory of dynamical systems or theory of optimization. Mechanics was a source of rise of individual scientific disciplines such as theory of control or theory of bifurcations and chaos.

Mechanics, in the sense of Gödel's approach, is a theory represented by the system hierarchically ordered. The highest level of the hierarchy is usually connected with a description of the whole "macro-body" with the use of time-varying parameters. On technical purposes, from mechanics one isolates mechanics of materials, solids mechanics and fluids mechanics.

Mechanics makes use of classical tools of fundamental physical laws, when relatively small number of material points is taken. While the number of these objects increases (in the case of molecules and atoms) then mechanics makes use of phenomenological theories. It turns out that there is no model, which describes transition on the basis of the knowledge concerning particular basic elements of structures such as atoms and molecules. Mechanics allows for explaining many biological or medical problems such as biological structures or hemostaza processes. The attention should be paid to biomechanics—a new and dynamically developing branch of mechanics.

It should be emphasized that examining any continuum medium is connected to the analyses of fields of displacements, deformations, temperature, velocity, acceleration and stress, but these fields vary in time randomly.

The outlined portrait of contemporary mechanics points the difficulties related to the theory of modelling. Moreover, modelling in this domain of science, supported by computational methods, gives hopes connected with solving of many courageous problems in the future such as: predictions based on computer simulations of properties of natural objects, working out new and complex models of continuum media, working out transition techniques from nano to micro scales, modelling of biological systems or yielding new materials of the desirable properties.

8.2 Mathematical Modelling

Let us start from the definition given by Gutenbaum [107]. A mathematical model is a collection of symbols and mathematical relations, and the rules how to operate them. The mentioned symbols and relations refer to specific elements of the modeled part of the reality. Interpretation of symbols and mathematical relations made a model come into being. Mathematical modelling is characterized by two essential features:

(i) *interdisciplinarity* (introducing the language of formal logic enables to examine many various scientific disciplines);
(ii) *universalism* (the same mathematical model can be applied in different branches of science).

Great scientists of eighteenth and nineteenth centuries pointed that applying the rules of mathematics and logic awaited empirical verification. Since mathematics and logic is a product of the human mind, and the latter undergoes evolutionary process according to the laws of the nature, then one can expect that the product of the mind is also consistent with the nature. Consequently, the product possesses a corresponding real element of the being.

There are many examples of application of outwardly pure abstract products, e.g. non-Euclidean geometry or Boole algebra. The aim of mathematical modelling is to describe our *reality*. An *experiment* is an ultimate verification of modelling correctness.

In general case, a sequence of events leading to appearance of a mathematical model is depicted in Fig. 8.1.

Mathematical modelling is successful and reaches sciences, which have been regarded as being resistant to mathematizing as e.g. history. Geometry

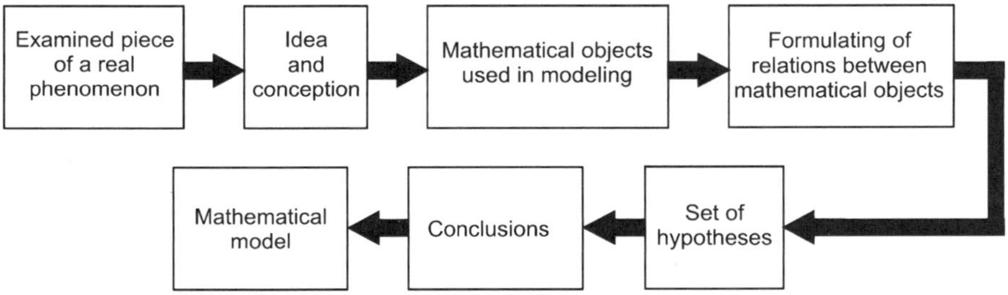

Fig. 8.1 Thinking process leading to the appearance of a mathematical model

undergoes unbelievable mathematical regime, as Spinoza (1632–1677) emphasized. He presented his philosophy in "a geometric manner" using definitions and axioms.

It was Galileo (1564–1642), who introduced mathematics into physics and Lavoisier (1743–1794) introduced mathematics into chemistry. In order to familiarize with the modelling procedure we will focus on mechanics.

8.3 Modelling in Mechanics

In order to examine an object or a phenomenon we are interested in, first we must build its physical model. A physical model is understood as a conception of physical phenomenon description and determination of parameters having the influence on its course.

The model should accurately represent processes proceeding in the real object. On the other hand, the model should not be too large. The next stage of modelling is the building of a mathematical model that is understood as mathematical formulation of physical laws. A mathematical model can be represented by algebraic equations, ordinary or partial differential equations, differential–integral equations, difference equations or difference equations with a delayed argument.

Although a mathematical model of a phenomenon should describe real conditions as accurately as possible (modern computational methods enable to analyse very complicated models with nonlinearity, stochasticity and various types of external interactions involved), it can happen that the costs connected with the analysis of so complicated problems are very high. Moreover, sometimes very simple models are sufficient to describe a phenomenon.

In the case of analysis of construction, one can treat the analysis as a system of coordinates, which are decision variables (constructional) and parameters from mathematical viewpoint. Moreover in the case of modelling of construction, one often makes use of a simplified model at the preliminary stage, while in final stages the model is more complicated.

The goal of mathematical description of construction is mathematical formulation of all constructional features, i.e. geometrical, material and dynamical ones. On the other hand, one needs to remember that constructional conditions make functional limitations. A complete mathematical model of construction covers determination of decision variables, determination of a set of good constructions and creating an optimization criterion—an objective function.

Mathematical models of construction can be classified as follows:

a) deterministic—all parameters are determined and each decision corresponds to one value of the objective function;
b) probabilistic—one or a few parameters are random variable with known probability distribution;
c) statistic—one or a few parameters are random variable with unknown probability distribution;
d) strategic—one or a few parameters can take on one of the determined values.

It is worth noticing that dynamical processes proceeding in a physical model are not accurate reflection of a real phenomenon but they rather present the actual knowledge about the examined phenomenon.

The differences between a real process and a process proceeding in a physical model are called disturbances. By the notion of disturbances we understand unknown or known influence of small changes of the system parameters on the process. This influence is intentionally ignored.

In order to understand the concept of disturbances better, let us consider an example of mechanism with ideally stiff links. Assumption of such a model is a kind of approximation, since in a real mechanism there are many (omitted) phenomena described by equations of theory of elasticity, rheology, fluid mechanics and others.

A mechanism with susceptible links can make the same movements as a mechanism with stiff links (classical), but additionally these movements can interfere with vibrations, which are connected with deformability of the constraints. The motion of a classical mechanism, called the base motion, is perturbed by additional motion, namely vibrations in this case. This additional motion will be called a disturbance of the base motion.

This observation also led to an original method applied in a new branch of mathematics and physics, namely the asymptotology. Due to the introduction of a naturally occurring quantity, namely a small parameter (perturbative), there exists a possibility of suitable separation of equations describing a modeled phenomenon into several systems. The systems describing processes, which differ in terms of only magnitude of some characteristics can be solved separately, that significantly simplifies the computing procedure. Next, the obtained particular solutions are combined into asymptotic series.

The next stage of modelling is the analysis of equations describing dynamics of machines. The analysis is performed in various ways, depending upon a type of the equations. Generally, one can isolate some essential steps applied at this stage. First, one seeks analytical solutions of such equations. One can do it if one deals with a linear mathematical model. It turns out that even very simple nonlinear systems do not possess unique solutions or it is very difficult to find solutions. Then, one uses analytical methods for finding approximated solutions based on elementary functions.

Such solutions allow for at least initial to familiarize with qualitative and often with sufficient accuracy quantitative dynamics of the system. However, so far numerical methods have been widely used in the analysis of the equations. On a present level of development of calculating machines (computers) and numerical methods, very many problems of dynamics of machines can be successfully solved.

Finally, the last stage of modelling covers the choice of system parameters so that one achieves the required dynamical properties. The choice is made within a framework of *synthesis* and *optimization*.

The next step, after synthesis and optimization, is a construction of a machine or a mechanism, which should be optimal with respect to the chosen optimization criteria. The detailed principles of constructing follows:

(a) functionality;
(b) reliability and durability;
(c) efficiency;
(d) lightness;
(e) cheapness and availability of materials;
(f) producibility;
(g) easiness of usage;
(h) ergonomicity;
(i) agreement with current norms and regulations;
(j) corrosion-proof and resistant to changes of temperature;
(k) aesthetics.

The optimization procedure can be defined as the choice of an element from a given set with regard to an order in this set [94]. Different variants of construction or control of particular processes can be included in the set. The set of solutions is usually bounded to the set of admissible solutions. Numbers (Euclidean space) or functions (Hilbert space) are components of a vector of decision variables.

The operator transforming an admissible set ϕ (or space of solutions E_x) into another set ϕ_q (or quality space E_q), which is called a set of attainable objectives is said to be an optimization criterion Q. The optimization criterion can be treated as a hypersurface located over the admissible set.

If $E_q = R^m (m = 2, 3, \ldots)$, then the operator $Q(q_1, \ldots, Q_m)$ associates each element belonging to the admissible set (ϕ) with m numbers characterizing the quality of the element. This is a job of poly-optimization, since m components of a vector of the function Q correspond m different optimization criteria. An element of a subset of m-dimensional Euclidean space $Q(x) = (q_1(x), \ldots, Q_m(x))$, called an *objective vector* for a fixed vector of decision variables, is a set of m numbers determining the value of particular quality coefficients. It is worth noticing that dimension n of a vector of decision variables does not depend on dimension of a vector of quality coefficients m.

From mathematical point of view, by optimization we understand finding such an element $x^* \in \phi$, that for $x \neq x^*$ the values of $Q(x)$ are not less than the values of a quality functional at the optimal point x^*, i.e. $(x^* \in \phi) : \wedge_{x \in \phi} Q(x) \geq Q(x^*)$.

The point $x^* = (x_1^*, \ldots, X_n^*)$ is said to be optimal, if the components $x_1^*, \ldots, x_n^*$ are optimal values of particular decision variables.

Depending on ϕ and Q we have to do with various optimization tasks. If an admissible set has a finite number of elements, then a problem of optimization is *discrete*. A problem of optimization is continuous if we have to do with continuous variation of decision variables. When we have to do with both mentioned tasks of optimization, then problems of optimization will be *mixed*.

Problems of optimization can be classified as problems with and without limitations, and static or dynamic problems. The case of static analysis in E^n belongs to a problem of linear or mathematical programming. If the dimension of Euclidean space is 1, then we have to do with minimization of a function of single or multi variables in one direction.

In the case of *dynamical* optimization, admissible solutions are elements of infinite-dimensional space, i.e. they are functions of independent variables. In practice, we mostly have to do with optimal choice of controlling, while an admissible set is given by equational and inequational constraints, and differential equations of a state.

Optimization problems can be classified with respect to the way they are solved. Generally, optimization problems can be classified as partial problems, and then we say about the *decomposition* of a problem. *Iterative* and *statistical* methods belong to the computational methods of optimization.

Besides the mentioned theoretical analysis, one should perform experimental research. The goal of such an examination is identification of elements of a physical model, identification of a mathematical model and verification of theoretical considerations. The experimental research is often performed in parallel with the theoretical research.

By the occasion, it is worth paying attention to significance of modern research techniques supported by reliable mathematical methods. As an example one can mention constructions consisting of non-uniform shells of various thickness. It turns out that the most reliable results can be obtained by using modern experimental methods.

There is a scheme of elements of modelling in dynamics of machines depicted in Fig. 8.2.

One distinguishes two classes of physical models. *Structural models*, belonging to the first class, whose organization is similar to organization of the analysed object and there exists correspondence between the elements of the object and model.

In the case of mechanical models, one distinguishes continuous and lumped models. The first ones are systems with continuously distributed parameters and they are usually described by means of partial differential equations, integral or differential-integral equations.

Lumped mechanical systems are described by ordinary differential equations and are simpler than continuous ones during the analysis. That is why continuous systems are often approximated by discrete systems. Moreover, one distinguishes linear and nonlinear models which are described by linear and nonlinear equations respectively. Usually, linear systems are approximations of nonlinear ones because the world outside is nonlinear.

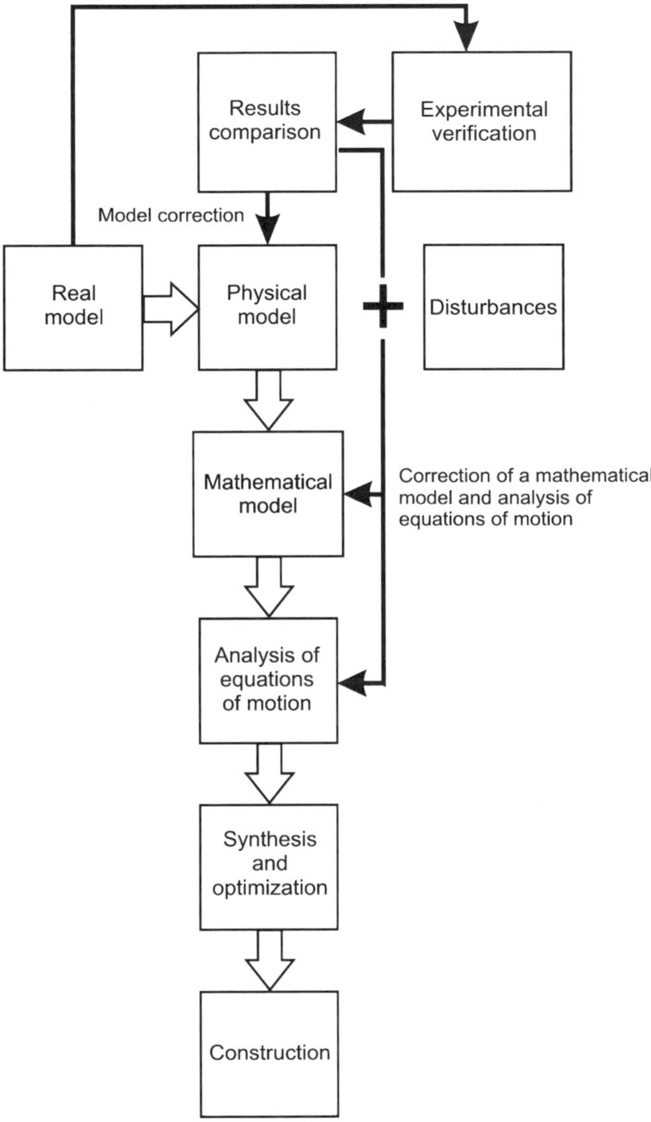

Fig. 8.2 Scheme of modelling stages in dynamics of machines

It is worth emphasizing that we will consider rather simplified mathematical models generated by problems related to nonlinear dynamics of machines. Moreover, we will show, even in regard to complicated systems of nonlinear equations, how to obtain much information, desired by engineers concerning the behaviour of the systems with the use of an associated simplified model, which is very often a system of linear differential equations.

Further considerations will be carried out on the basis of a mathematical model described by the equations

$$\frac{dy_s}{dt} = F_s(t, y_1, \ldots, y_n), \quad s = 1, 2, \ldots, n, \quad n \in N. \tag{8.1}$$

This is a very general formulation of equations. The processes proceeding between ideally stiff bodies, ones connected by a massless system of stiffness and damping can be brought to these equations. The system (8.1) needs more discussion. First of all, it is in normal form i.e. derivatives of the quantities are on the left-hand side of the equation. It can happen that the system $F_s(t, y_1, \ldots, y_n, \dot{y}_1, \ldots, \dot{y}_n) = 0$ is implicit, however such problems will not be discussed here. Most of methods and textbooks on numerical solutions of ordinary differential equations concern a system of the form (8.1). Moreover, the right-hand side of the system (8.1) is continuous, thus some of the processes proceeding in dynamical systems are not modelled by the system (8.1).

The techniques of transforming mechanics problems into a system of differential equations (8.1) are described in many textbooks on mechanics and will not be discussed here. With regard to a problem or needs of a user, on the medial stage leading to the system (8.1) one uses Lagrange's equations of the first or second kind, Lagrange's equation with multipliers, Boltzmann–Hamel or Maggi's equations (see [12, 13]).

The state of a dynamical system is described by $y_1, \ldots, y_n, t$, where t designates time, while $y_1, \ldots, y_n$ for each fixed $t \geq t_0 \geq 0$ take on real values. The Cauchy problem connected with the system (8.1) requires a "starting point", i.e. imposing initial conditions of the form $y(t_0) = y_0$, thus for the initial instant a state of the system is known, i.e. $y_{10}, y_{20}, \ldots, y_{n0}$.

Moreover, assume the function F is defined, continuous and its first derivative with respect to y is continuous as well for all $t \in I = t : t \geq t_0 \geq 0$ and $||y|| < \infty$, where $|| \cdot ||$ denotes the norm.

The mentioned assumptions ensure for each $t_0 \geq 0$ and $||y|| < \infty$ local existence and uniqueness of solutions as well as their continuous dependence on the initial instance y_0 in the finite time interval.

Moreover, we will deal with only such a class of equations for which solutions can be extended to $+\infty$, and this covers most of mechanical systems. Assume we know a particular solution to the system (8.1)

$$y_s = \varphi_s(t), \; s = 1, \ldots, n, \tag{8.2}$$

while $\varphi_s(t_0) = \varphi_{s0}$. This solution for $t > t_0$ means that for each value of t the instantaneous state of the system is generated by the initial state.

This raises a question: can one, on the basis of proximity of two states (proximity of two sets of initial conditions), come to conclusion about the distance between these two states as $t \to \infty$. Theory of *stability* of motion deals with such problems.

8.4 General Characteristics of Mathematical Modelling of Systems

One can distinguish the following types of models of systems connected with the goals of modelling:

(i) phenomenological model—describes and explains functioning of a system;
(ii) prognostic model—allows for predicting the behaviour of a model in future even when there are different conditions of environment interactions;
(iii) decision model—allows for a proper choice of input interactions satisfying required conditions;
(iv) normative model—allows for a proper choice of parameters and structure of a model for realization of particular tasks.

As one already mentioned, mathematical models should be compatible with the modeled system and easy in usage on the basis of the modelling procedure in mechanics. *Verification* and *validation* of a model plays a crucial role during the modelling procedure. Comparison of results obtained during the modelling procedure with a modeled object conserved (piece of the reality or more complex medial model) with respect of theory and experiment is called verification of a model.

The following criteria decide about consistency between an original and model:

(i) *Internal*, based on *formal* consistency (no logic and mathematical contradictions) and *algorithmic* (correctness of functioning of computational algorithms);
(ii) *External*, based on *heuristic* consistency (ability of interpretation of phenomena or verification of hypotheses and formulation new research task) and *pragmatic* (evaluating whether a built model is good).

In 1976 Zeigler [251] formulated the following types of consistency:

(i) *Replicative*—relies on evaluation of consistency with the same data as one used during identification of models;
(ii) *Predicative*—relies on evaluation of consistency with the modeled system under another conditions than during the performance of identification of a model;
(iii) *Structural*—relies on preservation of consistency also with respect to the structure of a model.

8.5 Modelling Control Theory

Now, we deal with modelling in automation and control theory which can be successfully applied in scientific disciplines such as: mechanics, electronics, physics, civil engineering, chemistry, chemical and process engineering, biochemistry and biomechanics, materials science and others. Moreover, many conceptions of the mentioned sciences can be easily transferred to another system, which is an object of considerations of cybernetics, psychology, linguistics or other arts. This chapter is based on the textbook [37].

Classification of control systems is performed on the basis of assumed criteria. One of the most important criterions is the one of required initial information (a priori) about a controlled system. According to this criterion, automatic control system can be divided into:

(i) ordinary control systems;
(ii) adaptive control systems;
(iii) distributed control systems.

Ordinary control systems require some initial information about a controlled process, i.e. detailed information about properties, equations, characteristics and parameters of the process before the start of controlling. Such systems are characterized by a constant structure and time-invariant values of parameters of particular elements during the system operation.

Ordinary control systems are widely discussed in literature and applied in practice. Since conditions variation of many physical processes is slow in time enough, it is sufficient to apply ordinary control systems (instead of more complicated control systems, namely adaptive ones).

In Fig. 8.3 one presented classification of ordinary control systems.

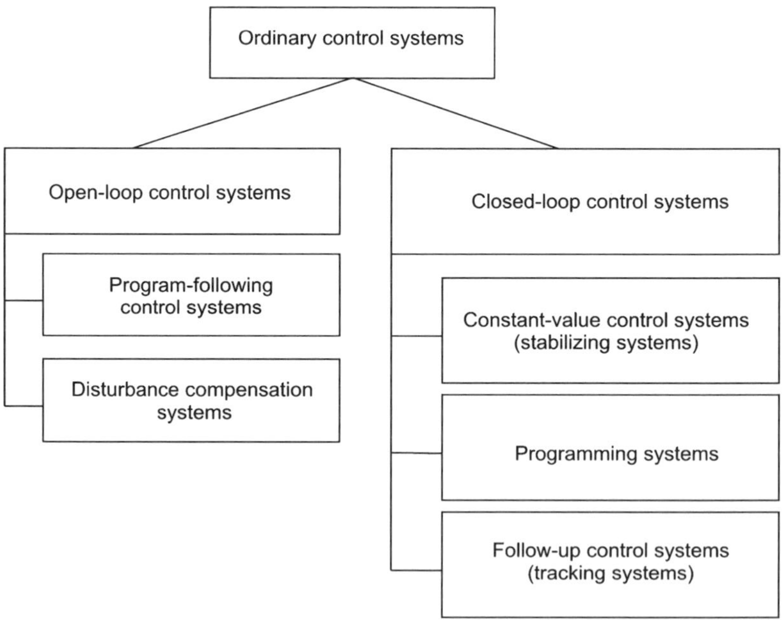

Fig. 8.3 Classification of ordinary control systems

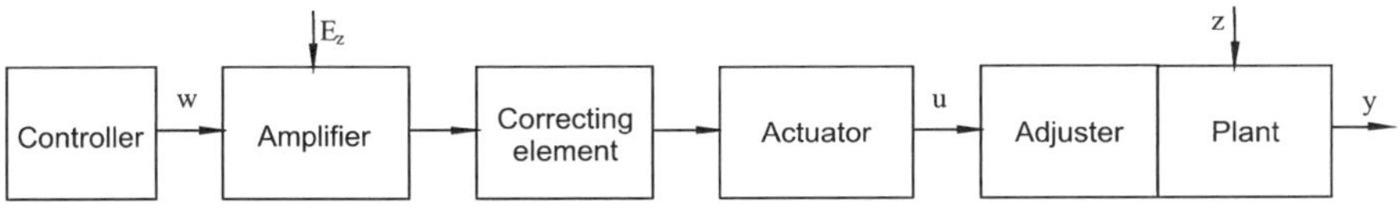

Fig. 8.4 Block diagram of an open-loop control

8.5.1 *Ordinary Automatic Control Systems*

Principles of operation and basic structure of control systems of different types will be presented in block diagrams.

8.5.1.1 Open-Loop Control Systems

Open-loop control systems constitute a group of systems, in which there is no feedback, which makes the input dependent on a selected output quantity. This type of systems requires complete initial information about a controlled system (process) with regard to the lack of feedback. The operating information is contained mainly in controlling and disturbing quantities. Open-loop control systems can be divided into open program-following control systems and open systems with disturbance compensation.

Principle of operation of an open-loop control system is depicted in Fig. 8.4.

Such systems perform jobs in order, preset by a controller, independently of the state of a controllable quantity y and disturbances acting on a plant (an object to be controlled). A correcting element in the system is necessary, when the response y_z of the object to the disturbance z differs from the response y_u of the object on the control signal u, and this can occur in both stationary and transient states. An amplifier is an element of the system which amplifies its output by means of the energy E_z taken from outside.

Open program-following control systems very often occur in processing industry in machines such as: numerically controlled machine, cyclic automata, etc. The information (program) is stored in controllers (memory devices), e.g. magnetic drums, perforated tapes or diskettes. The information in digital form is delivered to actuators and ensures the preset order and parameters of the procedure. In the similar way mechanical copy devices operate, in which a program is contained in a suitable type of cam mechanisms. Another example of an open-loop control system is control of the traffic lights.

A block diagram of an open system with disturbances compensation is depicted in Fig. 8.5.

Open systems with disturbance compensation are used to reduce undesirable influence of the environment (disturbances) on a process or object to be controlled by means of measurement of these interactions and compensation, which is performed by the additional inverse interaction on the object. The operating information is

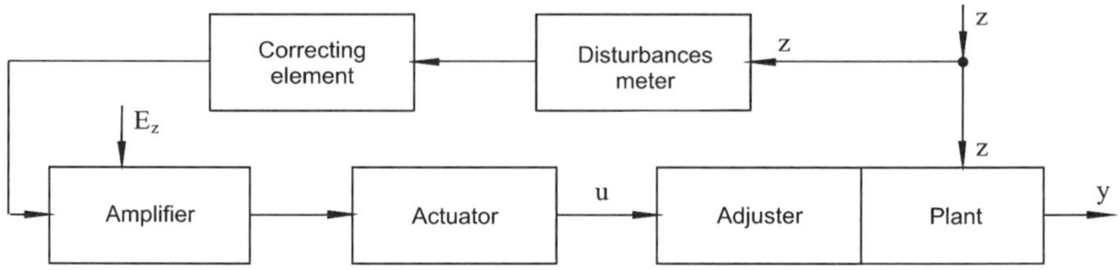

Fig. 8.5 Block diagram of open-loop control system with disturbance compensation

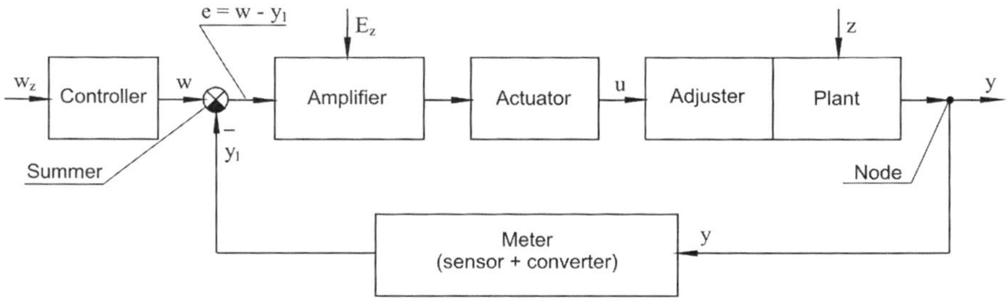

Fig. 8.6 Block diagram of a constant-value control system with a regulator of indirect operation

contained in disturbing quantities, but one needs to choose only those disturbances which influence the process most, since it is impossible to cover all disturbing interactions. As an example of this kind of systems, one can mention thermo-compensation systems in sensitive measurement devices.

8.5.1.2 Closed-Loop Control Systems (Automatic Control Systems)

Closed-loop control systems operate on the basis of the negative feedback principle, in other words measurement of the output quantity of an object (controllable quantity) and comparison of this result with the preset value of this quantity. Hence, a signal of control deviation arises. So the operating information is contained in the controlled signal itself. The initial information about a process to be controlled must be moderately complete, however less so than in open systems.

A control system with single controlled quantity and single feedback loop is called a one-dimensional system, while systems with many controlled quantities are called multi-dimensional.

Closed-loop control systems can be classified as follows: constant-value automatic control systems (stabilizing), program-following automatic control systems and follow-up control systems (tracking systems in other words).

A block diagram of a simple constant-value control system is depicted in Fig. 8.6.

As one can see in Fig. 8.6, the system possesses a backward feedback loop running through the meter of the controlled quantity y. Thus, it is a closed system.

In the main loop of the system there is an amplifier supplied with the energy E_z besides an actuator, adjuster and plant.

In consequence, the signal power from the controller is amplified and such systems are called systems with a controller of indirect operation. Otherwise, if a regulator does not amplify signals (there is no amplifier), then we have to do with simple systems with a regulator of direct action. A characteristic feature of constant-value control systems is constant value of the presetting quantity ($w(t) = \text{const.}$), which can be varied (if necessary) by means of various inputs w_z adjusted in the controller by means of handwheels. On the preset level w of the controlled value y the system will stabilize its course despite the undesirable disturbance z interaction on the object.

Control systems containing one loop of control are called one-loop systems. Such a system is depicted in Fig. 8.6. However, both one-dimensional and multi-dimensional systems can be multi-loop.

A set of elements of a simple regulator of indirect operation (see Fig. 8.6) is not always sufficient in practice. One introduces correcting serial and parallel elements into the system in order to ensure the necessary stability of operation, suitable quality and accuracy of control.

A block diagram of such a stabilizing one-dimensional system (which is then a one-loop system) is depicted in Fig. 8.7.

Constant-value control systems are the largest group of control systems, which are encountered in practice.

The second group of closed-loop control systems is different from stabilizing systems by that the presetting quantities $w(t)$ (preset values of controlled quantities) vary in time according to the program introduced by the service of the controller. Thus, these quantities are known, programmed functions of time ($w(t) = f(t)$).

A block diagram of the program-following control is similar to that in Fig. 8.7 but instead of an ordinary controller a program controller occurs. The controllable quantity y varies according to the program stored in a controller. Follow-up control systems do not possess a controller. The presetting quantity $w(t)$ is a random, previously unknown function of time ($w(t) = f_l(t)$). The controlled quantity y follows or tracks the course $w(t)$.

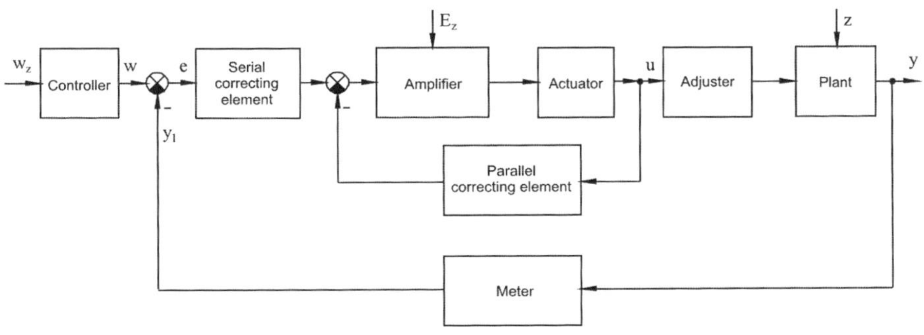

Fig. 8.7 Block diagram of a stabilizing system with a serial and parallel correcting element

In practice, one can also meet combined systems (closed-open), where feedback loop occurs with section of disturbances measurement.

8.5.1.3 Adaptive Automatic Control Systems

Adaptive control systems are such systems in which required initial information about a controlled process or object does not have to be complete. This means that in order to ensure the required accuracy and quality control, these systems require smaller range of the initial information than ordinary systems do. It follows from the fact that adaptive systems have an ability to adapt to such changes of the object operating conditions, which result in a change of its parameters or characteristics.

Activities connected with an object control in these systems rely on continuous or periodic examination of the object, and this results in completing the initial information.

Cars (changes of the tractive adhesion coefficient, changes of intensity of wind) and airplanes (changes of flight parameters according to altitude or weather conditions) are examples of objects, whose characteristics change as the operating conditions change.

Another type of objects of variable properties are objects with characteristics possessing extrema. The objective of an adaptive system is to keep an object operating point at maximal point of this characteristic. An example of such an object can be a radio receiver, whose circuit must be aligned to the frequency of a received electromagnetic wave in order to obtain a maximal value of intensity of the received signals. Another example is a water turbine, whose efficiency coefficient η under different loads Z_i varies according to curves possessing maxima according to the value of angle α of blades configuration (Fig. 8.8). Keeping a turbine operating point at maximal efficiency requires an adaptive system.

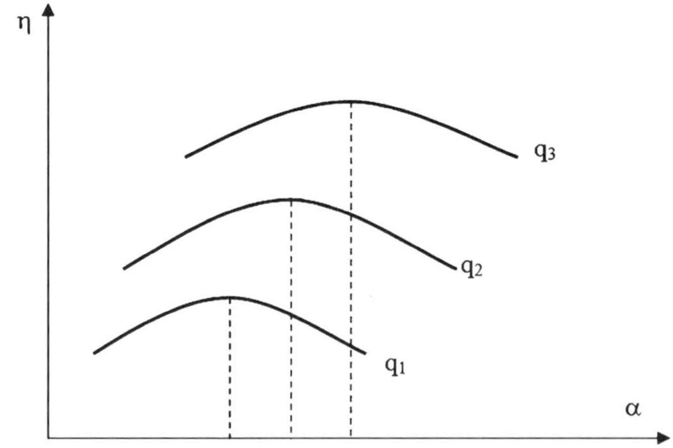

Fig. 8.8 Extremal characteristics of a water turbine

Adaptive systems can be divided into extremal control systems, control systems with self-adjustable correcting devices and self-optimizing systems. Extremal adaptive control systems refer to objects of characteristics, which possess extrema. The operating information in these systems is deviations from an extremum of a particular function of a single or multiple variable and it is not necessary to know precise position of the extremum at the beginning. In fact, it is sufficient to know a type of the extremal function and occurrence of the very extremum. The system gains the lacking information about position of the extremum during the seeking procedure.

Adaptive systems with self-adjustable devices allow to ensure the stability and required quality of control under incomplete knowledge of parameters and characteristics of a plant.

These systems are classified as follows: systems with open adjustment loops, closed adjustment loops and automatic control of object characteristics, and systems with extremal adjustment of correcting devices. Additional operating information in these systems is: information about disturbances influencing the parameters of a plant, information about deviation of transient processes in the main loop and information about deviation of quality control from the extremum of this quality.

Self-optimizing systems are a kind of control systems, which improves their characteristics during the control. The operating information, which concerns the existing interactions, and conclusions about necessary changes of the characteristics are gained by the system during the operation. Such processes as: prediction of occurrence of a random event, classification of complex situations, etc. are the basis of control algorithm of such systems. These systems are used in e.g. remote control and optimization of flight trajectories of rockets.

8.5.2 Distributed Automatic Control Systems

Distributed systems constitute a group of systems, which are meant to control motion of a large number of elements, means of transport or elements of communication network, etc.

Control in such systems is like playing game between two sides, from which one side is controlled by the system (computer) but the second side is not. The control is to ensure the most useful kind of motion of particular elements according to a criterion called a success function. This function can be e.g. required amount of transported goods in determined terms at the smallest costs of the transport. A system which is controlled is a set of means of transport, and the other side, the uncontrolled one, is a set of goods requirements, which pour in from various regions of a territory. In such systems, the required initial information about the uncontrolled side is poor. Only in successive game stages (control) the system gains the operating information about the other side.

A block diagram explaining the distributed system functioning is depicted in Fig. 8.9.

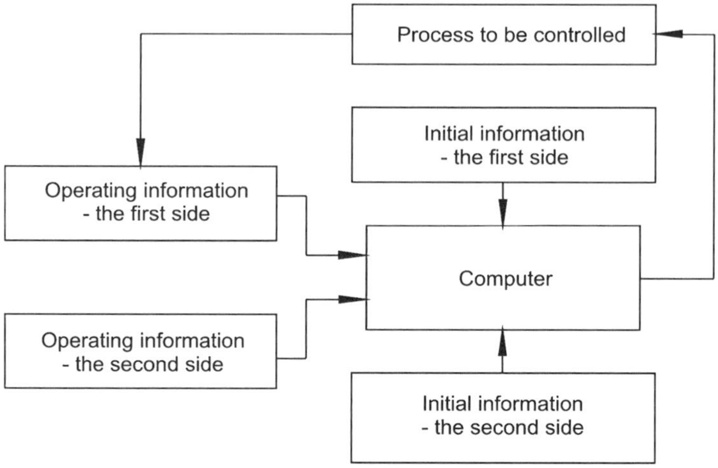

Fig. 8.9 Scheme of functioning of a distributed control system

8.5.3 Classification of Control Systems with Respect to Another Criteria

Besides the main criterion of classification of control systems with respect to initial information about a process to be controlled, it is advisable to discuss another classification criteria, which are used in theory and practice.

8.5.3.1 Classification with Respect to Purpose

It covers control systems or automatic regulation systems of: level, intensity, flow, power, voltage, angular velocity, pressure, temperature, humidity, etc.

8.5.3.2 Classification with Respect to Type of Energy of the Control Factor

It covers control systems such as: mechanical, electrical, electronical, hydraulic, pneumatic, electro-hydraulic, electro-pneumatic, chemical systems, etc.

8.5.3.3 Classification with Respect to Types of Signals Occurring in a System

We distinguish here: continuous systems, in which all occurring signals are continuous functions of time and discontinuous systems (discrete), in which at least one element generates a discrete quantized signal in time.

The following systems belong to the discrete systems:

(i) impulse systems, in which quantized in time signals occur, and this means that output quantities (coming from impulsators) appears only in pulsing instants (e.g. each 1 second), while the width of the pulse can be constant and the height can be variable or otherwise. One encounters impulse systems of variable period of pulsing;
(ii) relay systems, in which output signals of relays attain only two (e.g. zero–maximum) or three (minimum–zero–maximum) values, while transition from one to another level is done, when the input signal of a relay exceeds so-called switch point.

So-called finite automata belong to the class of discontinuous systems. These are systems, which take on (under influence of external signals and internal couplings) finite number of specific states, whose change obeys deterministic or stochastic rules (algorithms). Thus, a finite automaton will be a simple relay as well as digital mathematical machine. Complex finite automata can serve for modelling of neural networks of alive organisms.

8.5.3.4 Classification with Respect to Linearity of Elements

Linear systems, in which all elements have linear static characteristics (dependencies between inputs and outputs in stationary states) and are governed by linear differential, integral or algebraic equations. These are systems, for which one can apply the superposition principle, i.e. a complete output signal of the system with simultaneous action of several input signals is equal to the sum of output signals of the system.

Nonlinear systems are such systems in which there is at least one nonlinear element, and this makes that the whole system can be described by means of nonlinear equations, whose analysis is more difficult than the analysis of linear equations.

In the reality, one does not meet ideal linear elements. However with good approximation, one can often linearize for technical purposes nonlinear characteristic in neighbourhoods of operating points of an element.

8.5.3.5 Classification with Respect to Character of System Parameters

In this case we distinguish the following systems:

- **stationary**, whose parameters do not vary in time. In the case of stationary linear systems, they are described by linear equations of constant coefficients;
- **non-stationary (parametric)**, whose parameters clearly vary in time, mostly according to known functions of time. Equations describing these systems are so-called parametric equations, whose coefficients are known functions of time, e.g. harmonic functions.

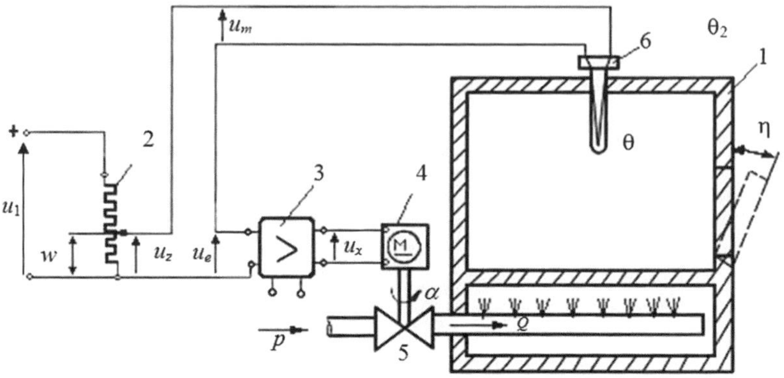

Fig. 8.10 Scheme of a temperature control system in a gas furnace

8.5.4 Examples of Control Systems and Their Block Diagrams

Classify elements occurring in control systems and make block diagrams (structural) of control systems, which are presented in schematic block diagrams and described with respect to their destination and operation. Moreover, determine the type of the considered control systems.

Example 8.1. Description of the system functioning.

The presented system (Fig. 8.10) serves for controlling the internal temperature of a furnace (1). It is achieved through the change of intensity Q of the heating medium (gas) delivered to a valve (5) under constant pressure p. The degree of valve opening is controlled by the electric engine (4), whose angle of rotation depends on the delivered voltage u_x, which is amplified in the amplifier (3) by the voltage u_e. Voltage u_e is a difference between the voltage u_z on the resistor (2) and the one u_m on the temperature sensor (6). So, the value of temperature θ can be changed by the service through adjusting position of the slider of the resistor (2), which is under constant voltage u_1.

After the analysis of the system functioning one can say that temperature control is carried on in a closed loop of signals, since the signal u_m from the measurement of the temperature θ backward-influences the value of this temperature, which in turn depends on the voltage u_z set on the resistor (2) by the service. The considered system is a constant-value automatic control system, in other words a stabilizing system.

Essential Elements in the System

1. Plant—a furnace (1), in which a process of temperature θ control proceeds by means of variation of inflow intensity Q of the heating medium.

2. Controller—a potentiometer (2); with its help the service can preset the required value of the temperature θ in the furnace (1) adjusting the potentiometer to the proper voltage value u_z.
3. Summer—an electric circuit of thermoelement connections (6), a potentiometer (2) and an amplifier (3).
4. Amplifier—an electrical amplifier (3), serves for amplifying a difference u_e voltage u_z and u_m to the value, which allows to start the engine (4).
5. Actuator—an electric engine (4), serves for changing the degree of valve opening (5). Elements mentioned in points 2–5 constitute a regulator.
6. Adjuster—a valve (5), serves for variation of inflow intensity Q of the heating medium.
7. Sensor—a thermoelement (6), serves for measurement of the temperature, which is transformed into voltage u_m.

Essential Quantities Occurring in the Considered Temperature Control System

1. Controllable quantity—temperature θ inside the furnace.
2. Input—position "w" of the potentiometer slider.
3. Presetting quantity—voltage u_z on the potentiometer.
4. Processed quantity (measuring)—voltage u_m on the thermo-element.
5. Control error—voltage u_e, which is a difference between the voltages u_z and u_m.
6. Controlling quantity (adjusting)—angle α of rotation of an electrical engine shaft (4) adjusting a valve (5).
7. Adjustable quantity—inflow intensity Q of the heating medium through the valve (5).
8. Disturbances—opening of the furnace (2), the external temperature θ_z, eventual changes of the supply pressure p.

Basing oneself on the analysis of signal courses in the considered system, one made its block diagram (Fig. 8.11).

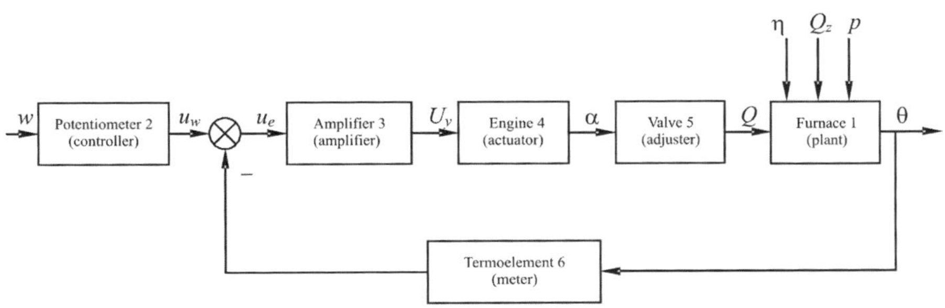

Fig. 8.11 Block diagram of the system from Fig. 8.10

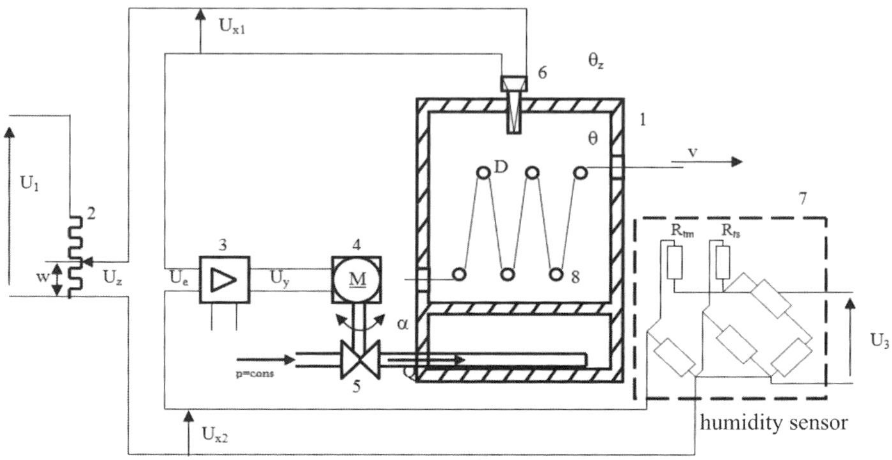

Fig. 8.12 Scheme of a system of humidity control of a material band

As one can see, the considered control system is a one-loop constant-value automatic control system (stabilizing), since there is a negative feedback and the presetting quantity U_z does not vary during the system operation.

Example 8.2. Description of the system functioning.

The presented system in Fig. 8.12 is meant for regulation of tapes humidity of material moving by means of system of rolls at velocity v. Functioning of this system is as follows: we insert the material (8) containing some amount of humidity into a drying chamber (1). The material containing some amount of humidity is subject to drying air flow of temperature θ measured by a sensor (6). The humidity D of the material is measured at the dryer output (1) by means of the humidity sensor (7). Signals from sensors (6) and (7) in a form of respective voltages U_{x1} and U_{x2} are subtracted from the adjusted voltage U_w on the potentiometer (2). The difference U_e amplified in the amplifier (3) arises, then it starts the engine (4), which change the degree of valve opening (5), and consequently the flow intensity Q. The value of temperature θ of the heating medium, which influences the temperature inside the dryer depends on the flow intensity Q. The temperature inside the dryer has an influence on the vaporization intensity of humidity in the material.

After the analysis of the system one can say that stabilization of humidity of the tape (8), coming out the dryer chamber (1) is made through the variation of the temperature θ inside the chamber. The signal coming from the temperature sensor (6) and the signal from the humidity sensor (7) are summed up with the input signal in the summer. Thus, we have to do with a double-loop system of humidity control.

Essential Elements in the System

1. Plant—a furnace (1) together with a material (8).
2. Controller—a potentiometer (2).
3. Summer—electric circuit of the thermoelement connections (6), a humidity sensor (7), a potentiometer (2) and an amplifier (3).
4. Amplifier—an electric amplifier (3).
5. Actuator—an electric engine (4).
6. Adjuster—a valve (5).
7. Sensor—a thermoelement (6) and a humidity sensor (7).

Essential Quantities Occurring in the Considered Temperature Control System

1. Controllable quantity—humidity D of a tape (8).
2. Input—position "w" of a slider on the potentiometer.
3. Presetting quantity—voltage U_z on a potentiometer (2).
4. Processed quantity—voltage U_{x1} and U_{x2} from the sensors.
5. Control error—input voltage U_e on an amplifier (3).
6. Adjusting quantity—angle α of rotation of a shaft of an engine (4).
7. Adjusted quantity—intensity Q of inflow of the heating medium.
8. Disturbances—opening (2) of the drying chamber, the outer temperature θ_z, variations of the velocity v of the tape displacement, variations of the tape humidity D_{wej} on the entry of the furnace chamber.

A block diagram of the considered system is depicted in Fig. 8.13

We can see from the block diagram that we have to do with a two-loop system. The first, internal loop concerns a temperature control and the second, external loop concerns the very humidity control. This is a constant-value control system of the humidity, whose value can be adjusted on the potentiometer (2). There are two negative feedbacks in the system, and the total control error U_e is a difference between the presetting quantity U_w and two measured quantities U_{x1} and U_{x2}.

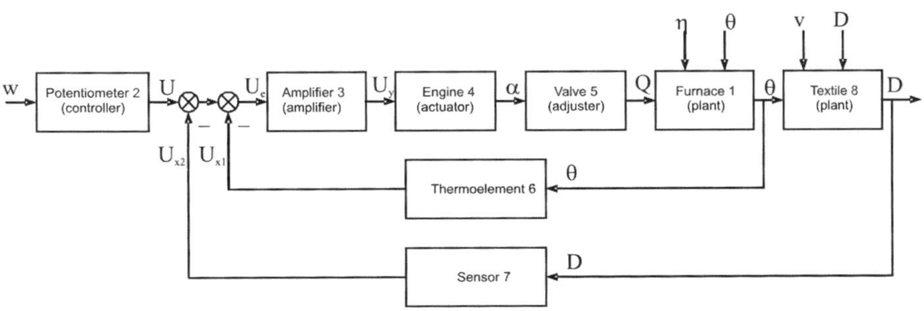

Fig. 8.13 Block diagram of the system from Fig. 8.12

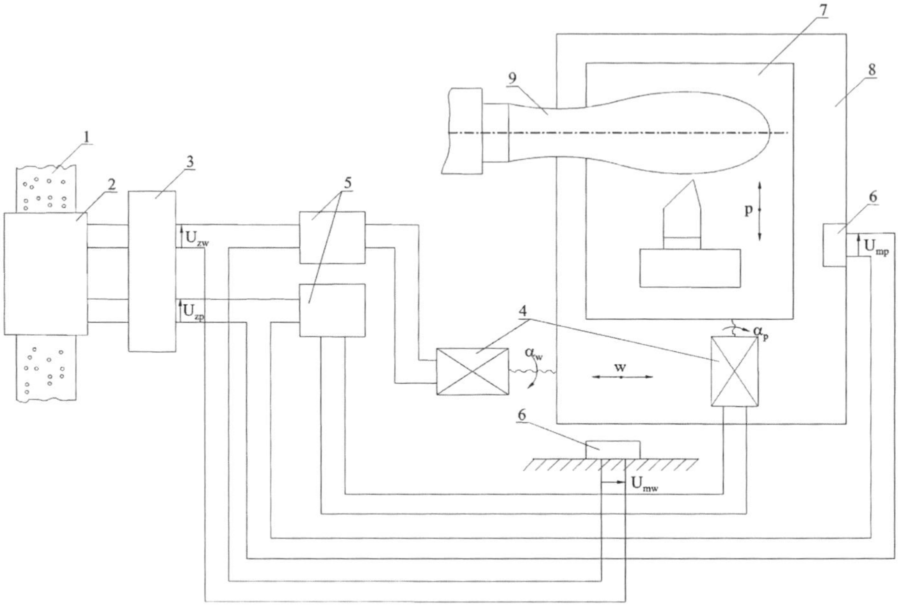

Fig. 8.14 Schematic diagram of the numerically controlled lathe

Example 8.3. Description of the system functioning.

Figure 8.14 presents a schematic block diagram of a numerically controlled lathe. The machining program of the item is saved on a perforated tape (1). Data read from the tape in the form of pulses sequence are changed into another kinds of pulses, from which each pulse corresponds to the precisely determined increment of the coordinates w and p. It is performed by an interpolator (3). Position of the board in arbitrary instant of time is measured by sensors (6). The information about this position is transmitted from the sensors in a form of signals, which come from the interpolator. Thus, the signals determine the required position of the board in a given instant of time. When there is a difference between these signals, the electric engines rotate and move the board by means of a gear, and this makes the difference vanish.

It follows from the above description that quantities presetting the position of the board in the form of signals U_{zw} and U_{zp} from the interpolator (3) are not constant, but they vary according to the treat program saved on a perforated tape (1). There are also two feedback loops reaching from the sensors (6) to the amplifiers (5). The considered system is a two-loop program-following control system.

Essential Elements in the System

1. Plant—a treated object (9) together with a cross slide (7) and carriage (8).
2. Controller—a perforated tape (1), a reader (2) and an interpolator (3).
3. Summer—an electric circuit of interpolator connections (3), sensors (6) and amplifiers (5).
4. Amplifier—an electric amplifiers (5).
5. Actuator—an electric engines (4).
6. Adjuster—gears moving the slide cross (7) and carriage (8).
7. Sensor—position sensors (6).

Essential Quantities Occurring in the System

1. Controllable quantity—coordinates y_w and y_p of positions of the slides.
2. Input—configuration of perforation on the tape (1) (w_z).
3. Presetting quantity—voltage U_{zw} and U_{zp} from an interpolator (3).
4. Processed quantity—voltage U_{mw} and U_{mp} from sensors (6).
5. Control error—input voltage of the amplifiers (5)
6. Adjusting quantity—angles α_w and α_p of rotation of shafts of electric engines (4).
7. Adjusted quantity—displacements of the slides on the gears
8. Disturbances—movements resistance of the slides, technological inaccuracy.

A block diagram of the considered system (Fig. 8.15) is as follows: the considered system is an example of a program-following control of two loops. The first loop serves for program-following control of the carriage position and the second loop serves for program-following control of the cross slide position; presetting

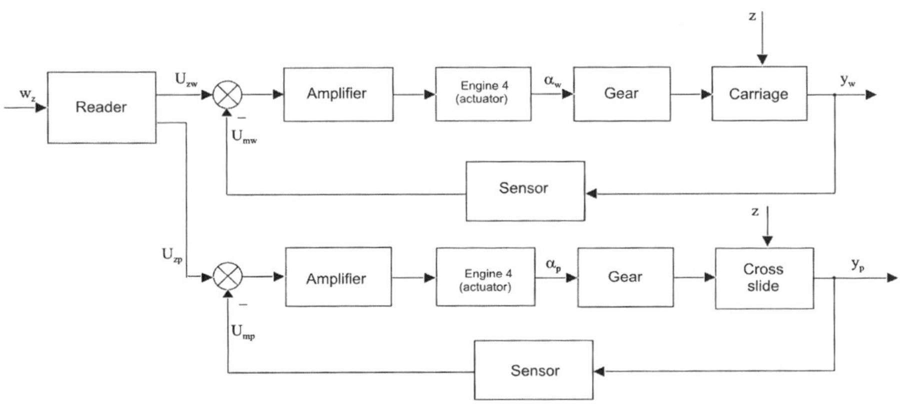

Fig. 8.15 Block diagram of the system in Fig. 8.14

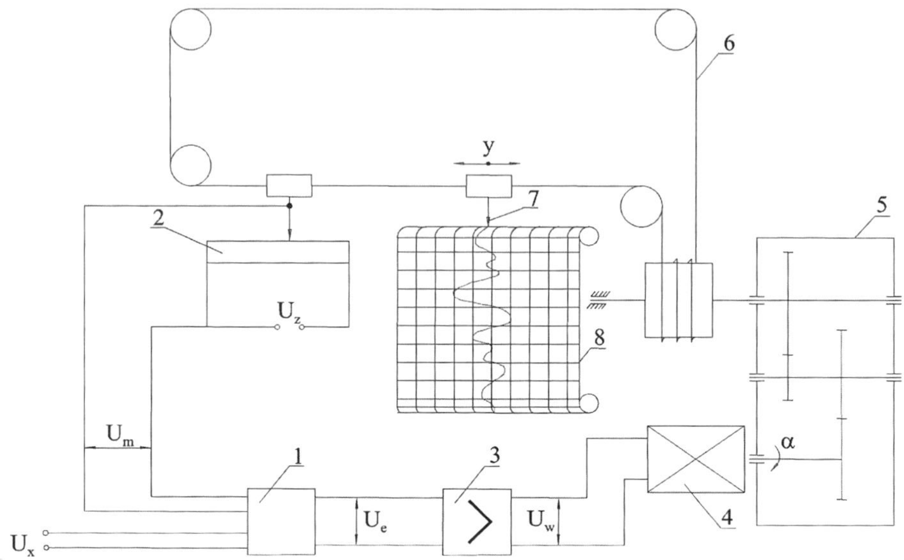

Fig. 8.16 Schematic block diagram of a register of voltage signals

quantities of these position are obtained from the controller consisting of a reader of the perforated tape (2) and interpolator (3).

Example 8.4. Description of the system functioning.

Figure 8.16 illustrates the principle of operation of a voltage signals register. A signal U_x, whose course is to be registered on a tape (8), is delivered to the input of a correcting element (1), where it is compared with the voltage U_m obtained from a slider of a potentiometer (2) supplied with a constant voltage U_z. Voltage $U_e = U_x - U_m$ is obtained on the output of the correcting element The voltage is amplified in an amplifier (3) and then it supplies the electric engine (4). The engine via the gear (5) and the strand (6) moves the slider of the potentiometer (2) in such a direction that the difference of voltage U_x and U_m decreases. The engine stops when the voltage U_e is zero. The strand (6) moves the pen (7). If the input signal U_x varies, then the slider of the potentiometer will be moved so that the voltage U_m follows the voltage U_x. The pen, which is displaced together with the slider saves on the tape (8) a graph of variation of the input signal U_x vs time.

It follows from the description of the system functioning that the input voltage U_x varies arbitrarily and independently of the service. This voltage is followed by the voltage U_m, whose value influences the position of the slider. The positions of the pen and the slider depend on the value of U_m. To conclude, the system depicted in Fig. 8.16 can be classified as a follow-up control system, tracking in other words, since the position y of a pen follows changes of the input voltage U_x and there is a closed cycle of signals in the system.

Essential Elements in the System

1. Plant—a pen (7) of the register.
2. Controller—it can be an electric device, whose voltage characteristics $U_x(t)$ is to be registered on the tape (8).
3. Summer—an electric comparison unit (1).
4. Amplifier—an electric amplifier (3).
5. Actuator—an electric engine (4).
6. Adjuster—a gear (5) with a strand (6).
7. Sensor—a potentiometer (2).

Essential Quantities Occurring in the System

1. Controllable quantity—position y of a pen of the register.
2. Presetting quantity—input voltage U_x.
3. Processed quantity—voltage $_m$ on the potentiometer.
4. Control error—voltage U_e on the correcting element (1).
5. Adjusting quantity—angle α of rotation of an engine shaft (4).
6. Adjusted quantity—displacement of the strand on the shaft.
7. Disturbances—resistance of the pen, elongation of the strand, variation of voltage U_z applied to the potentiometer, slides of the strand on a shaft.

A block diagram of the considered system has the following form.

In Fig. 8.17, one can see that this is a tracking system. Its structure is similar to the structure of a stabilizing system in Fig. 8.11, but the presetting quantity U_x varies arbitrarily and in a completely independent way, i.e. it can be neither adjusted nor programmed by the service.

Example 8.5. Description of the system functioning.

In Fig. 8.18 a schematic block diagram of a system is presented, which serves for keeping the flow intensity Q constant or in other words a constant discharge of the liquid in a common part of a pipeline. A discharge Q is, as one can see in Fig. 8.18, a sum of two discharges Q_1 and Q_2. The discharge Q_1 undergoes indeterminate changes, while the discharge Q_2 depends on the angular velocity of a

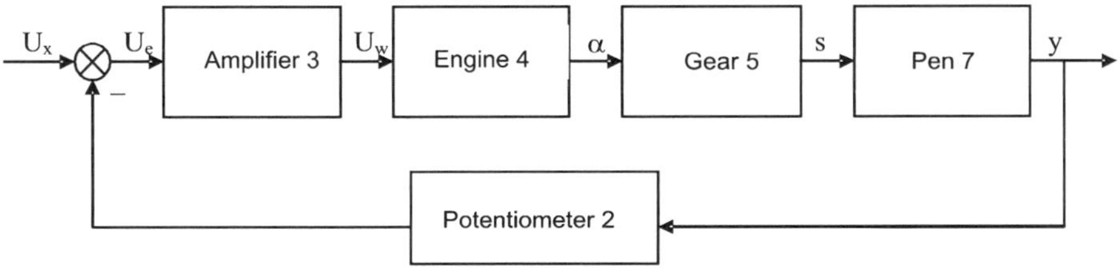

Fig. 8.17 Block diagram of a system in Fig. 8.17

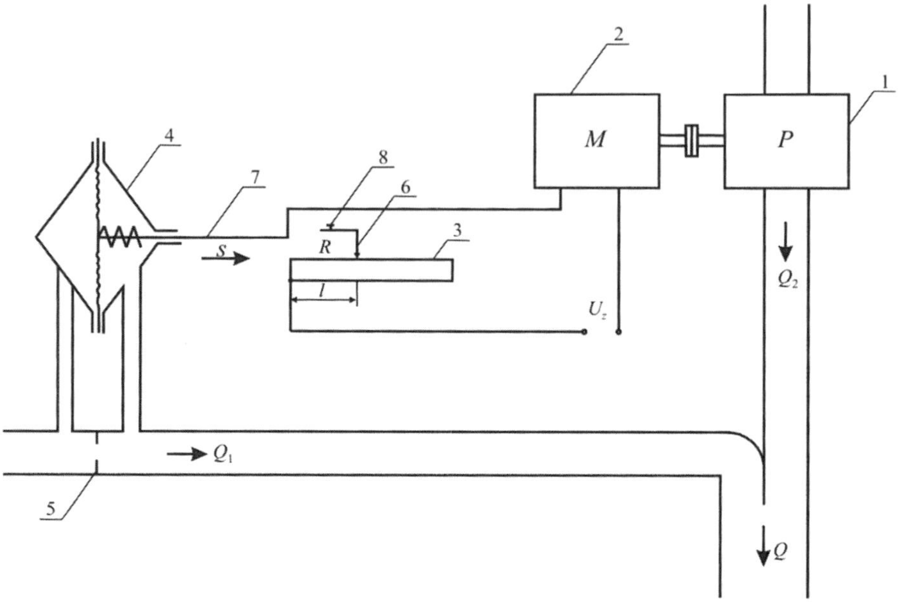

Fig. 8.18 Schematic block diagram of a system of keeping the flow Q constant

pump (1), which is driven by a direct-current motor (2). The motor is supplied with the constant voltage U_z, and its revolutions depend on the resistance R of a resistor (3), which in turn depends on a slider (6) position. The slider (6) position at fixed length of a mandrel (7) by means of a screw (8) depends on displacement of a sensor (4) membrane, which in turn depends on the pressure drop on a measuring orifice plate (5). This pressure drop depends on actual value of the flow intensity Q_1. As the flow intensity Q_1 grows, the resistance R grows as well, and in turn angular velocity of the engine (2) decreases. Hence, the discharge Q_2 of the pump (1) decreases. Thus, the total discharge Q is kept on a constant level. The service is able to change the value of the discharge Q by changing the mandrel (7) length.

It follows from the description of the system functioning that the system serves for keeping the discharge Q on a constant level. As follows from the schematic block diagram, the quantity Q is not measured. The quantity Q_1 is measured by means of the orifice (5) and the sensor (4). One needs to say that there is no classical feedback in the system. Thus, it is an open-loop automatic control system and not a closed-loop one. By further analysis we can conclude that it is an open-loop system with disturbance compensation. The disturbances are certain indeterminate changes of the flow intensity Q_1.

Essential Elements of the System

1. Plant—common part of the pipeline.
2. Controller—a mandrel (7) with a device (8) serving for adjusting of its initial position on a resistor (3).
3. Summer—a resistor (3).
4. Amplifier—there is none in this system.
5. Actuator—an electric engine (4).
6. Adjuster—a pump (1).
7. Disturbance sensor—an orifice (5) together with a membrane sensor (4).

Essential Quantities in the System

1. Controllable quantity—flow intensity Q.
2. Presetting quantity—initial position l of the slider on the resistor.
3. Processed quantity—displacement s of a sensor mandrel (4) corresponding to the flow intensity Q_1.
4. Control error—increment of the resistance se resulting from the displacement of the slider of the resistor with respect to the initial position l corresponding to the resistance R.
5. Adjusting quantity—angular velocity of an engine (2).
6. Adjusted quantity—flow intensity Q_2 from a pump (1).
7. Disturbances—flow intensity Q_1.

A block diagram of a system in Fig. 3.18 has the following form.

In Fig. 8.19 one can clearly see that it is an open-loop control system without feedback. In this system, there is a slotted line of disturbance influencing the controllable quantity.

Example 8.6. Description of the system functioning.

It follows from the schematic block diagram (see Fig. 8.20) that the system serves for controlling of a piston motion (3). It is performed with the use of a programmed motion of the slider (2) of a hydraulic divider, which guides the oil (which is

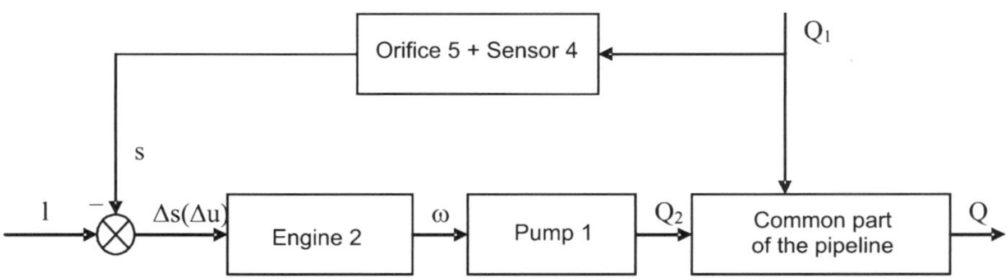

Fig. 8.19 Block diagram of the system in Fig. 8.18

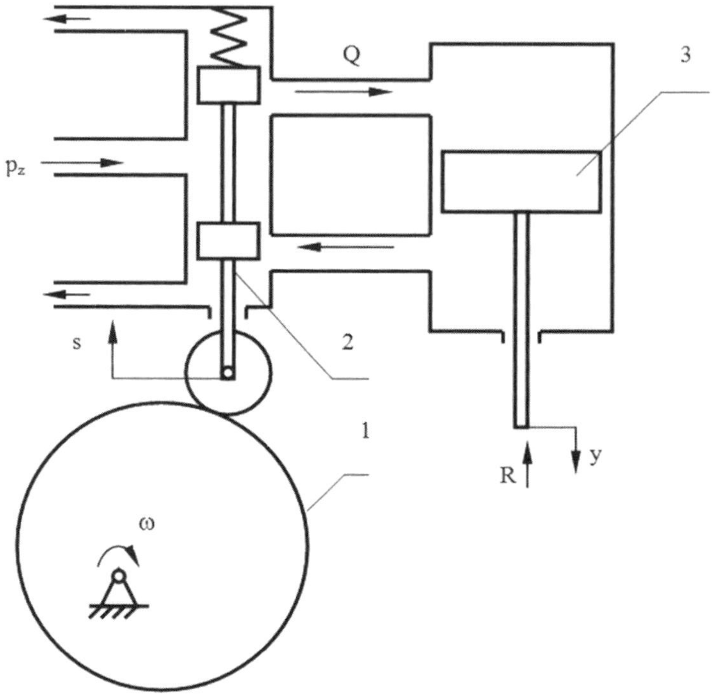

Fig. 8.20 Scheme of a piston motion control

delivered under constant pressure p_z) over or under the piston. The device presented in Fig. 8.20 consisting of a slider divider (2) and a piston, which moves in a cylinder is called a servo-mechanism. In this servo-mechanism slight movements of the slider make the piston move. Small movements of the slider requiring small forces to be applied result in large displacements of the piston (3), which can overcome significant resistance.

It follows from Fig. 8.20 that it is an open program-following control system since there is no feedback. This means that there is no connection between movement y of a piston (3) and movement s of a slider (2) of the hydraulic divider. The motion s, as can be seen, follows from the cam (1), which is a program of this motion, and motion of the piston (3) in consequence.

Essential Elements of the Considered System

1. Controller—a cam (1) of a particular angular velocity ω.
2. Amplifier—a hydraulic amplifier, which is a slider divider (2).
3. Actuator, adjuster and plant—a cylinder together with the piston (3).

A block diagram of a system in Fig. 8.20 has the following form (Fig. 8.21).

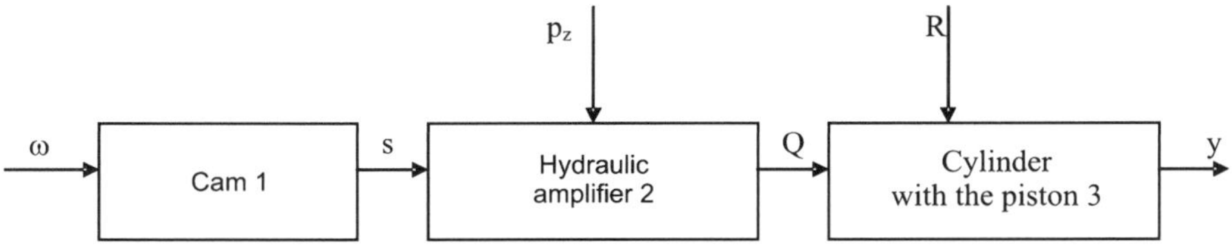

Fig. 8.21 Block diagram of the system in 8.20

Essential Quantities of the System

1. Controllable quantity—movements y of the piston (3).
2. Input—curvature of the cam (1).
3. Presetting quantity—position s of a divider slider.
4. Controlling quantity—degree of gaps opening s_1 delivering the oil to the engine.
5. Adjusted quantity—flow intensity Q of oil in the system.
6. Disturbances—resistance R of the piston movements, changes of the pressure p_z of oil supply.

Chapter 9
Phase Plane and Phase Space

9.1 Introduction

A dynamical state of an autonomous system is completely determined by the generalized coordinates $y_i(t)$ and the generalized velocities $\dot{y}_i(t)$ ($i = 1, 2, \ldots, n$, where n is the number of degrees of freedom). Treating time t as a parameter, a point of the coordinates $(y_i, \dot{y}_i)$ will be a point of $2n$-dimensional phase space. Motion of this point describes a phase trajectory as time increases. In the case of $n = 1$ a vibrating system has one degree-of-freedom and the phase space reduces to the phase plane. Then, a phase trajectory is a curve lying in the plane, and a set of all phase trajectories, corresponding to distinct initial conditions, form a phase portrait.

If the motion of one degree-of-freedom autonomous system (or two-dimensional system because it is governed by two first-order differential equations) is governed by the equation

$$\ddot{y} = F(y, \dot{y}), \tag{9.1}$$

then phase plane is said to be a plane with the rectangular coordinate system $(y, \dot{y} = v)$.

Equation (9.1) is transformed into a system of two first-order differential equations

$$\begin{aligned} \dot{y} &= v, \\ \dot{v} &= F(y, v). \end{aligned} \tag{9.2}$$

Equation (9.2) describes motion of a point $A(y, v)$ in the phase plane. Eliminating the time, we obtain an integral curve (phase trajectory) formula of the form

$$C(y, v) = 0. \tag{9.3}$$

J. Awrejcewicz, *Ordinary Differential Equations and Mechanical Systems*,
DOI 10.1007/978-3-319-07659-1_9

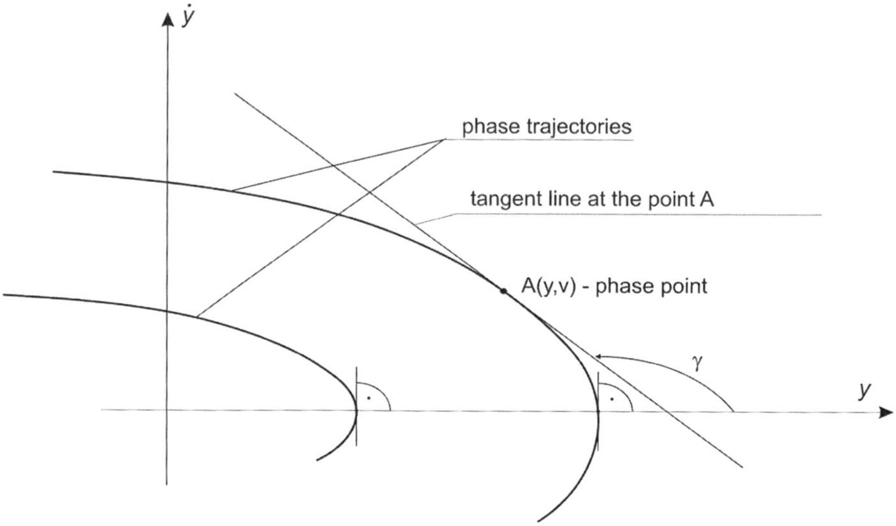

Fig. 9.1 Phase plane and phase trajectories

Dividing both sides of the system of Eq. (9.2) by themselves, we obtain

$$\frac{dv}{dy} = \frac{F(y, v)}{v} = \tan\gamma, \tag{9.4}$$

where γ is an angle between a phase trajectory and positive direction of the y-axis. There are a few phase trajectories depicted in Fig. 9.1 with the marked phase point and the angle γ.

Phase points, at which a tangent line is determined, will be called *ordinary* or *regular points*. Phase points, at which a tangent line is not determined, will be called *singular points*. The latter are equilibrium positions, determined from the equation

$$F(y, 0) = 0. \tag{9.5}$$

One can see in Fig. 9.1 that the phase trajectories intersect the y-axis at right angles. It turns out that each phase trajectory must pass through a regular point lying on the y-axis at right angle, since

$$\lim_{v\to 0} \frac{dv}{dy} = \lim_{v\to 0} \frac{F(y, v)}{v} = \infty, \tag{9.6}$$

and hence the value of γ at these points is $\pi/2$. A characteristic feature of nonlinear system follows from Eq. (9.5). These systems can possess one or several equilibrium positions depending on the character of the function $F(y, 0)$. Phase trajectories have some general properties, given below.

1. Direction of motion of the phase point $A(y, v)$ along the phase trajectory is such that the positive velocity v is in correspondence with increment of the displacement y according to positive direction of the y-axis, and the negative velocity v is in correspondence with the increment, which is opposite to the positive direction of the y-axis.
2. A phase trajectory cannot have a tangent line parallel to the v-axis at regular points, which do not lie on the y-axis. The phase trajectory cannot have a tangent line parallel to the v-axes at points, which do not lie in the v-axis.
3. If any continuous phase trajectory intersects the y-axis at two successive points, then there is at least one singular point between them.
4. In time interval, in which a continuous phase trajectory does not intersect the y-axis, the trajectory can intersect, at most, once any straight line parallel to the v-axis.
5. Closed curves in a phase plane correspond to periodic motions.

9.2 Phase Plain and Singular Points

A broader class of physical systems can be described by first-order differential equations of the form:

$$\frac{dy}{dt} = Q(y, v), \quad \frac{dv}{dt} = P(y, v). \tag{9.7}$$

Equation (9.4) is a particular case of Eq. (9.7). In what follows we analyse the linearized equation (9.7):

$$\frac{dv}{dy} = \frac{ay + bv}{cy + dv}. \tag{9.8}$$

In the dynamical system described by Eq. (9.8) there can be three types of phase trajectories, namely: a point, a closed (corresponds to a periodic solution) and open (corresponds to a non-periodic solution) curve. The aim of qualitative examination of the dynamical systems (9.8) is to determine a phase portrait and its topological structure. By a notion of topological structure we mean such properties of a phase portrait that remain unchanged under topological (i.e. unique and mutually continuous) mapping of a plane into itself. In order to perform such a qualitative analysis of the dynamical system (9.8), in most cases one can confine oneself to determining equilibrium positions, periodic trajectories and limit cycles, and phase semi-trajectories, which are curves separating qualitatively different phase trajectories in a neighbourhood of equilibrium position.

A limit cycle is said to be a closed phase curve, surrounded by a region completely filled with trajectories tending to the curve as $t \to +\infty$ or $t \to -\infty$. After Taylor expanding the functions $P(y, v)$ and $Q(y, v)$ about the analysed singular point we obtain

$$\frac{dv}{dy} = \frac{ay + bv + P'(y, v)}{cy + dv + Q'(y, v)}. \tag{9.9}$$

Ignoring the nonlinear terms, Eq. (9.9) takes the form

$$\begin{bmatrix} \dot{v} \\ \dot{y} \end{bmatrix} = \mathbf{A} \begin{bmatrix} v \\ y \end{bmatrix}, \tag{9.10}$$

$$\mathbf{A} = \begin{bmatrix} b & a \\ d & c \end{bmatrix}. \tag{9.11}$$

The point $y = v = 0$ is a *critical point* (or a trivial solution) of Eq. (9.10) and if $\det \mathbf{A} \neq 0$, then the system (9.10) is called simple [191]. Eigenvalues of the matrix $\mathbf{A}$ allow to determine the canonical basis in $\mathbf{R}^2$, where the matrix $\mathbf{A}$ takes a canonical form. A characteristic equation leading to determination of eigenvalues can be obtained by standard procedure, namely by assuming solutions of the form

$$\begin{aligned} v &= C_1 e^{\lambda t}, \\ y &= C_2 e^{\lambda t}, \end{aligned} \tag{9.12}$$

where C_1 and C_2 are constants. Substituting (9.12) into (9.10) we obtain

$$\begin{vmatrix} b - \lambda & a \\ d & c - \lambda \end{vmatrix} = 0, \tag{9.13}$$

and after expanding

$$\lambda^2 - (b + c)\lambda + bc - ad = 0. \tag{9.14}$$

By the above equation we find the discriminant

$$\Delta = (b - c)^2 + 4ad. \tag{9.15}$$

The above equation possesses the following roots

$$\lambda_{1,2} = \frac{1}{2}\left[(b + c) \pm \sqrt{(b - c)^2 + 4ad}\right]. \tag{9.16}$$

Considerations based on the phase plane (x, y) are transferred into the plane (ξ, η) and correspond to the canonical form of the matrix $\mathbf{A}$. After the transformation, corresponding curves in both planes are rotated and deformed but their qualitative features remain unchanged, e.g. a circle corresponds to an ellipse but both curves are closed. The character of the curves depends on the ratio λ_1/λ_2 and a constant C (see (9.20)).

We will consider the following cases:

1. Both roots are real and distinct, and of the same sign. We have such a situation when $\Delta > 0$, $ad - bc < 0$. Then, the matrix $\mathbf{A}$ in the canonical basis has the form

$$\mathbf{A} = \begin{bmatrix} \lambda_1 & 0 \\ 0 & \lambda_2 \end{bmatrix}, \tag{9.17}$$

where $\lambda_{1,2} \in \mathbf{R}$, and Eq. (9.10) takes the canonical form

$$\begin{aligned} \dot{\eta} &= \lambda_1 \eta, \\ \dot{\xi} &= \lambda_2 \xi. \end{aligned} \tag{9.18}$$

Equation (9.18) can be easily solved by separating the variables. Consequently, we obtain

$$\begin{aligned} \eta(t) &= C_1 e^{\lambda_1 t}, \\ \xi(t) &= C_2 e^{\lambda_2 t}. \end{aligned} \tag{9.19}$$

Next, we have

$$\ln \frac{\eta}{C_1} = \frac{\lambda_1}{\lambda_2} \ln \frac{\xi}{C_2},$$

thus

$$\ln \frac{\eta}{C_1} = \ln \left(\frac{\xi}{C_2} \right)^{\frac{\lambda_1}{\lambda_2}},$$

hence

$$\eta = C |\xi|^{\frac{\lambda_1}{\lambda_2}}, \tag{9.20}$$

where $C = (C_1/C_2)^{\frac{\lambda_1}{\lambda_2}}$. The singular point $(0, 0)$ is called a stable (unstable) node.

2. If $bc - ad = 0$ and $b + c < 0$, then by (9.14) we get

$$\lambda[\lambda - (b + c)] = 0, \tag{9.21}$$

and this implies $\lambda_2 = 0$ and $\lambda_1 = b + c < 0$. In this case, the analysed system is not simple. The matrix $\mathbf{A}$ has the following canonical form

$$\mathbf{A} = \begin{bmatrix} \lambda_1 & 0 \\ 0 & 0 \end{bmatrix}. \tag{9.22}$$

and its rank equals 1. Such a singular point is called a stable centre. It is noteworthy that when the rank of the matrix $\mathbf{A}$ equals 0, then the matrix is a zero matrix and each point of the phase plane is critical.

Critical points, called nodes, also occur when the discriminant of (9.14) $\Delta = 0$. Then $\lambda_1 = \lambda_2 = \lambda_0$ (double root) and if two linearly independent vectors are associated with a double eigenvalue, then canonical form of the matrix $\mathbf{A}$ reads:

$$\mathbf{A} = \begin{bmatrix} \lambda_0 & 0 \\ 0 & \lambda_0 \end{bmatrix}. \tag{9.23}$$

A critical point corresponding to this matrix is called a star-shaped node, which is stable, if $\lambda_0 < 0$ (and conversely). Only one eigenvector can be associated with a double eigenvalue. Then, a canonical form of the matrix $\mathbf{A}$ takes the form of the following Jordan block

$$\mathbf{A} = \begin{bmatrix} \lambda_0 & 0 \\ 1 & \lambda_0 \end{bmatrix}. \tag{9.24}$$

The differential equations (9.10) take the form

$$\begin{aligned} \dot{v} &= \lambda_0 v, \\ \dot{y} &= v + \lambda_0 y, \end{aligned} \tag{9.25}$$

and their solutions follow

$$\begin{aligned} v &= C_1 e^{\lambda_0 t}, \\ y &= (C_2 + C_1 t) e^{\lambda_0 t}. \end{aligned} \tag{9.26}$$

In this case, the critical point $(0, 0)$ is a *degenerate node*, which is stable for $\lambda_0 < 0$ and unstable for $\lambda_0 > 0$.

3. In this case both roots are real and have opposite signs. The orbits surround a singular point, which is called a *saddle*. Two orbits approach and move away from this point—these are axes of a coordinate system.
4. If the discriminant of (9.14) $\Delta < 0$ and $b + c \neq 0$, then the roots λ_1 and λ_2 are complex conjugate. Then, the critical point is a stable $b + c < 0$ or unstable $b + c > 0$ focus. Assume that $\lambda_{1,2} = \alpha \pm i\omega$, while $\alpha \neq 0$ and $\omega \neq 0$ (farther we will assume $\omega > 0$). In this case the canonical matrix has the form

$$\mathbf{A} = \begin{bmatrix} \alpha & -\omega \\ \omega & \alpha \end{bmatrix}, \tag{9.27}$$

and we will consider the following system of equations

$$\begin{aligned}\dot{v} &= \alpha v - \omega y,\\ \dot{y} &= \omega v + \alpha y.\end{aligned} \tag{9.28}$$

Parametric equations of orbits of the above system (its general solutions) are:

$$\begin{aligned}v(t) &= Ce^{\alpha t}\cos(\omega t + \varphi),\\ y(t) &= Ce^{\alpha t}\sin(\omega t + \varphi),\end{aligned} \tag{9.29}$$

where C and φ are any constants.

Orbits in neighbourhood of a focus can be also presented in the polar coordinates (ϱ, θ). Let us make a change of the variables

$$\begin{aligned}v &= \rho\cos\theta,\\ y &= \rho\sin\theta.\end{aligned} \tag{9.30}$$

By Eqs. (9.28) and (9.30) we get

$$\begin{aligned}\dot{\rho}\cos\theta - \rho\dot{\theta}\sin\theta &= \alpha\rho\cos\theta - \omega\rho\sin\theta,\\ \dot{\rho}\sin\theta + \rho\dot{\theta}\cos\theta &= \alpha\rho\cos\theta + \omega\rho\sin\theta.\end{aligned} \tag{9.31}$$

Multiplying (9.31) respectively by $\cos\theta$ and $\sin\theta$ (and by $\sin\theta$ and by $-\cos\theta$), and adding the equations we get

$$\begin{aligned}\dot{\rho} &= \alpha\rho,\\ \dot{\theta} &= \omega.\end{aligned} \tag{9.32}$$

The solution in the polar coordinates takes the form

$$\begin{aligned}\rho &= \rho_0 e^{\alpha t},\\ \theta &= \omega t + \theta_0,\end{aligned} \tag{9.33}$$

where ρ_0 and θ_0 define any initial conditions. The solutions (9.33) have simple physical interpretation. The argument θ grows linearly in time, while a ray originating from the focus and passing through the point $(y(t), v(t))$ rotate anticlockwise at angular velocity ω [rad/s].

By Eq. (9.33) after eliminating the time we obtain

$$\rho = \rho_0 e^{-\frac{\alpha}{\omega}\theta_0} e^{\frac{\alpha}{\omega}\theta}. \tag{9.34}$$

In this case the orbit is represented by a curve called a logarithmic spiral in the coordinates (ρ, θ). It is worth emphasizing that in the case of an unstable focus $\lim_{t\to+\infty} \rho(t) = +\infty$, and the shape of logarithmic spirals depends on the ratio $\frac{\alpha}{\omega}$. In the case when $b + c = 0$, then $\lambda_{1,2} = \pm i\omega$ $(\alpha = 0)$. Then Eq. (9.32) we get

$$\begin{aligned} \rho &= \rho_0 \equiv \text{const.}, \\ \theta &= \omega t + \theta_0. \end{aligned} \tag{9.35}$$

The above formulas represent a circle of radius ρ_0 in the polar coordinates (ρ, θ). While, by Eq. (9.28) we get

$$\begin{aligned} \dot{v} &= -\omega y, \\ \dot{y} &= \omega v, \end{aligned} \tag{9.36}$$

and eliminating the time we obtain

$$\frac{dy}{dv} = -\frac{v}{y}, \tag{9.37}$$

and hence $v^2 + y^2 = C^2$.

A critical point, in this case, is called a centre. The centre is a stable point but not asymptotically stable in Lyapunov's sense.

There is only one particular case left to discuss, namely the case of vanishing discriminant $\Delta = 0$, when zero is a double root, and the matrix $\mathbf{A}$ has the following canonical form

$$\mathbf{A} = \begin{bmatrix} 0 & 0 \\ 1 & 0 \end{bmatrix}. \tag{9.38}$$

A normal form of a system of differential equations takes the form

$$\begin{aligned} \dot{v} &= 0, \\ \dot{y} &= v, \end{aligned} \tag{9.39}$$

and their solutions are the following functions

$$\begin{aligned} v &= C_1, \\ y &= C_1 t + C_2. \end{aligned} \tag{9.40}$$

In Table 9.1 one classified phase portraits associated with critical points in two-dimensional space $\mathbf{R}^2$. Using linear transformation

$$\xi = \alpha y + \beta v, \quad \eta = \gamma y + \delta v. \tag{9.41}$$

Table 9.1 Phase portraits classification

Eigenvalues λ_1, λ_2	Eigenvectors (comment)	Name of a critical point
$0 < \lambda_1 < \lambda_2$		Unstable node
$\lambda_1 < \lambda_2 < 0$		Stable node
$\lambda_1 < 0 < \lambda_2$		Saddle
$0 = \lambda_1 < \lambda_2$		Unstable centre
$\lambda_1 < \lambda_2 = 0$		Stable centre
$\lambda_1 = \lambda_2 < 0$	Two eigenvectors	Stable star-shaped node
$\lambda_1 = \lambda_2 > 0$	Two eigenvectors	Unstable star-shaped node
$\lambda_1 = \lambda_2 < 0$	One eigenvector	Stable degenerate node
$\lambda_1 = \lambda_2 > 0$	One eigenvector	Unstable nondegenerate node
$\lambda_1 = \lambda_2 = 0$	One eigenvector	Degenerate centre
$\lambda_{1,2} = \alpha \pm i\omega$	$\alpha > 0, \omega \neq 0$	Unstable focus
$\lambda_{1,2} = \alpha \pm i\omega$	$\alpha < 0, \omega \neq 0$	Stable focus
$\lambda_{1,2} = \alpha \pm i\omega$	$\alpha = 0, \omega \neq 0$	Stable centre

one can transform Eq. (9.8) into the form of separated variables (see 9.18)

$$\frac{d\eta}{d\xi} = \frac{\lambda_1 \eta}{\lambda_2 \xi}. \tag{9.42}$$

Equation (9.41) yields

$$d\xi = \alpha dy + \beta dv, \quad d\eta = \gamma dy + \delta dv. \tag{9.43}$$

Inserting the nominator and denominator of the formula (9.8) instead of dy and dv, we obtain

$$\frac{d\eta}{d\xi} = \frac{\gamma(cy + dv) + \delta(ay + bv)}{\alpha(cy + dv) + \beta(ay + bv)}. \tag{9.44}$$

Comparing nominators and denominators of the above equation and of the formula (9.42), and using the linear transformation (9.41) we obtain the following system of equations

$$\begin{aligned} \gamma(cy + dv) + \delta(ay + bv) &= \lambda_1 \eta = \lambda_1(\gamma y + \delta v), \\ \alpha(cy + dv) + \beta(ay + bv) &= \lambda_2 \xi = \lambda_2(\alpha y + \beta v). \end{aligned} \tag{9.45}$$

In order to determine the constants γ and δ, for the first of the formulas (9.45), we equate the terms occurring by y and v. We obtain two algebraic equations of the form

$$\begin{aligned} \gamma(c-\lambda_1)+\delta a &= 0, \\ \gamma d+\delta(b-\lambda_1) &= 0. \end{aligned} \tag{9.46}$$

The algebraic equations for the second equation (9.45) have very similar structure and allow to determine the coefficients α and β

$$\begin{aligned} \alpha(c-\lambda_2)+\beta a &= 0, \\ \alpha d+\beta(b-\lambda_2) &= 0. \end{aligned} \tag{9.47}$$

This implies that λ_1 and λ_2 are roots of the same characteristic equation, which is formed by equating the characteristic determinant of the system of Eqs. (9.46) and (9.47) to zero, i.e.

$$\begin{vmatrix} c-\lambda & a \\ d & b-\lambda \end{vmatrix} = 0. \tag{9.48}$$

9.3 Analysis of Singular Points

Nowadays there are many softwares allowing to solve the differential equation (9.10) analytically and numerically. The obtained results are automatically plotted in a plane in the coordinates (y, v). The character of a singular point under consideration depends only on the coefficients a, b, c, d. The obtained phase trajectories are slightly deformed but it is possible to rotate them by solving the differential equation (9.10). This equation allows to obtain the rectified trajectories in the coordinate system (ξ, η). The shape of these graphs depends only on a ratio of the roots λ_1 and λ_2 of the characteristic equation (9.10).

9.3.1 Unstable Node

The first singular point $(0, 0)$ will be a node. In this case the roots λ_1 and λ_2 of the characteristic equation must be real and distinct, and have the same signs. These conditions will be satisfied when

$$(b-c)^2+4ad>0 \quad \text{and} \quad bc-ad>0. \tag{9.49}$$

The solution curves in the plane (ξ, η) will be parabolas passing through the point $(0, 0)$. If $b+c>0$, then a critical point is an unstable node, a phase point moves away from the origin as time increases. These conditions are satisfied for e.g.: $a=0, b=2, c=1, d=1$. This situation is presented in Figs. 9.2 and 9.3.

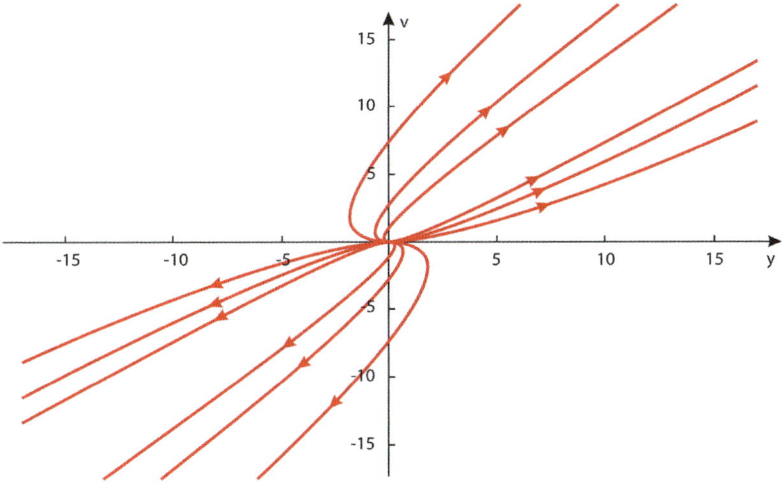

Fig. 9.2 The phase trajectories passing through the unstable node in the coordinates (y, v)

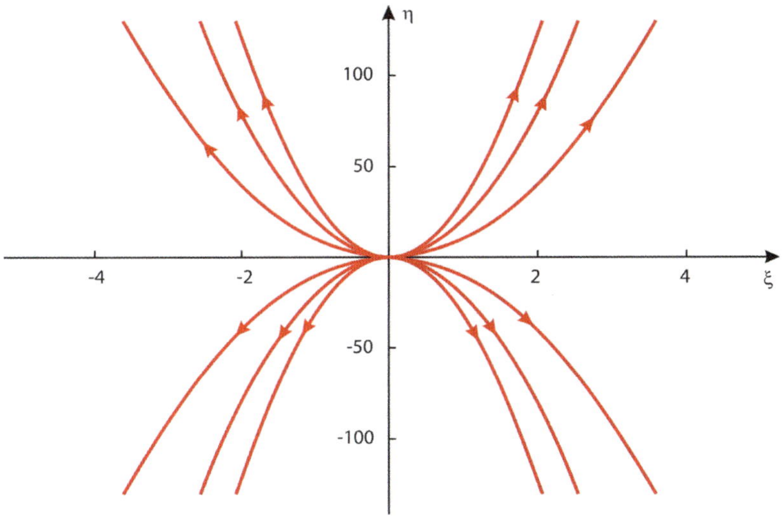

Fig. 9.3 Phase trajectories passing through the unstable node in the coordinates (ξ, η)

By the above graphs one can see that all the trajectories pass through the singular point $(0, 0)$, which is an unstable node because phase point move away from the node as time increases.

In Fig. 9.4 one can see the trajectory obtained numerically. This verifies the earlier obtained analytical solutions.

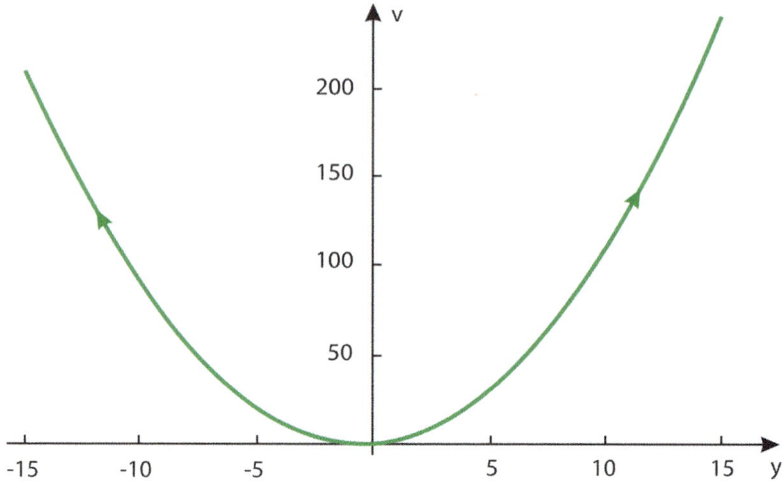

Fig. 9.4 Numerical solution for an unstable node in the coordinates (y, v)

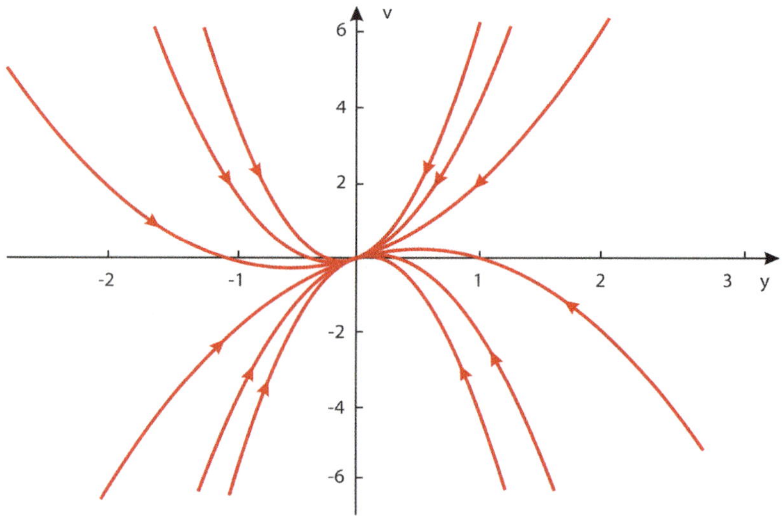

Fig. 9.5 Phase trajectories passing through the stable node in the coordinates (y, v)

9.3.2 Stable Node

If $b+c < 0$ then a phase point approaches to the singular point $(0, 0)$ as time grows. For instance, it takes place for $a = 1, b = -2, c = -1, d = 0$. These conditions are demonstrated in Figs. 9.5 and 9.6.

The phase trajectories are parabolas passing through the origin $(0, 0)$ of the coordinate system but the origin, which is a singular point is called a stable node,

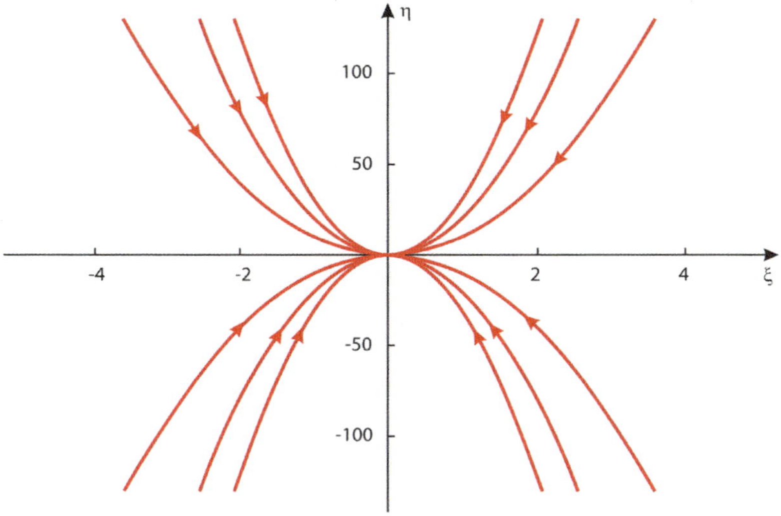

Fig. 9.6 Phase trajectories passing through the stable node in the coordinates (ξ, η)

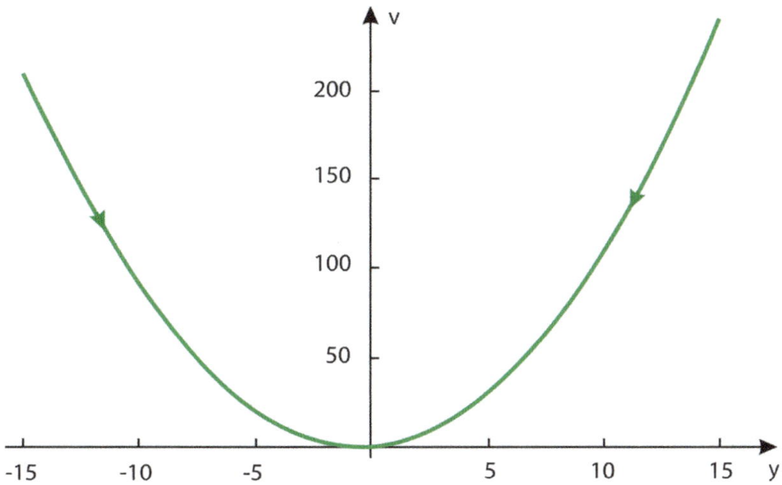

Fig. 9.7 Numerical solution for a stable node in the coordinates (y, v)

since phase points approach the point $(0, 0)$. Below in Fig. 9.7 one can see the numerical verification of the analytical solution.

If the roots λ_1 and λ_2 differ from each other significantly, then the phase trajectories change the direction more rapidly. Moreover, if one of the roots equals zero, then the curves are transformed into vertical straight lines.

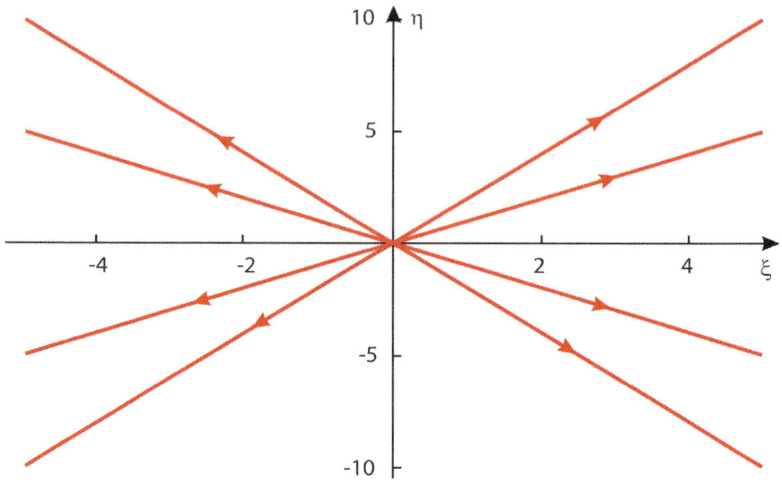

Fig. 9.8 Phase trajectories passing through the critical node in the coordinates (ξ, η)

9.3.3 *Critical Node*

When roots of the characteristic equation are real and $\lambda_1 = \lambda_2$, then a node is called a critical node. In this case we have

$$(b - c)^2 + 4ad = 0 \quad \text{and} \quad a = d = 0. \tag{9.50}$$

Let the coefficients be: $a = 0$, $b = 2$, $c = 2$, $d = 0$. If $b = c > 0$, then a phase point moves away from the origin as time increases. This situation is depicted in Fig. 9.8.

While $b = c < 0$, then the coefficients can be: $a = 0$, $b = -2$, $c = -2$, $d = 0$. Then phase points approach the point $(0, 0)$ as time increases. This situation is depicted in Fig. 9.9.

Figures 9.8 and 9.9 imply that the trajectories form a bunch of lines, on which a phase point approaches or moves away from the node $(0, 0)$, which is now called critical. Verification of the analytical solution is the numerical one depicted in Fig. 9.10.

9.3.4 *Degenerate Node*

We deal with a degenerate node if roots of the characteristic equation are equal and no special case occurs e.g.: $a = 2$, $b = 5$, $c = 1$, $d = -2$. Now, we have

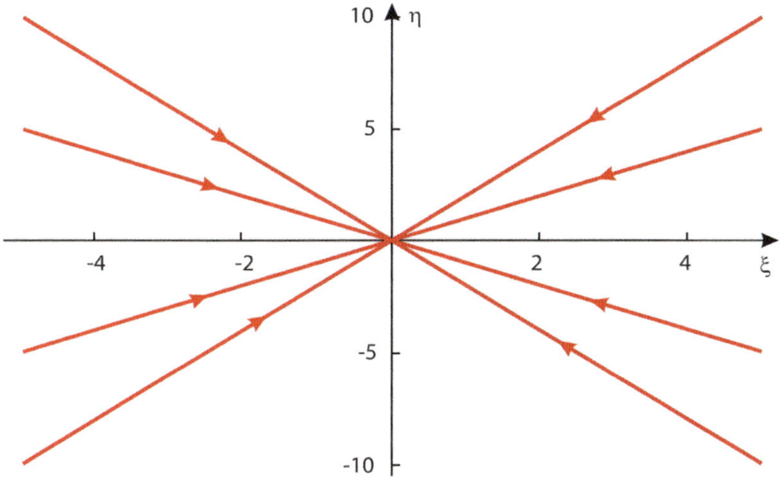

Fig. 9.9 Phase trajectories passing through the critical node in the coordinates (ξ, η)

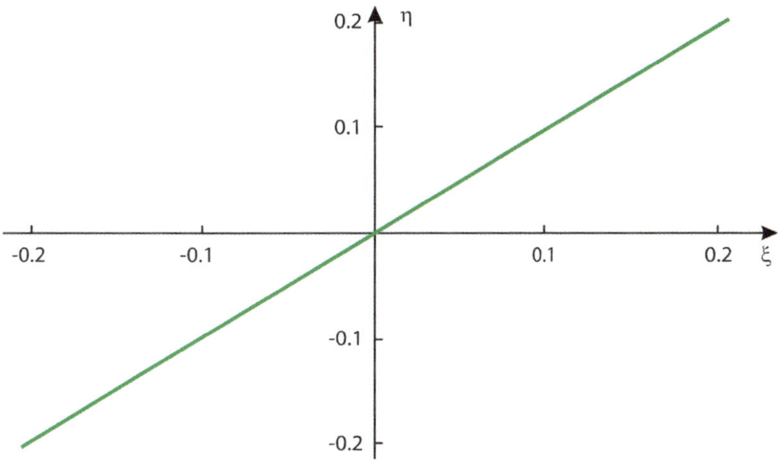

Fig. 9.10 Numerical solution for a critical node in the coordinates (ξ, η)

$b + c > 0$, and the singular point $(0, 0)$ is called a degenerate node and a phase point moves away from the origin of the coordinate system (Fig. 9.11).

When $a = 1$, $b = -2$, $c = -4$, $d = -1$, then $b + c < 0$, and a phase point approaches the origin and we also have to do with a degenerate node illustrated in Fig. 9.12. The numerical solution for a degenerate node is presented in Fig. 9.13.

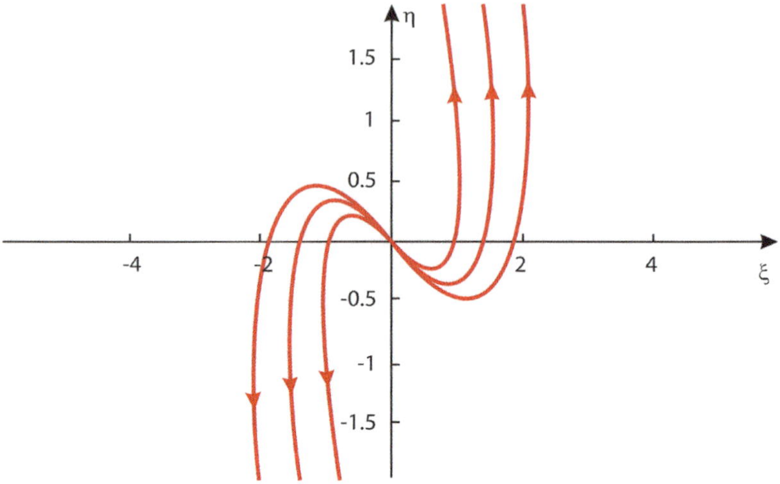

Fig. 9.11 Phase trajectories passing through the degenerated node in the coordinates (ξ, η)

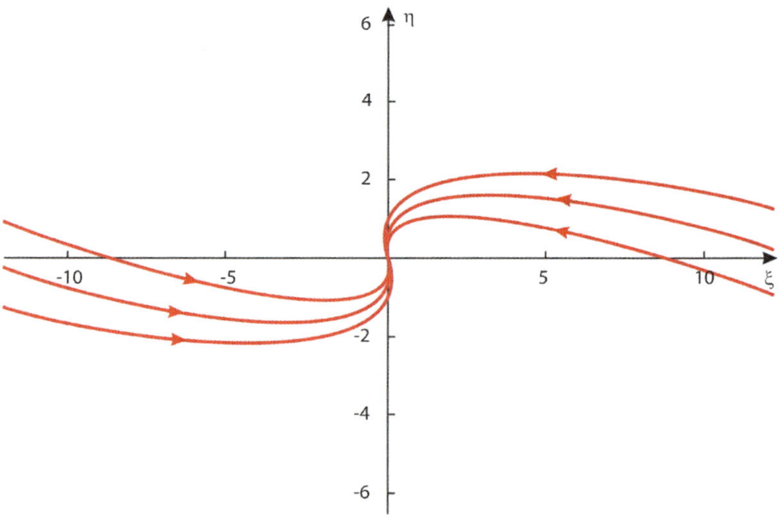

Fig. 9.12 Phase trajectories passing through the degenerated node in the coordinates (ξ, η)

9.3.5 Saddle

The second critical point is a saddle point, which is always unstable. In this case, the roots λ_1 and λ_2 are also real and distinct but they must be of opposite signs. This case occurs when the following conditions are satisfied:

$$(b-c)^2 = 4ad > 0 \quad \text{and} \quad bc - ad < 0. \tag{9.51}$$

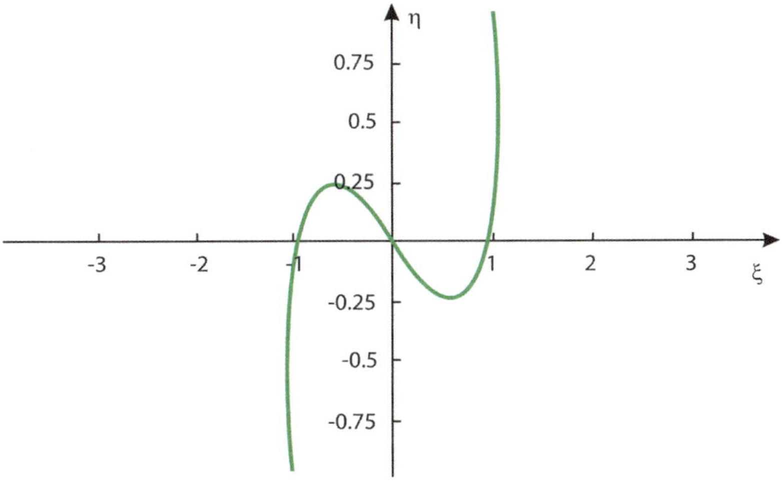

Fig. 9.13 A phase trajectory presenting a degenerated node obtained numerically in the coordinates (ξ, η)

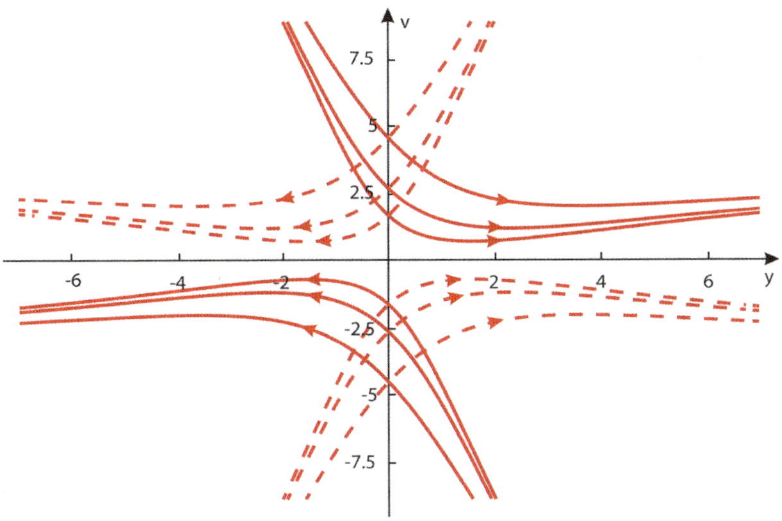

Fig. 9.14 Phase trajectories around the saddle in the coordinates (y, v)

The solution curves in the plane (ξ, η) are hyperbolae, which do not pass through a singular point. One of the roots (a positive one) is associated with the value growth of the solution as the time t increases, while the second solution tends to zero. In the plane (y, v) the curves will be deformed. Figure 9.14 illustrates this situation, where the coefficients: $a = 1, b = 2, c = -2, d = 1$ and then $\lambda_1 > \lambda_2$.

In order to "rectify" the phase trajectories, we transfer the solutions into the plane (ξ, η). This situation is illustrated in Fig. 9.15.

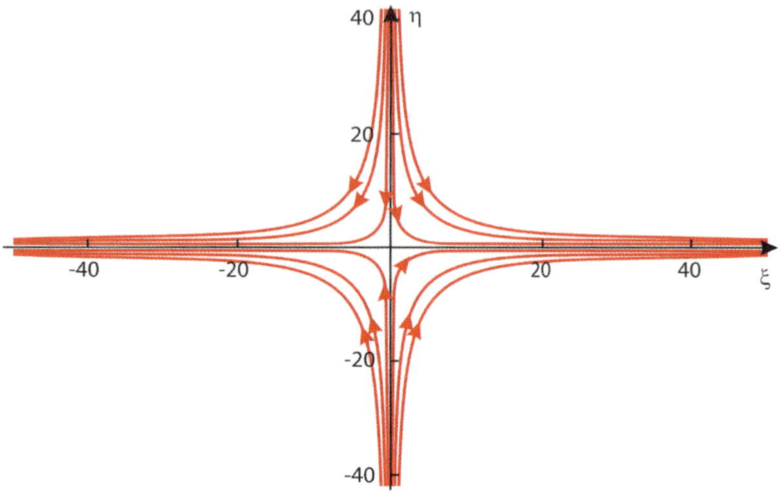

Fig. 9.15 Phase trajectories around the saddle in the coordinates (ξ, η)

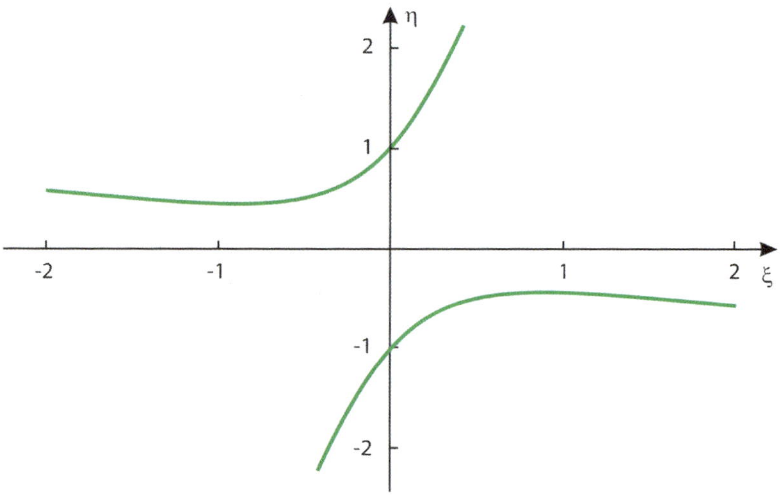

Fig. 9.16 Numerical solution in the case of a saddle in the coordinates (ξ, η)

Verification of the analytical solution is illustrated in Fig. 9.16, where there is numerically obtained singularity of saddle type.

All the analysed dynamical systems possessed real roots, which were solutions of the characteristic equation (9.14). This means that we did not have to do with any types of vibrations. Below, we characterize dynamical systems, whose roots of a characteristic equation are not real any more, i.e. there are no vibrations in these systems. For a stable and unstable focus damped oscillations appear, while in the case of a centre undamped oscillations appear.

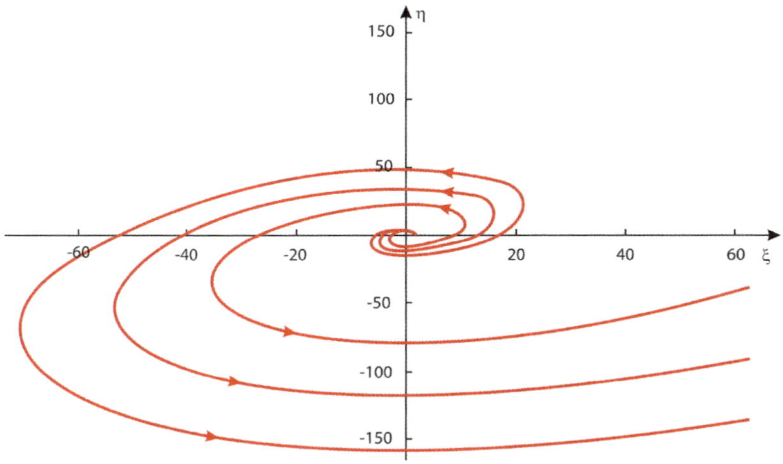

Fig. 9.17 Phase trajectories around the unstable focus in the coordinates (ξ, η)—numerical computations

9.3.6 *Unstable Focus*

The next analysed singular point appears, when

$$(b-c)^2 + 4ad < 0 \quad \text{and} \quad b + c \neq 0. \tag{9.52}$$

Then, roots of the characteristic equation (9.14) are complex conjugate, but any of the roots is neither real nor purely imaginary. It is possible for e.g. the following coefficients $a = 2, b = 0, c = 1, d = -1$ then $b + c > 0$, and a singularity of this type is called a non-stable focus, from which phase trajectories move away.

Figures 9.17 and 9.18 illustrate this situation.

9.3.7 *Stable Focus*

When $b + c < 0$ and the coefficients equal e.g. $a = -2, b = 0, c = -2, d = 2$ then we have a stable focus. Then, the phase trajectories approaches the origin of the coordinate system (Fig. 9.19).

We have also added the numerical solution of this problem (Fig. 9.20).

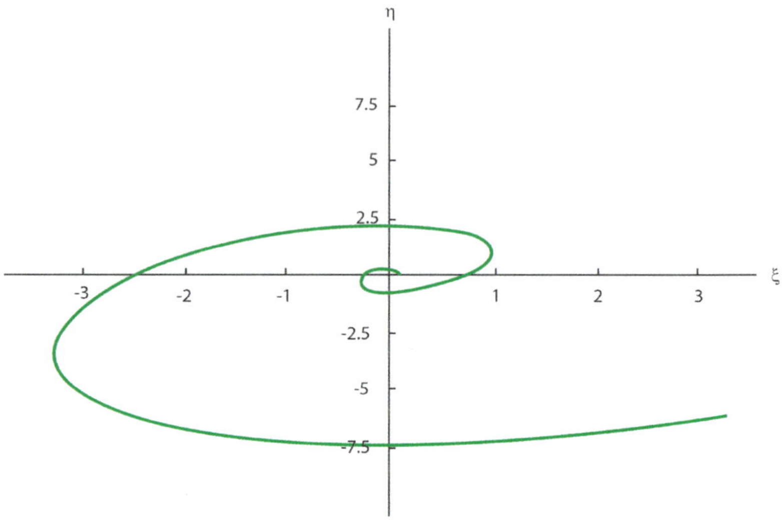

Fig. 9.18 Phase trajectories around the unstable focus in the coordinates (ξ, η)—numerical computations

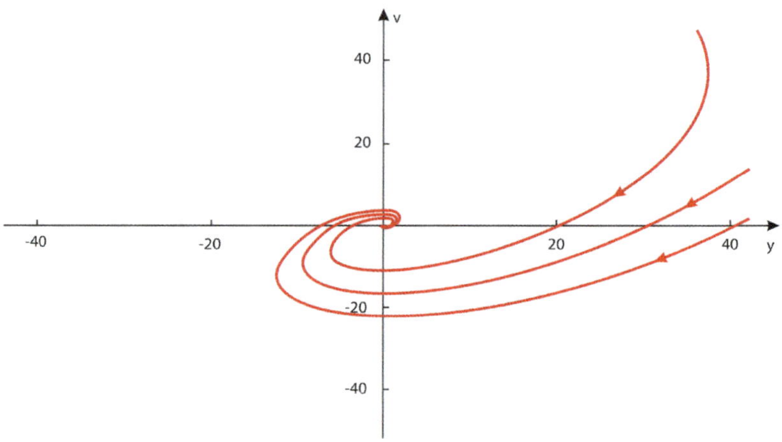

Fig. 9.19 Phase trajectories around the stable focus in the coordinates (y, v)

9.3.8 Centre

The last possible singularity, occurring in the origin is a centre point. The roots λ_1 and λ_2 are then complex conjugate and purely imaginary $\lambda_{1,2} = i\omega$, when

$$(b - c)^2 + 4ad < 0 \quad \text{and} \quad b + c = 0. \tag{9.53}$$

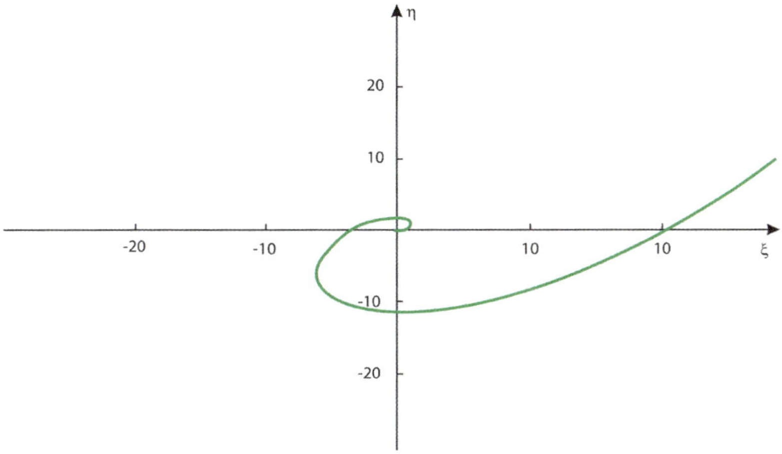

Fig. 9.20 The stable focus obtained numerically in the coordinates (ξ, η)

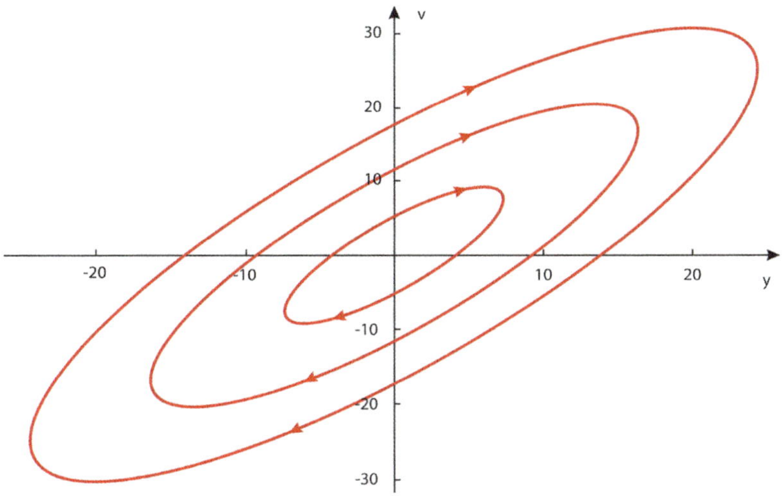

Fig. 9.21 Phase trajectories around the centre in the coordinates (y, v)

In the plane (y, v) the phase trajectories are deformed, but their character is left unchanged, thus they are closed curves surrounding the origin of the coordinate system. The coefficients can be selected in the following way: $a = -3$, $b = 2$, $c = -2$, $d = 2$. This is illustrated in Fig. 9.21.

In this case, the normal form of the equations differs from the previous one since we have to do with the case described by Eq. (9.37).

Figure 9.22 presents the phase trajectories around the origin of the coordinates (ξ, η), whereas the numerically obtained solution is shown in Fig. 9.23.

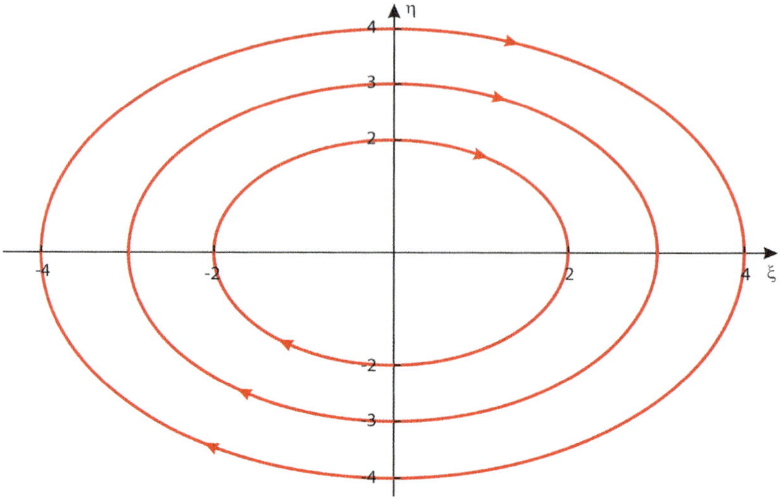

Fig. 9.22 Phase trajectories around the centre in the coordinates (ξ, η)

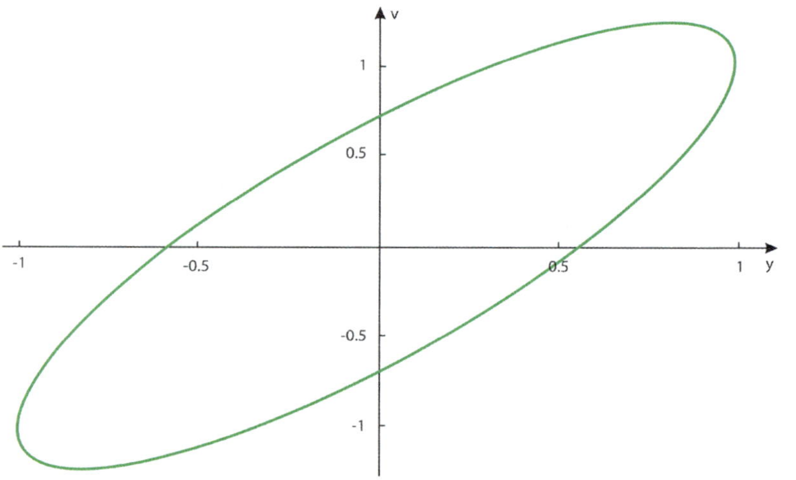

Fig. 9.23 The phase trajectory obtained numerically for a singularity of centre type in the coordinates (y, v)

It follows from the analysis performed in this subsection that the character of an equilibrium position and the shape of phase trajectories near the position depends only on the coefficients a, b, c, d. They have significant influence on the structure of the characteristic equation (9.14).

9.4 Analysis of Singular Points Governed by Three Differential Equations of First Order

In this section we deal with dynamical systems which are governed by three differential equations of first order. The obtained solutions will be presented by means of three-dimensional graphs of phase trajectories. Both analytical and numerical solutions will be plotted for properly selected values of all three constants C_1, C_2, C_3. Selecting in a proper way the values occurring in the equations, we will obtain singularities of special types. In Sect. 9.4.1, we will present the analysed system of equations and its characteristic equation, which will serve for determining proper matrices.

A given matrix will be characteristic for a specific type of a considered singularity. While, in Sect. 9.4.2, graphs of solutions of the corresponding system obtained numerically and analytically will be presented. These are solutions of a system of three first-order differential equations.

9.4.1 Theory Concerning the Solving a System of Differential Equations and Method for Determining Roots of a Polynomial of Third Degree

Considerations will be based on a system of three first-order differential equations. The analysed system of differential equations written in a form of rectangular coordinate system can be presented in the following way:

$$\begin{aligned}\frac{dx}{dt} &= ax + by + cz,\\ \frac{dy}{dt} &= dx + ey + fz,\\ \frac{dz}{dt} &= gx + hy + iz.\end{aligned} \tag{9.54}$$

In this system, the coefficients (characterizing the equations), i.e. a, b, c, d, e, f, g, h, i, can take on real as well as complex values. A solution of this system of equations, we will seek in the form

$$\begin{aligned}x &= C_1 \exp(\lambda t),\\ y &= C_2 \exp(\lambda t),\\ z &= C_3 \exp(\lambda t).\end{aligned} \tag{9.55}$$

The characteristic equation can be written in a matrix form

$$\begin{bmatrix} a & b & c \\ d & e & f \\ g & h & i \end{bmatrix} - \lambda \begin{bmatrix} 1 & 0 & 0 \\ 0 & 1 & 0 \\ 0 & 0 & 1 \end{bmatrix} = 0. \tag{9.56}$$

This equation has a trivial solution when we equate the determinant below to zero

$$\begin{vmatrix} a-\lambda & b & c \\ d & e-\lambda & f \\ g & h & i-\lambda \end{vmatrix} = 0. \tag{9.57}$$

Expanding the determinant we get the following characteristic equation

$$\lambda^3 - (a+e+i)\lambda^2 + (ai+ei+ae-cg-fh-bd)\lambda + bdi + afh + ceg - gbf - dhc - aei = 0. \tag{9.58}$$

The above properly selected coefficients allow to obtain singularities, we are interested in, in a three-dimensional space. The coefficients are responsible for the character of curves plotted after solving the system of differential equations (9.54).

9.4.2 Analysis of Singular Points Described by Three First-Order Differential Equations

Below, we consider and analyse different dynamical systems, in which we select and change the values, which we will write in the matrix form:

$$\mathbf{A} = \begin{bmatrix} a & b & c \\ d & e & f \\ g & h & i \end{bmatrix}. \tag{9.59}$$

9.4.2.1 Unstable Node

An unstable node will be the first analysed type of equilibrium in a three-dimensional phase space. We meet this type of singularity, when components of the matrix **A** are following

$$\mathbf{A} = \begin{bmatrix} 2 & 1 & 1 \\ 0 & 3 & 1 \\ 1 & -1 & 1 \end{bmatrix}.$$

Then the characteristic equation (9.58) possesses three roots. All of them are positive and real. They are: $\lambda_1 = 1$, $\lambda_2 = 2$, $\lambda_3 = 3$.

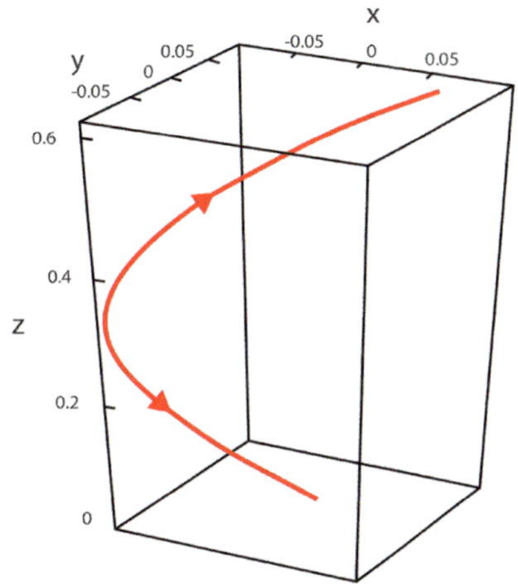

Fig. 9.24 The analytical solution, when an unstable node is an equilibrium position in the three-dimensional space

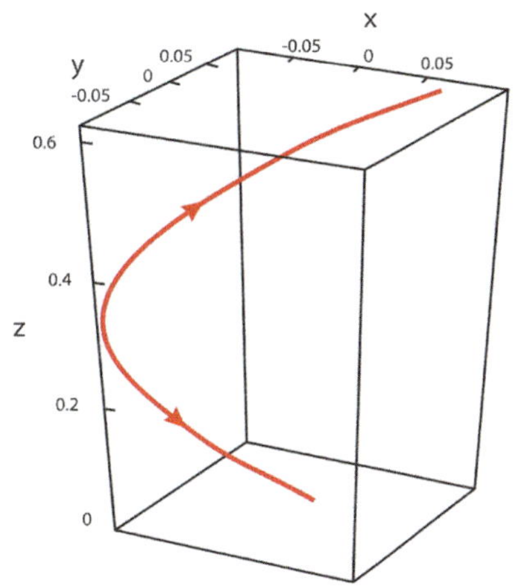

Fig. 9.25 The numerical solution, when an unstable node is an equilibrium position in the three-dimensional space

In Fig. 9.24, one can see that the solution is a parabola. One could have expected this, since we have obtained a similar graph during the analysis of equilibrium positions in the phase plane. Verification of this solution is a numerically obtained graph depicted in Fig. 9.25.

Another example of the matrix **A** (this matrix enabled to obtain equilibrium position of unstable node type), whose elements are

$$\mathbf{A} = \begin{bmatrix} 0.1 & 0 & 0 \\ 0 & 0.06 & 0 \\ 0 & 0 & 0.01 \end{bmatrix}.$$

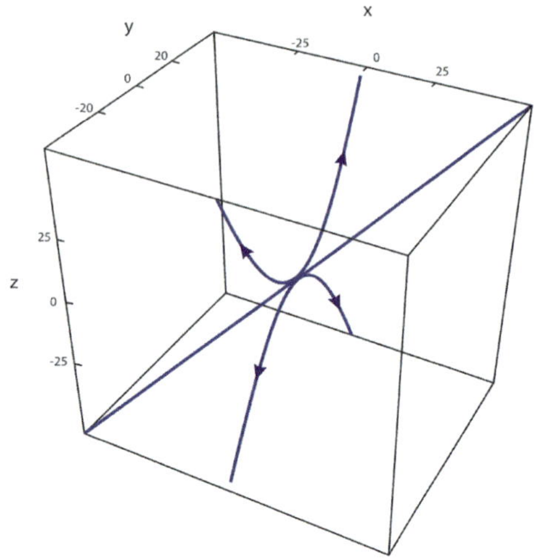

Fig. 9.26 The analytical solution, when an unstable node is an equilibrium position in three-dimensional space

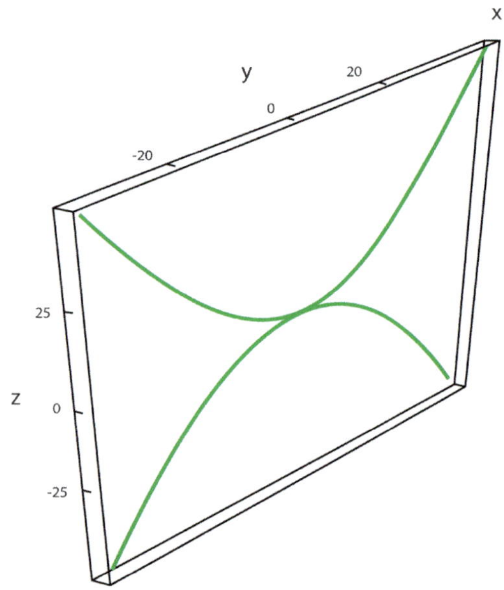

Fig. 9.27 The numerical solution, when an unstable node is an equilibrium position in three-dimensional space

Then roots of the characteristic equation are positive and real, but one of the roots is a double root. They are: $\lambda_1 = 0.06$, $\lambda_2 = 0.1$, $\lambda_3 = 0.1$. Below (Fig. 9.26), one can see the analytical solution graph (Fig. 9.27).

The graph presented in Fig. 9.26 was obtained with the use of symmetry principles and selecting the constants, which appear as a result of solving the system of differential equations (9.54).

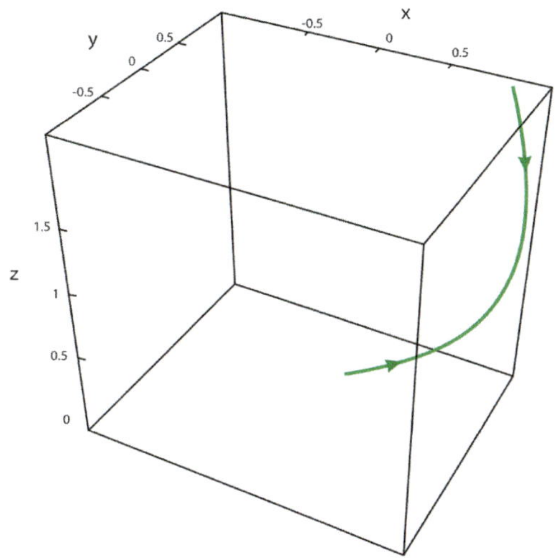

Fig. 9.28 The analytical solution, when a stable node is an equilibrium position in three-dimensional space

9.4.2.2 Stable Node

The second equilibrium—a stable node occurs, when the matrix **A** has the following components

$$\mathbf{A} = \begin{bmatrix} -2 & 1 & 1 \\ 0 & -1 & 1 \\ 1 & -1 & 3 \end{bmatrix}.$$

Then, the characteristic equations (9.58) possesses also three real roots but all of them are negative. They are: $\lambda_1 = -1$, $\lambda_2 = -2$, $\lambda_3 = -3$. Both analytical and numerical solutions of this example coincide and they are reported in Fig. 9.28.

9.4.2.3 Saddle

A next equilibrium position is a saddle point, which is always unstable. The matrix **A** has the following form then

$$\mathbf{A} = \begin{bmatrix} 0.1 & 0 & 0 \\ 0 & 0.09 & 0 \\ 0 & 0 & -0.09 \end{bmatrix}.$$

The characteristic equation (9.58) possesses then three real roots, but they are of opposite signs, i.e. two of them are positive and the last one is negative. They are: $\lambda_1 = -0.09$, $\lambda_2 = 0,09$, $\lambda_3 = 0.1$. The analytical solution is depicted in Fig. 9.29.

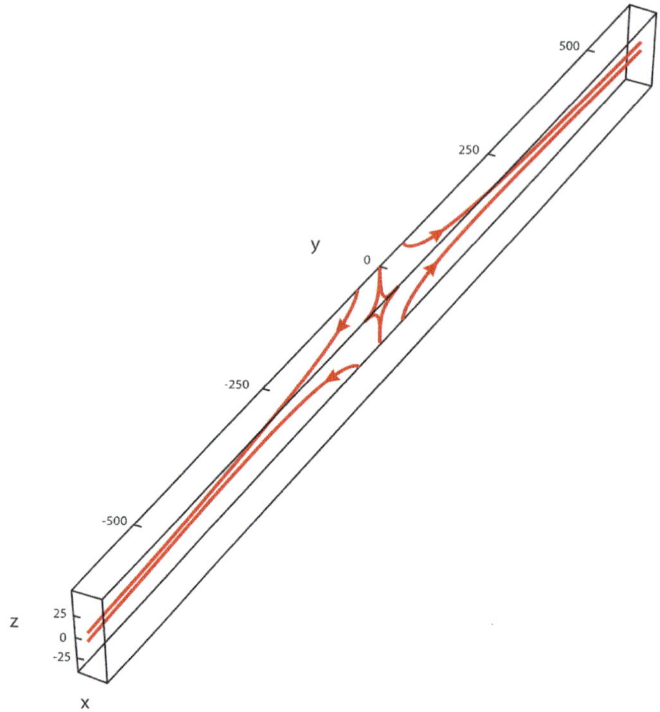

Fig. 9.29 The analytical solution, when a saddle is an equilibrium position in three-dimensional space

The solution is hyperbola, just like in the case of a two-dimensional saddle. In this case, when two roots of the characteristic equation are positive and one is negative, then the solution approaches the equilibrium position.

When a solution of Eq. (9.58) is three real roots of opposite signs, but two of them are negative and one is positive, then we also have to do with equilibrium position of saddle type. The matrix **A** has the following elements:

$$\mathbf{A} = \begin{bmatrix} 0.1 & 0 & 0 \\ 0 & -0.07 & 0 \\ 0 & 0 & -0.1 \end{bmatrix}.$$

Roots, as one can predict, are following: $\lambda_1 = 0.1$, $\lambda_2 = -0.07$, $\lambda_3 = -0.1$, and the analytical solution with the use of symmetry principles is depicted in Fig. 9.30.

In Fig. 9.30, similarly to Fig. 5.30, a hyperbola is a solution. This result differs from the previous one, since the phase trajectories move away from the equilibrium position. Roots of the characteristic equation have influence on this situation, since both of them are negative and previously were positive.

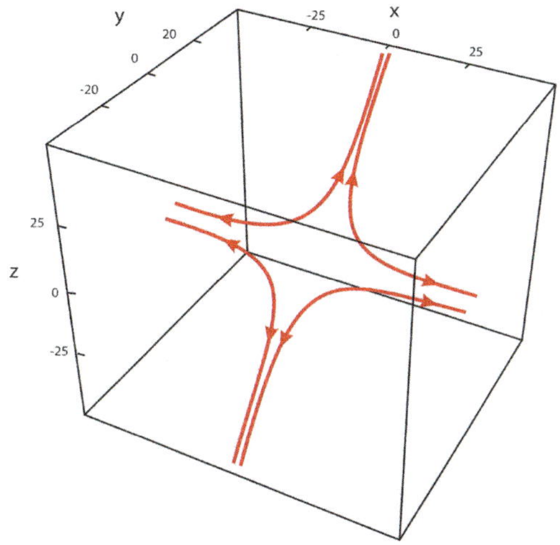

Fig. 9.30 The analytical solution, when a saddle is an equilibrium position in three-dimensional space

9.4.2.4 Unstable Focus

Now, we will consider the cases, when only one root of the characteristic equation (9.58) is real, while two remaining roots are complex conjugate.

When the matrix **A** has the following components

$$\mathbf{A} = \begin{bmatrix} 0.2 & 0 & 0 \\ 0 & 0.1 & -1 \\ 0 & 1 & 0.1 \end{bmatrix},$$

then roots of Eq. (9.58) equal: $\lambda_1 = 0.2$, $\lambda_2 = 0.1 + i$, $\lambda_3 = 0.1 - i$. Then we have to do with equilibrium position of unstable focus type. This situation occurs since the real root as well as the real parts of complex roots are positive. The obtained result is depicted in Fig. 9.31.

As a result, we obtained spirals stretching along the x-axis. As one can see a radius of these spirals grows and moves away from the equilibrium position. This type of singularity is called an unstable focus. Verification of this solution is a numerically obtained graph depicted in Fig. 9.32.

The spirals stretching is better seen for a similar matrix **A**, which also characterizes an unstable focus, namely:

$$\mathbf{A} = \begin{bmatrix} 0.05 & 0 & 0 \\ 0 & 0.1 & -1.3 \\ 0 & 1.3 & 0.1 \end{bmatrix}.$$

The roots are following: $\lambda_1 = 0.05$, $\lambda_2 = 0.1 + 1.3i$, $\lambda_3 = 0.1 - 1.3i$, and the solution is depicted in Fig. 9.33.

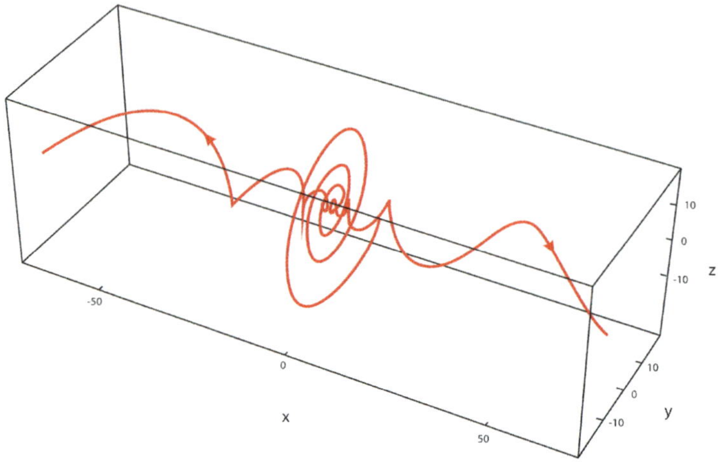

Fig. 9.31 The analytical solution, when an unstable focus is an equilibrium position in three-dimensional space

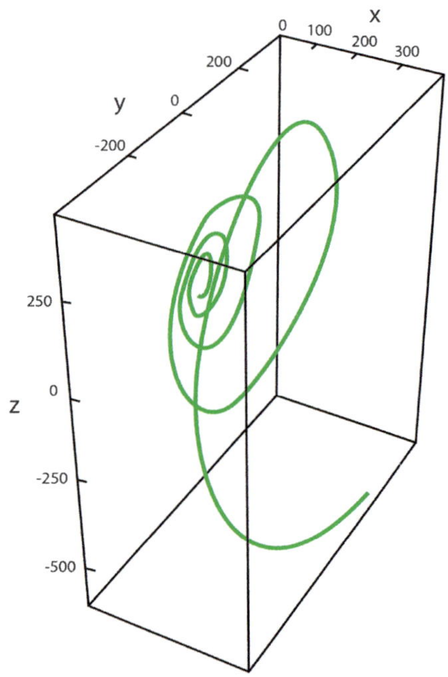

Fig. 9.32 The numerical solution, when an unstable focus is an equilibrium position in three-dimensional space

9.4.2.5 Stable Focus

Similar graphs, in which phase trajectories approach to the equilibrium position, occur in the case of a stable focus, which can be characterized by the following matrix

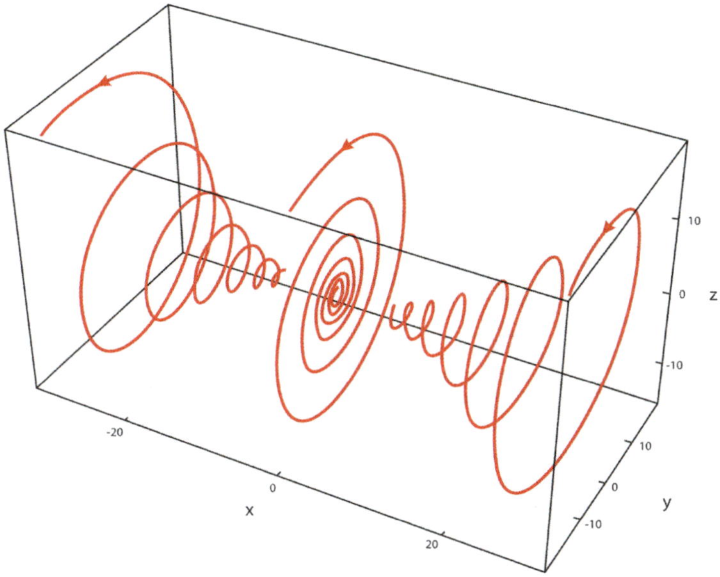

Fig. 9.33 The analytical solution, when an unstable focus is an equilibrium position in three-dimensional space

$$\mathbf{A} = \begin{bmatrix} -0.2 & 0 & 0 \\ 0 & -0.1 & -1 \\ 0 & 1 & -0.1 \end{bmatrix}.$$

In this case, a root of the characteristic equation (9.58) is negative. While complex conjugate roots have a negative real part: $\lambda_1 = -0.2$, $\lambda_2 = -0.1 + i$, $\lambda_3 = -0.1 - i$. Then, the solutions converge to the equilibrium position. This situation is illustrated in Fig. 9.34.

9.4.2.6 Saddle-Node

Similar graphs of phase trajectories can be obtained in the case of equilibrium position of saddle-node type. This singularity occurs when among three roots of Eq. (9.58), the real one is negative and real parts of the remaining complex conjugate roots are positive. It is possible when the matrix **A** has the following elements

$$\mathbf{A} = \begin{bmatrix} -0.1 & 0 & 0 \\ 0 & 0.09 & -0.5 \\ 0 & 0.5 & 0.09 \end{bmatrix}.$$

The roots are: $\lambda_1 = -0.1, \lambda_2 = 0.09+0.5i, \lambda_3 = 0.09-0.5i$, and the trajectories are presented in Fig. 9.35.

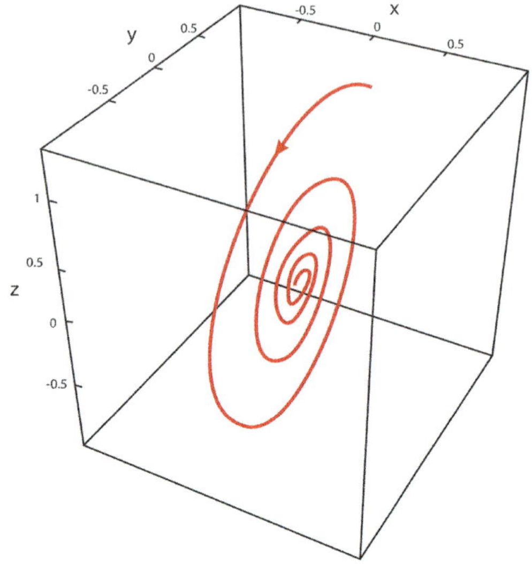

Fig. 9.34 The analytical solution, when a stable focus is an equilibrium position in three-dimensional space

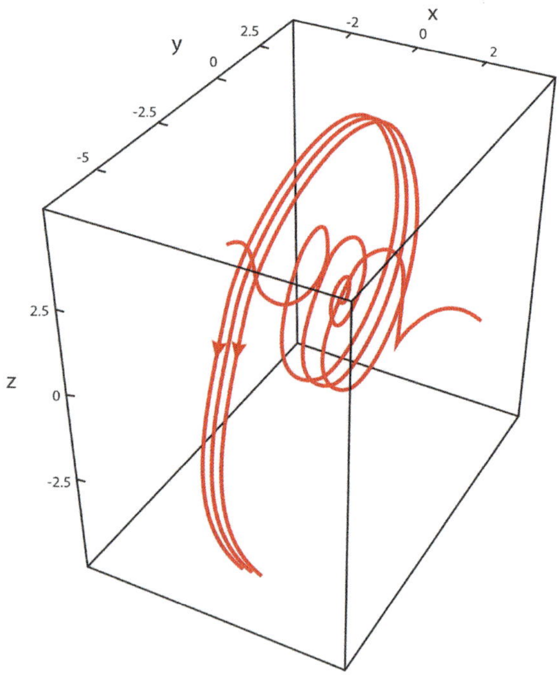

Fig. 9.35 The analytical solution, when a saddle-node is an equilibrium position in three-dimensional space

9.4.2.7 Saddle-Focus

An identical graph but flipped can be obtained if the matrix **A** has the form

$$\mathbf{A} = \begin{bmatrix} 0.1 & 0 & 0 \\ 0 & -0.1 & -0.5 \\ 0 & 0.5 & -0.1 \end{bmatrix}.$$

The roots are: $\lambda_1 = 0.1$, $\lambda_2 = -0.1+0.5i$, $\lambda_3 = -0.1-0.5i$. One can see that in this case a real root is positive and real parts of the complex roots are negative. The singularity of this type is called a saddle-focus, and the phase trajectories approach the equilibrium. This situation is illustrated in Fig. 9.36.

As one can see by the above considerations (just like in the case of phase plane), the graphs of three-dimensional phase trajectories corresponding to specific equilibrium points depend on coefficients occurring in the characteristic equation (9.58). Only they decide about the number of real roots of this equation and their values.

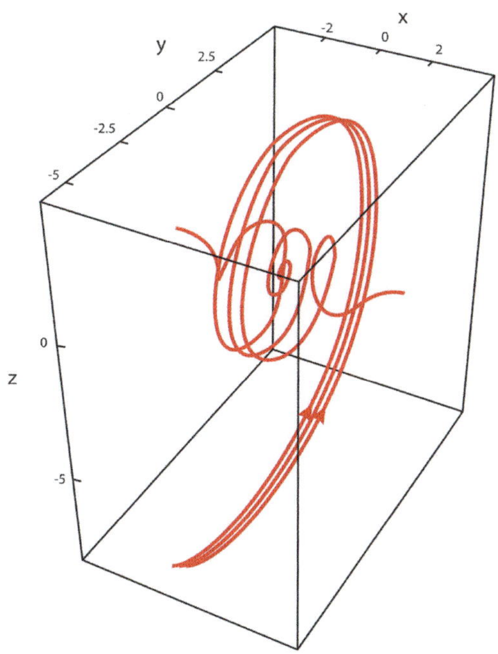

Fig. 9.36 The analytical solution, when a saddle-focus is an equilibrium position in three-dimensional space

Chapter 10
Stability

10.1 Introduction

If a dynamical system is governed by a system of equations

$$\frac{dy_s}{dt} = F_s(t, y_1, y_2, \ldots, y_n), \quad s = 1, 2, \ldots, n, \tag{10.1}$$

then a point $(y_1 \ldots y_n)$ will be called a phase point, and a space $y_1 \ldots y_n$ will be called a phase space. In this chapter, we will consider systems, for which the Cauchy theorem holds, which implies that for each point $(t_0, y_{10}, \ldots, y_{n0})$ there exists a single solution (10.1) satisfying the initial condition $y_s(t_0) = y_s(0), s = 1, 2, \ldots, n$. More detailed discussion on this problem, including discussion on critical points and curves, for which the Cauchy theorem holds can be found in [191]. The solution $y_s = y_s(t)$ is a phase trajectory and the time t is a parameter.

Definition 10.1 (Lyapunov Stability). A solution $y_s^* = y_s^*(t)$ of the system (10.1) is called stable in Lyapunov's sense, if for any numbers $\varepsilon > 0$ and t_0 there exists $\delta = \delta(\varepsilon, t_0) > 0$ such that all the solutions $y_s = y_s(t)$ of the system (10.1) along with the solutions $y_s^*(t)$, which satisfy the initial conditions

$$|y_s(t_0) - y_s^*(t_0)| < \delta, \tag{10.2}$$

also satisfy the inequality

$$|y_s(t) - y_s^*(t)| < \varepsilon, \tag{10.3}$$

for $t_0 \leqslant t < \infty$.

This means that the solution $y_s^*(t)$ is stable, if in any given instant of time t_0 all the solutions $y_s(t)$ sufficiently close to $y_s^*(t)$ lie within any small (but δ-dependent)

J. Awrejcewicz, *Ordinary Differential Equations and Mechanical Systems*,
DOI 10.1007/978-3-319-07659-1_10

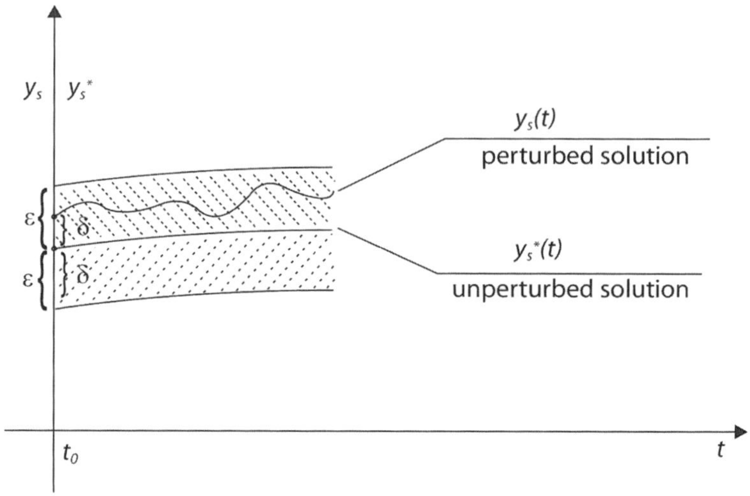

Fig. 10.1 Illustration of the stable solution in Lyapunov's sense

epsilon band. It is depicted in Fig. 10.1. It is worth emphasizing that stability of the solution $y_s^*(t)$ does not imply its boundedness, and the boundedness does not imply the stability. For example, solutions near resonance can be unbounded, yet stable.

On the other hand, chaotic orbits are characterized by the fact that when they start from any close initial conditions they move away from each other exponentially (they are unstable) but they stay bounded. Contradicting the definition (10.1) in the sense of logic, we obtain the following definition.

Definition 10.2. A solution $y_s^* = y_s^*(t)$ is called unstable in Lyapunov's sense, if for some $\varepsilon > 0$, t_0, and any $\delta > 0$ there exist at least one solution $y_s(t)$ and the instant $t_1 = t_1(\delta) > t_0$ such that

$$|y_s(t_0) - y_s^*(t_0)| < \delta \quad \text{and} \quad |y_s(t) - y_s^*(t)| \geq \varepsilon. \tag{10.4}$$

This situation is presented in Fig. 10.2.

Definition 10.3. A solution $y_s^* = y_s^*(t)$ is called asymptotically stable as $t \to +\infty$, if it is stable and the following condition is satisfied

$$\lim_{t \to \infty} |y_s(t) - y_s^*(t)| = 0. \tag{10.5}$$

Definition 10.4. If a solution $y_s^*(t)$ is asymptotically stable as $t \to +\infty$ for the whole considered region of initial conditions, then the solution $y_s^*(t)$ is called global asymptotically stable.

This means that a region of attraction of $y_s^*(t)$ is the whole considered space. One can show that it is sufficient to examine stability of a solution for any chosen

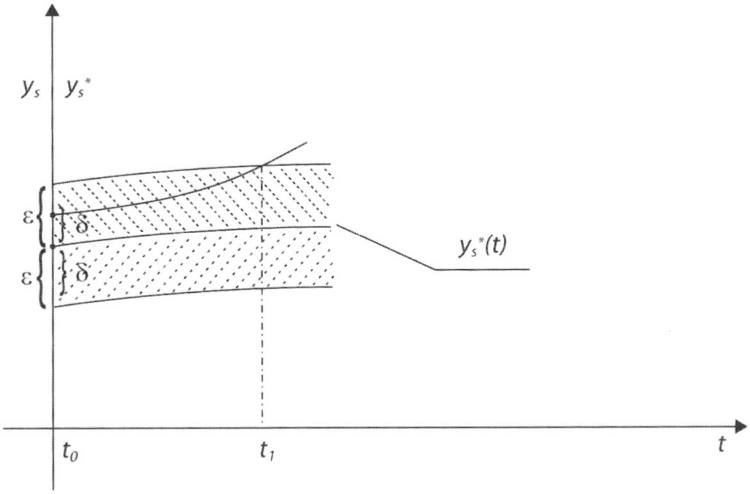

Fig. 10.2 Illustration of an unstable solution in Lyapunov's sense

instant of time t_0. If the examined solution is stable or unstable for $t = t_0$, then it is also stable or unstable respectively for each instant of time in the considered interval of variation of t [77]. Examination of stability by constantly acting perturbations is a very important problem, as will be clear hereafter.

Definition 10.5. A solution $y_s = y_s^*(t)$ of the system (10.1) is said to be orbital stable, if the solution trajectory $y_s(t)$ located sufficiently close to the solution $y_s^*(t)$ at instant t_0 stays at distance, no greater than any arbitrarily small number $\varepsilon > 0$, from this solution. Or equivalently, for each $\varepsilon > 0$ there exists such $\delta(\varepsilon, t_0)$, that if

$$|y_s(t_0) - y_s^*(t_0)| < \delta, \tag{10.6}$$

then

$$\rho(y_s(t), y_s^*(t)) < \varepsilon, \tag{10.7}$$

for $t \geq t_0$ and ρ is a distance between $y_s(t_0)$ and $y_s^*(t_0)$.

Definition 10.6. A solution $y^*(t)$ which is orbitally stable (or stable in Poincaré sense) is called asymptotically orbital stable, if there exists such $\delta > 0$, that for all solutions satisfying the condition

$$|y_s(t_0) - y_s^*(t_0)| < \delta, \tag{10.8}$$

the following condition is satisfied

$$\rho(y_s(t), y_s^*(t)) \to 0, \tag{10.9}$$

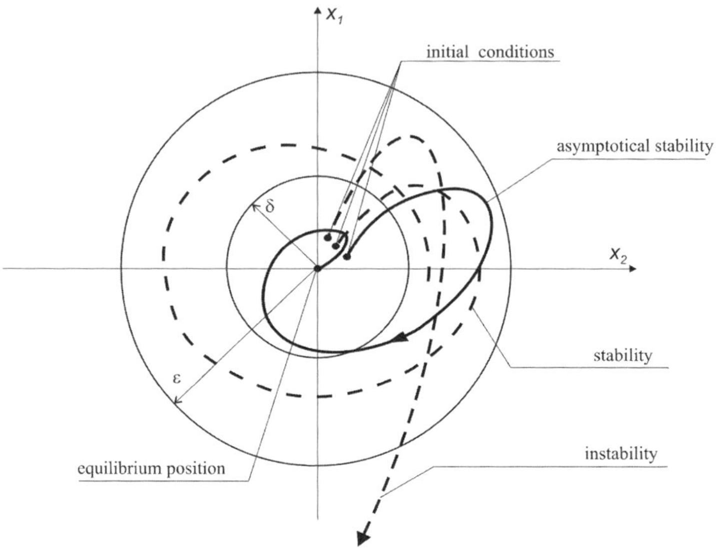

Fig. 10.3 Scheme of stability and instability of a solution represented by the origin of the coordinate system (equilibrium position)

as $t \to \infty$. This means that the trajectory $y_s(t)$ coincides with the trajectory $y_s^*(t)$ as $t \to \infty$.

It is worth emphasizing that a stable solution (asymptotically stable) in Lyapunov's sense will be orbitally stable (asymptotically orbital stable). While orbital stability does not imply Lyapunov stability.

If we consider only perturbations of the unperturbed solution $x(t)$, then we can introduce the following definitions of stability. An unperturbed solution is stable if for any real value of $\varepsilon > 0$ there exists real value of δ such that if $\|x(t_0)\| < \delta$, then $\|x(t)\| < \varepsilon$ for all $t > t_0$. A norm is defined by the formula

$$\|x\| = \sqrt{x^T x}. \tag{10.10}$$

An unperturbed solution is asymptotically stable, if $\lim_{t\to\infty} \|x(t)\| = 0$. Stability and instability of unperturbed solutions of equilibrium position on the basis of the analysis of perturbations are depicted in Fig. 10.3.

In the case of analysis of the unperturbed solution $y^*(t)$ the scheme of evolution of a perturbed and unperturbed trajectory is depicted in Fig. 10.4.

The above figure differs from Fig. 10.3 by that the origin is time-independent in the latter, while in Fig. 10.4 the unperturbed trajectory changes in time. A notion of stability in Poisson sense is rarely used. A phase trajectory is called stable in Poisson sense, when it is bounded and $\forall_{t_0>0}$ and $\delta > 0$ there exist (sufficiently large) values of time t_i, $0, 1, 2, \ldots$, for which the following inequality $\|x(t_i) - x(t_0)\| < \delta$, is satisfied, where δ is the neighbourhood of the mentioned singular point.

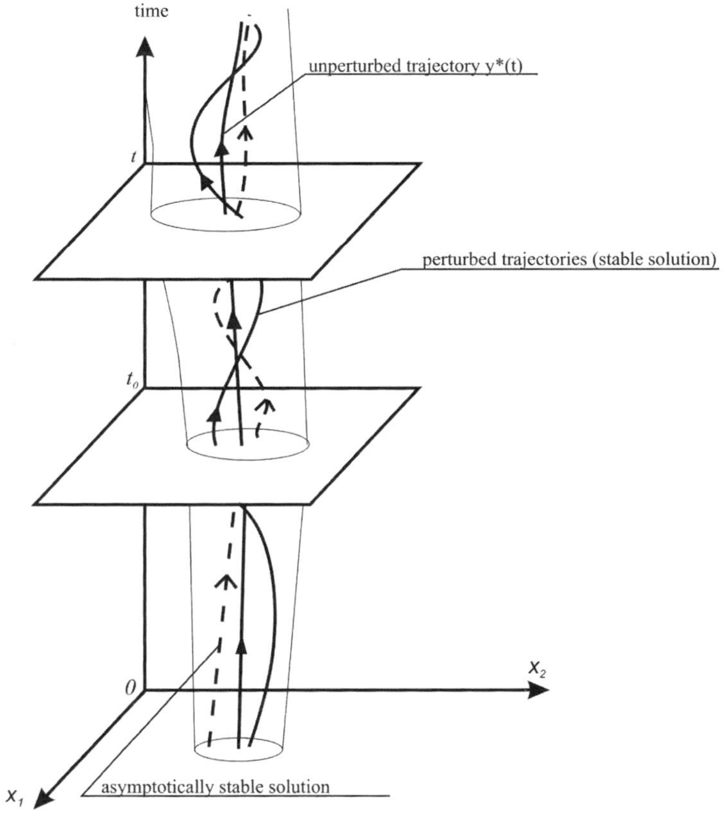

Fig. 10.4 Stable time-dependent solution

Definition 10.7 (Lagrange Stability). The system (10.1) is stable in Lagrange sense, if each of its solutions exists in time interval $t \in (t_0, \infty)$ and its norm is bounded in this interval.

It is worth emphasizing that Lagrange stability concerns stability of a system, thus stability of its all solutions (they are stable if they are bounded). As distinct from this case, the notion of stability in Lyapunov's sense concerns single phase trajectories and their stability is not connected with the notion of boundedness. Now, we give definitions of exponential stability.

Definition 10.8 (Exponential Stability). A trivial solution of the system (10.1) will be called exponentially stable as $t \to \infty$, if for any solution of this system $\{x(t)\} = \{t; t_0\{x_0\}\}$ originating from the region $\|(x_0)\| \leq \delta$ the following inequality holds

$$\|x(t)\| \leq L \|x(t_0)\| e^{-\alpha(t-t_0)}, \tag{10.11}$$

where L and α are some positive constants, which are independent of the choice of the solution $\{x(t)\}$.

Exponential stability of a trivial solution implies asymptotical stability. If we consider the exponent function $\exp(\beta t)$, $\beta \in \mathbb{R}$, then the coefficient β characterizes a function growth and we will call it a characteristic exponent of the function $\exp(\beta t)$. In the case of a complex function

$$F(t) = F_R(t) + iF_I(t), \tag{10.12}$$

of real variable t, then a number (or a symbol $\pm\infty$) defined by the formula

$$\chi[F] = \lim_{t\to\infty} (1/t) \ln |F(t)|, \tag{10.13}$$

will be called a characteristic exponent (a notion of a characteristic number of a function $F(t)$ has been introduced by Lyapunov), the number taken with the opposite sign is a characteristic exponent. It is worth emphasizing that by the inequality (10.11) a trivial solution is exponentially stable, if characteristic exponents $\{\chi[\{x(t)\}]\}$ of the solutions $\{x(t)\}$ of the system (10.1) close to the trivial solution satisfy the inequality

$$\chi[x(t)] \leq -\alpha < 0. \tag{10.14}$$

Definition 10.9 (Conditional Stability). A solution $x^*(t)$ of the system (10.1) is called conditionally stable as $t \to \infty$, if there exists such k-dimensional manifold (in the n-dimensional space) S_k $(k < n)$ that, for each solution satisfying the condition

$$x(t_0) \in S_k, \quad \|x(t_0) - x^*(t_0)\| < \delta(\varepsilon), \tag{10.15}$$

the following inequality holds

$$\|x(t) - x^*(t)\| < \varepsilon \quad for \quad t \geq t_0. \tag{10.16}$$

The above definition can be easily generalized on a case of conditional asymptotical stability.

A solution $x^*(t)$ will be called conditional asymptotically stable, if besides the mentioned conditions in the Definition 10.9, the following property holds

$$\lim_{t\to\infty} \|x(t) - x^*(t)\| = 0, \tag{10.17}$$

where $\|x(t_0) - x^*(t_0)\| < \delta$, while $\delta > 0$ is constant.

The sense of the above definition will be discussed in more detail with the use of an example of a system, which can be obtained from (10.1) by Taylor expansion and extraction of a linear part

$$\frac{dx}{dt} = Ax + F(t,x). \tag{10.18}$$

We assume that the constant matrix A possesses k eigenvalues of negative real parts and $n-k$ characteristic roots (eigenvalues) of non-negative real parts. Moreover, we assume that the function $F(t,x)$ satisfies Lipschitz condition, namely

$$\|F(t,x^*) - F(t,x)\| \le L\,\|x^* - x\|, \tag{10.19}$$

for all $t \in (-\infty, +\infty)$, and Lipschitz constant L is sufficiently small. Then, in some neighbourhood of the point $x = 0$ there exist manifolds S_k^+ and S_{n-k}^- of dimensions k and $n-k$, and such that for solutions of the system (10.18) the following boundary conditions are satisfied:

$$\begin{aligned} &\lim_{t\to+\infty} \|x(t)\| = 0, \text{ if } x(0) \in S_k^+, \\ &\lim_{t\to-\infty} \|x(t)\| = 0, \text{ if } x(0) \in S_{n-k}^-. \end{aligned} \tag{10.20}$$

It is worth emphasizing that the second boundary condition is sometimes used, when one seeks position of an unstable equilibrium point or a periodic orbit in the case of Poincaré map during numerical calculations.

The described notions of stability refer only to ideal systems governed by ordinary differential equations. In real systems it often happens that the observation time of the examined system is finite and sometimes very short. For such examples, it is impossible to examine asymptotical stability as $t \to +\infty$. Consequently, there exists a need for introduction a less rigorous definition of practical stability (*technical stability*).

Definition 10.10. Consider a system of equations of the form

$$\dot{x} = F(t,x) + R(t,x), \tag{10.21}$$

while $x(t_0) = a$. The function $R(t,x)$ is a perturbation constantly influencing the system, which is defined as sufficiently small, i.e. $\|R(t,x)\| \le r^2$, where r is a sufficiently small number.

If all the solutions of the system (10.21) which satisfy the initial condition, i.e. they are imposed with inaccuracies determined by the inequality

$$\|a - b\| = \sum_{i=1}^{n} (a_i - b_i)^2 < \delta^2 \tag{10.22}$$

and with constantly acting limited perturbations R satisfy the condition

$$\bigwedge_{t>t_0} \|x(t) - x^*(t)\| = \sum_{i=1}^{n} [x_i(t) - x_i^*(t)]^2 < \varepsilon^2, \tag{10.23}$$

then the solution $x^*(t)$ will be called technically stable with respect to limitations δ^2, ε^2 and r^2. A region determined by the number 2δ will be called a region of initial conditions ω, while a region determined by the number 2ε will be called a region of admissible solutions Ω.

We will point the differences between the above definition (referring to systems with constantly acting perturbations) and a definition in Lyapunov's sense. According to the definition in Lyapunov's sense, for each region of admissible solutions Ω there exists a region of initial conditions ω (can be selected), so that solutions starting from this region stay in the region Ω all the time. In the case of requirements determined by the definition of technical stability, notions of both regions are independent. They can be a result of requirements concerning the quality of technological, material conditions, etc.

In order to make a thorough study of this issue we will consider dynamics of a system governed by the first-order differential equation:

$$\frac{dy}{dt} = y\left(\alpha^2 - y^2\right), \quad \alpha > 0. \tag{10.24}$$

A singular solution (equilibrium positions) is obtained by equating the right-hand side of (10.24) to zero. The solutions are:

$$y_1^* = 0, \quad y_2^* = \alpha, \quad y_3 = -\alpha. \tag{10.25}$$

A general solution of the above equation has the form:

$$y = \frac{\alpha y_0 e^{\alpha^2(t-t_0)}}{\sqrt{(\alpha^2 - y_0^2) + y_0^2 e^{2\alpha^2(t-t_0)}}}, \tag{10.26}$$

where $y(t_0) = y0$

$$\lim_{t\to\infty} = y(t) = \pm\alpha.$$

For $\alpha = 1$, a solution for four different initial conditions is shown in Fig. 10.5.

Some singular points are stable when they attract solutions, which are within their neighbourhoods. And conversely, unstable singular points (or solutions) repel solutions, which are within their neighbourhoods. The former will be called attractors and the latter repellers [184].

With regard to detection of chaotic responses of simple dynamical system, but governed by at least a system of three second-order differential equations, it is not easy to define an attractor and repeller precisely and there exist several equivalent definitions.

Going back to our dynamical system (10.1) we can observe that its right-hand side generates a vector field, while the solution $x(t) = \Phi(t, t_0, x_0)$ will be called a

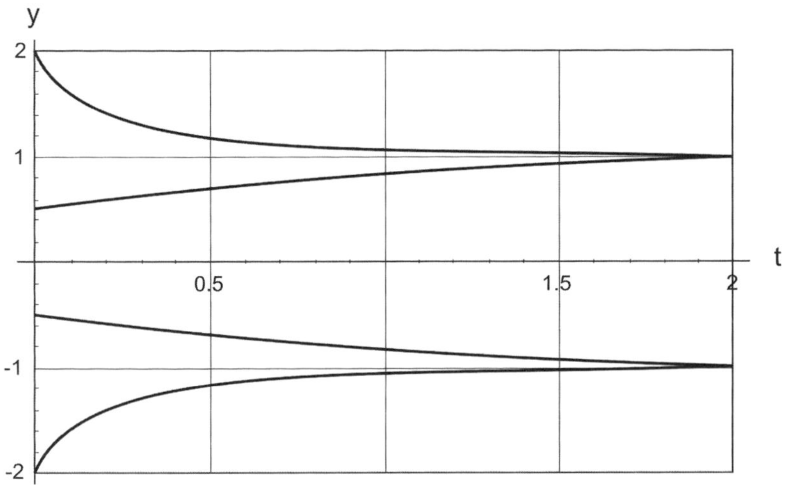

Fig. 10.5 Solution to Eq. (10.26)

vector state. The solution $x(t)$ is a vector, which has its origin in a point $x_0 = x(0)$ and whose end moves at velocity $f(x)$, which depends on the position.

Systems governed by differential equations belong to the class of dynamical systems with continuous time. If for every initial condition the Cauchy problem possesses one solution, then a dynamical system is unique.

As one can see by the last example, only a small class of differential equations possesses analytical solutions and most differential equations are solved numerically. With regard to applying of various numerical methods, it is necessary to introduce discrete time. Then, one introduces a notion of a dynamical system with discrete time (a cascade). A dynamical system (R^n, f) can present a cascade (discrete time) or *flow* (continuous time).

In the case of a cascade, a sequence of successive values of $\{\Phi^n(x)\}$, $n = 0, 1, 2, \ldots$ will be called a trajectory of the point x. If there exists a natural number $k \geq 2$ and a point x_0 satisfying the relationship $x_0 = \Phi^k(x_0)$, and $x_0 \neq \Phi^l(x_0)$ for $0 < l < k$, then x_0 will be called a periodic point of period k. A trajectory (periodic sequence) is connected with such a periodic point, whose k-element set $\{x_0, \Phi(x_0), \Phi^2(x_0) \ldots \Phi^{k-1}(x_0)^2\}$ will be called a *periodic orbit* associated with a point (x_0). Each point of this orbit is a periodic point of period k.

A point x^* will be called an ω-*limit* point of the trajectory $\{\Phi^n(x)\}$ if there exists a sequence such that $\lim_{n^* \to \infty} \Phi^{n^*}(x) = x^*$. The set of all ω-*limit* points will be called the Ω-*limit* of the trajectory $\{\Phi^n(x)\}$.

The compact subset X^* of the space R^n will be called an invariant set of a cascade (R^n, Φ) if $\Phi(x^*) = x^*$.

The mentioned fixed points, periodic points and sets Ω-*limit* are examples of invariant sets. Moreover we add quasi-periodic sets, which are one of Ω-*limit* sets. They can occur in a system of at least third order. The simplest case, which is often encountered in technical problems is a harmonically driven linear oscillator. Frequencies of free oscillations and excitation are not commensurable, e.g.: 1 and $\sqrt{2}$. In this case, the observation and registration of a stationary state of the system each period of the driving force allow to observe a course of a quasi-periodic orbit, which is a collection of points constituting the closed curve.

On the basis of the earlier introduced notions we can make a notion of an attractor and repeller clear. All of the mentioned sets, namely fixed points, periodic and quasiperiodic solutions can be attractors or repellers.

An attractor of a dynamical system (R^n, Φ) *is a bounded and closed invariant set* $A \subset R^n$ *if there exists such its neighbourhood* $O(A)$*, that for any* $x \in O(A)$ *the trajectory* $\{\Phi^n(x)\}$ *stays in* $O(A)$ *and moreover asymptotically approaches this point as* $n \to \infty$*. Moreover, a set of all values of* x*, for which the sequence* $\{\Phi^n(x)\}$ *tends towards the set* A *is called its set (basin) of attraction.*

A repeller of a dynamical system (R^n, Φ) *is a closed and bounded invariant set* $\overline{A} \subset R^n$ *if there exists such a neighbourhood* $O(\overline{A})$ *that if* $x \notin \overline{A}$ *and* $x \in O(\overline{A})$*, then no matter the neighbourhood we choose, the property holds such that for some* l *we have* $\Phi^k(x) \notin O(\overline{A})$ *for* $k > l$*.*

Recently, one encounters in the literature many examples of both strange attractors and repellers, whose basins of attraction or repulsion can possess very complicated character and properties [183–186, 189].

The next step in adaptation of definition for the purposes of real dynamical systems is to introduce a notion of technical stability in finite time. This problem has been stated and solved by Bogusz [52, 53].

Definition 10.11. Let a dynamical system be governed by Eq. (10.21) with the initial conditions $x(t_0) = x_0$, $x \in \omega$. Let $\omega \subset \Omega \subset E_n$, where Ω is a bounded and closed set, while each point $x \in \omega$ possesses a compact neighbourhood within Ω, i.e. $\underset{x\in\omega}{\forall} \underset{R>0}{\vee} K(x, R) \subset \Omega$, where $K(x, R)$ denotes a ball, described in a set E_n, with a centre at the point x and the radius R.

The dynamical system (10.21) will be called technically stable with respect to (ω, Ω) in the finite time $T > 0$, if each trajectory starting at instant t_0 from the point $x_0 \in \omega$ will not leave the set Ω within the time shorter that $t_0 + T$, $\underset{x_0\in\varpi}{\forall} x(t) \in \Omega$ for $t_0 < t < t_{0+T}$, and this illustrates Fig. 10.6.

Theorem 10.1. *Let a positive definite function* $V(x)$ *of class* C^1 *on* E_n *be given. If the following conditions are satisfied:*

$$(a)\ \frac{dV(x(t))}{dt} < 0,$$

for all $x(t) \in \Omega/\omega$ *and* $0 < t - t_0 \le T$*;*

$$(b)\ V(x_1, t_1) \le V(x_2, t_2),$$

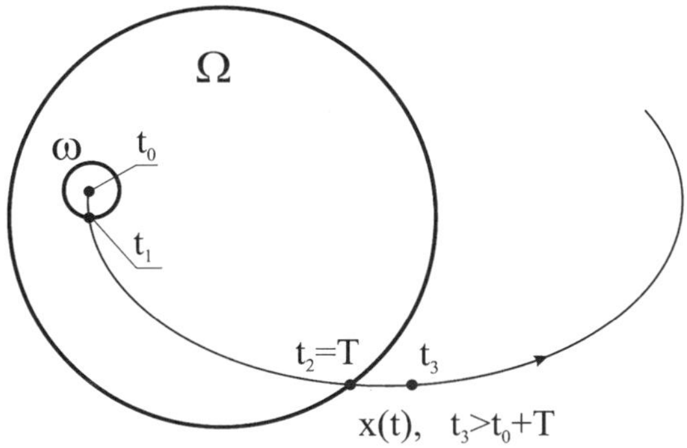

Fig. 10.6 Graphical interpretation of Definition 10.11

for all $x_1 \in \omega$, $x_2 \in E/\Omega$ *and* $0 < t_2 - t_1 < T$, *then the system* (10.1) *is technically stable in the time* T.

Proof. Let us denote by t_0, t_1, t_2 and t_3 the time of start of a trajectory from the point x_0, the time of leave of the set ω and set Ω, and the end of the observation when the trajectory is outside Ω (see 10.6). According to the theorem assumptions we have $0 < t_2 - t_1 < T$ and $V(x_1, t_1) \leq V(x_2, t_2)$. When a trajectory resides in the sets ω and Ω, the energy decreases along the trajectory as time increases, and this means that

$$V(t_2) - V(t_1) = \int_{t_1}^{t_2} \frac{\partial V(x(t))}{\partial t} dt < 0, \tag{10.27}$$

and $V(t_2) < V(t_1)$, which is in contradiction with the condition (b) of Theorem 10.1. □

Theorem 10.2. *Let a positive definite function V(x) be given of class* C^1 *onto* E_n. *If the following conditions are satisfied:*

(a) $\underset{x_1 \in \omega}{\forall}$ $V(x_1, t_1) < C_0$, $0 < t_1 - t_0 < T$;
(b) $V(x_2, t_2) \geq C_1$, $0 < t_2 - t_0 < T$, where $C_0 \leq C_1$;
(c) $\frac{dV(x(t))}{dt} < \frac{C_1 - C_0}{T}$

for $x(t) \in \Omega/\omega$ *and* $0 < t_2 - t_1 < T$, *then the system* (10.21) *is technically stable at the time* T.

More examples concerning this problem can be found in the mentioned papers of Bogusz as well as in [9]. A definition of stability in technical sense assumes that a region of admissible solutions Ω, with respect to technological processes, requirements, etc. includes a region of initial conditions ω. It often happens that a trajectory leaves the region of initial conditions $\omega \subset \Omega$, which come into the region (set) Ω and stay there. This means that one can assume that behaviour of the system is stable, when the system is in one of the equilibrium positions, from whose neighbourhood trajectories pass to a neighbourhood of another equilibrium position. The stability understood in that way comes forward the effective needs of technical stability investigations, which extend the range of Bogusz stability.

Definition 10.12. Consider the system (10.21) with initial conditions $x(0) = x_0$, while a function R is constantly acting perturbation, and $\|R(t,x)\| < r$, where r is a sufficiently small number. Let $\omega \subset E_n$ and $\Omega \subset E_n$ be two bounded and closed regions (Fig. 10.7), while $\omega \not\subset \Omega$. The system (10.21) is called technically stable with respect to (ω, Ω) at the time $t > T$, if each trajectory starting at the instant $t = 0$ at the point $x(0) \in \omega$ will not leave the region Ω within the time $t \geq T$, i.e. $x(0) \in \omega \Rightarrow \bigwedge_{t \geq T} x(t) \in \Omega$.

The above definition has been stated by Szpunar [225]. When one examines stability with the use of Szpunar's method, one can make use of the following theorem.

Theorem 10.3. *Assume that there exists a positive definite function $H(t-T,x)$ of class C^1 onto E^n. If the following conditions are satisfied: $H(0,x) = A(x)$, $H(t-T,x) \to -\infty$ as $|x| \to \infty$ in the interval $0 \leq t \leq T$, derivative of the function H along the solutions of the system* (10.21) *satisfies*

$$\frac{dH}{dt} < \frac{M-C^*}{T} \text{ for } t \geq 0,\ x \notin \Omega^* = \{x, A(x) \leq C^*\} \tag{10.28}$$

where: $M = \max\limits_{x(0)\in\omega_0} H[-T, x(0)]$, *and the following inequality holds for* C^*

$$M - m \leq C^* \leq M \text{ and } C^* < C \tag{10.29}$$

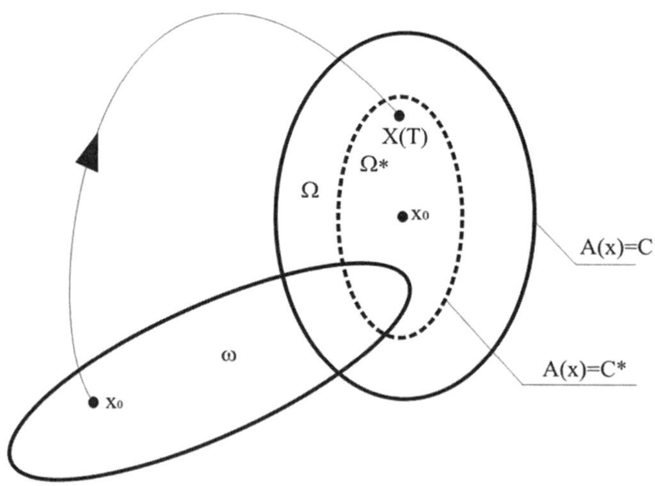

Fig. 10.7 Scheme of stability in Szpunar's sense with marked quantities occurring in Theorem 10.3

where: $m^* = \min_{x(0)\in\omega} H[-T, x(0)]$ *and the inequality* $H(t_1 - T, x_1) < H(t_2 - T, x_2)$ *for* $t_1 > t_2$*, is satisfied* $x_1 \in \Omega^*$, $x_2 \notin \Omega$ *then the system* (10.21) *is technically stable in Szpunar's sense.*

Radziszewski [203–206] and Sławiński [209–211] contributed to a discipline connected with stability of dynamical systems. These works are connected with so-called second Lyapunov's method and the choice of Lyapunov's function. Many recent results point that this method requires the least amount of assumptions, which weaken generality of considerations concerning a given system and that Lyapunov's function gives the best estimation of stability.

Consider the following nonlinear system [210]

$$\dot{x} = f(t, x) \tag{10.30}$$

where $f : I \times \Omega \to R^n$, $I = (\tau, \infty)$, $\tau \geq -\infty$ and Ω is a subset of R^n containing the point $(0, 0)$. We will assume that $f(t, 0) = 0$ for all $t \in I$. On the basis of the mentioned works we will show how one can estimate exponential stability, a set of attraction and the so-called exponential index of convergence for the equilibrium positions $(0, 0)$.

An exponential set of stability is defined by the relation:

$$\Omega_e(t_0) = \left\{ x_0 \in \Omega : \left(\bigvee_{\eta\geq 1}\right)\left(\bigvee_{\delta>0}\right)\left(\bigvee_{t\geq t_0}\right) \|x(t, t_0, x_0)\| < \eta \|x_0\| e^{-\sigma(t-t_0)} \right\}, \tag{10.31}$$

where $\|\cdot\|$ denotes a norm of a vector $\omega \in R^n$. The exponential index of stability is a limiting value of the derivative of Lyapunov's function.

$$V(x) = \|x\|_s = (x, Sx)^{1/2}, \tag{10.32}$$

where $x : R \to R^n$, $S \in M_n^+(R)$, and $M_n^+(R)$ is a set of symmetric positive definite $n \times n$ real matrices. Derivative of the function $V(x)$ reads

$$\dot{V}(t, x) = \frac{(x, Sf(t, x))}{\|x\|_S}. \tag{10.33}$$

Before we go further, let us consider a simple particular case

$$\dot{x} = Ax, \tag{10.34}$$

where $A \subset M_n(R)$. For this case, Eq. (10.33) takes the form:

$$\dot{V}(x) = \frac{(x, SAx)}{\|x\|_S}. \tag{10.35}$$

According to Lyapunov's definition the system (10.34) is exponentially stable if and only if there exists such γ that the following inequality holds

$$\dot{V}(x) \leq \gamma V(x), \tag{10.36}$$

for all $x \in R^n$. Assume that we choose γ in the following way

$$\gamma_s = \sup_{x \in R^n} \frac{\dot{V}(x)}{V(x)} = \sup \frac{(x, SAx)}{(x, Sx)}. \tag{10.37}$$

The system (10.34) will be called exponentially stable if there exists a positive definite symmetric matrix S, such that $\gamma < 0$. It is worth emphasizing that for the system (10.34) and matrices S the values of γ_s can be non-negative, in that case it is interesting to make such a choice of a matrix S that the value of γ_s is minimal, i.e. solve the equation

$$\tilde{\gamma} = \inf_{S \in M_n^+(R)} sup \frac{(x, SAx)}{(x, Sx)}. \tag{10.38}$$

A matrix S_0 satisfying (10.38) will be called a best matrix for the linear system (10.34). In general case, such a matrix may not exist. If the matrix A has a complex structure, then the matrix S_0 can be singular and $S_0 \notin M_n^+(R)$. One can prove that

$$\tilde{\gamma} = \max_j Re\lambda_j(A), \tag{10.39}$$

where $\lambda_j(A)$ is an eigenvalue of the matrix. The condition $\tilde{\gamma} < 0$ is a necessary and sufficient condition of exponential stability of a system. In the case of the considered linear system, a choice of the best matrix S is given by the following lemma

Lemma 10.1. *A best matrix S_0 for the linear system* (10.34) *is a matrix determined by the equations $S_0 = X^*X$, where X is a matrix of eigenvalues, which correspond to the eigenvalues of the analysed system, and $X^* = \overline{X}^T$.*

In what follows we consider a system described by differential equations of the form

$$\begin{bmatrix} \dot{x}_1 \\ \dot{x}_2 \end{bmatrix} = \begin{bmatrix} 0 & 1 \\ -1 & 2a \end{bmatrix} \begin{bmatrix} x_1 \\ x_2 \end{bmatrix}, \tag{10.40}$$

where $a \in R$ is some parameter. A determinant of the matrix occurring in (10.40) has the form

$$\begin{vmatrix} -\lambda & 1 \\ -1 & 2a - \lambda \end{vmatrix} = 0, \tag{10.41}$$

and this leads to the characteristic equation

$$\lambda^2 - 2a\lambda + 1 = 0 \tag{10.42}$$

Roots of the above equation are

$$\lambda_{1,2} = a \pm \sqrt{a^2 - 1}, \tag{10.43}$$

and the following equality holds $\lambda_1\lambda_2 = 1$.

According to the lemma, let us look for a matrix of eigenvalues corresponding the eigenvalues λ_1 and λ_2 which can take two forms. The former has the form

$$\begin{bmatrix} 1 & p \\ 1 & q \end{bmatrix} \begin{bmatrix} 0 & 1 \\ -1 & 2a \end{bmatrix} = \begin{bmatrix} \lambda_1 & 0 \\ 0 & \lambda_2 \end{bmatrix} \begin{bmatrix} 1 & p \\ 1 & q \end{bmatrix}, \tag{10.44}$$

and this leads to the following equations

$$\begin{aligned} p &= -\lambda_1, \ 1 + 2pa = \lambda_1 p, \\ q &= -\lambda_2, \ 1 + 2qa = \lambda_2 q. \end{aligned}$$

Consequently, the sought values $p = -\lambda_1$, $q = -\lambda_2$ and the above two equations (see the characteristic equation) are identically satisfied. A sought matrix X has the form

$$X_{(1)} = \begin{bmatrix} 1 & -\lambda_1 \\ 1 & -\lambda_2 \end{bmatrix}. \tag{10.45}$$

The following matrix equation presents another form of a matrix composed of eigenvectors

$$\begin{bmatrix} 1 & p \\ q & 1 \end{bmatrix} \begin{bmatrix} 0 & 1 \\ -1 & 2a \end{bmatrix} = \begin{bmatrix} \lambda_1 & 0 \\ 0 & \lambda_2 \end{bmatrix} \begin{bmatrix} 1 & p \\ q & 1 \end{bmatrix}, \tag{10.46}$$

and this leads to the following system of algebraic equations

$$\begin{aligned} p = -\lambda, \qquad & 1 + 2pa = \lambda_1 p, \\ q = -1/\lambda_2 \equiv -\lambda_1, \quad & q + 2a = \lambda_2. \end{aligned}$$

Finally, the sought matrix has the form

$$X_{(2)} = \begin{bmatrix} 1 & -\lambda_1 \\ -\lambda_1 & 1 \end{bmatrix}, \tag{10.47}$$

According to the lemma for $|a| > 1$ we have $\tilde{\gamma} = a + \sqrt{a^2 - 1}$ and the sought matrix S_0 has the following forms:

$$S_0^{(1)} = \begin{bmatrix} 1 & 1 \\ -\lambda_1 & -\lambda_2 \end{bmatrix} \begin{bmatrix} 1 & -\lambda_1 \\ 1 & -\lambda_2 \end{bmatrix} = 2 \begin{bmatrix} 1 & -\frac{(\lambda_1+\lambda_2)}{2} \\ -\frac{(\lambda_1+\lambda_2)}{2} & \frac{(\lambda_1^2+\lambda_2^2)}{2} \end{bmatrix}$$

and

$$S_0^{(2)} = \begin{bmatrix} 1 & -\lambda_1 \\ -\lambda_1 & 1 \end{bmatrix} \begin{bmatrix} 1 & -\lambda_1 \\ -\lambda_1 & 1 \end{bmatrix} = 2 \begin{bmatrix} \frac{1+\lambda_1^2}{2} & -\lambda_1 \\ -\lambda_1 & \frac{1+\lambda_1^2}{2} \end{bmatrix},$$

and taking into account the dependence $\lambda_{1,2}(a)$ we have:

$$\begin{aligned} S_0^{(1)} &= 2 \begin{bmatrix} 1 & -a \\ -a & 2a^2 - 1 \end{bmatrix}, \\ S_0^{(2)} &= 2k^2 \begin{bmatrix} 1 & -\frac{1}{a} \\ -\frac{1}{a} & 1 \end{bmatrix}, \end{aligned} \tag{10.48}$$

where $k^2 = a\lambda_1$. Substituting the obtained values of the matrix into (10.38) we get $\tilde{\gamma} = \lambda_1$.

Consider a limiting value of a derivative of Lyapunov's function of the form

$$\gamma_s(\rho) = \sup_{t \in I} \sup_{x \in B_\rho^s} \frac{(x, Sf(t, x))}{(x, Sx)}, \tag{10.49}$$

where $B_r^S = \{x \in R^n : \|x\|_S < \rho\}$.

According to Krasowski theorem concerning the exponential stability (see [138]), if there exists $S \in M_n^+(R)$ and $\rho > 0$ such that

$$\gamma_S(\rho) < 0, \tag{10.50}$$

then the trivial solution (10.34) is exponentially stable and $B_\rho^S \in \Omega_e(t_0)$ for each $t_0 \in I$. If there exists $S \in M_m^+(R)$ and $\rho > 0$ such that

$$\tilde{\gamma}(\rho) = \lim_{\rho \to 0} (\rho) < 0, \tag{10.51}$$

then a trivial solution of the system (10.34) is exponentially stable. In [210], the time duration of the transient process is given. The time needed to reaching the neighbourhood of the equilibrium position for any two sets $\Omega_i = \{x : \|x\|_S = \rho_i\}$ $i = 1, 2, \ldots$ such that $\rho_2 < \rho_1 < \rho$. If the condition (10.50) is satisfied, then time duration of a transient process can be estimated directly with the use of the introduced notion of a limiting value of derivative of Lyapunov's function in the following way:

$$T_{S,\rho}(\rho_1, \rho_2) \le \frac{1}{\gamma_S(\rho)} \ln \frac{\rho_2}{\rho_1}. \tag{10.52}$$

The time needed to overcome the distance between $x \in B_\rho^S$ and the equilibrium position decreases k-times in the sense of a norm $\| \cdot \|$ if

$$T_{S,\rho}^k \le -\frac{\ln k}{\gamma_S(\rho)}. \tag{10.53}$$

The value of the exponential index of convergence depends on the matrix S if for $\gamma_S(\rho) < 0$. The less value $\gamma_S(\rho)$, the better estimation is of $\Omega_e(t_0)$ and $T_{S,\rho}^k$. We demand the following minimal value of an exponential index of convergence with respect to the matrix $S \in M_n^+(R)$

$$\tilde{\gamma}_\rho = \inf_{S \in M_n^+} \gamma_S(\rho). \tag{10.54}$$

An analytical solution of such a stated problem can exist only for some systems of several degrees-of-freedom. In the case of nonlinear system the problem reduces to numerical determination of extremal values of the function

$$q(t,x) = \frac{(x, Sf(t,x))}{(x, Sx)}. \tag{10.55}$$

A dimension of such a stated problem of optimization (10.54) is $(n^2 + 3n)/2$, where n is a dimension of the considered phase space. If the matrix S is known, then the dimension of the problem is $n + 1$ and it is a linear function with respect to n. So in order to decrease the dimension of the optimization problem we assume the matrix S_0 (which is a linear part of the analysed system) as the best one of the matrices S occurring in (10.55). We will estimate the set of attraction and the exponential index of convergence for an equilibrium position after the linearization of the system

$$f(t,x) = Ax + h(t,x), \tag{10.56}$$

where: $A \in M_n(R)$ and $h : I \times R_n \to R_n$ is a continuous function at the point $(t, 0)$ and $\lim_{x \to 0} \frac{h(t,x)}{\|x\|} = 0$ with respect to the time t, where $\| \cdot \|$ is an arbitrarily chosen norm. Taking into account (10.49) and (10.50) we obtain

$$\tilde{\gamma}_S = \lim_{\rho \to 0} \sup_{t \in I} \sup_{x \in B_\rho^S} \frac{(x, SAx) + (x, Sh(t,x))}{(x, Sx)} = \max_i Re\lambda_i \left[\frac{1}{2}(A_S^T + A_S) \right], \tag{10.57}$$

where $A_S = S^{1/2}AS^{-1/2}$. If $S = S_0 \equiv X^*X$ and X satisfies the equation $XA = \Lambda X$, then

$$\tilde{\gamma}_{S_0} = \max_i Re\lambda_i(A) < 0. \quad (10.58)$$

The exponential index of convergence $\gamma_{S_0,\rho}$ is a non-decreasing function of ρ. Taking into account the condition (10.50) and performed calculations one can state the following theorem:

Theorem 10.4. *If* $\max_i \mathrm{Re}\,\lambda_i(A) < 0$ *and* $\lim_{x\to 0} \frac{h(t,x)}{\|x\|} = 0$ *with respect to* t*, then a trivial solution of such system* (10.32) *is exponentially stable and* $B_{\rho_0}^{S_0} \in \Omega_e(t_0)$*, where* $S_0 \equiv X^*X$ *and*

$$\rho_0 = \sup\{\rho : \gamma_s(\rho) < 0\}. \quad (10.59)$$

The above theorem is a good tool to estimate an exponential index of convergence and a set of attraction of the trivial solution. As can be seen by the mentioned definitions of stability, very serious difficulties arise by precise determination of many factors connected with real objects. Glendinning [99] states that there are about 60 different definitions of stability.

Despite the fact that discussion of these definitions is not a goal of this chapter, we will describe one more definition, which is one of the classical ones (besides Lyapunov [159], Poincaré [201, 202]). The definition has been given by Zhukovskiy [253]. Lyapunov's definition refers to analysis of a trajectory from neighbourhood of the examined one, whereas the Poincaré definition concerns a distance between neighbouring trajectories and the examined trajectory in the phase space. The Zhukovskiy definition seems to gather advantages of both of the mentioned definitions. In the case of the stability analysis of equilibrium positions all three definitions are equivalent.

With regard to detection of chaotic orbits, it turns out that applying two first definitions to evaluate the stability of chaotic orbits is not satisfactory. The main idea of Zhukovskiy's method is to estimate the distance between neighbouring trajectories on a piece of a surface, which is perpendicular to one of the trajectories [154], and this allows to introduce a completely new conception of linearization. In order to introduce a notion of Zhukovskiy stability we introduce the following class of a homeomorphic mapping. Let us recall a notion of homeomorphism: if X and Y are topological spaces, then a homeomorphism is called a continuous, bijective mapping between two spaces X and Y, and the inverse mapping is continuous as well. Two topological spaces will be called homeomorphic if there exists homeomorphism, which maps one space into another one.

$$\Gamma = \{\tau | \tau : [(0, +\infty) \to (0, +\infty), \tau(0) = 0]\}, \quad (10.60)$$

while the function Γ will play a role of a parameterizing factor of the trajectory $x(t)$.

Definition 10.13 (Zhukovskiy's Stability). A solution $x(t) = x(t, x_0)$ will be called stable in Zhukovskiy sense if $\forall_{\varepsilon>0} \vee_{\delta(\varepsilon)>0}$ for any $y_0 \in B_\delta(x_0)$ ($B_\delta(x)$ is an open ball of a radius δ and its centre x) one can find two parameterizing factors $\tau_1, \tau_2 \in \Gamma$ such that for all $t \geq 0$ the following inequality holds:

$$\|x(\tau_1(t), x_0) - x(\tau_2(t), y_0)\| < \varepsilon. \tag{10.61}$$

Next definitions concern asymptotical stability and non-stability in Zhukovskiy sense.

Definition 10.14. A solution $x(t) = x(t, x_0)$ is asymptotically stable in Zhukovski sense if one can find such $\delta_0 > 0$, that for all $y_0 \in B_{\delta_0}(x_0)$ there exist two parameterizing functions $\tau_1, t_2 \in \Gamma$, that the following condition is satisfied:

$$\lim_{t\to+\infty} \|x(\tau_1(t), x_0) - x(\tau_2(t), y_0)\| \to 0. \tag{10.62}$$

Definition 10.15. A solution $x = x(t, x_0)$ is called unstable in Zhukovskiy sense as $t \to +\infty$, if there exists such $e > 0$, that for any $\delta > 0$ one can select $y_0 \in K_\delta(x_0)$ in the way that for any two parameterizing function $\tau_1, \tau_2 \in M$ and for the first $\tau_1 \geq 0$ following inequality holds:

$$\|x(\tau_1(t), x_0) - x(\tau_2(t) y_0)\| \geq \varepsilon. \tag{10.63}$$

As was mentioned, in the case of stability analysis of equilibrium position Lyapunov's, Poincaré and Zhukovskiy's definitions are equivalent.

Consider a periodic solution $x = x(t, x_0)$, i.e. an image of this solution in the phase-space is a closed curve.

In this case, one can see the essential differences between definitions in Lyapunov and Poincaré sense, as was mentioned earlier. Assume we will consider an autonomous conservative harmonic oscillator. Then the perturbed solution with respect to the examined periodic solution will be also periodic. Thus, an arbitrary small perturbation makes a solution transfer from one closed curve to another one in the phase space and it is stable in Poincaré sense. However, it is not stable in Lyapunov sense because the periods of the mentioned orbits are distinct. For the considered case, Zhukovskiy's definition is equivalent to Poincaré but not to Lyapunov's one.

Consider a more general situation, when the analysed trajectory is neither periodic nor degenerate periodic orbit, i.e. an equilibrium position. In the considered case, a phase point belonging to the perturbed solution can leave neighbourhood of a segment of the analysed trajectory so as to find itself in ε neighbourhood of another segment of the trajectory. In this case, a trajectory is unstable in Zhukovskiy sense, though distances between phase points belonging to the perturbed and unperturbed orbit remain arbitrarily small. Thus, the solution is stable in Poincaré sense. Many other examples, which point out usefulness of applications of the Zhukovskiy definition (in particular, to analysis of stability of quasi-periodic or chaotic trajectories) have been discussed in [154].

10.2 Lyapunov's Functions and Second Lyapunov's Method

If

$$x_S(t) + y_S^*(t) = y_S(t), \tag{10.64}$$

then substituting (10.64) into Eq. (10.1) we get

$$\tfrac{dx_s}{dt} = X_s(t, x_1, \ldots, x_n),\ s = 1, \ldots, n, \tag{10.65}$$

where

$$X_S(t, x_1, \ldots, x_n) = F_S(t, x_1 + y_1^*, \ldots, x_n + y_n^*) - F_S(t, y_1^*, \ldots, y_n^*).$$

The quantities $x_S(t)$ are understood as perturbations of the analysed (unperturbed) solution $y_S^*(t)$. It is noteworthy that the solutions $x_S \equiv 0$ of Eq. (10.64) (which are called equations of perturbations hereafter) correspond to the solution $y_S^*(t)$. *Thus, stability analysis of the solution $y_S^*(t)$ reduces to the analysis of a trivial solution (equilibrium position) of equations of motion of a perturbed system which are also called reduced equations.*

Now, we will introduce some notions and definitions, which are necessary for further considerations.

Definition 10.16. A scalar, real and continuous function $V(t, x_1, \ldots, x_n)$ is called a weak definite (positive or negative) function, if $V(t, x_1, \ldots, x_n) \geq 0$ or $V(t, x_1, \ldots, x_n) \leq 0$, respectively.

Definition 10.17. A function $V(t, x_1, \ldots, x_n)$ is called positive definite if there exists a scalar function $W(x_1, \ldots, x_n)$ such that

$$V(t, x_1, \ldots, x_n) \geq W(x_1, \ldots, x_n) > 0, \tag{10.66}$$

$$V(t, 0 \ldots, 0) = W(0, \ldots, 0). \tag{10.67}$$

Similarly, we define a negative definite function V.

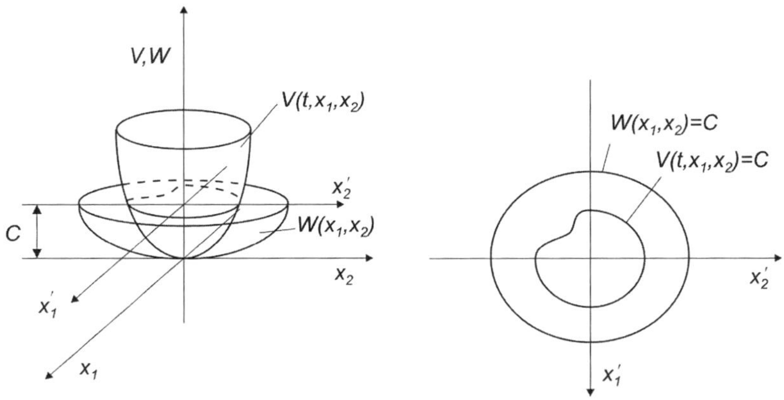

Fig. 10.8 Illustration of a positive definite function for a two-dimensional system

Definition 10.18. If there exists $W(x_1, \ldots, x_n)$ such that:

$$V(t, x_1, \ldots, x_n) \le W(x_1, \ldots, x_n) < 0, \tag{10.68}$$

$$V(t, 0 \ldots, 0) = W(0, \ldots, 0) = 0, \tag{10.69}$$

then the function $V(t, x_1, \ldots, x_n)$ is called negative definite.

A positive or negative definite function is called sign definite. One can show that if $V(t, x_1, \ldots, x_2) = V(x_1, \ldots, x_n)$ is a sign definite function, then there exists such a number $H > 0$, for which all the surfaces $V(x_1, \ldots x_n) = C$ are closed around the point $(0, \ldots, 0)$ for $|C| < H$. If there is time in the function V, then though the surfaces $V(t, x_1, \ldots, x_n)$ change as time flows, after a suitable selection of C the surface $V(t, x_1, \ldots, x_n)$ stays within the region surrounded by a closed surface $W(x_1, \ldots, x_n)$. For a system of two variables one can give a graphical illustration (see Fig. 10.8).

When one examines stability by means of Lyapunov's method one needs to relate a function $V(t, x_1, \ldots, x_n)$ to the reduced system (10.65). If a variable $(x_1, \ldots, x_n)$ of this system is the same variable occurring in the function $V(t, x_1 \ldots, x_n)$, then the function

$$\dot{V}(t, x_1, \ldots, x_n) = \frac{\partial V}{\partial t} + \sum_{i=1}^{n} \frac{\partial V}{\partial x_i} X_i(t, x_i, \ldots, x_n), \tag{10.70}$$

is called an exact derivative of the function $V(t, x_1, \ldots, x_n)$ with respect to time t, associated with the reduced system.

Theorem 10.5 (First Lyapunov's Theorem). *If for the reduced system* (10.64) *there exists a scalar function $V(t, x_1, \ldots, x_n)$, defined up to a sign which possesses* *a time derivative* $\dot{V}(t, x_1, \ldots, x_n)$ *and is associated with the system* (10.64)*, and is a function of a constant sign, which is opposite to the sign of* V *or is identically zero, then a trivial solution* $(x_1, \ldots, x_n) = (0, \ldots, 0)$ *of this system is stable in Lyapunov's sense as* $t \to +\infty$.

Theorem 10.6 (Second Lyapunov's Theorem). *Let for the reduced system* (10.64) *exist a function* $V(t, x_1, \ldots, x_n)$*, defined up to a sign which possesses a time derivative* $\dot{V}(t, x_1, \ldots, x_n)$*, defined up to a sign, which is opposite to the sign of* V*, then a trivial solution* $(x_1, \ldots, x_n) = (0, \ldots, 0)$ *of this system is asymptotically stable in Lyapunov's sense as* $t \to +\infty$.

Theorem 10.7 (The First Theorem on Instability). *Let for the reduced system* (10.64) *exist a function* $V(t, x_1, \ldots, x_n)$ *such that its derivative* $\dot{V}(t, x_1, \ldots, x_n)$ *with respect to* t*, associated with this system, is definite and can attain, in the neighbourhood of zero, values of the same sign as the function* $V(t, x_1, \ldots, x_n)$ *does, then a trivial solution (unperturbed solution) is unstable.*

Theorem 10.8 (The Second Theorem on Instability). *If for the reduced system* (10.64) *one can find a bounded function* $V(t, x_1, \ldots, x_n)$*, whose derivative can be transformed into the form*

$$\dot{V} = \lambda V + W, \tag{10.71}$$

where λ *is a particular constant positive number,* $W \equiv 0$ *or is a particular weak definite function and if the function* $V(t, x_1, \ldots, x_n)$ *is not weak definite of opposite sign to the function* W*, then a trivial solution is unstable.*

Two first Lyapunov's theorems and the first theorem on instability allow for a graphical illustration. If the function $V(t, x_1, \ldots, x_n)$ and its derivative $\dot{V}(x_1, \ldots, x_n)$ are definite and of opposite signs functions, then a phase point moving along the phase trajectory intersects each surface $V(x_1 \ldots, x_n) = C$ from outside, since the function V values should decrease. In this case, phase trajectories should approach the origin of the coordinate system (Fig. 10.9a).

Finally, in the case of fulfilment of conditions concerning the first theorem on instability, a phase point moving outward the curve (Fig. 10.9c) intersects the surface $V(t, x_1, \ldots, x_n) = C$. The case from Fig. 10.9b is particular, since $\dot{V} = 0$.

As can be noted, one assumed that the derivative $\dot{V}(t, x_1, \ldots, x_n)$, associated with the system, whose stability we want to determine, is weak definite at some neighbourhood of the origin. In order to show that a dynamical system is unstable it is sufficient to show that there exists at least one trajectory, which has its origin in the neighbourhood of an equilibrium position and the trajectory moves away from this point. It is not necessary to take into account the whole neighbourhood of the equilibrium position.

Theorem 10.9 (Chetayev Theorem). *Let for the reduced system* (10.65) *exist a function* $V(t, x_1, \ldots, x_n)$*, which possesses continuous partial derivatives of first order and whose projection of its cross-section on the plane* $(0, x_1, \ldots, x_n)$ *is not*

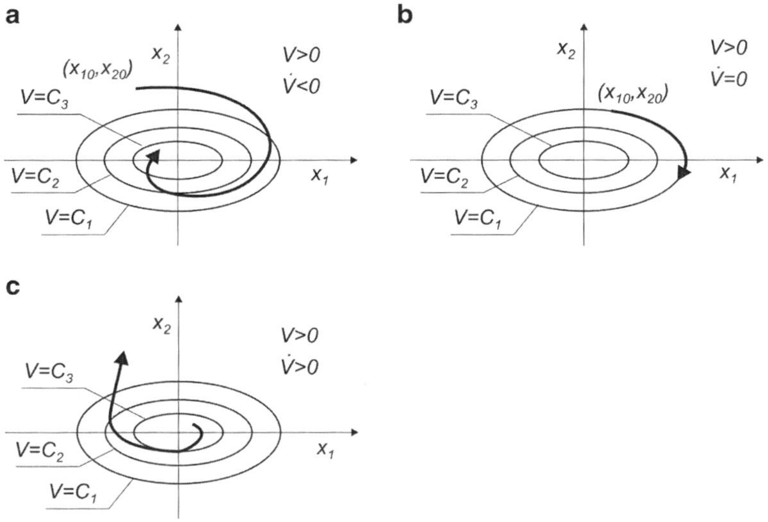

Fig. 10.9 Graphical illustration corresponding theorems on stability and instability

an empty and closed set, and its boundary includes the origin of the coordinate system and moreover on the boundary $V(t, x_1, \ldots, x_n) = 0$. *If the function* $V(t, x_1, \ldots, x_n)$ *is bounded in a region of projection and has a derivative in this region* $\dot{V}(t, x_1, \ldots, x_n)$ *associated with the system* (10.65), *provided that in each subregion of this region for which* $V(t, x_1, \ldots, x_n) \geq \alpha > 0$ *the inequality* $\dot{V}(t, x_1, \ldots, x_n) \geq \beta > 0$ *holds, where* $\beta = \beta(\alpha)$ *is a certain positive number, which depends on a positive number a, then a trivial solution of the system* (10.65) *is unstable in Lyapunov's sense as* $t \to +\infty$.

Projection of the cross-section of function $V(t, x_1, \ldots, x_n)$ on the plane $(0, x_1, \ldots, x_n)$ is depicted in Fig. 10.10. The origin belongs to the boundary of this projection. If there exists an internal point $(t_0, x_1^{(0)}, \ldots, x_n^{(0)})$ such that $0 < \|\{x^{(0)}\}\| < \delta$, and $V(t_0, x_1^{(0)}, \ldots, x_n^{(0)}) = \alpha > 0$, $\dot{V}(t, x_1(t), \ldots, x_n(t)) > 0$, then the projection of the trajectory, which has its origin in the cross-hatched region will leave this region.

Consider the reduced system (10.64) again. If a trivial solution of this system is global asymptotically stable, then it is asymptotically stable in Lyapunov's sense and its region of attraction is the whole space.

A function $V(t, x_1, \ldots, x_n)$ will be called infinitely large, if for an arbitrary positive number α there exist a positive number R such that the inequality $V(x_1, \ldots, x_n) > \alpha$ holds for all $x_1, \ldots, x_n$ lying inside a ball of radius R.

Theorem 10.10 (Barbashin–Krasovskiy Theorem). *If there exists a positive definite, infinite function* $V(x_1, \ldots, x_n)$ *and such that its derivative associated with the system* (10.65) *is a negative definite function (while* $dV/dt = 0$ *holds for*

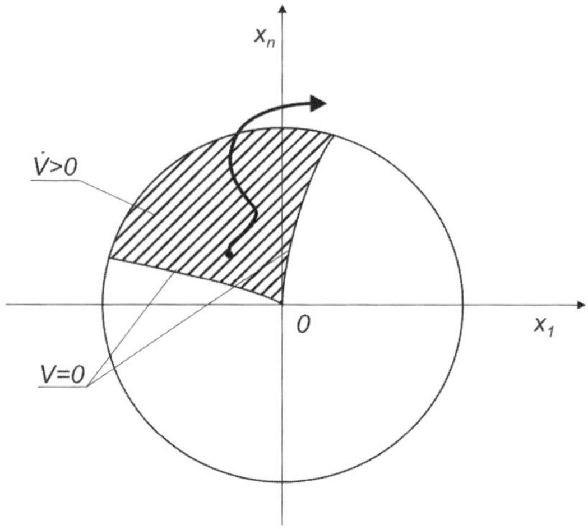

Fig. 10.10 Graphical illustration of Chetayev's theorem

$(x_1, \ldots, x_n) = (0, \ldots, 0)$), *then a trivial solution of the system* (10.65) *is global asymptotically stable.*

The above theorem concerns necessary and sufficient conditions of a system stability in Lagrange sense.

Theorem 10.11. *It is sufficient and necessary for the system* (10.1) *to be stable in Lagrange sense if there exists a function* $V(t, y_1, \ldots, y_n)$, *which satisfies the following conditions:*

(a) $V(t, y_1, \ldots, y_n) \geq W(y_1, \ldots, y_n)$, *and* $W(y_1, \ldots, y_n) \rightarrow \infty$ *for* $\|(y_1, \ldots, y_n)\| \rightarrow \infty$;
(b) for any expansion $y_s(t; t_0, y_{10}, \ldots, y_{n0})$, $s = 1, \ldots, n$ *a function* $V(t, y_s)$ *is non-increasing with respect to t.*

This last condition is equivalent to the condition $\dot{V}(t, y_1, \ldots, y_n) \leq 0$, where $\dot{V}$ is a derivative of a function associated with the analysed system.

Consider a very general nonlinear oscillator described by the following equation

$$\ddot{x} + p(t)\dot{x} + q(t)f(x) = 0, \tag{10.72}$$

with the following initial conditions: $p(t)$, $q(t)$ and $f(x) : 0 < q(t) < M$, $p(t) \geq -\dot{q}/2q$ and $\int_0^{\pm\infty} f(x)dx = +\infty$.

One can show that solutions and their derivatives of this equation are bounded for $t \in [t_0, \infty]$. Let us transform the system (10.72) into two first-order equations of the form

$$\frac{dx}{dt} = y, \tag{10.73}$$

$$\frac{dy}{dt} = -p(t)y - q(t)f(x). \tag{10.74}$$

We assume the following form of the function V:

$$V(t, x(t), y(t)) = \int_0^{x(t)} f(\zeta)d\zeta + \frac{y^2(t)}{2q(t)}, \tag{10.75}$$

and since $q(t) \leq M$, then we assume

$$W(x, y) = \int_0^{x} f(\zeta)d\zeta + \frac{y^2}{2M}, \tag{10.76}$$

where: $V(t, x, y) \geq W(x,y)$ and $W(x, y) \to +\infty$, when $x^2 + y^2 \to \infty$.

Next, we calculate the derivative associated with the system (10.72)

$$\frac{dV(t, x, y)}{dt} = f(x)y - \frac{y}{q}[py + qf] - \frac{y^2\dot{q}}{2q^2}, \tag{10.77}$$

and after transformations we obtain the inequality

$$\dot{V} = -\frac{y^2}{q}(p + \frac{\dot{q}}{2q}). \tag{10.78}$$

According to the inequality $p(t) \geqslant -\dot{q}/2q$, taking into account the assumption that $q > 0$, we obtain $\dot{V}[t, x(t), y(t)] \leq 0$, and this proves boundedness of $x(t)$ and $\dot{x}(t)$ in the interval $(0, \infty)$.

Now, we give a few theorems concerning characteristic exponents. Proofs of the theorems can be found in the monograph [77].

Theorem 10.12. *A characteristic exponent of finite sum of functions $f_m(t)$ ($m = 1, \ldots, M$) is not greater than the greatest one of the characteristic exponents of each of the sum components.*

Theorem 10.13. *A characteristic exponent of product of the finite number of functions $f_m(t)$($m = 1, \ldots, M$) is not greater than the sum of characteristic exponents of these functions.*

A characteristic exponent of a matrix $[F_{jk}(t)]$ is said to be a number or symbol ($\pm\infty$) defined by the formula

$$\chi[F] = \max_{j,k} \chi[F_{j,k}(t)]. \tag{10.79}$$

Theorem 10.14. *A characteristic exponent of the sum of finite number of matrices does not exceed the largest one from among characteristic exponents of these matrices.*

Theorem 10.15. *A characteristic exponent of the product of finite number of matrices is not greater than the sum of characteristic exponents of these matrices.*

A solution $\{y(t)\}$, which is not a trivial solution of the system (10.65) is exponentially stable, if the solution and close solution $x(t)$ for $t = t_0$ satisfy the inequality

$$\|x(t) - y(t)\| \le L \, \|x(t_0) - y(t_0)\| \, e^{-\beta(t-t_0)}, \tag{10.80}$$

for $t \ge t_0$ and for particular positive constants L and β.

For a homogeneous linear system of constant coefficients, the asymptotical stability of its trivial solution implies the exponential stability. In general it is not true, when the coefficients are variable.

Theorem 10.16. *If there exists a positive definite quadratic form*

$$V(x_1, \ldots, x_n) = (Ax, x), \tag{10.81}$$

whose derivative $\dot{V}(x_1, \ldots, x_n)$ associated with the reduced system (10.65) *satisfies the inequality*

$$\dot{V}(x_1, \ldots, x_n) \le W(x_1, \ldots, x_n), \tag{10.82}$$

while

$$W(x_1, \ldots, x_n) = -(Bx, x), \tag{10.83}$$

is a negative definite quadratic form, and matrices A and B are constant and symmetric, then a trivial solution of the system (10.65) *is exponentially stable as $t \to \infty$.*

Now, we will examine the stability with the use of Lyapunov's function.

Example 10.1. Examine the stability of equilibrium positions of a pendulum governed by the equation

$$\ddot{\phi} + \alpha^2 \sin\phi = 0.$$

The above equation is transformed into a system of two first-order differential equations

$$\frac{d\phi}{dt} = v,$$

$$\frac{dv}{dt} = -\alpha^2 \sin\phi.$$

First, we will consider a more general problem, namely a dynamical system of the form

$$\begin{aligned} \frac{d\phi}{dt} &= v, \\ \frac{dv}{dt} &= -F(\phi). \end{aligned} \tag{$*$}$$

We will show that if $F(\phi)$ is a function satisfying two conditions

$$F(0) = 0,$$

$$\phi F\,(\phi) > 0 \text{ for } \phi \neq 0$$

then the equilibrium position $\phi = v = 0$ is stable. One can observe that the system $(*)$ possesses a first integral of the form

$$\frac{v^2}{2} + \int_0^{\phi} F(\zeta) d\zeta = C.$$

Differentiating th above relation yields we get

$$v dv + F(\phi) d\phi = 0,$$

and this follows from Eq. $(*)$. In the interval $(0, \phi)$ the function $F(\phi) > 0$, and this means that

$$\int_0^{\phi} F(\zeta) d\zeta > 0$$

for $\phi \neq 0$. Hence, a function

$$V(\phi, v) = \frac{v^2}{2} + \int_0^{\phi} F(\zeta) d\zeta \tag{$**$}$$

is positive definite. Its derivative reads

$$\frac{dV}{dt} = \frac{\partial V}{\partial \phi} v - \frac{\partial V}{\partial v} F(\phi).$$

Therefore, we get

$$\frac{dV}{dt} = F(\phi) v - v F(\phi).$$

According to the first Lyapunov theorem (Theorem 10.1) the equilibrium position $\phi = v = 0$ is stable, since the function $\dot{V}$ is identically zero.

Now, let us go back to our studied system. The system possesses infinitely many equilibrium positions of the form $(\phi, v) = (k\pi, 0)$ for $k = 0, 1, 2, \ldots$. With regard to periodicity of a function $\sin\phi$, it is sufficient to examine only one top and bottom position of the pendulum, i.e. $(0, 0)$ and $(\pi, 0)$. Let (ϕ_0, v_0) be a particular solution we want to examine. Then, we obtain

$$\frac{d(\phi_0 + \phi)}{dt} = v_0 + v,$$

$$\frac{d(v_0 + v)}{dt} = -\alpha^2(\sin\phi_0 \cos\phi + \cos\phi_0 \sin\phi),$$

since for small ϕ we have $\sin\phi \approx \phi$ and $\cos\phi \approx 1$.

After linearization we have

$$\frac{d\phi}{dt} = v,$$

$$\frac{dv}{dt} = -\alpha^2\phi\cos\phi_0.$$

Solutions of the latter equations are sought in the form $\phi = Ae^{\lambda t}$, $v = Be^{\lambda t}$. We obtain the following characteristic equation

$$\begin{vmatrix} \lambda & -1 \\ \alpha^2\cos\phi_0 & \lambda \end{vmatrix} = \lambda^2 + \alpha^2\cos\phi_0 = 0.$$

For the top position, i.e. $(\pi, 0)$ we have $\lambda = \pm\alpha$ and since $\alpha > 0$, then the equilibrium position is unstable. For the bottom position, i.e. $(0, 0)$ we have $\lambda = \pm i\alpha$ and stability of the equilibrium position can be examined by means of Lyapunov's function. According to $(**)$ we have

$$V(\phi, v) = \frac{v^2}{2} + \int_0^\phi \alpha^2 \sin\varsigma d\varsigma = \frac{v^2}{2} + \alpha^2(1 - \cos\phi).$$

A function $V(\phi, v)$ is positive definite. Let us examine its derivative

$$\frac{dV(\phi, v)}{dt} = -v\alpha^2\sin\phi + \sin\phi\alpha^2 v \equiv 0.$$

Thus, according to Theorem 10.1 the equilibrium position $\phi = v = 0$ is stable.

□

Example 10.2. Examine stability of a solution $(0, 0)$ for a dynamical system

$$\frac{dx}{dt} = ax^5 + by,$$
$$\frac{dy}{dt} = cx + dy^3,$$

provided that $a < 0$ and $d < 0$.

By the linearized equation we have

$$\frac{dx}{dt} = by,$$
$$\frac{dy}{dt} = cx,$$

hence we get the characteristic equation

$$\begin{vmatrix} \lambda & b \\ c & \lambda \end{vmatrix} = \lambda^2 - bc = 0. \qquad (*)$$

If $bc > 0$, then the examined equilibrium position is unstable. If $bc < 0$, then by Eq. (*) we get purely imaginary complex conjugate roots $\lambda_{1,2} = \pm i\sqrt{bc}$, and this corresponds to the critical case. Examination of stability of an equilibrium position should be performed with the use of Lapunv's function.

We seek Lyapunov's function of the form

$$V(x, y) = A_1(x) + A_2(y),$$

where $A_1(0) = A_2(0) = 0$. Derivative of this function reads

$$\frac{dV}{dt} = \frac{dA_1}{dx}(ax^5 + by) + \frac{dA_2}{dy}(cx + dy^3).$$

We want function $\frac{dV}{dt}$ to have the following form

$$\frac{dA_1}{dx}by + \frac{dA_2}{dy}cx = 0.$$

By the latter equation we get

$$\frac{by}{\frac{dA_2}{dy}} = -\frac{cx}{\frac{dA_1}{dx}} = \frac{1}{2}, \qquad (**)$$

where one assumed that the ratios should equal $1/2$. By Eq. (**) we obtain

$$\frac{dA_2}{dy} = 2by,$$

$$\frac{dA_1}{dx} = -2cx,$$

and integrating we obtain

$$A_2 = by^2,$$

$$A_1 = -cx^2.$$

If these expressions are taken into account in $V(x, y)$, then we get the sought function

$$V(x, y) = by^2 - cx^2.$$

Its derivative is

$$\frac{dV}{dt} = 2bdy^4 - 2acx^6.$$

If the parameters b and c are of distinct signs, then $V(x, y)$ is a function determined up to a sign, since it is everywhere positive or negative except at the point $x = y = 0$. The function $\dot{V}$ has zeros different from $x = y = 0$, when the conditions are satisfied $bdy^4 = acx^6$. If $a < 0$, $d < 0$ and $bc < 0$, then the trivial solution $(0, 0)$ is asymptotically stable.

□

10.3 Classical Theories of Stability and Chaotic Dynamics

A natural question arises whether introduced classical theories of stability and instability can be applied to analysis of perturbed chaotic orbits? This problem was investigated by Leonov [154]. Firstly, let us introduce a definition of invariance of a set K.

Definition 10.19. We say that a set K is invariant, if $x(t, K) = K$ for any t, while

$$x(t, K) = \left\{ x(t, x_0)|_{x_0 \in K} \right\}.$$

An invariant set is generated by already described continuous systems (of continuous independent variable)

$$\frac{dx}{dt} = F(t), \quad t \in R^1, \tag{10.84}$$

or discrete ones (cascades with the independent variable $t = 0, 1, 2, \ldots$),

$$x(t+1) = F(x(t)), \tag{10.85}$$

where: $x \in R^n$ and $F(x)$ is a vector function.

In mathematical sense, if for any x_0 a trajectory $x(t, x_0)$ for $t \in (0, +\infty)$ is uniquely determined, then we say that Eqs. (10.84) and (10.85) govern a dynamical system.

Now, on the basis of Leonov's work [154], we extend classical theory of stability, referred not to single trajectories but to invariant sets.

Definition 10.20. We say that the invariant set K is locally attracting (attractor), if for some neighbourhood of e this set $K(\varepsilon)$, the following condition is satisfied

$$\lim_{t \to +\infty} \rho(K, x(t, x_0)) = 0, \quad \underset{x_0}{\forall} \in K(\varepsilon), \tag{10.86}$$

where ϱ is a distance between a point x and the set K. The distance is defined by the formula

$$\rho(K, x) = \inf_{z \in K} \|z - x\|,$$

where $\|\cdot\|$ is an Euclidean norm in R^n, and $K(\varepsilon)$ is a set of points x, for which $\rho(K, x) < \varepsilon$.

Definition 10.21. We say that the invariant set K is a globally attracting set if

$$\lim_{t \to +\infty} \rho(K, x(t, x_0)) = 0, \underset{x_0}{\forall} \in R^n.$$

Definition 10.22. We say that an invariant set K is locally uniform attracting, if for any number $\delta > 0$ and a bounded set $B \subset R^n$ there exists such a number $t(\delta, B)$, that the following relation holds

$$x(t, B) \subset K(\delta), \forall t \geq t(\delta, B).$$

A graphical interpretation of Definition 10.22 is illustrated in Fig. 10.11.

Note that if the neighbourhood $K(\varepsilon)$ grows and covers R^n, then $B \cap R^n = B$.

Definition 10.23. We say that the invariant set K is global uniformly attracting, if for any number $\delta > 0$ and bounded set $B \subset R^n$ there exists such a number $t(\delta, B) > 0$, that the following relation holds

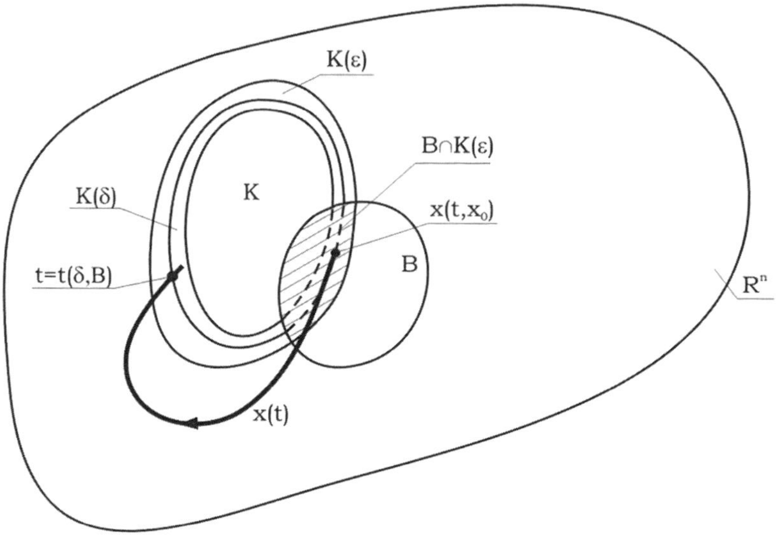

Fig. 10.11 A set K being *local uniformly attracting*

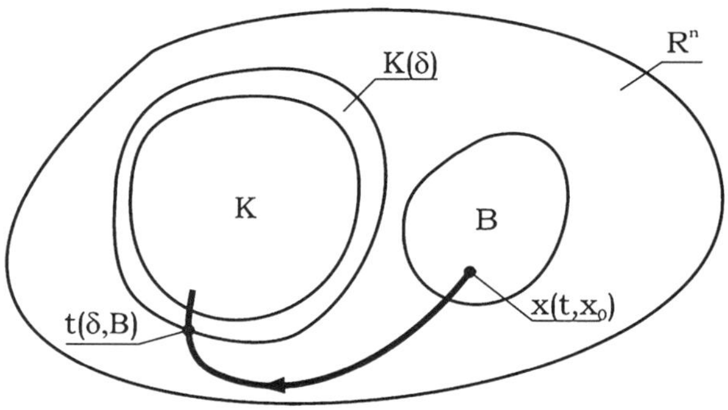

Fig. 10.12 A set K being *global uniformly attracting*

$$x(t, B) \subset K(\delta), \quad \forall t \geq t(\delta, B).$$

The above definition implies that a trajectory starting from any point of the set B will reach the set $K(\delta)$ and stay there (see Fig. 10.12).

Definition 10.24. We say that an invariant set K is stable in Lyapunov's sense if for any number $\varepsilon > 0$ one can find such a number $\delta > 0$, that the following relation holds

$$x(t, K(\delta)) \subset K(\varepsilon), \forall t \geq 0.$$

An illustration of this theorem is depicted in Fig. 10.13.

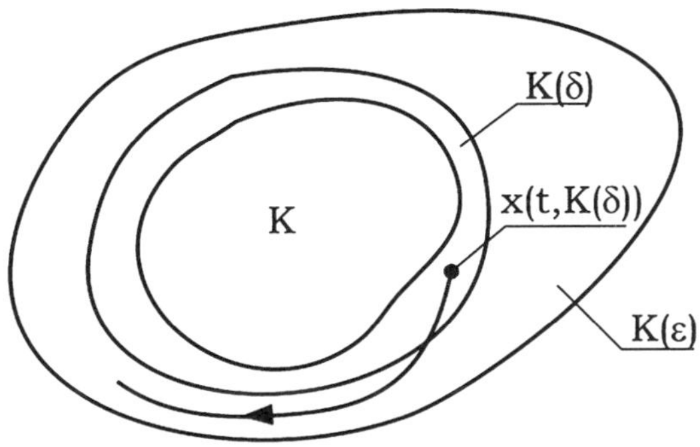

Fig. 10.13 A set K being stable in Lyapunov's sense

Referring to the earlier introduced notions of stability one can come to the following conclusions (proved in the monograph of Leonov [154]):

(i) For continuous dynamical systems, Lyapunov stability implies Zhukovskiy stability, and Zhukovskiy stability implies Poincaré stability.
(ii) For discrete dynamical systems, Lyapunov stability implies Poincaré stability.
(iii) In the case of equilibrium positions, all the introduced notions of stability are equivalent.
(iv) In the case of periodic orbits of discrete systems, the notions of Lyapunov and Poincaré stability are equivalent.
(v) In the case of periodic orbits of continuous systems, the notions of Poincaré and Zhukovskiy stability are equivalent.

It follows from the analysis of chaotic orbits that perturbed orbits are sensitive on perturbations and characterize of rapid mutual repelling. The considerations performed in monograph [154] imply that periodic orbits, in the case of continuous dynamical systems, can be stable in Poincaré and Zhukovskiy sense but unstable in Lyapunov's sense. On the other hand, orbits repelling each other (chaos) can be stable in Poincare sense. Moreover, Leonov comes to conclusion, on the basis of examples, that from among classical notions of instability the ones in Zhukovskiy and Lyapunov senses are the best choices for analysis of strange chaotic attractors, respectively for continuous and discrete systems.

Chapter 11
Modelling via Perturbation Methods

11.1 Introduction

Asymptotical methods, including perturbative ones, belong to the methods commonly applied in physics, applied mathematics and engineering sciences.

In this chapter, we will limit ourselves to analysis of only a few simple examples on the basis of modelling with the use of classical perturbative methods. Many monographs and thousands of scientific articles are devoted to this problem (see, for example, [7, 24]).

Before we start considering particular examples of modelling, we will briefly mention the advantages and disadvantages of asymptotic modelling.

The advantages include the following features of asymptotic methods:

(i) asymptotic methods are closely related to the examined physical process;
(ii) they enable to obtain solutions in analytic forms;
(iii) they enable to determine many important features of the sought solution;
(iv) they enable the unique mathematical approach to many various physical problems;
(v) they allow for initial estimation of possible solutions, which can be more thoroughly analysed via numerical methods;
(vi) they enable to construct solutions by means of combined methods, i.e. asymptotical and numerical ones.

The disadvantages include:

(i) it is difficult to estimate an error connected with a construction of an approximated solution;
(ii) the proof that an asymptotic series is convergent to a real solution is very laborious;
(iii) there are no general rules concerning the choice of perturbative parameters, so it depends on intuition and experience of a researcher.

J. Awrejcewicz, *Ordinary Differential Equations and Mechanical Systems*,
DOI 10.1007/978-3-319-07659-1_11

It is worth emphasizing that a small perturbation parameter, which plays a key role by evaluation of asymptotical convergence of series, can have a physical interpretation or can be introduced artificially.

The ideal for application of the asymptotic method is to introduce an artificial or small physical parameter ε, so that the problem can be reduced to the analytically solvable problem for $\varepsilon = 0$.

In spite of the standard classical approaches, there exists also a few new methods allowing to eliminate a small perturbation parameter, and hence to extend a validity of the obtained results. The mentioned approaches include the multiple scale methods [180] and its further modifications [134], max–min approach [118] and improved amplitude-formulation [117], homotopy perturbation method [116] and parameter-expanding method [119].

11.2 Selected Classical Perturbative Methods

11.2.1 Autonomous Systems

11.2.1.1 The Krylov Method

Consider a system governed by the equation

$$\ddot{y} + \alpha_0^2 y = \varepsilon Q(y). \tag{11.1}$$

For the sake of simplicity, we assume that $Q(0) = 0$, and this means that $y = 0$ is an equilibrium position. The function $Q(y)$ can be expanded around the equilibrium position, while $(dQ/dy)_{y=0} = 0$. A solution of Eq. (11.1) is sought in the form

$$y = y_0(t) + \sum_{k=1}^{K} \varepsilon^k y_k(t). \tag{11.2}$$

Additionally, in order to eliminate so-called *secular terms* (unbounded growth in time) one introduces the series

$$\alpha^2 = \alpha_0^2 + \sum_{k=1}^{K} \varepsilon^k \alpha_k. \tag{11.3}$$

Taking into account (11.2) and (11.3) in (11.1), we get

$$\sum_{k=0}^{K} \varepsilon^k \ddot{y}_k + \left(\alpha^2 - \sum_{k=1}^{K} \varepsilon^k \alpha_k\right) \sum_{k=0}^{K} \varepsilon^k y_k = \varepsilon\, Q\left(\sum_{k=0}^{K} \varepsilon^k y_k\right). \tag{11.4}$$

Next, we expand the right-hand side of (11.4) in a power series around $\varepsilon = 0$ and we get

$$\begin{aligned}
\varepsilon Q(y) = &\ \varepsilon Q\left(\sum_{k=0}^{K}\varepsilon^k y_k\right)\Bigg|_{\varepsilon=0} + \left\{Q\left(\sum_{k=0}^{K}\varepsilon^k y_k\right) + \varepsilon Q'\left(\sum_{k=0}^{K}\varepsilon^k y_k\right)\sum_{k=1}^{K} k\varepsilon^{k-1} y_k\right\}\Bigg|_{\varepsilon=0}\varepsilon \\
&+\left\{Q'\left(\sum_{k=0}^{K}\varepsilon^k y_k\right)\sum_{k=1}^{K} k\varepsilon^{k-1} y_k + Q'\left(\sum_{k=0}^{K}\varepsilon^k y_k\right)\sum_{k=1}^{K} k\varepsilon^{k-1} y_k + \varepsilon\, Q'\left(\sum_{k=1}^{K}\varepsilon^k y_k\right)\right. \\
&\left.\cdot\sum_{k=1}^{K} k\varepsilon^{k-1} y_k \sum_{k=1}^{K} k\varepsilon^{k-1} y_k + \varepsilon\, Q'\left(\sum_{k=1}^{K}\varepsilon^k y_k\right)\sum_{k=2}^{K} k(k-1)\varepsilon^{k-2} y_k\right\}\Bigg|_{\varepsilon=0}\frac{\varepsilon^2}{2} + \cdots \\
&= Q(y_0)\varepsilon + 2Q'(y_0)y_1\frac{\varepsilon^2}{2} + \cdots ,
\end{aligned} \tag{11.5}$$

where: $Q' = \frac{dQ}{dy}$, $Q'' = \frac{d^2Q}{dy^2}, \ldots$.

Taking into account (11.5) in (11.4) and equating terms standing by ε of the same powers in (11.4) we obtain:

$$\begin{aligned}
&\varepsilon^0 : \quad \ddot{y}_0 + \alpha^2 y_0 = 0, \\
&\varepsilon^1 : \quad \ddot{y}_1 + \alpha^2 y_1 = \alpha_1 y_0 + Q(y_0), \\
&\varepsilon^2 : \quad \ddot{y}_2 + \alpha^2 y_2 = \alpha_2 y_0 + \alpha_1 y_1 + y_1 Q'(y_0), \\
&\ldots \qquad\qquad \ldots
\end{aligned} \tag{11.6}$$

A solution of the first recurrent equation of a sequence of equations is a function

$$y_0 = a_0 \cos\psi, \tag{11.7}$$

where

$$\psi = \alpha t + \Theta_0. \tag{11.8}$$

Taking into account (11.7) in the second equation (11.6) we have

$$\ddot{y}_1 + \alpha^2 y_1 = \alpha_0 a_0 \cos\psi + Q(a_0\cos\psi). \tag{11.9}$$

The function $Q(a_0 \cos y)$ is even and periodic of period 2π, thus one can be expanded in Fourier series

$$Q(a_0\cos\psi) = \frac{1}{2}b_0 + \sum_{n=1}^{\infty} b_n \cos n\psi, \tag{11.10}$$

where

$$b_n = \frac{2}{\pi}\int_0^{\pi} Q(a_0\cos\psi)\cos n\psi\, d\psi. \tag{11.11}$$

Taking into account (11.10) in (11.9) we get

$$\ddot{y}_1 + \alpha^2 y_1 = \frac{1}{2} b_0 + (\alpha_1 a_0 + b_1 (a_0)) \cos\psi + \sum_{n=2}^{\infty} b_n \cos n\psi. \tag{11.12}$$

In order to obtain a periodic solution we need to eliminate a secular term in the solution $y_1(t)$, i.e. we assume that

$$\alpha_1 a_0 + b_1 (a_0) = 0. \tag{11.13}$$

By the above equation we determine $\alpha_1 = -\frac{b_1(a_0)}{a_0}$, so the first unknown component of the series (11.3). A solution of (11.12), including (11.13), is

$$y_1 = a_1 \cos (\alpha t + \Theta_1) + \frac{b_0}{2\alpha^2} + \sum_{n=2}^{\infty} \frac{b_n}{\alpha^2 - (n\alpha)^2} \cos n\psi. \tag{11.14}$$

Constants a_0 and Θ_0 are determined by the initial conditions, so we assume that $a_1 = \Theta_1 = 0$. This leads to determining a_0 and Θ_0 from Eq. (11.2), which can be troublesome in practice. That is why the constants a_0 and Θ_0 are often determined in a way that the solution y_0 satisfies the initial conditions.

In further considerations we assume $a_1 = \Theta_1 = 0$ and by (11.14) we get

$$y_1(t) = \frac{b_0}{2\alpha^2} + \sum_{n=2}^{\infty} \frac{b_n}{\alpha^2 - (n\alpha)^2} \cos n\psi. \tag{11.15}$$

Knowing y_1 we see that the right-hand side of the third equation of the system (11.6) is determined. By the condition of vanishing of the secular term in the solution $y_2(t)$ we get the coefficients α_2, and the solution y_2 is another component of the series (11.2). Confining ourselves to the first approximation $O(\varepsilon^2)$ we obtain:

$$y = a_0 \cos (\alpha t + \Theta_0) + \varepsilon \frac{1}{\alpha^2} \left[\frac{1}{2} b_0 + \sum_{n=2}^{\infty} \frac{b_n}{1 - n^2} \cos n (\alpha t + n\Theta_0) \right], \tag{11.16}$$

where

$$\alpha = \sqrt{\alpha_0^2 + \varepsilon \alpha_1 (a_0)} = \sqrt{\alpha_0^2 - \varepsilon \frac{b_1 (a_0)}{a_0}}. \tag{11.17}$$

A period of the sought solution reads

$$T = \frac{2\pi}{\sqrt{\alpha_0^2 - \varepsilon \frac{b_1(a_0)}{a_0}}}, \tag{11.18}$$

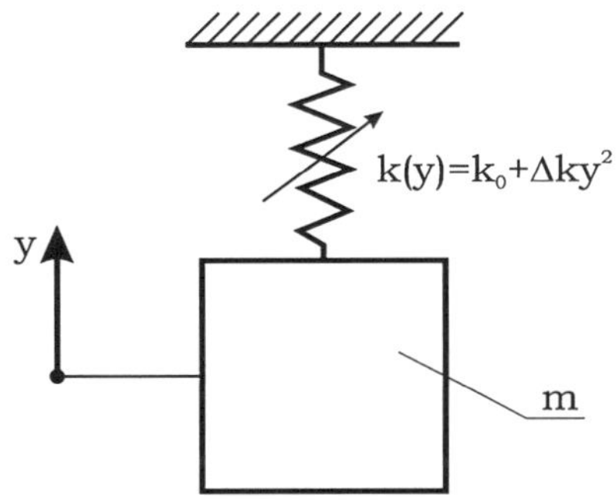

Fig. 11.1 Oscillator of a mass m and stiffness $k(y)$

and depends on the vibration amplitude.

Example 11.1. Determine equations of motion and period of free vibrations of a system depicted in Fig. 11.1. Assume that stiffness $k = k_0 + \Delta k y^2$.

The equation of motion has the form

$$\ddot{y} + \alpha_0^2 y = -\varepsilon y^3,$$

where

$$\alpha_0^2 = \frac{k_0}{m}, \quad \varepsilon = \frac{\Delta k}{m}.$$

The recurrent sequence of Eq. (11.6) takes the form

$$\ddot{y}_0 + \alpha^2 y_0 = 0,$$

$$\ddot{y}_1 + \alpha^2 y_1 = \alpha_1 y_0 - y_0^3,$$

$$\ddot{y}_2 + \alpha^2 y_2 = \alpha_2 y_0 - \alpha_1 y_1 - 3y_0^2 y_1.$$

A solution of the first equation of the recurrent sequence is a function

$$y_0 = a_0 \cos\psi, \quad \psi = \alpha t + \Theta_0.$$

Taking into account the second equation of the recurrent sequence and the following identity

$$\cos^3\psi = \frac{3}{4}\cos\psi + \frac{1}{4}\cos 3\psi,$$

we obtain

$$\ddot{y}_1 + \alpha^2 y_1 = a_0\left(\alpha_1 - \frac{3}{4}a_0^2\right)\cos\psi - \frac{1}{4}a_0^3\cos 3\psi.$$

The condition of vanishing of a secular term gives

$$\alpha_1 = \frac{3}{4} a_0^2.$$

Assuming $a_1 = \Theta_1 = 0$ we get

$$y_1 = \frac{a_0^3}{32\alpha^2} \cos 3\psi.$$

Therefore, we have

$$\ddot{y}_2 + \alpha^2 y_2 = \alpha_2 a_0 \cos\psi + \frac{3a_0^5}{128\alpha^2} \cos 3\psi - 3\frac{a_0^5}{32\alpha^2} \cos 3\psi \cos^2\psi.$$

Since

$$\cos^2\psi \cos 3\psi = \frac{1}{4} \left(\cos 5\psi + 2\cos 3\psi + \cos\psi\right),$$

we get

$$\ddot{y}_2 + \alpha^2 y_2 = \left(\alpha_2 a_0 - \frac{3a_0^5}{128\alpha^2}\right) \cos\psi - \frac{3a_0^5}{128\alpha^2} \cos 3\psi - \frac{3a_0^2}{128\alpha^2} \cos 5\psi.$$

Avoiding secularity yields

$$\alpha_2 = \frac{3a_0^4}{128\alpha^2},$$

and for $a_2 = Q_2 = 0$ we have

$$y_2 = \frac{3}{1024} \frac{a_0^5}{\alpha^4} \left(\cos 3\psi + \cos 5\psi\right).$$

Finally a solution (ignoring terms $O(\varepsilon^2)$) has the form

$$y = a_0 \cos\psi + \varepsilon \frac{a_0^3}{32\alpha^2} \cos 3\psi + \varepsilon^2 \frac{a_0^5}{1024\alpha^2} \left(3\cos 3\psi + \cos 5\psi\right), \qquad (*)$$

where

$$\alpha = \sqrt{\alpha_0^2 + \varepsilon \frac{3}{4} a_0^2 + \varepsilon^2 \frac{3}{128} \frac{a_0^4}{\alpha_0^2 + \varepsilon \frac{3}{4} a_0^2}}, \qquad \psi = \alpha t + \Theta_0.$$

The amplitude of oscillations, understood as the biggest displacement of the system from the equilibrium position, we obtain when

$$\psi = \alpha t + \Theta_0 = n\pi,$$

hence

$$t_n = \frac{n\pi - \Theta_0}{\alpha},$$

and we get from (∗)

$$A = a_0 + \varepsilon \frac{a_0^3}{32\alpha^2} + \varepsilon^2 \frac{a_0^5}{256\alpha^4},$$

The period of oscillations is

$$T = \frac{2\pi}{\sqrt{\alpha_0^2 + \varepsilon\frac{3}{4}a_0^2 + \varepsilon^2\frac{3}{128}\frac{a_0^4}{\alpha_0^2+\varepsilon\frac{3}{4}a_0^2}}}.$$

□

11.2.1.2 Krylov–Bogolubov–Mitropolskiy (KBM) Method

The Krylov method was developed and improved first by Bogolubov, and then by Mitropolskiy. It can be applied to second-order differential equations, which govern dynamics of mechanical systems. Its essential elements will be exposed here during the analysis of an autonomous system of a single degree-of-freedom of the form

$$\ddot{y} + \alpha_0^2 y = \varepsilon\, Q(y, \dot{y}). \tag{11.19}$$

Assume that $y = \dot{y} = 0$ is an equilibrium position and a function $Q(y, \dot{y})$ is analytic with respect to its variables. The main element which makes a difference between this method and the previous one is the assumption that the amplitude $a(t)$ and the phase $y(t)$ are functions of time. A solution of Eq. (11.19) is sought in the form

$$y(t) = a\cos\psi + \sum_{k=1}^{K} \varepsilon^k y_k\left[a(t), \psi(t)\right]. \tag{11.20}$$

In conservative systems, the amplitude a is constant so its derivative with respect to time reads

$$\frac{da}{dt} = 0. \tag{11.21}$$

Derivative y with respect to time is also constant for conservative systems. The derivative reads

$$\dot{\psi} = \alpha = \alpha_0 + \sum_{k=1}^{K} \varepsilon^k \alpha_k(a). \tag{11.22}$$

When we deal with nonconservative systems, we should complete the relationships (11.21) and (11.22) with time-dependent series, which are taken in the form

$$\dot{a} = \sum_{k=1}^{K} \varepsilon^k A_k\left[a(t)\right], \tag{11.23}$$

$$\dot{\psi} = \alpha_0 + \sum_{k=1}^{K} \varepsilon^k B_k\left[a(t)\right]. \tag{11.24}$$

In order to take into account (11.20) in (11.19) we calculate the first and second derivative of $y(t)$ with respect to time (the functions y_k are composed functions of two variables a and y) and we get

$$\dot{y} = \dot{a}\cos\psi - a\dot{\psi}\sin\psi + \sum_{k=1}^{K} \varepsilon^k \left(\frac{\partial y_k}{\partial a}\dot{a} + \frac{\partial y_k}{\partial \psi}\dot{\psi} \right), \tag{11.25}$$

$$\ddot{y} = \ddot{a}\cos\psi - \dot{a}\dot{\psi}\sin\psi - \dot{a}\dot{\psi}\sin\psi - a\ddot{\psi}\sin\psi - a\dot{\psi}^2\cos\psi$$
$$+ \sum_{k=1}^{K} \varepsilon^k \left(\frac{\partial^2 y_k}{\partial a^2}\dot{a}^2 + \frac{\partial^2 y_k}{\partial a \partial \psi}\dot{\psi}\dot{a} + \frac{\partial y_k}{\partial a}\ddot{a} + \frac{\partial^2 y_k}{\partial \psi \partial a}\dot{\psi}\dot{a} + \frac{\partial^2 y_k}{\partial \psi^2}\dot{\psi}^2 + \frac{\partial y_k}{\partial \psi}\ddot{\psi} \right). \tag{11.26}$$

Taking into account (11.20) and (11.26) on the left-hand side (L) of Eq. (11.19) we obtain

$$L = \left[\ddot{a} - \left(\dot{\psi}^2 - \alpha_0^2\right) a\right]\cos\psi - \left(2\dot{a}\dot{\psi} + a\ddot{\psi}\right)\sin\psi$$
$$+ \sum_{k=1}^{K} \varepsilon^k \left(\frac{\partial^2 y_k}{\partial a^2}\dot{a}^2 + \frac{\partial^2 y_k}{\partial \psi^2}\dot{\psi}^2 + 2\dot{a}\dot{\psi}\frac{\partial^2 y_k}{\partial a \partial \psi} + \frac{\partial y_k}{\partial a}\ddot{a} + \frac{\partial y_k}{\partial \psi}\ddot{\psi} + \alpha_0^2 y_k \right). \tag{11.27}$$

According to (11.23) and (11.24), we have

$$\ddot{a}=\sum_{k=1}^{K}\varepsilon^k\tfrac{d}{dt}A_k(a)=\sum_{k=1}^{K}\varepsilon^k\tfrac{dA_k}{da}\dot{a}=\sum_{k=1}^{K}\varepsilon^k\tfrac{dA_k}{da}\left(\sum_{k=1}^{K}\varepsilon^k A_k\right)=\varepsilon^2 A_1\tfrac{dA_1}{da}+O\left(\varepsilon^3\right),$$
$$\dot{\psi}^2-\alpha_0^2=\alpha_0^2+\left(\sum_{k=1}^{K}\varepsilon^k B_k\right)^2+2\alpha_0\sum_{k=1}^{K}\varepsilon^k B_k-\alpha_0^2$$
$$=\varepsilon 2\alpha_0 B_1+\varepsilon^2 2\alpha_0 B_2+\varepsilon^2 B_1^2+O\left(\varepsilon^3\right),$$
$$\dot{a}\,\dot{\psi}=\sum_{k=1}^{K}\varepsilon^k A_k\cdot\left(\alpha_0+\sum_{k=1}^{K}\varepsilon^k B_k\right)=\varepsilon\,\alpha_0 A_1+\varepsilon^2\alpha_0 A_2+\varepsilon^2 A_1 B_1+O\left(\varepsilon^3\right),$$
$$\ddot{\psi}=\sum_{k=1}^{K}\varepsilon^k\tfrac{dB_k}{da}\dot{a}=\sum_{k=1}^{K}\varepsilon^k\tfrac{dB_k}{da}\sum_{k=1}^{K}\varepsilon^k A_k=\varepsilon^2\tfrac{dB_1}{da}A_1+O\left(\varepsilon^3\right),$$
$$\dot{a}^2=\left(\sum_{k=1}^{K}\varepsilon^k A_k\right)^2=\varepsilon^2 A_1^2+O\left(\varepsilon^3\right),$$
$$\dot{\psi}^2=\left(\alpha_0+\sum_{k=1}^{K}\varepsilon^k B_k\right)^2=\alpha_0^2+\varepsilon 2\alpha_0 B_1+\varepsilon^2 2\alpha_0 B_2+\varepsilon^2 B_1^2+O\left(\varepsilon^3\right),\tag{11.28}$$

and ordering expressions standing by the same powers of ε, Eq. (11.27) takes the form

$$\begin{aligned}L=\varepsilon&\left[\alpha_0^2\left(\frac{\partial^2 y_1}{\partial\psi^2}+y_1\right)-2\alpha_0 B_1 a\cos\psi-2\alpha_0 A1\sin\psi\right]\\
+\varepsilon^2&\left[\alpha_0^2\left(\frac{\partial^2 y_2}{\partial\psi^2}+y_2\right)+2\alpha_0 B_1\frac{\partial^2 y_1}{\partial\psi^2}+2\alpha_0 A_1\frac{\partial^2 y_1}{\partial}a\partial\psi\right.\\
&+\left(A_1\frac{dA_1}{da}-\left(2\alpha_0 B_2+B_1^2\right)a\right)\cos\psi\\
&\left.-\left(2\alpha_0 A_2+2A_1 B_1+\frac{dB_1}{da}A_1 a\right)\sin\psi\right]+O\left(\varepsilon^2\right).\end{aligned}\tag{11.29}$$

The right-hand side P of Eq. (11.19) is expanded in the power series with respect to ε in the neighbourhood of $\varepsilon=0$. We obtain

$$\begin{aligned}P=\varepsilon Q\left(y,\dot{y}\right)&=\varepsilon\left\{Q\left(y,\dot{y}\right)|_{\varepsilon=0}+\left(\left.\frac{\partial Q\,\partial y}{\partial y\,\partial\varepsilon}\right|_{\varepsilon=0}+\left.\frac{\partial Q\,\partial\dot{y}}{\partial\dot{y}\,\partial\varepsilon}\right|_{\varepsilon=0}\right)+O\left(\varepsilon^2\right)\right\}\\
&=\varepsilon\,Q\left(a\cos\psi,-a\alpha_0\sin\psi\right)+\varepsilon^2\left[\frac{\partial Q}{\partial y}\left(a\cos\psi,-a\alpha_0\sin\psi\right)y_1\right.\\
&\left.+\frac{\partial Q}{\partial\dot{y}}\left(a\cos\psi,-a\alpha_0\sin\psi\right)+\left(A_1\cos\psi-aB_1\sin\psi+\alpha_0\frac{\partial y_1}{\partial\psi}\right)\right]+O\left(\varepsilon^2\right)\end{aligned}\tag{11.30}$$

Equating the expressions standing by ε with the same powers on the left-hand side of (11.29) and the right-hand side of (11.30) we obtain a sequence of differential equations

$$\frac{\partial^2 y_1}{\partial \psi^2} + y_1 = \frac{1}{\alpha_0^2}\left(f_0 + 2A_1\alpha_0 \sin\psi + 2B_1\alpha_0 a \cos\psi\right), \tag{11.31}$$

$$-\frac{\partial^2 y_2}{\partial \psi^2} + y_2 = \frac{1}{\alpha_0^2}\left(f_1 + 2A_2\alpha_0 \sin\psi + 2B_2\alpha_0 a \cos\psi\right), \tag{11.32}$$

$$\ldots$$

where

$$f_0 = Q\left(a\cos\psi, -a\alpha_0 \sin\psi\right), \tag{11.33}$$

$$\begin{aligned} f_1 = {} & -2B_1\alpha_0\frac{\partial^2 y_1}{\partial \psi^2} - 2A_1\alpha_0\frac{\partial^2 y_1}{\partial a \partial \psi} - \left(A_1\frac{dA_1}{da} - B_1^2 a\right)\cos\psi \\ & + \left(2A_1B_1 + aA_1\frac{dB_1}{da}\right)\sin\psi + \frac{\partial Q}{\partial y}\left(a\cos\psi, -a\alpha_0\sin\psi\right) y_1 \\ & + \frac{\partial Q}{\partial \dot{y}}\left(a\cos\psi, -a\alpha_0\sin\psi\right)\left(A_1\cos\psi - aB_1\sin\psi + \alpha_0\frac{\partial y_1}{\partial \psi}\right). \end{aligned} \tag{11.34}$$

A function f_0 determined by (11.33) is a periodic function of period 2π with respect to ψ, so it can be expanded into the Fourier series

$$f_0\left(a, \psi\right) = \frac{1}{2}b_{00}(a) + \sum_{n=1}^{\infty}\left(b_{0n}(a)\cos n\psi + c_{0n}(a)\sin n\psi\right), \tag{11.35}$$

where

$$b_{0n}(a) = \frac{1}{\pi}\int_0^{2\pi} Q\left(a\cos\psi, -a\alpha_0\sin\psi\right)\cos n\psi d\psi, \tag{11.36}$$

$$c_{0n}(a) = \frac{1}{\pi}\int_0^{2\pi} Q\left(a\cos\psi, -a\alpha_0\sin\psi\right)\sin n\psi d\psi, \quad n = 0, 1, 2, \ldots \tag{11.37}$$

Taking into account (11.35) in (11.31) and excluding from the series (11.35) those terms, which give in the solution of (11.31) secular terms, we have

$$\begin{aligned} \frac{\partial^2 y_1}{\partial \psi^2} + y_1 = {} & \frac{1}{\alpha_0^2}\left(c_{01} + 2A_1\alpha_0\right)\sin\psi + \frac{1}{\alpha_0^2}\left(b_{01} + 2B_1\alpha_0 a\right)\cos\psi \\ & + \frac{b_{00}}{2\alpha_0^2} + \frac{1}{\alpha_0^2}\sum_{n=2}^{\infty}\left(b_{0n}\cos n\psi + c_{0n}\sin n\psi\right). \end{aligned} \tag{11.38}$$

Two first components on the right-hand side of (11.38) give secular terms in the solution y_1. Their appearance in the solution is contradictory with the assumptions of asymptotic methods. Hence, the coefficients standing by $\sin\psi$ and $\cos\psi$ need to be equated to zero. This gives us relationships to determine unknown functions $A_1(a)$ and $B_1(a)$

$$A_1(a) = -\frac{c_{01}(a)}{2\alpha_0}, \tag{11.39}$$

$$B_1(a) = -\frac{b_{01}(a)}{2\alpha_0 a}. \tag{11.40}$$

Taking into account (11.39) and (11.40) in (11.38), a general solution of Eq. (11.38) is

$$y_1 = a_1 \cos(\psi + \Theta_1) + \frac{b_{00}}{2\alpha_0^2} + \frac{1}{\alpha_0^2}\sum_{n=2}^{\infty}\frac{1}{1-n^2}\left(b_{0n}\cos n\psi + c_{0n}\sin n\psi\right), \tag{11.41}$$

because by integration of Eqs. (11.23) and (11.24) two any constants appear. The constants can satisfy any initial conditions, imposed on Eq. (11.19) (we assume in further considerations $a_1 = 0$ and $\Theta_1 = 0$). Equation (11.41) takes the form

$$y_1 = \frac{b_{00}}{2\alpha_0^2} + \frac{1}{\alpha_0^2}\sum_{n=2}^{\infty}\left(b_{0n}\cos n\psi + c_{0n}\sin n\psi\right). \tag{11.42}$$

Note that the function f_1 determined by Eq. (11.34) and occurring in (11.32) is completely determined by the relationships (11.39), (11.41) and (11.42). It is easy to see that this function is periodic of period 2π with respect to y. It can be expanded into the Fourier series of the form

$$f_1(a,\psi) = \frac{1}{2}b_{10}(a) + \sum_{n=1}^{\infty}\left[b_{1n}(a)\cos n\psi + c_{1n}(a)\sin n\psi\right], \tag{11.43}$$

where

$$b_{1n}(a) = \frac{1}{\pi}\int_0^{2\pi} f_1(a,\psi)\cos n\psi d\psi, \tag{11.44}$$

$$c_{1n}(a) = \frac{1}{\pi}\int_0^{2\pi} f_1(a,\psi)\sin n\psi d\psi, \quad n = 0, 1, 2, \ldots \tag{11.45}$$

Doing analogously as with Eq. (11.31) we obtain the successive functions $A_2(a)$ and $B_2(a)$ of the series (11.23) and (11.24) in the form

$$A_2(a) = -\frac{c_{11}(a)}{2\alpha_0}, \tag{11.46}$$

$$B_2(a) = -\frac{b_{11}(a)}{2\alpha_0 a}, \tag{11.47}$$

and the function y_2 of the series (11.20)

$$y_2 = \frac{b_{10}}{2\alpha_0^2} + \frac{1}{\alpha_0^2}\sum_{n=2}^{\infty} \frac{1}{1-n^2}\left(b_{1n}\cos n\psi + c_{1n}\sin n\psi\right). \tag{11.48}$$

The unknown functions of time $a(t)$ and $y(t)$ occurring in the series (11.20) we will determine by solving the differential equations (11.23) and (11.24). These equations are easy to solve by means of separating of variables. Separating the variables in Eq. (11.23) we get

$$dt = \frac{da}{\sum\limits_{k=1}^{m} \varepsilon^k A_k(a)}. \tag{11.49}$$

Integrating (11.49) we have a function $a(t)$ in the implicit form

$$t = \int \frac{da}{\sum\limits_{k=1}^{m} \varepsilon^k A_k(a)} + a_0, \tag{11.50}$$

where a_0 is a constant dependent on initial conditions.

Even in the case, when the integral (11.49) can be expressed by elementary functions, one cannot obtain the explicit form of $a = a(t)$, which is necessary to solve Eq. (11.24). In the case when from (11.49) one can obtain an explicit form of $a(t)$ after separation of the variables, then from Eq. (11.24) we have

$$\psi = \int \left(\alpha_0 + \sum \varepsilon^k B_k\left[a(t)\right]\right) dt + \Theta_0. \tag{11.51}$$

Taking into account (11.42), (11.48), (11.50) and (11.51) in the series (11.15) we obtain a general solution of Eq. (11.20) with the constants a_0 and Θ_0 , dependent on the initial conditions.

Example 11.2. Determine oscillations of a system governed by the Rayleigh equation

$$\ddot{y} + \alpha_0^2 y = \left(2h - g\dot{y}^2\right)\dot{y}.$$

In order to make some transformations we introduce a parameter ε into the equation. It is seen that

$$Q(y, \dot{y}) = \left(2h - g\dot{y}^2\right)\dot{y}.$$

We seek a solution in the form

$$y(t) = a\cos\psi + \varepsilon y_1(a, \psi),$$

where

$$\dot{a} = \varepsilon A_1(a),$$

$$\dot{\psi} = \alpha_0 + \varepsilon B_1(a).$$

According to (11.31) we have

$$\frac{\partial^2 y_1}{\partial\psi^2} + y_1 = \frac{1}{\alpha_0^2}\left(f_0 + 2A_1\alpha_0\sin\psi + 2B_1\alpha_0 a\cos\psi\right),$$

where

$$f_0 = Q(a\cos\psi, -a\alpha_0\sin\psi) = a\alpha_0\left(\frac{3}{4}ga^2\alpha_0^2 - 2h\right)\sin\psi - \frac{1}{4}ga^3\alpha_0^3\sin 3\psi.$$

Taking two recently obtained formulas we get

$$\begin{aligned}\frac{\partial^2 y_1}{\partial\psi^2} + y_1 = \frac{1}{\alpha_0^2}\Bigg\{&\left[2A_1\alpha_0 + a\alpha_0\left(\frac{3}{4}ga^2\alpha_0^2 - 2h\right)\right]\sin\psi \\ &+ 2B_1\alpha_0 a\cos\psi - \frac{1}{4}ga^3\alpha_0^3\sin 3\psi\Bigg\}.\end{aligned} \tag{$*$}$$

By the condition of vanishing of secular terms we determine unknown coefficients A_1 and B_1, which read

$$A_1 = \frac{1}{2}a\left(2h - \frac{3}{4}ga^2\alpha_0^2\right),$$

$$B_1 = 0.$$

A solution of the equation ($*$), after taking into account the obtained A_1 and B_1, takes the form

$$y_1 = \frac{1}{32}ga^3\alpha_0\sin 3\psi.$$

We have

$$\dot{a} = \varepsilon a h \left(1 - K^2 a^2\right), \tag{$**$}$$

where

$$K^2 = \frac{3}{8} \frac{g\alpha_0^2}{h}.$$

If $K^2 > 0$, then after separation of variables, we obtain from ($**$)

$$\frac{da}{a\,(1 - Ka)\,(1 + Ka)} = \varepsilon h dt.$$

Decomposing the left-hand side of the later equation into partial fractions we obtain

$$\int \frac{da}{a} + \frac{K}{2} \int \frac{da}{(1 - Ka)} - \frac{K}{2} \int \frac{da}{(1 + Ka)} + \ln L = \varepsilon h t,$$

where L is an integration constant. After integration we have

$$\ln \frac{aL}{\sqrt{1 - K^2 a^2}} = \varepsilon h t,$$

and hence

$$a = \frac{a_0 e^{\varepsilon h t}}{\sqrt{1 + a_0^2 K^2 e^{2\varepsilon h t}}},$$

where $a_0 = L^{-1}$ is a constant, which does not depend on the initial conditions.

Below, we give a solution of Eq. (11.24), which takes the form

$$\psi = \alpha_0 t + \Theta_0.$$

Taking into account the obtained functions and assuming $\varepsilon = 1$ we find a general solution of the Rayleigh equation.

The constants a_0 and Θ_0 need to be determined so that the initial conditions must be satisfied.

□

11.2.1.3 Equivalent Linearization

Leaving only the first term in the solution (11.20) of Eq. (11.19) we have

$$y(t) = a \cos \psi, \tag{11.52}$$

while in the series (11.23) and (11.24) we leave the following terms

$$\dot{a} = \varepsilon A_1(a), \tag{11.53}$$

$$\dot{\psi} = \alpha_0 + \varepsilon B_1(a). \tag{11.54}$$

The solution (11.52) will be the first simplified approximation of a solution of Eq. (11.19). Below, we point out that the solution (11.52) of Eq. (11.19) satisfies, up to ε, the equation

$$\ddot{y} + 2h_e(a)\dot{y} + \alpha_e^2(a)y = O(\varepsilon). \tag{11.55}$$

The above equation is called an *equivalent linear equation* of the nonlinear equation (11.19). The parameters, the equivalent damping coefficient h_e and the equivalent eigenfrequency α_e occurring in this equation are determined below

$$h_e(a) = \frac{\varepsilon}{2\pi\alpha_0 a}\int_0^{2\pi} Q\left(a\cos\psi, -a\alpha_0\sin\psi\right)\sin\psi d\psi, \tag{11.56}$$

$$\alpha_e(a) = \alpha_0 - \frac{\varepsilon}{2\pi\alpha_0 a}\int_0^{2\pi} Q\left(a\cos\psi, -a\alpha_0\sin\psi\right)\cos\psi d\psi. \tag{11.57}$$

By Eq. (11.52) we have

$$\dot{y} = \dot{a}\cos\psi - a\dot{\psi}\sin\psi. \tag{11.58}$$

Taking into account (11.37) and (11.40) in (11.39), and including (11.56), Eq. (11.53) takes the form

$$\dot{a} = -ah_e(a). \tag{11.59}$$

Taking into account (11.36) and (11.41) in (11.40), and including (11.58), Eq. (11.54) takes the form

$$\dot{\psi} = \alpha_e(a). \tag{11.60}$$

Applying (11.59) to (11.60) in (11.58) we have

$$\dot{y} = -ah_e\cos\psi - a\alpha_e\sin\psi. \tag{11.61}$$

Differentiating (11.61) we get

$$\begin{aligned}\ddot{y} = -\dot{a}h_e\cos\psi - a\frac{dh_e}{da}\dot{a}\cos\psi + ah_e\dot{\psi}\sin\psi - \dot{a}\alpha_e\sin\psi -\\ -a\frac{d\alpha_e}{da}\dot{a}\sin\psi - a\alpha_e\dot{\psi}\cos\psi.\end{aligned} \tag{11.62}$$

Making use of (11.59) and (11.60) we have

$$\begin{aligned}\ddot{y} &= ah_e^2 \cos\psi + 2h_e a\alpha_e \sin\psi - a\alpha_e^2 \cos\psi + a^2 h_e \frac{dh_e}{da}\cos\psi \\ & a^2 h_e \frac{d\alpha_e}{da}\sin\psi - 2h_e^2 a\cos\psi - 2h_e\alpha_e \sin\psi + \alpha_e^2 a\cos\psi\end{aligned} \tag{11.63}$$

Introducing (11.59), (11.61) and (11.53) into (11.55) we get

$$\ddot{y} + 2h_e\dot{y} + \alpha_e^2 y = -ah_e^2\cos\psi + a^2 h_e \frac{dh_e}{da}\cos\psi + a^2 h_e \frac{d\alpha_e}{da}\sin\psi. \tag{11.64}$$

The right-hand side of (11.64) is of order of magnitude ε^2, since according to (11.56) and (11.57) the expressions h_e, dh_e/da, $d\alpha_e/da$ are of order of magnitude ε.

So, using Eqs. (11.56) and (11.57), we transform a nonlinear equation of type (11.19) into Eq. (11.55). Equation (11.55) will posses a periodic solution, if $h_e(a) = 0$, and roots of this nonlinear equation are amplitudes of approximated periodic solutions, determined by the equation

$$y_0 = a_0\cos\left[\alpha_e(a_0)\,t\right]. \tag{11.65}$$

While in the case, when $h_e(a) = 0$, then a solution is the function

$$y = A\cos\left(\alpha_e(a_0)\,t + \Theta\right), \tag{11.66}$$

where A and Θ are constants, which depend on the initial conditions.

Example 11.3. With the help of the equivalent linearization, determine the amplitude of vibrations of a system governed by the Rayleigh equation (see previous example).

According to the formulas (11.56) and (11.57), the equivalent damping coefficient and equivalent frequency read, respectively

$$h_e(a) = \frac{1}{2\pi\alpha_0 a}\int\limits_0^{2\pi}\left[2h\left(-a\alpha_0\sin\psi\right) - g\left(-a\alpha_0\sin\psi\right)^3\right]\sin\psi d\psi = -h + \frac{3}{8}ga^2\alpha_0^2,$$

$$\alpha_e(a) = \alpha_0 - \frac{1}{2\pi\alpha_0 a}\int\limits_0^{2\pi}\left[2h\left(-a\alpha_0\sin\psi\right) - g\left(-a\alpha_0\sin\psi\right)^3\right]\cos\psi d\psi = \alpha_0.$$

The equivalent equation takes the form

$$\ddot{y} + \left(-h + \frac{3}{8}ga^2\alpha_0^2\right)\dot{y} + \alpha_0^2 y = 0,$$

and the amplitude of periodic vibrations reads

$$a_0 = \frac{1}{\alpha_0}\sqrt{\frac{8h}{3g}}.$$

□

11.2.2 Nonautonomous Systems

11.2.2.1 Introduction

Now, we make use of the methods introduced in the preceding chapter for analysis of nonautonomous nonlinear systems governed by the equation

$$\ddot{y} + \alpha_0^2 y = \varepsilon\phi\ (y, \dot{y}, \omega t)\,. \tag{11.67}$$

Assume that a driving force is periodic, i.e. $\phi(\omega t + 2\pi) = \phi(\omega t)$, and expand this function in a Fourier series

$$\phi\ (y, \dot{y}, \omega t) = Q_{10}(y, \dot{y}) + \sum_{m=1}^{M} \{Q_{1m}(y, \dot{y}) \cos m\omega t + Q_{2m}(y, \dot{y}) \sin m\omega t\}\,. \tag{11.68}$$

We will assume that the functions Q_{10}, Q_{1m} and Q_{2m} are analytic in their variables, so we can expand them in a neighbourhood of the equilibrium position (for the sake of simplicity we assumed that $y = \dot{y} = 0$).

It is noteworthy to pay attention on variety of resonances occurring in a nonlinear system of one degree-of-freedom. In linear systems with periodic excitations, resonance could appear for the mth harmonic with the satisfied condition $m\omega = \alpha_0$, while for a nonlinear system the resonance appears for

$$m\omega = n\alpha_0, \tag{11.69}$$

where $m, n = 1, 2, 3, \ldots$. In dissipative systems during the resonance, one observes growth of the amplitude of vibrations.

Resonances can be classified as follows:

1. main resonance ($m = n = 1$);
2. subharmonic resonance ($m = 1, n > 1$);
3. ultraharmonic resonance ($m > 1, n = 1$);
4. ultrasubharmonic resonance ($m > 1, n > 1$).

11.2.2.2 Oscillations Away from Resonance

Consider a system

$$\ddot{y} + \alpha_0^2 y = \varepsilon\,[Q\ (y, \dot{y}) + P\ (\eta)]\,, \tag{11.70}$$

where ε is a small parameter and a driving force $P(\eta)$ is periodic with respect to $\eta = \omega t$ of period 2π. In order to analyse such a motion we will apply a simplified version of the Krylov–Bogolubov–Mitropolskiy method.

A solution of Eq. (11.70) is sought in the form

$$y = a\cos\psi + \varepsilon y_1(a, \psi, \eta), \tag{11.71}$$

where

$$\dot{a} = \varepsilon A_1(a), \tag{11.72}$$

$$\dot{\psi} = \alpha_0 + \varepsilon B_1(a), \tag{11.73}$$

while the terms of powers greater by one by ε have been ignored for the sake of simplicity. It also concerns further transformations. From Eq. (11.71) we get

$$\dot{y} = \dot{a}\cos\psi - a\dot{\psi}\sin\psi + \varepsilon\left(\frac{\partial y_1}{\partial a}\dot{a} + \frac{\partial y_1}{\partial \psi}\dot{\psi} + \frac{\partial y_1}{\partial \eta}\omega\right), \tag{11.74}$$

$$\begin{aligned}\ddot{y} = \ddot{a}\cos\psi - 2\dot{a}\dot{\psi}\sin\psi - a\ddot{\psi}\sin\psi - a\dot{\psi}^2\cos\psi + \varepsilon\left(\frac{\partial^2 y_1}{\partial a^2}\dot{a}^2 + 2\frac{\partial^2 y_1}{\partial a\partial\psi}\dot{a}\dot{\psi}\right.\\ \left. +2\frac{\partial^2 y_1}{\partial a\partial\eta}\dot{a}\omega + \frac{\partial y_1}{\partial a}\ddot{a} + \frac{\partial^2 y_1}{\partial \psi^2}\dot{\psi}^2 + 2\frac{\partial^2 y_1}{\partial\psi\partial\eta}\dot{\psi}\omega + \frac{\partial y_1}{\partial\psi}\ddot{\psi} + \frac{\partial^2 y_1}{\partial\eta^2}\omega^2\right).\end{aligned} \tag{11.75}$$

Taking into account (11.72) and (11.73) in (11.74),(11.75) we get

$$\dot{y} = -a\alpha_0\sin\psi + \varepsilon\left(A_1\cos\psi - aB_1\sin\psi + \frac{\partial y_1}{\partial\psi}\alpha_0 + \frac{\partial y_1}{\partial\eta}\omega\right), \tag{11.76}$$

$$\begin{aligned}\ddot{y} &= -a\alpha_0^2\cos\psi \\ &+\varepsilon\left(-2A_1\alpha_0\sin\psi - 2B_1\alpha_0 a\cos\psi + \frac{\partial^2 y_1}{\partial\psi^2}\alpha_0^2 + 2\frac{\partial^2 y_1}{\partial\psi\partial\eta}\alpha_0\omega + \frac{\partial^2 y_1}{\partial\eta^2}\omega_0^2\right).\end{aligned} \tag{11.77}$$

Since

$$\begin{aligned}\ddot{a} &= \frac{d}{dt}(\dot{a}) = \varepsilon^2 A_1\frac{dA_1}{da},\\ \ddot{\psi} &= \frac{d}{dt}(\dot{\psi}) = \varepsilon^2\frac{dB_1}{da}A_1,\\ \dot{\psi}^2 &= \alpha_0^2 + 2\varepsilon\alpha_0 B_1 + \varepsilon^2 B_1,\end{aligned} \tag{11.78}$$

the left-hand side L of Eq. (11.70) takes the form

$$L=\varepsilon\left(2A_1\alpha_0\sin\psi-2B_1\alpha_0 a\cos\psi+\frac{\partial^2 y_1}{\partial\psi^2}\alpha_0^2+2\frac{\partial^2 y_1}{\partial\psi\partial\eta}\alpha_0\omega+\frac{\partial^2 y_1}{\partial\eta^2}\omega^2\right)$$
$$+\varepsilon\alpha_0^2 y_1. \tag{11.79}$$

Expanding the right-hand side of Eq. (11.70) into a power series with respect to ε we have

$$\begin{aligned} P &= \varepsilon\left[Q\left(y,\dot{y}\right)+P\left(\eta\right)\right] \\ &= \varepsilon\left\{\left[Q\left(y,\dot{y}\right)+P\left(\eta\right)\right]\Big|_{\varepsilon=0}+\varepsilon\frac{1}{1!}\left[\frac{\partial Q}{\partial y}\frac{dy}{d\varepsilon}+\frac{\partial Q}{\partial y}\frac{d\dot{y}}{d\varepsilon}\right]\Big|_{\varepsilon=0}+\cdots\right\} \\ &= \varepsilon\left[Q\left(a\cos\psi,-a\alpha_0\sin\psi\right)+P\left(\eta\right)\right]+O\left(\varepsilon^2\right). \end{aligned} \tag{11.80}$$

The function $Q(\psi)=Q(\psi+2\pi)$, and $P(\omega t)=P(\omega(t+T))$, where $T=2\pi/\omega$. The right-hand side of Eq. (11.80) we expand into the Fourier series of the form

$$\begin{aligned} P = \varepsilon\Bigg[&\frac{1}{2}b_0(a)+\sum_{n=1}^{\infty}\left(b_n(a)\cos n\psi+c_n(a)\sin\psi\right)+\frac{1}{2}p_0 \\ &+\sum_{m=1}^{\infty}\left(p_m\cos m\eta+q_m\sin m\eta\right)\Bigg], \end{aligned} \tag{11.81}$$

where

$$\begin{aligned} b_n &= \frac{1}{\pi}\int_0^{2\pi} Q\left(a\cos\psi,-a\alpha_0\sin\psi\right)\cos n\psi d\psi, \\ c_n &= \frac{1}{\pi}\int_0^{2\pi} Q\left(a\cos\psi,-a\alpha_0\sin\psi\right)\sin n\psi d\psi, \\ p_m &= \frac{2}{T}\int_0^{T} P\left(\eta\right)\cos m\eta d\eta, \\ q_m &= \frac{2}{T}\int_0^{T} P\left(\eta\right)\sin m\eta d\eta, \end{aligned} \tag{11.82}$$

where $m,n=0,1,2,\ldots$.

Equating expressions occurring at the same powers ε in Eqs. (11.79) and (11.80) we get

$$\frac{\partial^2 y_1}{\partial \psi^2} + 2\frac{\omega}{\alpha_0}\frac{\partial^2 y_1}{\partial \psi \partial \eta} + \frac{\omega^2}{\alpha_0^2}\frac{\partial^2 y_1}{\partial \eta^2} + y_1 = \frac{1}{\alpha_0^2}\{[2A_1\alpha_0 + c_1(a)]\sin\psi$$

$$+ [2B_1\alpha_0 a + b_1(a)]\cos\psi + \frac{1}{2}b_0(a) + \sum_{n=2}^{\infty}(b_n(a)\cos n\psi + c_n(a)\sin\psi)$$

$$+\frac{1}{2}p_0 + \sum_{m=1}^{\infty}(p_m\cos m\eta + q_m\sin m\eta)\} . \tag{11.83}$$

The two first components on the right-hand side of Eq. (11.83) give secular terms. Equating them to zero we obtain

$$A_1 = -\frac{c_1(a)}{2\alpha_0} = -\frac{1}{2\pi\alpha_0}\int_0^{2\pi} Q\,(a\cos\psi, -a\alpha_0\sin\psi)\sin\psi d\psi, \tag{11.84}$$

$$B_1 = -\frac{b_1(a)}{2\alpha_0 a} = -\frac{1}{2a\pi\alpha_0}\int_0^{2\pi} Q\,(a\cos\psi, -a\alpha_0\sin\psi)\cos\psi d\psi. \tag{11.85}$$

Hence and by (11.83) we find

$$\frac{\partial^2 y_1}{\partial \psi^2} + 2\frac{\omega}{\alpha_0}\frac{\partial^2 y_1}{\partial \psi \partial \eta} + \frac{\omega^2}{\alpha_0^2}\frac{\partial^2 y_1}{\partial \eta^2} + y_1 = \frac{1}{\alpha_0^2}\left[\frac{1}{2}b_0(a) + \sum_{n=2}^{\infty}(b_n(a)\cos n\psi\right.$$

$$\left. + c_n(a)\sin\psi) + \frac{1}{2}p_0 + \sum_{m=1}^{\infty}(p_m\cos m\eta + q_m\sin m\eta)\right] . \tag{11.86}$$

We will require a solution of the above equation not to possess the first harmonic of free vibrations and thus

$$y_1 = \frac{1}{\alpha_0^2}\left[\frac{b_0(a)}{2} + \sum_{n=2}^{\infty}\left(\frac{b_n(a)}{1-n^2}\cos n\psi + \frac{c_n(a)}{1-n^2}\sin n\psi\right)\right.$$
$$\left. + \frac{p_0}{2} + \sum_{m=1}^{\infty}\left(\frac{p_m}{1-\left(\frac{\omega}{\alpha_0}m\right)^2}\cos m\eta + \frac{q_m}{1-\left(\frac{\omega}{\alpha_0}m\right)^2}\sin m\eta\right)\right] \tag{11.87}$$

Taking into account (11.87) in (11.71) we obtain a complete solution, exact to $O(\varepsilon^2)$. Next, according to (11.72), (11.73) and (11.84),(11.85), we obtain

$$\dot{a} = -\frac{\varepsilon}{2\pi\alpha_0}\int_0^{2\pi} Q\left(a\cos\psi, -a\alpha_0\sin\psi\right)\sin\psi d\psi, \tag{11.88}$$

$$\dot{\psi} = \alpha_0 - \frac{\varepsilon}{2\pi a\alpha_0}\int_0^{2\pi} Q\left(a\cos\psi, -a\alpha_0\sin\psi\right)\cos\psi d\psi. \tag{11.89}$$

These equations describe a transient state of the vibrating system. In the case of analysis of the steady state the problem is reduced to the analysis of the equation

$$\int_0^{2\pi} Q\left(a\cos\psi, -a\alpha_0\sin\psi\right)\sin\psi d\psi = 0, \tag{11.90}$$

by which we determine a. There may be several solutions since this equation is a nonlinear algebraic equation. We obtain corresponding phases from Eq. (11.79)

$$\psi = \int\left(\alpha_0 - \frac{\varepsilon}{2\pi a\alpha_0}\int_0^{2\pi} Q\left(a\cos\psi, -a\alpha_0\sin\psi\right)\cos\psi d\psi\right)dt + \Theta_0. \tag{11.91}$$

The obtained solution y consists of two parts. The first part contains the harmonic phases y and presents driven vibrations of the system, the second one (containing the harmonic phases of the driven force η) presents driven vibrations of the system. This is a result of the assumptions, made at the beginning. In a general case, it is not possible to separate free vibrations from driven ones. The obtained solution is undetermined for $\alpha_0 = m\omega$, i.e. for the frequency ω, at which resonance appears. The problem of determination of a solution in the vicinity of resonance will be analysed in the next section.

Example 11.4. Analyse the motion of a system, governed by the following differential equation

$$\ddot{y} + \alpha_0^2 y = \varepsilon\left[\left(2h - g\dot{y}^2\right)\dot{y} + p\cos\omega t\right]$$

far from the resonance.

Taking into account the studied equation, and according to (11.82) we obtain: $p_m = 0$ $(m \neq 1)$, $p_1 = p$, $q_m = 0$, $b_n = 0$, $c_2 = 0$, $c_3 - \frac{1}{4}ga^3\alpha_0^2$ and $c_n = 0$ for $n > 3$ and the solution

$$y = a\cos\psi + \varepsilon\left[\frac{1}{32}ga^3\alpha_0\sin 3\psi + \frac{p}{\alpha_0^2 - \omega^2}\cos\omega t\right].$$

The functions $a(t)$ are $y(t)$ solutions of Eqs. (11.88) and (11.89), which take the form in this case

$$\dot{a} = -\frac{\varepsilon}{2\pi\alpha_0}\int_0^{2\pi}\left[2h - g\left(-a\alpha_0 \sin\psi^2\right)\right](-a\alpha_0 \sin\psi)\sin\psi d\psi = ah(1 - K^2a^2),$$

$$\dot{\psi} = \alpha_0,$$

where

$$K^2 = \frac{3}{8}\frac{g\alpha_0^2}{h}.$$

For $K^2 > 0$, solutions of the first-order ODEs are, respectively

$$a = \frac{a_0 e^{\varepsilon ht}}{\sqrt{1 + a_0^2 K^2 e^{2\varepsilon ht}}},$$

$$\psi = \alpha_0 t + \Theta,$$

where a_0 and Θ are constants, which are depend on the initial conditions. Finally, we get

$$y = \frac{a_0 e^{\varepsilon ht}}{\sqrt{1 + a_0^2 K^2 e^{2\varepsilon ht}}}\cos(\alpha_0 t + \Theta)$$

$$+\frac{\varepsilon}{32}g\alpha_0\left[\frac{a_0 e^{\varepsilon ht}}{\sqrt{1 + a_0^2 K^2 e^{2\varepsilon ht}}}\right]^3 \sin(3\alpha_0 t + 3\Theta) + \frac{\varepsilon p}{\alpha_0^2 - \omega^2}\cos\omega t\,.$$

In the steady state $t \to \infty$ and finally we get

$$\lim_{t\to+\infty} y = \frac{1}{K}\cos(\alpha_0 t + \Theta) + \frac{\varepsilon}{32}\frac{g\alpha_0}{K^3}\sin(3\alpha_0 t + 3\Theta) + \frac{\varepsilon p}{\alpha_0^2 - \omega^2}\cos\omega t.$$

In the end, let us remind that according to the assumption, the obtained solution is valid for a small amplitude of the driving force and sufficiently far from resonance. □

11.2.2.3 Oscillations Near Resonance

Consider the system (11.70), but now we apply the method of equivalent linearization, discussed in Sect. 11.2.1.3, to solve the problem. We seek a solution in the form

$$y = a\cos\psi + \sum_{k=1}^{K} \varepsilon^k y_k(a,\psi,\eta)\,, \tag{11.92}$$

It turns out that derivatives of the amplitudes and the angle ψ are functions of the phase angle ϑ of the form

$$\dot{a} = \sum_{k=1}^{K} \varepsilon^k A_k(a,\vartheta)\,, \tag{11.93}$$

$$\dot{\psi} = \alpha_0 + \sum_{k=1}^{K} \varepsilon^k \bar{B}_k(a,\vartheta)\,. \tag{11.94}$$

In general $\vartheta(t)$ is related to the other phases by the equation

$$n\vartheta(t) = n\psi(t) - m\eta(t), \tag{11.95}$$

which allows to eliminate the variable y and obtain the equations

$$y = a\cos\left(\frac{m}{n}\eta + \vartheta\right) + \sum_{k=1}^{K} \varepsilon^k y_k(a,\vartheta,\eta)\,, \tag{11.96}$$

$$\dot{a} = \sum_{k=1}^{K} \varepsilon^k A_k(a,\vartheta)\,, \tag{11.97}$$

$$\dot{\vartheta} = \alpha_0 - \frac{m}{n}\omega + \sum_{k=1}^{K} \varepsilon^k \bar{B}_k(a,\vartheta)\,. \tag{11.98}$$

Further considerations will be carried out in a close neighbourhood of *ultrasubharmonic resonance*, which is exhibited by the equation

$$\alpha_0^2 - \left(\frac{m}{n}\omega\right)^2 = \varepsilon\,\Delta. \tag{11.99}$$

Taking into account (11.99) in (11.70) we get

$$\ddot{y} + \left(\frac{m}{n}\omega\right)^2 y = \varepsilon\left[Q\left(y, \dot{y}\right) + P\left(\eta\right) - \Delta y\right]. \tag{11.100}$$

By Eq. (11.99) we also obtain

$$\alpha_0 = \sqrt{\left(\frac{m}{n}\omega\right)^2 + \varepsilon\,\Delta} \cong \frac{m}{n}\omega + \frac{\Delta}{2\frac{m}{n}\omega}\varepsilon. \tag{11.101}$$

From Eqs. (11.96)–(11.98) we obtain (assuming that $O(\varepsilon^2)$)

$$y = a\cos\left(\frac{m}{n}\eta + \vartheta\right) + \varepsilon\, y_1\left(a, \vartheta, \eta\right), \tag{11.102}$$

$$\dot{a} = \varepsilon\, A_1\left(a, \vartheta\right), \tag{11.103}$$

$$\dot{\vartheta} = \varepsilon\left[\bar{B}_1\left(a, \vartheta\right) + \frac{\Delta}{2\frac{m}{n}\omega}\right] = \varepsilon\, B_1\left(a, \vartheta\right). \tag{11.104}$$

Differentiating (11.102) and taking into account (11.103) and (11.104) we obtain

$$\begin{aligned}\dot{y} = -a\frac{m}{n}\omega\sin\left(\frac{m}{n}\eta + \vartheta\right) + \varepsilon\Big[A_1\cos\left(\frac{m}{n}\eta + \vartheta\right) - \\ -aB_1\sin\left(\frac{m}{n}\eta + \vartheta\right) + \frac{\partial y_1}{\partial\eta}\omega\Big],\end{aligned} \tag{11.105}$$

$$\begin{aligned}\ddot{y} = -a\left(\frac{m}{n}\omega\right)^2\cos\left(\frac{m}{n}\eta + \vartheta\right) + \varepsilon\Big[-2a\frac{m}{n}\omega B_1\cos\left(\frac{m}{n}\eta + \vartheta\right) - \\ -2\frac{m}{n}\omega A_1\sin\left(\frac{m}{n}\eta + \vartheta\right) + \frac{\partial^2 y_1}{\partial\eta^2}\omega^2\Big].\end{aligned} \tag{11.106}$$

Taking into account the two last equations in (11.100) we obtain

$$\begin{aligned}&-a\left(\frac{m}{n}\eta\right)^2\cos\left(\frac{m}{n}\eta + \vartheta\right) + a\left(\frac{m}{n}\omega\right)^2\cos\left(\frac{m}{n}\eta + \vartheta\right)\\&+\varepsilon\Big[-2a\frac{m}{n}\omega\, B_1\cos\left(\frac{m}{n}\eta + \vartheta\right) - 2\frac{m}{n}\omega\, A_1\sin\left(\frac{m}{n}\eta + \vartheta\right) + \frac{\partial^2 y_1}{\partial\eta^2}\omega^2\\&+\left(\frac{m}{n}\omega\right)^2 y_1\Big] = \varepsilon\left\{Q\left[a\cos\left(\frac{m}{n}\eta + \vartheta\right), -a\frac{m}{n}\omega\sin\left(\frac{m}{n}\eta + \vartheta\right)\right]\right.\\&\left.+P(\eta) - \Delta a\cos\left(\frac{m}{n}\eta + \vartheta\right)\right\}.\end{aligned} \tag{11.107}$$

After equating the terms standing by ε we have obtained

$$\frac{\partial^2 y_1}{\partial \eta^2} + \left(\frac{m}{n}\right)^2 y_1 = 2\frac{m}{n}\frac{1}{\omega}A_1 \sin\left(\frac{m}{n}\eta + \vartheta\right)$$
$$+\left(2a\frac{m}{n}\frac{1}{\omega}B_1 - \frac{\Delta a}{\omega^2}\right)\cos\left(\frac{m}{n}\eta + \vartheta\right)$$
$$+\frac{1}{\omega^2}\left[\frac{1}{2}b_0(a) + \sum_{n'=1}^{\infty} b_{n'}(a)\cos n'\left(\frac{m}{n}\eta + \vartheta\right)\right.$$
$$\left. +c_{n'}(a)\sin n'\left(\frac{m}{n}\eta + \vartheta\right) + \frac{1}{2}p_0 + \sum_{m'=1}^{\infty} p_{m'}\cos m'\eta + q_{m'}\sin m'\eta\right], \tag{11.108}$$

where

$$b_{n'} = \frac{1}{\pi}\int_0^{2\pi} Q\left(a\cos\psi, -a\frac{m}{n}\omega\sin\psi\right)\cos n'\psi d\psi,$$
$$c_{n'} = \frac{1}{\pi}\int_0^{2\pi} Q\left(a\cos\psi, -a\frac{m}{n}\omega\sin\psi\right)\sin n'\psi d\psi,$$
$$p_{m'} = \frac{2}{T}\int_0^{T} P(\eta)\cos m'\eta d\eta, \tag{11.109}$$
$$q_{m'} = \frac{2}{T}\int_0^{T} P(\eta)\sin m'\eta d\eta, \quad m', n' = 0,\ 1,\ 2, \ldots$$

Further analysis will be carried out for the case $n = 1$, what follows from the fact that the influence of the phase angle ϑ is exposed then. By Eq. (11.108) we obtain

$$\frac{\partial^2 y_1}{\partial \eta^2} + m^2 y_1 = \left(2m\frac{1}{\omega}A_1 + \frac{1}{\omega^2}C_1\right)\sin\,(m\eta + \vartheta) + \frac{1}{\omega^2}q_m \sin m\eta$$
$$+\left(2am\frac{1}{\omega}B_1 - \frac{\Delta a}{\omega^2} + \frac{1}{\omega^2}b_1\right)\cos\,(m\eta + \vartheta) + \frac{1}{\omega^2}p_m\cos m\eta$$
$$+\frac{1}{\omega^2}\left[\frac{1}{2}b_0(a) + \sum_{n'=2}^{\infty} b_{n'}\cos n'\,(m\eta + \vartheta) + c_{n'}\sin n'\,(m\eta + \vartheta)\right. \tag{11.110}$$
$$\left. +\frac{1}{2}p_0 + \sum_{m'=1, m'\neq m}^{\infty}\left(p_{m'}\cos m'\eta + q_{m'}\sin m'\eta\right)\right].$$

Assume in the terms, giving the secular terms

$$2m\frac{1}{\omega}A_1 + \frac{1}{\omega^2}c_1 = \alpha, \qquad \frac{1}{\omega^2}q_m = \beta,$$
$$2am\frac{1}{\omega}B_1 - \frac{\Delta a}{\omega^2} + \frac{1}{\omega^2}b_1 = \gamma, \qquad \frac{1}{\omega^2}p_m = \delta, \tag{11.111}$$

then the terms can be transformed into the form

$$\alpha \sin\,(m\eta + \vartheta) + \beta \sin m\eta + \gamma \cos\,(m\eta + \vartheta) + \delta \cos m\eta$$
$$= (\alpha \cos\vartheta + \beta - \gamma \sin\vartheta) \sin m\eta + (\alpha \sin\vartheta + \gamma \cos\vartheta + \delta) \cos m\eta\,. \tag{11.112}$$

They will equal zero, when

$$\alpha \cos\vartheta + \beta - \gamma \sin\vartheta = 0\,,$$
$$\alpha \sin\vartheta + \delta + \gamma \cos\vartheta = 0\,. \tag{11.113}$$

Multiplying the first of Eq. (11.113) by $\cos\vartheta$, and the second one by $sin\vartheta$ and adding them we get

$$\alpha + \beta \cos\vartheta + \delta \sin\vartheta = 2m\frac{1}{\omega}A_1 + \frac{1}{\omega^2}C_1 + \frac{1}{\omega^2}q_m \cos\vartheta + \frac{1}{\omega^2}p_m \sin\vartheta = 0. \tag{11.114}$$

Multiplying the first of Eq. (11.113) by $\sin\vartheta$, and the second one by -$\cos\vartheta$ and adding them we get

$$\beta \sin\vartheta - \delta \cos\vartheta - \gamma = \frac{1}{\omega^2}q_m \sin\vartheta - \frac{1}{\omega^2}p_m \cos\vartheta - \tag{11.115}$$
$$-2am\frac{1}{\omega}B_1 + \frac{\Delta a}{\omega^2} - \frac{b_1}{\omega^2} = 0.$$

By the above equations we find

$$A_1 = -\frac{c_1}{2m\omega} - \frac{q_m}{2m\omega}\cos\vartheta - \frac{p_m}{2m\omega}\sin\vartheta,$$
$$B_1 = -\frac{\Delta}{2m\omega} - \frac{b_1}{2ma\omega} + \frac{q_m}{2ma\omega}\sin\vartheta - \frac{p_m}{2ma\omega}\cos\vartheta. \tag{11.116}$$

Taking into account (11.116) in (11.103) and (11.104) we find

$$\dot{a} = \varepsilon\left(-\frac{c_1}{2m\omega} - \frac{q_m}{2m\omega}\cos\vartheta - \frac{p_m}{2m\omega}\sin\vartheta\right), \tag{11.117}$$

$$\dot{\vartheta} = \alpha_0 - m\omega + \varepsilon\left(-\frac{b_1}{2ma\omega} + \frac{q_m}{2ma\omega}\sin\vartheta - \frac{p_m}{2ma\omega}\cos\vartheta\right). \tag{11.118}$$

In order to simplify the transformations, further considerations will be carried out for $m = 1$, namely we will seek a solution of the form

$$y = a\cos(\omega t + \vartheta). \tag{11.119}$$

By Eq. (11.101) we obtain

$$\varepsilon\Delta = 2\omega(\alpha_0 - \omega). \tag{11.120}$$

while by Eqs. (11.117) and (11.118) we obtain

$$\begin{aligned}\dot{a} &= \varepsilon\left(-\frac{c_1}{2\omega} - \frac{q_1}{2\omega}\sin\vartheta\right)\\ &= -\frac{\varepsilon}{2\pi\omega}\int_0^{2\pi} Q(a\cos\psi, -a\omega\sin\psi)\sin\psi d\psi - \frac{\varepsilon p}{2\omega}\sin\vartheta,\end{aligned} \tag{11.121}$$

$$\begin{aligned}\dot{\vartheta} &= \alpha_0 - \omega + \varepsilon\left(-\frac{b_1}{2a\omega} - \frac{p}{2a\omega}\cos\vartheta\right)\\ &= \alpha_0 - \omega - \frac{\varepsilon}{2\pi\omega a}\int_0^{2\pi} Q(a\cos\psi, -a\omega\sin\psi)\cos\psi d\psi - \frac{\varepsilon p}{2a\omega}\cos\vartheta.\end{aligned} \tag{11.122}$$

where $p = p_1$

In an analogous manner as in formulas (11.56) and (11.57) we introduce the following designations

$$h_e(a) = \frac{\varepsilon}{2\pi\alpha_0 a}\int_0^{2\pi} Q(a\cos\psi, -a\alpha_0\sin\psi)\sin\psi d\psi, \tag{11.123}$$

$$\alpha_e(a) = \alpha_0 - \frac{\varepsilon}{2\pi\alpha_0 a}\int_0^{2\pi} Q(a\cos\psi, -a\alpha_0\sin\psi)\cos\psi d\psi. \tag{11.124}$$

Now, we will show that the solution (11.119) satisfies, exact to ε the equivalent linear equation corresponding to the primary nonlinear equation of the form

$$\ddot{y} + 2h_e(a)\dot{y} + \alpha_e^2(a)y = \varepsilon p\cos\omega t. \tag{11.125}$$

Let us transform Eq. (11.121) in the following way

$$\begin{aligned}\dot{a} &= -ah_e + ah_e - \frac{\varepsilon}{2\pi\omega}\int_0^{2\pi} Q(a\cos\psi, -a\omega\sin\psi)\sin\psi d\psi - \frac{\varepsilon p}{2\omega}\sin\vartheta\\ &= -ah_e + \frac{\varepsilon}{2\pi\alpha_0}\int_0^{2\pi} Q(a\cos\psi, -a\alpha_0\sin\psi)\sin\psi d\psi\\ &\quad - \frac{\varepsilon}{2\pi\omega}\int_0^{2\pi} Q(a\cos\psi, -a\alpha_0\sin\psi)\sin\psi d\psi - \frac{\varepsilon p}{2\omega}\sin\vartheta.\end{aligned} \tag{11.126}$$

We eliminate α_0, in the above equation, on the basis of the relationship (11.120), and we get

$$\dot{a} = -ah_e + \frac{\varepsilon}{2\pi\left(\omega + \frac{\varepsilon\Delta}{2\omega}\right)} \int_0^{2\pi} Q\left(a\cos\psi, -a\left(\omega + \frac{\varepsilon\Delta}{2\omega}\right)\sin\psi\right)\sin\psi d\psi$$
$$-\frac{\varepsilon}{2\pi\omega}\int_0^{2\pi} Q\left(a\cos\psi, -a\omega\sin\psi\right)\sin\psi d\psi - \frac{\varepsilon p}{2\omega}\sin\vartheta. \tag{11.127}$$

After expansion of the second term of the right-hand side of Eq. (11.127) into a power series with respect to ε and ignoring terms with e we obtain

$$\dot{a} = -ah_e + \frac{\varepsilon}{2\pi\omega}\int_0^{2\pi} Q\left(a\cos\psi, -a\omega\sin\psi\right)\sin\psi d\psi$$
$$-\frac{\varepsilon}{2\pi\omega}\int_0^{2\pi} Q\left(a\cos\psi, -a\omega\sin\psi\right)\sin\psi d\psi - \frac{\varepsilon p}{2\omega}\sin\vartheta = -ah_e - \frac{\varepsilon p}{2\omega}\sin\vartheta. \tag{11.128}$$

Similarly, we obtain

$$\dot{\vartheta} = \alpha_e - \omega - \frac{\varepsilon p}{2a\omega}\cos\vartheta. \tag{11.129}$$

By Eq. (11.119) we obtain

$$\dot{y} = \dot{a}\cos\left(\omega t + \vartheta\right) - a\left(\omega + \dot{\vartheta}\right)\sin\left(\omega t + \vartheta\right), \tag{11.130}$$

$$\ddot{y} = \ddot{a}\cos(\omega t + \vartheta) - 2\dot{a}(\omega + \dot{\vartheta})\sin(\omega t + \vartheta) - a\ddot{\vartheta}\sin(\omega t + \vartheta)$$
$$-a(\omega + \dot{\vartheta})^2\cos(\omega t + \vartheta). \tag{11.131}$$

Taking into account the relationships (11.128) and (11.129) in the two first equations we have (exact to $O(\varepsilon^2)$)

$$\dot{y} = -ah_e\cos\left(\omega t + \vartheta\right) + \frac{\varepsilon p}{2\omega}\sin\omega t - a\alpha_e\sin\left(\omega t + \vartheta\right), \tag{11.132}$$

$$\ddot{y} = 2ah_e\alpha_e\sin\left(\omega t + \vartheta\right) + \varepsilon p\cos\omega t - a\alpha_e^2\cos\left(\omega t + \vartheta\right). \tag{11.133}$$

The left-hand side of (11.126), including (11.119), (11.130) and (11.131), is expressed in the form

$$L = \varepsilon p \cos \omega t - 2ah_e^2 \cos(\omega t + \vartheta) + h_e \frac{\varepsilon p}{\omega} \sin \omega t \cong \varepsilon p \cos \omega t. \tag{11.134}$$

Taking into account Eq. (11.134) and the right-hand side of (11.125) we see that the solution (11.119) satisfies, exact to ε, Eq. (11.125). Thus, the method of equivalent linearization allows to replace the nonlinear equation (11.70) with the equivalent linear equation (11.125) for vibrations near resonance. One needs to keep in mind that the unit equivalent damping coefficient $h_e(a)$ and the equivalent eigenfrequency $a_e(a)$ are functions of the amplitude of steady vibrations a. According to the theory of linear vibrations the amplitude of vibrations reads

$$a = \frac{\varepsilon p}{\sqrt{\left(\alpha_e^2(a) - \omega^2\right)^2 + 4h_e^2(a)\omega^2}}, \tag{11.135}$$

which allows to determine the frequency as a function of the amplitude

$$\omega = \sqrt{\left[\alpha_e^2(a) - 2h_e^2(a)\right] \pm \sqrt{4h_e^2(a)\left[h_e^2(a) - \alpha_e^2(a)\right] + \frac{\varepsilon^2 p^2}{\alpha^2}}}. \tag{11.136}$$

It turns out that one, two or no value of the circular frequency ω can exist satisfying Eq. (11.136). According to the theory of linear vibrations we obtain

$$\vartheta = \arctan \frac{-2h_e(a)\omega}{\alpha_e^2(a) - \omega^2}. \tag{11.137}$$

Thus, for every assumed amplitude a and value ω determined by Eq. (11.136) one can determine the corresponding phase angles ϑ.

It turns out that not every arcs (segments) of a resonance curve are stable, i.e. realizable physically. Moreover, there exist frequency intervals, whom correspond three or even five values of distinct amplitudes and phase angles. Now, we perform more thorough analysis of stability of steady vibrations, corresponding to particular segments of the resonance curve. Consider the equation of steady vibrations

$$y = a_0 \cos(\omega t + \vartheta_0), \tag{11.138}$$

where a_0 and ϑ_0 satisfy Eqs. (11.128) and (11.129). Hence, we have

$$\begin{aligned} -a_0 h_e(a_0) - \tfrac{\varepsilon p}{2\omega} \sin \vartheta_0 &= 0, \\ \alpha_e(a_0) - \omega - \tfrac{\varepsilon p}{2a_0\omega} \cos \vartheta_0 &= 0. \end{aligned} \tag{11.139}$$

In order to examine the stability of the solution of (11.138) one needs to analyse a solution, which is close to the solution

$$y = a \cos(\omega t + \vartheta), \tag{11.140}$$

where $a(t)$ and $\vartheta(t)$ are solutions of Eqs. (11.128) and (11.129). Hence

$$\begin{aligned} \dot{a} &= -ah_e(a) - \tfrac{\varepsilon p}{2\omega}\sin\vartheta = \varepsilon A\left[a(t), \vartheta(t), \omega\right], \\ \dot{\vartheta} &= \alpha_e(a) - \omega - \tfrac{\varepsilon p}{2a\omega}\cos\vartheta = \varepsilon B\left[a(t), \vartheta(t), \omega\right]. \end{aligned} \tag{11.141}$$

As was already mentioned, we will consider a solution, which is near the steady one, i.e.

$$\begin{aligned} a(t) &= a_0 + \delta_a(t), \\ \vartheta(t) &= \vartheta_0 + \delta_\vartheta(t), \end{aligned} \tag{11.142}$$

where $\delta(t)$ is sufficiently small. Taking into account (11.142) in (11.141) we get

$$\begin{aligned} \dot{\delta}_a &= \varepsilon A\left[(a_0 + \delta_a(t)), (\vartheta_0 + \delta_\vartheta(t)), \omega\right], \\ \dot{\delta}_\vartheta &= \varepsilon B\left[(a_0 + \delta_a(t)), (\vartheta_0 + \delta_\vartheta(t)), \omega\right], \end{aligned} \tag{11.143}$$

and next, the right-hand sides of (11.143) are expanded in power series with respect to δ_a and δ_ϑ around a point of the coordinates (a_0, ϑ_0), which yields (only the first powers of expansion have been included)

$$\begin{aligned} \dot{\delta}_a &= \varepsilon A\left[(a_0, \vartheta_0, \omega) + \frac{\partial A}{\partial a}(a_0, \vartheta_0)\,\delta_a + \frac{\partial A}{\partial \vartheta}(a_0, \vartheta_0)\,\delta_\vartheta\right], \\ \dot{\delta}_\vartheta &= \varepsilon B\left[(a_0, \vartheta_0, \omega) + \frac{\partial B}{\partial a}(a_0, \vartheta_0)\,\delta_a + \frac{\partial B}{\partial \vartheta}(a_0, \vartheta_0)\,\delta_\vartheta\right]. \end{aligned} \tag{11.144}$$

According to (11.139) we have

$$\begin{aligned} A(a_0, \vartheta_0, \omega) &= 0, \\ B(a_0, \vartheta_0, \omega) &= 0. \end{aligned} \tag{11.145}$$

We seek the following forms of the linear differential equations (11.144)

$$\begin{aligned} \delta_a &= D_a e^{rt}, \\ \delta_\vartheta &= D_\vartheta e^{rt}. \end{aligned} \tag{11.146}$$

Taking into account (11.146) in (11.144) by assumption of nonzero D_a and D_ϑ we obtain the following characteristic equation

$$\begin{aligned} &r^2 - \varepsilon r\left[\frac{\partial A}{\partial a}(a_0, \vartheta_0) + \frac{\partial B}{\partial \vartheta}(a_0, \vartheta_0)\right] \\ &+\varepsilon^2\left[\frac{\partial A}{\partial a}(a_0, \vartheta_0)\frac{\partial B}{\partial \vartheta}(a_0, \vartheta_0) - \frac{\partial A}{\partial \vartheta}(a_0, \vartheta_0)\frac{\partial B}{\partial a}(a_0, \vartheta_0)\right] = 0. \end{aligned} \tag{11.147}$$

The analysed steady solution will be stable, if the increments $\delta_a(t)$ and $\delta_\vartheta(t)$ tend to zero, as $t \to \infty$. It will take place when real parts of the roots of (11.147) are less than zero. According to the Viéta's equations, it takes place when

$$\frac{\partial A}{\partial a} + \frac{\partial B}{\partial \vartheta} < 0, \tag{11.148}$$

$$\frac{\partial A}{\partial a}\frac{\partial B}{\partial \vartheta} - \frac{\partial A}{\partial \vartheta}\frac{\partial B}{\partial a} > 0. \tag{11.149}$$

These conditions will be transformed into the form which enables the evaluation of stability of the solutions on the basis of the resonance curve. According to (11.141) we have

$$\begin{aligned} \frac{\partial A}{\partial a} &= -h_e(a_0) - a_0\frac{\partial h_e}{\partial a}(a_0), & \frac{\partial A}{\partial \vartheta} &= -\frac{\varepsilon p}{2\omega}\cos\vartheta_0, \\ \frac{\partial B}{\partial a} &= \frac{\partial \alpha_e}{\partial a}(a_0) + \frac{\varepsilon p}{2a_0^2\omega}\cos\vartheta_0, & \frac{\partial B}{\partial \vartheta} &= \frac{\varepsilon p}{2a_0\omega}\sin\vartheta_0. \end{aligned} \tag{11.150}$$

On the basis of the above relationships, the condition of stability (11.148) takes the form

$$-h_e(a_0) - a_0\frac{\partial h_e}{\partial a}(a_0) + \frac{\varepsilon p}{2\alpha_0\omega}\sin\vartheta_0 < 0. \tag{11.151}$$

making use of the first equation of (11.139) we obtain

$$-2h_e(a_0) - a_0\frac{\partial h_e}{\partial a}(a_0) < 0. \tag{11.152}$$

This condition is written in the form

$$\frac{d}{da_0}\left[a_0^2 h_e(a_0)\right] > 0, \quad \text{for} \quad a_0 > 0. \tag{11.153}$$

According to (11.139) we have

$$A\left[a_0(\omega), \vartheta_0(\omega), \omega\right] = 0, \tag{11.154}$$

$$B\left[a_0(\omega), \vartheta_0(\omega), \omega\right] = 0.$$

Differentiating Eq. (11.154) with respect to ω, we obtain

$$\begin{aligned} \frac{\partial A}{\partial a}\frac{da_0}{d\omega} + \frac{\partial A}{\partial \vartheta}\frac{d\vartheta_0}{d\omega} &= -\frac{dA}{d\omega}, \\ \frac{\partial B}{\partial a}\frac{da_0}{d\omega} + \frac{\partial B}{\partial \vartheta}\frac{d\vartheta_0}{d\omega} &= -\frac{dB}{d\omega}. \end{aligned} \tag{11.155}$$

Multiplying the first equations of the system (11.155) by $\partial B/\partial\vartheta$, and the second one by $-\partial A/\partial\vartheta$, next adding them up, we get

$$\frac{\partial A}{\partial a}\frac{\partial B}{\partial\vartheta}-\frac{\partial A}{\partial\vartheta}\frac{\partial B}{\partial a}=\left(\frac{da_0}{d\omega}\right)^{-1}\left(\frac{\partial A}{\partial\vartheta}\frac{dB}{d\omega}-\frac{\partial B}{\partial\vartheta}\frac{dA}{d\omega}\right). \tag{11.156}$$

Since

$$\begin{aligned}\frac{\partial A}{\partial\omega}&=\frac{\varepsilon p}{2\omega^2}\sin\vartheta_0,\\ \frac{\partial B}{\partial\omega}&=-1+\frac{\varepsilon p}{2a_0\omega^2}\cos\vartheta_0,\end{aligned} \tag{11.157}$$

and taking into account (11.150), according to (11.149), we obtain

$$-\left(\frac{da_0}{d\omega}\right)^{-1}\left[\frac{\varepsilon p}{2\omega}\cos\vartheta_0\left(-1+\frac{\varepsilon p}{2\alpha_0\omega^2}\cos\vartheta_0\right)+\frac{\varepsilon^2p^2}{4a_0\omega^3}\sin^2\vartheta_0\right]>0, \tag{11.158}$$

and taking only the terms standing by the first power of ε leads to the condition

$$\frac{\frac{\varepsilon p}{2\omega}\cos\vartheta_0}{\frac{da_0}{d\omega}}>0. \tag{11.159}$$

Since, the nominator of the inequality (11.159), according to the second equation (11.141), reads

$$\frac{\varepsilon p}{2\omega}\cos\vartheta_0=a_e\left[\alpha_0\left(a_0\right)-\omega\right], \tag{11.160}$$

for $a_0>0$ we have

$$\frac{\alpha_e\left(a_0\right)}{\frac{da_0}{d\omega}}>0. \tag{11.161}$$

It follows from the above inequality that the analysed steady solution is stable, if

$$\frac{da_0}{d\omega}>0\quad\text{and}\quad\alpha_e\left(a_0\right)>\omega\quad\text{or}\quad\frac{da_0}{d\omega}<0\quad\text{and}\quad\alpha_e\left(a_0\right)<\omega. \tag{11.162}$$

The performed analysis allows for complete determination of stability of the solutions on the basis of a resonance curve.

Chapter 12
Continualization and Discretization

12.1 Introduction

So far we considered oscillations of a single oscillator, or in a language of mechanics, a system of one degree-of-freedom. If there is a lot of oscillators, connected along a certain axis, then a problem of modelling can be reduced to continualization, i.e. there is a possibility of transition from many second-order ordinary differential equations to a single partial differential equation.

Usually, during the analysis of plates and shells governed by partial differential equations, the problem is reduced to a finite (large) number of second-order differential and algebraic equations by means of application of Bubnov–Galerkin method of higher order or the finite difference method [25]. However, in some cases a reversed process is required (see [169] and references therein).

Due to the modelling by means of continualization one can obtain relations characterizing micro and macro-properties of an analysed system.

12.2 One-Dimensional Chain of Coupled Oscillators

Consider vibrations of a single point mass (Fig. 12.1a) and two connected masses (Fig. 12.1b).

The equation of motion of a single oscillator has the form

$$m\ddot{y}_1 + 2ky_1 = 0, \tag{12.1}$$

or

$$\ddot{y} = -2\alpha^2 y_1, \tag{12.2}$$

J. Awrejcewicz, *Ordinary Differential Equations and Mechanical Systems*,
DOI 10.1007/978-3-319-07659-1_12

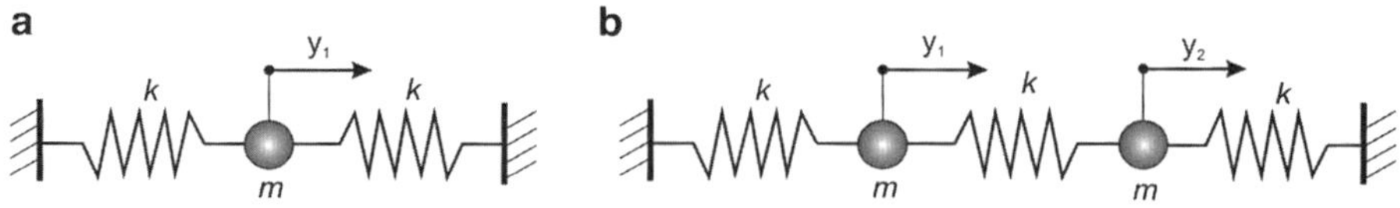

Fig. 12.1 System of one and two degrees of freedom (motion is horizontal and stiffness of the springs is the same everywhere)

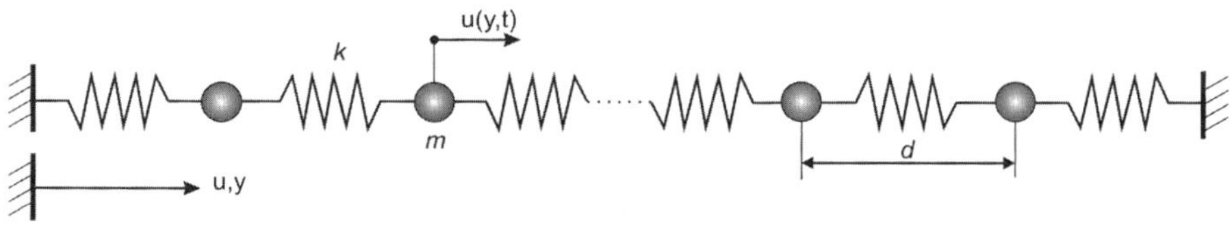

Fig. 12.2 Chain of point masses connected by springs of the same stiffness

where $\alpha^2 = k/m$. Note that the solution can be expressed by the function

$$y_1 = \sin\frac{\pi}{2}\left(A\cos\omega_1 t + B\sin\omega_1 t\right), \tag{12.3}$$

where $\omega_1 = \sqrt{2}\alpha$.

The equation of motion of two coupled oscillators has the form

$$\begin{aligned} m_1\ddot{y}_1 &= -ky_1 - k\left(y_1 - y_2\right), \\ m_2\ddot{y}_2 &= -ky_2 - k\left(y_2 - y_1\right), \end{aligned} \tag{12.4}$$

or after transformation

$$\begin{aligned} \ddot{y}_1 &= \alpha^2\left(y_2 - 2y_1\right), \\ \ddot{y}_2 &= \alpha^2\left(-2y_2 + y_1\right). \end{aligned} \tag{12.5}$$

Consider vibrations of a longitudinal chain of point masses (Fig. 12.2)

Let L be the number of masses (maximal number of frequencies) and i be the number of successive masses, whereas l denotes the number of successive oscillation modes.

In the case of a sequence of masses, we postulate the following form of the solution

$$y_i = X_i \cdot T_i(t), \tag{12.6}$$

$$\omega_l = 2\alpha\sin\frac{\pi l}{2(L+1)}. \tag{12.7}$$

where: $X_i = \sin\frac{\pi_i l}{L+1}$, $T_i(t) = A_i\cos\omega_l t + B_i\sin\omega_l t$.

Note that in the case from Fig. 12.1a we have $L = l = i = 1$, and by Eqs. (12.6), (12.7) we obtain: $\omega_l = 2\alpha \sin\frac{\pi}{4} = \sqrt{2}\alpha$, $y_1 = \sin\frac{\pi}{2}\,(A_1\cos\omega_1 t + B_1 \sin\omega_1 t)$.

In the case of two oscillators and the first mode ($l = 1$) we obtain

$$\begin{aligned} y_1 &= \sin\frac{\pi}{3}\,(A\cos\omega_1 t + B\sin\omega_1 t)\,, \\ y_2 &= \sin\frac{2\pi}{3}\,(C\cos\omega_1 t + D\sin\omega_1 t)\,. \end{aligned} \tag{12.8}$$

Substituting (12.8) into Eq. (12.5) we get terms *standing by* $\cos\omega_1 t$ *and* $\sin\omega_1 t$)

$$\begin{aligned} -A\sin\frac{\pi}{3}\omega_1^2 &= C\alpha^2\sin\frac{2\pi}{3} - 2A\alpha^2\sin\frac{\pi}{3}, \\ -C\sin\frac{2\pi}{3}\omega_1^2 &= A\alpha^2\sin\frac{\pi}{3} - 2C\alpha^2\sin\frac{2\pi}{3}, \\ -B\sin\frac{\pi}{3}\omega_1^2 &= D\alpha^2\sin\frac{2\pi}{3} - 2B\alpha^2\sin\frac{\pi}{3}, \\ -D\sin\frac{2\pi}{3}\omega_1^2 &= B\alpha^2\sin\frac{\pi}{3} - 2D\alpha^2\sin\frac{2\pi}{3}. \end{aligned} \tag{12.9}$$

Making use of the formula $\sin 2\alpha = 2\sin\alpha\cos\alpha$ we obtain the characteristic equation

$$\alpha^4 = 4\alpha^4 - 4\alpha^2\omega_1^2 + \omega_1^4. \tag{12.10}$$

By the frequency formula we have $\omega_1 = 2\alpha\sin\frac{\pi}{6} = \alpha$, and after the substitution $\omega_1 = \alpha$ into (12.10) we obtain an identity, which shows validity of (12.7). In the case of the second vibration mode $l = 2$ we have

$$\begin{aligned} y_1 &= \sin\frac{2\pi}{3}\,(A\cos\omega_2 t + B\sin\omega_2 t)\,, \\ y_2 &= \sin\frac{4\pi}{3}\,(C\cos\omega_2 t + D\sin\omega_2 t)\,. \end{aligned} \tag{12.11}$$

Similarly, one can show that the formula (12.2) is valid. In Fig. 12.3 X_i are depicted for the case of single and two coupled oscillators.

Going back to the chain of masses connected by springs of the same stiffness k, provided that the distances between the springs equal d, the motion of any oscillator i in the chain has the following form

$$\frac{d^2 y_i}{dt^2} = \alpha^2\,(y_{i+1} - 2y_i + y_{i-1})\,, \tag{12.12}$$

where $\alpha^2 = k/m$. It is easy to see that for $i = 1$, $L = 1$ we obtain (12.2), and for $i = 1$, $L = 2$, $i = 2$, $L = 2$ we get Eq. (12.5).

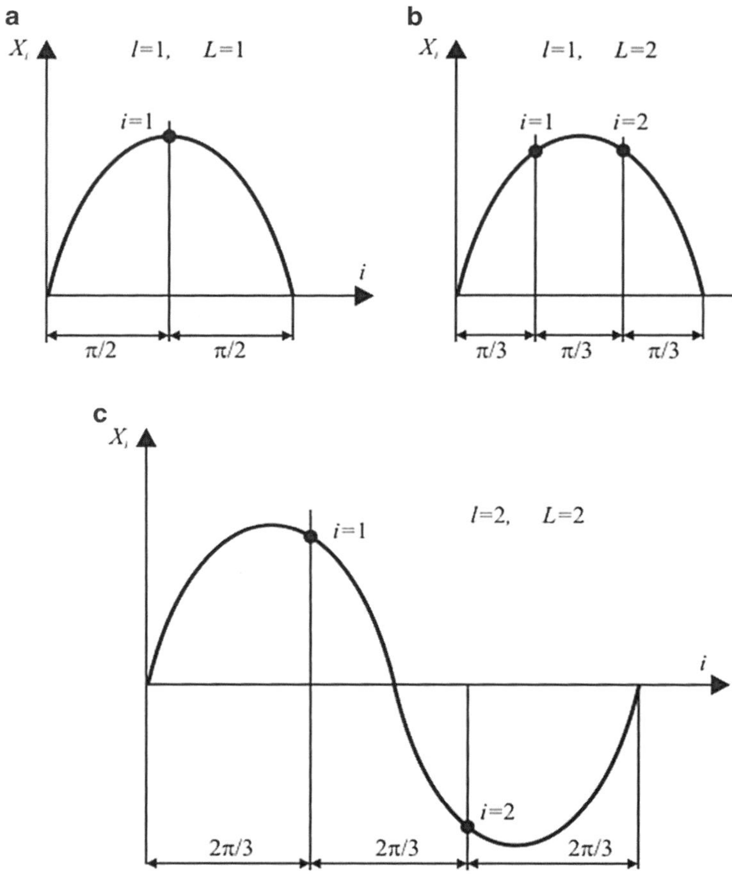

Fig. 12.3 Values of the functions X_i for the case of one (**a**) and two degrees-of-freedom for the first (**b**) and second (**c**) oscillation modes

Classical mathematical procedure of continualization relies on replacing the discrete displacement $y_i(t)$ with the continuous variable $u(y,t)$ of the form

$$y_i\,(t) = u\,(y,t)\,, \tag{12.13}$$

$$y_{i\pm1} = u\,(y \pm d, t) = \sum_{j=0}^{2J} (\pm 1)^j \,\frac{d^j}{j!}\,\frac{\partial^j u\,(y,t)}{\partial y^j} + 0\left(\left(d\,\frac{\partial u}{\partial y}\right)^{2J+1}\right), \tag{12.14}$$

where we assumed that the operator $d\frac{\partial u}{\partial y}$ is far less than 1, and this ensures convergence of the series. The obtained partial differential equation describes a wave of length significantly higher than d.

Substituting (12.13) and (12.14) into Eq. (12.12) we get

$$\frac{\partial^2 u(y,t)}{\partial t^2} - 2\alpha^2 \sum_{j=0}^{2J} \frac{d^{2j}}{(2j)!} \frac{\partial^{2j} u(y,t)}{\partial y^{2j}} = 0\left(\left(d\frac{\partial u}{\partial y}\right)^{2J+1}\right), \tag{12.15}$$

where the natural frequency of the chain masses $\alpha^2 = k/m$ and $2J$ denotes the number of terms in the series.

With regard to the assumed approximation of J we obtain different partial differential equations. For $J = 1$ by (12.15) we get

$$\frac{\partial^2 u(y,t)}{\partial t^2} - \alpha^2 d^2 \frac{\partial^2 u(y,t)}{\partial y^2} = 0\left(d^3\left(\frac{\partial u}{\partial y}\right)^3\right). \tag{12.16}$$

The obtained wave partial differential equation describes a propagation of waves of lengths, which are so large that one can neglect the structural non-homogenity of the chain. Taking into account the next term of the series we get

$$\frac{\partial^2 u(y,t)}{\partial t^2} - \alpha^2 d^2 \left(\frac{\partial^2 u(y,t)}{\partial y^2} + \frac{d^2}{12}\frac{\partial^4 u(y,t)}{\partial y^4}\right) = 0\left(d^5\left(\frac{\partial u}{\partial y}\right)^5\right). \tag{12.17}$$

The obtained partial differential equation describes propagation of waves of higher frequencies, including the internal microscale d.

Let us spend a while on physical interpretation of the obtained Eq. (12.17). Consider a harmonic wave of the form

$$u(y,t) = U \exp(i(\omega t - ky)), \tag{12.18}$$

where U is a complex amplitude of the wave, $i^2 = -1$, ω is the frequency and k is the number of the wave. Substituting (12.18) into (12.17) we obtain

$$\omega = \pm k d\alpha \sqrt{1 - \frac{d^2}{12}k^2}. \tag{12.19}$$

It follows from Eq. (12.19) that waves satisfying the inequality $k^2 > 12/d^2$ possess imaginary part of the form $\omega = \pm ikl\alpha\sqrt{\frac{l^2k^2}{12} - 1}$, which means that they grow exponentially to infinity, and it does not possess a physical interpretation (there is no extra source of energy in the system).

In [169] the appropriate procedure leading to verification of Eq. (12.17) has been proposed. The procedure allows to model a process of waves propagation.

12.3 Planar Hexagonal Net of Coupled Oscillators

Let us consider a modelling procedure of a plane hexagonal net of coupled oscillators shown in Fig. 12.4.

Equations of motion of the net will be derived on the basis of Lagrange equations of the second kind

$$\frac{d}{dt}\left(\frac{\partial T_{i,j}}{\partial \dot{q}_{i,j}}\right) + \frac{\partial V_{i,j}}{\partial q_{i,j}} = 0, \tag{12.20}$$

where $T_{i,j}$ and $V_{i,j}$ are kinetic and potential energy of the mass (i, j).

Consider the variation of length $\Delta s^{(i)}$ of the springs of stiffness k labeled by numbers $1, \ldots, 6$ (Fig. 12.4). In each step of derivation of the equations we assume that the hexagonal cells remain unchanged after the strain, namely they remain hexagonal.

For the spring 1 we have

$$\begin{gathered}\left(x^{i+2,j} + \Delta x^{i+2,j}\right) - \left(x^{i,j} + \Delta x^{i,j}\right) = s_*,\\ x^{i+2,j} - x^{i,j} = s,\end{gathered}$$

hence we get

$$\Delta s^{(1)} \equiv s_* - s = \Delta x^{i+2,j} - \Delta x^{i,j}. \tag{12.21}$$

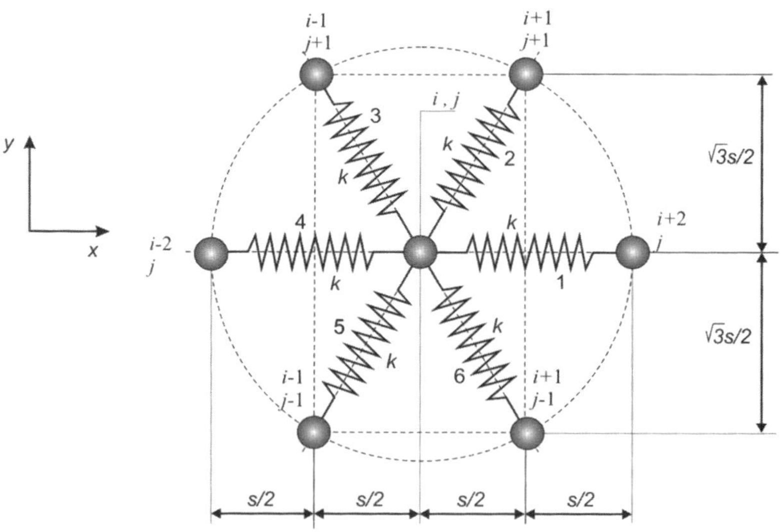

Fig. 12.4 Planar hexagonal net of oscillators

For the spring 2 we have

$$\left(x^{i+1,j+1}+\Delta x^{i+1,j+1}\right)-\left(x^{i,j}+\Delta x^{i,j}\right)=\frac{1}{2}s_*,$$

$$x^{i+1,j+1}-x^{i,j}=s^{(2)}=\frac{s}{2},$$

$$\left(y^{i+1,j+1}+\Delta y^{i+1,j+1}\right)-\left(y^{i,j}+\Delta y^{i,j}\right)=\frac{\sqrt{3}}{2}s_*,$$

$$y^{i+1,j+1}-y^{i,j}=\frac{\sqrt{3}}{2}s,$$

and hence we obtain

$$\Delta x^{i+1,j+1}-\Delta x^{i,j}=\frac{1}{2}\left(s_*-s\right)\ |\cdot\ \frac{1}{2},$$

$$\Delta y^{i+1,j+1}-\Delta y^{i,j}=\frac{\sqrt{3}}{2}\left(s_*-s\right)\ |\cdot\ \frac{\sqrt{3}}{2},$$

and finally

$$\Delta s^{(2)}=\frac{1}{2}\left(\Delta x^{i+1,j+1}-\Delta x^{i,j}\right)+\frac{\sqrt{3}}{2}\left(\Delta y^{i+1,j+1}-\Delta y^{i,j}\right). \tag{12.22}$$

For the spring 3 we have

$$-\left(x^{i-1,j+1}+\Delta x^{i-1,j+1}\right)-\left(x^{i,j}+\Delta x^{i,j}\right)=\frac{1}{2}s_*,$$

$$x^{i,j}-x^{i-1,j+1}=\frac{1}{2}s,$$

$$\left(y^{i-1,j+1}+\Delta y^{i-1,j+1}\right)-\left(y^{i,j}+\Delta y^{i,j}\right)=\frac{\sqrt{3}}{2}s_*,$$

$$-y^{i,j}+y^{i-1,j+1}=\frac{\sqrt{3}}{2}s,$$

hence

$$\Delta x^{i,j}-\Delta x^{i-1,j+1}=\tfrac{1}{2}\left(s_*-s\right)\ |\cdot\ \tfrac{1}{2},$$

$$-\Delta y^{i,j}+\Delta y^{i-1,j+1}=\tfrac{\sqrt{3}}{2}\left(s_*-s\right)\ |\cdot\ \tfrac{\sqrt{3}}{2},$$

and finally

$$\Delta s^{(3)}=\frac{1}{2}\left(\Delta x^{i,j}-\Delta x^{i-1,j+1}\right)+\frac{\sqrt{3}}{2}\left(-\Delta y^{i,j}+\Delta y^{i-1,j+1}\right). \tag{12.23}$$

For the spring 4 we have

$$-\left(x^{i-2,j}+\Delta x^{i-2,j}\right)+\left(x^{i,j}+\Delta x^{i,j}\right)=s_*,$$
$$x^{i,j}-x^{i-2,j}=s,$$

hence

$$\Delta s^{(4)}=\Delta x^{i,j}-\Delta x^{i-2,j}. \tag{12.24}$$

For the spring 5 we have

$$\left(x^{i,j}+\Delta x^{i,j}\right)-\left(x^{i-1,j-1}+\Delta x^{i-1,j-1}\right)=\frac{1}{2}s_*,$$
$$x^{i,j}-x^{i-1,j-1}=\frac{1}{2}s,$$
$$\left(y^{i,j}+\Delta y^{i,j}\right)-\left(y^{i-1,j-1}+\Delta y^{i-1,j-1}\right)=\frac{\sqrt{3}}{2}s_*,$$
$$y^{i,j}-y^{i-1,j-1}=\frac{\sqrt{3}}{2}s,$$

hence

$$\Delta x^{i,j}-\Delta x^{i-1,j-1}=\frac{1}{2}\left(s_*-s\right)\;|\cdot\frac{1}{2},$$
$$\Delta y^{i,j}-\Delta y^{i-1,j-1}=\frac{\sqrt{3}}{2}\left(s_*-s\right)\;|\cdot\frac{\sqrt{3}}{2},$$

and finally

$$\Delta s^{(5)}=\frac{1}{2}\left(\Delta x^{i,j}-\Delta x^{i-1,j-1}\right)+\frac{\sqrt{3}}{2}\left(\Delta y^{i,j}-\Delta y^{i-1,j-1}\right). \tag{12.25}$$

For the spring 6 we have

$$\left(x^{i+1,j-1}+\Delta x^{i+1,j-1}\right)-\left(x^{i,j}+\Delta x^{i,j}\right)=\frac{1}{2}s_*,$$
$$x^{i+1,j-1}-x^{i,j}=\frac{1}{2}s,$$
$$\left(y^{i,j}+\Delta y^{i,j}\right)-\left(y^{i+1,j-1}+\Delta y^{i+1,j-1}\right)=\frac{\sqrt{3}}{2}s_*,$$
$$y^{i,j}-y^{i+1,j-1}=\frac{\sqrt{3}}{2}s,$$

hence

$$\Delta x^{i+1,j-1} - \Delta x^{i,j} = \tfrac{1}{2}\left(s_* - s\right) \,|\cdot\, \tfrac{1}{2},$$
$$\Delta y^{i,j} - \Delta y^{i+1,j-1} = \tfrac{\sqrt{3}}{2}\left(s_* - s\right) \,|\cdot\, \tfrac{\sqrt{3}}{2},$$

and finally

$$\Delta s^{(6)} = \frac{1}{2}\left(\Delta x^{i+1,j-1} - \Delta x^{i,j}\right) + \frac{\sqrt{3}}{2}\left(\Delta y^{i,j} - \Delta y^{i+1,j-1}\right). \tag{12.26}$$

The kinetic and potential energy of the mass (i, j) read

$$\begin{aligned} T_{i,j} &= \frac{1}{2}m\left(\Delta \dot{x}^{i,j}\right)^2, \\ V_{i,j} &= \frac{1}{2}\sum_{i=1}^{6}\left(\Delta s^{(i)}\right)^2 k, \end{aligned} \tag{12.27}$$

and to the Lagrange equations (12.20) we need

$$\frac{\partial V_{i,j}}{\partial\left(\Delta x^{i,j}\right)} = k\left(\sum_{i=1}^{6}\Delta s^{(i)}\frac{\partial \Delta s^{(i)}}{\partial\left(\Delta x^{i,j}\right)}\right) = k\Bigg\{-\left(\Delta x^{i+2,j} - \Delta x^{i,j}\right)$$
$$+ -\frac{1}{4}\left(\Delta x^{i+1,j+1} - \Delta x^{i,j}\right) + -\frac{\sqrt{3}}{4}\left(\Delta y^{i+1,j+1} - \Delta y^{i,j}\right)$$
$$+\frac{1}{4}\left(\Delta x^{i,j} - \Delta x^{i-1,j+1}\right)$$
$$+\frac{\sqrt{3}}{4}\left(\Delta y^{i-1,j+1} - \Delta y^{i,j}\right) + \left(\Delta x^{i,j} - \Delta x^{i-1,j}\right) + \frac{1}{4}\left(\Delta x^{i,j} - \Delta x^{i-1,j-1}\right)$$
$$+\frac{\sqrt{3}}{4}\left(\Delta y^{i,j} - \Delta y^{i-1,j-1}\right) + \frac{1}{4}\left(\Delta x^{i+1,j-1} - \Delta x^{i,j}\right) - \frac{\sqrt{3}}{4}\left(\Delta y^{i,j} - \Delta y^{i+1,j-1}\right)\Bigg\}$$
$$= k\Bigg\{3\Delta x^{i,j} - \Delta x^{i+1,j} - \Delta x^{i-1,j} - \frac{1}{2}\Delta x^{i+1,j+1} - \frac{1}{2}\Delta x^{i-1,j+1} - \frac{1}{2}\Delta x^{i-1,j-1}$$
$$+\frac{1}{2}\Delta x^{i+1,j-1} - \frac{\sqrt{3}}{2}\Delta y^{i+1,j-1} + \frac{\sqrt{3}}{2}\Delta y^{i-1,j+1} - \frac{\sqrt{3}}{2}\Delta y^{i+1,j+1}\Bigg\}. \tag{12.28}$$

$$
\begin{aligned}
\frac{\partial V_{i,j}}{\partial\left(\Delta y^{i,j}\right)} &= k\left(\sum_{i=1}^{6}\Delta s^{(i)}\frac{\partial \Delta s^{(i)}}{\partial\left(\Delta y^{i,j}\right)}\right)\\
&= k\left\{\frac{\sqrt{3}}{2}\left[-\frac{1}{2}\left(\Delta x^{i+1,j+1}-\Delta x^{i,j}\right)-\frac{\sqrt{3}}{2}\left(\Delta y^{i+1,j+1}-\Delta y^{i,j}\right)\right]\right.\\
&\quad +-\frac{\sqrt{3}}{2}\left[\frac{1}{2}\left(\Delta x^{i,j}-\Delta x^{i-1,j+1}\right)-\frac{\sqrt{3}}{2}\left(\Delta y^{i-1,j+1}-\Delta y^{i,j}\right)\right]\\
&\quad +\frac{\sqrt{3}}{2}\left[\frac{1}{2}\left(\Delta x^{i,j}-\Delta x^{i-1,j-1}\right)+\frac{\sqrt{3}}{2}\left(\Delta y^{i,j}-\Delta y^{i-1,j-1}\right)\right]\\
&\quad \left.+\frac{\sqrt{3}}{2}\left[\frac{1}{2}\left(\Delta x^{i+1,j-1}-\Delta x^{i,j}\right)+\frac{\sqrt{3}}{2}\left(\Delta y^{i,j}-\Delta y^{i+1,j-1}\right)\right]\right\}\\
&= k\left\{\Delta x^{i,j}\left(\frac{\sqrt{3}}{4}-\frac{\sqrt{3}}{4}+\frac{\sqrt{3}}{4}-\frac{\sqrt{3}}{4}\right)-\frac{\sqrt{3}}{4}\Delta x^{i+1,j+1}\right.\\
&\quad +\frac{\sqrt{3}}{4}\Delta x^{i+1,j-1}+\frac{\sqrt{3}}{4}\Delta x^{i-1,j+1}-\frac{\sqrt{3}}{4}\Delta x^{i-1,j-1}\\
&\quad +\Delta y^{i,j}\left(\frac{3}{4}+\frac{3}{4}+\frac{3}{4}+\frac{3}{4}\right)\\
&\quad \left.-\frac{3}{4}\Delta y^{i+1,j+1}+-\frac{3}{4}\Delta y^{i-1,j+1}-\frac{3}{4}\Delta y^{i-1,j-1}-\frac{3}{4}\Delta y^{i+1,j-1}\right\}\\
&= \frac{k}{4}\left(-\sqrt{3}\Delta x^{i+1,j+1}+\sqrt{3}\Delta x^{i+1,j-1}+\sqrt{3}\Delta x^{i-1,j+1}\right.\\
&\quad -\sqrt{3}\Delta x^{i-1,j-1}+\\
&\quad \left.+12\Delta y^{i,j}-3\Delta y^{i+1,j+1}-3\Delta y^{i-1,j+1}-3\Delta y^{i-1,j-1}-\Delta y^{i+1,j-1}\right)\\
&= \frac{\sqrt{3}}{4}k\,(-\Delta x^{i+1,j+1}+\Delta x^{i+1,j-1}+\Delta x^{i-1,j+1}-\Delta x^{i-1,j-1}\\
&\quad +4\sqrt{3}\Delta y^{i,j}\\
&\quad +-\sqrt{3}\Delta y^{i+1,j+1}-\sqrt{3}\Delta y^{i-1,j+1}-\sqrt{3}\Delta y^{i-1,j-1}\\
&\quad -\sqrt{3}\Delta y^{i+1,j-1})\\
&= \frac{k}{4}\{12\Delta x^{i,j}-4\Delta x^{i+1,j}-4\Delta x^{i-1,j}-\Delta x^{i+1,j+1}
\end{aligned}
$$

$$
\begin{aligned}
&-\Delta x^{i-1,j+1} - \Delta x^{i-1,j-1} \\
&+-\Delta x^{i+1,j-1} + \sqrt{3}\Delta y^{i+1,j-1} + \sqrt{3}\Delta y^{i-1,j+1} \\
&-\sqrt{3}\Delta y^{i+1,j+1} - \sqrt{3}\Delta y^{i-1,j-1}\Big\}.
\end{aligned} \tag{12.29}
$$

Finally, the equation of motion takes the form

$$
\begin{aligned}
m\Delta \ddot{x}^{i,j} &= \frac{k}{4}\left(12\Delta x^{i,j} - 4\Delta x^{i+1,j} - 4\Delta x^{i-1,j} - \Delta x^{i+1,j+1} - \Delta x^{i-1,j+1} - \Delta x^{i-1,j-1}\right. \\
&\left. +-\Delta x^{i+1,j-1} + \sqrt{3}\Delta y^{i+1,j-1} + \sqrt{3}\Delta y^{i-1,j+1} - \sqrt{3}\Delta y^{i+1,j+1} - \sqrt{3}\Delta y^{i-1,j-1}\right); \\
m\Delta \ddot{y}^{i,j} &= \frac{\sqrt{3}}{4}k\left(-\Delta x^{i+1,j+1} + \Delta x^{i+1,j-1} + \Delta x^{i-1,j+1} - \Delta x^{i-1,j-1} + 4\sqrt{3}\Delta y^{i,j}\right. \\
&\left. +-\sqrt{3}\Delta y^{i+1,j+1} - \sqrt{3}\Delta y^{i-1,j+1} - \sqrt{3}\Delta y^{i-1,j-1} - \sqrt{3}\Delta y^{i+1,j-1}\right).
\end{aligned} \tag{12.30}
$$

Assuming that discrete displacements of mass (i, j) equal to the values of continuous displacements, i.e. $\Delta x^{i,j} = u$, $\Delta y^{i,j} = v$, the positions of masses surrounding the masses (i, j) are determined by means of expansion. For displacements in the x direction, we have

$$
\begin{aligned}
\Delta x^{(i\pm 1,j\pm 1)} &= u\left(\Delta x \pm \frac{1}{2}s, \Delta y \pm \frac{\sqrt{3}}{2}s\right) \\
&= u(\Delta x, \Delta y) + \left(\pm\frac{1}{2}s\right)\frac{\partial u}{\partial(\Delta x)}(\Delta x, \Delta y) + \left(\pm\frac{\sqrt{3}}{2}s\right)\frac{\partial u}{\partial(\Delta y)}(\Delta x, \Delta y) \\
&+\frac{1}{2!}\left(\pm\frac{1}{2}s\right)^2\frac{\partial^2 u}{\partial(\Delta x)^2}(\Delta x, \Delta y) + \frac{1}{2!}\left(\pm\frac{\sqrt{3}}{2}s\right)^2\frac{\partial^2 u}{\partial(\Delta y)^2}(\Delta x, \Delta y) \\
&+\left(\pm\frac{1}{2}s\right)\left(\pm\frac{\sqrt{3}}{2}s\right)\frac{\partial^2 u}{\partial(\Delta x)\partial(\Delta y)} + 0\left(s^3\right),
\end{aligned} \tag{12.31}
$$

$$
\Delta x^{(i\pm 1,j)} = u(\Delta x \pm s, \Delta y) = u(\Delta x, \Delta y) + (\pm s)\frac{\partial u}{\partial(\Delta x)} + \frac{1}{2!}(\pm s)^2\frac{\partial^2 u}{\partial(\Delta x)^2} + \cdots
$$

For displacements in the y direction we obtain

$$\Delta y^{(i\pm 1,j\pm 1)} = v\left(\Delta x \pm \frac{1}{2}s, \Delta y \pm \frac{\sqrt{3}}{2}s\right)$$

$$= v(\Delta x, \Delta y) + \left(\pm\frac{1}{2}s\right)\frac{\partial v}{\partial(\Delta x)} + \left(\pm\frac{\sqrt{3}}{2}s\right)\frac{\partial v}{\partial(\Delta y)} + \frac{1}{2!}\left(\frac{1}{2}s\right)^2\frac{\partial^2 v}{\partial(\Delta x)^2}$$

$$+\frac{1}{2!}\left(\pm\frac{\sqrt{3}}{2}s\right)^2\frac{\partial^2 v}{\partial(\Delta y)^2} + \left(\pm\frac{1}{2}s\right)\left(\pm\frac{\sqrt{3}}{2}s\right)\frac{\partial^2 v}{\partial(\Delta x)\,\partial(\Delta y)} + 0\left(s^3\right), \tag{12.32}$$

$$\Delta y^{(i\pm 1,j)} = v(\Delta x \pm s, \Delta y) = v(\Delta x, \Delta y) + (\pm s)\frac{\partial v}{\partial(\Delta y)}$$

$$+\frac{1}{2!}(\pm s)^2\frac{\partial^2 v}{\partial(\Delta y)^2} + 0\left(s^3\right).$$

Substituting (12.31), (12.32) into (12.30) we get

$$Fs\rho\frac{d^2u}{dt^2} = -\frac{3ks^2}{8}\left[3\frac{\partial^2 u}{\partial x^2} + \frac{\partial^2 u}{\partial y^2} + 2\frac{\partial^2 u}{\partial x \partial y}\right], \tag{12.33}$$

$$\rho\frac{d^2u}{dt^2} = -\frac{3}{8}E\left[3\frac{\partial^2 u}{\partial x^2} + \frac{\partial^2 u}{\partial y^2} + 2\frac{\partial^2 u}{\partial x \partial y}\right], \tag{12.34}$$

where $Fs\rho = m$ and $E = ksF^{-1}$.

Similarly we get

$$\rho\frac{d^2v}{dt^2} = -\frac{3}{8}E\left[3\frac{\partial^2 v}{\partial x^2} + \frac{\partial^2 v}{\partial y^2} + 2\frac{\partial^2 v}{\partial x \partial y}\right]. \tag{12.35}$$

More details on modelling by means of continualization, its defects, virtues and reliability can be found in [169, 170, 223].

12.4 Discretization

Models of mechanical systems such as rods, beam, plates or shells are made on the basis of partial differential equations and various types of theories (e.g. von Kármán, Donnell, Kirchhoff, Love or Mindlin theory, etc.). These problems are widely discussed in the literature. The mentioned elements of construction play an important role in civil engineering, aircraft and car industry, etc. On the other hand, it is known that during the modelling procedure one needs to introduce a number of simplifications but in such a way that preserves basic features of the analysed object. Basing oneself only on the analysis of shell constructions, one needs to emphasize that the constructions can be various, e.g. single and multi-layers, reinforced with ribs, of composite type or made of isotropic or anisotropic materials. Moreover, they can work not only in regions of Hooke's law validity but also in regions of plastic strains.

It is easy to come to conclusions that, as a result of modelling, one can obtain very complicated systems of equations, which are extremely difficult or even impossible to solve.

The goal of this section is to show how in the discretization procedure, understood as transition from partial differential equations, which describe nonlinear vibrations of cylindrical shells, to ordinary differential equations, the obtained model of the latter is not standard and fundamental questions arise concerning the way of its solution.

We analyse vibrations of a nonlinear shell with imperfections, governed by the so-called Donnell equations of the form

$$\frac{D}{h}\nabla^4 w_1 = L\left(w_1 + w_0, \Phi\right) + \frac{1}{R}\frac{\partial^2 \Phi}{\partial x^2} - \rho\frac{\partial^2 w_1}{\partial t^2}, \tag{12.36}$$

$$\frac{1}{E}\nabla^4 \Phi = -\frac{1}{2}L\left(w_1 + 2w_0, w_1\right) - \frac{1}{R}\frac{\partial^2 w_1}{\partial x^2}, \tag{12.37}$$

$$L(A, B) = \frac{\partial^2 A}{\partial x^2}\frac{\partial^2 B}{\partial y^2} + \frac{\partial^2 A}{\partial y^2}\frac{\partial^2 B}{\partial x^2} - 2\frac{\partial^2 A}{\partial x \partial y}\frac{\partial^2 B}{\partial x \partial y}, \quad D = \frac{Eh^3}{12(1-\nu^2)},$$

where E is the Young's modulus; ν is the Poisson coefficient; ρ is the density of shell; R, h are radius and thickness of the shell, respectively; x and y are longitudinal and circular coordinates; w_1 is a dynamic deflection; Θ is the Airy function; $w_0(x, y)$ are initial imperfections.

In 1974 Evensen [86] proposed the following solution

$$w_1 = f_1(t)\cos sy \sin rx + f_2(t)\sin sy \sin rx + f_3(t)\sin^2 rx, \tag{12.38}$$

where: $s = n/R$; $r = m\pi/l$; n is the number of waves in the circular direction, and m denotes the number of halfwaves in longitudinal direction; l is length of the shell; $f_1(t)$, $f_2(t)$, $f_3(t)$ are generalized coordinates.

Evensen assumed that the length of the central surface of the shell is constant during the vibrations, i.e.

$$\int_0^{2\pi R} \varepsilon_{22} dy = \int_0^{2\pi R} \left[\frac{\partial v}{\partial y} - \frac{w}{R} + \frac{R}{2}\left(\frac{\partial w}{\partial y}\right)^2\right] dy = 0, \tag{12.39}$$

which leads to the relationship

$$f_3 = \frac{(f_1^2 + f_2^2)n^2}{4R}. \tag{12.40}$$

The imperfections have the following form

$$w_0 = f_{10} \cos sy \sin rx + f_{20} \sin sy \sin rx, \tag{12.41}$$

where f_{10}, f_{20} are constants.

First, we substitute the expressions (12.38) and (12.41) into (12.36). One can show that a solution of Eq. (12.36) has the form

$$\begin{aligned} \Phi = \Phi_0 \cos 2rx + \Phi_1 \cos sy \sin rx + \Phi_2 \sin sy \sin rx + \Phi_3 \cos 2sy \\ + \Phi_4 \sin 2sy + \Phi_5 \cos sy \sin 3rx + \Phi_6 \sin sy \sin 3rx, \end{aligned} \tag{12.42}$$

where:

$$\begin{gathered} \Phi_0 = -\frac{Es^2}{32r^2}\left(f_{10}^2 + f_{20}^2\right), \\ \Phi_1 = \frac{E}{R}\left[\frac{r^2}{(r^2+s^2)^2}\left(f_1 - f_{10}\right) - \frac{s^2r^2n^2}{4(r^2+s^2)^2} f_1\left(f_1^2 + f_2^2\right)\right], \\ \Phi_2 = \frac{E}{R}\left[\frac{r^2}{(r^2+s^2)^2}\left(f_2 - f_{20}\right) - \frac{s^2r^2n^2}{4(r^2+s^2)^2} f_2\left(f_1^2 + f_2^2\right)\right], \\ \Phi_3 = \frac{Er^2}{32s^2}\left(f_2^2 - f_1^2 + f_{10}^2 - f_{20}^2\right), \quad \Phi_4 = \frac{Er^2}{16s^2}\left(f_{10}f_{20} - f_1 f_2\right), \\ \Phi_5 = \frac{Es^2r^2n^2}{4R(s^2+9r^2)^2} f_1\left(f_1^2 + f_2^2\right), \quad \Phi_6 = \frac{Es^2r^2n^2}{4R(s^2+9r^2)^2} f_2\left(f_1^2 + f_2^2\right). \end{gathered} \tag{12.43}$$

Below, we give ordinary differential equations obtained from Eq. (12.36) after application of the Galerkin method (c.f. [25]):

$$\begin{aligned} &\ddot{f}_1 + \omega_1^2 f_1 + \gamma f_2 + 2\chi f_1\left(\dot{f}_1^2 + f_1\ddot{f}_1 + \dot{f}_2^2 + f_2\ddot{f}_2\right) + \gamma_1 f_1\left(f_1^2 + f_2^2\right) \\ &\quad + g f_1\left(f_1^2 + f_2^2\right)^2 + \alpha_1 f_1 f_2 + \alpha_2 f_1^2 + \alpha_3 f_2^2 = \omega_0^2 f_{10}, \\ &\ddot{f}_2 + \omega_2^2 f_2 + \gamma f_1 + 2\chi f_2\left(\dot{f}_1^2 + f_1\ddot{f}_1 + \dot{f}_2^2 + f_2\ddot{f}_2\right) + \gamma_1 f_2\left(f_1^2 + f_2^2\right) \\ &\quad + g f_2\left(f_1^2 + f_2^2\right)^2 + \beta_1 f_1^2 + \beta_2 f_2^2 + \beta_3 f_1 f_2 = \omega_0^2 f_{20}, \end{aligned} \tag{12.44}$$

where:

$$2\chi = \frac{3}{2}\left(\frac{n^2}{2R}\right)^2, \quad \gamma = -\frac{Er^4}{8\rho} f_{10} f_{20}, \quad \omega_1^2 = \omega_0^2 - \frac{Er^4}{16\rho}\left(f_{10}^2 - f_{20}^2\right),$$

$$\omega_2^2 = \omega_0^2 + \frac{Er^4}{16\rho}\left(f_{10}^2 - f_{20}^2\right), \quad \omega_0^2 = \frac{1}{\rho}\left[\frac{D}{h}\left(s^2 + r^2\right)^2 + \frac{Er^4}{R^2\left(s^2 + r^2\right)^2}\right],$$

$$\gamma_1 = \frac{1}{\rho}\left[\frac{Er^4}{16} + \frac{D^2 r^4 n^4}{hR^2} - \frac{Er^4 s^4}{\left(s^2 + r^2\right)^2}\right],$$

$$g = \frac{3En^2 r^4 s^6}{16\rho}\left[\frac{1}{\left(s^2 + r^2\right)^2} + \frac{1}{\left(s^2 + 9r^2\right)^2}\right],$$

$$\alpha_1 = \frac{Er^4 s^4 f_{20}}{2\rho\left(s^2 + r^2\right)^2}, \quad \alpha_2 = \frac{3Er^4 s^4 f_{10}}{4\rho\left(s^2 + r^2\right)^2}, \quad \alpha_3 = \frac{Er^4 s^4 f_{10}}{4\rho\left(s^2 + r^2\right)^2},$$

$$\beta_1 = \frac{Er^4 s^4 f_{20}}{4\rho\left(s^2 + r^2\right)^2}, \quad \beta_2 = \frac{3Er^4 s^4 f_{20}}{4\rho\left(s^2 + r^2\right)^2}, \quad \beta_3 = \frac{Er^4 s^4 f_{10}}{2\rho\left(s^2 + r^2\right)^2}. \tag{12.45}$$

By (12.44) it is easy to see that a system of ordinary differential equations cannot be transformed into the standard form. Even its numerical attempt to solve it raises difficulties. A detailed description of derivation of the equations is given in [141].

By estimating the order of magnitude of some physical parameters and after introduction the small parameter $\varepsilon \ll 1$ and the following non-dimensional quantities

$$t_* = \omega_0 t, \quad x_* = f_1 h^{-1}, \quad y_* = f_2 h^{-1}, \quad \omega_1^* = \omega_1 \omega_0^{-1}, \quad \omega_2^* = \omega_2 \omega_0^{-1},$$

$$\varepsilon\gamma^* = \gamma\omega_0^{-2}, \quad \chi^* = \chi h^2, \quad \gamma_1^* = \gamma_1 \omega_0^{-2} h^2, \quad g^* = g\omega_0^{-2} h^4, \quad \varepsilon\alpha_1^* = \alpha_1 \omega_0^{-2} h,$$

$$\varepsilon\alpha_2^* = \alpha_2 \omega_0^{-2}, \quad \varepsilon\alpha_3^* = \alpha_3 \omega_0^{-2}, \quad f_{10}^* = f_{10} h^{-1}, \quad f_{20}^* = f_{20} h^{-1}, \quad \varepsilon\beta_1^* = \beta_1 \omega_0^{-2} h,$$

$$\varepsilon\beta_2^* = \beta_2 \omega_0^{-2} h, \quad \varepsilon\beta_3^* = \beta_3 \omega_0^{-2} h, \tag{12.46}$$

Equation (12.44) can be cast in the form ("*" over the non-dimensional symbols are ignored)

$$\begin{aligned}
&\ddot{x} + \omega_1^2 x + \varepsilon^2 \gamma y + 2\chi x\left(\dot{x}^2 + x\ddot{x} + \dot{y}^2 + y\ddot{y}\right) + \gamma_1 x\left(x^2 + y^2\right) \\
&\quad + gx\left(x^2 + y^2\right)^2 + \varepsilon\alpha_1 xy + \varepsilon\alpha_2 x^2 + \varepsilon\alpha_3 y^2 = f_{10}, \\
&\ddot{y} + \omega_2^2 y + \varepsilon^2 \gamma x + 2\chi y\left(\dot{y}^2 + y\ddot{y} + \dot{x}^2 + x\ddot{x}\right) + \gamma_1 y\left(x^2 + y^2\right) \\
&\quad + gy\left(x^2 + y^2\right)^2 + \varepsilon\beta_1 x^2 + \varepsilon\beta_2 y^2 + \varepsilon\beta_3 xy = f_{20}.
\end{aligned} \tag{12.47}$$

Introducing a small parameter opens a way to applications of perturbative methods in solving Eq. (12.47).

12.5 Modelling of Two-Dimensional Anisotropic Structures

In this section, we will show how a single amplitude equation can explain many nonlinear phenomena, experimentally observed in various branches of science [194]. In the case of so-called Reyleigh–Bernard fluid convection one experimentally observed the occurrence of both isotropic and anisotropic quasi-two-dimensional structures as well as so-called perpendicular and skew waves. In the case of examining some crystalline liquids one observed that convective instability leads to appearing normal, rectangular and chaotic vortices. These phenomena can be explained by a single amplitude equation.

Besides the mentioned examples, the equation under consideration can model a phenomenon of deflection of anisotropic, susceptible plate immersed in nonlinear medium of Duffing stiffness (see [69, 74, 121, 158, 194]).

If we consider transversal vibrations of the mentioned plate in the coordinate system x,y, then its equation of motion takes the form

$$m\ddot{u}+h\dot{u}+\gamma u^3 = -\left[\lambda_1\partial_x^4 + \lambda_2\partial_y^4 + 2\lambda_3\partial_x^2\partial_y^2 + \mu_1\partial_x^2 + \mu_2\partial_y^2 + \chi\right]u, \tag{12.48}$$

where: $\lambda_1, \lambda_2, \lambda_3$ are parameters (if $\lambda_1 = \lambda_2 = \lambda_3$, the problem is isotropic, if $\lambda_1 = \lambda_2 \neq \lambda_3$ then the system is symmetric); μ_1, μ_2 characterize the plate load; and γ i χ are stiffness coefficients of the medium; h is a viscous damping coefficient of the medium, and m is a mass of the plate. Let us rescale the above equation by introducing new variables

$$x = X\sqrt{\frac{2\lambda_1}{\mu_1}}, \qquad y = Y\frac{\sqrt{2\mu_2\lambda_1}}{\mu_1}. \tag{12.49}$$

Simple transformations lead to the relationships

$$\begin{aligned} \partial_x^2 &= \frac{\mu_1}{2\lambda_1}\partial_X^2, \qquad \partial_y^2 = \frac{\mu_1^2}{2\lambda_1\mu_2}\partial_Y^2, \\ \partial_x^4 &= \frac{\mu_1^2}{4\lambda_1^2}\partial_X^4, \qquad \partial_y^4 = \frac{\mu_1^4}{4\lambda_1^2\mu_2^2}\partial_Y^4, \end{aligned} \tag{12.50}$$

and substituting (12.49) into (12.48) we get

$$\begin{aligned} &\frac{4\lambda_1 m}{\mu_1^2}\ddot{u} + \frac{4\lambda_1 h}{\mu_1^2}\dot{u} + \frac{4\lambda_1\gamma}{\mu_1^2}u^3 \\ &= -\left[\partial_X^4 + \frac{\lambda_2\mu_1^2}{\lambda_1\mu_2^2}\partial_Y^4 + 2\frac{\lambda_3\mu_1}{\lambda_1\mu_2}\partial_X^2\partial_Y^2 + 2\partial_X^2 + 2\partial_Y^2 + \frac{4\lambda_1\chi}{\mu_1^2}\right]u. \end{aligned} \tag{12.51}$$

Introducing the operator $\nabla^2 = \partial_X^2 + \partial_Y^2$, we have

$$\left(1+\nabla^2\right)^2 = 1 + \partial_X^4 + \partial_Y^4 + 2\partial_X^2\partial_Y^2 + 2\partial_X^2\partial_Y^2 + 2\partial_X^2 + 2\partial_Y^2,$$

and the right-hand side P of Eq. (12.48), taking into account (12.50), takes the form

$$P = -\left\{\left(1+\nabla^2\right)^2 - 1 - \partial_Y^4 - 2\partial_X^2\partial_Y^2 + \frac{\lambda_2\mu_1^2}{\lambda_1\mu_2^2}\partial_Y^4 + \frac{2\lambda_3\mu_1}{\lambda_1\mu_2}\partial_X^2\partial_Y^2 + \frac{4\lambda_1\chi}{\mu_1^2}\right\}u,$$

and after transformation

$$P=\left\{-\left(1+\nabla^2\right)^2 - \left(\frac{\lambda_2\mu_1^2}{\lambda_1\mu_2^2}-1\right)\partial_Y^4 - 2\left(\frac{\lambda_3\mu_1}{\lambda_1\mu_2}-1\right)\partial_X^2\partial_Y^2 + \left(1-\frac{4\lambda_1\chi}{\mu_1^2}\right)\right\}u.$$

Introducing the following parameters we have

$$c = \frac{\lambda_2\mu_1^2}{\lambda_1\mu_2^2} - 1, \quad \eta = \frac{\lambda_3\mu_1}{\lambda_1\mu_2} - 1, \quad r = 1 - \frac{4\lambda_1\chi}{\mu_1^2},$$

hence

$$\frac{4\lambda_1 m}{\mu_1^2}\ddot{u} + \frac{4\lambda_1 h}{\mu_1^2}\dot{u} + \frac{4\lambda_1\gamma}{\mu_1^2}u^3 = \left[-\left(1+\nabla^2\right)^2 - c\partial_Y^2 - 2\eta\partial_X^2\partial_Y^2 + r\right]u.$$

The next step is to rescale the time t and the variable u according to the formulas

$$\tau = \frac{\chi}{h}t, \quad u = \sqrt{\frac{\chi}{\gamma}}v.$$

The analysed equation takes the form

$$\frac{4\lambda_1}{\mu_1^2}\left\{m\left(\frac{\chi^2}{h^2}\right)\sqrt{\frac{\chi}{\gamma}}v''+\chi\sqrt{\frac{\chi}{\gamma}}v'+\chi\sqrt{\frac{\chi}{\gamma}}v^3\right\} = \left\{-\left(1+\nabla^2\right)^2 - c\partial_Y^2 - 2\eta\partial_X^2\partial_Y^2 + r\right\}v.$$

Transforming the above equation we get

$$\frac{m\chi}{h^2}v'' + v' + v^3 = \left\{-\left(1+\nabla^2\right)^2 - c\partial_Y^2 - 2\eta\partial_X^2\partial_Y^2 + r\right\}v,$$

where one assumed $4\lambda_1\chi\mu_1^{-2} = 1$.

For large value of the coefficient h, the inertial term can be neglected and finally the equation takes the form

$$v' = \left\{-\left(1+\nabla^2\right)^2 - c\partial_y^2 - 2\eta\partial_x^2\partial_y^2 + r\right\}v - v^3. \tag{12.52}$$

For the sake of simplicity we go back to the variables x and y. Let us rescale the variables

$$X = \frac{1}{2}\varepsilon^{\frac{1}{2}}x, \quad Y = \frac{1}{2}\varepsilon^{\frac{1}{2}}y, \quad T = \varepsilon\tau, \tag{12.53}$$

where ε is a perturbation (small) parameter, and make the following substitution

$$\begin{aligned} \partial x &\to \partial x + \frac{1}{2}\varepsilon^{\frac{1}{2}}\partial_X, \\ \partial y &\to \partial y + \frac{1}{2}\varepsilon^{\frac{1}{2}}\partial_Y. \end{aligned} \tag{12.54}$$

By Eq. (12.54) we obtain

$$\begin{aligned}
\partial_x^2 &\to \partial_x^2 + \varepsilon^{\frac{1}{2}}\partial_x\partial_X + \frac{1}{4}\varepsilon\partial_X^2, \quad \partial_y^2 \to \partial_y^2 + \varepsilon^{\frac{1}{2}}\partial_y\partial_Y + \frac{1}{4}\varepsilon\partial_Y^2, \\
\partial_x^2\partial_y^2 &\to \partial_x^2\partial_y^2 + \varepsilon^{\frac{1}{2}}\partial_x\partial_y^2\partial_X + \frac{1}{4}\varepsilon\partial_X^2\partial_y^2 + \varepsilon^{\frac{1}{2}}\partial_y\partial_Y\partial_x^2 + \varepsilon\partial_x\partial_y\partial_X\partial_Y \\
&\quad + \frac{1}{4}\varepsilon^{\frac{3}{2}}\partial_y\partial_X^2\partial_Y + \frac{1}{4}\varepsilon\partial_x^2\partial_Y^2 + \frac{1}{4}\varepsilon^{\frac{3}{2}}\partial_x\partial_X\partial_Y^2 + \frac{1}{16}\varepsilon^2\partial_X^2\partial_Y^2, \\
\partial_y^4 &\to \partial_y^4 + \frac{3}{2}\partial_y^2\partial_Y^2 + \frac{1}{16}\varepsilon^2\partial_Y^2 + 2\partial_y^2\varepsilon^{\frac{1}{2}}\partial_y\partial_Y + \frac{1}{2}\varepsilon^{\frac{3}{2}}\partial_y\partial_Y^3, \\
1+\nabla^2 &= 1+\partial_x^2+\partial_y^2 \to \left(1+\partial_x^2+\partial_y^2\right)+\left[\varepsilon^{\frac{1}{2}}\left(\partial_x\partial_X+\partial_y\partial_Y\right)+\frac{1}{4}\varepsilon\left(\partial_X^2+\partial_Y^2\right)\right], \\
\left(1+\nabla^2\right)^2 &= \left(1+\partial_x^2+\partial_y^2\right)^2 + 2\left(1+\partial_x^2+\partial_y^2\right)\left[\varepsilon^{\frac{1}{2}}\left(\partial_x\partial_X+\partial_y\partial_Y\right)\right. \\
&\quad \left.+\frac{1}{4}\varepsilon\left(\partial_X^2+\partial_Y^2\right)\right] + \left[\varepsilon^{\frac{1}{2}}\left(\partial_x\partial_X+\partial_y\partial_Y\right)+\frac{1}{4}\varepsilon\left(\partial_X^2+\partial_Y^2\right)\right]^2 \\
&= \left(1+\partial_x^2+\partial_y^2\right)^2 + 2\varepsilon^{\frac{1}{2}}\left(1+\partial_x^2+\partial_y^2\right)\left(\partial_x\partial_X+\partial_y\partial_Y\right) \\
&\quad + \frac{1}{2}\varepsilon\left(1+\partial_x^2+\partial_y^2\right)\left(\partial_X^2+\partial_Y^2\right) + \varepsilon\left(\partial_x\partial_X+\partial_y\partial_Y\right)^2 \\
&\quad + \frac{1}{2}\varepsilon^{\frac{3}{2}}\left(\partial_x\partial_X+\partial_y\partial_Y\right)\left(\partial_X^2+\partial_Y^2\right) + \frac{1}{16}\varepsilon^2\left(\partial_X^2+\partial_Y^2\right)^2.
\end{aligned} \tag{12.55}$$

Solutions of Eq. (12.52) are sought in the form of the following series

$$v = \varepsilon^{\frac{1}{2}}\left(v_0 + \varepsilon^{\frac{1}{2}}v_1 + \varepsilon v_2 + \cdots\right). \tag{12.56}$$

Substituting (12.53)–(12.56) into (12.52) and equating terms occurring by the powers $\varepsilon^{\frac{1}{2}}$, ε i $\varepsilon^{\frac{3}{2}}$ we get a sequence of the following equations

$$
\begin{aligned}
\varepsilon^{\frac{1}{2}}: &\quad 0 = L_0 v_0, \\
\varepsilon: &\quad 0 = L_0 v_1 + L_1 v_0, \\
\varepsilon^{\frac{1}{2}}: &\quad \partial_T v_0 = L_0 v_2 + L_1 v_1 + L_2 v_0,
\end{aligned}
\tag{12.57}
$$

where

$$
\begin{aligned}
L_0 &= -\left(1+\nabla^2\right)^2 - c\partial_y^4 - 2\eta\partial_x^2\partial_y^2 + r_c, \\
L_1 &= -2\left[\partial_x\left(1+\partial_x^2\right)\partial_X + \partial_y\partial_Y + (1+c)\partial_y^3\partial_Y + (1+\eta)\left(\partial_y\partial_X + \partial_x\partial_Y\right)\partial_x\partial_y\right], \\
L_2 &= -\left\{\frac{3}{2}\,\partial_x^2\partial_X^2 + \frac{1}{2}\partial_X^2 + \frac{3}{2}(1+c)\partial_y^2\partial_Y^2 + \frac{1}{2}\partial_Y^2\right. \\
&\left. + \frac{1}{2}(1+\eta)\left(\partial_y^2\partial_X^2 + \partial_x^2\partial_Y^2 + 4\partial_x\partial_y\partial_X\partial_Y\right) - 1 + u_0^2\right\}.
\end{aligned}
\tag{12.58}
$$

Let us go back to Eq. (12.52) and consider only its stationary solution $v_0 = 0$. Let us check how the perturbation Δv behaves around the equilibrium position. By Eq. (12.52) after linearization we obtain

$$
\partial_T \Delta v = \left\{-\left(1+\nabla^2\right)^2 - c\partial_y^2 - 2\eta\partial_x^2\partial_y^2 + r\right\}\Delta v. \tag{12.59}
$$

Its solutions are sought in the form

$$
\Delta v = e^{\sigma T} e^{i(qx+py)}. \tag{12.60}
$$

We obtain

$$
\begin{aligned}
&\partial_T \Delta v = \sigma e^{\sigma T} e^{i(qx+py)}, \quad \nabla^2(\Delta v) = \frac{\partial^2}{\partial x^2}\Delta v + \frac{\partial^2}{\partial y^2}\Delta v = -(q^2+p^2)e^{\sigma T}e^{i(qx+py)}, \\
&\left(1+\nabla^2\right)^2 = \left[1-(q^2+p^2)\right]^2 e^{\sigma T}e^{i(qx+py)}, \quad \frac{\partial^2(\Delta v)}{\partial x^2} = -q^2 e^{\sigma T}e^{i(qx+py)}, \\
&\frac{\partial^2(\Delta v)}{\partial y^2} = -p^2 e^{\sigma T}e^{i(qx+py)}, \quad \frac{\partial^4(\Delta v)}{\partial y^4} = p^4 e^{\sigma T}e^{i(qx+py)},
\end{aligned}
\tag{12.61}
$$

and substituting (12.60) and (12.61) into (12.59) we have

$$
\sigma = -\left[1-\left(q^2+p^2\right)\right] - cp^4 - 2\eta q^2 p^2 + r. \tag{12.62}
$$

The loss of stability will be related to the algebraic condition $\sigma = \sigma(r; p, q) = 0$. As was shown in [194] the critical values r_c, p_c, q_c, in this case, read

$$q_c^2 = 1, \quad p_c^2 = 0, \quad r_c = 0, \tag{12.63}$$

$$q_c^2 = 0, \quad p_c^2 = \frac{1}{1+c}, \quad r_c = \frac{c}{1+c}, \tag{12.64}$$

$$q_c^2 = \frac{c-\eta}{c-2\eta-\eta^2}, \quad p_c^2 = \frac{-\eta}{c-2\eta-\eta^2}, \quad r_c = \frac{-\eta^2}{c-2\eta-\eta^2}, \quad \eta < 0. \tag{12.65}$$

The critical values (12.63) and (12.64) define periodic normal structures in directions x and y, respectively. The values determined by the relationships (12.65) are responsible for occurrence of mixed structures (skew), where $q_c \neq 0$ and $p_c \neq 0$. As a small parameter ε one can assume $\varepsilon = r - r_c$.

Solutions of a recurrent sequence of Eq. (12.57) are sought for the following generating solution (for $\eta > 0$)

$$v_0 = \frac{A_0 e^{ix} + \bar{A}_0 e^{-ix}}{\sqrt{3}}, \tag{12.66}$$

where $\bar{A}_0$ is complex conjugate of A_0. We find

$$\begin{aligned} \frac{\partial v_0}{\partial x} &= \frac{iA_0 e^{ix} - i\bar{A}_0 e^{-ix}}{\sqrt{3}}, \quad \frac{\partial^2 v_0}{\partial x^2} = \frac{-A_0 e^{ix} - \bar{A}_0 e^{-ix}}{\sqrt{3}}, \\ \frac{\partial^3 v_0}{\partial x^3} &= \frac{-iA_0 e^{ix} + i\bar{A}_0 e^{-ix}}{\sqrt{3}}, \quad \frac{\partial^4 v_0}{\partial x^4} = \frac{A_0 e^{ix} + \bar{A}_0 e^{-ix}}{\sqrt{3}}. \end{aligned} \tag{12.67}$$

Third of Eq. (12.57) generates a so-called amplitude equation of the form

$$\frac{\partial A_0}{\partial T} = \left\{ \partial_X^2 + \frac{1}{2}\eta\partial_Y^2 + 1 - |A_0|^2 \right\} A_0 \tag{12.68}$$

for $\eta > 0$. In the case of $\eta < 0$ a generating solution is sought in the form

$$v_0 = \frac{A_0(X, Y, T)e^{i(q_c x + p_c y)} + \bar{A}_0(X, Y, T)e^{-i(q_c x + p_c y)}}{\sqrt{3}}, \tag{12.69}$$

which leads to the following amplitude equation

$$\frac{\partial A_0}{\partial T} = \left\{ (q_c\partial_X + p_c\partial_Y)^2 + cp_c^2\partial_Y^2 + 2\eta q_c p_c \partial_X \partial_Y + 1 - |A_0|^2 \right\} A_0. \tag{12.70}$$

During derivation of Eq. (12.68) (and similarly (12.70)) one made use of the transformation

$$\begin{aligned} u_0^3 &= \left(A_0^2 e^{2ix} + 2A_0\bar{A}_0 + \bar{A}_0^2 e^{-2ix}\right)\left(A_0 e^{ix} + \bar{A}_0 e^{-ix}\right) \\ &= A_0^3 e^{3ix} + 2A_0^2\bar{A}_0 e^{ix} + \bar{A}_0^2 A_0 e^{-ix} + A_0^2\bar{A}_0 e^{ix} + 2A_0\bar{A}_0^2 e^{-ix} + A_0^3\bar{A}_0 e^{-3ix} \\ &\cong 3A_0^2\bar{A}_0 e^{ix} + 3A_0\bar{A}_0^2 e^{-ix}. \end{aligned}$$

In further considerations, we will confine ourselves to the case $\eta > 0$. We seek a periodic solution in the form

$$A = Fe^{i(QX+PY)}, \tag{12.71}$$

Substituting (12.71) into (12.68) we find the value

$$F = \sqrt{1 - Q^2 - \frac{1}{2}\eta P^2}. \tag{12.72}$$

Let us examine the stability of the obtained solution [194]. Substituting

$$A = (F + f)e^{i(QX+PY)}, \tag{12.73}$$

into (12.68) (where f is a complex perturbation) and performing the linearization procedure we obtain

$$\frac{\partial f}{\partial T} = \left[\partial_X^2 + \frac{1}{2}\eta\partial_Y^2 + 2iQ\partial_X + i\eta P\partial_Y - |F|^2\right] f - F^2\bar{f}. \tag{12.74}$$

A solution of the above equation is sought in the form

$$f = \left[f_1 e^{i(KX+LY)} + f_2 e^{-i(KX+LY)}\right] e^{\sigma T}, \tag{12.75}$$

where:

$$\begin{aligned} &\sigma^2 + 2\left[F^2 + K^2 + \tfrac{1}{2}\eta L^2\right]\sigma + D = 0, \\ &D = 2\left(F^2 - 2Q^2\right)K^2 + \eta\left(F^2 - \eta P^2\right)L^2 - 4\eta QPKL + \left(K^2 + \tfrac{1}{2}\eta L^2\right)^2. \end{aligned} \tag{12.76}$$

The instability ($\sigma > 0$) appears when $D < 0$.

Chapter 13
Bifurcations

13.1 Introduction

We consider the following system of ordinary differential equations

$$\frac{dx}{dt} = \tilde{F}(x, \lambda), \tag{13.1}$$

where: $x \in \mathbb{R}^n$, $\lambda \in \mathbb{R}^k$, $\tilde{F} : \mathbb{R}^n \times \mathbb{R}^n \to \mathbb{R}^n$. The evolutionary system (13.1) can be represented by the vector field x_λ. A solution of the system (13.1) is defined by the phase flow $\Phi_\lambda : \mathbb{R}^k \times \mathbb{R}^n \to \mathbb{R}^n$, where $\Phi_\lambda(x, t) = x_\lambda(t)$ with the attached initial conditions $x = x_\lambda(0)$.

One can use a terminology introduced by Arnold [11]. An object depending on parameters is said to be a *family*. A small change of parameters leads to the object *deformations*. It appears that very often an analysis of all potential deformations is reduced to analysis of a representative one, further referred as a *versal deformation*. The latter can be found using a procedure of reducing a linear problem to that with a Jordan form matrix.

Each set of parameters λ is related to a special configuration of the phase space of the considered dynamical system. It can happen that for different values of $\lambda \in \mathbb{R}^k$, the system behaves qualitatively different. The hyperplanes separating different subspaces of the investigated phase space correspond to the bifurcation sets of parameters. It may happen also that these separating hyperplanes can possess a very complicated structure.

We have already considered a matrix with multiple eigenvalues and we have shown, how to reduce it to a Jordan canonical form. As it has been pointed by Arnold [11], in many engineering oriented sciences, when a matrix is approximately known the obtained results may be qualitatively different from expectations. It is caused by a fact that even a slight perturbation can easily destroy a Jordan canonical form.

J. Awrejcewicz, *Ordinary Differential Equations and Mechanical Systems*,
DOI 10.1007/978-3-319-07659-1_13

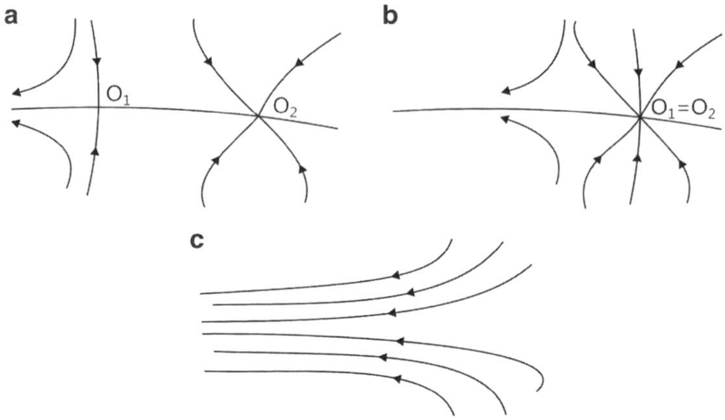

Fig. 13.1 Saddle-node bifurcation. The points O_1 and O_2 approach each other (**a**), overlap (**b**) and vanish (**c**)

However, in the case of parameterized families of matrices their perturbation does not change the multiple eigenvalue matrix form from the family. The problem defined by Arnold [11] is focused on a construction of the simplest form and to determine the minimum number of parameters to which a considered family can be reduced. A versal deformation is called *universal* if the change of the introduced transformation is determined uniquely. A versal deformation is *miniversal* one if the dimension of the parameter space is the smallest required to realize a versal deformation. The questions concerned a construction of miniversal deformations (normal forms) of matrices with multiple eigenvalues and the minimal number of parameters are also addressed in the monograph [243].

The main results are summed up in the following theorem.

Theorem 13.1. *Every matrix A possesses a miniversal deformation and the number of its parameters is equal to the codimension of the orbit of A.*

The smallest number of parameters of a versal deformation of the matrix A can be formally found following the steps given by Gantmacher [97] and Arnold [11].

To introduce a background of dynamical system bifurcations we briefly follow Neimark [181], who analysed some properties of two- and three-dimensional phase space bifurcations. In Fig. 13.1 three steps of the phase plain (portrait) changes are shown which refer to the saddle-node bifurcations. In Fig. 13.2 a situation when a stable focus changes lead to an occurrence of a periodic orbit is shown. In Fig. 13.3 three successive steps leading to occurrence of a stable and an unstable periodic orbits are shown. Note that doubled limit cycle creates the so-called critical orbit. In Fig. 13.4 the successive steps of an occurrence of a stable periodic orbit associated with a stable type separatrice is shown. In Fig. 13.5 three successive steps of a bifurcation changing the separatrices associated with two saddles are shown.

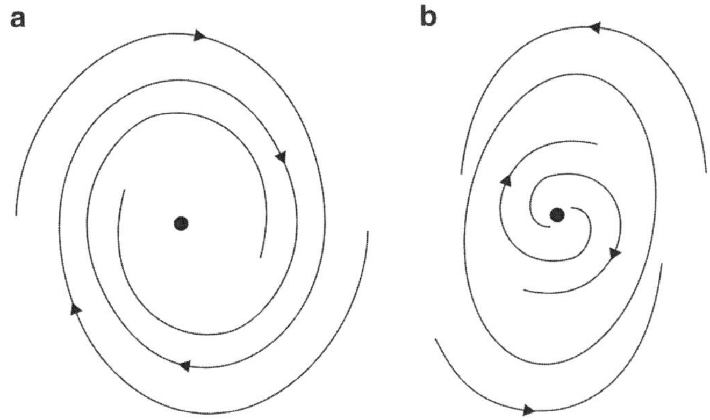

Fig. 13.2 Focus—periodic orbit bifurcation. A stable focus (**a**) becomes unstable and a stable periodic orbit is born (**b**)

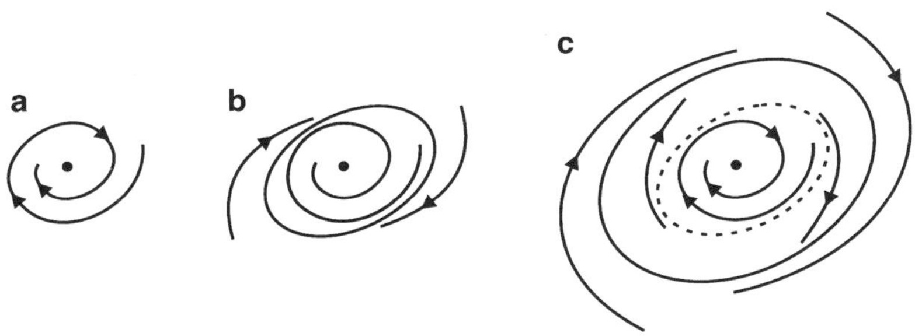

Fig. 13.3 Bifurcation leading to occurrence of two periodic orbits. A stable focus (**a**) becomes unstable and a stable periodic orbit is born (**b**), which eventually becomes also unstable and a second (stable) periodic orbit appears (**c**)

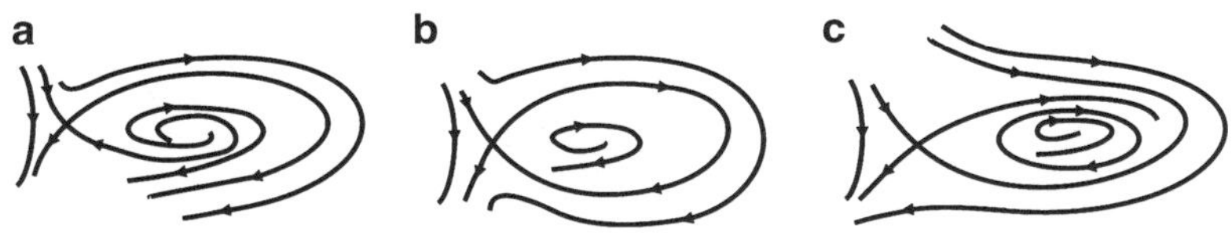

Fig. 13.4 Stable and unstable manifolds of a saddle (**a**) become closed (**b**) eventually leading to occurrence of a stable periodic orbit without a saddle point (**c**)

The dashed area corresponds to significant changes of the phase flow. Generally, bifurcations can be separated into two classes: static and dynamical bifurcations. *Static* bifurcations are related to equilibrium, whereas *dynamic* bifurcations are related to other objects of a phase space.

Recall that $\lambda \in \mathbb{R}^k$ and assume that the eigenvalues σ_i, related to a being investigated locally equilibrium (singular point) depend on $k-1$ *passive* parameters (scalars) and one *active* parameter λ^*. It is clear that for fixed passive parameters

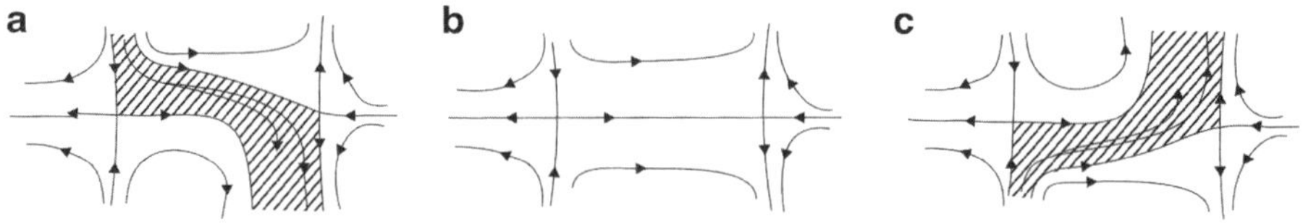

Fig. 13.5 Bifurcation changing separatrices of two saddles ((**a**) a trajectory before (**b**) in a critical state (**c**) and after bifurcation)

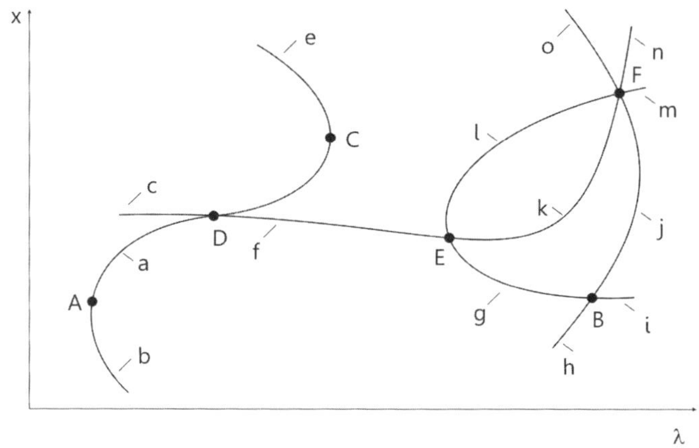

Fig. 13.6 Global bifurcation diagram

and for quasi-static changes of an active parameter one gets curves in the space $(\mathrm{Re}\,\sigma_i, \mathrm{Im}\,\sigma_i, \lambda^*)$, further called trajectories corresponding to the eigenvalues σ_i.

Suppose that increasing λ^* one of the eigenvalues crosses origin (an investigated equilibrium) and other eigenvalues remain either in left-hand side plane (LHP) or right-hand side plane (RHP). A small change of $\pm\Delta\lambda$ close to λ^* results in $\lambda^* \pm \Delta\lambda$. Two situations are possible. If a being investigated equilibrium changes its stability when λ^* changes from $\lambda^* - \Delta\lambda$ to $\lambda^* + \Delta\lambda$ and the leading eigenvalue remains always real then this bifurcation is called *divergence*.

The Hopf (or *flatter*) bifurcation occurs when a pair of complex conjugate eigenvalues crosses (with nonzero velocity) the imaginary axis of the plane $(\mathrm{Re}\,\sigma, \mathrm{Im}\,\sigma)$. The previously stable equilibrium becomes unstable and a new periodic orbit is born. A divergence belongs to one-dimensional bifurcation whereas Hopf bifurcation is two-dimensional one.

Note that although Hopf [124] stated the theorem valid for n-dimensional vector field, the sources related to this problem can be found in the work of Poincaré [202], and the first study of two-dimensional vector fields including a theorem formulation belongs to Andronov [8]. Hence some authors (see [243] for example) call this bifurcation as the Poincaré–Andronov–Hopf one.

In order to introduce a fundamental background of bifurcations we follow the diagram shown in Fig. 13.6 (see also Iooss and Joseph [127]).

The global bifurcation diagram includes branches of solutions (small letters) and branching points (capital letters). A *solution branch* corresponds to the uniqueness of the dependence $x(\lambda)$. However there may exist points where the uniqueness dependence is violated. They are called branching points. One also distinguishes *primary* branching points (A—limiting point, B—bifurcation point) and *secondary* bifurcation points (D—tangent point; C—limiting point; E—bifurcation point; F—multiple bifurcation point).

Consider two solutions $x(\lambda_0)$ and $x(\lambda_0) + \varepsilon$ of Eq. (13.1). Both of them satisfy the equations

$$\begin{aligned} \frac{dx(\lambda_0)}{dt} &= \tilde{F}(\lambda_0, x(\lambda_0)), \\ \frac{d(x(\lambda_0) + \varepsilon)}{dt} &= \tilde{F}(\lambda_0, x(\lambda_0) + \varepsilon). \end{aligned} \tag{13.2}$$

Hence in order to analyse equilibria states we obtain

$$F(\lambda, \varepsilon) = 0, \tag{13.3}$$

where $F(\lambda, \varepsilon) = \tilde{F}(\lambda_0, x(\lambda_0) + \varepsilon) - \tilde{F}(\lambda_0, x(\lambda_0))$.

The isolated solutions of Eq. (13.3) can be classified in the following way [127]:

1. *Regular* point. In this point derivative $F_\lambda \neq 0$ or $F_\varepsilon \neq 0$ and from the implicit function theorem one can find either a curve $\lambda(\varepsilon)$ or $\varepsilon(\lambda)$.
2. *Regular limiting* point. In this point $\lambda_\varepsilon(\varepsilon)$ changes its sign and $F_\lambda(\lambda, \varepsilon) \neq 0$.
3. *Singular* point. In this point $F_\lambda = F_\varepsilon = 0$
4. *Double bifurcation* point. In this point two curves with different tangents intersect each other.
5. *Bifurcation-limiting* point. In this double point the derivative λ_ε changes its sign.
6. *Tangent* point. This is a common point of two curves with the same tangent.
7. *Higher order singular* point. In this point first- and second-order derivatives are equal to zero.

13.2 Singular Points in 1D and 2D Vector Fields

13.2.1 1D Vector Fields

Our attention is focused on a first-order ordinary differential equation with one parameter λ of the form (13.1). For real values of both x and λ values the singular points (equilibria) are defined by the algebraic equation (13.3).

We are going to analyse an existence and uniqueness of singular points qualitatively including a construction of bifurcating solutions together with their stability estimation. Note that (in general) considering an implicit function (13.3) one can get either the function $x(\lambda)$ or $\lambda(x)$, or isolated points, or it cannot be explicitly described. Here, following the earlier introduced classification given in Sect. 13.1, one deals either with *regular* or *singular* points. A regular point belongs to only one curve, whereas a singular point can be either isolated or it can belong to a few curves (branches). An order of singularity defines number of the associated branches. Here we would like briefly to explain a word "singular". It appears first as the name of equilibria of the dynamical systems. However, it appears second time when a nonlinear algebraic equation is considered and it concerns a classification of the roots of the algebraic equation.

The classification of equilibria results from the implicit function theorem.

Theorem 13.2. *Given the function $f(x, y)$ with continuous partial derivatives in the neighbourhood of (x_0, y_0), where $f(x_0, y_0) = 0$. If either $f_x(x_0, y_0) \neq 0$ or $f_y(x_0, y_0) \neq 0$, then:*

(i) There exists such α and β that for $x_0 - \alpha < x < x_0 + \alpha$, (or for $y_0 - \beta < y < y_0 + \beta$) we have a unique solution $y = y(x)$ (or $x = x(y)$);
(ii) The function $y = y(x)$ (or $x = x(y)$) is differentiable in the neighbourhood $|x - x_0| < \alpha$ (or $|y - y_0| < \beta$), and $\frac{dy}{dx}(x) = -\frac{f_x(x,y(x))}{f_y(x,y(x))}$ (or $\frac{dx}{dy}(y) = -\frac{f_y(x(y),y)}{f_x(x(y),y)}$).

Now we can define the singular points more precisely. A point (λ_0, x_0) is said to be *regular*, if both partial derivatives of the function $F(\lambda, x)$ are not simultaneously equal to zero, i.e.

$$F_\lambda^2(\lambda_0, x_0) + F_x^2(\lambda_0, x_0) \neq 0 \tag{13.4}$$

A regular point for which $F_x(\lambda_0, x_0) = 0$ is called the *extremal* point (it can be either minimum or maximum). A point for which (13.4) does not hold is said to be *singular*. A point (λ_0, x_0) is said to be singular of nth order, if the associated with this point derivatives up to the order $n - 1$ are equal to zero and at least one of the nth order derivatives is different from zero.

The solution branches are defined by the Taylor series. Introducing the new function $v = \frac{\lambda-\lambda_0}{x-x_0}$ and $w = \frac{x-x_0}{\lambda-\lambda_0}$ and dividing the Taylor series representation of $F(x, \lambda) = 0$ by $\frac{(x-x_0)^n}{n!}$ and by $\frac{(\lambda-\lambda_0)^n}{n!}$, we obtain the following nth order algebraic equations,

$$\begin{aligned} &A_0 v^n + A_1 v^{n-1} + \cdots + A_{n-1} v + A_n + O(\lambda - \lambda_0) = 0, \\ &A_0 + A_1 w + \cdots + A_{n-1} w^{n-1} + A_n w^n + O(x - x_0) = 0, \end{aligned} \tag{13.5}$$

where: $A_i = \frac{\partial^n F}{\partial x^i \partial \lambda^{n-i}}(\lambda_0, x_0),\ i = 1, 2, \ldots, n.$

In a limit case $(\lambda, x) \to (\lambda_0, x_0)$ the algebraic equations with $O(\lambda - \lambda_0) = O(x - x_0) = 0$ serve to find a tangent of a slope to a solution branch to the axis x or λ, respectively. An existence of n solutions to Eq. (13.5) yields n different branches (some of them, however, can be degenerated). Real distinct solutions correspond to distinct intersecting branches, whereas multiple solutions correspond to same tangent values of different solutions in the point (λ_0, x_0) or overlapping of multiple branches. Complex solutions correspond to the so-called degenerated branches (points) [144].

A point (λ_0, x_0) is called *degenerated* if it belongs simultaneously to degenerated branches (solutions).

An order of degeneracy of a singular point (λ_0, x_0) is defined by a number of complex roots of Eq. (13.5). In words, degeneracy order corresponds to a number of degenerated branches. Following the approach given in monograph [144] consider the first equation of the system (13.5), further referred as $H(\nu, \varepsilon) = 0$, where $\varepsilon = \lambda - \lambda_0$. Let ν_{0i} be solutions to the first equation of (13.5). A unique solution with the tangent ν_{0i} is obtained when $H_\nu(\nu_{0i}, 0) \neq 0$. This holds when ν_{0i} is a simple root of $H(\nu_{0i}, \varepsilon) = 0$. Note that in order to distinguish different tangent and identical branches passing through the point (λ_0, x_0) one needs to calculate higher order derivatives, which for instance define curvatures of the branches.

If $A_0, A_1, \ldots, A_{n-1}$ are equal to zero, then the first equation of (13.5) has $n - k$ roots, whereas the second possesses k roots equal to zero. Geometrically, it means that k branches have a tangent parallel to the axis λ in the point (λ_0, x_0). Now we briefly outline a construction of a solution branch in vicinity of the singular point (λ_0, x_0), for which

$$F(\lambda_0, x_0) = 0. \tag{13.6}$$

Let us approximate function F by its double Taylor series of the form (we follow here the approach given in monograph [144]):

$$\begin{aligned}
F(\lambda, x) &= \alpha(\lambda) + a(\lambda)(x - x_0) + \frac{1}{2!}b(\lambda)(x - x_0)^2 + \frac{1}{3!}c(\lambda)(x - x_0)^3 + \cdots \\
&= \alpha_0 + \alpha_1(\lambda - \lambda_0) + \frac{1}{2!}\alpha_2(\lambda - \lambda_0)^2 + \frac{1}{3!}\alpha_3(\lambda - \lambda_0)^3 + \cdots \\
&\quad + [a_0 + a_1(\lambda - \lambda_0) + \frac{1}{2!}a_2(\lambda - \lambda_0)^2 + \frac{1}{3!}a_3(\lambda - \lambda_0)^3 + \cdots](\lambda - \lambda_0) \\
&\quad + \frac{1}{2!}[b_0 + b_1(\lambda - \lambda_0) + \frac{1}{2!}b_2(\lambda - \lambda_0)^2 \\
&\quad + \frac{1}{3!}b_3(\lambda - \lambda_0)^3 + \cdots](\lambda - \lambda_0)^2 + \frac{1}{3!}[c_0 + c_1(\lambda - \lambda_0) + \frac{1}{2!}c_2(\lambda - \lambda_0)^2 \\
&\quad + \frac{1}{3!}c_3(\lambda - \lambda_0)^3 + \cdots](\lambda - \lambda_0)^3 + \cdots,
\end{aligned} \tag{13.7}$$

where: $\alpha_i = \frac{\partial^i F}{\partial \lambda^i}(\lambda_0, x_0)$, $a_i = \frac{\partial^{i+1} F}{\partial x \partial \lambda^i}(\lambda_0, x_0)$, $b_i = \frac{\partial^{i+2} F}{\partial x^2 \partial \lambda^i}(\lambda_0, x_0)$, $c_i = \frac{\partial^{i+3} F}{\partial x^3 \partial \lambda^i}(\lambda_0, x_0)$, and so on. Note that $\alpha_0 = 0$. If $F(\lambda, x_0) = 0$ (independently of λ) then $x = x_0$ is a solution. In addition, if (λ_0, x_0) is a regular point then $x = x_0$ is only one solution. However, if (λ_0, x_0) is a singular point then in spite of $x = x_0$ there may exist also additional branches of solutions. If $x - x_0 = \varepsilon$, then a solution can be parameterized by

$$\lambda - \lambda_0 = \varphi(\varepsilon), \tag{13.8}$$

where $\varphi(\varepsilon)$ can be approximated by the series

$$\varphi(\varepsilon) = \mu_0 + \mu_1\varepsilon + \mu_2\varepsilon^2 + \cdots \tag{13.9}$$

It can be easily shown that $\lambda_0 = 0$. Substituting (13.8) and (13.9) into (13.7), and comparing terms standing by the same powers of ε the following equations are obtained

$$\begin{aligned}
\varepsilon &: \alpha_1\mu_1 + a_0 = 0,\\
\varepsilon^2 &: \frac{1}{2}\alpha_1\mu_1^2 + a_1\mu_1 + \frac{1}{2}b_0 + \alpha_1\mu_2 = 0,\\
\varepsilon^3 &: \alpha_0\mu_3 + \frac{1}{2}\alpha_2\mu_1\mu_2 + \frac{1}{6}\alpha_3\mu_1^3 + a_1\mu_2 + \frac{1}{2}a_2\mu_1^2 + \frac{1}{2}b_1\mu_1 + \frac{1}{6}c_0 = 0,\\
&\dots
\end{aligned} \tag{13.10}$$

which define $\mu_1, \mu_2, \mu_3, \ldots$ Hence a solution in the neighbourhood of (λ_0, x_0) has the following parameterized form

$$\begin{aligned}
x - x_0 &= \varepsilon,\\
\lambda - \lambda_0 &= \lambda_1\varepsilon + \lambda_2\varepsilon^2 + O(\varepsilon^3).
\end{aligned} \tag{13.11}$$

Note that if $\alpha_1 = F_\lambda(\lambda_0, x_0) = 0$, then one should use another parametrization, i.e. $x - x_0 = \psi(\varepsilon)$, $\lambda - \lambda_0 = \varepsilon$. Note also that if (λ_0, x_0) is singular then $\alpha_1 = a_0 = 0$ and we get

$$\alpha_2\mu_1^2 + 2a_1\mu_1 + 2b_0 = 0. \tag{13.12}$$

Therefore, if

(i) $a_1^2 > 2b_0\alpha_2$, then we have two distinct roots for μ;
(ii) $a_1^2 = 2b_0\alpha_2$, then we have double real solution (two branches of solution are tangent);
(iii) $a_1^2 < 2b_0\alpha_2$, then the roots are complex (the point (λ_0, x_0) is singular).

If $b_0 = a_1 = \alpha_2 = 0$, then (λ_0, x_0) is the triple point and first non-trivial algebraic equation is defined by the terms standing by ε^3. Many other examples are given in the mentioned monograph [144].

When various branches of solutions are found, the next step is focused on analysis of their stability. Consider one-dimensional problem governed by the equations

$$\begin{aligned}
\dot{x} &= F(\lambda, x),\\
\dot{v} &= F_\lambda(\lambda, x)v,
\end{aligned} \tag{13.13}$$

where the second equation governs behaviour of perturbations related to a being investigated singular points (equilibria). The discussed earlier Lyapunov's theorems can be used to estimate stability. Since we deal with one-dimensional case, then $F_\lambda(\lambda, x)$ defines the characteristic exponent of (13.13). If $\sigma = F_\lambda(\lambda, x)$ is negative, then a constant solution x of the original nonlinear differential equation is stable.

Observe that during stability investigation in the neighbourhood of the point (λ_0, x_0) we deal with the functions:

(i) $\sigma = \sigma(\lambda) = F_x(\lambda, x(\lambda))$ if a being investigated branch has the form $x = x(\lambda)$;
(ii) $\sigma = \sigma(\varepsilon) = F_x(\lambda(\varepsilon), x(\varepsilon))$ if a being investigated branch is parameterized by the equations $x - x_0 = \varepsilon$, $\lambda - \lambda_0 = \varphi(\varepsilon)$.

We give two examples studied in monograph [144].

Example 13.1. Consider the first-order differential equation with the right-hand side $F(\lambda, x) = -1-2\lambda+2\lambda x+\lambda^2+3x^2-2x^3$. Display existence and classification of singular points and build the corresponding bifurcation diagram.

Following the steps described earlier we obtain

$$F_x = 2\lambda + 6x - 6x^2,$$
$$F_\lambda = -2 + 2\lambda + 2x.$$

Solving two equations $F_x = 0$ and $F_\lambda = 0$, we obtain (0,1) as the singular point, which is doubled ($F_x(0,1) = 0$, $F_{xx}(0,1) \neq 0$). Introducing the parametrization

$$\varepsilon = x - 1,$$
$$\lambda = \mu_1\varepsilon + \mu_2\varepsilon^2 + O(\varepsilon^3),$$

we obtain:

$$\varepsilon^2: \quad \mu_1^2 + 2\mu_1 - 3 = 0$$
$$\varepsilon^3: \quad \mu_2 + \mu_1\mu_2 - 1 = 0.$$

Solving above algebraic equations we get: $\mu_{11} = 1$, $\mu_{12} = -3$, $\mu_{21} = 0,5$, $\mu_{22} = -0,5$. Hence, the following branches of solutions are found:

$$x = \varepsilon,$$
$$\lambda_1(\varepsilon) = 2\varepsilon - \varepsilon^2 + O(\varepsilon^3).$$

and

$$x = \varepsilon,$$
$$\lambda_2(\varepsilon) = -\varepsilon + O(\varepsilon^3).$$

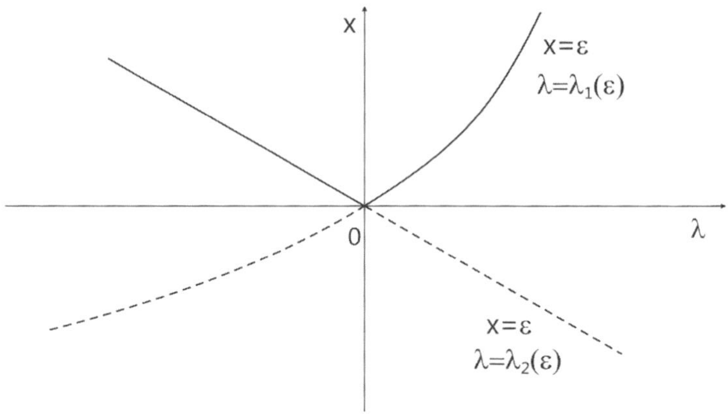

Fig. 13.7 Stable (*solid line*) and unstable (*dashed line*) branches of solutions in a vicinity of (0,0)

Stability of the obtained branches are defined by the exponents

$$\sigma_1(\varepsilon) = F_x[\lambda_1(\varepsilon), \varepsilon] = -6\varepsilon + O(\varepsilon^2),$$
$$\sigma_2(\varepsilon) = F_\lambda[\lambda_2(\varepsilon), \varepsilon] = -3\varepsilon + O(\varepsilon^2).$$

The corresponding bifurcation diagram is reported in Fig. 13.7. For $\varepsilon > 0$ ($\varepsilon < 0$) we have stable (unstable) branches.

□

Example 13.2. Consider a vertical slim rod of length l and a buckling caused by its gravity.

The equilibrium conditions of the rod element are:

(i) $\rho A\ddot{w}dx = dN \sin\alpha + dT \cos\alpha$ (transversal motion);
(ii) $dN \cos\alpha = qdx + dT \sin\alpha$ (longitudinal static condition);
(iii) $\frac{dM}{dx} = T$ (static condition),

where: ρ is the material density; A is the area of rod cross section; N is the normal force; T is the transversal force; M is the bending moment; α defines buckling angle and $(\cdot) = d/dt$.

From (iii), taking into account (i) and (ii), we get:

$$\rho\frac{Ad^2\ddot{w}}{dt^2} = \frac{\partial^2 M}{\partial x^2} + q\tan\alpha,$$

where

$$\tan\alpha = \frac{\partial w}{\partial x}, \quad \sqrt{1 + (\frac{dw}{dx})^2} = \frac{1}{\cos\alpha}.$$

The bending moment and a rod curvature are linked via relation

$$M = -EI\left(1 + \frac{dw}{dx}\right)^{-\frac{3}{2}},$$

where EI denotes the rod stiffness. Since we are going to consider rather large rod deflections then we include two nonlinear terms of the Taylor series of $M(w)$, and hence

$$M = -EI\frac{\partial^2 w}{\partial x^2}\left[1 - \frac{3}{2}\left(\frac{\partial w}{\partial x}\right)^2 + \frac{8}{15}\left(\frac{\partial w}{\partial x}\right)^4\right].$$

In addition, we approximate $(\cos\alpha)^{-1}$ by

$$\frac{1}{\cos\alpha} = 1 + \frac{1}{2}\left(\frac{\partial w}{\partial x}\right)^2.$$

Taking into account the above relations we obtain the partial differential equation governing slim vertical rod dynamics

$$\frac{d^2 w}{dt^2} = g\frac{\partial w}{\partial x} - \frac{EI}{\rho A}\left[\frac{\partial^4 w}{\partial x^4}\left(1 - \left(\frac{\partial w}{\partial x}\right)^2 + \frac{9}{8}\left(\frac{\partial w}{\partial x}\right)^4\right) - 9\frac{\partial w}{\partial x}\frac{\partial^2 w}{\partial x^2}\frac{\partial^3 w}{\partial x^3}\right.$$
$$\left. + - 3\frac{\partial^2 w}{\partial x^2} + 18\left(\frac{\partial w}{\partial x}\right)^3\frac{\partial^2 w}{\partial x^2}\frac{\partial^3 w}{\partial x^3} + 21\left(\frac{\partial w}{\partial x}\right)^2\left(\frac{\partial^2 w}{\partial x^2}\right)\right].$$

The boundary conditions for $w(x,t)$ include:

(a) geometrical

$$w(0,t) = 0, \quad \frac{dw}{dx}(0,t) = 0;$$

(b) approximate mechanical

$$\frac{d^2 w}{dx^2}(l,t) = 0, \quad \frac{d^3 w}{dx^3}(l,t) = 0.$$

In order to obtain an ordinary differential equation we separate the variables

$$w(x,t) = u(t)h(x).$$

Taking $h(x)$ as the fourth-order polynomial, and after orthogonalization procedure, we get

$$\ddot{u} = \left(\frac{A}{\lambda} - \frac{B}{\lambda^4}\right)u + \frac{C}{\lambda^6}u^3 - \frac{D}{\lambda^8}u^5,$$

where A, B, C, D are passive parameters, and $\lambda = l$ is the active parameter. Note that although we have second-order differential equation, we are going to investigate equilibria which are governed by

$$F(\lambda, u) = \left(\frac{A}{\lambda} - \frac{B}{\lambda^4}\right)u + \frac{C}{\lambda^6}u^3 - \frac{D}{\lambda^8}u^5 = 0$$

in a way similar to that of first-order differential equation. The trivial solution $u = 0$ corresponds to a straight form of the rod. The singular points are defined by

$$F_u(\lambda, u) = \frac{A}{\lambda} - \frac{B}{\lambda^4} + \frac{3C}{\lambda^6}u^2 - \frac{D}{\lambda^8}u^4 = 0,$$

$$F_\lambda(\lambda, u) = \left(-\frac{A}{\lambda^4} + \frac{4B}{\lambda^5}\right)u - \frac{2C}{\lambda^7}u^3 + \frac{4D}{\lambda^9}u^5 = 0.$$

For $u = 0$ from $F_\lambda = 0$ we get $\lambda_0 = \left(\frac{B}{A}\right)^{\frac{1}{3}}$. The point $(\lambda_0, u_0) = (\lambda_0, 0)$ is the singular point. Because $F_{\lambda u}(\lambda_0, 0) \neq 0$ than this is second-order singular point. The horizontal line $u = 0$ is one of the solutions crossing by this singular point. The second branch can be parameterized in the following way

$$u = \varepsilon,$$

$$\lambda - \lambda_0 = \mu_1\varepsilon + \mu_2\varepsilon^2 + \mu_3\varepsilon^3 + \mu_4\varepsilon^4 + O(\varepsilon^5).$$

The being sought numbers μ_1, μ_2, μ_3 and μ_4 are found from the equations:

$$\varepsilon: \; a_1\mu_1 + \frac{1}{2}b_0 = 0;$$

(where $a_1 = F_{\lambda u}(\lambda_0, 0) = \frac{3A}{\lambda_0^2} > 0$, $b_0 = F_{uu}(\lambda_0, 0) = 0$ and hence $\mu_1 = 0$)

$$\varepsilon^2: \; a_1\mu_2 + \frac{1}{6}c_0 = 0;$$

(where $c_0 = F_{uuu}(\lambda_0, 0) = \frac{6C}{\lambda_0^6} > 0$, and hence $\lambda_2 = -\frac{C}{3A\lambda_0^4} < 0$; it can be easy checked that also $\lambda_3 = 0$)

$$\varepsilon^4: \; a_1\mu_4 + \frac{1}{2}a_2\mu_2^2 + \frac{1}{6}c_1\mu_2 + \frac{1}{5!}c_0 = 0;$$

(where $a_2 = F_{\lambda\lambda u}(\lambda_0, 0) = -\frac{18A^2}{B} < 0$, $c_1 = F_{\lambda uuu}(\lambda_0, 0) = \frac{36C}{\lambda_0^3} > 0$, $c_0 = F_{uuuuu}(\lambda_0, 0) = -\frac{5D}{\lambda_0^8} < 0$.) In result, we have found the following parameterized branches of non-trivial solution

$$u = \varepsilon$$
$$\lambda - \lambda_0 = \mu_2\varepsilon^2 + \mu_4\varepsilon^4 + O(\varepsilon^6),$$

where $\lambda_2 < 0$ and $\lambda_4 > 0$. Now let us investigate stability. In a case of the trivial solution

$$\sigma_1 = F_u(\lambda, 0) = \frac{A}{\lambda}\left(1 - \frac{\lambda_0^3}{\lambda^3}\right).$$

Hence $\sigma_1 > 0$, ($\sigma_1 < 0$) when $\lambda > \lambda_0$ ($\lambda < \lambda_0$). In the case of the second branch we have

$$\sigma_2(\varepsilon) = F_u\left(\lambda(\varepsilon), \varepsilon\right) = \frac{A}{\lambda(\varepsilon)} - \frac{B}{\lambda^4(\varepsilon)} - \frac{3C}{\lambda^6(\varepsilon)}\varepsilon^2 + \frac{5D}{\lambda^8(\varepsilon)}\varepsilon^4,$$

and consequently

$$\sigma_2(0) = 0; \quad \frac{d\sigma_2}{d\varepsilon}(0) = 0, \quad \frac{d^2\sigma_2}{d\varepsilon^2}(0) = -\frac{6C}{\lambda_0^6} < 0.$$

The exponent σ_2 can be approximated by

$$\sigma_2(\varepsilon) = -\frac{3C}{\lambda_0^6}\varepsilon^2 + O(\varepsilon^3).$$

In the extremum of this branch, which is equal to $\varepsilon_{ex} = \pm\left(\frac{\mu_2}{\mu_4}\right)^{\frac{1}{2}}$, a change of stability occurs.

The results are shown in Fig. 13.8, where the hysteresis loop $E1, E2, E3, E4$ is remarkable. □

13.2.2 Two-Dimensional Vector Fields

Consider how two nonlinear algebraic equations with one parameter λ of the form

$$F_i\left(x_1, x_2, \lambda\right) = 0, \quad i = 1, 2 \tag{13.14}$$

and $x_{10}, x_{20}, \lambda_0$ is a solution of (13.14).

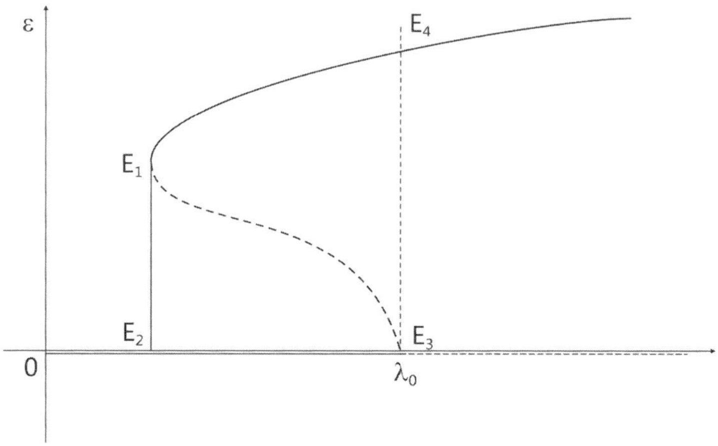

Fig. 13.8 Bifurcation diagram of a slim rod buckling

Theorem 13.3 (Implicit Function Theorem). *If the Jacobian associated with (13.14)* $\frac{\partial F_i}{\partial x_i}(x_{10}, x_{20}, \lambda_0) \neq 0$ *then there is a neighbourhood of the point* $(x_{10}, x_{20}, \lambda_0)$*, where the curve defined by (13.14), is unique, i.e. there exist such functions* $x_i(\lambda)$ *that*

$$F_i\left[x_1(\lambda), x_2(\lambda), \lambda\right] = 0, \; i = 1, 2. \tag{13.15}$$

In words, if the associated Jacobian with the investigated point differs from zero then there are no other solutions in a neighbourhood of the point $(x_{10}, x_{20}, \lambda_0)$. A necessary condition for bifurcation is defined by $\frac{\partial f_i}{\partial x_i}(x_{10}, x_{20}, \lambda_0) = 0, \; i = 1, 2.$

Now we are going to describe briefly a construction of a bifurcation solution of the system (13.14). We consider a trivial case, i.e. we assume that $x_{10} = x_{20} = 0$ is the solution of Eq. (13.14), and $\lambda = 0$ corresponds to a critical state of the system.

Since

$$F_i(0, 0, \lambda) = 0, \quad i = 1, 2, \tag{13.16}$$

hence

$$\begin{aligned} F_1(x_1, x_2, \lambda) &= a(\lambda)x_1 + b(\lambda)x_2 + \alpha_1(\lambda)x_1^2 + 2\beta_1(\lambda)x_1x_2 \\ &\quad + \gamma(\lambda)x_2^2 + O\left(\|x\|^3\right), \\ F_2(x_1, x_2, \lambda) &= c(\lambda)x_1 + d(\lambda)x_2 + \alpha_2(\lambda)x_1^2 + 2\beta_2(\lambda)x_1x_2 \\ &\quad + \gamma(\lambda)x_2^2 + O\left(\|x\|^3\right), \end{aligned} \tag{13.17}$$

where $a, b, c, d, \alpha_i, \beta_i$ and γ_i (i=1, 2) depend on the parameter λ and $\|\cdot\|$ denotes a norm in two-dimensional Euclidian space. The associated variational (perturbational) equations have the form

$$\begin{bmatrix} \dot{x}_1 \\ \dot{x}_2 \end{bmatrix} = \left(\begin{bmatrix} a_0 & b_0 \\ c_0 & d_0 \end{bmatrix} + \begin{bmatrix} a_1 & b_1 \\ c_1 & d_1 \end{bmatrix} \lambda + O\left(\lambda^2\right) \right) \begin{bmatrix} x_1 \\ x_2 \end{bmatrix}. \tag{13.18}$$

A bifurcation condition leads to the equation

$$\det\left[A_0\right] = \begin{vmatrix} a_0 & b_0 \\ c_0 & d_0 \end{vmatrix} = 0. \tag{13.19}$$

It is clear that $\det\left[A_0\right] = 0$, if at least one of the eigenvalues of the matrix A_0 is equal to zero. Knowing that $(0, 0, 0)$ is the bifurcation point one needs to find a number of bifurcating solutions (from this point) occurring in a small vicinity of $\lambda = 0$. Taking into account the small increments of the variables and functions in (13.16) and dividing the obtained linear equations by $d\lambda$ one obtains

$$\begin{aligned} \frac{dF_1}{d\lambda} &= a_0 \frac{dx_1}{d\lambda} + b_0 \frac{dx_2}{d\lambda}, \\ \frac{dF_2}{d\lambda} &= c_0 \frac{dx_1}{d\lambda} + d_0 \frac{dx_2}{d\lambda}. \end{aligned} \tag{13.20}$$

This linear approximation cannot be used to find uniquely higher order derivatives. One should include the nonlinear terms. For the case, when only quadratic terms are taken into account, we get

$$\begin{aligned} a_0 \frac{dx_1}{d\lambda} + b_0 \frac{dx_2}{d\lambda} + \alpha_{10} \left(\frac{dx_1}{d\lambda} \right)^2 + 2\beta_{10} \frac{dx_1}{d\lambda} \frac{dx_2}{d\lambda} + \gamma_{10} \left(\frac{dx_2}{d\lambda} \right)^2 &= 0, \\ c_0 \frac{dx_1}{d\lambda} + d_0 \frac{dx_2}{d\lambda} + \alpha_{20} \left(\frac{dx_1}{d\lambda} \right)^2 + 2\beta_{20} \frac{dx_1}{d\lambda} \frac{dx_2}{d\lambda} + \gamma_{20} \left(\frac{dx_2}{d\lambda} \right)^2 &= 0. \end{aligned} \tag{13.21}$$

A number of bifurcating solutions is defined by a number of intersection points of two conical curves represented by Eq. (13.21). One can have 1, 2 or 3 solutions, in spite of the trivial one.

In general we have three different cases to be considered in two-dimensional vector fields:

(i) one eigenvalue is equal to zero;
(ii) two eigenvalues are zero with degeneracy order 1;
(iii) two eigenvalues are zero with a degeneracy order 2.

We consider only the case (i) and we follow the steps studied in the monograph [144].

Let us introduce the following parameterization

$$x_1 = \varepsilon,\ x_2 = \varepsilon y(\varepsilon),\ \lambda = \varepsilon\mu(\varepsilon), \tag{13.22}$$

or

$$x_1 = \varepsilon x(\varepsilon),\ x_2 = \varepsilon,\ \lambda = \varepsilon\mu(\varepsilon). \tag{13.23}$$

The functions $x(\varepsilon)$, $y(\varepsilon)$ and $\lambda(\varepsilon)$ are polynomials and can be found in a way described earlier. Using parameterization (13.22) the being analysed Eq. (13.14) can be presented in the form

$$\varepsilon g_i\left[\mu(\varepsilon), y(\varepsilon), \varepsilon\right] = 0, \quad i = 1, 2, \tag{13.24}$$

where

$$\begin{aligned} g_1 &= a_0 + a_1\varepsilon\mu + (b_0 + b_1\varepsilon\mu)\,y + \alpha_1\varepsilon + 2\beta_1\varepsilon y + \gamma_1\varepsilon y^2 + O\left(\varepsilon^2\right), \\ g_2 &= c_0 + c_1\varepsilon\mu + (d_0 + d_1\varepsilon\mu)\,y + \alpha_2\varepsilon + 2\beta_2\varepsilon y + \gamma_2\varepsilon y^2 + O\left(\varepsilon^2\right). \end{aligned} \tag{13.25}$$

In fact, we are going to find the functions $y(\varepsilon)$ and $\mu(\varepsilon)$ which satisfy the equations

$$\begin{aligned} a_0 + b_0 y + \varepsilon\left[(a_1 + b_1 y)\,\mu + \alpha_1 + 2\beta_1 y + \gamma_1 y^2\right] + O\left(\varepsilon^2\right) = 0, \\ c_0 + d_0 y + \varepsilon\left[(c_1 + d_1 y)\,\mu + \alpha_2 + 2\beta_2 y + \gamma_2 y^2\right] + O\left(\varepsilon^2\right) = 0. \end{aligned} \tag{13.26}$$

For $\varepsilon = 0$ we obtain two dependent equations and hence $y_0 = y(0) = -\frac{a_0}{b_0} = -\frac{c_0}{d_0}$ for $b_0 \neq 0$ or $d_0 \neq 0$.

In the case $b_0 = d_0 = 0$ one needs to apply the following parametrization $x_1 = \varepsilon x(\varepsilon), x_2 = \varepsilon, \lambda = \varepsilon\mu(\varepsilon)$. In a similar way, one obtains $x_0 = x(0) = -\frac{b_0}{a_0} = -\frac{d_0}{c_0}$ for either $a_0 \neq 0$ or $c_0 \neq 0$. Let

$$y(\varepsilon) = y_0 + \varepsilon\tilde{y}(\varepsilon) \tag{13.27}$$

Substituting (13.27) into (13.26) we get

$$\varepsilon h_i(\lambda, \tilde{y}, \varepsilon) = 0, \quad i = 1, 2, \tag{13.28}$$

where

$$\begin{aligned} h_1 &= b_0\tilde{y} + \mu(a_1 + b_1 y) + \alpha_1 + 2\beta_1 y_0 + \gamma_1 y_0^2 + O(\varepsilon), \\ h_2 &= d_0\tilde{y} + \mu(c_1 + d_1 y) + \alpha_2 + 2\beta_2 y_0 + \gamma_2 y_0^2 + O(\varepsilon). \end{aligned} \tag{13.29}$$

Since for $\varepsilon = 0$ we have $h_i = 0$, hence we get two linear equations of the form

$$\begin{aligned} b_0\tilde{y_0} + \mu_0 (a_1 + b_1 y_0) + \alpha_{10} + 2\beta_{10} y_0 + \gamma_{10} y_0^2 = 0, \\ d_0\tilde{y_0} + \mu_0 (c_1 + d_1 y_0) + \alpha_{20} + 2\beta_{20} y_0 + \gamma_{20} y_0^2 = 0, \end{aligned} \tag{13.30}$$

which serve to find $(y_0,\tilde{\ }\mu_0)$. The first (linear) approximation to the being sought bifurcated solutions has the form

$$x_1 = \varepsilon, x_2 = \varepsilon y_0, \lambda = \varepsilon \mu_0. \tag{13.31}$$

Note that the solution $(\tilde{y_0}, \mu_0)$ is the only one, and $\varepsilon = \sqrt{x_1^2 + \left(\frac{x_2}{y_0}\right)^2}$ measures a distance between the bifurcational and trivial solutions. One can include more terms in order to get nonlinear functions $\varepsilon(\lambda)$ and $\mu(\varepsilon)$. The associated linear variational equation defines a local phase flow in a vicinity of a being investigated non-trivial solution. Introducing the local variables $v_1 = u_1 - \varepsilon$ and $v_2 = u_2 - \varepsilon y_0$ one obtains

$$\begin{bmatrix} \dot{v}_1 \\ \dot{v}_2 \end{bmatrix} = \begin{bmatrix} a_0 + \varepsilon(\lambda_0 a_1 + 2\alpha_{10} + 2\beta_{10}) & b_0 + \varepsilon(\lambda_0 b_1 + 2\beta_{10} + 2\gamma_{10}) \\ c_0 + \varepsilon(\lambda_0 c_1 + 2\alpha_{20} + 2\beta_{20}) & d_0 + \varepsilon(\lambda_0 d_1 + 2\beta_{20} + 2\gamma_{20}) \end{bmatrix} \begin{bmatrix} v_1 \\ v_2 \end{bmatrix} \tag{13.32}$$

The bifurcating solution (13.31) is unstable in Lyapunov sense if at least one eigenvalue $\sigma_i(\varepsilon), i = 1, 2$ has a positive real part. For a case $a_0 = b_0 = c_0 = 0$ and $d_0 < 0$ from (13.32) one gets

$$\begin{bmatrix} \dot{v}_1 \\ \dot{v}_2 \end{bmatrix} = \begin{bmatrix} c_1\varepsilon & c_2\varepsilon \\ c_3\varepsilon & d_0 + c_4\varepsilon \end{bmatrix} \begin{bmatrix} v_1 \\ v_2 \end{bmatrix}, \tag{13.33}$$

where c_i are real values. The associated eigenvalues are easily found

$$\sigma_{1,2}(\varepsilon) = \frac{d_0 + (c_1 + c_2)\varepsilon}{2} \pm \frac{1}{2}\sqrt{d_0^2 + 2d_0(c_4 - c_1)\varepsilon}, \tag{13.34}$$

and for enough small ε they can be represented by their linear part only

$$\sigma_1(\varepsilon) = c_1\varepsilon, \sigma_2(\varepsilon) = d_0 + c_4\varepsilon. \tag{13.35}$$

The corresponding bifurcational diagram is shown in Fig. 13.9. Unstable solutions are marked by a dashed line.

Example 13.3. Display a bifurcation diagram and investigate stability of all branches of solutions occurred in the system

$$\begin{aligned} \dot{x_1} &= px_1 - px_2 - x_1^2 + x_2^2 + x_1^3, \\ \dot{x_2} &= px_2 + x_1x_2 + 2x_1^3. \end{aligned}$$

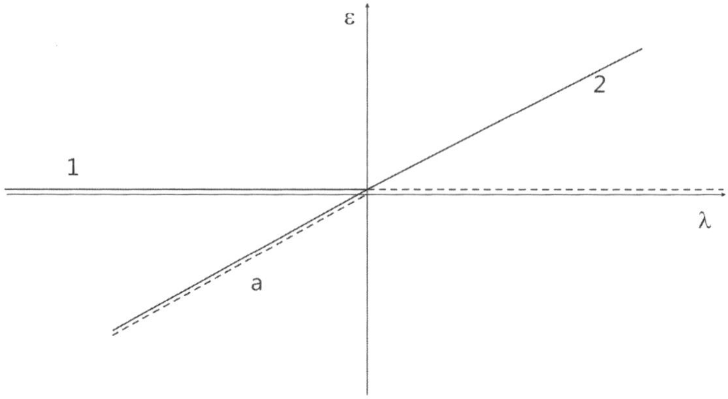

Fig. 13.9 Bifurcational diagram corresponding to simple zero eigenvalue (*1* primary (trivial) solution; *2* secondary (bifurcated) solution)

First we observe that a matrix associated with the origin does not have the terms independent on p, hence $A_0 = 0$. We have double zero eigenvalue for $p = 0$, with the Riesz index $\nu = 1$ and degeneracy order 2. The eigenvalues of the matrix A are $\sigma_{1,2} = p$.

We introduce the following parametrization

$$\begin{aligned} x_1 &= \varepsilon, \\ x_2 &= \varepsilon y_0 + \varepsilon^2 y_1 + O\left(\varepsilon^3\right), \\ p &= \varepsilon \mu_0 + \varepsilon^2 \mu_1 + O\left(\varepsilon^3\right), \end{aligned}$$

and we substitute the above equations to the analysed differential equations. The following algebraic equations are obtained

$$\begin{aligned} \varepsilon^2 : \; & \begin{aligned} \mu_0 - \mu_0 y_0 - 1 + y_0^2 &= 0, \\ (\mu_0 + 1)\, y_0 &= 0; \end{aligned} \\ \varepsilon^3 : \; & \begin{aligned} \mu_1 (1 - y_0) + y_1 (2y - \mu_0) &= -1, \\ \mu_1 y_0 + y_1 (1 + \mu_0) &= -2. \end{aligned} \end{aligned} \qquad (*)$$

Their solutions are

$$y_0^{(1)} = 0, \quad y_0^{(2)} = 1, \quad y_0^{(3)} = -2,$$
$$\mu_0^{(1)} = 1, \quad \mu_0^{(2)} = -1, \quad \mu_0^{(3)} = -1.$$

Since for each of three above pairs (y_0, μ_0) the main determinant of $(*)$ is not equal to zero, there are three pairs of solutions $\left(y_1^{(k)}, \mu_1^{(k)}\right)$ corresponding to three pairs $\left(y_0^{(k)}, \mu_0^{(k)}\right)$ for $k = 1, 2, 3$:

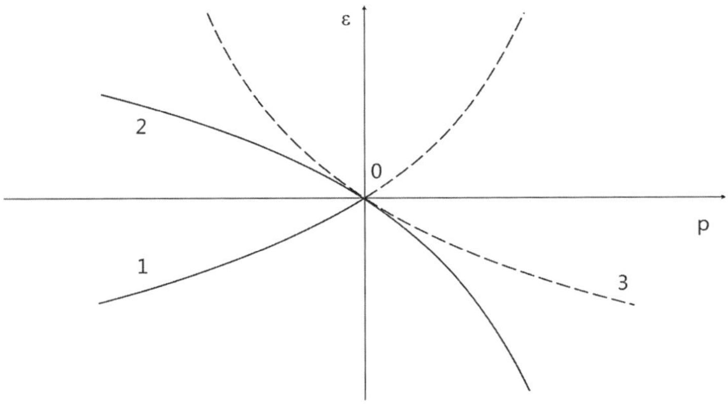

Fig. 13.10 Four branches of solutions (*dashed curves* correspond to unstable ones)

$$y_1^{(1)} = -1, \quad y_1^{(2)} = -\frac{1}{3}, \quad y_1^{(3)} = \frac{4}{3},$$
$$\mu_1^{(1)} = -1, \quad \mu_1^{(2)} = -2, \quad \mu_1^{(3)} = 1.$$

Therefore we have the following three branches of bifurcating solutions (see Fig. 13.10)

$$\begin{aligned} x_1^{(1)} &= \varepsilon, & x_1^{(2)} &= \varepsilon, & x_1^{(3)} &= \varepsilon, \\ x_2^{(1)} &= -\varepsilon^2 + O\left(\varepsilon^3\right), & x_2^{(2)} &= \varepsilon - \tfrac{1}{3}\varepsilon^3 + O\left(\varepsilon^2\right), & x_2^{(3)} &= -2\varepsilon - \tfrac{4}{3}\varepsilon^2 + O\left(\varepsilon^2\right), \\ p^{(1)} &= \varepsilon - \varepsilon^2 + O\left(\varepsilon^3\right), & p^{(2)} &= -\varepsilon - 2\varepsilon^2, & p^{(3)} &= -\varepsilon + \varepsilon^2 + O\left(\varepsilon^3\right). \end{aligned}$$

A stability of each bifurcated solutions depends on the eigenvalues of the matrix calculated in the point $x_1 = \varepsilon$, $x_2 = \varepsilon y_0$, $p = \varepsilon\mu_0$

$$B(\varepsilon) = \begin{bmatrix} \frac{\partial F_1}{\partial x_1} & \frac{\partial F_1}{\partial x_2} \\ \frac{\delta F_2}{\partial x_1} & \frac{\partial F_2}{\partial x_2} \end{bmatrix} = \varepsilon^2 \begin{bmatrix} \mu_0 - 2 & 2y_0 - \mu_0 \\ y_0 & 1 + \mu_0 \end{bmatrix} = \varepsilon_2 B_0.$$

Now, taking three pairs (y_0, μ_0) we obtain

$$B_0^{(1)} = \begin{bmatrix} -1 & -1 \\ 0 & 2 \end{bmatrix}, \; B_0^{(2)} = \begin{bmatrix} -3 & 3 \\ 1 & 0 \end{bmatrix}, \; B_0^{(3)} = \begin{bmatrix} -3 & -3 \\ -2 & 0 \end{bmatrix},$$

and the corresponding eigenvalues read

$$\sigma_{01}^{(1)} = -1, \; \sigma_{02}^{(1)} = 2, \; \sigma_{01,2}^{(2)} = -\tfrac{3}{2} \pm \tfrac{1}{2}\sqrt{21}, \; \sigma_{01,2}^{(3)} = -\tfrac{3}{2} \pm \tfrac{1}{2}\sqrt{33}.$$

□

All of the so far presented examples are taken from the monograph [144].

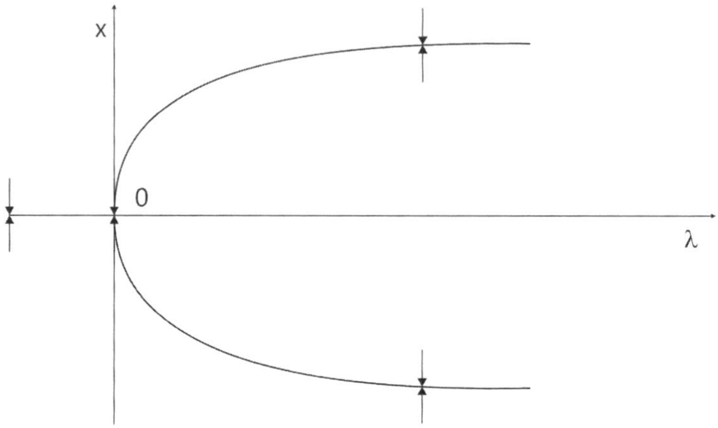

Fig. 13.11 Saddle-node bifurcation

13.2.3 Local Bifurcation of Hyperbolic Fixed Points

In the previous sections we have shown a general approach to analyse bifurcation appeared in one- and two-dimensional vector fields. Here we present local bifurcations of hyperbolic fixed points. A reader can easily apply the described earlier method to construct the corresponding bifurcation diagrams.

(i) *A saddle-node bifurcation*

This bifurcation is governed by the equation

$$\dot{x} = \lambda - x^2 \tag{13.36}$$

with the corresponding bifurcation diagram shown in Fig. 13.11.

(ii) A *transcritical bifurcation.*

This bifurcation is characterized by vector field

$$\dot{x} = \lambda x - x^2 \tag{13.37}$$

and its associated bifurcation diagram is shown in Fig. 13.12.

(iii) A *pitchfork bifurcation.*

This bifurcation occurs in the one-dimensional system with a cubic type non-liberality and is governed by the vector field

$$\dot{x} = \lambda x - x^3. \tag{13.38}$$

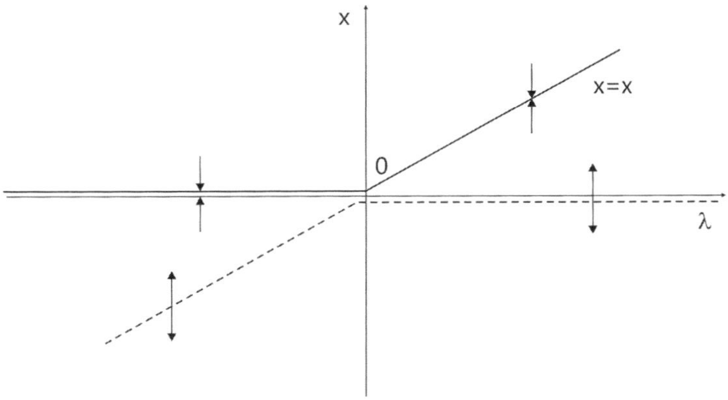

Fig. 13.12 A transcritical bifurcation

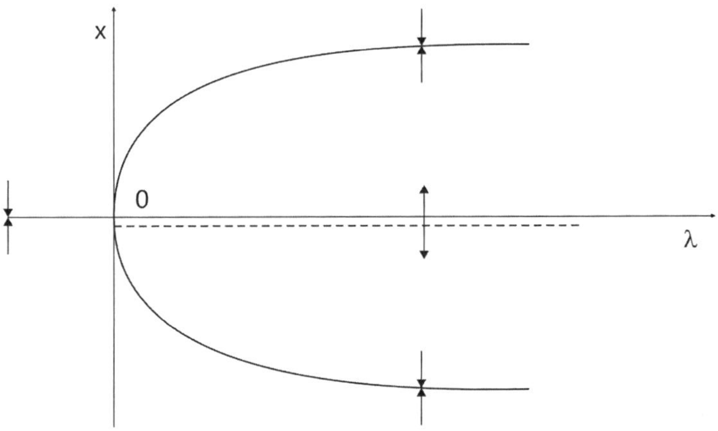

Fig. 13.13 A pitchfork bifurcation

The bifurcation diagram is shown in Fig. 13.13.

Note that for $\lambda < 0$ we have one branch of solutions, whereas for $\lambda > 0$ there are three branches of solution (two stable and one unstable).

There is another important question related to the so-called normal forms of the classical local bifurcations. In words, having a Taylor expansion around a trivial non-hyperbolic fixed point of a general parameter family of one-dimensional vector fields one can characterize the different geometry of the curves passing through origin by an appropriate truncation of the series. For example by adding the signs "$\pm$" instead of the sign before two terms on the right-hand side of (13.36), (13.37) and (13.38) we get the normal form of the mentioned local bifurcations.

13.2.4 Bifurcation of a Non-hyperbolic Fixed Point (Hopf Bifurcation)

The Hopf bifurcation plays a role of a door between static and dynamics and is very important in engineering. In spite of the original work of [124] a problem when a previously stable equilibrium becomes unstable and in a critical bifurcation point periodic orbit appear has been analysed also by Andronov et al. in the thirtieth (see [8,43]). Then in seventieth this central problem for dynamics has been reconsidered by many researchers like Chow and Mallet-Paret [64], Crandall and Rabinowitz [73], Hale [112], Hale and Oliveria [113], Hassard et al. [115], Holmes [123], Marsden and McCracken [164], Golubitsky and Schaeffer [100], Golubistky et al. [101], and others [125]. Here we briefly follow the approach described by Hassard [115].

Consider the following differential equations

$$\dot{x} = F(p, x), \qquad x \in \mathbb{R}^n. \tag{13.39}$$

We assumed that p is the bifurcation parameter and for $p_{cr} = 0$ we have $F(p_{cr}, 0) = 0$. We say that the system (13.39) has a family of periodic solutions with the parameter $\varepsilon \in (0, \varepsilon_0)$, and the amplitude of periodic solutions tends to zero when the formally introduced parameter $\varepsilon \to 0$.

Theorem 13.4 (Hopf). *Given*

(i) $F(p, 0) = 0$ *for every* p *from an open interval including* $p = 0$ *and* $0 \in \mathbb{R}^n$ *is the isolated fixed point of (13.39);*
(ii) The function F *is analytical with regard to* p *and* x *in a certain neighbourhood of* $(0, 0) \in \mathbb{R}^n \times \mathbb{R}^1$;
(iii) The matrix $A(p)$ *of the linearized system in a vicinity of zero solution has a pair of conjugated eigenvalues* σ *and* $\bar{\sigma}$, *where*

$$\sigma(p) = \xi(p) + i\eta(p), \tag{13.40}$$

and

$$\xi(0) = 0, \qquad \frac{d\xi}{dp}(0) \neq 0, \qquad \eta(0) = \eta_0 > 0; \tag{13.41}$$

(iv) Other eigenvalues of the matrix $A(0)$ *possess negative real parts.*

Then the system (13.39) has a family of periodic solutions. In addition, there is a certain ε^H *and an analytical function*

$$p^H(\varepsilon) = \sum_{i=2}^{\infty} p_i \varepsilon^i, \qquad (0 < \varepsilon < \varepsilon^H), \tag{13.42}$$

such that for every $\varepsilon \in (0, \varepsilon^H)$ *the system (13.39) for* $p = p^H(\varepsilon)$ *has a periodic solution* $x_\varepsilon(t)$.

The period of periodic solution $x_\varepsilon(t)$ is the analytical function

$$T^H(\varepsilon) = \frac{2\pi}{\eta_0}(1 + \sum_{i=2}^{\infty} \tau_i \varepsilon^i), \qquad (0 < \varepsilon < \varepsilon^H). \tag{13.43}$$

For every $L > 2\pi/p_0$ there is a neighbourhood R of the point $x = 0$ ($x \in \mathbb{R}^n$) and an open interval J including 0, such that for every $p \in J$ the periodic solution to the system (13.39) which lie in R and have a period smaller then T^H are the members of the family $x_{\varepsilon(t)}$ for which $p^H(\varepsilon) = p, \varepsilon \in (0, \varepsilon^H)$. The solutions which differ by an initial phase and corresponding to the same $x_\varepsilon(t)$ cannot be distinguishable. If $p^H(\varepsilon)$ is not identically equal to zero, then a first nonzero coefficient p_i has an even index. There exists such $\varepsilon^1 \in (0, \varepsilon^H)$ that $p^H(\varepsilon)$ is either positive or negative for $\varepsilon \in (0, \varepsilon^1)$. Two Floquet exponents of the solution $x_\varepsilon(t)$ tend to zero for $\varepsilon \to 0$. One of them is identically equal to zero for $\varepsilon \in (0, \varepsilon^H)$, whereas second one

$$\sigma(\varepsilon) = \sum_{i=2}^{\infty} \sigma_i \varepsilon^i, \qquad 0 < \varepsilon < \varepsilon^H. \tag{13.44}$$

The periodic solution $x_\varepsilon(t)$ is orbitally stable (unstable) with the asymptotic phase, if $\sigma(\varepsilon) < 0 \quad (\sigma(\varepsilon) > 0)$.

The proof of the Hopf theorem is here omitted, but it can be found in the work of Marsden and McCracken [164]. Although the Hopf theorem in its source version is related to the systems with analytical right-hand sides, but the similar like theorems have been formulated later for the right-hand sides being differentiable (see [115]). We have used the same reference to formulate the Hopf theorem. The main idea of the proof is focused on reduction of initial n-dimensional system to two-dimensional Poincaré form with use of the central manifold theory. A construction of a bifurcated solution relies on application of the normal Poincaré form. Another alternative approach has been presented by Iooss and Joseph [127] (see also Kurnik [145]). We present the later one, since it seems to be more economical leading to estimate bifurcating solutions and their stability.

Assume that self-excited oscillations equation of a mechanical system with lumped parameters is governed by the equation

$$\dot{U} = F(\nu, U; Q), \tag{13.45}$$

where $U \in \mathbb{R}^n$ and $F : \mathbb{R}^n \times \mathbb{R}^n \to \mathbb{R}^n$, ν is the parameter governing self-excitation and further referred as the bifurcation parameter; Q is another parameter.

Assume that F is analytical with regard to U and ν in a vicinity of ν_{cr}. Let $U^* = U^*(\nu, Q)$ be a constant solution. Note that for each Q the constant solution depends on bifurcation parameter, and hence we have a family of constant solutions. On the other hand, for a given ν and Q, one may have many solutions $U_1^*, U_2^*, \ldots$.

Dealing with a Hopf bifurcation we consider one of the U_i^* solutions. The corresponding critical value of ν is found investigating stability of the being analysed constant solution $U^*(\nu, Q)$. Assuming that $A(\nu, Q)$ is the matrix corresponding to the linearized system, then the characteristic equation is

$$\det(A - \sigma I) = 0, \tag{13.46}$$

where I is the identity $n \times n$ matrix.

The critical point $\nu = \nu_{cr}(Q)$ is defined by $\mathrm{Re}\{\sigma_1(\nu_{cr}, Q)\} = 0$, where $\sigma_1 > \sigma_i$, $i = 2, \ldots, n$. Then the system (13.45) is transformed to its local form by introducing the variable $u = U - U^*(\nu, Q)$, and the parameter $\omega = \nu - \omega_{cr}$. Equation (13.45) is recast to the following form

$$\dot{u} = f(\omega, u), \tag{13.47}$$

where: $f(\omega, u) = F(\omega + \omega_{cr}, u + U^*; Q) - F(\omega + \omega_{cr}, U^*; Q)$, $Q, \omega \in \mathbb{R}^+, u, U^* \in \mathbb{R}^n$.

Observe that $f(\omega, 0) = 0$. The parameter Q will be further omitted to simplify our considerations. Therefore the problem is reduced to consideration of the following system

$$\dot{u} = A(\omega)u + N(\omega, u), \tag{13.48}$$

where $A(\omega, Q) = A(\omega)$, $A_{ij} = [\frac{\partial f_i}{\partial u_j}]_{u=0}$.

Let $\sigma_1(\omega), \ldots, \sigma_n(\omega)$ be the eigenvalues of $A(\omega)$. Claiming also that the assumptions of Hopf theorem are satisfied in a vicinity of the critical point:

$$\begin{aligned} \mathrm{Re}\,\sigma_1(0) = \xi(0) = 0, \\ \mathrm{Im}\,\sigma_2(0) = \eta(0) = \Omega_0, \\ \frac{d\xi}{d\omega}(0) \neq 0, \end{aligned} \tag{13.49}$$

where Ω_0 is the positive number and $\sigma_1, \sigma_2 = \bar{\sigma}_1$ are simple eigenvalues.

Let q and q^* be the eigenvectors of the matrix $A(\omega)$ and $A^*(\omega) = A^T(\omega)$, respectively. They are associated with the imaginary eigenvalues in the critical point. These are found from the equations

$$\begin{aligned} (A(0) - \sigma_1(0)I)q = 0, \\ (A^T(0) - \bar{\sigma}_1(0)I)q^* = 0. \end{aligned} \tag{13.50}$$

The required uniqueness is achieved by introduction of the normalization procedure

$$\langle q, q^* \rangle = 1, \tag{13.51}$$

where $\langle q, q^* \rangle = \sum_{i=1}^{n} a_i \bar{b}_i$ is the Hermitean scalar product in complex n-dimensional Euclidean space and the vectors q and $\bar{q}^*$ are orthogonal, i.e.

$$\langle q, \bar{q}^* \rangle = \sum_{i=1}^{n} q_i q_i^* = 0. \tag{13.52}$$

They are linearly independent and they will play a role of skeleton of being sought periodic solutions. The vector q^* will be used during implementation of the orthogonal condition within the Fredholm alternative.

The being sought bifurcated solution has the form

$$u(s, \varepsilon) = \sum_{i=1}^{\infty} \frac{1}{n!} u^{(n)}(s) \varepsilon^n, \tag{13.53}$$

where $s = \Omega(\varepsilon)t$. We take

$$\begin{aligned} \Omega(\varepsilon) &= \Omega_0 + \sum_{i=1}^{\infty} \frac{1}{n!} \varepsilon^n \Omega_n, \\ \omega(\varepsilon) &= \omega_{cr} + \sum_{i=1}^{\infty} \frac{1}{n!} \varepsilon^n \omega_n, \quad (\omega_{cr} = 0), \end{aligned} \tag{13.54}$$

where Ω is the frequency of the sought periodic solution; $\Omega_0 = \operatorname{Im} \sigma_1(0)$; ω_n and Ω_n are the series terms; $u^{(n)}(s)$ is the series of $2\pi-$ periodic continuous and differentiable functions; ε is the parameter measuring the distance between a trivial, and periodic solutions in a sense of the applied norm.

Note that an existence of bifurcated solutions in the form (13.53) is guaranteed by Hopf theorem. Let $P_{2\pi}$ is the space of complex and $2\pi-$periodic functions, continuous and differentiable, where the following scalar product is defined

$$[a(s), b(s)] \stackrel{\text{df}}{=} \frac{1}{2\pi} \int_0^{2\pi} < a(s), \bar{b}(s) > ds, \tag{13.55}$$

and we take the norm

$$\|a\| = \sqrt{[\bar{a}, a]}. \tag{13.56}$$

In addition, following Iooss and Joseph ([127]), we introduce the Maclaurin series of the function $f(u, t; \omega, p)$ with regard to u. Since $f(0, t; \omega, p) \equiv 0$, hence

$$f(u, t; \omega, p) = \sum_{i=1}^{\infty} \frac{1}{n!} f_{uu\ldots u}(\omega, p, t, 0|u|u| \ldots |u), \tag{13.57}$$

where $f_{uu\dots u}(\omega, p, t, 0|u|u|\dots|u)$ is nth-linear operator acting on the vector u in neighbourhood of $u = 0$.

In general, an nth-linear operator of the vector field $f(u, t, \omega, p)$ acting on the vectors $a_1, a_2, \dots, a_n$ in an arbitrary point u_0 is defined by

$$f_{uu\dots u}(\omega, p, t, u_0|a_1|a_2|\dots|a_n) = \lim_{h_1, h_2, \dots, h_n \to 0} \frac{\partial^{(n)} f(u_0 + h_1 a_1 + \dots + h_n a_n, t; \omega, p)}{\partial h_1 \partial h_2 \dots \partial h_n} \tag{13.58}$$

The right-hand side of (13.47) can be developed into the Maclaurin series

$$f(\omega, u) = \sum_{i=1}^{\infty} \frac{1}{n!} f_{uu\dots u}(\omega, 0|u|u|\dots|u). \tag{13.59}$$

On the other hand, each term of the series (13.59) is developed into the Maclaurin series with regard to ω:

$$\begin{aligned} f(\omega, u) &= f_u(0, 0|u) + f_{u\omega}(0, 0|u)\omega + \frac{1}{2!} f_{u\omega\omega}(0, 0|u)\omega^2 + \cdots \\ &\dots + \frac{1}{2!}\left\{ f_{uu}(0, 0|u|u) + f_{uu\omega}(0, 0|u|u)\omega + \frac{1}{2!} f_{uu\omega\omega}(0, 0|u|u)\omega^2 + \cdots \right\} \\ &+ \frac{1}{3!}\left\{ f_{uuu}(0, 0|u|u|u) + f_{uuu\omega}(0, 0|u|u|u)\omega + \frac{1}{2!} f_{uuu\omega\omega}(0, 0|u|u|u)\omega^2 + \cdots \right\} + \cdots \end{aligned} \tag{13.60}$$

and hence

$$\begin{aligned} f(\omega, u) &= \sum_{i=1}^{\infty} \frac{\varepsilon^i}{i!} f_u(0, 0|u^{(i)}) \\ &+ \sum_{i=1}^{\infty}\sum_{j=1}^{\infty} \frac{\varepsilon^{i+j}}{i!j!} f_{u\omega}(0, 0|u^{(i)})\omega_j + \frac{1}{2!} f_{uu\omega}(0, 0|u^{(i)}|u^{(j)}) \\ &+ \sum_{i=1}^{\infty}\sum_{j=1}^{\infty}\sum_{k=1}^{\infty} \frac{\varepsilon^{i+j+k}}{i!j!k!} \{ f_{u\omega\omega}(0, 0|u^{(i)})\omega_j\omega_k + \frac{1}{2!} f_{uu\omega}(0, 0|u^{(i)}|u^{(j)})\omega_k \\ &+ \frac{1}{3!}\{ f_{uuu}(0, 0|u^{(i)}|u^{(j)}|u^{(k)})\} + \sum_{i=1}^{\infty}\sum_{j=1}^{\infty}\sum_{k=1}^{\infty}\sum_{l=1}^{\infty} \frac{\varepsilon^{i+j+k+l}}{i!j!k!l!}\{\dots\} + \cdots \end{aligned} \tag{13.61}$$

The left-hand side of (13.47) has the form

$$\dot{u} = \Omega_0 \sum_{i=1}^{\infty} \frac{\varepsilon^i}{i!} \frac{du^{(i)}}{ds} + \sum_{i=1}^{\infty}\sum_{j=1}^{\infty} \frac{\varepsilon^{i+j}}{i!j!} \Omega_j \frac{du^{(i)}}{ds}. \tag{13.62}$$

Comparing the same terms by the same powers of ε in (13.61) and (13.62) we get

$$-\Omega_0 \frac{du^{(1)}}{ds} + f_u(0,0|u^{(1)}) = 0 \quad (\varepsilon^1)$$

$$-\Omega_0 \frac{du^{(n)}}{ds} + f_u(0,0|u^{(n)}) = \sum_{i+j=n} \frac{n!}{i!j!}\{\Omega_1 \frac{du^{(j)}}{ds} - f_{u\omega}(0,0|u^{(i)})\omega_j +$$

$$-\frac{1}{2!} f_{uu}(0,0|u^{(i)}|0^{(j)})\} - \sum_{i+j+k=n} \frac{n!}{i!j!k!} \cdots + \cdots (\varepsilon^n \quad \text{for} \quad n > 1). \tag{13.63}$$

Introducing the operator

$$J_0(\cdot) \overset{\text{df}}{=} -\Omega_0 \frac{d(\cdot)}{ds} + f_u(0,0|(\cdot)), \tag{13.64}$$

Equation (13.63) takes the following form

$$\begin{cases} J_0 u^{(1)} = 0, \\ J_0 u^{(n)} = g_n(s), \end{cases} \tag{13.65}$$

where

$$g_n(s) = g_n(s + 2\pi) = n\Omega_{n-1} \frac{du^{(1)}}{ds} - n\omega_{n-1} f_{u\omega}(0,0|u^{(1)}) - R_{n-1},$$

$$R_{n-1} = \sum_{i+j=n} \sum_{i,j \geq 1} \binom{n}{i} \frac{1}{2!} f_{uu}(0,0|u^{(i)}|u^{(j)}) + \sum_{i+j=n} \sum_{i \geq 2, j \geq 1} \frac{n!}{i!j!} \left\{ -\Omega_j \frac{du^{(i)}}{ds} \right.$$

$$\left. + \omega_j f_{u\omega}(0,0|u^{(i)}) \right\} + \sum_{i+j+k=n} \sum_{i,j,k \geq 1} \frac{n!}{i!j!k!} \left\{ \frac{1}{2!} f_{u\omega\omega}(0,0|u^{(i)})\omega_j\omega_k \right.$$

$$\left. + \frac{1}{2!} f_{uu\omega}(0,0|u^{(i)}|u^{(j)})\omega_k + \frac{1}{3!} f_{uuu}(0,0|u^{(i)}|u^{(j)}|u^{(j)}) \right\} + \cdots \tag{13.66}$$

Let us introduce the following harmonic functions

$$z = qe^{is}, \quad z^* = q^* e^{is}. \tag{13.67}$$

The following properties hold

$$J_0 z = J_0 \bar{z} = 0, \tag{13.68}$$

and

$$J_0^* \stackrel{\text{df}}{=} \Omega_0 \frac{d(\cdot)}{ds} + f_n(0, 0|(\cdot)), \tag{13.69}$$

where: $f_u^* = A^*(0)u = A^T(0)u$.

The function z and $\bar{z}$ are linearly independent, and we assume the following orthogonality condition

$$[u, z^*] = \varepsilon. \tag{13.70}$$

This condition implies a chain of conditions

$$[u^{(1)}, z^*] = 1, \quad [u^{(n)}, z^*] = 0 \qquad for \quad n > 1. \tag{13.71}$$

In other words it means that the fundamental harmonic $e^{is}(e^{-is})$ appears only in $u^{(1)}$. The condition (13.70) gives an iteration for ε as the amplitude of the bifurcated solution. In order to solve (13.65), we assume

$$u^{(1)} = cz + \bar{c}\bar{z}, \tag{13.72}$$

where c is the complex constant.

The orthogonalization condition gives

$$[u^{(1)}, z^*] = c[z, z^*] = \bar{c}[z, z^*], \tag{13.73}$$

and hence $c = 1$ and

$$u^{(1)} = [z + z^*]. \tag{13.74}$$

The being sought periodic solution of (13.65) exists if the following orthogonality condition is satisfied (Fredholm alternative).

$$[g_n, z^*] = 0. \tag{13.75}$$

This condition eliminates the occurring secular terms from $g_n(s)$. Taking into account (13.72) and (13.74) in (13.75) we obtain the following complex equation

$$n\Omega_{n-1}i - n\omega_{n-1} < f_{u\omega}(0, 0|q|q^*) > -[R_{n-1}, z^*] = 0 \tag{13.76}$$

with two unknowns Ω_{n-1} and ω_{n-1}. If $\sigma(\omega) = \xi(\omega) + i\eta(\omega)$ is the eigenvalue associated with the vector $q(\omega)$, then

$$\sigma q = f_u(\omega, 0|q). \tag{13.77}$$

Differentiating both sides of (13.75) with regard to ω and taking $\omega = 0$ we obtain

$$\{A(0) + i\Omega_0 I\}q_\omega(0) = \sigma_\omega(0) - f_{u\omega}(0, 0|q(0)), \tag{13.78}$$

where $q_\omega(0)$ is unknown. The Fredholm alternative applied to (13.78) gives

$$\langle \sigma_\omega(0)q(0) - f_{u\omega}(0, 0|q(0)), q^*\rangle = 0. \tag{13.79}$$

Taking into account the earlier introduced notation $q(0) = q$ and $q^*(0) = q^*$ ($\langle q, \bar{q}^*\rangle = 1$) we get

$$\sigma_\omega(0) =< f_{u\omega}(0, 0|q(0)), q^* > . \tag{13.80}$$

A solvability condition for $n > 1$ with regards to $u^{(n)} \in P_{2\pi}$ yields the equation

$$-n\Omega_{n-1}i + n\omega_{n-1}r_\omega(0) + [R_{n-1}, z^*] = 0. \tag{13.81}$$

Separating real and imaginary parts we obtain

$$\omega_{n-1} = -\frac{\mathrm{Re}[R_{n-1}, z^*]}{n\xi_\omega(0)} \tag{13.82}$$

and

$$\Omega_{n-1} = -\mathrm{Re}[R_{n-1}, z^*]\frac{\eta_{\omega(0)}}{\xi_\omega(0)} + \frac{1}{n}\mathrm{Im}[R_{n-1}, z^*]. \tag{13.83}$$

The obtained dependencies (13.82) and (13.83) allow to find the unknown coefficients. Taking $n = 2$, we get $R_1 = f_{uu}(0, 0|u^{(1)}|u^{(1)})$ and $[R_1, z^*] = 0$, which implies that $\omega_1 = \Omega_1 = 0$. It can be shown that

$$\omega_{2k-1} = \Omega_{2k-1} = 0 \qquad k \in \mathbb{N}. \tag{13.84}$$

It means that the functions $\omega(\varepsilon)$ i $\Omega(\varepsilon)$ are even.

In order to obtain ω_2 and Ω_2 we take $n = 3$, and we get

$$R_2 = \frac{3}{2}f_{uu}(0, 0|u^{(1)}|u^{(2)}) + f_{uuu}(0, 0|u^{(1)}|u^{(1)}|u^{(1)}). \tag{13.85}$$

To find $[R_2, z^*]$ one needs to solve (13.66) for $n = 2$ ($u^{(2)}(s)$). The right-hand side of $g_2(s)$ can be presented by

$$g_2 = -R_1 = -f_{uu}(0, 0|q|q)e^{i2s} - 2f_{uu}(0, 0|q|\bar{q}) - f_{uu}(0, 0|\bar{q}|\bar{q})e^{-i2s}, \tag{13.86}$$

where the condition $[g_2, z^*] = 0$ is always satisfied. Let us denote

$$-f_{uu}(0,0|q|q) = P, \; -2f_{uu}(0,0|q|q) = S, \tag{13.87}$$

and hence

$$g_2 = S + Pe^{i2s} + \bar{P}e^{-i2s} \tag{13.88}$$

The solution $u^{(2)}(s)$ of (13.65) is a sought in the form

$$u^{(2)}(s) = K + Le^{i2s} + \bar{L}e^{-i2s} \tag{13.89}$$

where K, L are real and complex vectors of the form

$$K = A^{-1}(0)S, L = \{A(0) - 2i\Omega_0 I\}^{-1}P, \tag{13.90}$$

where $A(0)$ and $A(0) - 2i\Omega_0 I$ are nonsingular.
Denoting

$$Le^{i2s} = y, \tag{13.91}$$

we obtain

$$\begin{aligned} f_{uu}(0,0|u^{(1)}|u^{(2)}) &= f_{uu}(0,0|z+\bar{z}|K+y+\bar{y}) = f_{uu}(0,0|z|K) \\ +f_{uu}(0,0|z|y) + f_{uu}(0,0|z|\bar{y}) &+ f_{uu}(0,0|\bar{z}|K) + f_{uu}(0,0|\bar{z}|y) + f_{uu}(0,0|\bar{z}|\bar{y}). \end{aligned} \tag{13.92}$$

Applying orthogonalization to each term of (13.72) with z^* we obtain

$$\begin{aligned} [f_{uu}(0,0|u^{(1)}|u^{(2)}), z^*] &= \langle (f_{uu}(0,0|q|K) + f_{uu}(0,0|q|L)), q^* \rangle \\ &= \sum_{i=1}^{n}\sum_{j=1}^{n}\sum_{k=1}^{n} b_{ijk}\bar{q}_i^*(q_j K_k + \bar{q}_j L_k), \end{aligned} \tag{13.93}$$

where

$$b_{ijk} = \frac{\partial^2 f_i}{\partial u_j \partial u_k}|_{u=0,\omega=0}. \tag{13.94}$$

Noting that

$$\begin{aligned} [f_{uuu}(0,0|u^{(1)}|u^{(1)}|u^{(1)}), z^*] &= 3\langle f_{uuu}(0,0|q|q|\bar{q}) \rangle \\ &= \sum_{i=1}^{n}\sum_{j=1}^{n}\sum_{k=1}^{n}\sum_{l=1}^{n} c_{ijkl}\bar{q}_i^* q_j q_k q_l, \end{aligned} \tag{13.95}$$

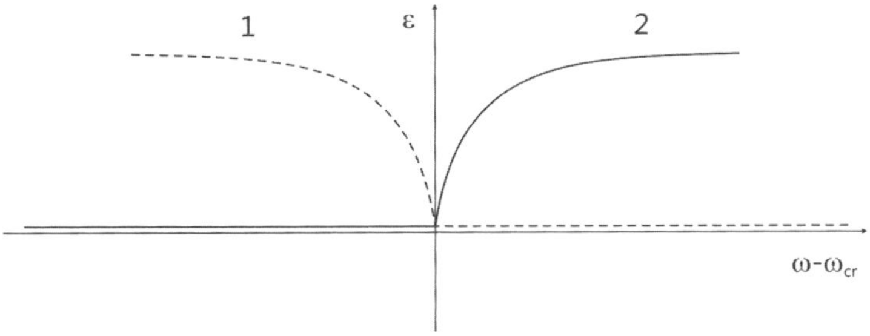

Fig. 13.14 Amplitude of the periodic solution versus the bifurcation parameter for sub-critical (*1*) and super-critical (*2*) bifurcation

where:

$$c_{ijkl} = \frac{\partial^3 f_i}{\partial u_j \partial u_k \partial u_l}|_{u=0,\omega=0}, \tag{13.96}$$

we obtain

$$\begin{aligned}[R_2, z^*] = &\frac{3}{2}\langle f_{uu}(0,0|q|K) + f_{uu}(0,0|\bar{q}|L), q^*\rangle \\ &+3\langle f_{uuu}(0,0|q|q|\bar{q}), q^*\rangle.\end{aligned} \tag{13.97}$$

The first-order approximation of the bifurcated solution is

$$u(s,\varepsilon) = \varepsilon u^{(1)}(s) = \varepsilon(z+\bar{z}) = 2\varepsilon\text{Re}\{q^{is}\} = 2\varepsilon\{\text{Re q}\cos\Omega(\varepsilon)t - \text{Im q}\sin\Omega(\varepsilon)t\}, \tag{13.98}$$

where:

$$\Omega = \Omega_0 + \frac{1}{2}\varepsilon^2\Omega_2 \tag{13.99}$$

and

$$\omega = \omega_{cr} + \frac{1}{2}\varepsilon^2\omega_2. \tag{13.100}$$

The obtained dependence $\omega(\varepsilon)$ is shown in Fig. 13.14.

For $\omega_2 > 0$ a super-critical bifurcation occurs, whereas for $\omega_2 < 0$ a subcritical bifurcation appears. In the case of $\omega_2 = 0$ one has to calculate ω_4. Eliminating ε from (13.99) and (13.100) we obtain the dependence of self-excited frequency versus the bifurcation parameter

$$\Omega = \Omega_0 + \frac{\Omega_2}{\omega_2}(\omega - \omega_{cr}). \tag{13.101}$$

The second-order approximation to the bifurcated solution has the form

$$u(s,\varepsilon)=\varepsilon u^{(1)}+\frac{1}{2}\varepsilon^2u^{(2)}=2\varepsilon\mathrm{Re}\{qe^{is}\}+\frac{1}{2}\varepsilon^2K+\varepsilon^2\mathrm{Re}\{Le^{i2s}\}. \tag{13.102}$$

It is seen that it contains harmonic and superharmonic parts as well as the constant parts. The latter causes a shift of oscillation origin.

One can proceed in a similar way to get the successive approximations for $n=3,4,\ldots$.

For the first-order approximation we get

$$\begin{aligned}\|u(s,\varepsilon)\| &= \sqrt{[u,u]}=\left\{\frac{1}{2\pi}\int_0^{2\pi}<u,u>ds\right\}^{\frac{1}{2}}\\ &=\varepsilon\left\{\frac{1}{2\pi}\int_0^{2\pi}\sum_{i=1}^{n}(q_ie^{is}+\bar{q}_ie^{-is})(\bar{q}_ie^{-is}+q_ie^{is})ds\right\}^{\frac{1}{2}}\\ &=\varepsilon\left\{\frac{1}{2\pi}\int_0^{2\pi}\sum_{i=1}^{n}(q_i^2e^{i2s}+2q_i\bar{q}_i+(\bar{q}_i{}^2e^{-i2s})ds\right\}^{\frac{1}{2}}\\ &=\varepsilon\left\{2\sum_{i=1}^{n}2q_i\bar{q}_i\right\}^{\frac{1}{2}}=\varepsilon\left\{2\sum_{i=1}^{n}|q_i|^2\right\}^{\frac{1}{2}}=\varepsilon\sqrt{2}\|q\|. \end{aligned}\tag{13.103}$$

In the second-order approximation we have

$$u(t,\omega)=2\sqrt{\frac{2\omega}{\omega_2}}\mathrm{Re}\{qe^{i\Omega t}\}+\frac{\omega}{\omega_2}K+\frac{2\omega}{\omega_2}\mathrm{Re}\{Le^{i2\Omega t}\}. \tag{13.104}$$

The $\omega(\varepsilon)$ approximation using nth order polynomial

$$\omega(\varepsilon)=\omega_{cr}+\frac{1}{2}\omega_2\varepsilon^2+\frac{1}{24}\omega_4\varepsilon^4+O(\varepsilon^6) \tag{13.105}$$

yields the picture shown in Fig. 13.15.

The term ω_4 is defined by

$$\omega_4=-\frac{\mathrm{Re}[R_4,z^*]}{5\xi_\omega(\omega_{cr})}, \tag{13.106}$$

where:

$$\begin{aligned}R_4 &= 30\omega_2\langle f_{uu}(q|K)+f_{uu\omega}(\bar{q}|L)+f_{uuu\omega}(q|q|q),q^*\rangle\\ &+15\langle f_{uuu}(q|K|K)+2f_{uuu}(\bar{q}|L|\bar{L})+2f_{uuu}(\bar{q}|K|L),q^*\rangle.\end{aligned}\tag{13.107}$$

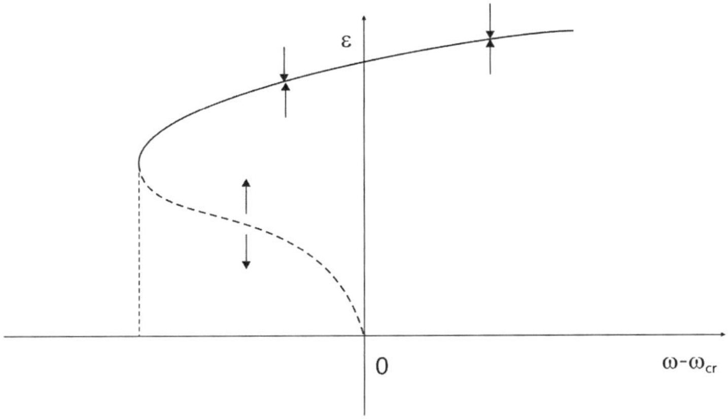

Fig. 13.15 Hopf bifurcation diagram

Now we briefly describe a stability estimation of the periodic bifurcated solution $V(s,\varepsilon)$ of Eq. (13.47). Considering v as the perturbation to the investigated solution V, i.e. by substituting

$$v = u - V \tag{13.108}$$

to (13.47) we get

$$\dot{v} = g(\omega, v, s, \varepsilon), \tag{13.109}$$

where

$$g(\omega, v, s, \varepsilon) = f(\omega, V + v) - f(\omega, V). \tag{13.110}$$

The linearized equation (13.109) has the form

$$\dot{v} = g_v(\omega, 0|v), \tag{13.111}$$

where

$$g_v(\omega, 0|v) = f_n(\omega, v(s,\varepsilon)|v). \tag{13.112}$$

Equation (13.111) has periodic coefficients. A stability of $v = 0$ depends on the eigenvalues of the monodromy matrix. One of the eigenvalues (Floquet exponents) is equal to zero, whereas the second depends analytically on ε, i.e. $\sigma = \sigma(\varepsilon)$ and $\sigma(0) = 0$.

Theorem 13.5 (Orbital Stability). *A limit cycle is asymptotical orbitally stable if all Floquet exponents have negative real parts.*

The next theorem gives hints how to find real Floquet exponents.

Theorem 13.6. *A real Floquet exponent* $\sigma(\varepsilon)$ *can be presented in the form*

$$\sigma(\varepsilon) = \hat{\sigma}(\varepsilon)\frac{d\omega}{d\varepsilon}, \tag{13.113}$$

where: $\hat{\sigma}(\varepsilon)$ *is the smooth function in neighbourhood of* $\varepsilon = 0$ *such that* $\frac{d\hat{\sigma}(\varepsilon)}{d\varepsilon}(0) = -\xi(0)$, *and* $\frac{\hat{\sigma}(\varepsilon)}{\varepsilon}$ *is an even function. Since*

$$\frac{d\omega}{d\varepsilon} = \varepsilon\omega_2 + O(\varepsilon^4) \tag{13.114}$$

and

$$\hat{\sigma}(\varepsilon) = -\xi_\omega(0)\varepsilon + O(\varepsilon^3), \tag{13.115}$$

therefore

$$\sigma(\varepsilon) = -\xi_\omega(0)\omega_2\varepsilon^2 + O(\varepsilon^4). \tag{13.116}$$

To conclude, we have two following cases:

(i) $\omega\xi_\omega(0) > 0$, *and* $\sigma(\varepsilon) < 0$. *Then the solution is orbitally asymptotic stable;*
(ii) $\omega\xi_\omega(0) < 0$, *and* $\sigma(\varepsilon) < 0$. *Then the solution is orbitally unstable.*

Example 13.4 (See [144]). We consider the Van der Pol equation of the following form

$$\dot{x_1} = x_2,$$
$$\dot{x_2} = -x_1 + \omega x_2 - x_1^2 x_2.$$

The matrix of the linearized system in the vicinity of $(0, 0)$ is

$$A(\omega) = \begin{bmatrix} 0 & 1 \\ -1 & \omega \end{bmatrix}.$$

The eigenvalues of $A(\omega)$ are defined by the equation

$$\sigma^2 - \omega\sigma + 1 = 0.$$

Hence, for $|\omega| < 2$ there exist complex conjugate roots, whereas for $|\omega| \geq 2$ we have real eigenvalues $\sigma_2 << \sigma_1 < 0$ for $\omega < -2$, and $\sigma_1 > \sigma_2 > 0$ for $\omega > 2$. The complex conjugate roots are

$$\sigma_{1,2} = \frac{\omega}{2} \pm \frac{i}{2}\sqrt{4 - \omega^2}.$$

The trajectories of the eigenvalues σ_1 and σ_2 form the up and down half-circle of the complex plane $(\mathrm{Re}\,\sigma)^2 + (\mathrm{Im}\,\sigma)^2 = 1$. Depending on ω we have the following types of the singular points

(1) $\omega < -2$ (stable node);
(2) $\omega = -2$ (stable node degenerated);
(3) $-2 < \omega < 0$ (stable focus);
(4) $\omega = 0$ (asymptotically stable focus);
(5) $0 < \omega < 2$ (unstable focus);
(6) $\omega = 2$ (unstable node degenerated);
(7) $\omega > 2$ (unstable node).

Let us check the Hopf bifurcation theorem assumptions:

(i) $f(\omega, 0)$ for every ω, $x = 0$ is isolated equilibrium;
(ii) the function f is analytical for $(x, \omega) \in \mathbb{R}^2 \times \mathbb{R}$;
(iii) the matrix $A(\omega)$ has a pair of complex eigenvalues such that $\mathrm{Re}\sigma_1(0) = \mathrm{Re}\sigma_2(0) = 0$, $\quad \mathrm{Im}(\sigma_1) = 1 > 0$ and $\frac{d(\mathrm{Re}\sigma_1)}{d\omega}(0) = \frac{1}{2} \neq 0$.

The bifurcated solutions and the eigenvectors are defined by the equations that can be found using the Iooss and Joseph [127] method, and they are reported in [144]. □

13.2.5 Double Hopf Bifurcation

One may extend a concept of analysis of Hopf bifurcation into a case when a few pairs of purely imaginary eigenvalues cross an imaginary axis of the complex plane with nonzero velocities. Although there exist many papers devoted to this problem [73, 100, 164], but we follow here Dei Yu [250] research results.

If the Jacobian of higher-dimensional dynamical system possesses two pairs of purely imaginary eigenvalues, then the so-called double Hopf bifurcation may be exhibited. If the ratio of the two eigenvalues is not a rational number, then associated bifurcation in non-resonant.

Following Yu [250], consider the following dynamical system

$$\dot{\mathbf{x}} = Ax + F(x). \tag{13.117}$$

where: $x \in \mathbb{R}^n$, $F : \mathbb{R}^n \to \mathbb{R}^n$, $F(0) = F'(0) = 0$, and

$$A = \begin{bmatrix} 0 & \omega_{1c} & 0 & 0 & 0 \\ -\omega_{1c} & 0 & 0 & 0 & 0 \\ 0 & 0 & 0 & \omega_{2c} & 0 \\ 0 & 0 & -\omega_{2c} & 0 & 0 \\ 0 & 0 & 0 & 0 & B \end{bmatrix}. \tag{13.118}$$

Notice that B matrix is of order $(n-4)\times(n-4)$ and it is hyperbolic one, which means that its associated eigenvalues have no zero real parts. We assume also that $\omega_{1c}/\omega_{2c} \neq k/l$, where $k, l \in \mathbb{C}$.

In an explicit form Eq. (13.117) reads

$$
\begin{aligned}
\dot{x}_1 &= \omega_{1c}x_2 + F_1(x), \quad \dot{x}_2 = -\omega_{1c}x_1 + F_2(x),\\
\dot{x}_3 &= \omega_{2c}x_4 + F_3(x), \quad \dot{x}_4 = -\omega_{2c}x_3 + F_4(x),\\
\dot{x}_p &= -\alpha_p x_p + F_p(x), \quad p = 5, 6, \ldots, m_1 + 4,\\
\dot{x}_q &= -\alpha_q x_q + \omega_q x_{q+1} + F_q(x), \quad \dot{x}_{q+1} = -\omega_q x_q - \alpha_q x_{q+1} + F_{q+1}(x),\\
& q = m_1 + 5, m_1 + 7, \ldots, n-1
\end{aligned} \tag{13.119}
$$

and $n = 4 + m_1 + 2m_2$.

In what follows, the multiple time scale method is used assuming that $t = t(T_0, T_1, T_2, \ldots)$, and $T_0 = t$, $T_1 = \varepsilon t$, $T_2 = \varepsilon^2 t$, and so on. Therefore

$$
\begin{aligned}
\frac{d}{dt} &= \frac{\partial}{\partial T_0}\frac{\partial T_0}{\partial t} + \frac{\partial}{\partial T_1}\frac{\partial T_1}{\partial t} + \frac{\partial}{\partial T_2}\frac{\partial T_2}{\partial t} + \cdots\\
&= D_0 + \varepsilon D_1 + \varepsilon^2 D_2 + \cdots .
\end{aligned} \tag{13.120}
$$

A being sought solution is expanded into the power series with respect to ε

$$
x_i(t;\varepsilon) = \varepsilon x_{i1}(T_0, T_1, T_2, \ldots) + \varepsilon^2 x_{i2}(T_0, T_1, T_2, \ldots) + \varepsilon^3 x_{i3}(T_0, T_1, T_2, \ldots) + \cdots \tag{13.121}
$$

Substituting (13.121) into (13.2.5), and accounting (13.120) the following sequence of perturbation equations is obtained

$$
\begin{aligned}
\varepsilon: D_0x_{11} &= \omega_{1c}x_{21},\\
D_0x_{21} &= -\omega_{1c}x_{11},\\
D_0x_{31} &= \omega_{2c}x_{41},\\
D_0x_{41} &= -\omega_{2c}x_{31},\\
D_0x_{p1} &= -\alpha_p x_{p1},\\
D_0x_{q1} &= -\alpha_q x_{q1} + \omega_q x_{(q+1)1},\\
D_0x_{(q+1)1} &= -\omega_q x_{q1} + \alpha_q x_{(q+1)1};
\end{aligned} \tag{13.122}
$$

$$
\begin{aligned}
\varepsilon^2: D_0x_{12} &= \omega_{1c}x_{22} - D_1x_{11} + F_{12}(x_1),\\
D_0x_{22} &= -\omega_{1c}x_{12} - D_1x_{21} + F_{22}(x_1),\\
D_0x_{32} &= \omega_{2c}x_{42} - D_1x_{31} + F_{32}(x_1),\\
D_0x_{42} &= -\omega_{2c}x_{32} - D_1x_{41} + F_{42}(x_1),\\
D_0x_{p2} &= -\alpha_p x_{p2} + F_{p2}(x_1),\\
D_0x_{q2} &= -\alpha_q x_{q2} + \omega_q x_{(q+1)2} + F_{q2}(x_1),\\
D_0x_{(q+1)2} &= -\omega_q x_{q2} + \alpha_q x_{(q+1)2} + F_{(q+1)2}x_1),
\end{aligned} \tag{13.123}
$$

where x_1 corresponds to the first-order approximation and $f_{i2} = d^2[F_i(x)/\varepsilon]/d\varepsilon^2)$ $_{\varepsilon=0}$ are the functions of x_1.

Differentiating the first of Eq. (13.122) and substituting the second equation into the resulting equation yields

$$D_0^2 x_{11} + \omega_{1c}^2 x_{12} = 0, \tag{13.124}$$

which has the following solution

$$x_{11} = r_1(T_1, T_2, \ldots) cos[\omega_{1c} T_0 + \Phi_1((T_1, T_2, \ldots)] \equiv r_1 cos\Theta_1, \tag{13.125}$$

where r_1 is the amplitude and Φ_1 is the phase. Knowing x_{11} it is easy to obtain x_{21} from the third equation of (13.122). Proceeding in a similar way one finds

$$x_{31} = r_2(T_1, T_2, \ldots) cos[\omega_{2c} T_0 + \Phi_2(T_1, T_2, \ldots)] \equiv r_2 cos\Theta_2, \tag{13.126}$$

and then x_{41} is defined by the fourth equation of (13.122). The other variables $x_{j1} = 0, \quad j = 5, 6, \ldots, n$.

Observe also that

$$D_0 r_1 = D_0 r_2 = 0, \qquad D_0\Phi_1 = D_0\Phi_2 = 0. \tag{13.127}$$

From Eq. (13.123) we obtain

$$D_0^2 x_{12} + \omega_{1c}^2 x_{12} = -D_1 D_0 x_{11} - \omega_{1c} D_1 x_{21} + D_0 F_{12} + \omega_{1c} F_{22}. \tag{13.128}$$

The condition for avoiding secular terms determine $D_1 r_i$ and $D_1\Phi_i, \quad i = 1, 2$. In the next step one finds the particular solution of (13.128). Having obtained x_{12} one easily finds x_{22} from the second equation of (13.123).

Using Eqs. (13.125) and (13.126) the following first-order differential equations are obtained

$$\frac{dr_i}{dt} = D_0 r_i + \varepsilon D_1 r_i + \varepsilon^2 D_2 r_i + \varepsilon^3 D_3 r_i + \cdots \tag{13.129}$$

$$\frac{d\Phi_i}{dt} = D_0\Phi_i + \varepsilon D_1\Phi_i + \varepsilon^2 D_2\Phi_i + \varepsilon^3 D_3\Phi_i + \cdots \tag{13.130}$$

Recall that we analyse non-resonant double Hopf bifurcation, and hence we get $D_{2k+1} r_i = D_{2k+1}\Theta_i = 0$ for $k = 0, 1, 2, \ldots$, whereas

$$\begin{aligned} D_{2k} r_1 &= r_1[a_{2k0} r_1^{2k} + a_{(2k-2)2} r_1^{2k-1} r_2^2 + \cdots + a_{2(2k-2)} r_1^{2k-1} + a_{02k} r_2^{2k}], \\ D_{2k} r_2 &= r_2[b_{2k0} r_1^{2k} + b_{(2k-2)2} r_1^{2k-1} r_2^2 + \cdots + b_{2(2k-2)} r_1^2 r_2^{2k-2} + b_{02k} r_2^{2k}] \end{aligned} \tag{13.131}$$

and $D_{2k}\Phi_i$ have similar form.

Introducing a back scaling $\varepsilon x_i \to x_i, \quad \varepsilon r_i \to r_i$, Eqs. (13.129), (13.130) have the new form

$$\frac{dr_i}{dt} = D_2 r_i + D_4 r_i + D_6 r_i + \cdots \tag{13.132}$$

$$\frac{d\Phi_i}{dt} = D_2 \Phi_i + D_4 \Phi_i + D_6 \Phi_i + \cdots \tag{13.133}$$

and

$$\frac{d\Theta_i}{dt} = \omega_{ic} + \frac{d\Phi_i}{dt}, \tag{13.134}$$

where: $\Theta_i = \omega_{ic} T_0 + \Phi_i \equiv \omega_{ic} t + \Phi_i$, and $i = 1, 2, \ldots$.

It is worth noticing that Eqs. (13.132), 13.133) and (13.134) are called *normal forms*. A reason is that $D_{2k} r_1$, $D_{2k} r_2$, $D_{2k} \Phi_1$ and $D_{2k} \Phi_2$ are obtained by eliminating *secular* terms (i.e. resonant terms). The *resonant* terms are actually applied in Poincaré normal form theory.

Furthermore, the obtained periodic solution given by (13.124)–(13.126) and (13.132)–(13.134) represents both asymptotic and transient solution for the *critical* variables (modes) $x_i, \quad i = 1, 2, 3, 4 \quad (r_i = r_i(t)$, and $\Phi_i = \Phi_i(t))$. The *non-critical* variables (modes) $x_j, \quad j = 5, 6, \ldots$ are found from Eq. (13.121) (they are excited by critical variables),

Observe also that the periodic solution (13.121) can be treated as the nonlinear transformation between (13.117) and the normal forms (13.132)–(13.134). This observation is supported by the following consideration. The periodic solution can be written as

$$\begin{aligned}
x_1 &= r_1 \cos\Theta_1 + h_1(r_1\cos\Theta_1, r_1\sin\Theta_1, r_2\cos\Theta_2, r_2\sin\Theta_2),\\
x_2 &= -r_1 \sin\Theta_1 + h_2(r_1\cos\Theta_1, r_1\sin\Theta_1, r_2\cos\Theta_2, r_2\sin\Theta_2),\\
x_3 &= r_2 \cos\Theta_2 + h_3(r_1\cos\Theta_1, r_1\sin\Theta_1, r_2\cos\Theta_2, r_2\sin\Theta_2),\\
x_4 &= -r_2 \sin\Theta_2 + h_4(r_1\cos\Theta_1, r_1\sin\Theta_1, r_2\cos\Theta_2, r_2\sin\Theta_2),\\
x_i &= h_i(r_1\cos\Theta_1, r_1\sin\Theta_1, r_2\cos\Theta_2, r_2\sin\Theta_2), \quad i = 5, 6, \ldots, n.
\end{aligned} \tag{13.135}$$

Introducing the new variables

$$y_1 = r_1\cos\Theta_1, \quad y_2 = -r_2\sin\Theta_1, \quad y_3 = r_2\cos\Theta_2, \quad y_4 = -r_2\sin\Theta_2, \tag{13.136}$$

from (13.135) one gets

$$x_i = y_i + h_i(y_1, y_2, y_3, y_4), \quad i=1,2,3,4, \quad x_i = h_j(y_1, y_2, y_3, y_4), \quad i=5,6,\ldots,n. \tag{13.137}$$

The obtained equation, which is the same as (13.121), manifests a transformation between $x_i, \quad i = 1, 2, \ldots, n$ and $y_j, \quad j = 1, 2, 3, 4$. In other words, the first four equations can be considered as a nonlinear transformations between the coordinates x_1, x_2, x_3, x_4 and y_1, y_2, y_3, y_4. The remaining equations represent the projection of the original system to the four-dimensional centre manifold governed by the critical variables y_1, y_2, y_3, y_4. Using the Cartesian coordinates the normal form governed by (13.132)–(13.134) reads

$$\begin{aligned}
\dot{y}_1 &= \omega_{1c} y_2 + g_1(y_1, y_2, y_3, y_4), \\
\dot{y}_2 &= -\omega_{1c} y_1 + g_2(y_1, y_2, y_3, y_4), \\
\dot{y}_3 &= \omega_{2c} y_4 + g_3(y_1, y_2, y_3, y_4), \\
\dot{y}_4 &= -\omega_{2c} y_3 + g_4(y_1, y_2, y_3, y_4).
\end{aligned} \tag{13.138}$$

To sum up, the nonlinear transformation (13.138), which is equivalent to (13.121), represents a transition between the original system (13.117) and the normal form (13.138).

Now both stability and bifurcation analysis may be carried out using Eqs. (13.132)–(13.134) or (13.138).

Following Yu [250], in order to outline a general bifurcational analysis, the normal form is presented explicitly up to third order

$$\dot{r}_1 = r_1[\alpha_{11}\mu_1 + \alpha_{12}\mu_2 + a_{20}r_1^2 + a_{02}r_2^2], \dot{r}_2 = r_2[\alpha_{21}\mu_1 + \alpha_{22}\mu_2 + b_{20}r_1^2 + b_{02}r_2^2], \tag{13.139}$$

and similarly

$$\dot{\Theta}_1 = \omega_{1c} + \beta_{11}\mu_1 + \beta_{12}\mu_2 + c_{20}r_1^2 + c_{02}r_2^2, \dot{\Theta}_2 = \omega_{2c} + \beta_{21}\mu_1 + \beta_{22}\mu_2 + d_{20}r_1^2 + d_{02}r_2^2, \tag{13.140}$$

For the convenience two parameter variables $\alpha_{11}\mu_1 + \alpha_{12}\mu_2, \quad \alpha_{21}\mu_1 + \alpha_{22}\mu_2$ are used, where μ_1, μ_2 are the perturbation parameters.

The tracked behaviour follows:

1. Equilibria (E):

$$r_1 = r_2 = 0 \tag{13.141}$$

2. Hopf bifurcation with frequency ω_1 $(H(\omega_1))$:

$$r_1^2 = -\frac{1}{a_{20}}(\alpha_{11}\mu_1 + \alpha_{12}\mu_2), r_2 = 0, \omega_1 = \omega_{1c} + \beta_{11}\mu_1 + \beta_{12}\mu_2 + c_{20}r_1^2. \tag{13.142}$$

3. Hopf bifurcation with frequency ω_2 $(H(\omega_2))$:

$$r_2^2 = -\frac{1}{b_{02}}(\alpha_{21}\mu_1 + \alpha_{22}\mu_2), r_1 = 0, \omega_2 = \omega_{2c} + \beta_{21}\mu_1 + \beta_{22}\mu_2 + d_{02}r_2^2. \tag{13.143}$$

4. 2-D tori with frequencies ω_1, ω_2 (2-D tori):

$$
\begin{aligned}
r_1^2 &= -\frac{a_{02}(\alpha_{21}\mu_1 + \alpha_{22}\mu_2) - b_{02}(\alpha_{11}\mu_1 + \alpha_{12}\mu_2)}{a_{20}b_{02} - a_{02}b_{20}},\\
r_2^2 &= -\frac{b_{20}(\alpha_{21}\mu_1 + \alpha_{12}\mu_2) - a_{20}(\alpha_{21}\mu_1 + \alpha_{22}\mu_2)}{a_{20}b_{02} - a_{02}b_{20}},\\
\omega_1 &= \omega_{1c} + \beta_{11}\mu_1 + \beta_{12}\mu_2 + c_{20}r_1^2 + c_{02}r_2^2,\\
\omega_2 &= \omega_{2c} + \beta_{21}\mu_1 + \beta_{22}\mu_2 + d_{20}r_1^2 + d_{02}r_2^2.
\end{aligned}
\tag{13.144}
$$

Evaluating the Jacobian of Eq. (13.139) on the equilibrium (13.141) yields two critical lines

$$
L_1: \qquad \alpha_{11}\mu_1 + \alpha_{12}\mu_2 = 0 \quad (\alpha_{21}\mu_1 + \alpha_{22}\mu_2 < 0), \tag{13.145}
$$

$$
L_2: \qquad \alpha_{21}u_1 + \alpha_{22}u_2 = 0 \quad (\alpha_{11}u_1 + \alpha_{12}u_2 < 0), \tag{13.146}
$$

and L_1 (L_2) corresponds to occurrence of a family of periodic solutions after Hopf bifurcation with ω_1 (ω_2). The solutions are stable if the following inequalities are satisfied

$$
\alpha_{11}\mu_1 + \alpha_{12}\mu_2 > 0 \quad \text{and} \quad \alpha_{21}\mu_1 + \alpha_{22}\mu_2 < 0. \tag{13.147}
$$

Now, evaluating the Jacobian of Eq. (13.139) on the Hopf bifurcation solution (13.142), yields the stability conditions

$$
\alpha_{11}\mu_1 + \alpha_{22}\mu_2 > 0 \quad \text{and} \quad \alpha_{21}\mu_1 + \alpha_{22}\mu_2 - \frac{b_{20}}{a_{20}}(\alpha_{11}\mu_1 + \alpha_{12}\mu_2) < 0. \tag{13.148}
$$

One may check that the $H(\omega_1)$ periodic solution exists, when $a_{20} < 0$.
The second inequality in (13.148) yields the critical line

$$
L_3: \quad (\alpha_{21} - \frac{b_{20}}{a_{20}}\alpha_{11})\mu_1 + (\alpha_{22} - \frac{b_{20}}{a_{20}}\alpha_{12})\mu_2 = 0 \quad (\alpha_{11}\mu_1 + \alpha_{12}\mu_2) > 0). \tag{13.149}
$$

The L_3 line represents a secondary Hopf bifurcation with frequency ω_2 from the limit cycle produced by $H(\omega_1)$ (i.e. a 2-D torus is created).

Similarly, the (13.143) solution associated with $H(\omega_2)$ is stable when the inequalities hold

$$
\alpha_{21}\mu_1 + \alpha_{22}\mu_2 > 0 \quad \text{and} \quad \alpha_{11}\mu_1 + \alpha_{12}\mu_2 - \frac{a_{02}}{b_{02}}(\alpha_{21}\mu_1 + \alpha_{22}\mu_2) < 0 \tag{13.150}
$$

and it exists when $b_{02} < 0$.

A critical line L_4 is defined by the second inequality of (13.150):

$$L_4: \quad (\alpha_{11} - \frac{a_{02}}{b_{02}}\alpha_{21})\mu_1 + (\alpha_{12} - \frac{a_{02}}{b_{02}}\alpha_{22})\mu_2 = 0 \quad (\alpha_{11}\mu_1 + \alpha_{22}\mu_2) > 0). \tag{13.151}$$

Crossing L_4 a secondary Hopf bifurcation with frequency ω_1 occurs from the periodic orbits born after $H(\omega_2)$. Again the 2-D tori is produced, with solutions, governed by Eq. (13.144).

In [250] a family of solutions lying on 2-D tori is traced via evaluation the Jacobian on Eq. (13.144) yielding

$$\mathbf{J_T} = \begin{bmatrix} 2a_{20}r_1^2 & 2a_{02}r_1r_2 \\ 2b_{20}r_1r_2 & 2a_{02}r_2^2 \end{bmatrix}. \tag{13.152}$$

The stability of the quasi-periodic motion is defined by trace (Tr < 0) and determinant (Det > 0) of the Jacobian $\mathbf{J_T}$.

The mentioned stability conditions are supplemented by the following existence conditions

$$\begin{aligned} &a_{02}(\alpha_{21}\mu_1 + \alpha_{22}\mu_2) - b_{02}(\alpha_{21}\mu_1 + \alpha_{12}\mu_2) > 0, \\ &b_{20}(\alpha_{11}\mu_1 + \alpha_{12}\mu_2) - a_{20}(\alpha_{21}\mu_1 + \alpha_{22}\mu_2) > 0. \end{aligned} \tag{13.153}$$

Observe that the existence region boundaries (13.153) are defined by the critical lines L_3 and L_4, i.e. the periodic solution associated with $H(\omega_1)$ ($H(\omega_2)$) bifurcates from the critical line L_3 (L_4) into a quasi-periodic solution with the stability boundary L_3 (L_4).

Note that $r_1 > 0$ and $r_2 > 0$ guarantee satisfaction of conditions (13.153), and $Det > 0$ yields

$$a_{20}b_{02} - a_{02}b_{20} > 0, \tag{13.154}$$

and hence the condition $Tr < 0$ gives

$$a_{20}(a_{02} - b_{02})(\alpha_{21}\mu_1 + \alpha_{22}\mu_2) - b_{02}(a_{22} - b_{20})(\alpha_{11}\mu_1 + \alpha_{12}\mu_2) < 0. \tag{13.155}$$

From (13.155) one gets

$$\begin{aligned} &L_5: [a_{20}(a_{02} - b_{02})\alpha_{21} - b_{02}(a_{22} - b_{20})\alpha_{11}]\mu_1 \\ &+[a_{20}(a_{02} - b_{02})\alpha_{22} - b_{02}(a_{22} - b_{20})\alpha_{12}]\mu_2 = 0. \end{aligned} \tag{13.156}$$

On L_5 a quasi periodic solution may loose its stability, and bifurcate to a 3-D torus with frequencies $(\omega_1, \omega_2, \omega_3)$.

A combination of perturbation approaches and harmonic balance technique to analyse various Hopf type bifurcations is presented in works [18–21, 30] and examples of numerical technique to trace dynamical behaviour using a path following method taken from mechanics and biomechanics are reported in [15–17].

13.3 Fixed Points of Maps

There are three distinct situations possible for bifurcations of fixed points of maps.

(i) $D_x f(x_0, \lambda_0)$ *has a single eigenvalue equal to 1.*
In this case the problem is reduced to a study of one-dimensional centre manifold

$$x \mapsto f(x, \lambda), \quad x \in \mathbb{R}^1,\ \lambda \in \mathbb{R}^1. \tag{13.157}$$

Fixed point can be transformed to the origin, where $f(0,0) = 0$ and $\frac{\partial f}{\partial x}(0,0) = 1$. In this case three situations are possible.
A *saddle-node bifurcation* at $(x, \lambda) = (0,0)$ takes place if

$$\begin{aligned} f(0,0) &= 0, \\ \tfrac{\partial f}{\partial \lambda}(0,0) &= 1, \end{aligned} \tag{13.158}$$

and

$$\begin{aligned} &\frac{\partial f}{\partial \lambda}(0,0) \neq 0, \\ &\frac{\partial^2 f}{\partial x^2}(0,0) \neq 0. \end{aligned} \tag{13.159}$$

The map

$$x \mapsto f(x, \lambda) = x + \lambda \pm x^2, \quad x \in \mathbb{R}^1, \quad \lambda \in \mathbb{R}^1 \tag{13.160}$$

can serve as a normal form for the saddle-node bifurcations for maps.
A *transcritical bifurcation* can be represented by the map

$$x \mapsto f(x, \lambda) = x + \lambda x \pm x^2, \quad x \in \mathbb{R}^1, \quad \lambda \in \mathbb{R}^1. \tag{13.161}$$

Note that $(x, \lambda) = (0,0)$ is a non-hyperbolic fixed point with eigenvalue 1 ($f(0,0) = 0$, $\frac{\partial f}{\partial x}(0,0) = 1$). A transcritical bifurcation appears if

$$\frac{\partial f}{\partial \lambda}(0,0) = 0, \qquad \frac{\partial^2 f}{\partial x \partial \lambda}(0,0) \neq 0, \qquad \frac{\partial^2 f}{\partial x^2}(0,0) \neq 0. \tag{13.162}$$

A *pitchfork bifurcation* occurs in a one parameter family of smooth one-dimensional maps (13.157) with a non-hyperbolic fixed point ($f(0,0) = 0$, $\frac{\partial f}{\partial x}(0,0) = 1$) if

$$\frac{\partial f}{\partial \lambda}(0,0) = 0, \quad \frac{\partial^2 f}{\partial x^2}(0,0) = 0, \quad \frac{\partial^2 f}{\partial x \partial \lambda}(0,0) \neq 0, \quad \frac{\partial^3 f}{\partial x^3}(0,0) \neq 0, \tag{13.163}$$

and its normal form is given by

$$x \mapsto f(x,\lambda) = x + \lambda x \pm x^3, \quad x \in \mathbb{R}^1, \quad \lambda \in \mathbb{R}^1. \tag{13.164}$$

(ii) $D_x f(x_0, \lambda_0)$ *has a single eigenvalue equal to* -1.
In contrary to the bifurcation (i) this bifurcation does not have an analog with one-dimensional dynamics of vector fields, since it refers to period-doubling bifurcation at $(x,\lambda) = (0,0)$. It occurs when

$$\begin{aligned} &f(0,0) = 0, \quad \frac{\partial f}{\partial x}(0,0) = -1, \quad \frac{\partial f}{\partial \lambda}(0,0) = 0, \\ &\frac{\partial^2 f}{\partial x^2}(0,0) = 0, \quad \frac{\partial^2 f}{\partial x \partial \lambda}(0,0) \neq 0, \quad \frac{\partial^3 f}{\partial x^3}(0,0) \neq 0. \end{aligned} \tag{13.165}$$

One can display the period-doubling bifurcation using the map

$$x \mapsto f(x,\lambda) = -x - \lambda x + x^3. \tag{13.166}$$

(iii) $D_x f(x_0, \lambda_0)$ *possesses two complex conjugate eigenvalues having modulus 1.*
This situation corresponds to the Neimark–Sacker bifurcation or sometimes it is referred as a secondary Hopf bifurcation ([106, 243]).

Consider the map (13.157) but for $x \in \mathbb{R}^2$, and let us again introduce a suitable transformation that $(x,\lambda) = (0,0)$ and $f(0,0)$. The associated matrix $D_x f(0,0)$ possesses two complex conjugate eigenvalues $\lambda(0)$ and $\bar{\lambda}(0)$, with $|\lambda(0)| = 1$. In addition, we require that $\lambda^n(0) \neq 1$ for $n = 1,2,3,4$.

It can be shown [243] that the normal form of the Neimark–Sacker bifurcation is governed by the complex map

$$z \mapsto \mu(\lambda) z + c(\lambda) z^2 \bar{z} + O(4), \quad z \in \mathbb{C},\ \lambda \in \mathbb{R}^1. \tag{13.167}$$

We change the variables letting

$$z = r e^{2\pi i \Theta}, \tag{13.168}$$

and we get

$$\begin{aligned} &r \mapsto |\mu(\lambda)| \left(r + \left(\mathrm{Re}\left(\frac{c(\lambda)}{\mu(\lambda)} \right) \right) r^3 + O(r^4) \right), \\ &\Theta \mapsto \Theta + \phi(\lambda) + \frac{1}{2\pi} \left(\mathrm{Im}\left(\frac{c(\lambda)}{\mu(\lambda)} \right) \right) r^2 + O(r^3), \end{aligned} \tag{13.169}$$

where

$$\phi(\lambda) \equiv \frac{1}{2\pi}\tan^{-1}\frac{\omega(\lambda)}{\alpha(\lambda)},$$
$$c(\lambda) = \alpha(\lambda) + i\omega(\lambda). \tag{13.170}$$

The Taylor expansion of (13.169) in $\lambda = 0$ gives

$$r \mapsto \left(1 + \frac{d}{d\lambda}|\mu(\lambda)|\,|_{\lambda=0}\lambda\right) r + \left(\mathrm{Re}\left(\frac{c(0)}{\mu(0)}\right)\right) r^3 + O(\lambda^2 r, \lambda r^3, r^4),$$
$$\Theta \mapsto \Theta + \phi(0) + \frac{d}{d\lambda}\phi(\lambda)|_{\lambda=0}\lambda + \frac{1}{2\pi}\left(\mathrm{Im}\frac{c(0)}{\mu(0)}\right) r^2. \tag{13.171}$$

The truncated normal form has the form

$$r \mapsto r + (d\lambda + ar^2)r,$$
$$\Theta \mapsto \Theta + \phi_0 + \phi_1\lambda + br^2, \tag{13.172}$$

where

$$d=\frac{d}{d\lambda}|\mu(\lambda)|_{\lambda=0},\ a=\mathrm{Re}\left(\frac{c(0)}{\mu(0)}\right),\ \phi_0=\phi(0),\ \phi_1=\frac{d}{dx}(\phi(\lambda)),\ b=\frac{1}{2\pi}\mathrm{Im}\frac{c(0)}{\lambda(0)}. \tag{13.173}$$

Following [243] there are four potential cases for the bifurcation of an invariant circle from a fixed point.

(a) $d > 0$, $a > 0$. The origin is an unstable fixed point for $\lambda > 0$ and an asymptotically stable fixed point for $\lambda < 0$ with an unstable invariant circle for $\lambda < 0$.
(b) $d > 0$, $a < 0$. The origin is an unstable fixed point for $\lambda > 0$ and an asymptotically stable fixed point for $\lambda < 0$ with an asymptotically stable invariant circle for $\lambda > 0$.
(c) $d > 0$, $a < 0$. The origin is an asymptotically stable fixed point for $\lambda > 0$ and an unstable fixed point for$\lambda < 0$ with an unstable invariant circle for $\lambda > 0$.
(d) The origin is an asymptotically stable fixed point for $\lambda > 0$, and an unstable fixed point for $\lambda < 0$, with an asymptotically stable invariant circle for $\lambda < 0$.

Note that here the bifurcation consists of a circle which has many different orbits. Hence one should start with the initial condition laying on this circle. Since in this case $r = (-\frac{\lambda d}{a})^{\frac{1}{2}}$, then the associated circle map has the form

$$\Theta \mapsto \Theta + \phi_0 + \left(\phi_1 + \frac{d}{a}\right)\lambda. \tag{13.174}$$

For $\phi_0 + (\phi_1 + \frac{d}{a})\lambda$ rational (irrational) all orbits on invariant circle are periodic (quasiperiodic, i.e. densely fill the circle).

13.4 Continuation Technique

The introduced earlier considerations related to local bifurcations can be used to follow the dynamics in a systematic way. In a case of vector fields, one can construct a local Poincaré map, and then to reduce the problem of one order to study a bifurcation of a fixed point of the map. In a case of a k periodic orbit one can take into account the bifurcation of a fixed point of the kth iterate of the map.

Coming back to the differential equation (13.1) and taking into account the initial condition $x(t_0) = x_0$ (for a given specific numerical value of λ_0) one can trace a periodic orbit occurred via Hopf bifurcation. We integrate numerically (13.1) during the time equal to exact (or estimated) period of a new periodic orbit corresponding to the parameter $\lambda_0 + \Delta\lambda$, where $\|\Delta\lambda\| \ll 1$.

In a case of second-order differential equation the situation is shown in Fig. 13.16.

In a case of periodic solution with the period T_0 we have the following boundary condition

$$x(T_0, x_0) - x_0 = 0. \tag{13.175}$$

Equation (13.175) yields x_0 using, for instance, Newton's method. Using Taylor expansion around kth order approximation of $x_0^{(k)}$ we take only a linear term, and we obtain a linear correction $\Delta x_0^{(k)}$. A Jacobian of the Newton method is defined by

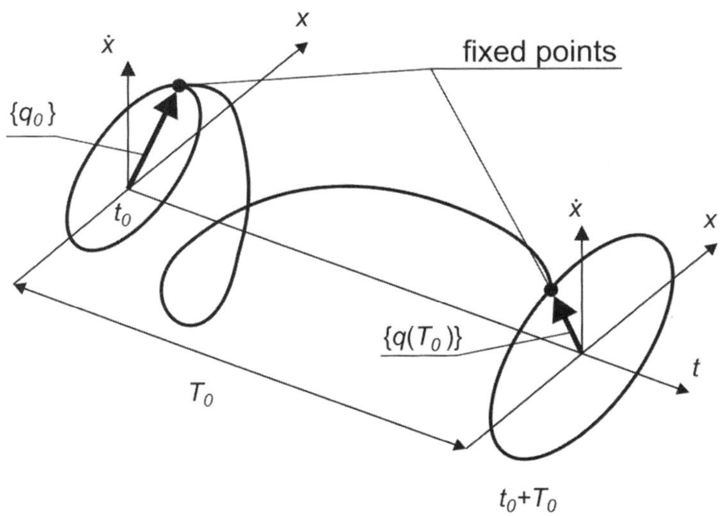

Fig. 13.16 Transformation of the state vector x_0

$$J = \frac{\partial x(T_0, x_0)}{\partial x_0} - I = N - I, \tag{13.176}$$

where I is the identity matrix. After computation with the required accuracy we use the matrix N obtained in the last computational step as the monodromy matrix. The associated eigenvalues decide about stability and bifurcation of the analysed periodic orbit (or equivalently, a fixed point of the maps). This method is called shooting and has already found a wide treatment in literature (see, for instance Seydel [214], Awrejcewicz [14]).

Another equivalent approach is based on the Galerkin approximation [234]. It is well known that any periodic solution (function) can be represented by its Fourier expansion

$$y_k = a_0 + \sum_{k=1}^{K} (a_{c_k} \cos k\omega t + a_{s_k} \sin k\omega t), \tag{13.177}$$

where K is the number of a highest harmonics, and ω is the fundamental frequency. The assumed solution is substituted to the system of the second-order differential equations governing dynamics of oscillating systems. Since from the assumption we have the kth order Fourier approximation to a periodic solution, then between left- and right-hand sides a difference occurs (a residual vector), which satisfies the equation

$$r(a_0, a_{c_k}, a_{s_k}, t) = r(a_0, a_{c_k}, a_{s_k}, t + T). \tag{13.178}$$

In general, when $K \to \infty$, then $r \to 0$. The residual vector is also expanded into Fourier series

$$r = s_0 + \sum_{k=1}^{K} (s_{c_k} \cos k\omega t + s_{s_k} \sin k\omega t). \tag{13.179}$$

The condition $r = 0$ implies that $s_0 = s_{c_k} = s_{s_k} = 0$, and hence

$$\begin{aligned}
&\frac{1}{T}\int_0^T r(a_0, a_{c_k}, a_{s_k}, t)dt = 0,\\
&\frac{2}{T}\int_0^T r(a_0, a_{c_k}, a_{s_k}, t)\cos n\omega t dt = 0,\\
&\frac{2}{T}\int_0^T r(a_0, a_{c_k}, a_{s_k}, t)\sin n\omega t dt = 0, \quad k, n = 1, 2, 3, \ldots
\end{aligned} \tag{13.180}$$

Introducing the vectors

$$s = \begin{pmatrix} s_0 \\ s_{c_k} \\ s_{s_k} \end{pmatrix}, \quad a = \begin{pmatrix} a_0 \\ a_{c_k} \\ a_{s_k} \end{pmatrix}, \tag{13.181}$$

equation $r = 0$ can be substituted by

$$s(a) = 0, \tag{13.182}$$

and can be solved using the iterational Newton method. In practice the Fast Fourier Transformation (FFT) is used during computation of s [49,71]. After an appropriate choice of a starting point for the vector **a** we compute

$$\begin{aligned} \dot{y_k} &= \sum_{k=1}^{K} (k\omega a_{s_k} \cos k\omega t - k\omega a_{c_k} \sin k\omega t), \\ \ddot{y_k} &= \sum_{k=1}^{K} (-k^2\omega^2 a_{c_k} \cos k\omega t - k^2\omega^2 a_{s_k} \sin k\omega t). \end{aligned} \tag{13.183}$$

Applying $(\mathrm{FFT})^{-1}$ we can find (13.183), then we substitute (13.183) into the governing equations to get (13.182). The FFT is especially economic when a number of samples N_{FFT} satisfies $N_{\mathrm{FFT}} = 2^M > 4K$. The error of the introduced estimation is equal to

$$\Delta = \left(\sum_{n=k+1}^{\frac{N_{\mathrm{FFT}}}{2}} (s_{c_{n_k}}^2 + s_{s_{n_k}}^2) \right)^{\frac{1}{2}}. \tag{13.184}$$

A stability of the found solution is estimated by the variational equations.

Example 13.5. Consider a model of human vocal chords oscillations (see more details in [15–17]). The human lungs produces the air pressure required for larynx to be opened. The vocal chords start to continue to open because of inertia, and then their elastic properties cause their closing. The air stream leaves the larynx, and then Bernoulli suction effect appears. Next the described cycle repeats. Display a bifurcation diagram corresponding to the mechanical model of human vocal chords shown in Fig. 13.17.

A point mass can move in the directions x and y. Since the vocal chords cannot touch each other, the artificial damping c_s and stiffness k_s have been introduced. The hyperbolic type stiffness associated with the coefficient k_s approaches infinity when the mass approaches origin. The larynx has been modelled as a reservoir with stiff walls and real elastic properties have been included in the modified air

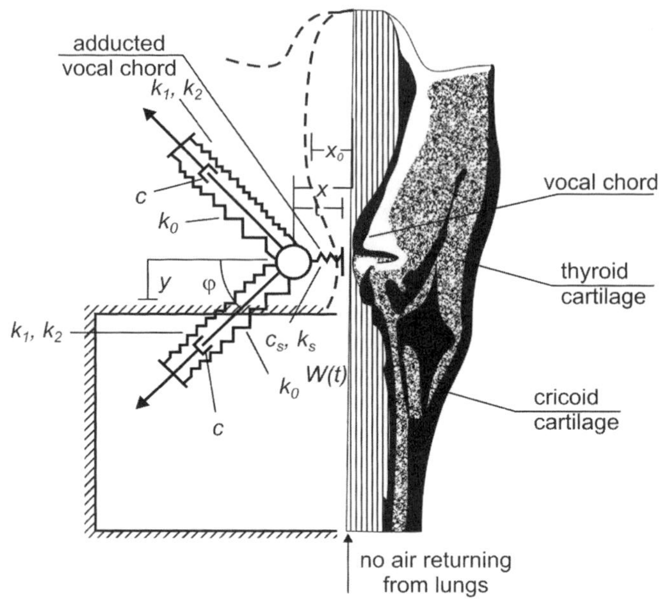

Fig. 13.17 Mechanical model of a human vocal chord

parameters. It is assumed that the lungs pressure generates equal forces acting on the vocal chords, which move symmetrically. This is a reason that we consider only one vocal chord oscillations. In a case of some pathological cases (for instance caused by cancer) the symmetry is violated and the number of equations is doubled.

The ordinary non-dimensional differential equations have the form (see [17] for more details)

$$\ddot{X} + C\dot{X} + \{K_x + K_D[(X - X_0)^2 + Y^2]\}(X - X_0) - K_{xy}Y -$$
$$-K_s X^{-s}(1 - C_s\dot{X}) = EP,$$
$$\ddot{Y} + C\dot{Y} + \{K_y + K_D[(X - X_0)^2 + Y^2]\}Y - K_{xy}(X - X_0) = EP,$$
$$\dot{P} = Q - \begin{cases} (X - 1)P^{0,5}, & \text{for } X > 1, \\ 0, & \text{for } x \le 1. \end{cases}$$

where: C corresponds to damping properties of the vocal chords ($C < 1$); K_y represents a vertical stiffness of a vocal chord ($K_y \in (0,7;0,9)$), $K_x = 1$; K_{xy} is the stiffness coefficient coupling the vocal chord displacements in two directions $K_{xy} \in (0,3;0,5)$); K_0 is a Duffing (cubic) type stiffness coefficient ($K_0 < 1$); K_s is a coefficient of a hyperbolic type stiffness ($K_s < 0,1$); $s = 4$; C_s represents damping ($C_s < 1$); X_0 is the horizontal equilibrium position of the vocal chord; E is the average surface of the vocal chord ($E \in (0,1;0,0)$) and Q is outlay of the air stream ($Q \in (0,0;100,0)$).

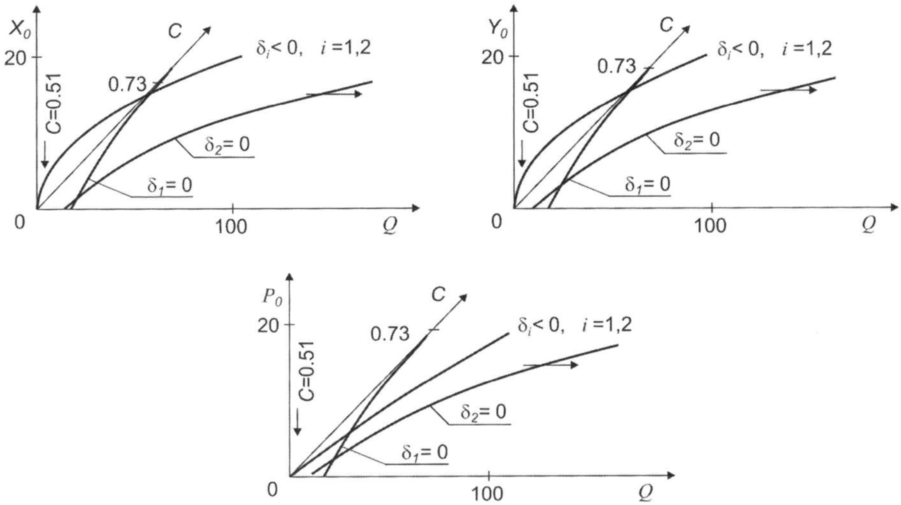

Fig. 13.18 Equilibrium and stability

The analysed system of differential equations is strongly nonlinear especially in a vicinity of origin (see the term responsible for hyperbolic type stiffness). We take the following fixed parameters: $K_y = O.9$; $K_{xy} = 0.3$; $K_0 = 0.001$; $K_s = 0,001$; $C_s = 0.5$; $E = 1$. First we calculate equilibria positions (see Fig. 13.18).

The solid curves in the planes (X_0, Q), (Y_0, Q) and (P_0, Q) correspond to the equilibrium positions. They do not change with the change of C, but their stability depends on C. The solid curves located in the plane (C, Q) correspond to stability loss boundaries.

Denote the eigenvalues associated with an equilibrium by $\lambda_{1,2} = \delta_1 \pm i\omega_1$, $\lambda_{3,4} = \delta_2 \pm i\omega_2$ (it can be proved that the fifth eigenvalue is real and negative). If $\delta_i = 0$ $(i = 1, 2)$, then for a fixed C value the Hopf bifurcation occurs with an increase or decrease of Q (see directions of vertical arrows). A point of intersection of both curves can be interpreted as a meeting point of two frequencies ω_1 and ω_2. If they are irrational then a quasi-periodic orbit appears.

The bifurcation diagram for $K_y = 0.3$; $K_D = K_s = 0.001$; $D_s = 0.5$; $K_{xy} = 0.3$; $X_0 = E = 0.4$; $Q = 7$ has been reported in Fig. 13.19, and the calculations have been carried out using shooting method.

Damping coefficient C has been taken as an active parameter. Its decrease causes occurrence of the Hopf bifurcation and a periodic orbit appears (this is example of self-exciting oscillations). A slight further decrease of C yields an increase of the oscillation amplitude represented by the branch $1'$. In the point $PD1'$ we have period doubling bifurcation and a new subharmonic solution appears (branch $3'$). This solution is further traced numerically along this branch and in the point Q it changes its stability. As the numerical analysis shown in the vicinity of Q a quasi-periodic solution appears.

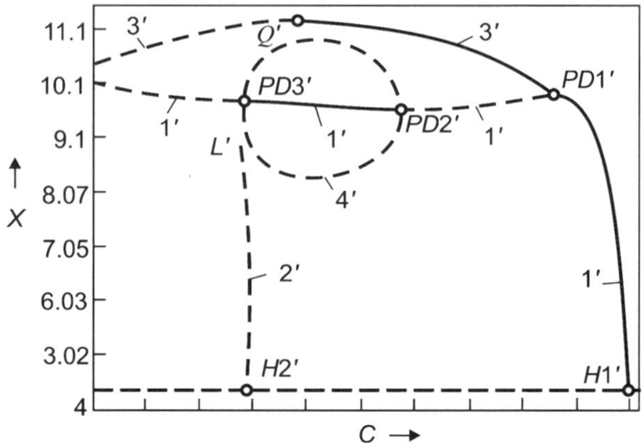

Fig. 13.19 Bifurcation diagram of the human vocal chords

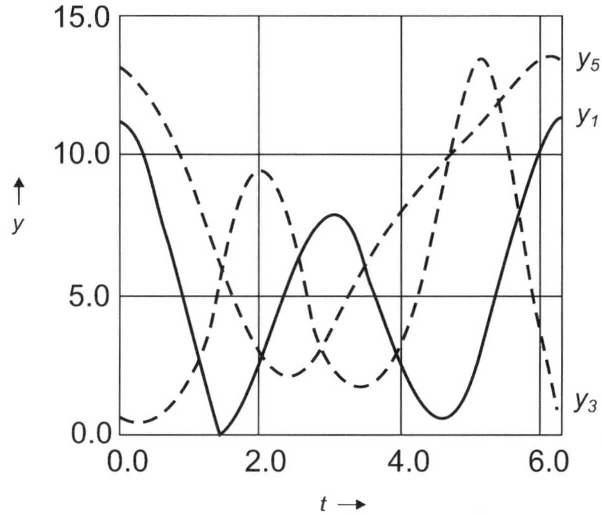

Fig. 13.20 Periodic oscillation for vocal chords

The branch 1 becomes unstable in the point $PD1'$. However between $PD2'$ and $PD3$ we have stable periodic solution. In the point $H2'$ another periodic solution branch appears which is unstable.

It is clear that for each point of the bifurcation curve one can easily obtain phase trajectories or time histories. For instance, the latter are shown in Fig. 13.20 for $C = 0.16$.

During calculations the period has been normalized to 2π and (from this figure) it is seen that this a subharmonic stable solution of the branch $3'$. A cusp of y_i is visible, which is in agreement with our hyperbolic type stiffness assumption. It must be emphasized that in this case the shooting method is much more economical in comparison to the Urabe–Reiter method. In the latter case one needs to take high number of harmonics which extremely extends computation time. The shooting method does not have the mentioned drawback. □

13.5 Global Bifurcations

As it has been pointed out in monographs [216,217], sometimes a periodic orbit does not exist on the stability boundary and Poincaré map is not defined i.e. it cannot be analysed using a local approach.

First we can consider *bifurcation of a homoclinic loop to a saddle-node equilibrium*, which is displayed by the equations

$$\dot{x_1} = \mu + \lambda_2 x^2 + G(x_1, \mu),$$
$$\dot{x_2} = [A + h(x_1, x_2, \mu)]x_2, \tag{13.185}$$

where the eigenvalues of A lie in LHP (see Fig. 13.21).

For $\mu < 0$ saddle O_1 and a node O_2 are distinct. They approach each other when $\mu \to 0^+$. In the critical point $\mu = 0$ the saddle-node equilibrium with the homoclinic loop disappears and a stable periodic orbit P_μ is born with a period of $\pi / \sqrt{\mu\lambda_2}$. In fact this observation can be formulated as a theorem and proved (see [216]).

We give one more example taken from [85], where two mutually coupled oscillators are considered:

$$\dot{x}_1 = x_2,$$
$$\dot{x}_2 = -\varepsilon(1 - \beta {x_1}^2 + {x_1}^4)x_2 - x_1 + \alpha x_3,$$
$$\dot{x}_3 = x_4,$$
$$\dot{x}_4 = -\varepsilon(1 - \beta {x_3}^2 + {x_3}^4)x_4 - x_3 + \alpha x_1, \tag{13.186}$$

where $0 < \alpha < 1$ is a coupling factor, β controls amplitude and ε is a control parameter. It has been shown that two symmetric solutions disappear simultaneously for $\varepsilon = 0.448$ via saddle-node bifurcation. Just before this value a switching between two symmetric solutions occurred. The situation is displayed in Fig. 13.22.

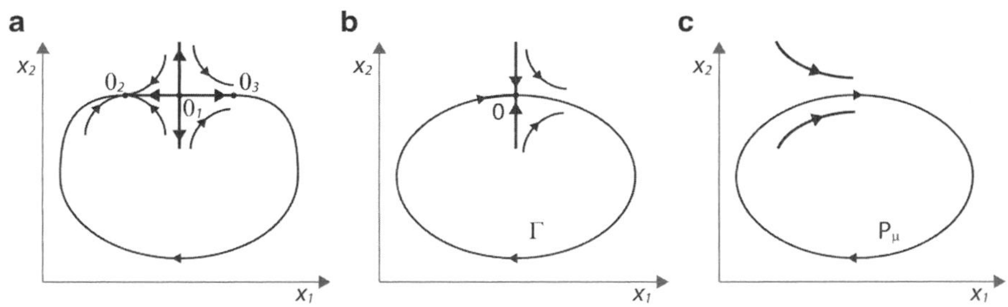

Fig. 13.21 Successive steps of a bifurcation of a saddle-node equilibrium with a homoclinic trajectory

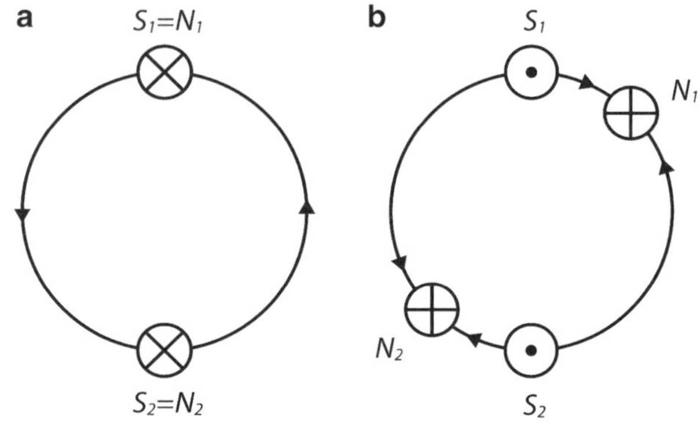

Fig. 13.22 A heteroclinic cycle at (**a**) and after (**b**) saddle-node bifurcation

The authors using Poincaré map, traced the unstable manifold W_n of the saddle in a vicinity of the saddle-node bifurcation and they observed the following scenario (see Fig. 13.22). One W_n of S_1 goes in N_1, whereas the other W_n goes in N_2. One of W_n of S_2 goes in N_1 and the other W_n of it goes in N_2. A node and a saddle coalesce at the saddle-node bifurcation point creating a degenerate saddle. A heteroclinic cycle links two degenerate saddles at this point. Two stable N_1 and N_2 are connected by unstable manifolds even after disappearance of the synchronized solution via saddle-node bifurcation and nodes are replaced by their traces. A flow stays traced of N_1 and N_2 for a relatively long time and then quickly moves along a heteroclinic cycle linking two traces.

Another important global bifurcation can lead to occurrence of an *invariant torus* or *Klein bottle*, which has been analysed by Afraimovich and Shilnikov ([1, 2]).

Theorem 13.7 (See [216]). *If the global unstable set of the saddle-node is a smooth compact manifold (a torus or a Klein bottle) at $\mu = 0$, then a smooth closed attractive invariant manifold T_μ (a torus or a Klein bottle, respectively) exists for all small μ.*

With a continuous change of μ, the invariant manifold will change continuously. For $\mu < 0$ we have a set composed of the unstable manifold of the saddle periodic orbit $P^-(\mu)$ with the stable periodic orbit $P^+(\mu)$ (by $P^\pm(\mu)$ we denote periodic orbits occurred after the saddle-node bifurcations). The invariant manifold for $\mu = 0$ is represented by w_p^u. For $\mu > 0$ the Poincaré rotation number approaches zero for $\mu \to 0^+$ (in case of torus). Hence, on the axis μ there are infinitely many phase-locking (resonant) zones (periodic orbits) as well as infinitely many zones composed of irrational values of μ (quasiperiodic orbits).

More detailed analysis of the briefly mentioned global bifurcations is given in the monograph [216] and is beyond of this book. There are also many references which include examples of various bifurcations in mechanics (see, for example [38, 145, 153, 190, 212, 231]).

Finally the basic phenomena of bifurcations exhibited by continuous dynamical systems, as well as explicit procedures for application of mathematical theorems to particular real-world problems are widely reported in monograph [146].

13.6 Piece-Wise Smooth Dynamical System

13.6.1 Introduction

The piece-wise smooth dynamical systems (PSDS) are governed by the equation:

$$\dot{x} = F(x; \lambda), \tag{13.187}$$

where: $x = x(t) \in \mathbb{R}^n$ represents the system state in time instant t; $\lambda \in \mathbb{R}^m$ is the vector of parameters, whereas the transformation $F : \mathbb{R}^n \times \mathbb{R}^m \to \mathbb{R}^n$ is the piece-wise smooth function. In other words, the phase space $D \in \mathbb{R}^n$ is divided into finite number of subspaces V_i, where the function F is smooth. The subspaces are separated by $(n-1)$ hyperplanes Σ_{ij}, where the 'discontinuities' are exhibited.

The PSDS governed by (13.187) can be classified in the following manner:

(i) Systems with discontinuous Jacobian DF, with continuous but non-smooth vector field F, and with smooth system's state $x(t)$;
(ii) Systems with discontinuous vector field F and with non-smooth but continuous $x(t)$;
(iii) Systems with discontinuous $x(t)$. In this case, whenever the system is in contact with Σ_{ij}, its state undergoes a jump described by $x^+ = g(x^-; \lambda)$, where $x^-(x^+)$ describes the system state just before (after) a contact.

Dynamical systems belonging to class (i) are often called Filippov systems [93, 152]. In mechanical and electrical engineering there exist many various systems with piece-wise linear characteristics (see for example [46, 47, 161, 256]). Non-smooth mechanical systems with impact and/or friction have very long history and are described in many books (see, for instance [29] and references therein). Systems with Coulomb friction can be treated as those with a jump of a damping characteristics (class (ii))—see [95, 151, 152], or they exhibit a stick-slip movement. A stick takes place, when a resulting force acting on a moving body is less than the associated Coulomb force. Often the authors use differential inclusions in order to attack this problem more rigorously (see [143]). Three simple examples of mechanical systems with piece-wise nonlinearities are shown in Fig. 13.23.

In the first case (Fig. 13.23a) the system is governed by the equation

$$m\ddot{x} + F(x) = 0, \tag{13.188}$$

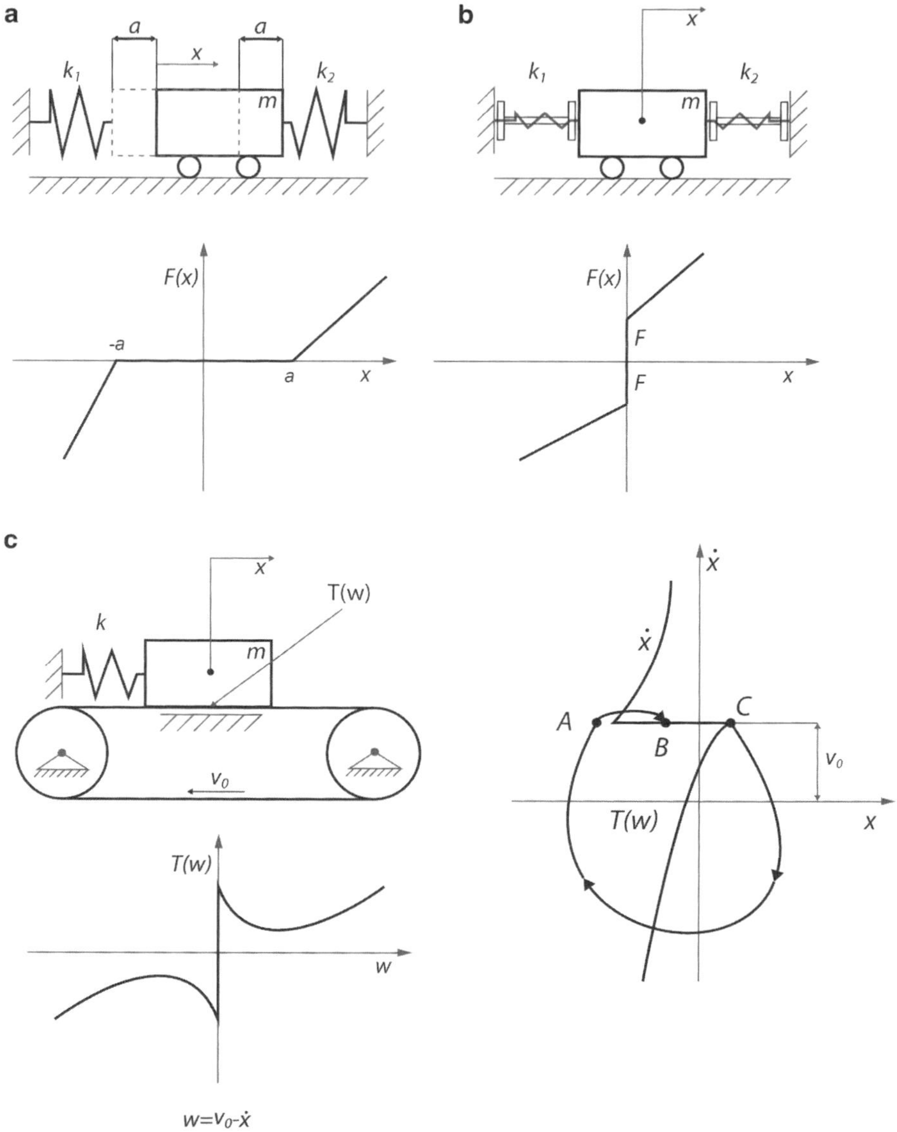

Fig. 13.23 One-degree-of-freedom system with a gap (**a**), discontinuous force (**b**), and stick-slip periodic motion (**c**)

where

$$F(x) = \begin{cases} 0 & \text{for } -a \leq x \leq a, \\ k_2(x-a) & \text{for } x > a, \\ k_1(x+a) & \text{for } x < a. \end{cases} \tag{13.189}$$

In the points $\pm a$ the function $F(x)$ is not differentiable. In the second case (Fig. 13.23b) the springs are initially stretched and governing differential equation is the same as (13.188), but

$$F(x) = \begin{cases} k_2 x + F_0 & \text{for } x > 0, \\ 0 & \text{for } x = 0, \\ k_1 x - F_0 & \text{for } x < 0. \end{cases} \tag{13.190}$$

In the third case the system is self-exited and is governed by equation

$$m\ddot{x} + kx = T(w) - T(0), \tag{13.191}$$

where:

$$T(w) = mg[\mu_0 \cdot sgnw - \alpha w + \beta w^3], \quad \omega = v_0 - \dot{x}. \tag{13.192}$$

In the above μ_0, α, β are coefficients describing friction. Note that the periodic orbit is not differentiable in the points A, B, C and the interval BC corresponds to stick, i.e. the mass m does not move in relation to tape. The system shown in Fig. 13.23a belongs to class (i), the system presented in Fig. 13.23b belongs to class (ii), and system given in Fig. 13.23c belongs to class (iii). The mechanical systems with impacts can be either modelled as a system with the sudden stiffness change (class (ii)), or as systems belonging to class (iii). In the latter case, when a surface $\Sigma_{i,j}$ is achieved, a sudden change of velocity occurs owing to the Newton's law and a definition of the restitution coefficient. In this case the system can be considered as that of one sided constraints and its behaviour is governed by algebraic inequalities. The described situation is equivalent also to a Dirac impulse of the function F on one of two sides of hyperplane $\Sigma_{i,j}$.

It is clear that in mechanical systems various classes of discontinuities may appear simultaneously. Although often friction and impact are independent ([238]) but more realistic are situations, where impact and friction appear together.

13.6.2 Stability

Here we consider a simple situation when on $\Sigma_{i,j}$ discontinuities do not occur. Recall also that when an analysed orbit belongs to any subspace V_i, than situation is clear, since stability concepts of smooth systems can be applied. Therefore, a stability devoted to PSDS is more general and when $\Sigma_{i,j}$ vanishes it is reduced to classical (smooth) stability concepts.

Assume that a periodic orbit $x(t)$ of a PSDS intersects the hyperplanes $\Sigma_{i,j}$ finite times in a periodic manner, and the intersections are non-degenerated, i.e. an orbit intersects $\Sigma_{i,j}$ transversally, a contact time with $\Sigma_{i,j}$ is infinitely short and all $\Sigma_{i,j}$ are smooth in contact points.

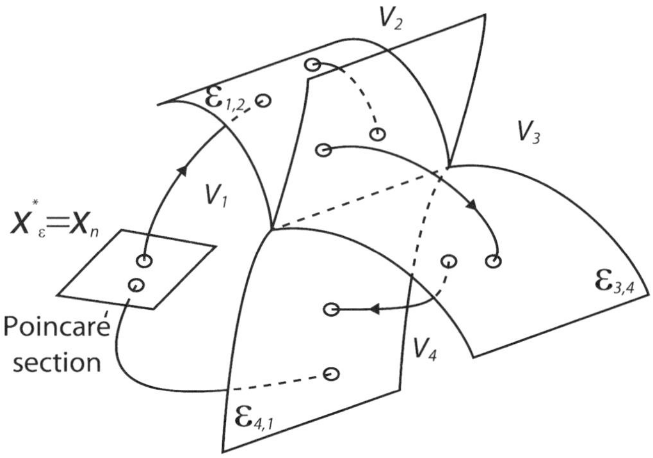

Fig. 13.24 A piece-wise smooth periodic orbit and a Poincaré section Σ

One may introduce a Poincaré section Σ which do not overlap with any $\Sigma_{i,j}$. In what follows we are going to analyse locally an intersection point of x(t) with Σ (further referred as x^*_Σ). To estimate stability one has to define a monodromy matrix ϕ^* owing to our earlier considerations for completely smooth systems. In order to obtain ϕ^* corresponding to piece-wise smooth periodic orbit, consider the matrix of fundamental solutions $\phi(t, t_0)$ satisfying the following linear differential equations

$$\dot{\phi}(t, t_0) = DF_i(x(t))\phi(t, t_0), \quad \phi(t, t_0) = I, \tag{13.193}$$

where D is the differential operator and DF_i denotes Jacobian. Observe that in time intervals $D_i = \{t \in \mathbb{R}; t_{i-1} < t < t_i\}$, for $i = 1, 2, \ldots, k+1$, the orbit $x(t)$ belongs to subspace V_i, where $F = F_i$ is smooth, and $F_{k+1} = F_1$ (see also Fig. 13.24 for $k = 4$). In each of the intersection (discontinuous) points defined by $t = t_i$ the orbit undergoes sudden changes, which means that the fundamental matrix also changes in these points. These changes can be formally expressed in the following way

$$\phi(t_i^+, t_0) = S_i\phi(t_i^-, t_0), \qquad i = 1, 2, \ldots, k. \tag{13.194}$$

In the above the superscript (−) denotes time instant just before a discontinuity, whereas the superscript (+) denotes time instant just after a discontinuity, and S_i is called saltation matrix.

In order to obtain explicitly a saltation matrix, one has to trace perturbations of two neighbourhood orbits $x(t)$ and $\overline{x}(t)$ (see Fig. 13.25).

The described method has been introduced for class (ii) by Aizerman–Gantmakher [3], and has been extended to the systems with discontinuous vector state [175].

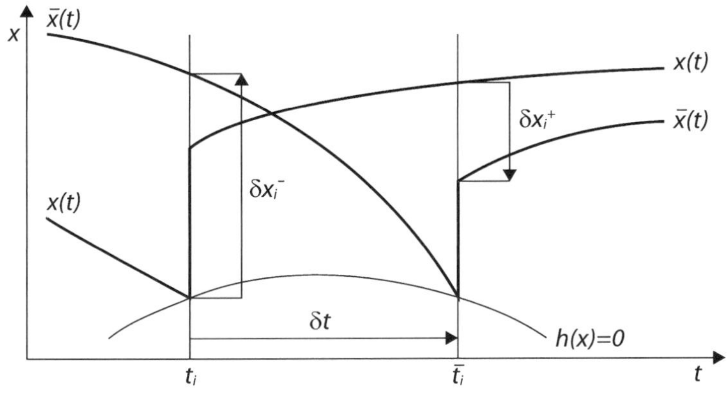

Fig. 13.25 Perturbed $\bar{x}(t)$ and unperturbed $x(t)$ orbits and the associated perturbations

Let us trace the orbit $x(t)$ in the time intervals D_i and D_{i+1}:

$$\begin{aligned} \dot{x} &= F_i(x), && \text{for } t_{i-1} < t < t_i, \\ 0 &= h_i(x_i^-), && \text{for } \quad t = t_i, \\ x_i^+ &= g_i(x_i^-), && \text{for } \quad t = t_i, \\ \dot{x} &= F_{i+1}(x), && \text{for } t_i < t < t_{i+1}, \end{aligned} \tag{13.195}$$

where: $x_i^- = \lim_{\substack{t \to t_i \\ t < t_i}} x(t)$, $x_i^+ = \lim_{\substack{t \to t_i \\ t > t_i}} x(t)$. Note that in the time instant $t = t_i$ defined by zero of smooth function $h_i(x)$ in V_i a discontinuity appears and in general the system state can exhibit a jump defined by a smooth function $g_i(x)$. In addition, it can happen that $F_i \neq F_{i+1}$. The function $h_i(x)$ can be scalar (for example during impacts in mechanical systems), or it can be a vector.

The perturbed solution $\bar{x}(t) = x(t) + \delta x(t)$ touches the constraint in time instant

$$\bar{t_i} = t_i + \delta t_i, \tag{13.196}$$

and it satisfies Eq. (13.195), where now the bars over x have to be added.

Assuming $\delta t_i > 0$ and following the introduced notations shown in Fig. 13.25, one obtains

$$\delta x_i^- = \bar{x}(t_i) - x_i^-, \delta x_i^+ = \bar{x}_i^+ - x(\bar{t}_i). \tag{13.197}$$

Since the perturbed orbit $\bar{x}(t)$ for $t = \bar{t_i}$ satisfies (13.195), the Taylor series expansions is used and the following manipulations hold:

$$\begin{aligned} 0 &= h_i(\bar{x}_i^-) = h_i\left(\bar{x}(t_i + \delta t_i)\right) \\ &\approx h_i(\bar{x}(t_i) + \frac{d\bar{x}}{dt}\,|_{t_i}\,\delta t_i) \end{aligned}$$

$$
\begin{aligned}
&\approx h_i(\bar{x}(t_i) + F_i(\bar{x}(t_i))\delta t_i) \\
&\approx h_i(x_i^- + \delta x_i^- + F_i(x_i^- + \delta x_i^-)\delta t_i) \\
&\approx h_i(x_i^- + \delta x_i^- + F_i(x_i^-)\delta t_i + DF_i(x_i^-)\delta x_i^- \delta t_i) \\
&\approx h_i(x_i^- + \delta x_i^- + F_i(x_i^-)\delta t_i) \\
&\approx h_i(x_i^-) + Dh_i(x_i^-)\delta x_i^- + Dh_i(x_i^-)F_i(x_i^-)\delta t_i.
\end{aligned} \tag{13.198}
$$

Finally, the following equation is obtained

$$
Dh_i(x_i^-)[\delta x_i^- + F_i(x_i^-)\delta t_i] = 0. \tag{13.199}
$$

In order to realize a non-degenerated contact between the orbit and discontinuous hyperplane $\Sigma_{i,j}$ the following condition should be assumed

$$
\mathrm{rank} Dh_i(x_i^-)F_i(x_i^-) = \mathrm{rank}[Dh_i(x_i^-)F_i(x_i^-), Dh_i(x_i^-)\delta x_i^-]. \tag{13.200}
$$

From (13.199) one obtains

$$
\delta t_i = -\frac{Dh_i(x_i^-)\delta x_i^-}{Dh_i(x_i^-)F_i(x_i^-)}. \tag{13.201}
$$

Now the Taylor series is applied to Eq. (13.195) and the manipulations similar to those in (13.198) are carried out:

$$
\begin{aligned}
\overline{x}_i^+ = g_i(\bar{x}_i^-) &\approx g_i(x_i^- + \delta x_i^- + F_i(x_i^-)\delta t_i) \\
&\approx g_i(x_i^-) + Dg_i(x_i^-)[\delta x_i^- + F_i(x_i^-)\delta t_i] \\
&\approx x_i^+ + Dg_i(x_i^-)[\delta x_i^- + F_i(x_i^-)\delta t_i].
\end{aligned} \tag{13.202}
$$

From (13.195), (13.197) and (13.202) one gets

$$
\begin{aligned}
\delta x_i^+ &= \overline{x}_i^+ - x(\bar{t}_i) \\
&\approx x_i^+ + Dg_i(x_i^-)[\delta x_i^- + F_i(x_i^-)\delta t_i] - (x_i^+ - \delta x_i^+) \\
&\approx x_i^+ + Dg_i(x_i^-)[\delta x_i^- + F_i(x_i^-)\delta t_i] - (x_i^+ + \dot{x}_i^+\delta t_i) \\
&\approx x_i^+ + Dg_i(x_i^-)[\delta x_i^- + F_i(x_i^-)\delta t_i] - (x_i^+ + F_{i+1}(x_i^+)\delta t_i),
\end{aligned} \tag{13.203}
$$

which finally yields

$$
\delta x_i^+ = Dg_i(x_i^-)\delta x_i^- + [Dg_i(x_i^-)F_i(x_i^-) - F_{i+1}(x_i^+)]\delta t_i, \tag{13.204}
$$

and δt_i is defined by (13.201). Recall that the saltation matrix transforms the perturbation δx_i^- just before a discontinuity point into the perturbation δx_i^+ just after the discontinuity point via formula

$$\delta x_i^+ = S_i \delta x_i^- . \tag{13.205}$$

Hence, accounting (13.204) and (13.201), the saltation matrix is found

$$S_i = Dg_i(x_i^-) + [F_{i+1}(x_i^+) - Dg_i(x_i^-)F_i(x_i^-)]\frac{Dh_i(x_i^-)}{Dh_i(x_i^-)F_i(x_i^-)}. \tag{13.206}$$

Note that similar considerations can be repeated for $\delta t_i < 0$.

The introduced theory can be used as an extension of earlier one to trace through the described continuation technique the piece wise smooth periodic orbits and also classical algorithms for computations of Lyapunov exponents in smooth systems [89, 244, 245] (one has to include jumps of $\delta x(t)$ in each of discontinuity points) can be applied.

Example 13.6. Derive a saltation matrix using the Aizerman–Gantmakher theory in one-degree-of-freedom mechanical system with an impact.

The following second-order differential equation governs dynamics with impacts of the oscillator

$$\ddot{q} = F_q(q, \dot{q}, t), \qquad q \le q_{\max},$$

where $q_{\max}$ defines a barrier position. Assume that in the time instant $t = t_k$, $q(t_k) = q_{\max}$ and an impact modelled within Newton's hypothesis

$$\dot{q}^+ = -e\dot{q}^-$$

occurs, where: $\dot{q}^- = \dot{q}^-(t_k) = \lim_{\substack{t \to t_k \\ t < t_k}} \dot{q}(t)$ is the oscillator velocity just before impact, whereas $\dot{q}^+ = \dot{q}^+(t_k) = \lim_{\substack{t \to t_k \\ t > t_k}} \dot{q}(t)$ is the oscillator velocity just after impact, and e is the restitution coefficient. The following phase coordinates are introduced

$$x = \mathrm{col}\{x_1, x_2, x_3\} = \mathrm{col}\{q, \dot{q}, t\}.$$

Hence, we have

$$\dot{x} = F(x), \qquad x \in \mathbb{R}^3,$$

where

$$F(x) = \text{col}\{x_2, F_q(x), 1\} = \text{col}\{\dot{q}, F_q, 1\}.$$

The phase space configuration is defined by the inequality

$$h(x) \geq 0,$$

where $h(x) = q_{\max} - x_1 = q_{\max} - q$.

In the time instant $t = t_k$ defined by $h(x^-(t_k)) = 0$, we get

$$x^+ = g(x^-),$$

where: $x^- = x^-(t_k) = \lim_{\substack{t \to t_k \\ t < t_k}} x(t)$, $x^+ = x^+(t_k) = \lim_{\substack{t \to t_k \\ t > t_k}} x(t)$. The function defining a jump of the state vector in the point of discontinuity has the form

$$g(x) = \text{col}\{x_1, -ex_2, x_3\} = \text{col}\{q, -e\dot{q}, t\}.$$

The associated Jacobians with $h(x)$ and $g(x)$ have the form

$$Dh(x) = \begin{bmatrix} -1 & 0 & 0 \end{bmatrix},$$

$$Dg(x) = \begin{bmatrix} 1 & 0 & 0 \\ 0 & -e & 0 \\ 0 & 0 & 1 \end{bmatrix}.$$

From (13.201) one gets

$$\delta t = -\frac{Dh(x^-)\delta x^-}{Dh(x^-)F(x^-)} = -\frac{\begin{bmatrix} -1 & 0 & 0 \end{bmatrix} \begin{Bmatrix} \delta x_1^- \\ \delta x_2^- \\ \delta x_3^- \end{Bmatrix}}{\begin{bmatrix} -1 & 0 & 0 \end{bmatrix} \begin{Bmatrix} x_2^- \\ F_q^- \\ 1 \end{Bmatrix}} = -\frac{\delta x_1^-}{x_2^-} = -\frac{\delta q^-}{\dot{q}^-},$$

where $F_q^- = F_q(x^-)$.

Observe that for a degenerated impact, when $\dot{q}^- \to 0$, the time δt approaches infinity. The saltation matrix is obtained from (13.206):

$$S = Dg(x^-) + [F(x^+) - Dg(x^-)F(x^-)]\frac{Dh(x^-)}{Dh(x^-)F(x^-)}$$

$$= \begin{bmatrix} 1 & 0 & 0 \\ 0 & -e & 0 \\ 0 & 0 & 1 \end{bmatrix} + \left(\begin{Bmatrix} \dot{q}^+ \\ F_q^+ \\ 1 \end{Bmatrix} - \begin{bmatrix} 1 & 0 & 0 \\ 0 & -e & 0 \\ 0 & 0 & 1 \end{bmatrix} \begin{Bmatrix} \dot{q}^- \\ F_q^- \\ 1 \end{Bmatrix} \right) \frac{[-1\ 0\ 0]}{[-1\ 0\ 0] \begin{Bmatrix} \dot{q}^- \\ F_q^- \\ 1 \end{Bmatrix}}$$

$$= \begin{bmatrix} 1 & 0 & 0 \\ 0 & -e & 0 \\ 0 & 0 & 1 \end{bmatrix} + \left(\begin{Bmatrix} \dot{q}^+ \\ F_q^+ \\ 1 \end{Bmatrix} - \begin{Bmatrix} \dot{q}^- \\ -eF_q^- \\ 1 \end{Bmatrix} \right) \frac{[-1\ 0\ 0]}{(-\dot{q}^-)}$$

$$= \begin{bmatrix} 1 & 0 & 0 \\ 0 & -e & 0 \\ 0 & 0 & 1 \end{bmatrix} + \begin{bmatrix} -(\dot{q}^+ - \dot{q}^-) & 0 & 0 \\ -(F_q^+ + eF_q^-) & 0 & 0 \\ 0 & 0 & 0 \end{bmatrix} (-\dot{q}^-)^{-1}$$

$$= \begin{bmatrix} \dot{q}^+/\dot{q}^- & 0 & 0 \\ (F_q^+ + eF_q^-)/\dot{q}^- & -e & 0 \\ 0 & 0 & 1 \end{bmatrix},$$

where $F_q^+ = F_q(x^+)$, and it can be cast to the following form

$$S = \begin{bmatrix} -e & 0 & 0 \\ (F_q^+ + eF_q^-)/\dot{q}^- & -e & 0 \\ 0 & 0 & 1 \end{bmatrix}.$$

The obtained matrix transforms a perturbation δx^- just before a barrier into the perturbation just after the barrier δx^+. It is worth noticing that in the case of grazing bifurcation (tangent to the barrier surface), the velocity $\dot{q}^- \to 0$, and possesses matrix element $S_{21} \to \infty$.

13.6.3 Orbits Exhibiting Degenerated Contact with Discontinuity Surfaces

In the previous section the eigenvalues of Jacobian changed smoothly with a smooth change of a bifurcation parameter. Hence, the analysed piecewise smooth orbits can exhibit all bifurcations described earlier and associated with smooth systems.

Here, we are going to analyse the cases, where a Poincaré map (or a vector field during analysis of stationary points), its Jacobian and the associated eigenvalues exhibit discontinuities, which are associated with smooth changes of a bifurcation parameter. In this case either the obtained bifurcations are qualitatively similar to those in smooth systems, or they are completely new.

The most explored bifurcation in PSDS is a so-called grazing bifurcation. It occurs, when a part of trajectory becomes tangent to one of the discontinuity surfaces while changing a bifurcation parameter smoothly. It cannot be predicted by

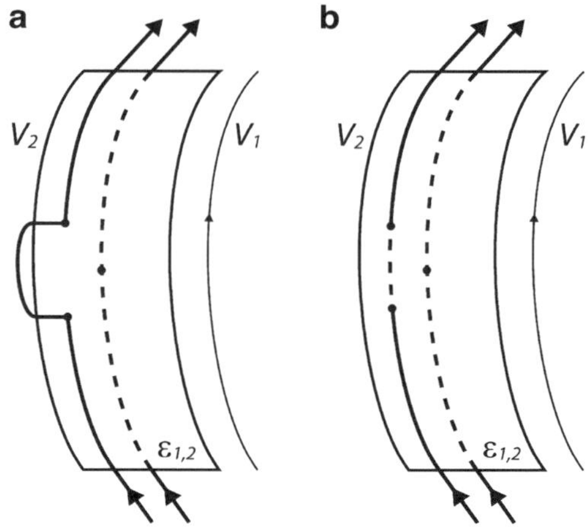

Fig. 13.26 Grazing bifurcations exhibited by Filippov's system (**a**) and a system with discontinuous vector state (**b**)

tracing a Jacobian behaviour. This type of bifurcation has been extensively studied by Feigin ([89–92]), although it is known in Russian literature as C-bifurcation.

However, to be more precise the C-bifurcation is typical for Filippov's systems [80], whereas grazing bifurcation is more practically oriented one and appears in systems with impacts. During investigations of maps a so-called border-collision bifurcation may appear, which is related to grazing bifurcation of a vector field ([41, 79, 80, 183, 185, 186]). Namely, it characterizes a collision of a mapping point with a discontinuity surface.

In Fig. 13.26 a grazing bifurcation is schematically shown for two different PSDS. Namely, a system with continuous vector state (Fig. 13.26a), and a system with a jump of a vector state (Fig. 13.26b) are displayed. The letter case is typical for impacting systems.

Although the classical grazing bifurcation assumes that a discontinuous surface in a contact place is smooth, but in practise very often it can be non-smooth. Imagine that a surface $\Sigma_{i,j}$ is composed of two smooth parts $\Sigma_{i,j}^{(1)}$ and $\Sigma_{i,j}^{(2)}$, which intersection creates a set C with dimension $(n-2)$, where n is dimension of phase space (see Fig. 13.27).

Two first bifurcations (Fig. 13.27a, b) belong to corner-collision bifurcation, whereas the third one Fig. 13.27c is more complicated.

Another important class of bifurcations exhibited by PSDS is associated with a sliding motion along $\Sigma_{i,j}$ (a trajectory remains on $\Sigma_{i,j}$ on a finite time interval). In the Filippov systems the vector field either forces the trajectories to move into $\Sigma_{i,j}$ from its both sides (attraction sliding mode) or to move away from it (repulsion sliding mode). More details are given in [46, 152].

In our example reported in Fig. 13.23c a real stick phase of the mass (which moves together with tape) can be referred as the attraction sliding mode.

When PSDS moves along a discontinuous surface a sliding type bifurcation can occur [46, 82]. In this case an orbit interacts with a part $\overline{\Sigma}$ of a discontinuity hyperplane Σ. Four different sliding bifurcations in Filippov's systems are reported in Fig. 13.28.

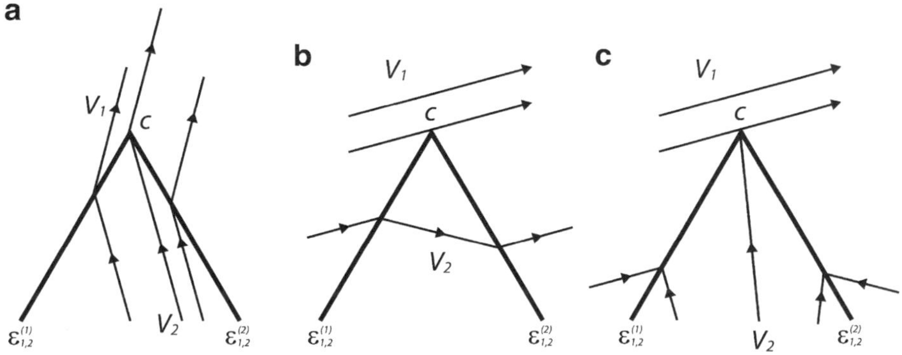

Fig. 13.27 Internal (**a**) and external (**b**) corner—collision bifurcation and bifurcation with sliding on $\Sigma^{(i)}_{1,2}$, $i = 1, 2$ (see [78])

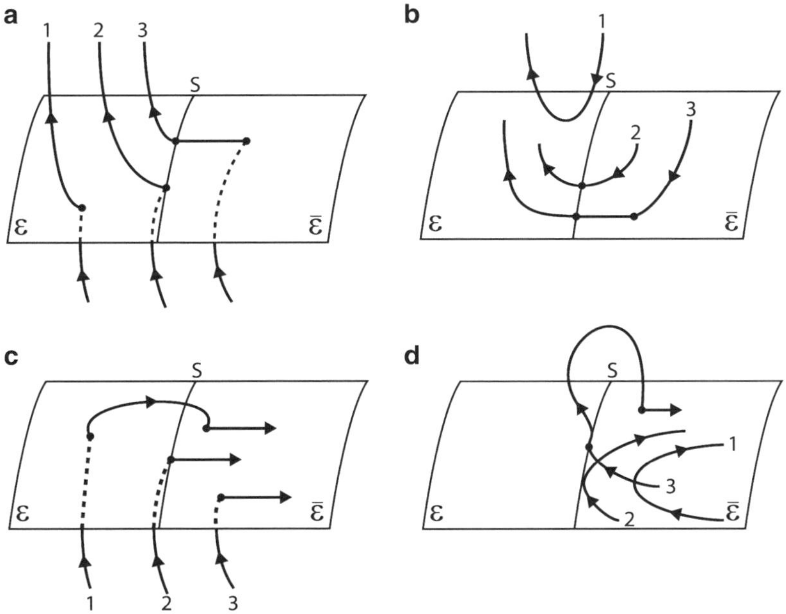

Fig. 13.28 Four different sliding bifurcations in a Filippov's system

Type I sliding bifurcation is shown in Fig. 13.28a. Increasing a bifurcation parameter the orbit first intersects transversally Σ, and then it moves to right successively touching the S-border between Σ and $\overline{\Sigma}$ (trajectory 2) and begins to slide. Any trajectory leaves $\overline{\Sigma}$ tangently.

In Fig. 13.28b the trajectory 1 approaches Σ and $\overline{\Sigma}$, it touches Σ and $\overline{\Sigma}$ simultaneously (orbit 2). Increasing further a bifurcation parameter a sliding part on $\overline{\Sigma}$ occurs (see orbit 3). This bifurcation is said to be grazing-sliding and it generalizes a grazing bifurcation concept.

The third example (Fig. 13.28c) is referred as a sliding bifurcation of type II (sometimes it is also called switching-sliding).

First, the orbit 1 intersects transversally Σ. Two other orbits remain in $\overline{\Sigma}$. All of the orbits are associated with sliding out of the border S.

The last (Fig. 13.28d) type of sliding bifurcation is called multisliding. All orbits belong to $\overline{\Sigma}$. A change of a bifurcation parameter yields to partition of the orbit into two parts (one of it lies on $\overline{\Sigma}$, and the other one lies outside of $\overline{\Sigma}$). The described multisliding bifurcation is a member of a sliding adding scenario, which yields an occurrence of periodic orbits with increasing number of sliding intervals [81].

13.6.4 Bifurcations in Filippov's Systems

Filippov's systems belong to classes (i) and (ii) and they are characterized by systems with at least smooth vector state.

In [151, 152] bifurcations of periodic solutions in systems with discontinuous vector fields are analysed. Although it is assumed that the Poincaré section in a bifurcation point is continuous but not necessarily smooth. Assume that a being analysed periodic solution is in contact with a discontinuous surface Σ. If the perturbed orbit intersects Σ and moves into another part of the phase space, where it stays infinitely short time, and then it returns to the previous phase domain, then the associated Poincaré map is continuous. It can be proved that if the analysed periodic orbit in tangent to a smooth discontinuity surface, then the Poincaré map is smooth and its Jacobian is continuous. When the orbit goes through the point of discontinuity surface which is non-smooth, then the associated Jacobian is non-continuous. To omit the occurred problem a concept of a generalized Clarke's derivatives can be applied. It yields definitions of both generalized Jacobian and generalized fundamental matrix. It can be shown also that a generalized Jacobian can be obtained via linear approximation of a non-smooth vector field F in a thin phase space $\overline{\Sigma}$ with ϵ thickness including a surface of discontinuity Σ. In words, a non-continuous vector field F is substituted by continuous but non-smooth vector field $\tilde{F}$, when $\epsilon \to \infty$.

13.6.5 Bifurcations of Stationary Points

Consider first codimension 1 bifurcations of a fixed point $x^* = x^*_{\Sigma_{i,j}}$ of the vector field F lying on the non-smooth surface $\Sigma_{i,j}$, being a border between subspaces V_i and V_j, where F is smooth.

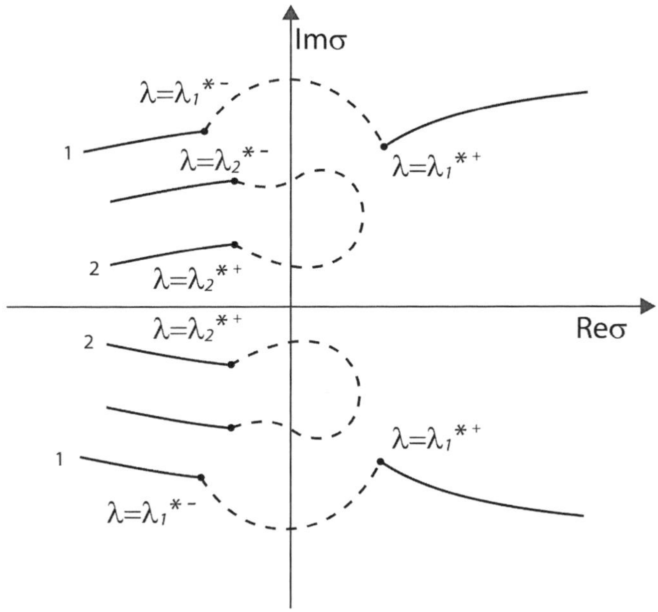

Fig. 13.29 Two different path of eigenvalues σ displaying discontinuous bifurcations

Since the point $x^*_{\Sigma_{i,j}}$ belongs simultaneously to V_i and V_j, it possesses two different values of a $DF_i(x^*_{\Sigma_{i,j}}; \lambda_{\Sigma_{i,j}})$ and $DF_j(x^*_{\Sigma_{i,j}}; \lambda_{\Sigma_{i,j}})$ on each of two sides of $\Sigma_{i,j}$. In what follows the following generalized Jacobian is defined in the point $x^*_{\Sigma_{i,j}}$:

$$\tilde{D}F(x^*_{\Sigma_{i,j}}; \lambda_{\Sigma_{i,j}}) = (1-q)DF_i(x^*_{\Sigma_{i,j}}; \lambda_{\Sigma_{i,j}}) + qDF_j(x^*_{\Sigma_{i,j}}; \lambda_{\Sigma_{i,j}}), \tag{13.207}$$

where $0 \leq q \leq 1$.

Formula (13.207) yields possible values of Jacobian in the point $x^*_{\Sigma_{i,j}}$ and it represents the smallest convex set possessing two values of Jacobian on two sides of $\Sigma_{i,j}$. Note that for $q = 0$ we have $\tilde{D}F(x^*_{\Sigma_{i,j}}; \lambda_{\Sigma_{i,j}}) = DF_i(x^*_{\Sigma_{i,j}}; \lambda_{\Sigma_{i,j}})$, whereas for $q = 1$ we get $\tilde{D}F(x^*_{\Sigma_{i,j}}; \lambda_{\Sigma_{i,j}}) = DF_j(x^*_{\Sigma_{i,j}}; \lambda_{\Sigma_{i,j}})$.

The mentioned set does not only define eigenvalues of Jacobians in a discontinuous point, but also defines a path of their jumps, when x^* goes through $\Sigma_{i,j}$. Now, if during such a jump the imaginary axis of the phase plane is crossed, then the associated bifurcation is called discontinuous one (see Fig. 13.29).

In the first case (curve 1) two complex conjugate eigenvalues cross once the imaginary axis, whereas in the second case (curve 2) the eigenvalues paths intersect the imaginary axis two times.

It is worth noticing that for each classical local bifurcation there exist also corresponding non-smooth bifurcations.

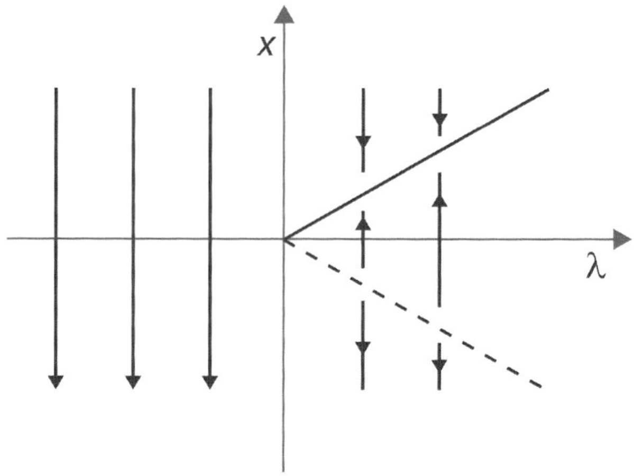

Fig. 13.30 Discontinuous saddle-node bifurcation

For instance, classical smooth saddle-node bifurcation can be displayed by the following equation

$$\dot{x} = F(x;\lambda) = \lambda - \mid x \mid . \tag{13.208}$$

The PSDS (13.208) does not contain fixed points for $\lambda < 0$, whereas for $\lambda > 0$ it has two fixed points $x^* = \pm\lambda$ (one of them is stable (solid curve), and one unstable (clashed curve)—see Fig. 13.30).

The generalized Jacobian computed in (0;0) is equal to $\tilde{D}F(0;0) = -2q + 1$ for $0 \leq q \leq 1$. The associated eigenvalue in the bifurcation point $\sigma = [-1, 1]$. For $q = 0 (q = 1)$ we have $\sigma = 1$ $(\sigma = -1)$, whereas for $q = \frac{1}{2}$ we have $\sigma = 0$.

Although it is not difficult to construct the corresponding discontinuous partners to the classical smooth bifurcations, we consider only one more bifurcation of the following non-smooth system

$$\dot{x} = F(x;\lambda) = \pm x + \left| x + \frac{1}{2}\lambda \right| - \left| x - \frac{1}{2}\lambda \right| . \tag{13.209}$$

The system has three stationary points $x^* = 0$ (stable for $\lambda < 0$, and unstable for $\lambda > 0$), and $x^* = \pm\lambda$ with the marked stability in Fig. 13.31.

Since two non-smooth vector field surfaces appear in the point (0;0), two parameters q_i, i=1,2 are needed to define a generalized Jacobian $\tilde{D}F(0;0) = 2(q_2 - q_1) - 1$, where $0 \leq q_i \leq 1$. For $q_2 = q_1 + \frac{1}{2}$ the associated eigenvalue crosses the imaginary axis yielding the discontinuous pitchfork bifurcation.

One may also construct a discontinuous Hopf bifurcation, which is governed by two following non-smooth equations

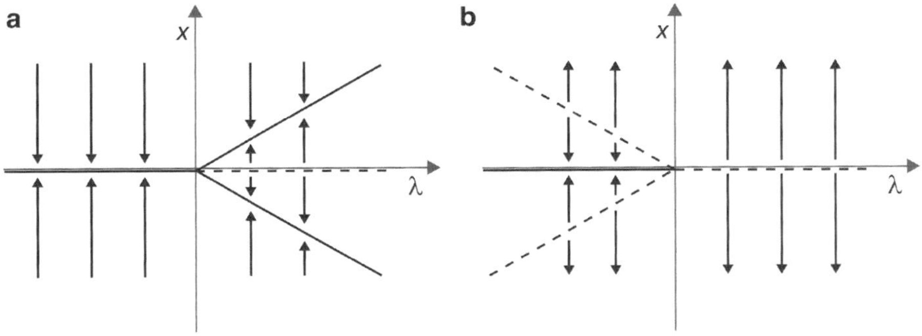

Fig. 13.31 Discontinuous pitchfork bifurcation (supercritical (**a**) and subcritical (**b**))

$$\dot{x}_1 = -x_1 - \omega x_2 + \frac{x_1}{\sqrt{x_1^2 + x_2^2}}\left(\left|\sqrt{x_1^2 + x_2^2} + \frac{1}{2}\lambda\right| - \left|\sqrt{x_1^2 + x_2^2} - \frac{1}{2}\lambda\right|\right),$$
$$\dot{x}_2 = \omega x_1 - x_2 + \frac{x_2}{\sqrt{x_1^2 + x_2^2}}\left(\left|\sqrt{x_1^2 + x_2^2} + \frac{1}{2}\lambda\right| - \left|\sqrt{x_1^2 + x_2^2} - \frac{1}{2}\lambda\right|\right). \tag{13.210}$$

Introducing the transformation

$$x_1 = r\cos\Theta, \qquad x_2 = r\sin\Theta \tag{13.211}$$

one gets

$$\begin{aligned} \dot{r} &= -r + \left| r + \tfrac{1}{2}\lambda \right| - \left| r - \tfrac{1}{2}\lambda \right|, \\ \dot{\Theta} &= \omega. \end{aligned} \tag{13.212}$$

For $\lambda < 0$ the stationary point $[x_1, x_2]^T = [0; 0]^T$ is stable, whereas for $\lambda > 0$ it becomes unstable, and a new stable periodic solution with the amplitude $r = \sqrt{x_1^2 + x_2^2}$ appears (a supercritical bifurcation).

13.6.6 Bifurcations of Periodic Orbits

A similar idea of generalized derivative can be used to trace periodic orbits yielded by a non-smooth vector field. A so-called generalized fundamental matrix is defined via relation

$$\tilde{\phi} = (1 - q)\phi^- + q\phi^+, \tag{13.213}$$

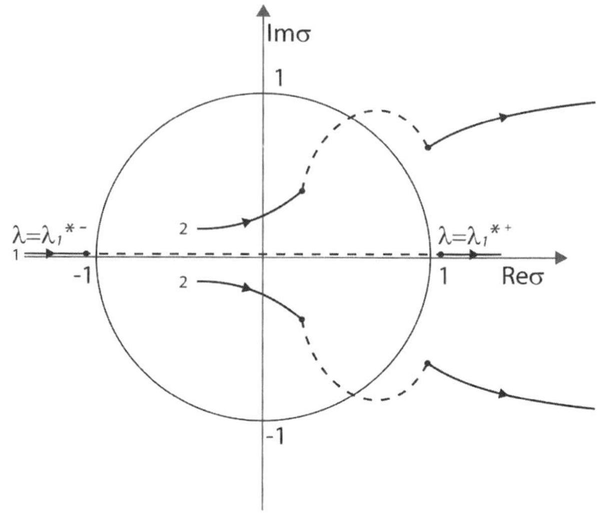

Fig. 13.32 Two paths of eigenvalues during fold-flip (*2*) and Neimark–Sacker (*1*) discontinuous bifurcations

where $0 \leq q \leq 1$. For $\lambda = \lambda^*$ we have two values $\phi = \phi^-$ or $\phi = \phi^+$ ($\phi^- = \lim_{\substack{\lambda \to \lambda^* \\ \lambda < \lambda^*}} \phi$; $\phi^+ = \lim_{\substack{\lambda \to \lambda^* \\ \lambda > \lambda^*}} \phi$). Two typical discontinuous bifurcations, i.e. Neimark–Sacker and fold-flip are shown in Fig. 13.32.

The unit circle of complex eigenvalues plane is crossed via a jump by a pair of complex conjugated eigenvalues and it is referred as the discontinuous Neimark–Sacker bifurcation. A real eigenvalue can jump from inside of the unit circle through either -1 (discontinuous period doubling or flip bifurcation) or through +1 (fold bifurcation). However, a real eigenvalue $\sigma < -1$ can also jump over the circle getting new value $\sigma > 1$. This discontinuous bifurcation is called fold-flip (path 1 in Fig. 13.32).

The described simple smooth and they corresponding non-smooth bifurcations have relatively simple geometrical interpretation (see Fig. 13.33).

One-dimensional Poincaré map in the vicinity of smooth saddle-node bifurcation (Fig. 13.33a) does not intersect the line $x_{i+1} = x_i$ for $\lambda < \lambda^*$, is tangent to this line for $\lambda = \lambda^*$, and possesses two intersection points for $\lambda > \lambda^*$ (the last case corresponds to two periodic orbits of an associated dynamical system). The Poincaré map of non-smooth saddle-node bifurcation touches $x_{i+1} = x_i$ in non-smooth point (cusp), and hence it has different left-hand side and right-hand side derivatives.

In Fig. 13.33b for smooth period doubling bifurcation one-dimensional Poincaré map intersects $x_{i+1} = x_i$ in the point, where the associated derivative is equal to -1. On the other hand, in the case of non-smooth period doubling bifurcation (dashed lines) the associated Poincaré map intersects $x_{i+1} = x_i$ in the cusp and it has two different values of a derivative.

To sum up, the situation of classification of possible non-smooth bifurcations is far to be completed. As it has been already mentioned, such bifurcation may have their analogues in classical smooth bifurcations or they can be combinations

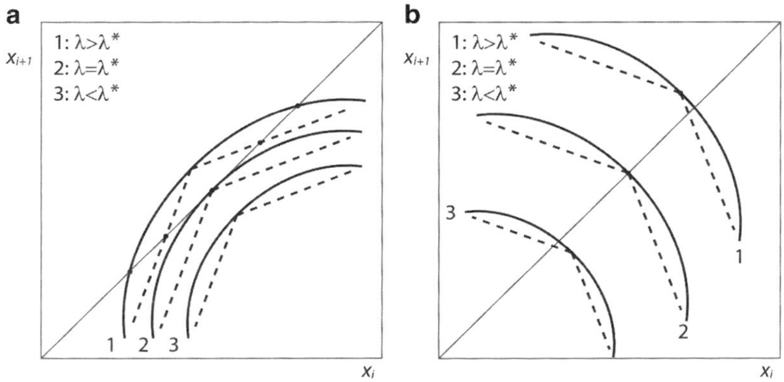

Fig. 13.33 Poincaré maps displaying saddle-node (**a**) and period doubling (**b**) smooth (*solid curve*) and non-smooth (*dashed lines*) bifurcations

of a sequence of such smooth bifurcations. However, there are also non-smooth bifurcations which have not their analogy in smooth systems.

The earlier briefly described C-bifurcations or border collision bifurcation are also locally analysed using piecewise smooth local modes in [78–80, 82, 92]. The popular grazing bifurcation is classified through estimation of real eigenvalues of a map, which are less than −1 and larger than +1 on both sides of a discontinuity. However, the Neimark–Sacker bifurcation is not discussed.

Almost nothing is done in the field of global non-smooth bifurcations and local non-smooth bifurcations occurred in high dimensional systems.

Chapter 14
Optimization of Systems

14.1 Introduction

A phenomenon of optimization occurs in both the nature and human activity. Stems are built optimally, a sunflower is filled with grains optimally and also hunting dogs chase a fox optimally. For the first time, the problems of optimization were met by the Greeks, who knew that:

(i) a circle is a figure of the smallest circumference with the fixed area;
(ii) a ball is a figure of the largest volume with the fixed area;
(iii) an equilateral triangle is of the biggest area of triangles with fixed circumference.

Foundations of contemporary optimization are connected with variational calculus and the brachistochrone problem, stated by Bernoulli in 1969 which was solved by him 1 year later. In a gravitational field, one needs to find the curve joining two points (not lying on a vertical line) along which a particle falling from rest accelerated by gravity travels in the least time (we neglect friction). In physics, this problem is connected with the Fermat principle: a light ray passes through two points along such a curve that the time of motion along the curve is shortest.

Euler is regarded as a creator of the variational calculus, and Lagrange is regarded as a main representative. Presently, variational principles are commonly known and applied in various domains of science, such as:

(i) biology and optics (the Fermat principle);
(ii) mechanics, electrotechnics and elektronics (the Hamilton–Lagrange, Jacobi and Dirichlet principles);
(iii) mechanics and strength of materials (the Castigliano–Menabrei principle);
(iv) geometrical optics (the Hilbert–Malus principle);

J. Awrejcewicz, *Ordinary Differential Equations and Mechanical Systems*,
DOI 10.1007/978-3-319-07659-1_14

(v) Maxwell's equations in electrodynamics (the Nöther and Bessel–Hagen principles);
(vi) electrostatics (the Friedrichs principle).

14.2 Simple Examples of Optimization in Approximative Problems

Consider a problem solved by Steinhaus [222] (Fig. 14.1).

For a given convex function $F(x)$ in the interval $[a,b]$ we seek a linear function

$$\sigma(x) = mx + n \tag{14.1}$$

which minimizes the following functional

$$J = \int_a^b |F(x) - G(x)|\, dx. \tag{14.2}$$

This means that a hatched area in Fig. 14.1 should have the smallest value. Steinhaus showed that the sought points are as follows: $u = (3a + b)/4$ and $v = (3b + a)/4$.

Substituting (14.1) into (14.2) we get

$$\begin{aligned} J = \int_a^b (F(x) - mx - n)\, dx = \int_a^u (F(x) - mx - n)\, dx \\ + \int_u^v (mx + n - F(x))\, dx + \int_v^b (F(x) - mx - n)\, dx. \end{aligned} \tag{14.3}$$

Next, we seek the minimum of (14.3) with respect to m and n. We obtain

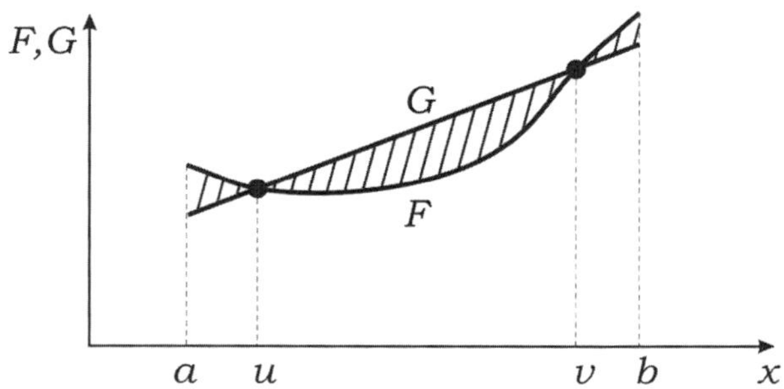

Fig. 14.1 Linear approximation σ of the convex curve F

$$\frac{dJ}{dm} = \int_a^u (-x)dx + (F(u) - mu - n)\frac{\partial u}{\partial m}$$

$$+ \int_v^b (-x)dx - (F(v) - mv - n)\frac{\partial v}{\partial m}$$

$$+ - \int_u^v (-x)dx - (F(v) - mv - n)\frac{\partial v}{\partial m} + (F(u) - mu - n)\frac{\partial u}{\partial m}$$

$$= -\frac{1}{2}\left(u^2 - a^2\right) - \frac{1}{2}\left(b^2 - v^2\right) + \frac{1}{2}\left(v^2 - u^2\right) = 0,$$

since $F(u) = mu + n$ and $F(v) = mv + n$.

After transformation we get

$$u^2 - v^2 = \frac{a^2 - b^2}{2}. \tag{14.4}$$

Similarly, we find

$$\frac{dJ}{dn} = -\int_a^u dx + \int_u^v dx - \int_v^b dx$$

$$= -(u - a) + (v - u) - (b - v) = 0.$$

or

$$u - v = \frac{a - b}{2}. \tag{14.5}$$

Comparing the expressions (14.4) and (14.5), we get the following relationship

$$u + v = a + b. \tag{14.6}$$

Adding up and subtracting corresponding sides of (14.5) and (14.6), we get abscissae of the points given by Steinhaus.

Markov pointed out that if the function $F(x)$ is continuous in the interval $[a,b]$, and if the difference $F(x) - P_n(x)$, where $P_n(x) = \alpha_0 + \alpha_1 x + \cdots + \alpha_n x^n$, changes the sign only at the points

$$x_k = \frac{a + b}{2} - \frac{b - a}{2}\cos\frac{k\pi}{n + 2}, \quad k = 1, \ldots, n + 1, \tag{14.7}$$

then the following inequality holds

$$\int_a^b |F(x) - P_n(x)|\, dx \le \int_a^b |F(x) - Q_n(x)|\, dx, \tag{14.8}$$

where $Q_n(x)$ is any polynomial of at most n degree, and the sign in the equality (14.8) appears for $Q_n(x) \equiv P_n(x)$. For a particular case, it follows from (14.8) that

$$P_n(x_k) = F(x_k),\ k = 1, \ldots, n+1. \tag{14.9}$$

This means that knowing (e.g. from measurements) values of the functions $F(x_1), \ldots, F(x_{n+1})$, we can determine a polynomial $P_n(x)$. On the other hand, Lagrange interpolation formula allows to evaluate $P_n(x)$. Having the abscissae x_k determined from the formula (14.7), we build a polynomial

$$w(x) = (x - x_1)(x - x_2)\ldots(x - x_{n+1}). \tag{14.10}$$

The polynomial $P_n(x)$ takes the form

$$P_n(x) = \sum_{k=1}^{n+1} F(x_k)\frac{w(x)}{(x - x_k)w(x_k)}. \tag{14.11}$$

Necessity of applying the interpolating formula (14.11) follows from the fact that we do not know if the polynomial (14.9) is the best approximation.

Let the function $F(x)$ be approximated by a parabola of the second degree. According to Eq. (14.7) we have the following abscissae (nodes)

$$\begin{aligned} x_1 &= \frac{a+b}{2} - \frac{b-a}{2}\cos\frac{\pi}{4} = \frac{a+b}{2} - \frac{b-a}{2}\frac{\sqrt{2}}{2}, \\ x_2 &= \frac{a+b}{2} - \frac{b-a}{2}\cos\frac{\pi}{2} = \frac{a+b}{2}, \\ x_3 &= \frac{a+b}{2} - \frac{b-a}{2}\cos\frac{3\pi}{4} = \frac{a+b}{2} + \frac{b-a}{a}\frac{\sqrt{2}}{2}. \end{aligned} \tag{14.12}$$

14.3 Conditional Extrema

In most cases encountered in the nature and technics, additional constraints are imposed on problems, which deals with seeking of extrema of functions within their domain of definition. This problem reduces to finding a minimum of a real function $F(u)$ of a domain of definition $D \subset R^n$ and with imposed constraints of the form

$$G_i(u) = 0, \quad i = 1, \ldots, m, \quad m < n. \tag{14.13}$$

Suppose that using m Eq. (14.13) we determine the following relationships

$$\begin{aligned} u_1 &= g_1(u_{m+1}, \ldots, u_n) \\ &\ldots \\ u_m &= g_m(u_{m+1}, \ldots, u_n), \end{aligned} \tag{14.14}$$

i.e. we have determined m variables u_i as functions of remaining $n - m$ variables. Although, previously our problem was connected with the minimalization of the function

$$F(u_1, u_2, \ldots, u_m, u_{m+1}, \ldots, u_n) \tag{14.15}$$

with the constraints (14.13), then now according to (14.14) our problem boils down to minimization without constraints of the function

$$F(g_1, g_2, \ldots, g_m, u_{m+1}, \ldots, u_n) = 0. \tag{14.16}$$

It turns out that it is possible only for very simple cases. In the case of algebraic equations of degree $n \geq 5$ and transcendental equations such a procedure is not possible because we cannot "resolve" the constraints (14.13).

One can see this clearly in the case $u \in R^2$. The function $G_1(u_1, u_2)$ can possess three following forms:

$$G_1(u_1, u_2) = u_1^2 + u_2^2 = 0, \tag{14.17}$$

and a solution is one point $u_1 = u_2 = 0$; or

$$G_1(u_1, u_2) = u_1^2 + u_2^2 + 2 = 0, \tag{14.18}$$

and solutions of this equation do not exist; or finally

$$G_1(u_1, u_2) = u_1^2 + u_2^2 - 2 = 0, \tag{14.19}$$

and a solution is infinitely many points lying on a circle.

As one can see by the above examples, significant and difficult questions arise. Is there any continuous curve $u_2 = g_1(u_1)$, whose all points satisfy $G_1(u_1, u_2) = 0$, and is it unique? Thus, the problem concerns estimation of local character of the function $g_1(u_1, u_2)$ in the neighbourhood of the point (u_1^0, u_2^0) satisfying $g(u_1^0, u_2^0) = 0$. One makes use of the Implicit Function Theorem given below.

Theorem 14.1 (Implicit Function Theorem). *If the function* $g(u_1, u_2)$ *have continuous partial derivatives* g_{u_1} and g_{u_2} *in a neighbourhood of the point* (u_1^0, u_2^0) *and if*

$$g_{u_1}(u_1^0, u_2^0) = 0 \text{ and } g_{u_2}(u_1^0, u_2^0) \neq 0, \tag{14.20}$$

then

(i) there exists such $\delta > 0$ *that for each sufficiently small* $\varepsilon > 0$, *and for each value of* $u_1^0 - \delta < u_1 < u_1^0 + \delta$ *there exists a unique function* $u_2(u_1)$, *which is a solution to the equation* $g = 0$ *in the interval* $u_2^0 - \varepsilon < u_2 < u_2^0 + \varepsilon$;
(ii) the function $u_2(u_1)$ *is of class* C^1 *in the interval* $(u_1^0 - \delta, u_2^0 + \delta)$, *i.e. it is continuous and has first continuous derivative, which is determined by the formula*

$$u_2'(u_1) = -\frac{g_{u_1}(u_1, u_2)}{g_{u_2}(u_1, u_2)}, \tag{14.21}$$

while $u_2 = u_2(u_1)$.

The above theorem can be easily generalized into a system of implicit functions.

Note that if there is no constraint G_1, then a stationary point (u_{10}, u_{20}) of a function

$$F(u_1, u_2) = 0, \tag{14.22}$$

is determined by the equations

$$\begin{aligned} \frac{\partial F(u_1, u_2)}{\partial u_1} &= 0, \\ \frac{\partial F(u_1, u_2)}{\partial u_2} &= 0, \end{aligned} \tag{14.23}$$

since their solution allows to obtain u_{10}, u_{20}.

Since the increments u_1 and u_2 are arbitrary, the exact differential also vanishes

$$dF = \left.\frac{\partial F}{\partial u_1}\right|_{u_1 = u_{10}} du_1 + \left.\frac{\partial F}{\partial u_2}\right|_{u_2 = u_{20}} du_2 = 0. \tag{14.24}$$

Let us introduce a constraint

$$G_1(u_{10}, u_{20}) = 0. \tag{14.25}$$

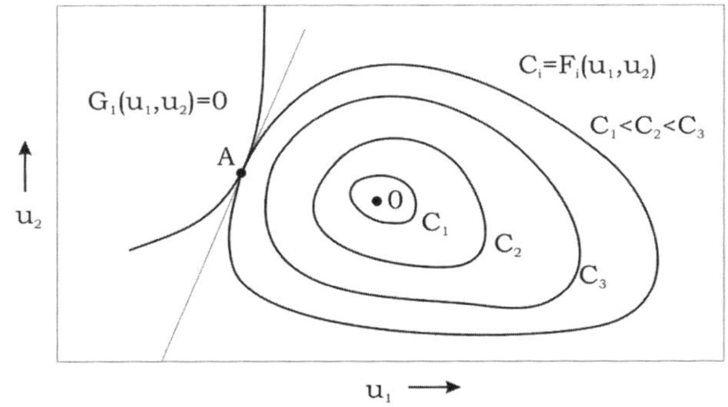

Fig. 14.2 Graphical illustration of existence of the constraint function extremum

Note that also Eq. (14.24) will stay valid but not for all differential increments du_1 and du_2. These increments must satisfy the additional equation

$$dG_1 = \left.\frac{\partial G_1}{\partial u_1}\right|_{u_{10}} du_1 + \left.\frac{\partial G_1}{\partial u_2}\right|_{u_{20}} du_2 = 0. \tag{14.26}$$

The Jacobian of the homogeneous equations (14.24) and (14.26) reads

$$J\left(\frac{F, G_1}{u_1, u_2}\right) = \left|\begin{matrix} \frac{\partial F}{\partial u_1} & \frac{\partial F}{\partial u_2} \\ \frac{\partial G_1}{\partial u_1} & \frac{\partial G_1}{\partial u_2} \end{matrix}\right|_{u_{10},\, u_{20}} = 0 \tag{14.27}$$

We mention here the geometric interpretation of the considered problem given in [102] (see Fig. 14.2).

The function $F(u) = C_i$, $i = 1, 2, \ldots$, $C_i < C_{i+1}$, possesses a minimum at the point 0. On the other hand, a graph of the constraint function $G_1(u_1, u_2) = 0$ is represented by a solid curve. Increasing the value of C_i for some $i = I$, the curve $F(u_1, u_2)C_I$ will possess a tangent point A with the curve $G(u_1, u_2) = 0$. This point determines the minimal value of the function $F(u)$ with constraints. The straight line in the figure passing through the point A is tangent to both the curve $G(u) = 0$ and $F = \text{const}\, C_I$.

In the case of $F = \text{const}$ by Eq. (14.24) we get

$$\frac{du_{20}}{du_{10}} = -\frac{\partial F/\partial u_{10}}{\partial F/\partial u_{20}}, \tag{14.28}$$

whereas in the case of $G(u_1, u_2)$ by Eq. (14.26) we find

$$\frac{\partial u_{20}}{\partial u_{10}} = -\frac{\partial G/\partial u_{10}}{\partial G/\partial u_{20}}. \tag{14.29}$$

By Eqs. (14.28) and (14.29) we obtain

$$\frac{\partial F}{\partial u_{10}}\frac{\partial G}{\partial u_{20}} - \frac{\partial F}{\partial u_{20}}\frac{\partial G}{\partial u_{10}} = 0, \tag{14.30}$$

what follows from (14.27).

This method can be easily generalized to the case of seeking the extremum of a function of several variables at many constraints [102], but it turns out that evaluation of determinants of high degree is laborious.

We will briefly describe more often applied method, namely the so-called method of Lagrange's multipliers.

In the considered two-dimensional case, Eq. (14.30) can be written in the form

$$\frac{\partial F/\partial u_{10}}{\partial G/\partial u_{10}} = \frac{\partial F/\partial u_{20}}{\partial G/\partial u_{20}} = -\lambda, \tag{14.31}$$

where the constant λ is called the Lagrange multiplier. By Eq. (14.31) we get

$$\begin{aligned} \frac{\partial F}{\partial u_{10}} + \lambda\frac{\partial G}{\partial u_{10}} &= 0, \\ \frac{\partial F}{\partial u_{20}} + \lambda\frac{\partial G}{\partial u_{20}} &= 0. \end{aligned} \tag{14.32}$$

In the considered stationary point $\mathrm{A}(u_{10}, u_{20})$ gradients of the functions F and G are colinear and of opposite signs. The multiplier λ represents sensitivity of the function F on change of the constraints of G at the point (u_{10}, u_{20}), if $\partial G/\partial u_{10} \neq 0$ and $\partial G/\partial u_{20} \neq 0$.

Note that after introducing the notion of the Lagrange multipliers, the functions (14.32) can be understood as existence conditions of a stationary point of the function $F + \lambda G$ without constraints. The initial problem, which relies on finding a stationary point (u_{10}, u_{20}) of the function $F(u_1, u_2)$ with the constraining condition $g(u_1, u_2) = 0$ has been reduced to the analysis of the function $F + \lambda G$ of three unknowns u_{10}, u_{20} and λ_0, but without constraints.

The performed reasoning can be easily repeated in the case of seeking the extremum of the function $F(u_1, \ldots, u_m, v_1, \ldots, v_n)$ with the constraints
$G_i(u_1, \ldots, u_m, v_1, \ldots, v_n)$, $i = 1, \ldots, n$.

The Lagrange idea (multipliers) comes from considerations concerning mechanical systems.

14.4 Static Optimization

The *static optimization* is used in various problems concerning design which leads to e.g. optimization of construction with respect to various criteria as well as in

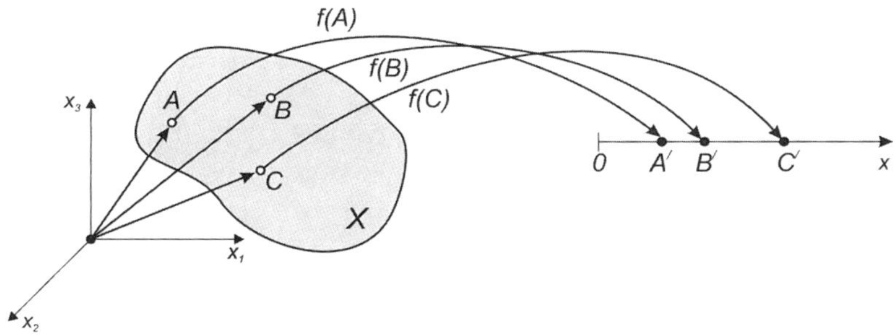

Fig. 14.3 An objective function $f : X \to R^1$

problems concerning organization of undertakings in various branches of economy and management. A problem of static optimization occurs, when both objective functions and limitations have the form of algebraic relationships. Problems of dynamic optimization are reduced to static optimization by discretization of time and spatial variables. It is commonly applied in e.g. dynamical analysis of plates and shells [25].

Decision variables, in problems of static optimization, are real numbers and form a vector of the decision variables $x = [x_1, \ldots, x_n]^T$. A set X, where

$$x \in X \subseteq R^n, \tag{14.33}$$

is called a set of admissible decisions, and is bounded and closed (compact).

An *objective function* is called a mapping $f : X \to R^1$ *ordering* the set X (Fig. 14.3).

The mapping f should have the following property:

$$\forall\,(x_1, x_2) \in X, \quad x_1 = x_2 \to f(x_1) = f(x_2). \tag{14.34}$$

Ordering the set X with the function f is defined by means of *equivalence relation*

$$f(x_1) \not\succ f(x_2) \Leftrightarrow x_1 \leq x_2, \tag{14.35}$$

a symbol "$\not\succ$" is understood as "x_1 is not better than x_2"

A *strict order* of the set X we obtain for an injective function, i.e. such that

$$\forall\,(x_1, x_2) \in X, \;\; x_1 \neq x_2 \Leftrightarrow f(x_1) \neq f(x_2) \tag{14.36}$$

and it is defined as follows

$$\forall\,(x_1, x_2) \in X, \;\; f(x_1) < f(x_2) \Leftrightarrow x_1 < x_2. \tag{14.37}$$

A global maximum (minimum) of a function $f(x)$ in a set X is called an element $x^* \in X$ such that there does not exist any better element than x, i.e.

$$\forall x \in X, \quad f(x^*) \not\geq f(x) \tag{14.38}$$

and respectively

$$\forall x \in X, \quad f(x^*) \not> f(x). \tag{14.39}$$

The problem of static optimization reduces to the determination of $x^* \in X$. It is enough to consider a problem of maximalization of f, since a problem of minimalization can be easily reduced to the problem of maximalization by taking $-f$, i.e.

$$\begin{aligned} &\max_{x \in X} f(x) = \min_{x \in X}[-f(x)], \\ &x^* = \arg\max_{x \in X} f(x) = \arg\min_{x \in X}[-f(x)]. \end{aligned} \tag{14.40}$$

The function $f(x)$ can possess in the considered set X one or many global maxima, or none (the reader can easily find such cases).

The element x^* is a strict global maximum in the set X if the following strict inequality is satisfied

$$\forall x \in X, \quad f(x^*) > f(x). \tag{14.41}$$

Theorem 14.2 (Weierstrass). *If X is a compact set and the function $f(x)$, $\forall x \in X$ is continuous, then a global maximum of $f(x)$ exist and can occur inside or on the boundary of the set X.*

We skip the proof of this theorem (see e.g. [107]).

An element ${}^*x \in X$ is called a *local maximum* of a function $f(x)$ if in the ε -neighbourhood of the element ${}^*x$ the function $f(x)$ does not attain values greater than $f({}^*x)$, i.e.

$$\begin{aligned} &\varepsilon > 0 : f\left({}^*x\right) \geq f\left(x\right), \quad \forall x \in [X \cap N_\varepsilon\left({}^*x\right)], \\ &N_\varepsilon\left({}^*x\right) = \left\{ x \in R^n : \sqrt{\sum_{i=1}^{N}\left({}^*x_i - x_i^2\right)} < \varepsilon \right\}. \end{aligned} \tag{14.42}$$

Note that in the case of a function $f(x)$, $x \in R^1$, the problem of minimal value reduces to equating the first derivative $f^{|}(x^*) = 0$ to zero and $f^{||}(x^*) \leq 0$.

14.4.1 Local Approximation of a Function

Since we have already mentioned about local extrema of function, we need to emphasize that problems of optimization are connected with a local approximation of function.

Taylor expansion in a neighbourhood of a point x_0 has the form

$$f(x) = \sum_{n=0}^{\infty} \frac{f^{(n)}(x_0)}{n!}(x - x_0)^n, \tag{14.43}$$

where $f^{(n)} = \frac{\partial f}{\partial x^n}(x_0)$. Not every function can be Taylor expanded. If we want to take only N terms in the series (14.43), then

$$f(x) = f(x_0) + \sum_{n=1}^{N} \frac{f^n(x)}{n!}(x - x_0)^n + O\left(|x - x_0|^N\right), \tag{14.44}$$

where the symbol $O\left(|x - x_0|^N\right)$ means that terms of indices $n > N$ will decrease faster than $|x - x_0|^N$ (they are small in comparison with $|x - x_0|^N$).

The first-order approximation has the form

$$f(x) = f(x_0) + \frac{df(x_0)}{dx}(x - x_0), \tag{14.45}$$

and the second-order approximation has the form

$$f(x) = f(x_0) + \frac{df(x_0)}{dx}(x - x_0) + \frac{1}{2}\frac{d^2 f(x_0)}{dx^2}(x - x_0)^2. \tag{14.46}$$

Consider a vector $x \in R^n$ and introduce Euclidean norm of the form

$$\|x\| = \sqrt{\sum x_i^2}. \tag{14.47}$$

The quadratic approximation of $f(x)$ around the point x_0 has the form

$$\begin{aligned} f(x) = f(x_0) + \sum_{i=1}^{n} \frac{\partial f(x_0)}{\partial x_i}(x_i - x_{i0}) \\ + \frac{1}{2}\sum_{i=1}^{n}\sum_{j=1}^{n} \frac{\partial^2 f(x_0)}{\partial x_i \partial x_j}(x_i - x_{i0})(x_j - x_{j0}) + O\left(\|x - x_0\|^2\right), \end{aligned} \tag{14.48}$$

where: $x = (x_1, \ldots, x_n)^T$, $x_0 = (x_{10}, \ldots, x_{n0})^T$. Let us introduce a notion of gradient of a function f, namely

$$\nabla f = \left(\frac{\partial f}{\partial x_1}, \frac{\partial f}{\partial x_2}, \ldots, \frac{\partial f}{\partial x_n} \right), \tag{14.49}$$

Next, we introduce a matrix called a hessian

$$\nabla^2 f \equiv H = \begin{bmatrix} \frac{\partial^2 f}{\partial x_1^2} & \cdots & \frac{\partial^2 f}{\partial x_1 x_n} \\ \vdots & & \vdots \\ \frac{\partial^2 f}{\partial x_n x_1} & \cdots & \frac{\partial^2 f}{\partial x_n^2} \end{bmatrix}. \tag{14.50}$$

We define a vector of perturbations

$$\partial x = x - x_0 \tag{14.51}$$

and a function of perturbations

$$\partial f = f(x) - f(x_0). \tag{14.52}$$

Equation (14.48) can be written in the form

$$\partial f = \nabla f(x) \partial x + \frac{1}{2} \partial x^T H(x_0) \partial x + O\left(\|\partial x\|^2 \right), \tag{14.53}$$

where H is a square and symmetric matrix.

First, we considered $x \in R^1$, next $x \in R^n$ for the case of one function f, and now our considerations can be generalized onto a vectorial function $f(x) = [f_1(x), f_2(x), \ldots, f_m(x)]^T$. In this case, gradient of the function (vector) is replaced with a matrix, called Jacobian, of the form

$$\frac{\partial f}{\partial x} \equiv J = \begin{bmatrix} \frac{\partial f_1}{\partial x_1} & \cdots & \frac{\partial f_1}{\partial x_n} \\ \vdots & & \vdots \\ \frac{\partial f_m}{\partial x_1} & \cdots & \frac{\partial f_m}{\partial x_n} \end{bmatrix} \equiv [\nabla f_1, \nabla f_2, \ldots, \nabla f_m]^T. \tag{14.54}$$

14.4.2 Stationary Points and Quadratic Forms

Assume that a function $f(x)$ possesses a minimum at the point $f^*(x^*) = \min$. The necessary condition of existence of the internal minimum is

$$f(x^*) = 0. \tag{14.55}$$

Points x^* satisfying this condition are called stationary. This condition is necessary since it can be satisfied also at points, which are not local minima. In the case of a singular point, by Eq. (14.53) we get

$$\partial f_* = \frac{1}{2}\partial x^T H(x_*)\partial x, \tag{14.56}$$

and this allows to conclude that:

If the necessary condition (14.55) *is satisfied and Hessian of the function* $f(x)$ *is positive definite at the stationary point* x_**, then this point is a local minimum.*

In practice, positive definiteness of $\nabla^2 f$ is reduced to a diagonalization condition of Hessian. A function

$$f(x) = x^T H x \tag{14.57}$$

is called a quadratic form. The quadratic form (14.57) is called positive (negative) definite, if $\forall x \in R^n$ and $x \neq 0$ attain positive values (negative).

A quadratic form is called *semi-positive (semi-negative)* definite, if $\forall x \in R^n$ it has *non-negative (non-positive)* values.

A quadratic form possessing any values is called sign-indefinite.

Similarly, we define Hessian H. Character of a quadratic form is determined by means of eigenvalues of the matrix H, i.e.

$$\det(\sigma I - H) = 0, \tag{14.58}$$

where $I_{n\times n}$ is a unit matrix.

If the matrix H is symmetric and their elements are real, then its eigenvalues are real numbers. The classification of stationary points and quadratic forms (function $f(x)$) was given in Table 14.1.

One can show, on the basis of the definition of concave function (see further considerations), that a semi-negative (semi-positive) definite quadratic form is a convex (concave) function. A semi-negative (positive) definite quadratic form is a strictly concave (strictly convex) function.

In order to evaluate the character of the matrix H one needs to evaluate the values of n major subdeterminants of the matrix **H** called the *Sylvester determinants.*

If determinants of the Sylvester matrix H are positive (non-negative), then the matrix H is positive (semipositive) definite. However, if we want to know if H is negative or seminegative definite, then we need to reduce the problem to the earlier described one by introducing the matrix-H.

14.5 Convexity of Sets of Functions

In order to introduce a notion of a convex function we will consider a function $f(x)$ illustrated in Fig. 14.4.

Table 14.1 Classification of quadratic forms [107, 192]

Eigenvalues σ_i	det H	$f(x)$	H	Types of functions and stationary points
$\sigma_i < 0$	$\det H \neq 0$	$x^T H x < 0$	$H < 0$	Strictly concave function; strict global maximum
$\sigma_i > 0$	$\det H \neq 0$	$x^T H x > 0$	$H > 0$	Strictly convex function; strict global minimum
$\sigma_i \leq 0$ $V_i : \sigma_i = 0$	$\det H = 0$	$x^T H x \leq 0$	$H \leq 0$	Concave function; global maximum; peaks
$\sigma_i \geq 0$ $V_i : \sigma_i = 0$	$\det H = 0$	$x^T H x \geq 0$	$H \geq 0$	Convex function; global minimum; valley
$\sigma_i \neq 0$	$\det H \neq 0$	$x^T H x \in R^1$		Regular saddle point
$V_i : \sigma_i = 0$	$\det H = 0$	$x^T H x \in R^1$		Singular saddle point

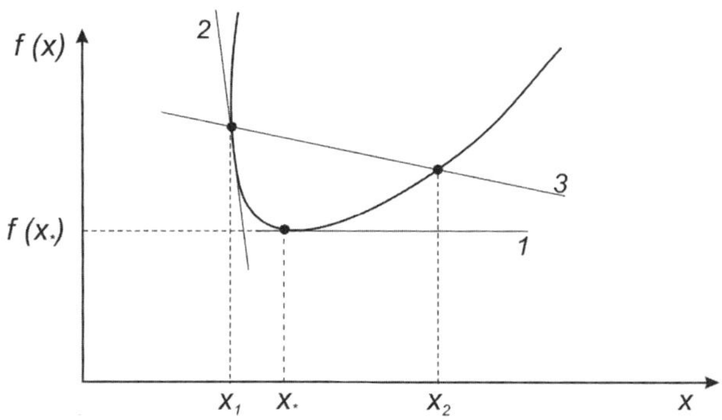

Fig. 14.4 Geometrical interpretation of convexity of a function $f(x)$

The following properties of a convex function follow from Fig. 14.4

(i) tangent 2 at any point of the function $f(x)$ will never intersect the graph of the function;
(ii) secant 3 is located above the function $f(x_1) \leq f(x) \leq f(x_2)$.

The property of *local convexity* ensures the existence of minimum, while the property of *global convexity* ensures the existence of global minimum.

A set X is convex is for any two elements $(x_1, x_2) \in X$ all the elements belonging to a segment of the line connecting x_1 and x_2 belong to X (Fig. 14.5a).

A set $S \subseteq R^n$ is convex if for all two points x_1 and x_2 belonging to S a point (see Fig. 14.5b)

$$x(\lambda) = \lambda x_1 + (1 - \lambda)x_2, \quad \forall \lambda \in [0, 1]. \tag{14.59}$$

Definition 14.1 (Of a Convex Function). A function $f(x) = f(x_1, \ldots, x_n)$ defined on a convex set $S \subseteq R^n$ is called convex, if for any two points $x_1, x_2 \in S$ and for all $\lambda \in [0, 1]$ the inequality holds

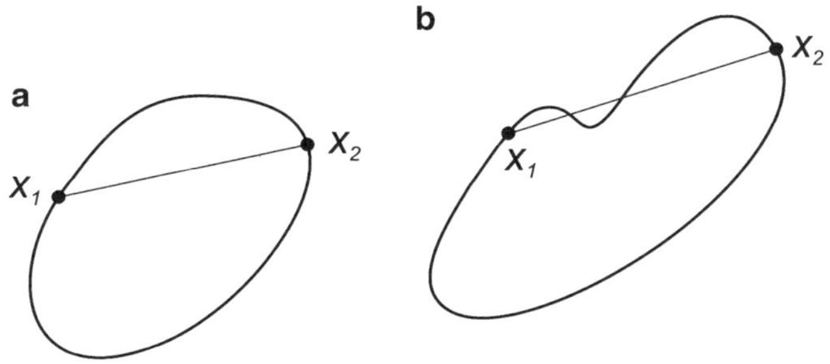

Fig. 14.5 Convex set (**a**) and nonconvex (**b**)

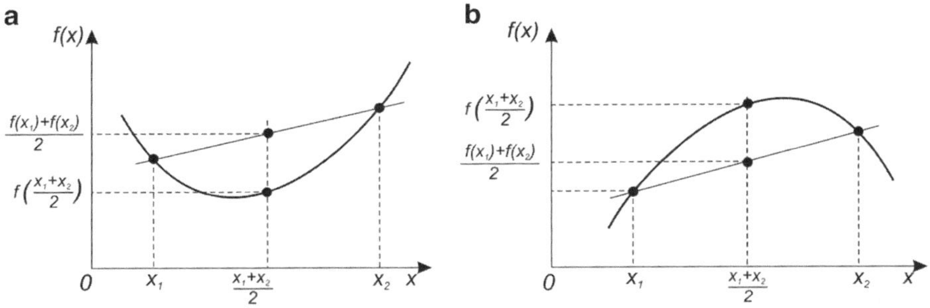

Fig. 14.6 Convex (**a**) and concave (**b**) functions

$$f[\lambda x_1 + (1-\lambda)x_2] \leq \lambda f(x_1) + (1-\lambda) f(x_2). \tag{14.60}$$

This means that if the function $f(x)$ is convex, then a segment connecting any two points of $f(x)$ lies on or over $f(x)$.

Definition 14.2 (A Concave Function). A function $f(x)$ defined on a concave set $S \subset R^n$ is called concave, if for any $x_1, x_2 \in S$ and for all $\lambda \in [0, 1]$ the inequality holds

$$f[\lambda x_1 + (1-\lambda)x_2] \geq \lambda f(x_1) + (1-\lambda) f(x_2). \tag{14.61}$$

If $f(x)$ is concave, then every segment connecting any two points lies on or under the surface $f(x)$. Convex and concave functions for $\lambda = 1/2$ and $x \in R^1$ are illustrated in Fig. 14.6.

If in the inequalities (14.60) and (14.61), only strict inequalities occur, then we obtain a definition of a strictly convex and concave function.

In Fig. 14.7 one presented a graph of a function $f(x)$, which is neither convex nor concave in the interval $[x_1,x_4]$. However, in the intervals $[x_1,x_2]$ and $[x_3,x_4]$ the function is concave, and in the interval $[x_2,x_3]$ the function is convex.

Below, we give properties of convexity of a set and function.

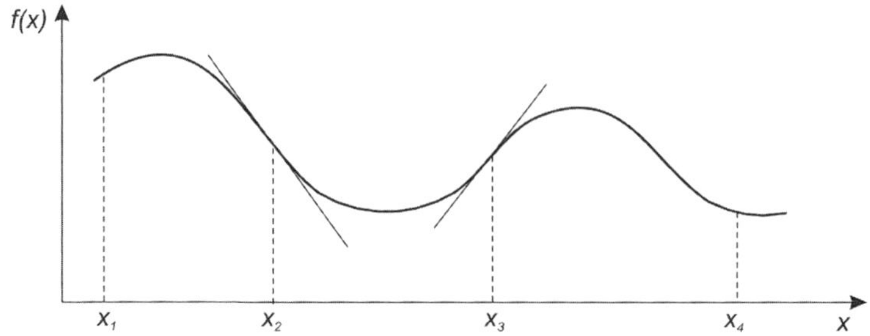

Fig. 14.7 Function piecewise convex and concave

(i) If S is a convex set and $\alpha \in R$, then a set αS is also convex.
(ii) If S_1, S_2 are convex sets, then a set $S_1 + S_2$ is also convex.
(iii) Intersection of two convex sets is a convex set.
(iv) If f_1, f_2 are convex functions on a set S, then $f_1 + f_2$ is also a convex function on the set S.

Theorem 14.3 (A Convex Optimization). *If $f(x)$ is a concave function on a convex set X, then each local maximum is global in the set X.*

Proof of this theorem is given in [107].

Theorem 14.4 (A Differentiable Function). *A differentiable function is convex if and only if its hessian is semi-positive definite in a whole region of convexity.*

14.6 Problems of Optimization Without Constraints

This problem relies on determination of $x^* \in R^n$ (an admissible set is the whole R^n) such that

$$f\left(x^*\right) = \max f\left(x\right), \ \forall x \in R, \tag{14.62}$$

where: $f : R^n \to R$ and the function is of class C^2, i.e. is continuous and twice differentiable with respect to x.

According to definition (14.49) a gradient of a function ∇f determines a direction of the largest growth of the function $f\left(x\right)$ at given point (see Fig. 14.8).

Now, we will show what condition the point $x = x^*$ being a local maximum of the function $f\left(x\right)$ in R^n must satisfy (see Fig. 14.9 for R^3).

For a local maximum $x = x^*$ we have

$$\forall r_\varepsilon > 0 : f\left(x^*\right) \geq f\left(x^* + \varepsilon r\right), \tag{14.63}$$

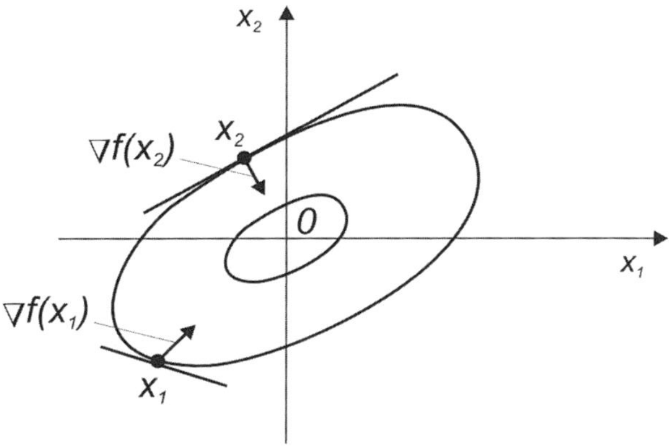

Fig. 14.8 Direction of the gradient for $x \in R^2$

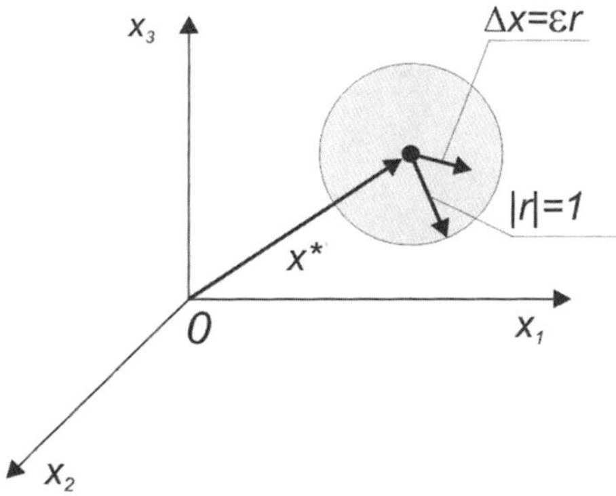

Fig. 14.9 Increment of the vector $\Delta x = \varepsilon r$ in the neighbourhood of a point x^*

and after Taylor expansion of the function in the neighbourhood of $x = x^*$ we obtain

$$f\left(x^* + \varepsilon r\right) = f\left(x^*\right) + \varepsilon \nabla f\left(x^*\right) r + \frac{1}{2}\varepsilon^2 r^T H\left(x^*\right) r + O\left(\varepsilon^3\right) \tag{14.64}$$

Substituting (14.64) into (14.63) we get

$$\nabla f\left(x^*\right) r + \frac{1}{2}\varepsilon^2 r^T \nabla f\left(x^*\right) r + O\left(\varepsilon^3\right) \leq 0. \tag{14.65}$$

With the condition $\varepsilon \to 0$ by (14.65) we obtain

$$\nabla f\left(x^*\right) r \leq 0, \tag{14.66}$$

and for arbitrarily chosen direction r this condition can be satisfied only if

$$\nabla f\left(x^*\right) = 0. \tag{14.67}$$

Table 14.2 Conditions of local optimality

First-order necessary conditions	Second-order necessary conditions	Sufficient conditions	Character of the point
$\nabla f(x^*) = 0^T$	$H(x^*) \le 0$	$H(x^*) < 0$	Local maximum
	$H(x^*) \ge 0$	$H(x^*) > 0$	Local minimum
	$H(x)$ is sign-indefinite		Saddle
	$H(x) = 0$		One needs to take higher terms of the Taylor series

The obtained condition (14.67) is called a necessary condition of the first order for local maxima at the point x^*. Hessian, evaluated at the stationary point, decides about the character of the stationary point $x = x^*$.

If $x = x^*$ is a stationary point, and moreover

$$H\left(x^*\right) \le 0, \tag{14.68}$$

then at some directions of the perturbations Δx of the point x^* the function $f(x)$ can decay, and at the other directions, no changes are observed. In this case we cannot distinguish between maximum and saddle, hence the condition (14.68) is only so-called necessary condition of the second order for maximum of a function.

The sufficient condition of maximum at the point $x = x^*$ has the form

$$H\left(x^*\right) < 0, \tag{14.69}$$

i.e. the hessian at the point x^* should be negative definite.

In Table 14.2 of [107], one gave the necessary and sufficient conditions of local optimality.

The sufficient conditions determine a stationary point as a local strict minimum or maximum. If $f(x)$ is a concave function, the fact that the point x^* is stationary is a necessary and sufficient condition for x^* to be a global maximum (minimum).

If $f(x)$ is a strictly concave function (strictly convex), then it possesses exactly one stationary point, which is a global strict maximum (minimum).

14.7 Optimality Conditions of a Quadratic Form

Consider a quadratic form

$$f(x) = \frac{1}{2}x^T Hx + a^T x + b, \tag{14.70}$$

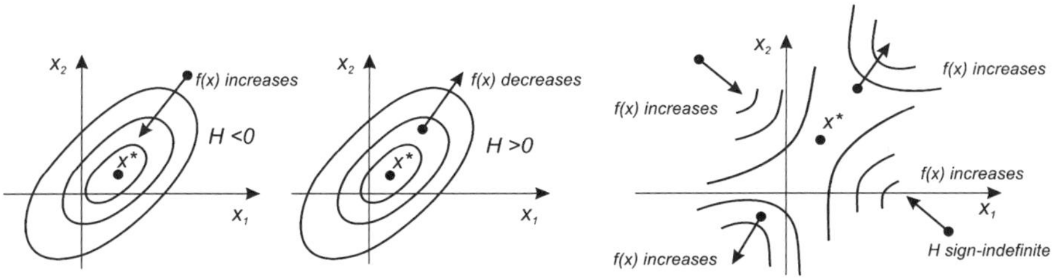

Fig. 14.10 Isometric curves and directions of growth of the function $f(x)$

where $H_{n \times n}$ is symmetric and $H \not\equiv 0$; a is a vector of dimension n; H is a hessian of the form $f(x)$, and b is a scalar.

Differentiating (14.70) we get

$$\nabla f(x) = (Hx + a)^T, \tag{14.71}$$

and the condition of stationary (14.67) has the form

$$Hx + a = 0. \tag{14.72}$$

The above linear vectorial equation possesses a solution if and only if

$$H = \text{rank}(H, a) \tag{14.73}$$

If (rank H) $= n$, then det $H \neq 0$ and Eq. (14.72) possesses a unique solution

$$x^* = -H^{-1}a. \tag{14.74}$$

If (rank H) $< n$ and (14.73) is satisfied, then Eq. (14.72) possesses infinitely many solutions, which can have maxima, minima and singular saddles.

For the case $x \in R^2$ and for $H < 0$, $H > 0$ and H, sign indefinite one showed graphic illustrations of *isometric curves* (Fig. 14.10), i.e. $f(x) = \text{const}$ in Fig. 14.8, which was taken from the work [107].

14.8 Equivalence Constraints

A problem of optimization with equivalence constraints relies on determination $x^* \in X$, satisfying

$$\forall x^* \in X, f\left(x^*\right) = \max f(x), \tag{14.75}$$

where the admissible set X has the form

$$X = \{x \in R^n : \quad a(x) = b\}, \tag{14.76}$$

while

$$f : R^n \rightarrow R^1; \quad a : R^n \rightarrow R^m, m < n, \quad b \in R^m.$$

In order to illustrate the problem we will consider the case $n = 2, m = 1$. In this case we have

$$a(x_1, x_2) = b, \tag{14.77}$$

and assume that the above function can be presented in the form

$$x_2 = g(x_1). \tag{14.78}$$

Substituting (14.78) into the *objective function* (14.75), the problem reduces to the determining

$$f\left[x_1^*, g\left(x_1^*\right)\right] = \max_{x_1 \in R^1} f\left[x_1, g\left(x_1\right)\right]. \tag{14.79}$$

Having x_1^* (14.78) we find

$$x_2^* = h\left(x_1^*\right). \tag{14.80}$$

The sufficient condition of local optimality has the form

$$\frac{df}{dx_1} = \frac{\partial f}{\partial x_1} + \frac{\partial f}{\partial g}\frac{dg}{dx_1} = 0, \tag{14.81}$$

or including (14.78)

$$\frac{\partial f}{\partial x_1} + \frac{\partial f}{\partial x_2}\frac{dx_2}{dx_1} = 0. \tag{14.82}$$

Let us take into account the constraining equation (14.77), which implies

$$da = \frac{\partial a}{\partial x_1}dx_1 + \frac{\partial a}{\partial x_2}dx_2 = 0. \tag{14.83}$$

We will further assume that a point $\left(x_1^*, x_2^*\right)$ is a regular point. Let in its neighbourhood

$$\frac{\partial a}{\partial x_2} \neq 0. \tag{14.84}$$

Hence by (14.83) we obtain

$$\frac{dx_2}{dx_1} = -\left.\frac{\partial a/\partial x_1}{\partial a/\partial x_2}\right|_{(x_1^*, x_2^*)} . \tag{14.85}$$

The necessary conditions of optimality are determined by Eqs. (14.82) and (14.84) which imply

$$\left.\left(\frac{\partial f}{\partial x_k} - z\frac{\partial a}{\partial x_k}\right)\right|_{(x_1^*, x_2^*)} = 0, \ k = 1, 2 \tag{14.86}$$

where: $z = \frac{\partial f/\partial x_2}{\partial a/\partial x_2}$. As we can see, we have two equations (14.86) and the constraining equation (14.77) to our disposal, and three unknowns, i.e. x_1, x_2 and z. Equation (14.86) can be written in the equivalent form

$$\left.\frac{\partial f/\partial x_1}{\partial f/\partial x_2}\right|_{(x_1^*, x_2^*)} = -\left.\frac{\partial a/\partial x_1}{\partial a/\partial x_2}\right|_{(x_1^*, x_2^*)} . \tag{14.87}$$

Note that the left-hand side of the above equation equals $-dx_2/dx_1$ (see (14.82) for $df = 0$, i.e. $f = \text{const}$). The right-hand side of (14.87) equals dx_2/dx_1 (see (14.82) for $da = 0$). Finally, Eq. (14.87) takes the form

$$\left.\frac{dx_2}{dx_1}\right|_{\substack{f = \text{const} \\ (x_1^*, x_2^*)}} = -\left.\frac{dx_2}{dx_1}\right|_{a(x_1^*, x_2^*) = b} , \tag{14.88}$$

which allows for geometric presentation of the local optimality condition (Fig. 14.11).

Figure 14.11 shows that at the point (x_1^*, x_2^*) satisfying the necessary conditions of optimality, the curve $a(x_1, x_2) = b$ is tangent to the isometric curve $f = \text{const}$.

14.9 The Lagrange Function and Multipliers

Let us introduce Lagrange's function of the following form

$$L(x_1, x_2, z) = f(x_1, x_2) + z[b - a(x_1, x_2)] . \tag{14.89}$$

For x_1^*, x_2^* this function must satisfy the stationarity conditions of the form

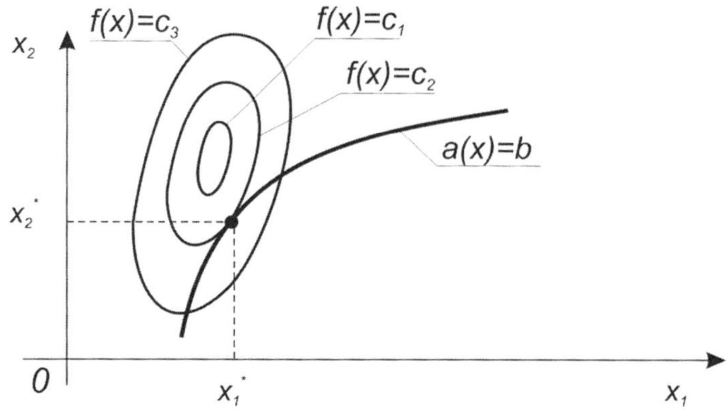

Fig. 14.11 Geometric interpretation of Eq. (14.88)

$$\frac{\partial L}{\partial x_k} = \frac{\partial f}{\partial x_k} - z\frac{\partial a}{\partial x_k} = 0, \tag{14.90}$$

$$\frac{\partial L}{\partial z} = b - a\left(x_1, x_2\right). \tag{14.91}$$

As one can see, the conditions (14.90) and (14.91) are in accordance with the earlier obtained conditions (14.86) and (14.77). A new introduced variable z is the Lagrange multiplier of interpretation, determined by the equation $z = \frac{\partial f/\partial x_2}{\partial a/\partial x_2}$. Now, let us consider a general case $n \geq 2, m \geq 1$ and $m < n$, and apply the direct method. From m constraining equations

$$b - a\left(x\right) = 0 \tag{14.92}$$

we determine m variables $x^m = \left(x_1, \ldots, x_m\right)^T$ as a function of the remaining $n - m$ variables $x^{n-m} = \left(x_{m+1}, \ldots, x_n\right)^T$, i.e. let

$$x^m = g\left(x^{n-m}\right), \quad g : R^{n-m} \to R^m. \tag{14.93}$$

We obtain the necessary conditions of local optimality

$$\begin{gathered}\frac{\partial f}{\partial x} = -z^T G\left(x\right) = 0^T, \\ b - a\left(x\right) = 0,\end{gathered} \tag{14.94}$$

where

$$z^T = \frac{\partial f}{\partial x^m}\left[G_m\left(x\right)\right]^{-1}, \tag{14.95}$$

$$G(x) = \begin{bmatrix} \frac{\partial a_1(x)}{\partial x_1} & \cdots & \frac{\partial a_1(x)}{\partial x_n} \\ \cdots & \cdots & \cdots \\ \frac{\partial a_m(x)}{\partial x_1} & \cdots & \frac{\partial a_m(x)}{\partial x_n} \end{bmatrix}, \tag{14.96}$$

$$G_m(x) = \begin{bmatrix} \frac{\partial a_1(x)}{\partial x_1} & \cdots & \frac{\partial a_1(x)}{\partial x_m} \\ \cdots & \cdots & \cdots \\ \frac{\partial a_m(x)}{\partial x_1} & \cdots & \frac{\partial a_m(x)}{\partial x_m} \end{bmatrix}, \tag{14.97}$$

where $G(x)$ is the Jacobi matrix. Note that a system of Eq. (14.94) consists of $n+m$ equations of $n+m$ variables $(x_1, \ldots, x_n, z_1, \ldots, z_m)$. In order to determine $x = x^*$, the vector z should be uniquely determined by Eq. (14.95), which holds if and only if

$$\operatorname{rank} G(x) = m. \tag{14.98}$$

The above condition ensures that we have m linearly independent columns of the matrix $G(x)$, so the matrix $[G_m(x)]^{-1}$ occurring in (14.95) exists, and the vector z is uniquely determined by the relationship (14.95).

In this case one can also build the Lagrange function

$$L(x, z) = f(x) + z^T [b - a(x)], \tag{14.99}$$

where $z = (z_1, \ldots, z_m)^T$ is a vector of the Lagrange multipliers. Below, there are stationarity conditions of the Lagrange function

$$\begin{aligned} \frac{\partial L}{\partial x} &= \frac{\partial f}{\partial x} - z^T G(x) = 0^T, \\ \left(\frac{\partial L}{\partial y}\right)^T &= b - a(x) = 0, \end{aligned} \tag{14.100}$$

which are identical to the conditions (14.95).

We will not deal with the second-order necessary conditions, i.e. sufficient conditions for existence of a global maximum (minimum) at the point x^* in the case with constraints (see e.g. [107]). Below, we give *geometric interpretation of the Lagrange multipliers*. We will write the first equation of the system (14.100) in the expanded form

$$\frac{\partial f(x^*, z^*)}{\partial x_k} = \sum_{k=1}^{m} z_i^* \frac{\partial a_i(x^*)}{\partial x_k}, \quad k \in \{1, \ldots, n\}. \tag{14.101}$$

At the stationary point, particular components of gradient of the objective function are equal to the weighted sum of elements of the appropriate column of the Jacobi matrix (14.96), while the Lagrange multipliers are the weighted coefficients.

14.10 Inequivalence Constraints

Mostly, problems of static optimization concerning sets of decision variables are given in the form of constraints of inequivalent type.

Our task is determination $x^* \in X$ such that

$$f\left(x^*\right) = \max_{x \in X} f\left(x\right), \tag{14.102}$$

where:

$$\begin{aligned} &x = \{x \in R^n : \quad x \geq 0, \quad a\left(x\right) \leq b\}, \\ &f : \quad R^n \to R^1; \quad a : R^n \to R^m, \quad b \in R^m. \end{aligned} \tag{14.103}$$

We will assume that f is of class C^2, and the functions $a_i, i = 1, \ldots, m$ are of class C^1. We do not assume any relation between n and m. We will solve the problem in two stages. First, we will consider the case $m = 0$, i.e. when the decision variables are not negative

$$x = \left(x_1, \ldots, x_n\right)^T \geq 0, \; x \in R^n. \tag{14.104}$$

Let $x^* \in X$ be a solution of the maximalization problem of $f\left(x\right)$ in the set X. Each component x_k^* of the vector x^* is characterized by the following properties:

(i) An optimal point x_k^* lies within the set X (see Fig. 14.12)

$$x_k^* > 0 \text{ while } \frac{\partial f(x^*)}{\partial x_k} = 0; \tag{14.105}$$

(ii) An optimal point x_k^* lies on the boundary of the set X (see Fig. 14.12), i.e.

$$x_k^* = 0 \text{ while } \frac{\partial f(x^*)}{\partial x_k} \leq 0; \tag{14.106}$$

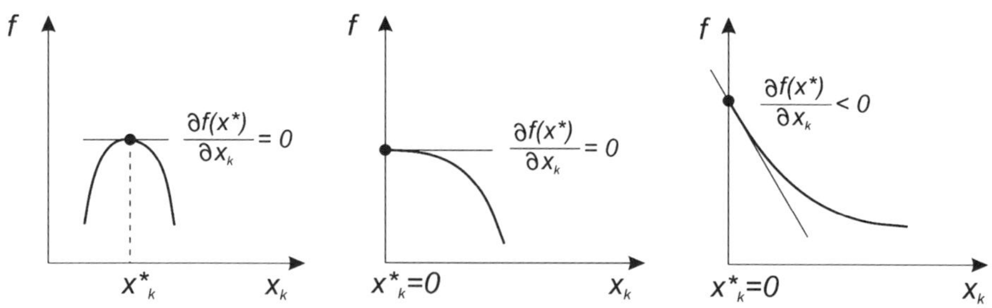

Fig. 14.12 Possible positions of optimal points

Cases (14.105) and (14.106) can be written in the following way

$$\forall j \in \{1, \ldots, n\}, \frac{\partial f(x^*)}{\partial x_k} \leq 0, \quad \frac{\partial f(x^*)}{\partial x_k} x_k^* = 0, \tag{14.107}$$

or in the vector form

$$\nabla f(x) \leq 0^T, \nabla f\left(x^*\right) x^* = 0, x^* \geq 0. \tag{14.108}$$

Now, we enter the second stage, i.e. $X = \{x \in R^n, x > 0, a(x) \leq b\}$. Firstly, let us introduce additional variables, called *complementing variables* of the form

$$s = b - a(x), \tag{14.109}$$

where: $s = [s_1, \ldots, s_m]^T$. Note that $a(x) \leq b$ is equivalent to the inequality

$$s \geq 0. \tag{14.110}$$

Our new task is determination $x^* \in X$ in such a way that

$$\begin{gathered} f\left(x^*\right) = \max_{x \in X} f(x), \\ X = \{x \in R^n : \quad x \geq 0, s \in R^m : s \geq 0, b - a(x) - s = 0\}. \end{gathered} \tag{14.111}$$

In such a new formulated problem, the equivalence constraints occur

$$b - a(x) - s = 0, \tag{14.112}$$

and constraints on non-negativeness $n + m$ variables of the form

$$x \geq 0, \quad s \geq 0. \tag{14.113}$$

Let us introduce the Lagrange function of the form

$$L(x, z, s) : \quad f(x) + z^T [b - a(x) - s], \tag{14.114}$$

where z is an m-element vector of the Lagrange multipliers. Constraints on non-negativeness of x and s we introduce on the basis of the conditions (14.108) applied to the Lagrange function (14.114). Necessary conditions of existence of the maximum at the point $x^* \in X$ are:

$$\frac{\partial L(x, z, s)}{\partial x} \leq 0^T, \quad \frac{\partial L(x, z, s)}{\partial s} \leq 0^T, \quad \frac{\partial L(x, z, s)}{\partial z} = 0^T, \tag{14.115}$$

$$\frac{\partial L(x,z,s)}{\partial x}x = 0, \quad \frac{\partial L(x,z,s)}{\partial s}s = 0, \tag{14.116}$$

$$x \geq 0, \quad s \geq 0. \tag{14.117}$$

Differentiating (14.114) with respect to x, z and s, and eliminating the vector s according to (14.109) we obtain

$$\frac{\partial L}{\partial x} = \frac{\partial f}{\partial x} - z^T \nabla f(x), \quad \frac{\partial L}{\partial a} = -z^T, \quad \frac{\partial L}{\partial x} = 0^T. \tag{14.118}$$

At the maximum point, according to (14.115)–(14.118), the following necessary conditions are satisfied

$$\frac{\partial f(x)}{\partial x} - z^T \nabla f(x) \leq 0^T, \tag{14.119}$$

$$\left[\frac{\partial f(x)}{\partial x} - z^T \nabla f(x)\right] x = 0, \tag{14.120}$$

$$[b - a(x)]^T z = 0; \; b - a(x) \geq 0, \tag{14.121}$$

$$x \geq 0, \quad z \geq 0. \tag{14.122}$$

The obtained conditions (14.119)–(14.122) are called the Kuhn–Tucker conditions. Note that the inequality (14.122) goes into equality, when there are no non-negativity of decision variables. If as the Lagrange function we use the following one

$$L(x,z): \quad f(x) + z^T [b - a(x)], \tag{14.123}$$

then the Kuhn–Tucker conditions at the point of maximum x^* are as follows

$$\frac{\partial L(x,z)}{\partial x} \leq 0^T, \quad \frac{\partial L(x,z)}{\partial z} \geq 0^T, \tag{14.124}$$

$$\frac{\partial L(x,z)}{\partial x}x = 0, \quad \frac{\partial L(x,z)}{\partial z}z = 0, \tag{14.125}$$

$$x \geq 0, \quad z \geq 0. \tag{14.126}$$

The Kuhn–Tucker conditions (14.124)–(14.126) are sufficient for existence of a *global maximum*, if $f(x)$ is concave and functions $a_i(x)$, $i \in \{1, \ldots, m\}$ are convex for $x \in X$. If the mentioned conditions of concavity of $f(x)$ and convexity of $a_i(x)$ are connected with the neighbourhood ε of the point x^*, then (14.124)–(14.126) are sufficient conditions of *local optimality*.

The considered problem of optimization with inequivalence constraints and constraints on non-negativeness of decision variables is called a *saddle problem* [109].

It turns out that the point (x^*, z^*) is a saddle point of Lagrange's function, which attains maximum with respect to x and minimum with respect to z

$$L\left(x, z^*\right) \leq L\left(x^*, z^*\right) \leq L\left(x^*, z\right) \tag{14.127}$$

for all $x \geq 0, z \geq 0$. It follows from (14.124) that in the case of a convex problem, changes of x, (z) from the point $\left(x^*, z^*\right)$ satisfying the Kuhn–Tucker conditions cannot make the Lagrange function increase (decrease). During the iterations one seeks the maximum with respect to x at z constant, and next the minimum with respect to z at x constant, determined in the preceding iteration till the assumed convergence is reached, i.e. saddle point.

14.11 Shock Absorber Work Optimization

14.11.1 Introduction

In this section we will illustrate computational problems occurring solving of the optimization problems even for the simple linear oscillator with one degree of freedom (e.g. [102]).

14.11.2 Optimization Example

Consider the autonomous system with one degree of freedom serving as a shock absorber with mass m, stiffness k and damping c.

Dimensional motion equation governing dynamics of the one degree-of-freedom mechanical system has the following form

$$m\ddot{x} + c\dot{x} + kx = 0. \tag{14.128}$$

After the Laplace transformation we obtain

$$L\left[\ddot{x}\right] = \int_0^\infty \ddot{x}e^{-st}dt = \dot{x}e^{-st}\Big|_0^\infty + s\int_0^\infty \dot{x}e^{-st}dt$$

$$= -\dot{x}_0 + s\, xe^{-st}\Big|_0^\infty + s^2\int_0^\infty xe^{-st}dt = -\dot{x}_0 - sx_0 + s^2X\left(s\right);$$

$$L\left[\dot{x}\right] = \int_0^\infty \dot{x}e^{-st}dt = xe^{-st}\Big|_0^\infty + s\int_0^\infty e^{-st}dt = -x_0 + sX\left(s\right),$$

and after substitution into (14.128) we get

$$X(s) = \frac{m x_0 s + m\dot{x}_0 + c x_0}{m s^2 + c s + k}. \tag{14.129}$$

Poles of the function $X(s)$ are determined by the characteristic equation

$$m s^2 + c s + k = 0,$$

roots of which are

$$s_{1,2} = \frac{-c \pm \sqrt{c^2 - 4km}}{2m},$$

and therefore Eq. (14.129) can assume the form

$$X(s) = \frac{x_0\left(s + \frac{c}{m} + y\right)}{(s - s_1)(s - s_2)}, \quad y = \frac{\dot{x}_0}{x_0}. \tag{14.130}$$

Make distribution (14.130) into simple fractions

$$\frac{x_0\left(s + \frac{c}{m} + y\right)}{(s - s_1)(s - s_2)} = \frac{A}{s - s_1} + \frac{B}{s - s_2},$$

and further

$$x_0\left(s + \frac{c}{m} + y\right) = A(s - s_2) + B(s - s_1).$$

From the above equation we get a system of two equations

$$x_0 = A + B,$$
$$x_0\left(\frac{c}{m} + y\right) = -A s_2 - B s_1,$$

which yields

$$A = \frac{-x_0\left(\frac{c}{m} + y + s_1\right)}{s_2 - s_1},$$
$$B = \frac{x_0\left(\frac{c}{m} + y + s_2\right)}{s_2 - s_1}.$$

Finally, we get

$$X(s) = -\frac{x_0}{s_2 - s_1}\frac{\frac{c}{m} + y + s_1}{s - s_1} + \frac{x_0}{s_2 - s_1}\frac{\frac{c}{m} + y + s_2}{s - s_2}$$
$$= \frac{x_0}{s_2 - s_1}\left(-\frac{\frac{c}{m} + y + s_1}{s - s_1} + \frac{\frac{c}{m} + y + s_2}{s - s_2}\right). \tag{14.131}$$

It should be noted that

$$s_1 + \frac{c}{m} = -\frac{c}{2m} + \frac{\sqrt{c^2 - 4mk}}{2m} + \frac{c}{m} = \frac{c}{2m} + \frac{\sqrt{c^2 - 4mk}}{2m} = -s_2,$$
$$s_2 + \frac{c}{m} = -\frac{c}{2m} - \frac{\sqrt{c^2 - 4mk}}{2m} + \frac{c}{m} = \frac{c}{2m} - \frac{\sqrt{c^2 - 4mk}}{2m} = -s_1,$$

and from Eq. (14.131) we get

$$X(s) = \frac{x_0}{s_2 - s_1}\left(\frac{s_2 - y}{s - s_1} - \frac{s_1 - y}{s - s_2}\right), \tag{14.132}$$

and after coming back to the time domain we obtain sought solution

$$x(t) = \frac{x_0(s_2 - y)}{s_2 - s_1}e^{s_1 t} - \frac{x_0(s_1 - y)}{s_2 - s_1}e^{s_2 t}. \tag{14.133}$$

In the case of a double root of the characteristic equation we

$$s_1 = s_2 = s_0 = -\frac{c}{2m}, \tag{14.134}$$

and applying transformations, we get

$$X(s) = \frac{x_0\left(s + \frac{c}{m} + y\right)}{(s - s_0)^2} = \frac{x_0(s - s_0) + x_0\left(s_0 + \frac{c}{m} + y\right)}{(s - s_0)^2}$$
$$= \frac{x_0}{s - s_0} + \frac{x_0\left(s_0 + \frac{c}{m} + y\right)}{(s - s_0)^2}.$$

Because

$$s_0 + \frac{c}{m} = \frac{c}{2m} = -s_0,$$

so

$$X(s) = \frac{x_0}{s - s_0} - \frac{x_0(s_0 - y)}{(s - s_0)^2}, \tag{14.135}$$

which leads to the solution in the time-domain of the form

$$x(t) = x_0 e^{s_0 t} - x_0 (s_0 - y) t e^{s_0 t} = x_0 (1 - s_0 t) e^{s_0 t} + \dot{x}_0 t e^{s_0 t}. \tag{14.136}$$

For the roots $s_1 \neq s_2$ of Eq. (14.133), the following solution is obtained

$$x(t) = \frac{x_0 s_2 - \dot{x}_0}{s_2 - s_1} e^{s_1 t} - \frac{x_0 s_1 - \dot{x}_0}{s_2 - s_1} e^{s_2 t}. \tag{14.137}$$

Our goal is to determine the minimum time needed for extreme $x(t)$ to appear. It will depend on the roots s_1 and s_2, what means that it depends on the choice of the m, c and k parameters. After a comparison of the first derivative (14.137) to zero (necessary condition of extreme) we have

$$\dot{x}(t) = \frac{x_0 s_2 - \dot{x}_0}{s_2 - s_1} s_1 e^{s_1 t} - \frac{x_0 s_1 - \dot{x}_0}{s_2 - s_1} s_2 e^{s_2 t} = 0, \tag{14.138}$$

and hence

$$t_e = \frac{1}{s_2 - s_1} \ln \left[\frac{(x_0 s_2 - \dot{x}_0) s_1}{(x_0 s_1 - \dot{x}_0) s_2} \right]. \tag{14.139}$$

Let us differentiate t_e sequentially in respect to s_1 and s_2. We obtain

$$\frac{dt_e}{ds_1} = \frac{1}{(s_2 - s_1)^2} \ln \left[\frac{(x_0 s_2 - \dot{x}_0) s_1}{(x_0 s_1 - \dot{x}_0) s_2} \right]$$

$$+ \frac{1}{s_2 - s_1} \frac{(x_0 s_1 - \dot{x}_0) s_2}{(x_0 s_2 - \dot{x}_0) s_1} \left[\frac{(x_0 s_2 - \dot{x}_0)(x_0 s_1 - \dot{x}_0) s_2 - x_0 s_2 (x_0 s_2 - \dot{x}_0) s_1}{(x_0 s_1 - \dot{x}_0)^2 s_2^2} \right]$$

$$= \frac{1}{(s_2 - s_1)^2} \ln \left[\frac{(x_0 s_2 - \dot{x}_0) s_1}{(x_0 s_1 - \dot{x}_0) s_2} \right] + \frac{-x_0}{s_1 (s_2 - s_1)(x_0 s_1 - \dot{x}_0)};$$

$$\frac{dt_e}{ds_2} = -\frac{1}{(s_2 - s_1)^2} \ln \left[\frac{(x_0 s_2 - \dot{x}_0) s_1}{(x_0 s_1 - \dot{x}_0) s_2} \right]$$

$$+ \frac{1}{s_2 - s_1} \frac{(x_0 s_1 - \dot{x}_0) s_2}{(x_0 s_2 - \dot{x}_0) s_1} \left[\frac{s_1 x_0 s_2 (x_0 s_1 - \dot{x}_0) - s_1 (x_0 s_1 - \dot{x}_0)(x_0 s_2 - \dot{x}_0)}{(x_0 s_1 - \dot{x}_0)^2 s_2^2} \right]$$

$$= -\frac{1}{(s_2 - s_1)^2} \ln \left[\frac{(x_0 s_2 - \dot{x}_0) s_1}{(x_0 s_1 - \dot{x}_0) s_2} \right] + \frac{\dot{x}_0}{s_2 (s_2 - s_1)(x_0 s_2 - \dot{x}_0)}.$$

In order to determine the $t_e$minimum value due to s_1 and s_2, we obtain

$$\frac{1}{(s_2 - s_1)^2} \ln \left[\frac{(x_0 s_2 - \dot{x}_0)\, s_1}{(x_0 s_1 - \dot{x}_0)\, s_2} \right] - \frac{\dot{x}_0}{s_1 (s_2 - s_1)(x_0 s_1 - \dot{x}_0)} = 0,$$

$$\frac{1}{(s_2 - s_1)^2} \ln \left[\frac{(x_0 s_2 - \dot{x}_0)\, s_1}{(x_0 s_1 - \dot{x}_0)\, s_2} \right] - \frac{\dot{x}_0}{s_2 (s_2 - s_1)(x_0 s_2 - \dot{x}_0)} = 0,$$

and therefore

$$\begin{aligned} &\ln \left[\frac{s_1 (x_0 s_2 - \dot{x}_0)}{s_2 (x_0 s_1 - \dot{x}_0)} \right] - \frac{\dot{x}_0 (s_2 - s_1)}{s_1 (x_0 s_1 - \dot{x}_0)} = 0, \\ &\ln \left[\frac{s_1 (x_0 s_2 - \dot{x}_0)}{s_2 (x_0 s_1 - \dot{x}_0)} \right] - \frac{\dot{x}_0 (s_2 - s_1)}{s_2 (x_0 s_2 - \dot{x}_0)} = 0. \end{aligned} \tag{14.140}$$

From the above equations it follows that

$$\frac{\dot{x}_0 (s_2 - s_1)}{s_1 (x_0 s_1 - \dot{x}_0)} = \frac{\dot{x}_0 (s_2 - s_1)}{s_2 (x_0 s_2 - \dot{x}_0)}$$

or otherwise

$$s_2 (x_0 s_2 - \dot{x}_0) = s_1 (x_0 s_1 - \dot{x}_0)\,. \tag{14.141}$$

After dividing by x_0 the above equation we get

$$s_2 (s_2 - y) = s_1 (s_1 - y)\,, \tag{14.142}$$

where $y = \dot{x}_0 / x_0$. Applying the transformation (14.141), we have

$$y (s_2 - s_1) = s_2^2 - s_1^2,$$

or

$$y \equiv \frac{\dot{x}_0}{x_0} = s_1 + s_2 = -\frac{c}{m}. \tag{14.143}$$

Equation (14.141) yields

$$\frac{s_1}{s_2} = \frac{x_0 s_2 - \dot{x}_0}{x_0 s_1 - \dot{x}_0}. \tag{14.144}$$

After applying of (14.144) in (14.139), we obtain

$$t_e = \frac{1}{s_2 - s_1} \ln \left(\frac{s_1}{s_2} \cdot \frac{s_1}{s_2} \right) = \frac{2}{s_2 - s_1} \ln \left| \frac{s_1}{s_2} \right|. \tag{14.145}$$

The first equation of the system (14.140) implies that

$$\ln\left(\frac{s_1\left(x_0 s_2-\dot{x}_0\right)}{s_2\left(x_0 s_1-\dot{x}_0\right)}\right)=\frac{\dot{x}_0\left(s_2-s_1\right)}{s_1\left(x_0 s_1-\dot{x}_0\right)},$$

and taking into account (14.144), we obtain

$$\ln\left(\frac{s_1}{s_2}\right)^2=\frac{\frac{\dot{x}_0}{x_0}\left(s_2-s_1\right)}{s_1\left(s_1-\frac{\dot{x}_0}{x_0}\right)}$$

or otherwise

$$\ln\left(\frac{s_1}{s_2}\right)^2=\frac{y\left(s_2-s_1\right)}{s_1\left(s_1-y\right)}. \tag{14.146}$$

After taking into account (14.146) in (14.145), we obtain

$$t_e=\frac{1}{s_2-s_1}\ln\left(\frac{s_1}{s_2}\right)^2=\frac{y}{s_1\left(s_1-y\right)}.$$

Bearing in mind that $y=-c/m$, and $s_1=-\frac{c+\sqrt{c^2-4km}}{2m}$, we get

$$\begin{aligned} t_e&=\frac{-\frac{c}{m}}{s_1\left(\frac{-c+\sqrt{c^2-4km}+2c}{2m}\right)}=\frac{-2c}{s_1\left(c+\sqrt{c^2-4km}\right)}\\ &=\frac{-4cm}{\left(-c+\sqrt{c^2-4km}\right)\left(c+\sqrt{c^2-4km}\right)}\\ &=\frac{-4cm}{c^2-4km-c^2}=\frac{c}{k}. \end{aligned} \tag{14.147}$$

According to (14.133), we have

$$X_e\equiv X\left(t_e\right)=\frac{x_0}{s_2-s_1}e^{s_1 t_e}\left(\left(s_2-y\right)-\left(s_1-y\right)e^{\left(s_2-s_1\right)t_e}\right), \tag{14.148}$$

and according to (14.145), we obtain

$$\left(s_2-s_1\right)t_e=\ln\left(\frac{s_1}{s_2}\right)^2. \tag{14.149}$$

After substituting (14.149) into (14.148) we have

$$
\begin{aligned}
X_e &= \frac{x_0 e^{s_1 t_e}}{s_2 - s_1} \left((s_2 - y) - (s_1 - y)\, e^{\ln\left(\frac{s_1}{s_2}\right)^2} \right) \\
&= \frac{x_0 e^{s_1 t_e}}{s_2 - s_1} \left((s_2 - y) - \left(\frac{s_1}{s_2}\right)^2 (s_1 - y) \right) \\
&= \frac{x_0 e^{s_1 t_e}}{s_2^2 (s_2 - s_1)} \left(s_2^2 (s_2 - y) - s_1^2 (s_1 - y) \right),
\end{aligned} \tag{14.150}
$$

and after taking into account (14.142), we obtain

$$
\begin{aligned}
X_e &= \frac{x_0 e^{s_1 t_e}}{s_2^2 (s_2 - s_1)} \left(s_2 s_1 (s_1 - y) - s_1^2 (s_1 - y) \right) \\
&= \frac{x_0 e^{s_1 t_e}}{s_2^2 (s_2 - s_1)} s_1 (s_1 - y)(s_2 - s_1) = \frac{x_0 s_1 (s_1 - y)}{s_2^2} e^{s_1 t_e}.
\end{aligned} \tag{14.151}
$$

On the other hand, according to (14.133) yields

$$
\begin{aligned}
X_e &= \frac{x_0}{s_2 - s_1} \left((s_2 - y)\, e^{s_1 t_e} - (s_1 - y)\, e^{s_2 t_e} \right) \\
&= \frac{x_0}{s_2 - s_1} e^{s_2 t_e} \left((s_2 - y)\, e^{-(s_2 - s_1) t_1} - (s_1 - y) \right) \\
&= \frac{x_0}{s_2 - s_1} e^{s_2 t_e} \left((s_2 - y)\, e^{\ln\left(\frac{s_1}{s_2}\right)^{-2}} - (s_1 - y) \right) \\
&= \frac{x_0}{s_2 - s_1} e^{s_2 t_e} \left(\frac{s_2^2}{s_1^2} (s_2 - y) - (s_1 - y) \right) \\
&= \frac{x_0}{s_1^2 (s_2 - s_1)} e^{s_2 t_e} \left(s_2^2 (s_2 - y) - s_1^2 (s_1 - y) \right).
\end{aligned} \tag{14.152}
$$

Then repeated application of the (14.142), we obtain

$$
\begin{aligned}
X_e &= \frac{x_0 e^{s_2 t_e}}{s_1^2 (s_2 - s_1)} \left(s_2 s_1 (s_1 - y) - s_1^2 (s_1 - y) \right) \\
&= \frac{x_0 e^{s_2 t_e}}{s_1^2 (s_2 - s_1)} s_1 (s_1 - y)(s_2 - s_1) \\
&= \frac{x_0 e^{s_2 t_e} (s_1 - y)}{s_1}.
\end{aligned} \tag{14.153}
$$

After applying (14.151) and (14.153) we have

$$X_e^2 = X_e \cdot X_e = \frac{x_0 s_1 (s_1 - y)}{s_2^2} \cdot \frac{x_0 (s_1 - y)}{s_1} e^{(s_1+s_2)t_e}, \tag{14.154}$$

and since

$$s_1 + s_2 = -\frac{c}{m},$$

then

$$X_e^2 = \frac{x_0^2 (s_1 - y)^2}{s_2^2} e^{-\frac{c}{m} t_e}. \tag{14.155}$$

Note that

$$\frac{s_1 - y}{s_2} = \frac{-\frac{c}{2m} + \frac{\sqrt{c^2-4km}}{2m} + \frac{c}{m}}{-\frac{c}{2m} - \frac{\sqrt{c^2-4km}}{2m}} = \frac{c + \sqrt{c^2 - 4km}}{-c - \sqrt{c^2 - 4km}} = -1,$$

and hence, due to (14.155) and (14.147), we obtain

$$X_e^2 = X_0^2 e^{-\frac{c^2}{km}},$$

and finally the following extreme value is found

$$X_e = \pm X_0 e^{-\frac{c^2}{2km}}, \tag{14.156}$$

It remains to consider the case of a double root

$$s_1 = s_2 = -\frac{c}{2m}.$$

After differentiating (14.136) and aligning the derivative to zero, we obtain

$$\begin{aligned} \dot{x}(t) &= -x_0 s e^{st} + x_0 s (1 - st) e^{st} + \dot{x}_0 s e^{st} + \dot{x}_0 s t e^{st} \\ &= -x_0 s^2 t e^{st} + \dot{x}_0 (1 + st) e^{st} = 0, \end{aligned} \tag{14.157}$$

that is

$$t \left(\dot{x}_0 s - x_0 s^2 \right) = -\dot{x}_0$$

or in another form

$$t_e = \frac{\dot{x}_0}{x_0 s^2 - \dot{x}_0 s} = \frac{\frac{\dot{x}_0}{x_0}}{s^2 - \frac{\dot{x}_0}{x_0} s} = \frac{\tilde{y}}{s^2 - \tilde{y} s}, \tag{14.158}$$

where $\tilde{y} = \frac{\dot{x}_0}{x_0}$.

Let us look for such s that

$$\frac{dt_e}{ds} = -\frac{\tilde{y}\,(2s - \tilde{y})}{\left(s^2 - \tilde{y}s\right)^2} = 0, \tag{14.159}$$

what leads to the finding

$$s = \frac{1}{2}\tilde{y} = \frac{\dot{x}_0}{2x_0}. \tag{14.160}$$

After taking into account (14.160) in (14.158), we obtain

$$t_e = \frac{\tilde{y}}{\frac{1}{4}\tilde{y}^2 - \frac{1}{2}\tilde{y}^2} = -\frac{4}{\tilde{y}} = -\frac{2}{s} = \frac{4m}{c}, \tag{14.161}$$

because $t_e = \frac{4m}{c}$. Reached extreme value of the shock absorber inclination is

$$\begin{aligned} X_e \equiv X\,(t_e) &= x_0 e^{st_e}\left[1 - st_e + \tilde{y}t_e\right] \\ &= x_0 e^{-\frac{c}{2m}\frac{4m}{c}}\left[1 - \frac{c}{2m}\cdot\frac{4m}{c} - \frac{c}{m}\cdot\frac{4m}{c}\right] \\ &= -x_0 e^{-2}. \end{aligned} \tag{14.162}$$

If the initial conditions $x_0 = 0$, then

$$x\,(t) = \dot{x}_0 t e^{st}, \tag{14.163}$$

while $\dot{x}\,(t) = 0$ leads to the equation

$$\dot{x}_0 e^{st} + \dot{x}_0 t s e^{st} = 0,$$

and hence

$$t_e = -\frac{1}{s} = \frac{2m}{c}. \tag{14.164}$$

In this case, the extreme is

$$X_e = \dot{x}_0 t_e e^{st_e} = \dot{x}_0 \frac{2m}{c} e^{-1} = \frac{2m\dot{x}_0}{ec}. \tag{14.165}$$

If the condition (14.160) is true for any time t, we get a perfectly working damper without vibrations. This is due to the following reasoning.

From Eq. (14.160) in the form

$$\frac{\dot{x}(t)}{x(t)} = -\frac{c}{m} \tag{14.166}$$

we eliminate the constant c and substitute it to the equation of motion (14.128). This operation gives

$$\ddot{x} - \frac{\dot{x}^2}{x} + \alpha^2 x = 0, \tag{14.167}$$

where $\alpha^2 = k/m$. Then let us multiply both sides of (14.167) by x to obtain

$$\ddot{x}x - \dot{x}^2 + \alpha^2 x^2 = 0$$

or

$$\frac{\ddot{x}x - \dot{x}^2}{x^2} + \alpha^2 = 0.$$

Write the above equation in the equivalent form

$$\frac{d}{dt}\left(\frac{\dot{x}}{x}\right) + \alpha^2 = 0. \tag{14.168}$$

After the first integration of Eq. (14.168) we have

$$\frac{\dot{x}}{x} = -\alpha^2 t + \frac{\dot{x}_0}{x_0},$$

and the second one yields

$$\ln\frac{x(t)}{x_0} = -\alpha^2\frac{t^2}{2} + \frac{\dot{x}_0}{x_0}t,$$

and hence

$$x(t) = x_0 e^{-\alpha^2\frac{t^2}{2} + \frac{\dot{x}_0}{x_0}t},$$

or

$$x(t) = x_0 e^{-\frac{k}{m}\frac{t^2}{2} - \frac{c}{m}t}. \tag{14.169}$$

It can be seen that in this case $x(t)$ disappears faster than is the case for an exponential decay.

At the end let us consider the case, where the roots of the characteristic equation are complex conjugate of the form

$$s_{1,2} = \alpha \pm i\omega.$$

According to (14.138) we have

$$\dot{x}(t) = \frac{x_0}{s_2 - s_1}\left[(s_2 - y)\, s_1 e^{s_1 t} - (s_1 - y)\, s_2 e^{s_2 t}\right] = 0,$$

that is

$$(s_2 - y)\, s_1 e^{i\omega t} = (s_1 - y)\, s_2 e^{-i\omega t},$$

and so

$$e^{i2\omega t} = \frac{(s_1 - y)\, s_2}{(s_2 - y)\, s} \equiv z. \tag{14.170}$$

Therefore we have

$$i2\omega t = \ln z. \tag{14.171}$$

On the other hand,

$$\ln z = \ln|z| + i \arg z,$$

and because$|z| = 1$ we get

$$\ln z = i \arg z,$$

$$\arg z = \arctan\left(\frac{\operatorname{Im} z}{\operatorname{Re} z}\right).$$

Let us transform (14.170) to obtain

$$z = \frac{(s_1 - y)\, s_2}{(s_2 - y)\, s_1} = \frac{(s_1 - y)\, \bar{s}_1}{(\bar{s}_1 - y)\, s_1} = \frac{|s|^2 - y\bar{s}}{|s|^2 - ys}, \tag{14.172}$$

where $s_1 = \bar{s}_2 \equiv s$.

Further transformation of Eq. (14.172) yields

$$\begin{aligned} z &= \frac{\left(|s|^2 - y\bar{s}\right)^2}{\left(|s|^2 - ys\right)\left(|s|^2 - y\bar{s}\right)} = \frac{|s|^4 - 2y\,|s|^2\,\bar{s} + y^2\bar{s}^2}{|s|^4 + y^2\,|s|^2 - y\,|s|^2\,(s + \bar{s})} \\ &= \frac{|s|^4 - 2y\,|s|^2\,\alpha + y^2\left(\alpha^2 - \omega^2\right) + i\left(2y\omega\,|s|^2 - y^2 2\alpha\omega\right)}{|s|^2\left(|s|^2 + y^2 - y\,(s + \bar{s})\right)}. \end{aligned} \tag{14.173}$$

From the above equations it follows that

$$\frac{\operatorname{Im} z}{\operatorname{Re} z}=\frac{2y\omega\left|s\right|^{2}-2\alpha\omega y^{2}}{\left|s\right|^{4}-2y\left|s\right|^{2}\alpha+y^{2}\left(\alpha^{2}+\omega^{2}\right)}=\frac{2\omega\left(\left|s\right|^{2}-\alpha y\right)y}{\left(\left|s\right|^{2}-y\alpha\right)^{2}-y^{2}\omega^{2}}. \tag{14.174}$$

From Eq. (14.171) after taking into account (14.174) we have

$$i2\omega t_{e}=i\arctan\left[\frac{2\omega\left(\left|s\right|^{2}-\alpha y\right)y}{\left(\left|s\right|^{2}-y\alpha\right)^{2}-y^{2}\omega^{2}}\right]-i2k\pi,$$

and hence

$$t_{e}=\frac{1}{2\omega}\left[\arctan\left(\frac{2\omega\left[\left(\alpha^{2}+\omega^{2}\right)-\alpha y\right]y}{\left[\alpha^{2}+\omega^{2}-\alpha y\right]^{2}-y^{2}\omega^{2}}\right)\right]-2k\pi. \tag{14.175}$$

Let us look for extremes of the equation

$$\frac{\partial te}{\partial\alpha}=0, \tag{14.176}$$

and let us introduce the variable

$$u\left(\alpha\right)=\left(\alpha^{2}+\omega^{2}-y\alpha\right)^{2}-y^{2}\omega^{2}. \tag{14.177}$$

After taking into account (14.175) and (14.177), we obtain from Eq. (14.176)

$$\frac{\partial t_{e}}{\partial\alpha}=\frac{1}{2\omega\left(1+u^{2}\right)}\frac{\partial u}{\partial\alpha}=0, \tag{14.178}$$

$$\frac{\partial u}{\partial\alpha}=\frac{2\omega y}{\left[\left(\alpha^{2}+\omega^{2}-y\alpha\right)^{2}-y^{2}\omega^{2}\right]^{2}}\left(\left(2\alpha-y\right)\left[\left(\alpha^{2}+\omega^{2}-y\alpha\right)^{2}-y^{2}\omega^{2}\right]\right.$$
$$\left.+-\left(\alpha^{2}+\omega^{2}-y\alpha\right)\left[2\left(\alpha^{2}+\omega^{2}-y\alpha\right)\left(2\alpha-y\right)\right]\right)=0,$$

and finally from (14.178), we obtain

$$\left(2\alpha-y\right)\left[\left(\alpha^{2}+\omega^{2}-y\alpha\right)^{2}-y^{2}\omega^{2}-2\left(\alpha^{2}+\omega^{2}-y\alpha\right)^{2}\right]=0,$$

this leads to the description

$$\alpha=\frac{1}{2}y=\frac{1}{2}\frac{\dot{x}_{0}}{x_{0}}. \tag{14.179}$$

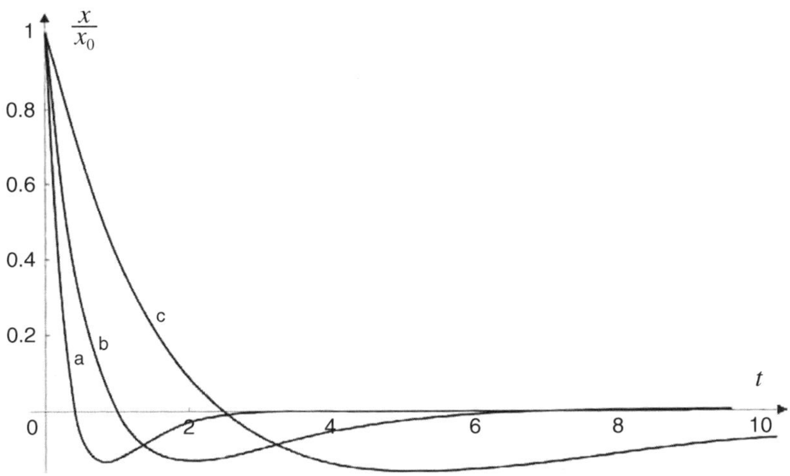

Fig. 14.13 Timing diagram (described by Eq. (14.183)) modified for optimal shock absorber $\frac{c}{m}=5$ (**a**); 2 (**b**); 0.8 (**c**)

On the other side, we compute

$$\frac{dt_e}{d\omega}=-\frac{1}{2\omega^2}(\arctan u-2k\pi)+\frac{1}{2\omega}\frac{1}{1+u^2}\frac{du}{d\omega}$$
$$=-\frac{1}{\omega}t_e+\frac{1}{2\omega}\frac{1}{1+u^2}\frac{du}{d\omega}=0,$$

and therefore

$$\frac{1}{2}\cdot\frac{1}{1+u^2}\cdot\frac{du}{d\omega}-t_e=0. \tag{14.180}$$

In accordance with the considerations carried out in the book [102] solution of (14.180) is performed to find

$$\omega_*=2.26\left(\frac{1}{2}\frac{\dot{x}_0}{x_0}\right)=2.26\alpha_*. \tag{14.181}$$

You can see that the time t_e reaches minimum due to α, and due to the ω it reaches the maximum in the point (α_*,ω_*) and is equal to

$$t_e=-0.65\frac{x_0}{\dot{x}_0}. \tag{14.182}$$

Designated critical point is a saddle (for $t > 0$ there have to be $\frac{x_0}{\dot{x}_0} < 0$). In Fig. 14.13 shows an example of function graphs

$$\frac{x(t)}{x_0} = \left(1 - \frac{ct}{2m}\right) c^{-\frac{ct}{2m}}, \tag{14.183}$$

for designed optimal shock absorber.

Chapter 15
Chaos and Synchronization

15.1 Introduction

New point of view, introduced into known definitions in mathematic and empiric sciences by developments in nonlinear dynamic, provides novel interpretations of one of many philosophical trends based on determinism and indeterminism. Until now those two concepts were treated as mutually exclusive; however examples of chaotic motions appearing in a simple physical, chemical or biological systems indicate possibility that relationship between them exists. Even though, theoretically, the determination of motion trajectory is possible by the introduction of the highly accurate initial conditions, obtaining sufficient accuracy is impossible in practice. This issue has much wider range, and as every real state of system is described with a certain inaccuracy, it should be described as probability distribution and not as numbers. This is the reason why in determined system we expect typical for stochastic systems dynamics (it will be described and illustrated for a simple mapping and ordinary differential equations further). This type of deterministic systems motion, contrary to random variable systems, is called deterministic chaos.

In traditional physics and mechanics discrete and continuous systems can be distinguished. The former are described by ordinary differential equations, and the latter are described by partial differential equations. Proponents of differential equations argue that ordinary differential equations yield sufficient accuracy for the partial equation approximation. At the same time proponents of partial differential equations give a lot of counter-examples and show that it is possible to study the partial differential equations without the need to build on ODEs.

In mechanics there is more compromise to the situation, as often in this approach continuous systems are approximated by discrete systems, even if the method is based on rigid finite elements and its variations. For example, the beam can be approximated as system of point masses connected by mass-less springs, also with regard to attenuation. This kind of approximation gives very good results that are sufficient for the needs of applications. Also the reverse approaches are often

J. Awrejcewicz, *Ordinary Differential Equations and Mechanical Systems*,
DOI 10.1007/978-3-319-07659-1_15

applied. As example can be given a system of many oscillators connected in series with susceptible elements and performing planar movement. When there are large number of masses, it can be treated as the continuous system and we can get a full solution to the problem by finding a solution to partial differential equations describing the vibrations of such modeled beam. This example shows the relativity of concepts such as continuous and discrete systems and the ability to transition from one to the other, which may be dictated by the needs of the researcher. And here also, in the process of "continualization" that is in simulating continuous system by increasing the number of masses and springs, ideal is not required, obtained for example by increasing the number of masses tending to infinity as it is in impossible in practice (see Chap. 12).

There appears deeper reflection on the basis of the above considerations. Classical mathematical ideal was based on finding accurate solutions by any means. Nature tells us, however, that this idealization is not only very expensive, but sometimes even unattainable. That is why we should adapt nature's guidelines and try not to perform the computer simulation of infinite accuracy of the initial conditions to find the "true" trajectory. Instead of this we should use tools in advance adapted to assumed inaccuracies in the initial conditions. Moreover, such an approach should not be seen as a painful abandonment of the pursuit of the ideal, but as a new competitive face of mathematics in relation to the classical approach. Absolute accuracy is an unattainable utopia for many aspects of nonlinear dynamics.

This is somehow the opposition to the classical position and can be clearly seen in a new branch of mathematics, represented by asymptology. In this science achieving ideals is deliberately dispensed. Absolute accuracy is also possible to achieve, but only when the asymptotic series are coinciding. The main tools of asymptology are based on the fact that these series do not have to be consistent. In short, this idea can be interpreted as follows: convergent series describes the function $y(t) = y_0$ for $t \to \infty$, and asymptotic series are described function for $t = t_0$ for $y \to y_0$. Just a few dozen years ago, the idea of using the description of the phenomena using divergent asymptotic series seemed ridiculous.

Here we quote one more argument against achieving ideal at any cost. Even if we have a few (completely accurate) particular solutions of an analysed dynamical system, we cannot take full advantage of them. It is not applicable for nonlinear superposition principle system and we cannot find a general solution by adding the special arrangements.

Firstly let us summarize the main idea, which will continue to scroll through the pages of this chapter. Absolute accuracy is welcome, but not at any price. Failure in obtaining absolute accuracy often becomes basis/ground for creation of the new mathematics, physics or mechanics significantly extending the scope of their traditional approaches. Secondly, we should not cling to the definitions of determinism–indeterminism, stochasticity–regularity, big–small, precise–inaccurate, discrete–continuous, etc., while remaining within their framework. It turns out that there are legitimate transitions between them, leading to a deeper understanding of the Nature.

The second major goal in mind and aim of this chapter is an indication of the existence of certain universal rules and laws of dynamical systems. In physics and mechanics there are known systems of repeating similarities in structures on different levels. If you succeed in finding those structures by the application of mathematical methods such as renormalization, they will have universal characteristic independent of the type the describing equations or their projections point. Physical properties are self-similar and repeated in decreasing scale. For example, M. Feigenbaum noted that the shape of the researched by him "fig tree" that was obtained by the analysis of doubling of the period occurring in logistics mapping $x \to Ax(1-x)(A \in [0,4], x \in [0,1])$ is self-similar. Twig of this tree has a shape similar to the shape of the whole tree, and an approximation (scaling factor) increases with decreasing branches with a view to the "magic" number of 4,669 ... This number is also appeared in the analysis of trigonometric mapping $x \to A \sin x$. It turned out that with the scaling factor (which is independent of the equation), and knowing the rules for the construction of such a "tree" it can be quickly reconstructed (and created). There are two reasons for it: law, rule or pattern of conduct and the number (scaling factor).

Many years spent by the ancient Greeks on the analysis of geometric figures and numbers were not vain [139]. For example, the main power of Pythagoreans (sixth century BC) was their mathematical knowledge. They were striving to build knowledge of numbers construction and relationships between them. Pythagoras dared even to say that "all things are numbers". It was a great aesthetic experience for him to establish the link between the tones in the music, and numbers. Numbers revealed themselves to Pythagoras not only in the field of listening experiences, but also in aesthetic experiences like shapes and colors. Following the cards of this book reader can certainly see that many of these insights are reflected in the development of modern nonlinear dynamics, chaos theory and fractals.

According to Aristotle, the numbers were beautiful, and this beauty was manifested through sound or visual form. Summarizing the activities of the Pythagoreans Aristotle wrote that all things are, imitate or reproduce the numbers, and the elements of numbers the elements of all things (for example trades of parity and oddity are numbers elements). One is the base of everything and it is the cause of the creation of two, and both of these numbers are the reason for the creation of all other numbers. And further, the numbers create the points, points form lines, lines form spaces, etc. For them non-elastic point was the link between the geometric and arithmetic form of the world. Every natural number is finite in itself, and only a series of numbers extends to infinity. Number ten was for them an ideal and $\sqrt{2}$ insulted the "holy" majesty of numbers (the existence of this number was kept secret by the Pythagoreans for a long time). According to the Pythagorean school whole material and spiritual worlds are under the rule of the natural numbers. Each irrational number can be approximated with high accuracy to the rational number, and what is more, Bernoulli representation consisting of the numbers 1 and 2 is a precursor of chaos. Could it be that after so many years the secret of the Pythagorean philosophy is still relevant? In present there again can be noticed a return of interest for numbers and number sequences, but this time inspired by the development of modern nonlinear dynamics. Revived the role and significance of the numbers, including rational and irrational numbers, and their relative position (i.e. on the interval $[1, 0]$), strings Frobenius and Dedekind steps, approximations by rational numbers through irrational numbers, zero-one approximation of real numbers, and so-called Bernoulli shift. There is no need to convince anyone about the importance of the number π and of the golden ratio. This time, however, start a discussion took place form a completely different point. It started from considerations relating to differential equations, which are mathematical models of some systems real (physical) or from analysis of some projections, which can be obtained from the differential equations. In both, equations and some projections, observed doubling in the period of solutions with a change in the parameter effectively lead to chaotic motion in accordance with scaling factor equal to 4.669... No one disputes the role of irrational numbers and the existence of quasi-periodic solutions lying on the two- or multi-dimensional torus, while the rational numbers associated are with a periodic solutions. And here again appears the analogy to the earlier discussion. For some (critical) parameters of the system torus disappears and appears periodic solution what lead to placing of the rational numbers in role of the irrational numbers. It turns out that in the interval [0,1], almost all real numbers have decimal representation, which are random, and it means that a sequence of consecutive digits in the decimal representation of the number repeats. Almost every number in that range will have a different number of digits, and therefore randomly selected number will also be random decimal representation. In the "language" of numbers, in practical numerical calculations, periodic motion detection in decimal representation means that starting from a given number, that is from the long decimal representation, after some number of digits (equal to the period) again appears this number (to be precise, with the same extension of the decimal representation). You can prove that such feature occurs when the number is a rational number. In the neighbourhood of any two given rational numbers, there is a whole "ocean" of infinitely many rational and irrational numbers, which are mixed with each other. It can be shown that for a certain types of mapping very close numbers (in terms of decimal expansions), after multiple projections, are different from each other.

A careful reader will see that after all the real system time (independent variable) "floats" on a continuous basis and goes through all mixed together rational and irrational numbers, while numerical procedures based are on discrete models and discrete dynamics. Fortunately, however, there are close links between the discrete and continuous dynamics.

Kingdom of numbers extends even further. Fractional numbers play a key role in the dynamics of the so-called fractals through so-called fractal dimensions introduced by Hausdorff. Fractals, contrary to the intention of their creators, were somehow geometric method of the study of irregular dynamics, also of the deterministic chaos. There are some, though far analogies between fractal and chaotic dynamics. Scenario of transition to chaos through bifurcations of doubling period of newly emerging orbits leads in effect to the coexistence of infinitely many periodic and unstable orbits and also creates a qualitatively new geometric structure known as a strange attractor. Similar situation is in the case of simple geometric shapes such as triangles, lines and ellipses, and arranging them in a certain scale, according to a certain pattern (low), what leads to a qualitatively new structure, in which the decrease in its basic elements (circle, triangle, etc.) leads to the loss of their shapes creating the new structure. This last one has a new property—self-similarity. It shows the similarity between the different elements in respect to the size. As mentioned above, this structure is characterized by a dimension that is a fraction and not an integer. An example may be a snowflake, which has a very complicated boundary line. This line is not a one-dimensional curve (dimension 1) and is not filling the two-dimensional surface (dimension 2). Thus, the dimension is searched in the range of $1 < w < 2$.

Chaotic dynamics and synchronization are two opposing processes that can be observed in physical, biological or chemical systems. The first one is expressed in a tendency for the disorder, while the second one tends to the simplicity and regularity. By changing the parameters and the initial conditions it is possible to move from one process to another. While synchronization processes have been observed for a long time, and were subjects for a scientific analysis since the seventeenth century, starting from the works devoted to the analysis of synchronization of clocks ticking, and later also test of the synchronization in sound tuning fork generators or movement of the planets, however the phenomenon of chaos in deterministic systems have been detected recently. They are related to the pioneering works of Poincaré, Lorenz and Ueda. In this chapter, among others, we are going to discuss these two extreme processes on the examples of one- and two-dimensional mappings and on the dynamic systems. As a tool of analysis used are analytical and numerical methods.

15.2 Modelling and Identification of Chaos

First, we will consider a so-called Poincaré maps. They are widely used for the analysis of dynamic systems. They are based on the introduction of the plane (or hyperplane), which crosses the returning phase flow without any contact with the trajectory of this flow (Fig. 15.1).

Γ trajectory of the flow lies in three-dimensional space, and the points of intersection with the plane belong to it, the Poincaré mapping represents a mapping of the plane into itself. Undeniable advantage of this mapping is reduction of the dimension of the design space by one (obtained points lie in a plane). In the case of two-dimensional flow (that means lying on the plane) mapping points are arranged along a line (so-called one-dimensional mapping).

In practice, we introduce the secant plane in a relatively simple way, as will be discussed on the example of the non-autonomous system with one degree of freedom described by the following differential equation

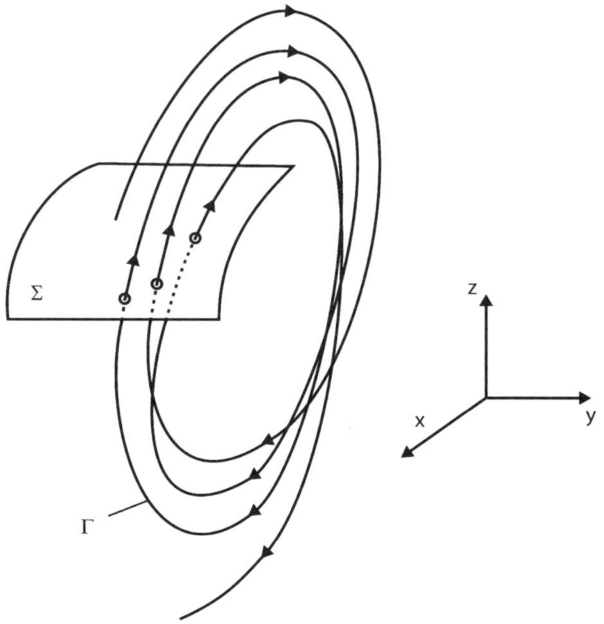

Fig. 15.1 Poincaré map schematic representation

$$\frac{d^2x}{dt^2} + F\left(x, \frac{dx}{dt}\right) = F_0 \cos \omega t. \tag{15.1}$$

The period of the exciting force is $T = 2 = \pi/\omega$ and the discrete value of time

$$t_n = t_0 n T, \quad n = 0, 1, 2, \ldots \tag{15.2}$$

we record the speed and movement

$$\begin{aligned} v_n &= \frac{dx}{dt}(t_n), \\ x_n &= x(t_n). \end{aligned} \tag{15.3}$$

If

$$\begin{aligned} x(t_0) &= x_0, \\ \nu(t_0) &= \nu_0, \end{aligned} \tag{15.4}$$

then points

$$\begin{aligned} x_n &= x(t_n, x_0, \nu_0), \\ \nu_n &= \nu(t_n, x_0, \nu_0), \end{aligned} \tag{15.5}$$

form Poincaré map.

In this case, the Poincaré map is a two-dimensional one. For the autonomous system ($F_0 = 0$) and assuming that the force $F\left(x, \frac{dx}{dt}\right)$ causes the self-excitation of the system, we will obtain one-dimensional map, for observation (or measurement) value of $x(t_n)$ will be made in moments of time t_n for which $v(t_n) = \frac{dx}{dt}(t_n) = 0$. Periodic solution (periodic orbit) will be presented as a fixed point of this mapping. If it will be stable in the sense of Lyapunov (to this stability theory devoted is a vast literature [44, 187, 244]), then a sequence of points prior to it will be going to this point. This is known as attractor. If a fixed point mapping is unstable, then the sequence of points will be "running away" from this point. In this case singular point x_0 is unstable in the sense of Lyapunov and can be called *repiler* [106, 214, 224].

These simple considerations can be widely generalized. Let the solution to a system of n ordinary differential equations in normal form to have the form $x = X(t, x_0)$, where $x_0 = X(t_0, x_0)$. If the Cauchy problem has a single solution set by said initial condition, the following compounds occur

$$(t_2, X(t_1, x_0)) = X(t_1 + t_2, x_0), \tag{15.6}$$

what is easy to check, taking into account the fact that

$$X(t_1 + t_2, x_0) = x_2, X(t_1, x_0) = x_1, X(t_2, x_1) = x_2. \tag{15.7}$$

Family of mappings $X_t(x) = X(t, x)$ that is defined by the solution $x(t)$ defines a dynamical system in the space $\mathbb{R}^n$ what is noted as $(\mathbb{R}^n, X_t)$. If $t \in [0, \infty)$, then dynamical system is continuous (and we are calling it the flow), and if $t \in N$, the dynamical system is discrete (and a cascade is meant).

Returning to the secant plane geometric interpretation, it appears that there is some functional relationship between the coordinates of two successive points of intersection of the trajectory with the plane of the form

$$x_{k+1} = F(x_k), \tag{15.8}$$

what sometimes can be expressed in an analytical form. Difference equation (15.8) describes the cascade.

Point x_0 is called a fixed point of the cascade (15.8) if

$$F(x_0) = x_0. \tag{15.9}$$

For each point x n-fold application of the operation leads to a point F

$$x(n) = F^n(x). \tag{15.10}$$

Trajectory passing through any point x in the phase space $\mathbb{R}^n$ will be called finite or infinite set of iteration (15.8). We say that the point x is periodic with a period k, if

$$F^k(x) = x, \tag{15.11}$$

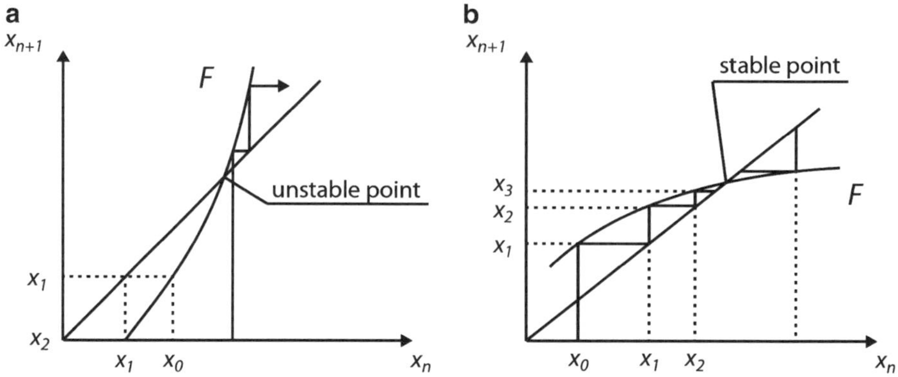

Fig. 15.2 Stable (**b**) and unstable (**a**) fixed point of the map (15.8)

k is the smallest integer with this property. If you choose x_0 close to the tested point x, and if the distance $\|x\ (k) - x\| \to 0$, when the $n \to \infty$, then x_0 is asymptotically stable.

Based on Eq. (15.8) and Fig. 15.2 is easy to imagine the stability criteria for mapping of the fixed point.

Based on the graphical construction of the mapping (15.8) shown in Fig. 15.2, it can be concluded that the mapping fixed point is stable (unstable) if

$$\left| \frac{dx_{k+1}}{dx_k} \right| < 1(> 1). \tag{15.12}$$

Finally, also worth mentioning here are advantages of the Poincaré mapping. In general, studying dynamics of systems governed by ordinary differential equations we deal mainly with the analysis of equilibrium positions and of periodic solutions. In the case of periodic solutions, the stability condition means that for a considered fixed point (and consequently also periodic orbit) to be stable, the monodromy matrices eigenvalues (multipliers) should lie in a unit circle of radius 1. Often this method is used for the practical pre-determining of the stability of the orbit found numerically, for example, using the "shooting" method or Urabe–Reiter method (see Chap. 13).

Now, some basic concepts and definitions will be introduced, basing on the works of Kudrewicz [142] and Samoilenko [207].

Definition 15.1. String formed of successive mapping points $\{F^k(x)\}$ for $k = 0, 1, 2, \ldots$ will be called the trajectory (orbit) of the point x.

Comment.
The set $x_1, F(x_1), F^2(x), F^3(x), \ldots$ be the orbit of a point x. As an example, consider the mapping $F : w[0, 1] \to [0, 1]$ of the following form

$$F(x) = x(1 - x), \tag{15.13}$$

then for $x = \frac{1}{2}$ orbit is determined by the points $\frac{1}{2}, \frac{1}{4}, \frac{3}{16}, \ldots$.

Definition 15.2. If there exist a sequence $\{k_n\}$ of natural numbers and

$$\lim_{k_n \to \infty} F^{k_n}(x) = x_\Omega, \tag{15.14}$$

then a point x_Ω will be called Ω-border point of the $\{F^k(x)\}$ trajectory. The set of all such points is called the set of Ω-border of the $\{F^k(x)\}$ trajectories, and we denote it by $\Omega(x)$.

Definition 15.3. Invariant set of cascade Z is a set satisfying the condition $X(Z) = Z$. Most of invariant sets are equilibrium points or periodic trajectories.

There is also need for comment of Ω-border sets that can be divided into attractors and repliers.

Definition 15.4. The closed and bounded invariant set A is called attractor. If there exists its neighbourhood $O(A)$ such that for any $x \in O(A)$ the trajectory $\{F^k(x)\}$ tends to A for $k \to \infty$, the set of all x satisfying this condition is called the attracting set of the attractor A.

The following comments hold:

1. Often, while defining the attractor, it is said that the set that fulfills conditions in Definition 15.4 does not contain in itself a different set that satisfies these conditions.
2. Chaotic attractors are those attractors that contain at least one chaotic trajectory. Trajectory is called chaotic if at least one of the Lyapunov exponents associated with it is positive.
3. Strange attractors are called attractors, which have a complex geometric structure.
4. Typically these two terms are used interchangeably. However, they can exist independently (see Grebogi et al. [104] and Jacobson [128]).
5. Dynamical systems may possess several coexisting attractors. One of the main tasks in this case is to define the initial conditions, for which the phase flow will be attracted by the individual attractors. It turns out that the boundaries between the different attractor pools can have very complicated shapes, for example they may be fractals [165].

Definition 15.5. Fixed point x_0 of the map (15.8) for $\mathbb{R}^2$ is called hyperbolic, if this point derivative DF has eigenvalues different from 1. If this point is hyperbolic and has two real eigenvalues, λ_i $(i = 1, 2)$ and at the same time $|\lambda_1| < 1$ and $|\lambda_2| > 1$, then the point is a saddle. Manifolds

$$\begin{aligned} W^S(x_0, F) &= \left\{x : \left\|F^n x - W^S(x_0, F)\right\| \to 0, \quad \text{for} \quad n \to +\infty\right\}, \\ W^n(x_0, F) &= \{x : \|F^n x - W^n(x_0, F)\| \to 0, \quad \text{for} \quad n \to -\infty\}, \end{aligned} \tag{15.15}$$

are called the stable and unstable respectively, but they are invariant, that means

$$\begin{aligned} F\left(W^S(x_0, F)\right) &= W^S(x_0, F), \\ F\left(W^n(x_0, F)\right) &= W^n(x_0, F). \end{aligned} \tag{15.16}$$

If the given dynamical system described by system of ordinary differential equations meets Lipschitz conditions and there exists a solution to the Cauchy problem, the solution is unambiguous and accurately determined by the initial conditions. It is analogous to the train travelling on tracks, movements of which can be determined at any moment in time. And yet discovered strange chaotic attractors of Lorenz, Ueda, Hénon and others, seem to deny those obvious facts. Some uncertainty is intuitively understood particularly for the complex physical systems, where a small change in phases can lead to large changes in the systems dynamics in the intervals of the independent variable, i.e. time. We will explain this with an example of the system considered by Landau. For this we will consider two extreme cases of a dynamic system: chaos and synchronization. Without going into detail, the synchronization will be understood as tendency of subsystems of the complex dynamic system to perform "similar" dynamics, such as manifested by subsystems periodic motions within the same periods, and consequently causing the synchronization, that is the periodic movement with the same period for the entire system. This phenomenon has already been observed by Huygens during the analysis of the clocks ticking synchronization. Currently, the phenomenon has broader understanding and refers to the mutual organization of the biological systems, and in the mechanics this issue appears when considered are issues of the rotor vibration synchronizations and in the stabilization [50,104,142,207]. Consider first oscillating system fully synchronized, that is one that the vibration frequencies $\omega_1, \omega_2, \ldots \omega_k$ appearing in it satisfy the condition

$$l_1\omega_1 + l_2\omega_2 + \cdots + l_k\omega_k = 0, \tag{15.17}$$

where $\{l_1, \ldots l_k\} \in C$ and C is the set of integers. We say then that the system is in full resonance, and it reveals in the increases of the characteristic vibration amplitude for each of the ω_i in the discrete set $\omega = \{\omega_1, \ldots, \omega_k\}$. However, if each subset will vibrate independently from the others that are with its own (independent from the others) period then the system is not synchronized. In practice, the lack of synchronization is associated with the existence of irrational numbers, for example for $k = 2$, $\omega_1 = 1$ and $\omega_2 = \sqrt{2}$ solution $x = x(\varphi_1, \varphi_2)$, where $\varphi_1 = \omega_1 t$ and $\varphi_2 = \omega_2 t$ is in steady state on a two-dimensional manifold (torus), and the solution

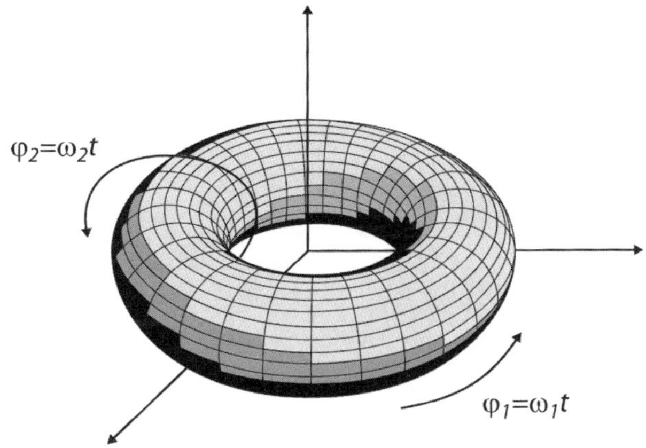

Fig. 15.3 Torus and two incommensurate frequencies ω_1 and ω_2

$x = x(\varphi_1, \varphi_2)$, where time is the parameter, is called the quasi-periodic solution. An example of such a two-dimensional manifold is shown in Fig. 15.3.

The issue of quasi-periodic orbits requires a deep study and it is only briefly addressed in this work. It contains more problems that are not completely solved. This applies mainly to determining the stability of multi-dimensional attractors - tori, tracking changes of the quasi-periodic orbits with the change of the parameters and their bifurcation [65, 83, 131, 156, 228].

One may imagine that if phases $\varphi_1^0, \ldots, \varphi_k^0$ are additionally changing even in minor range, the response of the system $x(\omega_1 t + \varphi_1^0, \ldots, \omega_k t + \varphi_k^0)$ can be subjected to significant changes over time and as a result lead to the appearance of chaotic motion.

Now we will point out another possible appearance of chaos in simple dynamical systems. For autonomous oscillators with single degree of freedom and with limited trajectory (performing recurrent motion) only positions of equilibrium or periodic orbits may be attractors. However, the situation is drastically changed for three-dimensional systems, specifically for systems with $1^1/_2$ degree of freedom or those oscillators with an external forcing.

It appeared therefore that trajectories in the three-dimensional systems may be present in a sub-phase space $\mathbb{R}^3$, but they can constantly wander between the positions of unstable equilibrium states and unstable periodic orbits. Although basing on the fact of existence of the limits for such subspace, we know that there is a time after which such trajectory will be arbitrarily close to the start point lying on the attractor, but it is impossible to obtain information when it will return there.

Flow of phase trajectories can be imagined on the example of a liquid that even in very small volume consists of a large number of particles. If in such a volume there is attractor which is a periodic orbit, then introducing a small drop of colored liquid at any of origin point in the volume, you will find that after a while the color will be determining the orbit. However, in the case when attractor is a chaotic strange attractor, the trajectory lying on such attractor starts to wander along the entire

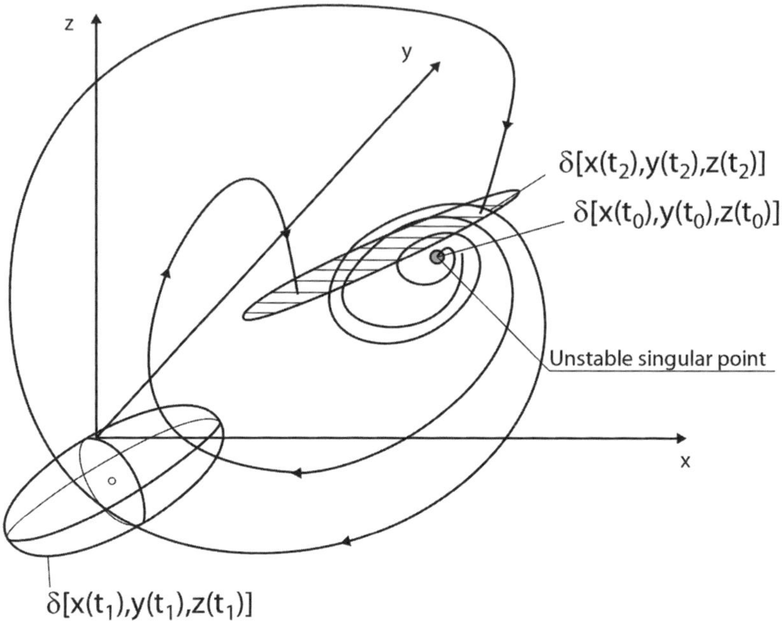

Fig. 15.4 An example of an unstable trajectory wandering in the limited area $\mathbb{R}^3$

volume of the liquid and the liquid becomes colored. Colored and colorless particles will be mixed. This process takes place in a relatively short time and therefore is not the result of diffusion, but rather is related to the turbulent movement of liquid molecules. This analogy is even deeper. The intensity of the color will indicate the probability of finding a phase point in this area and it does not depend on the initial position (the starting point).

Armed with the knowledge of the instability role, let us consider now evolution (change over time) of an area initial conditions taken from the plane x, y, that is, such that $z = 0$ and $\delta = \delta(x, y)$ which has been hatched in Fig. 15.4.

Area $\delta(t_0)$ is a very close neighbourhood of the unstable strange point. Two distinct trajectories in this figure exponentially flee from each other and initially tiny set of initial conditions $\delta(t_0)$ is transforms into a volume $\delta(t_1)$, and because the trajectories are limited, they must "turn around" and as a result for $z = 0$ the two points, which lie very close together (for $t = t_0$) after time $t = t_2$ are found far apart. The question arises, how far apart, or in rather, how small should be distance between them at the starting point so they could be found close to each other once again.

15.3 Lyapunov Exponents

Before answering this question, let us return to the old theory of characteristic numbers introduced by Lyapunov that, when with opposite sign, define a so-called Lyapunov exponents (details are given in Demidovich [77]). Earlier described was and exponential divergence of trajectories and Lyapunov exponents are a measure of such discrepancy.

According to (15.10) cascade (15.8) can be presented in the form of

$$\mathbf{x}(k+1) = F(\mathbf{x}(k)). \tag{15.18}$$

With each point $\mathbf{x}$ in the phase space can be associated array $DF^k(\mathbf{x})$ called the mapping Jacobian F^k, which is formed from the local linearization, which in practice amounts to calculating the kth iteration derivative for point $\mathbf{x}$. Starting from the point $\mathbf{x}(0)$ for the kth iteration, the matrix is expressed with the relationship

$$\mathbf{J}(k) = DF^k(\mathbf{x}(0)), \tag{15.19}$$

wherein $DF^k(\mathbf{x}(0))$ can be obtained as the product of

$$DF^k(DF(\mathbf{x}(k-1)) \cdot \ldots \cdot DF(\mathbf{x}(0))) = DF^k(\mathbf{x}(0)). \tag{15.20}$$

Having calculated $\mathbf{J}(k)$ for small increments we get

$$\mathbf{x}(k+1) \approx DF(\mathbf{x}(k))\,\mathbf{x}(0) = DF^k(\mathbf{x}(0))\,\mathbf{x}(0) = \mathbf{J}(k)\,\mathbf{x}(0). \tag{15.21}$$

Coming back to the discussion related to Fig. 15.4, assume that in a neighbourhood of the unstable point we select two points x_1 and x_2, which after k iterations (in accordance with the previous considerations) will evolve into the points $x_1(k)$ and $x_2(k)$, defined by the relationship

$$\mathbf{x}_1(k) - \mathbf{x}_2(k) \approx \mathbf{J}(k)(\mathbf{x}_1(0) - \mathbf{x}_2(0)). \tag{15.22}$$

We can take the whole flow generated by the initial conditions, which are located for example in the sphere. In the general case, however, it will be n-dimensional sphere $K_0 = K(0)$, which after the k iterations it becomes ellipsoid $K(k)$ (such ellipsoid when $n = 3$ is indicated in Fig. 15.4 as $\delta(t_2)$). If now, instead of a single point x_0 we take the sphere K_0, and respectively, instead of $x(k+1)$ we take ellipsoid K_{k+1}, then, according to (15.21) we obtain

$$K_{k+1} = DF(x(k))\,K_k. \tag{15.23}$$

Ellipsoids K_{k+1} and K_0 possess n principal axes and the system has n related with them Lyapunov exponents. For the system to be chaotic it is enough if one of the exponents (the largest) is positive. Lyapunov exponents is defined by the formula

$$\lambda_j = \lim_{k \to \infty} \frac{1}{k} \log \left(\alpha_{j,k} \right), \tag{15.24}$$

where $\alpha_{j,k}$ is the length of the jth axis of the ellipsoid K_k.

It turns out that for a very wide class of mappings F there exists the limit defined by (15.24), which for almost all $x(0)$ does not depend on $x(0)$, which means that it also is independent from the initial conditions. Then λ is a measure of changes in the initial conditions and if the error in the determination of the initial conditions is $\Delta x(0)$ then kth iteration will have the value

$$\Delta x_k \approx 10^{\lambda k} \Delta x \left(0\right), \tag{15.25}$$

for sufficiently small $\Delta x(0)$ and large enough k. Let now the maximum Lyapunov exponent to have the value $\lambda = 0.1$, which is not too high requirement for dynamical systems and let $k_* = 10^2$ and $\Delta x\left(0\right) = 10^{-5}$. According to (15.25) for the k_* iteration calculate $\Delta x(k_*) = 10^5$, and such accuracy of the calculations cannot be accepted. On the other hand we want to obtain k_* iterations error was $\Delta x(k_*) = 10^{-5}$. Thus, there appears the question problem, what the accuracy of the initial conditions definition we should apply. Using the formula (15.25), we calculate that $\Delta x(0)$ is 10^{-15}, and preserving so high accuracy of the calculations is extremely difficult. Of course, uncertainty increases with the iterations increase.

So the problem comes down to setting infinite precision to the initial conditions. According to the principle, each state of the real physical system can only be determined with reasonable accuracy and is determined rather by a probability distribution instead of a number. If the trajectory is considered stable, then the initial error rapidly decreases with time, and if it is unstable, it is growing rapidly with the increase of iterations resulting in the unpredictability of its behaviour, that is chaos in the determined system.

Now, let us discuss for a while the possibility of recursion (return) of trajectories, in such a way so it will lay on the attractor. We have already mentioned that the trajectory remaining in limited surface must have the possibility of returning in arbitrary close neighbourhood of the starting point. It turns out that it is a common characteristic for the phase surface. We are talking about a singularity point called the saddle. The precursor of chaotic motion is the cross-section of the stable and unstable saddle point variety. This is possible for at least a three-dimensional system.

15.4 Frequency Spectrum

In engineering calculations, both in the computer simulation and in the analysis of the real object in the laboratory, one of the most popular methods of analysis is a technique based on the analysis of the frequency spectrum. In what follows we apply the Fourier Fast Transformation (FFT). The transformation of the waveform from the time domain to frequency domain can be described by the relation

$$W(\omega) = \lim_{T\to\infty} \frac{1}{T} \left| \int_{-T}^{T} x(t)\, e^{-i\omega t} dt \right|. \tag{15.26}$$

In the case of regular movements (periodic and quasi-periodic) frequency spectrum consists of discrete components, while the continuous frequency spectrum corresponds to the chaotic trajectory $x(t)$.

15.5 Function of Autocorrelation

The autocorrelation function is a competitive tool to the Lyapunov exponents. It is widely described in the literature, particularly in respect to the differential equations. It is determined by the relationship

$$A(t) = \lim_{T\to\infty} \frac{1}{T} \int_{0}^{T} F[x(t+\tau)]\, F(x(\tau))\, d\tau, \tag{15.27}$$

assuming that the analysed system is ergodic. If the $A(t)$ include periodic or quasi-periodic components, then also in the researched system exists the periodic or quasi-periodic orbit. If the two trajectories lying close to each, separate and over time move independently, then $A(t)$ quickly approaches zero. This corresponds to a situation where at least one of the Lyapunov exponents is positive.

It is worth to mention some of the characteristics of the autocorrelation function.

1. It is a real and even function with the point of maximum in $t = 0$, which can assume both positive and negative value.
2. In the case stochastic process study with a mean meaning $\langle x(t)\rangle = 0$, it has the shape of the sharply outlined pulse.
3. For a stochastic process—white noise, the function $A(t)$ has the shape of a δ function.
4. If (on average,) slope of the autocorrelation function is has approximately exponential character, the dynamic state of the phenomenon is associated with the beginning of the chaotic motion.

If for the mapping (15.18) we define the mean value $\langle x(k)\rangle$ dependent on $x(0)$ as

$$\langle x_k \rangle = \lim_{k \to \infty} \frac{1}{k} \sum_{k=0}^{K} x(k), \tag{15.28}$$

then the autocorrelation function is given by

$$A(\ell) = \lim_{K \to \infty} \frac{1}{K} \sum_{k=0}^{K} (x(k) - \langle x_k \rangle)(x(k+\ell) - \langle x(k) \rangle), \tag{15.29}$$

where $\ell = 0, 1, 2, \ldots$.

15.6 Modelling of Nonlinear Discrete Systems

15.6.1 Introduction

In the vast majority of cases, the dynamics of physical systems is governed by partial or by ordinary differential equations. The former is often replaced by a variety of the methods reducing through the ordinary differential equations systems. The next step leading to further problem reduction is replacing differential equations with mappings. This method of proceeding is based on the use of the analytical methods.

Another, independent, method of research is based on the analysis of the simplest mappings various types of dynamics, in this case the one-dimensional ones, and in particular on trying to obtain deepest possible understanding of the chaotic dynamics basing on those mappings.

If you have a wide range of knowledge about the dynamics of one-dimensional representations, then the dynamics (even complex) systems described by differential equations can sometimes be understood by one-dimensional mappings.

Analytical methods face a number of limitations in the analysis of nonlinear dynamics, therefore, in most cases carried out are numerical analysis. In practice, this means replacing the continuous dynamics (in the equation the independent variable that is time, is a constant) by the discrete dynamics (in numerical methods time variables have the discrete values). It turns out that there are deep connections between the "continuous dynamic" and "discrete dynamics". In the numerical analysis used is the Poincaré mapping method. Links between points obtained on the Poincaré surface are described by differential equations. With the introduction of such a representation is not only reduced dimensionality of the dynamics, but also in the analysis of chaotic dynamics the introduction of the Poincaré surface led to the elimination of periodic movements of the points, what allows to focus attention on the chaotic dynamics. A further extension of the method of discretization is states discretization, for example, by assigning to the numbers only two values, zeros or ones.

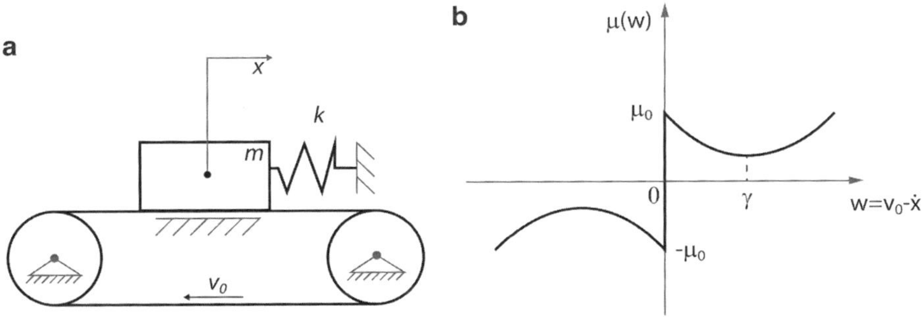

Fig. 15.5 Mechanical system with one degree of freedom: a block lying on a belt moving at a speed of v_0 (**a**) and the coefficient friction characteristics (**b**)

It may happen that for the analysis of two-dimensional mappings, there is also the possibility of their reduction to one-dimensional mappings. This occurs when the mapping in one direction is highly tensile, and in the other one it is strongly compressive. This will make the points along the one or two lines, and can be considered a one-dimensional mapping of one line into the other.

Now we will consider examples of the dynamics of simple physical systems that can be reduced to one-dimensional mappings analysis. In the nonlinear systems with friction self-exciting vibration can appear [149]. Figure 15.5 shows such classic case.

The body of mass m (block) is located on the tape with a coefficient of friction depending on the velocity relative to the body and the tape with characteristics shown in Fig. 15.5b. It turns out that the range of the relative speed $0 < w < \gamma$ block equilibrium position becomes unstable. There appear vibrations that are beginning to grow reaching a limit cycle (periodic orbit). The equation of motion has the form

$$m\ddot{x} + kx = mg\left[\mu_0 - \alpha w + \beta w^3\right], \tag{15.30}$$

where on the right side described analytically are friction forces. Now assume that for $x = 0$ the crash occurs, when $\dot{x} \geq a$—the rapid change in velocity, and furthermore we will assume that the dynamic is related to the sloping part coefficient of friction [181]. Then the dynamics of the considered system can be approximated by the equation

$$\ddot{x} + 2h\dot{x} + \alpha^2 x = 0, \tag{15.31}$$

$$\dot{x}_+ - \dot{x}_- = -b, \tag{15.32}$$

where $\dot{x}_+$ and $\dot{x}_-$ is the speed before and after the impact of the amplitude b. Dynamics described by Eqs. (15.31) and (15.32) can be represented by mapping

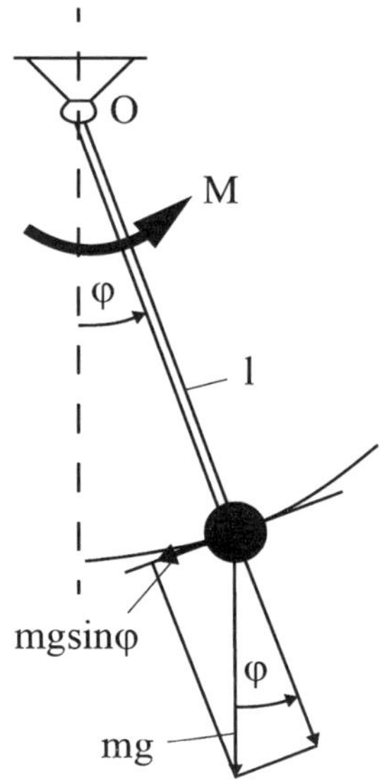

Fig. 15.6 Flat pendulum excited by $M = M_0 + M_1 \cos \omega t$

$$\bar{y} = qy + \begin{cases} 0 & \text{for } 0 \le y < \frac{a}{q}, \\ -b & \text{for } y \ge \frac{a}{q}, \end{cases} \tag{15.33}$$

where

$$\bar{y} = \dot{\bar{x}}, \quad y = \dot{x} \ge 0, \quad q = e^{2h\sqrt{\alpha^2 - h}} > 1. \tag{15.34}$$

Position $y_* = b/(q-1) > a$ is an unstable fixed point of this mapping. For $y < y_*$ chaotic vibrations appear in the $a - b \le y \le a$ range of changes. The example above was connected with the analytical method, while the following example refers to the numerical methods.

Consider the motion of a pendulum of length l, mass m and moment of mass inertia B (Fig. 15.6).

The equation of motion is:

$$B\ddot{\varphi} + c\dot{\varphi} + mgl \sin \varphi = M_0 + M_1 \cos \omega t, \tag{15.35}$$

where c is the viscous environment damping coefficient. Assuming $mgl = B$, this equation can be reduced to the form

$$\ddot{\varphi} + h\dot{\varphi} + \sin \varphi = M_1 + M_2 \cos \omega t, \tag{15.36}$$

where

$$h = \frac{c}{B}, \qquad M_1 = \frac{M_0}{B}, \qquad M_2 = \frac{M_1}{B}. \tag{15.37}$$

We will approximate

$$\dot{\varphi} = \frac{\varphi_n - \varphi_{n-1}}{t_n - t_{n-1}}, \tag{15.38}$$

$$\ddot{\varphi} = \frac{\varphi_{n+1} - 2\varphi_n + \varphi_{n-1}}{(t_{n+1} - t_n)(t_n - t_{n-1})}. \tag{15.39}$$

After taking into account (15.38) and (15.39) Eq. (15.36) takes the form (see [221])

$$\varphi_{n+1} - 2\varphi_n + \varphi_{n-1} + hT(\varphi_n - \varphi_{n-1}) + T^2 \sin\varphi_n = T^2(M_1 + M_2), \tag{15.40}$$

where $t_n = \frac{2\pi}{\omega}n$.

Assuming

$$R_1 = \frac{T(M_1 + M_2)}{h}, \qquad hT = 1 - b, \qquad R_2 = T^2, \tag{15.41}$$

from (15.40), we obtain

$$\varphi_{n+1} - 2\varphi_n + \varphi_{n-1} + (1-b)(\varphi - \varphi_{n-1}) + R_2 \sin\varphi_n = (1-b)R_1. \tag{15.42}$$

Equation (15.42) can be represented as an equivalent

$$\begin{aligned} r_{n+1} &= br_n - R_2 \sin\varphi_n, \\ \varphi_{n+1} &= \varphi_n + R_1 - R_2 \sin\varphi_n + br_n, \end{aligned} \tag{15.43}$$

where:

$$r_n = \varphi_n - \varphi_{n-1} - R_1. \tag{15.44}$$

We are still dealing here with a two-dimensional representation, but for very high damping such that $hT = 1$, we get one-dimensional representation of a circle into a circle (which will be discussed later).

15.6.2 Bernoulli's Map

Let us consider the mapping F carrying out a unit vector into itself, that is $[0, 1) \to [0, 1)$, in the form

$$x_{k+1} = F(x_k), \tag{15.45}$$

$$F(x_k) = 2x_k \bmod 1, \tag{15.46}$$

for $k = 0, 1, 2, \ldots$, while the modulo function limits the range of the obtained results to the unit (take only the rest after dividing by 1). This mapping is called the Bernoulli map and can also be written in the form of difference equation

$$x(k+1) = \begin{cases} 2x(k) & \text{for } 0 \le x(k) < 0.5, \\ 2x(k) - 1 & \text{for } 0.5 \le x(k) < 1. \end{cases} \tag{15.47}$$

This mapping has only one fixed point $x_0 = 0$, which is unstable.

Let us consider how this mapping will behave for $x(0) = {}^1/_{11}$, the rational number. We obtained the following sequence of numbers

$$\begin{gathered} x(0) = \frac{1}{11}, \quad x(1) = \frac{2}{11}, \quad x(2) = \frac{4}{11}, \quad x(3) = \frac{8}{11}, \\ x(4) = \frac{5}{11}, \quad x(5) = \frac{10}{11}, \quad x(6) = \frac{9}{11}, \quad x(7) = \frac{7}{11}, \\ x(8) = \frac{3}{11}, \quad x(9) = \frac{6}{11}, \quad x(10) = \frac{1}{11}, \end{gathered} \tag{15.48}$$

this represents a periodic orbit with a period equal 10. For $x(0) = \frac{1}{5}$ we get

$$x(0) = \frac{1}{5}, \quad x(1) = \frac{2}{5}, \quad x(2) = \frac{4}{5}, \quad x(3) = \frac{3}{5}, \quad x(4) = \frac{1}{5}. \tag{15.49}$$

So again we get a periodic orbit, but this time the period equals 4. It turns out that for all rational numbers in the considered unit interval the iteration results are in the form of periodic orbits. However, the situation is quite different if we choose as a starting point irrational number. To each point of the set [0,1], we can assign an infinite sequence $\{a_0, a_1, a_2, \ldots\}$, called the address, in such a way that

$$a_0 = 0, \qquad x(0) = \frac{1}{2}a_1 + \frac{1}{2^2}a_2 + \frac{1}{2^3}a_3 + \frac{1}{2^4}a_4 + \cdots \tag{15.50}$$

Numbers a_i can take only the values 0 or 1. Therefore, the $x(0)$ can be written as an infinite sequence of zeros and ones of the form

$$x(0) = \{a_1, a_2, a_3, a_4, \ldots\}. \tag{15.51}$$

It turns out that with such a representation of a real number, we can see an important property of the mapping (15.47). Let us consider it on the example of $x(0) = 0.32$. Reader is able to quickly perform calculations (for example using a basic calculator) finding sequence of a_i values, which are given below

$$x(0) \equiv 0.32 = \{0,\ 1,\ 0,\ 1,\ 0,\ 0,\ 0,\ 1,\ 1,\ 1, \ldots\}. \qquad (15.52)$$

After the first iteration we get the number of $x(1) = 0.64$ which has the following address

$$x(1) \equiv 0.64 = \{1,\ 0,\ 1,\ 0,\ 0,\ 0,\ 1,\ 1,\ 1, \ldots\}. \qquad (15.53)$$

A careful reader will see a pattern. Address (15.53) was obtained by shifting by one digit to the left of the address (15.52). It turns out that this regularity is the place for all the numbers from the interval [0,1]. The following iteration is associated with a shift by one digit to the left of the previous one address. This shift is called the Bernoulli shift. The second note concerns the finite and the infinite characteristic: the finite rational number is represented here by infinite series. In the interval [0,1], most of the numbers are irrational. These figures have random decimal representations or in other words, almost all the numbers from the interval [0,1] have random decimal representations.

Let us reflect on other analogies given by Schuster [213]. We assign the number to zero the head, and the number one tails and consider a coin toss. Tossing a coin repeatedly we receive following address $\{0, R, R, 0, \ldots\}$, which corresponds to the exactly one real number from the interval [0,1].

Now consider the following oddity. Take two numbers $x^{(1)}(0)$ and $x^{(2)}(0)$ that have for example 10^{16} the same decimal digits, so in the calculations are identified as the same. By subjecting these numbers to 10^{16} iterations (15.47) we come to the seventeenth place in the numbers addresses, and so to the places where they differ. Further iterations will already represent these different numbers. This raises a very clear parallel to the observed phenomenon of the deterministic chaos, i.e. in each subsequent realization of the same process, starting with a theoretically the same initial conditions, the response is always different because of the inevitable, albeit very small differences in their realizations.

The second property of the irrational numbers and of the Bemoulli shift is that any finite subset of the infinite set represents the number of repeats it in this set infinitely many times, and shift Bemoulli tries to move the subsequences to the left an infinite number of times.

Bernoulli mapping has one more feature typical of chaos. It is associated with the operation of stretching and folding. If the numbers subjected to iterations are in the range $[0, {}^1/_2)$, the projection extends corresponding sections (it is multiplies them by 2). If starting from some iteration, they are in the range of numbers larger than ${}^1/_2$, then in the following iterations their results are decreasing and the numbers are returning into the [0,1] interval.

At the end let us mention one more characteristic trait of this mapping, which is also typical of the chaos. We have shown that starting from the rational number received periodic orbits. They are unstable. Since the interval [0,1] there are infinitely many rational numbers, there is also an infinite number of unstable periodic orbits in this range.

15.6.3 Logistic Map

The logistic mapping received relatively detailed analysis

$$x(k+1) = px(k)(1 - x(k)), \tag{15.54}$$

wherein the parameter $p \in [0, 4]$. The main feature of this mapping is the section stretching or compression, and then folding it in half. It turns out that for a fixed value of the parameter p mapping will "wrap" the output section and place it in the range $[0, p/4]$. To illustrate, let us consider the case of $p = 1$ and let us start with numbers in the range [0,0.5]. Number zero becomes zero, and 0.5 change into 0.25, any other numbers are in the range [0, 0.25]. Considering the interval (0.5,1] it can be noticed that the number 1 becomes zero. The number of 0.7 becomes 0.21, 0.9 changes into 0.09. Basing on these trivial examples, we can see that also the numbers range (0.5,1) change range [0,0.25], with the numbers lying closer to the 1 are mapped into lying closer to zero. For parameter p greater than 4, almost all sequences $\{x(k)\}$ diverge to infinity. For the boundary value $p = 4$ the solution of Eq. (15.54) can be expressed in an analytical form

$$x(k) = \frac{1}{2}\left(1 - \cos\left[2^k \text{arc}\cos(1 - 2x(0))\right]\right). \tag{15.55}$$

Let us conduct now analysis of the typical nonlinear dynamics. Let us find fixed points of the mapping (15.54) and then examine their stability. Fixed points we find from the equations

$$px_*(1 - x_*) = x_*. \tag{15.56}$$

Obtained are the following two points:

$$x_*^{(1)} = 0, \qquad x_*^{(2)} = \frac{p-1}{p}. \tag{15.57}$$

Each of these solutions is stable when

$$\left|\left(\frac{df(x)}{dx}\right)_{x=x_*}\right| < 1, \tag{15.58}$$

where: $f(x) = px(1 - x)$.

Simple calculation shows that

$$\begin{aligned} \left|\left(\frac{df(x)}{dx}\right)_{x=x_*^{(1)}}\right| &= p, \\ \left|\left(\frac{df(x)}{dx}\right)_{x=x_*^{(2)}}\right| &= 2 - p, \end{aligned} \tag{15.59}$$

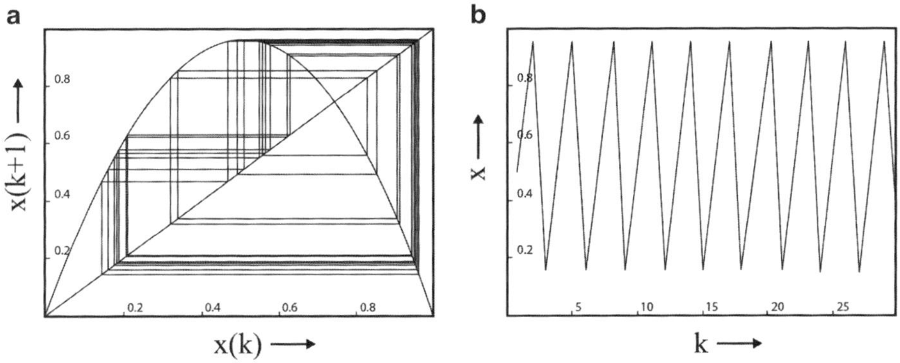

Fig. 15.7 Web chart for logistics mapping and for $p = 3.83$ (**a**) and the periodic course (**b**) corresponding to a closed curve in figure (**a**)

and the first solution is stable for $|p| < 1$, while the other one for $|2 - p| < 1$. Now, let us consider a few numerical examples of the logistic mapping. Figure 15.7 is an example of "web chart" for $p = 3.83$. As is clear from the preceding discussion, this parameter value both fixed mapping points are unstable. In the $x(k), x(k+1)$ coordinate system drawn were the function $f(x)$ and the diagonal. They are used for a simple determining of the next mapping points after the successive iterations. As you can see from the figure, the initial condition $x(0) = 0.3$ trajectory mapping tends to periodic orbit.

If we consider the mapping described of the function $x(k+3) = f^3(x(k))$, and on the vertical axis we take every third iteration point, that is $x(k+3)$, then we get web chart shown in Fig. 15.8. As can be seen from this figure, depending on the initial conditions of the trajectories, they are attracted by one of the three points at which the curve $f(x)$ is tangent to the diagonal of the pictures frame.

For every fifth iteration $x(k+5) = f^5(x(k))$, the chart of the curve $f(x)$ is more complicated (Fig. 15.9). Trajectory relatively quickly reaches a stable periodic orbit.

Now let us examine the behaviour of this mapping when changing parameter $p \in [2, 4]$. According to earlier solutions, a fixed mapping point equal to zero is unstable in the considered range of parameter changes. The second fixed point is stable when $p \in [2, 3]$. For the point $p = 3$ doubling period bifurcation occurs. Previously stable point now becomes unstable.

However, there is a new stable solution in the range for a period 4. When changing the parameter again, its stability is lost, and there is an orbit with a period of $2^3 = 8$, and so on, until it reaches the orbits with period 2^k. When $k \to \infty$ parameter p reaches a limit equal to $p_g = 3.5699$. It turns out that in the $p \in [p_g, 4]$ a similar bifurcation cascade can be observed for a period orbits 3 and 4, that is 3^k and 4^k, where $k = 1, 2, 3, \ldots$. They are called periodicity windows that correspond to the specific compartments of parameter p. That means that chaos is observed for some nowhere dense subsets of parameter p that have positive value.

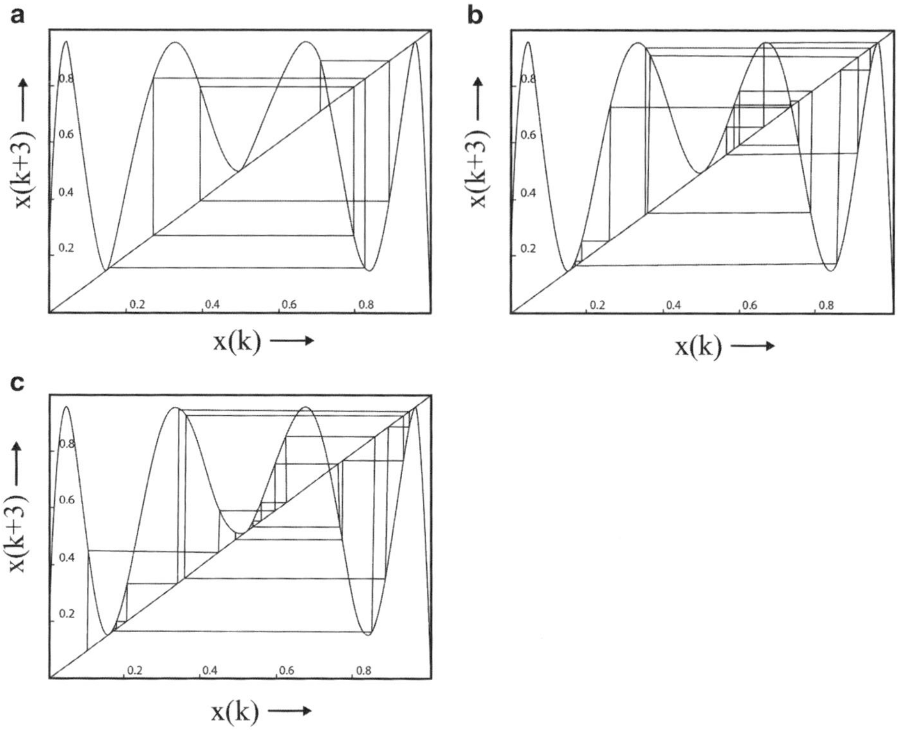

Fig. 15.8 Web chart for logistic mapping and for $p = 3.83$ in the coordinate system $x(k)$ and $x(k+3)$ for different initial conditions: (**a**) $x(0) = 0.7$; (**b**) $x(0) = 0.35$; (**c**) $x(0) = 0.1$

Fig. 15.9 Web chart for logistic mapping and for every fifth iteration

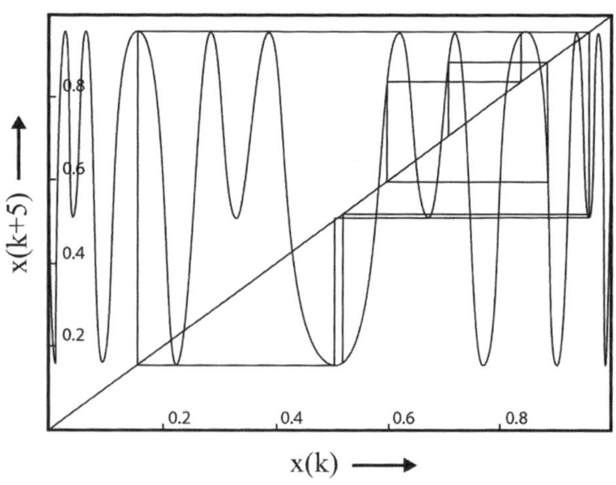

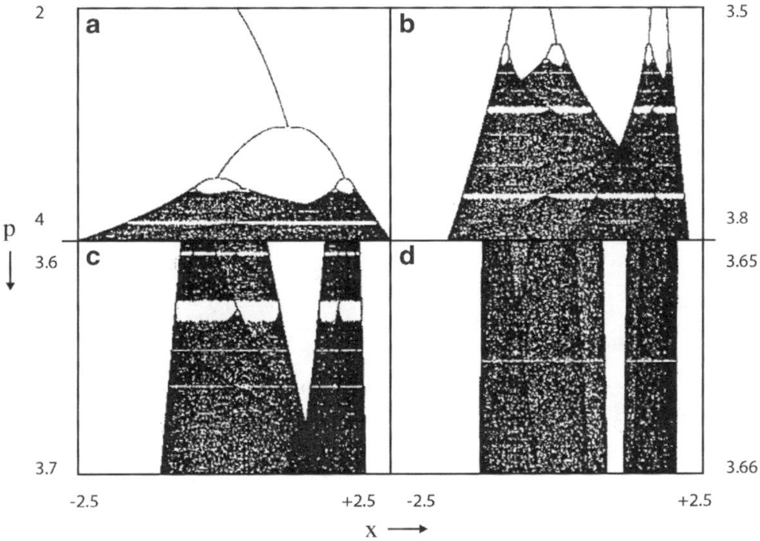

Fig. 15.10 Bifurcation chart of the logistic mapping for different ranges of the changes in the control parameter p: (**a**) $p = [2, 4]$; (**b**) $p \in [3.5, 3.8]$; (**c**) $p \in [3.6, 3.7]$; (**d**) $p \in [3.56, 3.66]$

Figure 15.10 shows the so-called bifurcation chart and the following drawings were created as a result of the enlargement of the previous one for a specific range of parameter p. Bifurcation cascade doubling period, chaotic movements and windows of periodicity are shown clearly.

Figure 15.11 shows the logistic mapping for $p = 3.7$ for investigations of the chaotic mapping dynamics process. After about four million iterations, and as you can see from the chart of chaotic attractor is a part of the segment [0,1] and is defined by the projection of the parabola marked with a thick line onto the horizontal axis. Further points obtained by iteration are arranged along this stretch in a completely unpredictable (chaotic) way.

Autocorrelation functions $A(l)$ for $p = 4$ is determined by the formula (15.29). According to (15.28) for almost all initial conditions we get

$$\langle x_k \rangle = \frac{1}{2}, \tag{15.60}$$

then

$$A(l) = \begin{cases} 1/8 \text{ for } l = 0 \\ 0 \quad \text{for } l \neq 0 \end{cases}, \tag{15.61}$$

for almost all initial conditions.

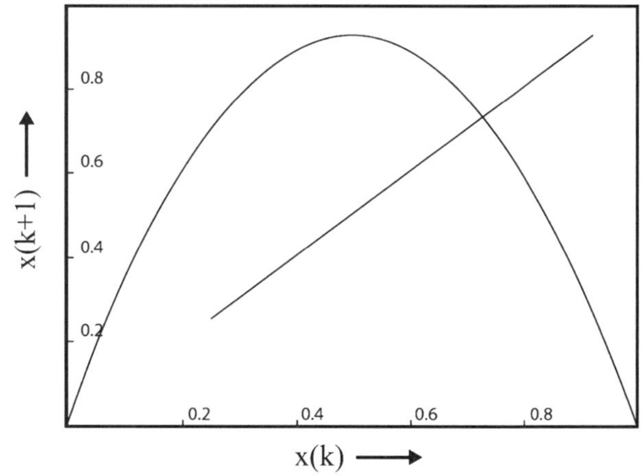

Fig. 15.11 Chaotic logistic map for $p = 3.7$

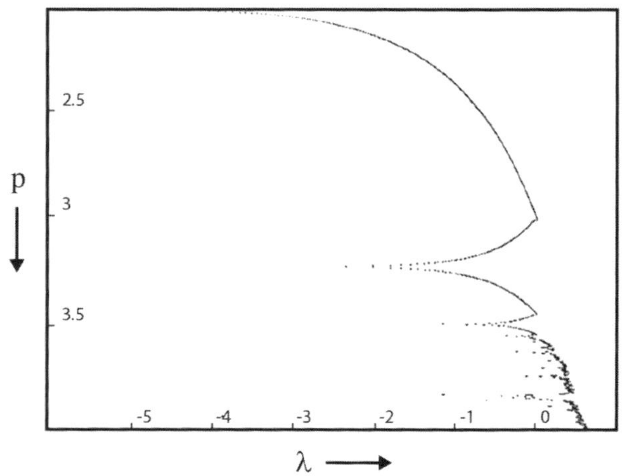

Fig. 15.12 λ exponent changes accompanying changes in the parameter p in the range

Since the analytical solution lo the logistic mapping is important for $p = 4$ we compute the associated Lyapunov exponent

$$\lambda = \lim_{k \to \infty} \frac{1}{k} \log \left| \frac{x(k)}{x(0)} \right| = \log 2. \tag{15.62}$$

For this parameter, the exponent value $\lambda_n = 0.693144$ is calculated numerically for 206 000 iterations, what yields to the error value $\delta = |\lambda - \lambda_n| = 0.00000318$. Lyapunov exponents' values for $p \in [2, 4]$ are shown in Fig. 15.12.

Analytical form of solutions for $p = 4$ (it is worth noting that for the value of the parameter number $1/2$ maps into 1, while in the following iteration 1 becomes zero) allows for the transformation

$$y = \frac{2}{\pi} \text{arc sin} \sqrt{x}, \tag{15.63}$$

reducing logistic map to the Bernoulli map (15.47).

15.6.4 Map of a Circle into a Circle

This is another one-dimensional representation, which we will analyse. Mapping of a circle into a circle is described by the equation

$$\phi(k+1) = F(\phi(k)) \equiv \phi(k) + R_1 + R_2 \sin\phi(k) \mod 2\pi. \tag{15.64}$$

This mapping depends on two parameters R_1 and R_2 and may represent the nonlinear oscillator phase transition, wherein the parameter value for R_1 describes two frequencies ratio, and R_2 is the nonlinear enhancement effects coefficient [213, 221]. This simple representation shows many interesting features of the nonlinear dynamics, namely the periodic, quasi-periodic and chaotic dynamics.

It is worth to point out some basic properties of (15.64) mapping [213]:

(a) The function F has the characteristic

$$F(\phi + 2\pi) = \phi + 2\pi + R_1 + R_2 \sin\phi = 2\pi + F(\phi). \tag{15.65}$$

(b) For $|R_2| < 1$ a $F(\phi)$ map exists and is differentiable (a diffeomorphism).
(c) For $R_2 = -1$ reversed mapping F^{-1} becomes non-differentiable, while for $|R_2| > 1$ it is ambiguous.

Figure 15.13 presents the $F(\phi)$ map for $R_1 = 0.4$ and different values of R_2, what confirms the previously mentioned property. For all iteration the value characterizing the average displacement by an angle ϕ is defined by the formula

$$w = 2\pi w^* = \lim_{N\to\infty} \frac{F^N(\phi_0) - \phi_0}{N}. \tag{15.66}$$

The average period shift is defined as $T_w = 2\pi/w$, where w is the angular frequency of rotation (winding number) while, the rotation frequency as $w^* = 1/w$. These relations are similar to the concept of the frequency of a periodic circular orbit that is not lying on the torus, and the frequency. It turns out [213, 221] that for $R_2 < 1$ the limit of the formula (15.66) always exists, but can be represented either as rational or irrational number. If it is a rational number, then range of the parameters R_1, R_2, for which $w = p/q$, $p, q \in N$ with respect to the mapping (15.64) is called tongues.

Consider now in more detail the dynamics of trajectories lying on a two-dimensional torus (Fig. 15.14).

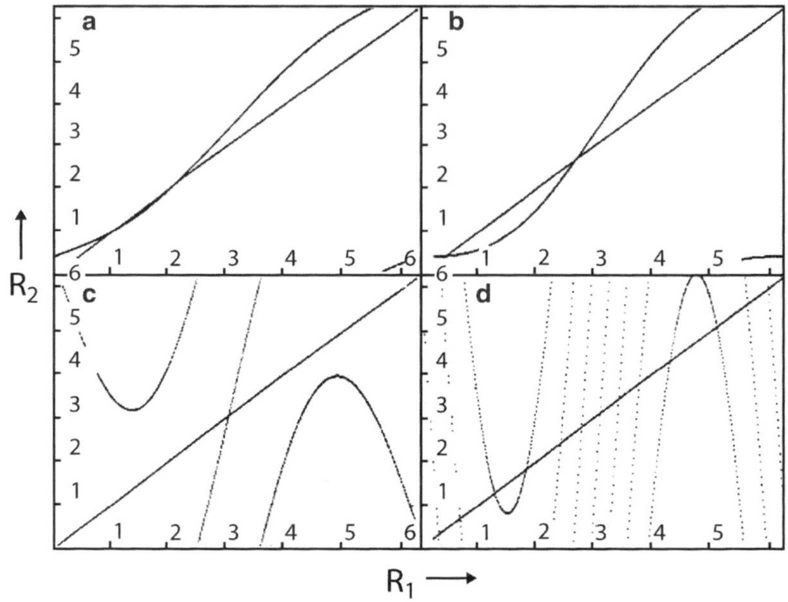

Fig. 15.13 The map (15.64) for $R_1 = 0.4$ and different values of R_2: (**a**) -0.5; (**b**) -1; (**c**) -5; (**d**) -20

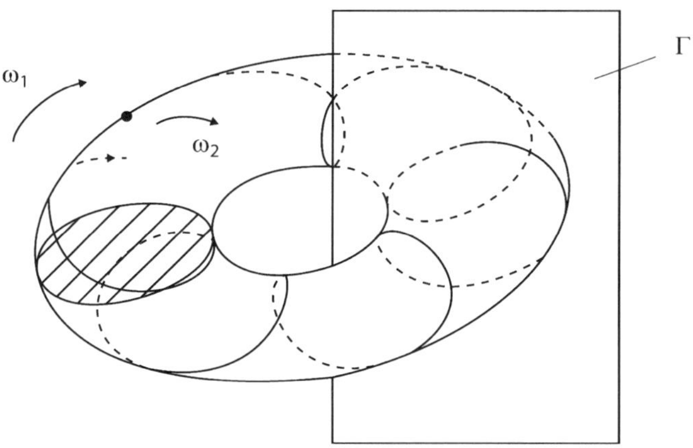

Fig. 15.14 Poincaré map—cross-section of the torus by plane Γ

Let, for example,

$$w = \frac{\omega_1}{\omega_2} = \frac{p}{q} = \frac{3}{5}. \tag{15.67}$$

where ω_1 and ω_2 are frequencies marked in Fig. 15.14. Let us consider the journey of the point starting from the plane Γ. This point will cross the plane again after the

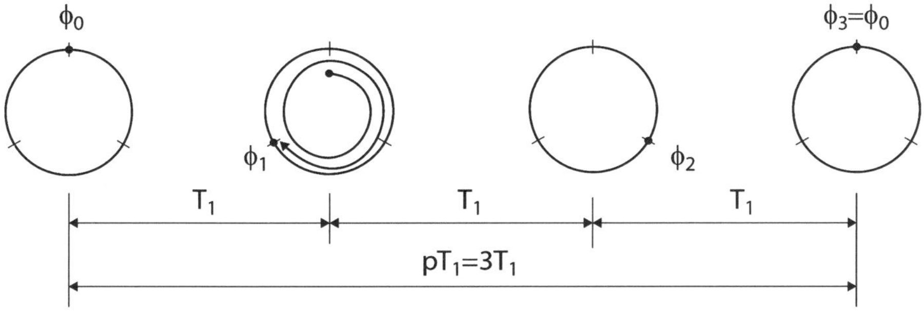

Fig. 15.15 Point motion within the plane Γ observed in T_1 intervals. Between successive positions performs point $5/3$ turn, what means that in time $3T_1$ point will do five turns and then the movement will be repeated

time $T_1 = 2\pi/\omega_1$. Figure 15.15 shows a picture of stroboscopic photos distant from each other in the time by T_1.

Let us consider the mapping

$$\phi_{n+1} = \phi_n + w \cdot 2\pi. \tag{15.68}$$

For $w = \frac{3}{5}$ we obtain successively

$$\begin{aligned} \phi_1 &= \phi_0 + \frac{3}{5}2\pi, \\ \phi_2 &= \phi_1 + \frac{3}{5}2\pi = \phi_0 + 2 \cdot \frac{3}{5}2\pi, \\ &\vdots \\ \phi_5 &= \phi_0 + 3 \cdot 2\pi, \end{aligned} \tag{15.69}$$

which means that $\phi_5 = \phi_0 \bmod 2\pi$, and in the general case

$$\phi_q = \phi_0 + p \cdot 2\pi. \tag{15.70}$$

For N-turns we obtain the definition of the circular rotation defined by (15.66).

The plane Γ we get three fixed points, while in the mapping plane (15.68) there are five fixed points (in the plane perpendicular to the Γ there are also five fixed points). According to (15.69) in the plane (ϕ_{n-1}, ϕ_n) we get five fixed points $\phi_1^*, \phi_2^*, \ldots \phi_5^*$. If the point ϕ_i^* belongs to on the q-periodic orbit generated by the mapping (15.64), then according to (15.70) we have

$$F^q_{R_1,R_2}\left(\phi_i^*\right) = \phi_i^* + 2\pi p \equiv \phi_i \,(\bmod 2\pi), \tag{15.71}$$

where $i = 1, 2, \ldots, q$, and F_{R_1,R_2} means that this function is dependent on the parameters R_1 and R_2. It also means that starting from the point ϕ_i^* we are coming back to it through q iterations, or after moving by the angle $2\pi p$. On this occasion, it is good to come back to the interpretation related to Fig. 15.15. At the same mapping point we will be back after q rotations (with a frequency ω_2) or after moving by the angle of $2\pi p$. According to (15.65), we have

$$F\left(\phi_i^*\right) = \phi_i^* + R_1 + R_2 \sin \phi_i^*, \tag{15.72}$$

and we calculate

$$\frac{dF\left(\phi_i^*\right)}{d\phi_i^*} = 1 + R_2 \cos \phi_i^*. \tag{15.73}$$

Complete orbit consisting of points, ϕ_i^*, $i = 1, 2, \ldots, q$ is stable if each of the points ϕ_i^* is stable, that is:

$$\prod_{i=1}^{q} |1 + R_2 \cos \phi_i^*| < 1. \tag{15.74}$$

We will consider now the simplest case where $w = 1$, so $p = q = 1$. According to (15.72), we obtain

$$\phi_0^* = \phi_0^* + R_1 + R_2 \sin \phi_0^*$$

and

$$R_1 = -R_2 \sin \phi_0^*.$$

However, from the condition (15.74) we have

$$|1 + R_2 \cos \phi_0^*| < 1. \tag{15.75}$$

For $R_2 < 1$ the loss of the stability limits are reached when the

$$R_2 \cos \phi_0^* = 0, \tag{15.76}$$

that is for $\phi_0^* = \pm(\pi/2)$. Therefore, the width of the first tongue is

$$R_1 = -R_2 \sin \phi_0^* = \pm R_2,$$

what is confirmed by the observation of the area in the vicinity of 0 and 2π in Fig. 15.15.

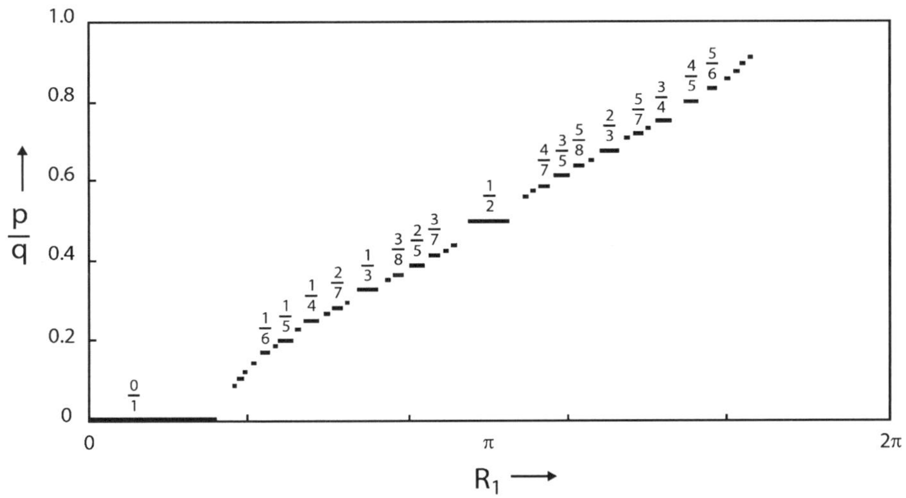

Fig. 15.16 The structure of a circle within a circle mapping for $R_2 = -1$ (the so-called devil's stairs)

15.6.5 Devil's Stairs, Farey Tree and Fibonacci Numbers

In [39, 40, 129] work a similar analysis was preformed for a previously considered circle within a circle mapping for different values of the rotation number $w = p/q$ and for $R_2 < 1$. It turned out that for each rational value of w and for each of the R_2 in considered interval the q-periodic orbit is stable over some a range of parameter $\Delta R_1(w, R_2)$. However, for $|R_2| = -1$, it turned out that the sum of all those intervals for all rational numbers is 2π, as shown in Fig. 15.16 and the graph is called the devil's stairs.

The other characteristics can be observed in the structure shown in Fig. 15.16: (a) the length of the intervals corresponding to the values of p/q increases with the decrease of q; (b) if we take two numbers $w_1 = p_1/q_1$ and $w_2 = p_2/q_2$, then there is a rational number $w = (p_1 + p_2)/(q_1 + q_2)$ between them. For example when taking $w_1 = \frac{1}{3}$ and $w_2 = \frac{2}{5}$, we receive $w = \frac{3}{8}$ and it is a value, which corresponds to the length of the interval shown in Fig. 15.16 between w_1 and w_2, and at the same time it is a rational number with the smallest denominator lying between w_1 and w_2. This construction allows for the creation of so-called. Farey tree, as shown in Fig. 15.17.

Physical interpretation of the results from Fig. 15.16 is as follows: there is such a systems synchronization that changes in the parameter R_1 (in a real system frequencies ω_1 and ω_2) in a certain range do not lead to changes in the parameters p and q, and thus to the change of the frequency (period) of the periodic orbit. Now we will discuss the possibility of approximation of quasi-periodic dynamics by periodic dynamics, which is connected with the possibility of approximation of

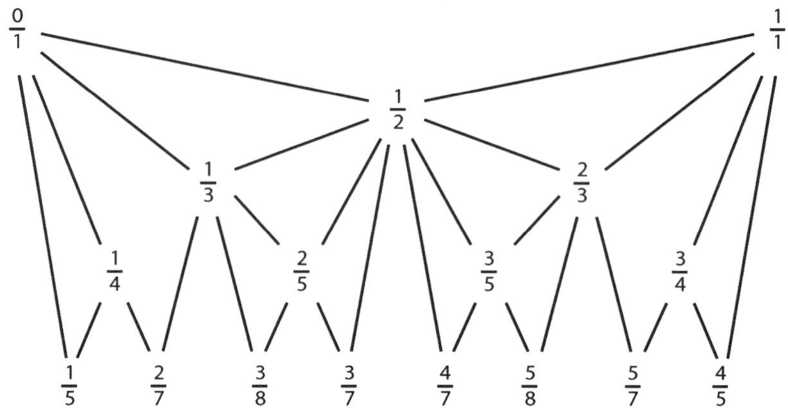

Fig. 15.17 Farey Tree enabling arrangement of the numbers in the interval [0,1]

irrational numbers by a sequence of rational numbers [135]. In general, any real number "a" may be represented by a continued fraction $[a_0, a_1, a_2, \ldots]$ of the form

$$a = a_0 + \cfrac{1}{a_1 + \cfrac{1}{a_2 + \cfrac{1}{a_3 + \cfrac{1}{a_4 + \cdots}}}}, \tag{15.77}$$

where a_i belong to the set of natural numbers.

For numbers that are rational continued fraction is finite, and for irrational numbers it is infinite. In practice, the appearance of a large value a_i in relation (15.77) results in a rapid convergence of a fraction of that number. Slowest convergence fraction is characterized by the number $w = (\sqrt{5} - 1)/2$, which is a number corresponding to the golden division. It corresponds to the division of the section of length L into two parts l and $L - l$ such that $w = l/L = (L - l)/l$. This number plays an important role in the chaotic dynamics and fractal theory, and its continued fraction is an infinite set consisting only of the 1 with the exception for $a_0 = 1$.

For $0 < w < 1$ the number of "w" can be approximated with continued fraction

$$w \cong \frac{r_k}{s_k} = [a_1, a_2, \ldots a_k], \tag{15.78}$$

where r_k and s_k are natural numbers calculated from the formulas

$$r_k = a_k r_{k-1} + r_{k-2}, \quad k = 2, 3, \ldots \tag{15.79}$$

$$s_k = a_k s_{k-1} + s_{k-2}, \quad k = 2, 3, \ldots \tag{15.80}$$

where $r_1 = 1$, $r_0 = 0$, $s_0 = 1$, $s_1 = a_1$.

We will consider as an example number $1/\sqrt{2}$, for which $w \cong 0.7071068\ldots$. Successively computing $a_1 = \mathrm{INT}(1/w) = \mathrm{INT}(1.4142\ldots) = 1$ (here we take the integer part of the obtained number). Then we calculate

$$a_2 = \mathrm{INT}\left[\frac{w}{1 - wa_1}\right] = \mathrm{INT}\,(2.414215365) = 2, \tag{15.81}$$

and a_3 as

$$a_3 = \mathrm{INT}\left[\frac{1 - wa_1}{w\,(1 + a_1a_2) - a_2}\right] = \mathrm{INT}\,(2.414211) = 2, \tag{15.82}$$

$$a_4 = \mathrm{INT}\left[\frac{w\,(1 + a_1a_2) - a_2}{1 + a_2a_3 - w\,[a_1\,(1 + a_2a_3) + a_3]}\right] = \mathrm{INT}\,(2.414213489) = 2 \tag{15.83}$$

and thus we can continue this process of calculations. Using the formulas (15.79) and (15.80) we get $r_2 = 2$, $s_2 = 3$, $r_3 = 5$, $s_3 = 7, \ldots$, $r_6 = 29$, $s_6 = 41$, $r_7 = 70$, $s_7 = 99$. Ending calculation on the seventh word it is noticeable that $1/\sqrt{2}$ can be approximated by the value

$$w \cong \frac{r_7}{s_7} = 0.707070707, \tag{15.84}$$

this gives an error about 0.000036. In general, the correct is inequality

$$\left|w - \frac{r_k}{s_k}\right| \leq \frac{1}{s_k s_{k-1}}. \tag{15.85}$$

For the golden ratio have $a_k = 1$

$$s_k = s_{k-1} + s_{k-2}, \quad k = 2, 3, 4, \ldots \tag{15.86}$$

where: $r_1 = 1$, $r_0 = 0$, $s_0 = 1$, $s_1 = 1$. Then we calculate the sequence

$$\begin{aligned} r_2 = r_1 = 1; \quad r_3 = r_2 + r_1 = 2; \quad r_4 = r_3 + r_2 = 3; \quad r_5 = 5; \ldots \\ s_2 = 2; \quad s_3 = s_2 + s_1 = 3; \quad s_4 = s_3 + s_2 = 5; \quad s_5 = 8 \ldots \end{aligned} \tag{15.87}$$

and the obtained results can be generalized as

$$r_k = r_{k-1} + r_{k-2}, \tag{15.88}$$

$$s_k = r_{k+1} = r_k + r_{k-1}. \tag{15.89}$$

The next sequence of numbers approximating the terms w_* is defined as

$$w_k = \frac{r_k}{s_k} = \frac{r_k}{r_{k+1}} = \frac{r_k}{r_k + r_{k-1}} = \frac{1}{1 + \frac{r_{k-1}}{r_k}}, \tag{15.90}$$

which are similar to the values

$$w_* = \lim_{k \to \infty} w_k, \tag{15.91}$$

with the strings (15.79) and (15.80) being the Fibonacci sequences. According to (15.90) and (15.91) we get the equation

$$w_* = \frac{1}{1 + w_*}, \tag{15.92}$$

One of its elements is actually the $\left(\sqrt{5} - 1\right)/2$.

15.6.6 Hénon Map

With such a map we have met already in the previous section in the analysis of the pendulum flat motion that was treated with the time-varying torque.

Another two-dimensional representation, which we will devote more attention, is the Hénon map [120], which can be regarded as an extension of the earlier discussed logistic map. It is governed by the equation

$$\begin{aligned} x_{n+1} &= r - a x_n^2 + y_n, \\ y_{n+1} &= b x_n, \end{aligned} \tag{15.93}$$

or

$$(x, y) \to \left(r - a x^2 + y, bx\right), \tag{15.94}$$

where a, b and r serve as a bifurcation parameters. Mappings Jacobian (15.93) is

$$\det \begin{vmatrix} -2a x_n & 1 \\ b & 0 \end{vmatrix} = -b, \tag{15.95}$$

and therefore the system is dissipative for $|b| < 1$. It turns out that for $0 < b < 1$, $r = 1$ and $a > 0$, the mapping has two fixed points defined by the equation

$$x_{1,2} = \frac{-(1-b) \pm \sqrt{(1-b)^2 + 4a}}{2a}, \quad y_{1,2} = -b x_{1,2}. \tag{15.96}$$

If $a > (1-b)^2/4$ both points are real numbers and one of them is always unstable, while the other is unstable for $a > 0.75(1-b)^2$. This mapping is the basic for considerations of many interesting elements in nonlinear dynamics.

1. Let $r = 2.1$, $a = 1$, $b = -0.3$. In Fig. 15.18 on the plane (x, y) shown is a Hénon strange chaotic attractor. Furthermore, in the following figures from "a" to "d" are marked with crossed respectively periods 1, 2, 5 and 10 periodic orbits. The method of searching for such orbits is based on the use of Newton's method or its variants [184,219]. If x_* is a fixed point of the mapping $F(x, p)$ dependent on the parameter p, then satisfied is equation:

$$x_* = F(x_*, p). \tag{15.97}$$

Let the point x be placed near the point x_*. Introduce the matrix N

$$N = D_x F(x, p), \tag{15.98}$$

which elements are the partial derivatives with respect to x. Performing linearization around the point x we get

$$(x, p) + N dx = x + dx, \tag{15.99}$$

where we have

$$dx = (N - I)^{-1}(x - F(x, p)), \tag{15.100}$$

and the I above is the identity matrix. The expression $x - F(x, p) = E$ express an error of calculation, which for $x = x_*$ equals zero (this is the exact value). It turns out that Newton's method does not always make it possible to reduce the error in the next step of the calculation. Modified Newton's method allows you to choose such increase dx that the convergence is maintained.

In the case of periodic orbits marked with crosses in Fig. 15.18 starting points for the modified Newton's method were selected at random. Two points were found for the period one (a), four points with period two (b), three different orbits with period five (c), and fifteen different orbits with period ten. In the last case, as starting points for the modified Newton method 9031 random points were chosen. It is worth noting that many of the found periodic orbits do not belong to chaotic attractor.
2. The next example involves a bifurcation curve. On the vertical axis we put the parameter b, while on the horizontal axis x. In fact, it is the mapping of the family of attractors depending on the parameter b in the plane (b, x), $0.1 \leq b \leq 0.3$ (see Fig. 15.19). For $r = 1.3$ the chaotic dynamic of mapping is interrupted windows of periodicity for some values of b (there are infinitely many of them), however when reducing of the r and $b \approx 0.26$ bifurcation occurs, chaotic motion

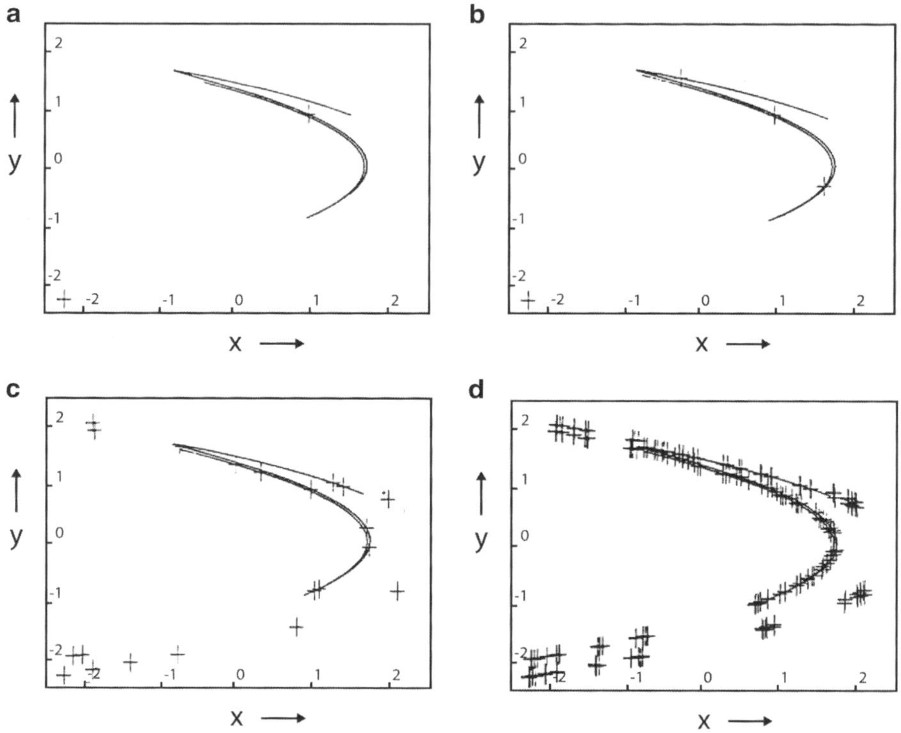

Fig. 15.18 Hénon strange chaotic attractor and periodic points of Hénon mapping (marked with *crosses*) with the following periods: (**a**) 1, (**b**) 1, 2, (**c**) 1, 2, 5, (**d**) 1, 2, 5, 10

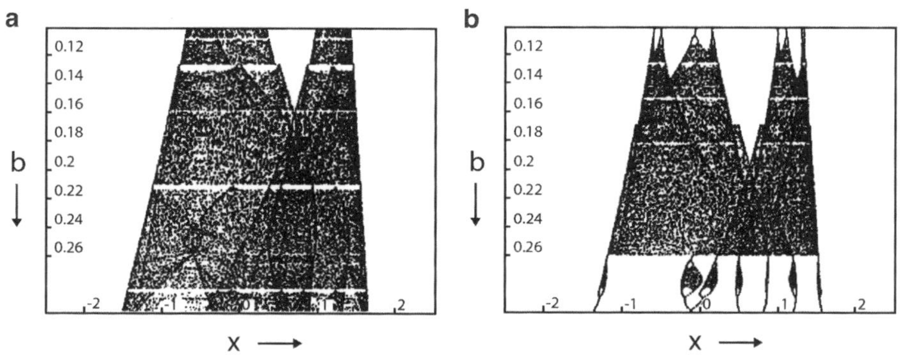

Fig. 15.19 Bifurcation curve of the Hénon map and the parameters (**a**) $r = 1.3$; (**b**) $r = 1.25$

disappears and periodic motion appears. Then, each of the "branches" doubles and with further reduction of b formed are the so-called bubbles of chaotic motion (Fig. 15.19b).

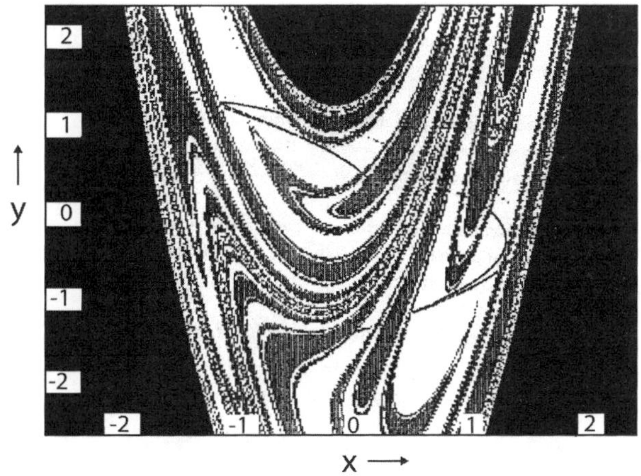

Fig. 15.20 Pools of attraction for the Hénon map and for $a = 1, r = 1, b = 0.48$

3. Basing on Hénon map we will discuss the concept of attractor attraction pools. By the attractor attraction pool will be defined the set of all initial conditions in phase space, which will be "attracted" by the attractor, that is after "start" of each of these initial conditions trajectories over time will be on the attractor. These pools of attraction for the Hénon mapping are shown in Fig. 15.20. For a set of parameters, as shown, there are three different attractors. One of them is ∞ (black area), stable periodic orbit with a period of eight (a gray area), and a strange chaotic attractor, in the figure consisting of two parts, which "attracts" the initial conditions from the white area.
4. Now we will discuss the puzzling similarities between the Hénon attractor and the unstable variety of the fixed point lying within the attractor. Through a stable variety of the mapping fixed point we understand a set of points leading up to this point with the number of iterations tending to infinity defining the Hénon map. However, the concept of unstable variety of the mapping fixed point we mean a set of points which are attracted by the iterations with the opposite direction (or repelled by applying the initial iterations). Figure 15.21a shows the Hénon chaotic attractor, while Fig. 15.21b shows the set of points attracted by the reverse iteration by an unstable fixed point with coordinates $(0.855, 0.898)$. It is striking similarity here between the two sets. It is believed that these sets are identical, but has not been proofed as accurate (see for example [219]).
5. Now we will turn our attention to the similarity between the Hénon attractor attraction pool and attractor which is infinity (∞) (this will be a set of points that the iterations tending to infinity "escape" to infinity (Fig. 15.22a)), and a stable fixed point attraction pool within the Hénon attractor (Fig. 15.22b). The calculations were performed assuming the parameters: $a = 1$, $b = -0.32$, $r = 2.10$. You can see that approximately the pool of attraction of a stable point of the Hénon attractor fits in the Hénon chaotic attractor attraction pool.
6. There is also the possibility of the chaotic trajectory contained in a limited area in the phased space, however all other trajectories situated in the neighbourhood "escape" to infinity, so they are not bounded. Such invariant set will be called

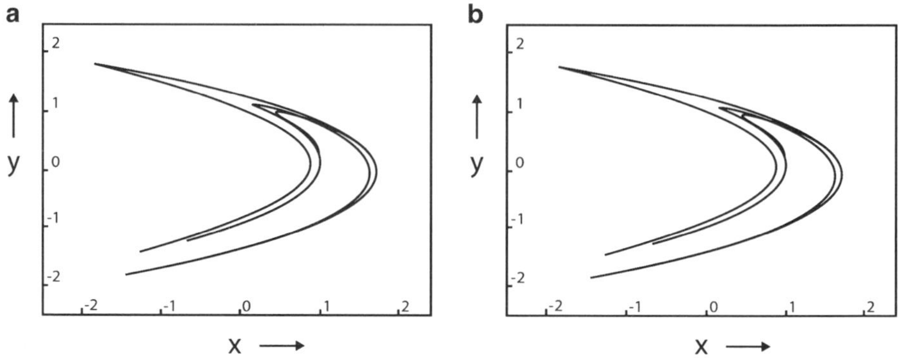

Fig. 15.21 Hénon attractor for $a = 1$, $r = 1.38$ and $b = 0.32$ (**a**) and unstable variety of the fixed point with coordinates $(x, y) = (0.855, 0.898)$ (**b**)

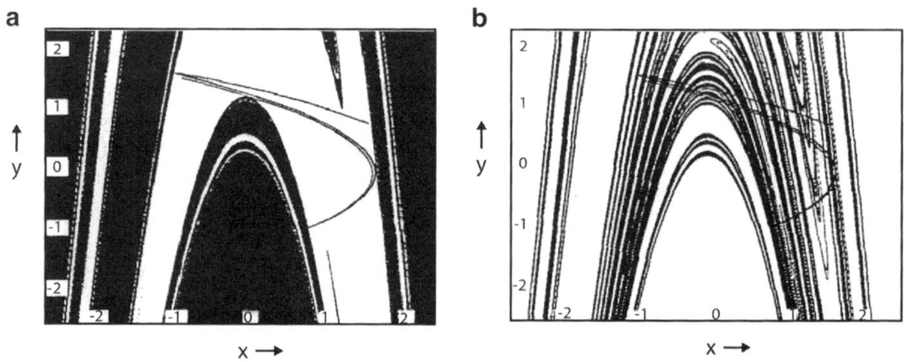

Fig. 15.22 Attraction pool for infinity (∞), marked with a *black* and the Hénon attractor attraction pool (**a**) and attracting pool of stable fixed point lying within the Hénon map, which is approximately (0.907, 0.966) (**b**). In both figures (**a**) and (**b**) marked is also Hénon attractor

Hénon mapping chaotic saddle. This invariant and compact set is unstable, so almost all trajectories of the neighbourhood will be distancing themselves, and in the considered case, they will "escape" to infinity.

Figure 15.23 shows an example of an unstable set that is invariant and compact, on which lays the chaotic trajectory. Calculations were performed for $a = 1$, $b = 0.4$ and $r = 4$.

7. From Fig. 15.23 we can conclude that for certain Hénon map parameters, there are two attractors which are attracting sets of the initial conditions. The first one is a pool of initial conditions attracted by Hénon chaotic attractor, and the second is a pool of initial conditions that in time are "fleeing" to infinity, or are attracted by infinity. There are also points belonging to the boundaries of the two pools, and the initial conditions are not attracted by any of these attractors [184]. This limited trajectory is shown in Fig. 15.24 for $a = 1$, $b = -0.3$, $r = 2.12$.

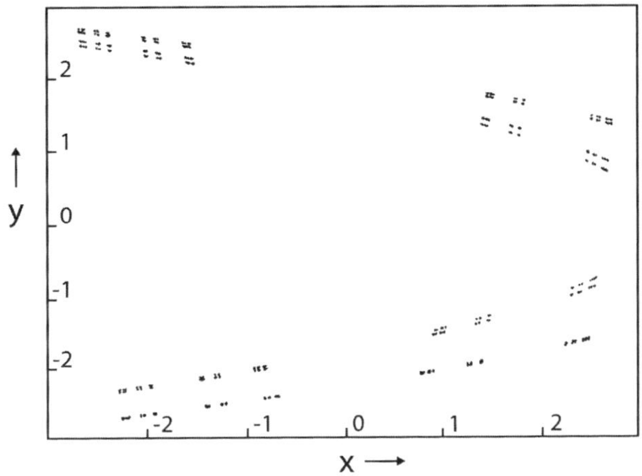

Fig. 15.23 Unstable invariant set containing a chaotic trajectory for Hénon map

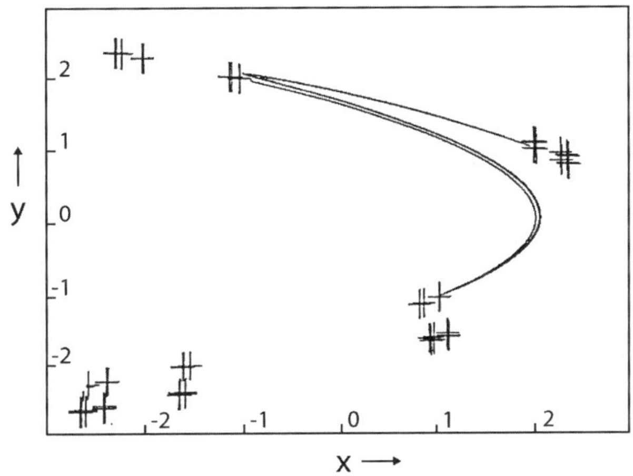

Fig. 15.24 Border trajectory (marked with *crosses*) belonging to the boundaries of the Hénon attractor attraction pools (marked with *points*) and attractor lying in an "infinity"

8. In this example, basing on the Hénon map illustrated is a way leading to chaos by doubling of the period. The calculation results are shown in Fig. 15.25a–c.

   In the first one you can see the way that leads to chaotic motion by successive doubling period of vibration. Basing on Fig. 15.25a, b can be calculated the relations between lengths of the subsequent curves in between the points of the bifurcation, which are: $d_2/d_4 = 4.33$, $d_4/d_8 = 4.42$, $d_8/d_{16} = 4.54$ and apparently they tend to the Feigenbaum constant (approximately 4.67). Figure 15.25c shows the graph of changes in Lyapunov exponent λ in relation to Fig. 15.23b, that is for the same range of changes in parameter r. Where it is positive, there is chaos.

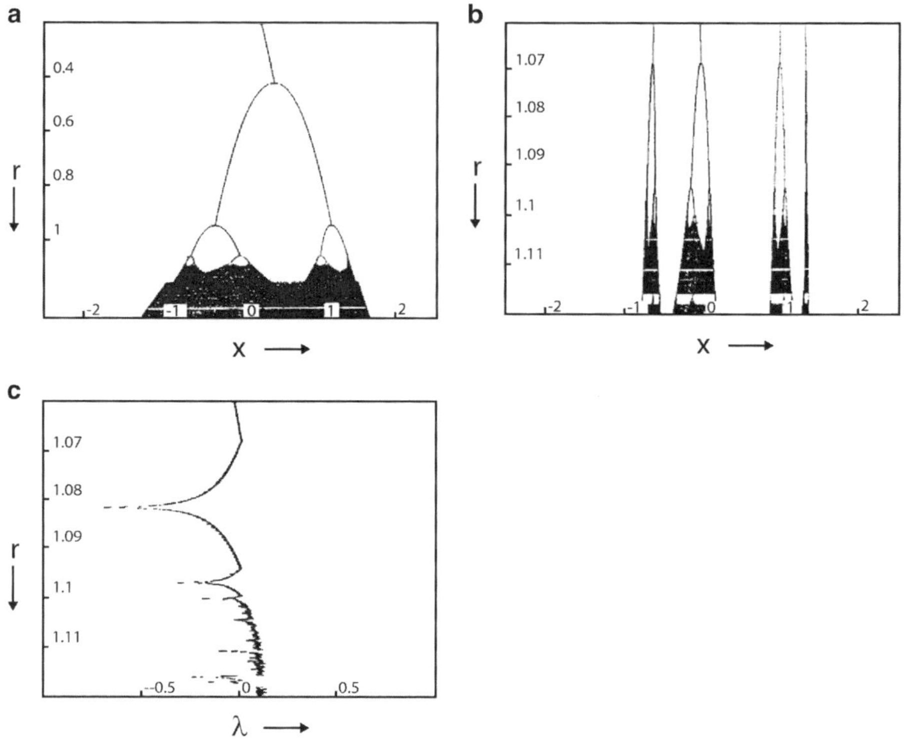

Fig. 15.25 Bifurcation graph illustrating the period doubling cascade leading to chaos (**a**) of the graph window (**a**) for $1.06 \leq r \leq 1.12$ (**b**) and the Lyapunov exponent corresponding (**b**) to the figure (**c**)

15.6.7 Ikeda Map

Ikeda map is described by the equation

$$z \to \rho + c_2 z \exp\left[i\left(c_1 - c_3\left(1 + z^2\right)^{-1}\right)\right], \quad z = x + iy, x, y \in R, i^2 = -1. \tag{15.101}$$

Figure 15.26 shows three successive iterations of an ellipse located in the upper right-hand corners of the pictures for the following parameters: $\rho = 0.5$, $c_1 = 0.4$, $c_2 = 0.9$, $c_3 = 6$. Chaotic dynamic is more visible with each of the iterations.

Let us now consider the dynamics of Ikeda map (15.101) for the same parameters as before, but now iterated ellipse is shifted to the left compared to the one in the previous case (Fig. 15.27). As can be seen from this figure chaotic dynamics is revealed here much earlier.

Figure 15.28 presented is only the first iteration of the ellipse, but in this case it is lying along the $y = 0.5$ and for different values of the control parameter ρ. Increase of parameter ρ from 0.5 to 1.0 affects the deepening of the dynamics of Ikeda chaotic mappings.

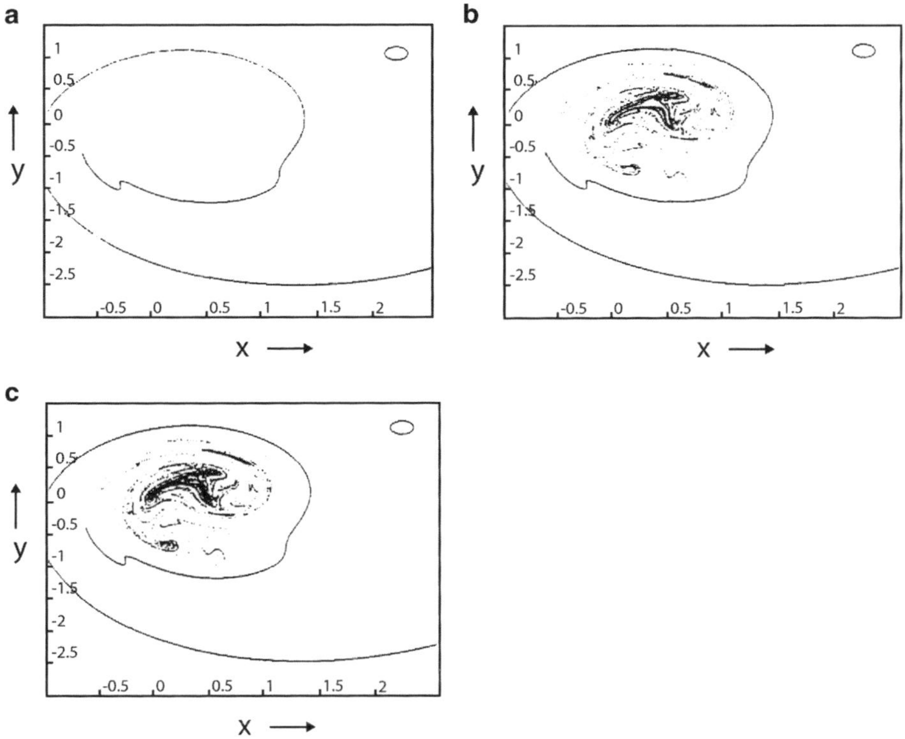

Fig. 15.26 The first (**a**), second (**b**) and third (**c**) iteration of the ellipse shown in the upper right corners of the drawing for mapping Ikeda

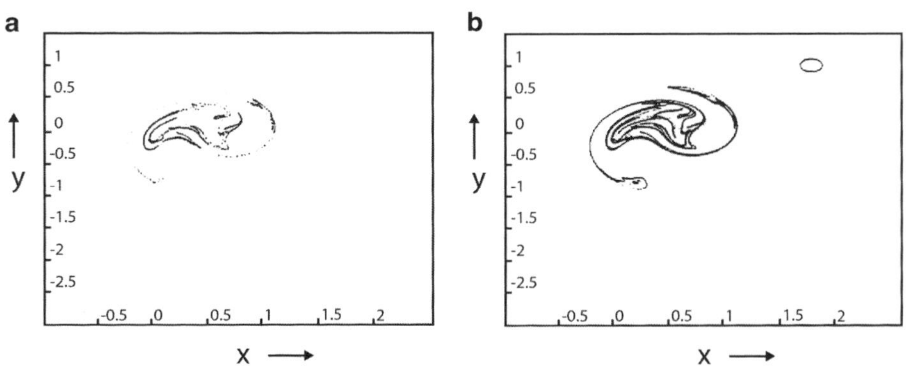

Fig. 15.27 The first (**a**) and second (**b**) iteration of the ellipse for the Ikeda map

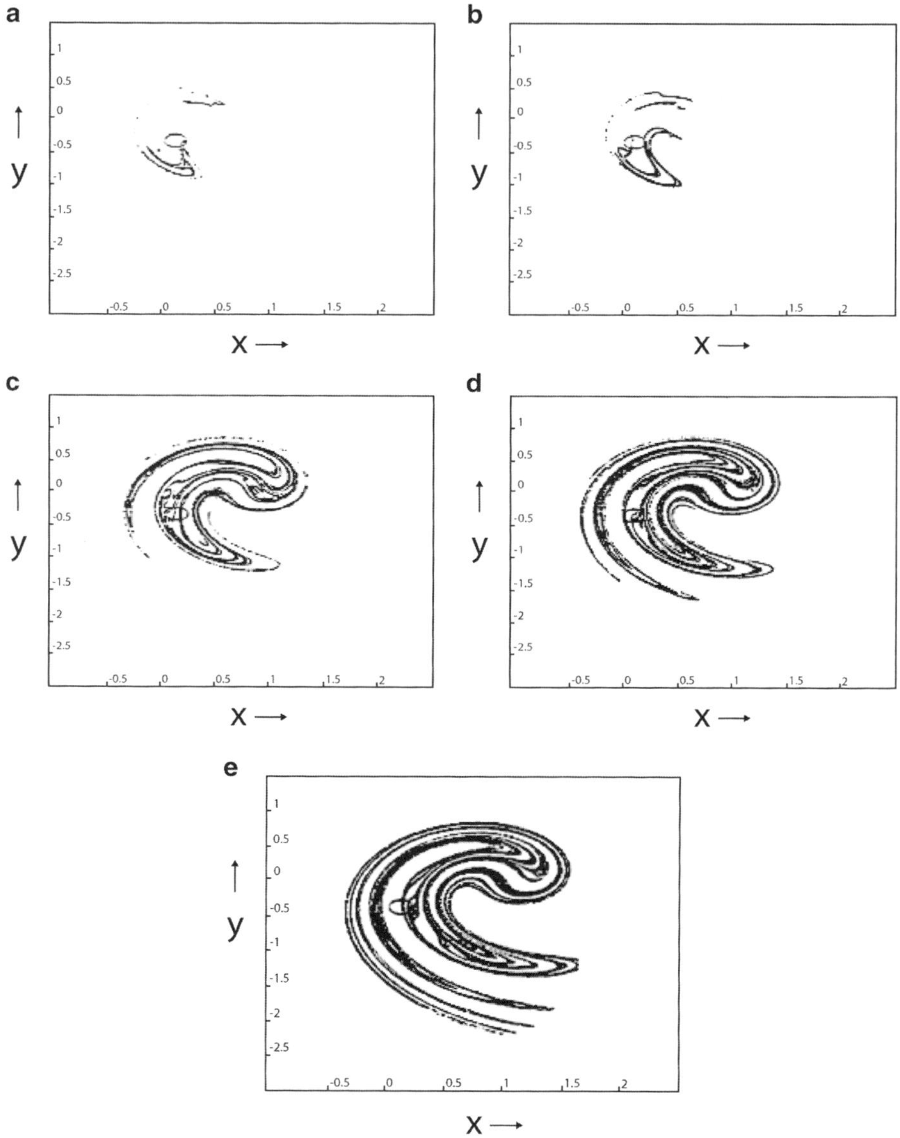

Fig. 15.28 The first iteration of the ellipse for different values of the parameter ρ: (**a**) 0.6; (**b**) 0.7; (**c**) 0.8; (**d**) 0.9; (**e**) 1

15.7 Modelling of Nonlinear Ordinary Differential Equations

15.7.1 Introduction

In order to determine time evolution of the natural processes we should have the knowledge of the functional dependencies between the function that describes this

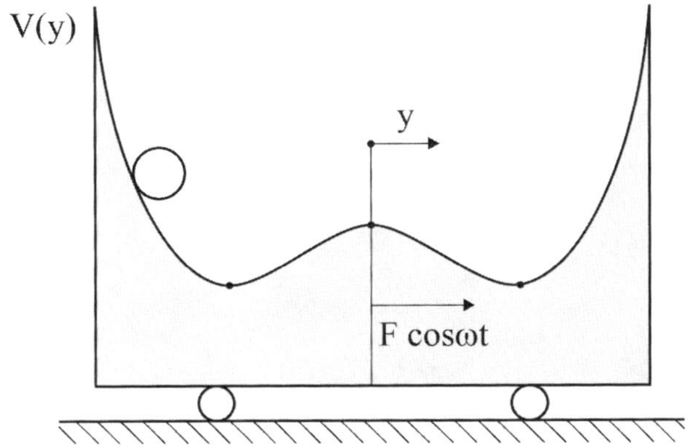

Fig. 15.29 Ball movement in the vessel along the potential $V(y)$

process and its derivative (or derivatives) and in addition we have to know the initial conditions. As it has been already mentioned, the relationship between an unknown function and its derivative is called a differential equation. Nowadays it is very difficult to imagine the development in many fields of science without knowledge of the differential equations theory. There are many directions of development in modern theory of differential equations and various methods of teaching depending on the needs of the designated public. This section deals only with a few examples of systems of differential equations describing the dynamics of simple physical systems in terms of chaotic dynamics (see also the monograph [224]).

15.7.2 Non-autonomous Oscillator with Different Potentials

Imagine that ball (material point) is in the vessel with the cross-section indicated in Fig. 15.29.

The equation of the ball motion is:

$$\ddot{y} + c\dot{y} + \frac{dV(y)}{dy} = F\cos\omega t. \tag{15.102}$$

The potential of $V(y)$ may have two minima and one maximum, as is shown in Fig. 15.29, or it may assume other shapes (Fig. 15.30).

If we describe the potential with equation

$$V(y) = \frac{\alpha y^2}{2} + \frac{\beta y^4}{4}, \tag{15.103}$$

the case of Fig. 15.29 corresponds to the potential of $\alpha < 0$ and $\beta > 0$, and for the potential of Fig. 15.30a we have $\alpha > 0$ and $\beta > 0$, and for the potential shown in Fig. 15.30b we have $\alpha > 0$ and $\beta < 0$.

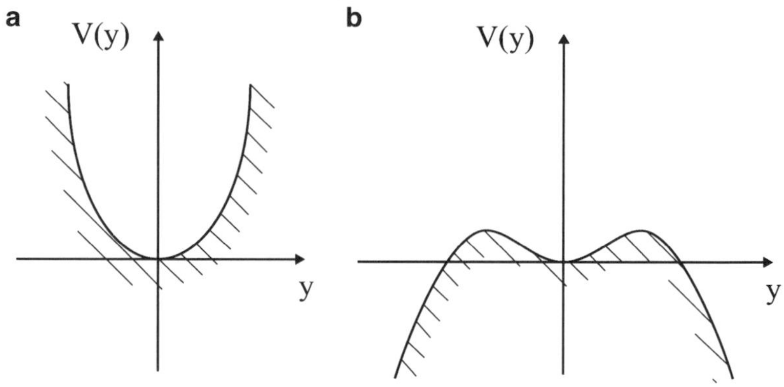

Fig. 15.30 Typical shapes of the potential $V(y)$

Consider first the case of the system with no force and no damping. Then the dynamics of the system is described by the equations

$$\dot{y} = x,$$
$$\dot{x} = -\alpha y - \beta y^3. \tag{15.104}$$

Let us find a balance ball positions. In this case from $\dot{y} = \dot{x} = 0$ and Eq. (15.104), we obtain

$$x = 0,$$
$$y\left(\alpha + \beta y^2\right) = 0, \tag{15.105}$$

which allows you to find three equilibrium positions $(y_0, x_0) = (0, 0)$ and $(x_0, y_0) = \left(\pm\sqrt{\frac{-\alpha}{\beta}}, 0\right)$. Let us examine the stability of each of the found balance positions. For this purpose, assume that δ_x and δ_y are small perturbations respectively for x_0 and y_0, and we have

$$x = x_0 + \delta x,$$
$$y = y_0 + \delta y, \tag{15.106}$$

which together with (15.104) leads after the linearization (that is leaving only the linear segments because of δ_x and δ_y) of the equations

$$\delta\dot{y} = +\delta x,$$
$$\delta\dot{x} = -\alpha\delta y - 3\beta y_0^2\delta y. \tag{15.107}$$

We will look for solutions of (15.107) in the following form:

$$\delta x = Xe^{\lambda t}, \qquad \delta y = Ye^{\lambda t}, \tag{15.108}$$

what after substituting into (15.107) yields the characteristic equation

$$\begin{vmatrix} -1 & \lambda \\ \lambda & \alpha + 3\beta y_0^2 \end{vmatrix} = 0, \tag{15.109}$$

from which we determine the following roots

$$\lambda_{1,2} = \pm\sqrt{-\alpha - 3\beta y_0^2}. \tag{15.110}$$

Next let us consider the case shown in Fig. 15.31. Then for (0,0) we have $\lambda_{1,2} = \pm\sqrt{-\alpha}$, and since $\alpha < 0$, the roots are real and of opposite signs. Location (0,0) is a saddle. Two remaining equilibrium positions correspond to the eigenvalues

$$\lambda_{1,2} = \pm i\sqrt{-2\alpha}, \tag{15.111}$$

that are imaginary values. Those positions of equilibrium are variety points of middle type. Location (0,0) is called hyperbolic, and the remaining equilibrium positions are elliptic. Phase trajectories with three equilibria are shown in Fig. 15.31.

Particularly noteworthy are two phase trajectories the shape of loop locked into eight. Trajectories coming out of the saddle-point 0 and returning to it is called the homoclinic trajectory (orbit). Homoclinic orbits can be described analytically in the form of the following two equations

$$y_H(t) = \sqrt{\frac{-2\alpha}{\beta}}\sec h\left(\pm\sqrt{-\alpha}\,(t - t_0)\right),$$

$$x_H(t) = -\alpha\sqrt{\frac{2}{\beta}}\sec h\left(\pm\sqrt{-\alpha}\,(t - t_0)\right)\tanh\left(\pm\sqrt{-\alpha}\,(t - t_0)\right), \tag{15.112}$$

where t is the time parameter.

15.7.3 Melnikov Function and Chaos

The basic idea of the Melnikov method [22, 106] is to use a solution of the uninterrupted integrable system of two differential equations to solve the disturbed system of equations. Let the dynamics of the system to be described by the equations:

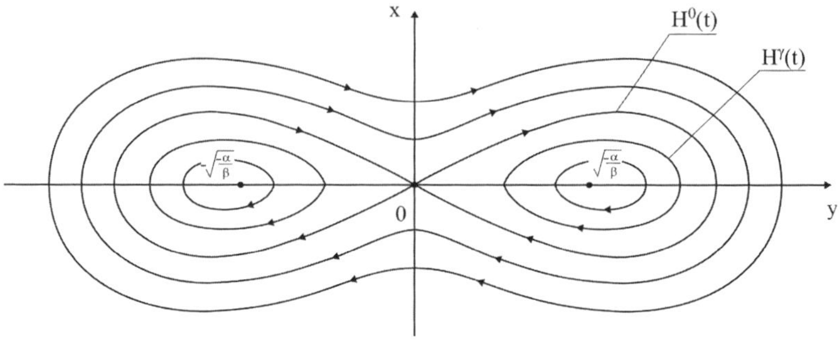

Fig. 15.31 Three equilibria and the surrounding them phase trajectories

$$\dot{x} = f_1(x, y) + \varepsilon g_1(x, y, t),$$
$$\dot{y} = f_2(x, y) + \varepsilon g_2(x, y, t) \quad (15.113)$$

Parameter $\varepsilon > 0$ is a value $\varepsilon \ll 1$ and is called the small perturbation parameter. It emphasizes the "smallness" of time-dependent disorders $g_i (i = 1, 2)$. In such a system chaotic motion may appear, and a set of parameters for which it appears can be determined with the method described below.

For $\varepsilon = 0$ undisturbed system has two homoclinic orbits $H^0(t)$ to the saddle point (0,0). The core of homoclinic orbits is filled with one-parameter family of periodic orbits $H^\gamma(t)$ with period T^γ dependent on parameter $\gamma \in (1, 0)$—see Fig. 15.31.

If in the system (15.113) forcing functions g_i $(i = 1, 2)$ are periodic in time, while the functions f_i $(i = 1, 2)$ have homoclinic orbit (as in Fig. 15.31), then the Melnikov function as follows

$$M(t_0) = \varepsilon \int_{-\infty}^{\infty} \{(f_1 [x_{H^0(t-t_0)}, y_{H^0(t-t_0)}] g_2 [x_{H^0(t-t_0)}, y_{H^0(t-t_0)}, t])$$
$$- f_2 [x_{H^0(t-t_0)}, y_{H^0(t-t_0)}] g_1 [x_{H^0(t-t_0)}, y_{H^0(t-t_0)}, t]\} \, dt. \quad (15.114)$$

If the function $M(t_0)$ does not yield zero values, then the stable and unstable manifolds do not intersect anywhere beyond the saddle point. If the equation $M(t_0) = 0$ has a solution, then additional intersection occurs. Let us now return to Eq. (15.102) and potential (15.103).

The equation of motion of the oscillator with such a choice of the potential takes the form

$$\dot{y} = x,$$
$$\dot{x} = -\alpha y - \beta y^3 - \varepsilon c x + \varepsilon F \cos \omega t, \quad (15.115)$$

where εc and εF highlight the "smallness" of the distinguished parameters.

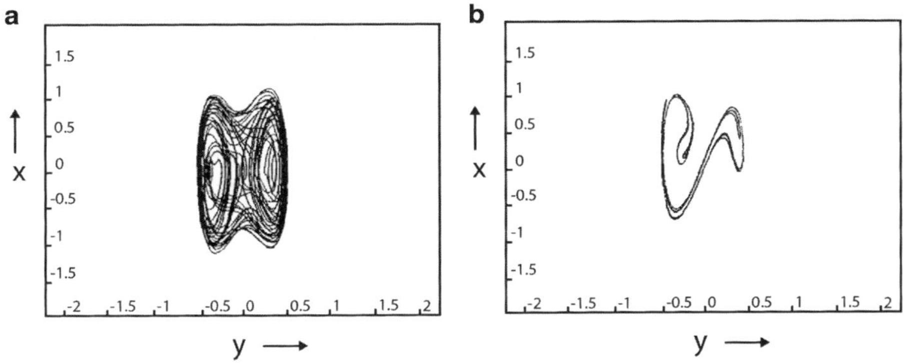

Fig. 15.32 Phase trajectory (**a**) and Poincaré map (**b**) for the Duffing oscillator governed by Eq. (15.115)

Assuming

$$f_1 = x, \quad f_2 = -\alpha y - \beta y^3,$$
$$g_1 = 0, \quad g_2 = \epsilon(F \cos \omega t - cx), \tag{15.116}$$

and using (15.114), we obtain

$$M\left(t_0\right) = \frac{4c\sqrt{(-\alpha)^3}}{3\beta} + \pi F \omega \sqrt{\frac{2}{\beta}} \frac{\sin \omega t_0}{\cosh \frac{\pi \omega}{2\sqrt{-\alpha}}}. \tag{15.117}$$

The function $M(t_0)$ changes sign for the following relationship between the parameters

$$F = \frac{4c\sqrt{(-\alpha)^3}}{3\pi\omega\sqrt{2\beta}} \cosh\left[\frac{\pi\omega}{2\sqrt{-\alpha}}\right]. \tag{15.118}$$

Let us take into consideration the following parameters: $c = 0.8$, $\alpha = -12$, $\beta = 100$, $\omega = 3.3$. The value of the last parameter is calculated from the formula (15.118) obtaining $F = 1.3295$. Equations (15.115) for given parameters were solved numerically and the numerical simulation results are shown in Fig. 15.32.

The phase trajectory "jumps" in a random way between two points corresponding to a minimum of two wells of the potential $V(y)$. Figure 15.32b shows the strange chaotic attractor on the plane in the form of an infinite set of points, while the distance in time between two successive points is $T = 2\pi/\omega$ (the Poincaré map).

In Fig. 15.33 as a control parameter taken was the amplitude of the exciting force F (other parameters unchanged) and plotted the maximum value of the Lyapunov

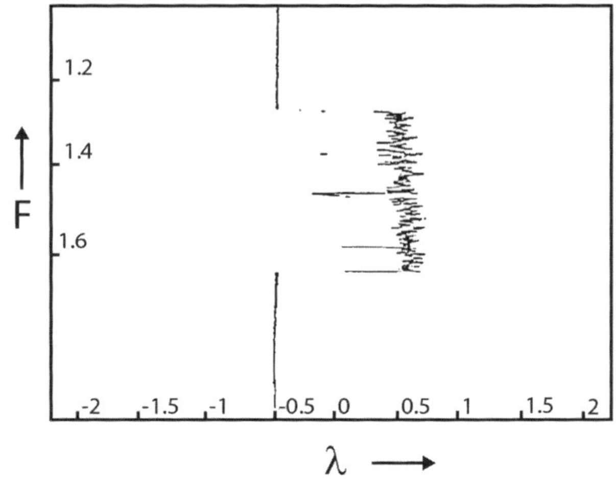

Fig. 15.33 Changes of Lyapunov exponent as a function of the parameter F

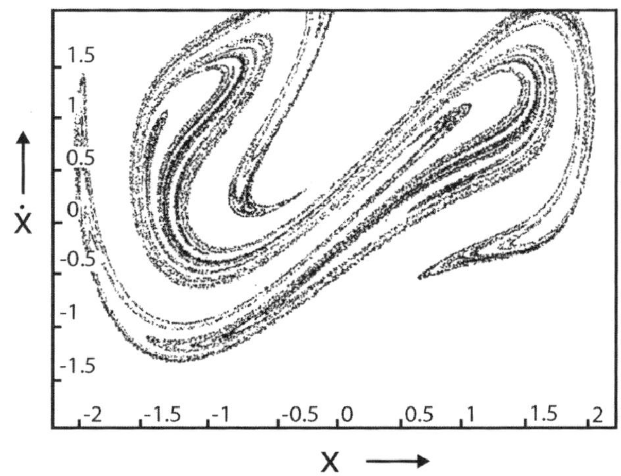

Fig. 15.34 Strange chaotic attractor discovered by Ueda

exponent for $1 \leq F \leq 2$. You can see that chaos appears for $F = 1.33$, and then disappears in the vicinity of $F \cong 1.62$.

Consider the case of the potential when $\alpha = 0$ and $\beta > 0$. This case was analysed by Ueda [233]. Presented strange chaotic attractor is often referred to as *Japanese attractor*.

Vibrations of many simple physical systems can be simplified to the Duffing equation. The equation of motion of the plane pendulum of inertia mass moment equal $B = ml^2$, with air resistance coefficient c forced by the moment $M = M_1 \cos \omega t$ has the form (see Fig. 15.34)

$$B\ddot{\varphi} = -mgl \sin \varphi - c_0\dot{\varphi} + M_1 \cos \omega t. \tag{15.119}$$

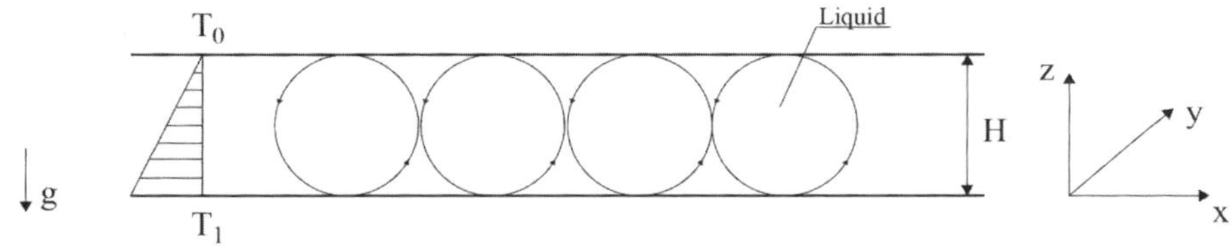

Fig. 15.35 Scheme of the process of Rayleigh–Benard convection used to derive equations Lorenz

After dividing by B we get

$$\ddot{\varphi} + c\dot{\varphi} + \beta \sin\varphi = F \cos\omega t, \tag{15.120}$$

where: $c = c_0/B, \beta = mgl/B, F = M_1/B$.

Chaotic dynamics of the pendulum takes a place for $\omega = 1$, $F = 2.4$, $c = 0.2$, $\beta = 1$.

15.7.4 Lorenz Attractor

Lorenz model is a system of three nonlinear ordinary differential equations of the first order [157]. Now we will derive those equations basing on the old problem of Rayleigh–Benard (reading of this induction of equations process can be omitted without problems in further analysis of the chaotic dynamics).

Let between two infinitely long plates with H distance be a liquid (Fig. 15.35). The liquid is heated from the bottom. Let u to be the velocity of liquid particles, let T_s to be the temperature surface, ρ_s to be density surface and pressure p_s, where T_0 corresponds ρ_0 and g is the acceleration due to gravity. Temperature, pressure and density are changed according to the following formulas (for $u = 0$), ΔT is the linear increase of the temperature.

$$\begin{aligned} T_s(z) &= T_0 + \Delta T - \left(\frac{z}{H}\right)\Delta T, \\ \rho_s(z) &= \rho_0[1 - \alpha(T_s(z) - T_0)], \\ \nabla p_s(z) &= -\rho_s(z) g \bar{z}, \end{aligned} \tag{15.121}$$

where $\bar{z}$ is the normal vector in the z direction. Firstly (that is with the provision of the low thermal energy) occurs laminar convection. Subsequently, stable vortices are formed, wherein the temperature increase is nonlinear described by

$$\Theta(x, y, z, t) = T(x, y, z, t) - T_s(z). \tag{15.122}$$

Speed u changes in time and the dynamics of the flow is described following system of partial differential equations

$$\frac{\partial u}{\partial t} + (u \cdot \nabla) u = \alpha \Theta g \bar{z} - \left(\frac{1}{\rho_0} \right) \nabla \delta p + \nu \nabla^2 u,$$
$$\frac{\partial \Theta}{\partial t} + (u \cdot \nabla) \Theta = \chi \nabla^2 \Theta - u_z \left(\frac{\Delta T}{H} \right),$$
$$\nabla u = 0. \tag{15.123}$$

Here δp is the pressure change proportional to the convection state, ν is the liquid kinetic viscosity coefficient, χ is a constant thermal diffusion process, and ∇^2 is the Laplace operator. Since $u_y = 0$, then the remaining components of the velocity vector can be obtained from the equations:

$$u_x = -\frac{\partial \psi}{\partial z}, \quad u_z = \frac{\partial \psi}{\partial x}. \tag{15.124}$$

In addition, Lorenz introduced the following boundary conditions:

$$\Theta(0) = \Theta(H) = \psi(0) = \psi(H) = \nabla^2 \psi(0) = \nabla^2 \psi(H) = 0. \tag{15.125}$$

Function of temperature dispersion Θ and flow function ψ can be found in the form of the following Fourier series:

$$\Theta(x, y, z, t) = \sum_{j=1}^{J} \sin(j\pi z) \Theta_j(x, t),$$
$$\psi(x, y, z, t) = \sum_{j=1}^{J} \sin(j\pi z) \psi_j(x, t). \tag{15.126}$$

Further Lorenz limited his considerations to only three basic solutions, and taking into account the boundary conditions he obtained

$$\Theta(x, z, t) = \frac{\Delta T R_c}{R_a \pi} \left[\sqrt{2} Y(t) \cos\left(\frac{\pi a x}{H}\right) \sin\left(\frac{\pi z}{H}\right) - z(t) \sin\left(\frac{2\pi z}{H}\right) \right],$$
$$\psi(x, z, t) = \frac{\sqrt{2}\left(1 + a^2\right)\chi}{a} X(t) \sin\left(\frac{\pi a x}{H}\right) \sin\left(\frac{\pi z}{H}\right), \tag{15.127}$$

where R_a is the Rayleigh number with the critical value R_c:

$$R_a = \frac{\alpha g H^3 \Delta T}{\chi \nu}, \quad R_c = \frac{\pi^4 \left(1 + a^2\right)^3}{a^2}. \tag{15.128}$$

X, Z and Z are amplitudes of three successive forms of the assumed solution that are dependent on time. Lorenz equations we obtain by substituting (15.127) into Eq. (15.123). They have the following form

$$\frac{dX}{d\tau} = \sigma (Y - X),$$
$$\frac{dY}{d\tau} = -XZ + rX - Y,$$
$$\frac{dZ}{d\tau} = XY - bZ, \tag{15.129}$$

where

$$\tau = \frac{\pi \left(1 + a^2\right) \chi t}{\chi^2}, \qquad \sigma = \frac{\mu}{\chi}, \qquad b = \frac{4}{1 + a^2}, \qquad r = \frac{R_a}{R_c}. \tag{15.130}$$

During computer simulations of these equations Lorenz noticed irregular oscillations for certain parameters of this strongly simplified version of a physical model. He also noted in the plane (X, Y) a geometric shape somewhat resembling a human kidney. Phase point wandered around the left or right kidney, while the jumps between them were random and impossible to predict.

Lorenz equations system is an autonomous system (without acting external force). Let us try to determine the equilibrium position of the system and investigate their stability. For $b > 0$, $\sigma > 0$ and $r > 0$, Eq. (15.129), we obtain

$$\sigma (Y - Z) = 0,$$
$$-XY + rX - Y = 0,$$
$$XY - bZ = 0. \tag{15.131}$$

As one can easily verify $(X, Y, Z) = (0, 0, 0)$. Disturbing the equilibrium position and limiting the discussion to the differential equations of the linear disorders we obtain the characteristic equation of the form

$$[\lambda + b] \left[\lambda^2 + (\sigma + 1) \lambda + \sigma (1 - r)\right] = 0. \tag{15.132}$$

For $0 < r < 1$ Eq. (15.132) has three real roots neither of which is negative. This means that considered equilibrium position is a stable. For $r = 1$ occurs solutions branching–bifurcation. For $r > 1$ we have the following solution set (the equilibrium)

$$X = Y = \pm \sqrt{b (r - 1)},$$
$$Z = r - 1. \tag{15.133}$$

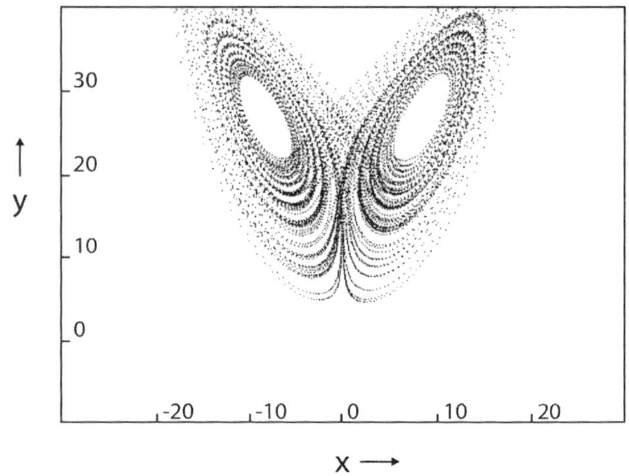

Fig. 15.36 Lorenz attractor projection on the plane (X, Y) for $\sigma = 10$, $r = 28$, $b = 2.67$

The previous solution (0,0,0) still exists, but is unstable. Characteristic equation in the considered case takes the form

$$\lambda^3 + (\sigma + b + 1)\,\lambda^2 + (r + \sigma)\,b\lambda + 2\sigma b\,(r - 1) = 0. \tag{15.134}$$

Both the equilibrium positions defined by Eq. (15.133) lose their stability after exceeding the critical value of the parameter.

$$r_c = \frac{\sigma\,(\sigma + b + 3)}{\sigma - b - 1}. \tag{15.135}$$

Lorenz attractor projection is shown in Fig. 15.36. Numerically calculated attractor dimension equals $d = 1.768 \pm 0.071$. Figure 15.37 shows a graph of the Lyapunov exponent λ in dependence of the control parameter r. For big values it assumes positive values, what reflects the chaotic traffic.

15.8 Synchronization Phenomena of Coupled Triple Pendulums

15.8.1 Mathematical Model

The investigated system consists of N identical triple pendulums [27]. Each triple pendulum, exhibited in Fig. 15.38a is a plane subsystem of three rigid links, rotationally coupled in points O_j $(j = 1, 2, 3)$, with viscous damping of the coefficients $\bar{c}_j$ $(i = 1, 2, \ldots, N)$, respectively. The position of the system is defined by $3N$ angles $\psi_{i,j}$ $(i = 1, 2, \ldots, N,\ j = 1, 2, 3)$. Masses of the corresponding

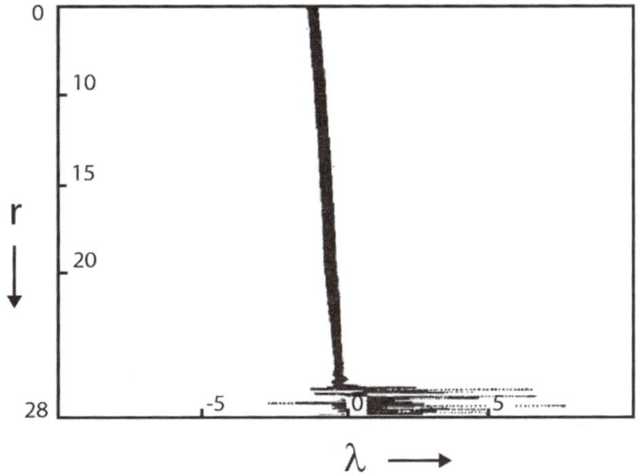

Fig. 15.37 λ Lyapunov exponent as a function of the parameter r

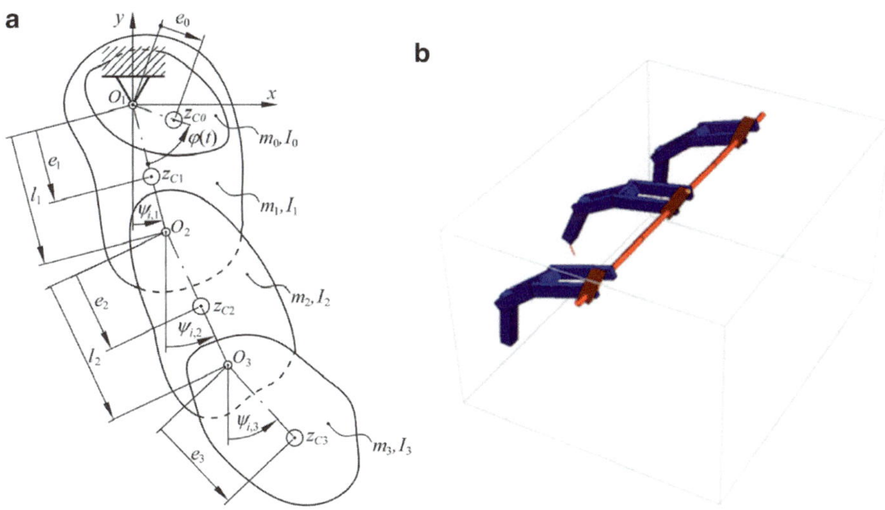

Fig. 15.38 The ith triple pendulum (**a**) and example of three coupled sets of pendulums (**b**)

links are denoted by m_j $(j = 1, 2, 3)$, while I_j denote mass moments of inertia of the corresponding bodies with respect to axes z_{cj} $(j = 1, 2, 3)$ - principal central axes perpendicular to the motion plane. It is assumed that the mass centers (axes z_{cj}) of the links lie on the lines including the corresponding joints (O_j, $j = 1, 2, 3$). The first link of each pendulum is forced by the external and common signal $\varphi(\tau) = \bar{\omega}\tau$ (where τ is time), realized by relative rotation of additional body (of mass m_0 and inertia moment I_0) connected to the first link in the joint O_1. Other geometric parameters of the system are visible in Fig. 15.38a. The pendulum sets are situated along a line perpendicular to the motion plane and they are coupled by viscous and elastic connections (with angular viscous damping coefficient $\bar{k}_s\bar{c}_{rs}$ and angular stiffness $\bar{k}_s$) between the first links of neighbouring sets of triple pendulums. Figure 15.38b exhibits an example of three coupled sets of triple pendulums.

The system is governed by the following set of differential equations in the Lagrange formalism

$$\frac{d}{d\tau}\left(\frac{\partial \bar{T}_i}{\partial \dot{\psi}_{i,j}}\right) - \frac{\partial \bar{T}_i}{\partial \psi_{i,j}} + \frac{\partial \bar{V}_i}{\partial \psi_{i,j}} = \bar{Q}_{i,j} \quad i = 1, \ldots, N, \quad j = 1, 2, 3, \tag{15.136}$$

where $(\dot{\ldots})$—denotes derivative with respect to real time τ, $\bar{T}_i$, $\bar{V}_i$—real kinetic and potential energy of the ith pendulum, $\bar{Q}_{i,j}$—real generalized forces acting in the system. The reader may find more material devoted to numerical and experimental investigations of the triple pendulum set in [27, 28, 36].

Real kinetic energy of the ith pendulum follows

$$\begin{aligned} \bar{T}_i = {} & A_i(\dot{\psi}_{i,1}) + \frac{1}{2}B_1\dot{\psi}^2_{i,1} + \frac{1}{2}B_2\dot{\psi}^2_{i,2} + \frac{1}{2}B_3\dot{\psi}^2_{i,3} \\ & + N_{12}\dot{\psi}_{i,1}\dot{\psi}_{i,2}\cos(\psi_{i,1} - \psi_{i,2}) + N_{13}\dot{\psi}_{i,1}\dot{\psi}_{i,3}\cos(\psi_{i,1} - \psi_{i,3}) \\ & + N_{23}\dot{\psi}_{i,2}\dot{\psi}_{i,3}\cos(\psi_{i,2} - \psi_{i,3}) \end{aligned} \tag{15.137}$$

where:

$$\begin{aligned} B_1 &= I_0 + I_1 + e_1^2 m_1 + e_0^2 m_0 + l_1^2 (m_2 + m_3), \\ B_2 &= I_2 + e_2^2 m_2 + l_2^2 m_3, \\ B_3 &= I_3 + e_3^2 m_3, \\ N_{12} &= m_2 e_2 l_1 + m_3 l_1 l_2, \\ N_{13} &= m_3 e_3 l_1, \\ N_{23} &= m_3 e_3 l_2, \end{aligned} \tag{15.138}$$

and where

$$A_i(\dot{\psi}_{i,1}) = \frac{1}{2}\bar{\omega}_i\left[I_0\left(\bar{\omega}_i - 2\dot{\psi}_{i,1}\right) + e_0^2 m_0\left(\bar{\omega}_i + 2\dot{\psi}_{i,1}\right)\right]. \tag{15.139}$$

Real potential energy of gravitational forces for the i-th pendulum is as follows

$$\bar{V}_i = -M_0\cos(\psi_{i,1} + \bar{\omega}_i\tau) - M_1\cos\psi_{i,1} - M_2\cos\psi_{i,2} - M_3\cos\psi_{i,3}, \tag{15.140}$$

where

$$
\begin{aligned}
M_0 = m_0 g\, e_0, \quad M_1 = m_1 g\, e_1 + (m_2 + m_3)\, g\, l_1,\\
M_2 = m_2 g\, e_2 + m_3 g\, l_2, \quad M_3 = m_3 g\, e_3,
\end{aligned}
\tag{15.141}
$$

and g is the gravitational acceleration.

The coupling between the adjacent sets of triple pendulums is modelled by following generalized forces

$$
\begin{aligned}
\bar{Q}_{i,1} &= -\bar{c}_1 \dot{\psi}_{i,1} - \bar{c}_2 \left(\dot{\psi}_{i,1} - \dot{\psi}_{i,2} \right) \\
&\quad + -\bar{k}_s \Big[(\psi_{i+1,1} - \psi_{i,1}) + (\psi_{i-1,1} - \psi_{i,1}) + \bar{c}_{rs} \left(\dot{\psi}_{i+1,1} - \dot{\psi}_{i,1} \right) \\
&\quad + \bar{c}_{rs} \left(\dot{\psi}_{i-1,1} - \dot{\psi}_{i,1} \right) \Big], \\
\bar{Q}_{i,2} &= -\bar{c}_2 \left(\dot{\psi}_{i,2} - \dot{\psi}_{i,1} \right) - \bar{c}_3 \left(\dot{\psi}_{i,2} - \dot{\psi}_{i,3} \right), \\
\bar{Q}_{i,3} &= -\bar{c}_3 \left(\dot{\psi}_{i,3} - \dot{\psi}_{i,2} \right), \quad \text{where} \quad i = 1, \ldots, N
\end{aligned}
\tag{15.142}
$$

and where we assume that $\psi_{0,1} = \psi_{1,1}$, $\dot{\psi}_{0,1} = \dot{\psi}_{1,1}$, $\psi_{N+1,1} = \psi_{N,1}$, $\dot{\psi}_{N+1,1} = \dot{\psi}_{N,1}$.

Then we introduce the non-dimensional time t

$$
t = \alpha_1 \tau, \tag{15.143}
$$

where

$$
\alpha_1 = \left(M_1 B_1^{-1} \right)^{\frac{1}{2}}. \tag{15.144}
$$

Furthermore, we take

$$
\frac{d\,(\ldots)}{d\tau} = \alpha_1 \frac{d\,(\ldots)}{dt} \tag{15.145}
$$

and therefore

$$\begin{aligned}
\overset{|}{\psi}_{i,j} &= \alpha_1 \dot{\psi}_{i,j}, \\
\bar{\omega} &= \alpha_1 \omega, \\
\frac{\partial(\ldots)}{\partial \overset{|}{\psi}_{i,j}} &= \frac{\partial(\ldots)}{\partial \dot{\psi}_{i,j}} \frac{\partial \dot{\psi}_{i,j}(\overset{|}{\psi}_j)}{\partial \overset{|}{\psi}_{i,j}} = \frac{1}{\alpha_1} \frac{\partial(\ldots)}{\partial \dot{\psi}_j},
\end{aligned} \tag{15.146}$$

where $i = 1, \ldots, N$, $j = 1, 2, 3$, ω denotes the non-dimensional counterpart of real angular frequency $\bar{\omega}$ and $(\ldots)$ is derivative with respect to non-dimensional time t.

Dividing both sides of the real equations (15.136) by $2E_1$, where

$$E_1 = \frac{1}{2}\alpha_1^2 B_1 = \frac{1}{2} M_1, \tag{15.147}$$

the following non-dimensional Lagrange formulation of the governing equations is obtained

$$\frac{d}{dt}\left(\frac{\partial T_i}{\partial \dot{\psi}_{i,j}}\right) - \frac{\partial T_i}{\partial \psi_{i,j}} + \frac{\partial V_i}{\partial \psi_{i,j}} = Q_{i,j}, \quad i = 1, \ldots, N \quad j = 1, 2, 3. \tag{15.148}$$

Non-dimensional kinetic energy of the i-th pendulum is as follows

$$\begin{aligned}
T_i = \frac{\bar{T}_i}{2E_1} &= a_i(\dot{\psi}_{i,1}) + \frac{1}{2}\dot{\psi}_{i,1}^2 + \frac{1}{2}\beta_2 \dot{\psi}_{i,2}^2 + \frac{1}{2}\beta_3 \dot{\psi}_{i,3}^2 \\
&+ \nu_{12}\dot{\psi}_{i,1}\dot{\psi}_{i,2}\cos(\psi_{i,1} - \psi_{i,2}) + \nu_{13}\dot{\psi}_{i,1}\dot{\psi}_{i,3}\cos(\psi_{i,1} - \psi_{i,3}) \\
&+ \nu_{23}\dot{\psi}_{i,2}\dot{\psi}_{i,3}\cos(\psi_{i,2} - \psi_{i,3})
\end{aligned} \tag{15.149}$$

where

$$\beta_2 = \frac{B_2}{B_1}, \beta_3 = \frac{B_3}{B_1}, \nu_{12} = \frac{N_{12}}{B_1}, \nu_{13} = \frac{N_{13}}{B_1}, \tag{15.150}$$

and

$$a_i(\dot{\psi}_{i,1}) = \frac{A_i(\overset{|}{\psi}_{i,1})}{2E_1} = \frac{1}{2B_1}\omega_i \left[I_0\left(\omega_i - 2\dot{\psi}_{i,1}\right) + e_0^2 m_0 \left(\omega_i + 2\dot{\psi}_{i,1}\right)\right]. \tag{15.151}$$

Non-dimensional potential energy of gravitational forces for the i-th pendulum has the form

$$V_i = \frac{\bar{V}_i}{2E_1} = -\mu_0 \cos(\psi_{i,1} + \omega_i t) - \cos\psi_{i,1} - \mu_2 \cos\psi_{i,2} - \mu_3 \cos\psi_{i,3} \tag{15.152}$$

where

$$\mu_0 = \frac{M_0}{M_1}, \mu_2 = \frac{M_2}{M_1}, \mu_3 = \frac{M_3}{M_1}. \tag{15.153}$$

Non-dimensional generalized forces have the following form

$$\begin{aligned}
Q_{i,1} &= \frac{\bar{Q}_{i,1}}{2E_1} = -c_1\dot{\psi}_{i,1} - c_2\left(\dot{\psi}_{i,1} - \dot{\psi}_{i,2}\right) \\
&\quad + -k_s\left[\left(\psi_{i+1,1} - \psi_{i,1}\right) + \left(\psi_{i-1,1} - \psi_{i,1}\right) + c_{rs}\left(\dot{\psi}_{i+1,1} - \dot{\psi}_{i,1}\right)\right. \\
&\quad \left.+ c_{rs}\left(\dot{\psi}_{i-1,1} - \dot{\psi}_{i,1}\right)\right], \\
Q_{i,2} &= \frac{\bar{Q}_{i,2}}{2E_1} = -c_2\left(\dot{\psi}_{i,2} - \dot{\psi}_{i,1}\right) - c_3\left(\dot{\psi}_{i,2} - \dot{\psi}_{i,3}\right), \\
Q_{i,3} &= \frac{\bar{Q}_{i,3}}{2E_1} = -c_3\left(\dot{\psi}_{i,3} - \dot{\psi}_{i,2}\right), \quad i = 1, 2, \ldots, N,
\end{aligned} \tag{15.154}$$

where

$$\begin{aligned}
c_j &= \frac{\bar{c}_j}{\sqrt{M_1 B_1}}, \quad \text{for} \quad j = 1, 2, 3, \\
k_s &= \frac{\bar{k}_s}{M_1}, \quad c_{rs} = \bar{c}_{rs}\sqrt{\frac{M_1}{B_1}}
\end{aligned} \tag{15.155}$$

and where we take $\psi_{0,1} = \psi_{1,1}$, $\dot{\psi}_{0,1} = \dot{\psi}_{1,1}$, $\psi_{N+1,1} = \psi_{N,1}$, $\dot{\psi}_{N+1,1} = \dot{\psi}_{N,1}$.

Finally, the non-dimensional governing equations can be written as follows

$$\begin{aligned}
\mathbf{M}(\boldsymbol{\psi}_i)\,\ddot{\boldsymbol{\psi}}_i + \mathbf{N}(\boldsymbol{\psi}_i)\,\dot{\boldsymbol{\psi}}_i^2 + \mathbf{C}\,\dot{\boldsymbol{\psi}}_i + \mathbf{p}(\boldsymbol{\psi}_i, t) &= \mathbf{f}\left(\boldsymbol{\psi}_{i-1}, \boldsymbol{\psi}_i, \boldsymbol{\psi}_{i+1}, \dot{\boldsymbol{\psi}}_{i-1}, \dot{\boldsymbol{\psi}}_i, \dot{\boldsymbol{\psi}}_{i+1}\right) \\
i &= 1, 2, \ldots, N
\end{aligned} \tag{15.156}$$

where

$$\boldsymbol{\psi}_i = \begin{Bmatrix} \psi_{i,1} \\ \psi_{i,2} \\ \psi_{i,3} \end{Bmatrix}, \quad \dot{\boldsymbol{\psi}}_i = \begin{Bmatrix} \dot{\psi}_{i,1} \\ \dot{\psi}_{i,2} \\ \dot{\psi}_{i,3} \end{Bmatrix}, \quad \dot{\boldsymbol{\psi}}_i^2 = \begin{Bmatrix} \dot{\psi}_{i,1}^2 \\ \dot{\psi}_{i,2}^2 \\ \dot{\psi}_{i,3}^2 \end{Bmatrix}, \quad \ddot{\boldsymbol{\psi}}_i = \begin{Bmatrix} \ddot{\psi}_{i,1} \\ \ddot{\psi}_{i,2} \\ \ddot{\psi}_{i,3} \end{Bmatrix},$$

$$\mathbf{M}(\boldsymbol{\psi}_i) = \begin{bmatrix} 1 & \nu_{12}\cos\left(\psi_{i,1} - \psi_{i,2}\right) & \nu_{13}\cos\left(\psi_{i,1} - \psi_{i,3}\right) \\ \nu_{12}\cos\left(\psi_{i,1} - \psi_{i,2}\right) & \beta_2 & \nu_{23}\cos\left(\psi_{i,2} - \psi_{i,3}\right) \\ \nu_{13}\cos\left(\psi_{i,1} - \psi_{i,3}\right) & \nu_{23}\cos\left(\psi_{i,2} - \psi_{i,3}\right) & \beta_3 \end{bmatrix},$$

$$\mathbf{N}(\boldsymbol{\psi}_i) = \begin{bmatrix} 0 & \nu_{12}\sin(\psi_{i,1}-\psi_{i,2}) & \nu_{13}\sin(\psi_{i,1}-\psi_{i,3}) \\ -\nu_{12}\sin(\psi_{i,1}-\psi_{i,2}) & 0 & \nu_{23}\sin(\psi_{i,2}-\psi_{i,3}) \\ -\nu_{13}\sin(\psi_{i,1}-\psi_{i,3}) & -\nu_{23}\sin(\psi_{i,2}-\psi_{i,3}) & 0 \end{bmatrix},$$

$$\mathbf{C}=\begin{bmatrix} c_1+c_2 & -c_2 & 0 \\ -c_2 & c_2+c_3 & -c_3 \\ 0 & -c_3 & c_3 \end{bmatrix}, \quad \mathbf{p}(\psi_i,t) = \begin{Bmatrix} \sin\psi_{i,1}+\mu_0\sin(\psi_{i,1}+\omega t) \\ \mu_2\sin\psi_{i,2} \\ \mu_2\frac{\nu_{12}}{\nu_{13}}\sin\psi_{i,3} \end{Bmatrix},$$

$$\mathbf{f} = k_s \begin{Bmatrix} \psi_{i-1}-2\psi_i+\psi_{i+1}+c_{rs}\left(\dot{\psi}_{i-1}-2\dot{\psi}_i+\dot{\psi}_{i+1}\right) \\ 0 \\ 0 \end{Bmatrix}. \tag{15.157}$$

15.8.2 Numerical Simulations

The following set of real parameters is constant during the numerical simulations presented in the current section

$$\begin{aligned} & m_0 = m_1 = m_2 = m_3 = 1\,\text{kg}, \\ & I_0 = I_1 = I_2 = I_3 = \frac{1}{12}\,\text{kg}\cdot\text{m}^2, \\ & e_1 = 0, \quad e_0 = e_2 = e_3 = \frac{1}{2}\,\text{m}, \quad l_1 = l_2 = 1\,\text{m}, \\ & \bar{c}_1 = \bar{c}_2 = \bar{c}_3 = 0.1\,\text{N m s}, \quad \bar{c}_{rs} = 0.347611\,\text{s}, \quad g = 10\,\text{m/s}^2. \end{aligned} \tag{15.158}$$

The set of real quantities (15.158) leads to the following non-dimensional parameters

$$\begin{aligned} & \beta_2 = 0.5517, \quad \beta_3 = 0.1379, \\ & \mu_0 = 0.25, \quad \mu_2 = 0.75, \\ & \nu_{23} = 0.2068, \quad \nu_{12} = 0.6207, \quad \nu_{13} = 0.2068, \\ & c_1 = c_2 = c_3 = 0.01438, \quad c_{rs} = 1. \end{aligned} \tag{15.159}$$

Figure 15.39a exhibits bifurcational diagram for one ($N = 1$) or for uncoupled triple pendulums ($k_s = 0$), with the excitation angular frequency ω as a bifurcational parameter. In Fig 15.39b–d there are presented two exemplary orbits: the periodic (b) for $\omega = 0.68$ and the chaotic one for $\omega = 0.72$ (c, d). Then, for (chaotic behaviour of uncoupled systems), we present bifurcational diagrams (Fig. 15.40) of dynamical behaviour of three coupled pendulums ($N = 3$), with coupling coefficient k_s as a control parameter. The first Poincaré section (for $k_s = 0$) of each bifurcational diagram is performed by the use of the following set of initial conditions

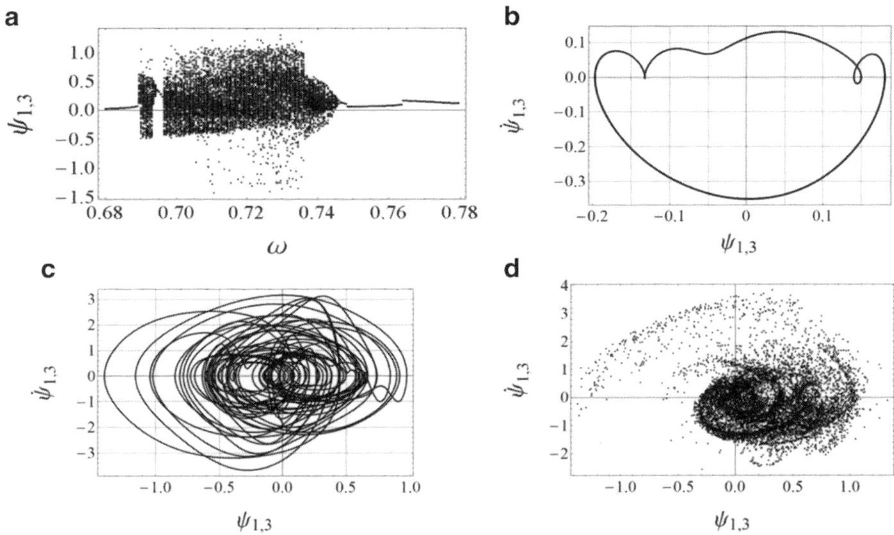

Fig. 15.39 Bifurcational diagram for one (N =1) triple pendulum (**a**) and the corresponding periodic solution for ω =0.68 (**b**) and chaotic attractor for ω =0.72 [(**c**) trajectory, (**d**) Poincaré section]

$$\psi_{i,j}(0) = 0, \quad \dot{\psi}_{i,j}(0) = 10^{-5}i \quad \text{where} \quad i = 1, 2, \ldots, N, \quad j = 1, 2, 3, \tag{15.160}$$

so the pendulums start from closely located, but different states. During a jump to the next Poincaré section (the change of control parameter), see Fig. 15.40a, the system state preserves continuity or is restarted to the initial conditions (15.160)—see Fig. 15.40b. Figure 15.40 exhibits rich spectrum of synchronization phenomena governed by the investigated system (the associated Poincaré maps are reported in Fig. 15.41). In particular, we have observed the intervals of chaotic and periodic behaviour of the system, or even regions of coexistence of chaotic and periodic attractors. We have also found the intervals of exact synchronization between chaotic behaviour of all three pendulums and the zones of exact synchronization between irregular motion of the first pendulum and the third one, while the second pendulum moves non-synchronically on chaotic attractor. We can also observe other kinds of non-exact synchronization, usually between periodic motions of the pendulums.

To conclude, in this section the preliminary research results of the system of coupled triple pendulums are presented. We have identified and shown examples of rich dynamics exhibited by the investigated system, including many different kinds of synchrony and opening the route to more deep and general view of synchronization phenomenon. Since there is a direct mechanical interpretation of the proposed model, the experimental verification is potentially possible. There are

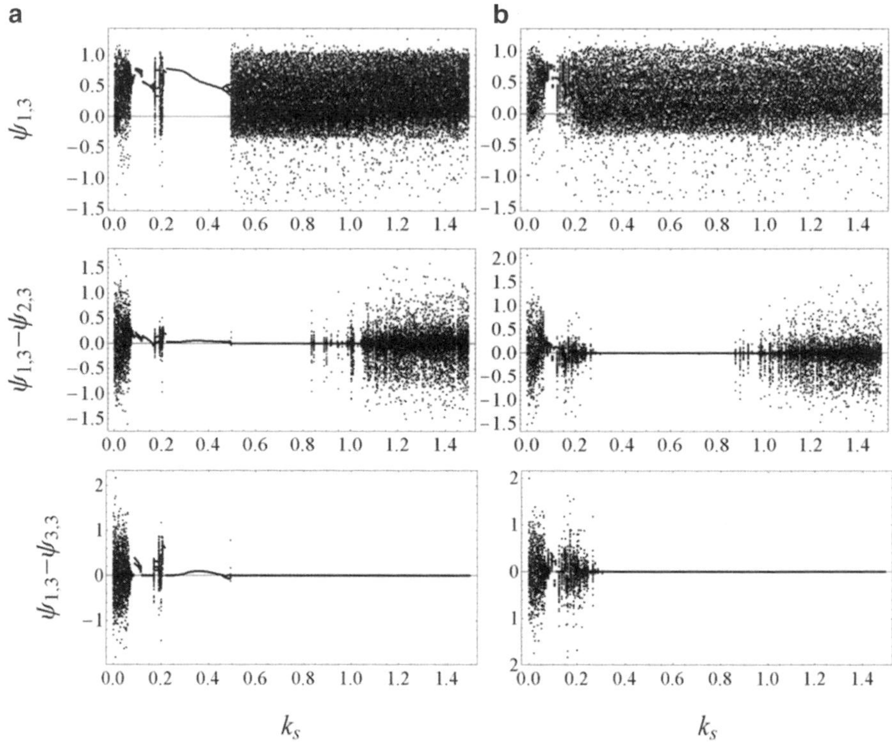

Fig. 15.40 Bifurcational diagrams for three (N =3) coupled triple pendulums

many possibilities of further research of the system, e.g. investigations of larger number of coupled subsystems of pendulums consisting of larger or smaller number of links.

15.9 Chaos and Synchronization Phenomena Exhibited by Plates and Shells

15.9.1 Introduction

In the past two decades a key role of the theory of bifurcation and chaos has been exhibited in the studies on high-dimensional nonlinear systems, and in particular structural members like beams, plates and shells. On the other hand, the mentioned structural members are widely applied in civil aerospace and mechanical engineering, including space stations, satellite solar panels, precision micromachines and instruments, and so on. In engineering an attempt to fabricate light-weight

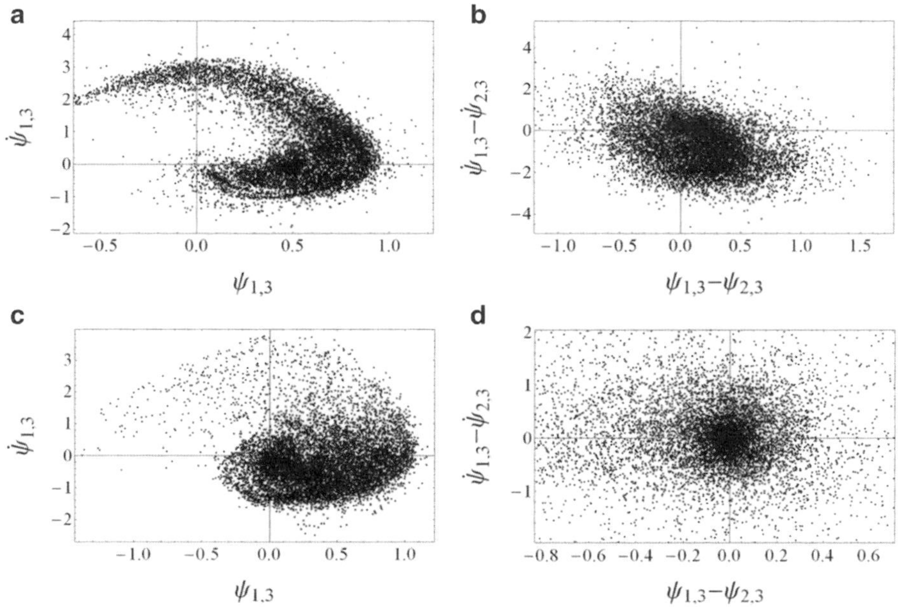

Fig. 15.41 Poincaré sections corresponding to Fig. 15.40a, for $k_s = 0.05$ (**a**, **b**) and $k_s = 1.4$ (**c**, **d**)

high-speed and energy-saving structures by simultaneously keeping large structural flexibility and stability, even for relatively large vibration deformation, is observed. In order to satisfy the engineering expectations, novel mathematical models are needed, supported by development of the theory of bifurcation and chaos as well as novel theoretical/numerical tools aimed at solving the governing partial differential equations are highly required. Below, a brief state of the art validating the mentioned remarks is given.

An averaging method was applied by Yang and Sethna [246, 247] to detect and analyse local and global bifurcations in parametrically excited nearly squared plates for symmetric and anti-symmetric cases. They formulated analytical conditions for the Shilnikov-type homoclinic orbits and deterministic chaos. A double mode approach to predict chaotic vibrations of a large deflection plate utilizing the Melnikov method was proposed by Shu et al. [218]. Lyapunov exponents, bifurcation diagrams and fractal dimension concepts were applied by Yeh et al. [249] to study chaotic and bifurcation vibrations of a simply supported thermo-elastic circular plate in large deflection.

Nagai et al. reported analytical results for a shallow cylindrical panel with a concentrated mass under periodic excitation [178] as well as experimental results of a shallow cylindrical shell-panel [177]. Amabili [5, 6] analysed the transition to chaotic vibrations for circular cylindrical shell and doubly curved panels in the vicinity of the fundamental frequency.

Ye et al. [248] analysed chaotic vibrations of antisymmetric cross-ply laminated composite rectangular thin plate under parametric excitation.

Wang et al. [241] studied chaotic vibrations of a bimetallic shallow shell of revolution under time-varying temperature excitation using the Melnikov functions, Poincaré maps, phase portraits, Lyapunov exponents and Lyapunov dimensions. They reported the onset of chaos, transient chaos, direct and reversed period-doubling scenario, jump phenomena and interior crisis. Nonlinear dynamics and chaos of a simply supported functionally orthotropic gradient material rectangular plate in thermal environment subjected to parametric and external excitations was studied by Zhang et al. [252]. The governing partial differential equations were reduced to ordinary differential equations modelling the truncated three degree-of-freedom nonlinear mechanical system.

Touzé et al. [232] studied von Kármán equation for thin plates which exhibit large amplitude vibrations putting emphasis on the transition from periodic to chaotic vibrations in free-edge, perfect and imperfect circular plates. The bifurcation diagrams, Lyapunov exponents and Fourier spectra were applied to analyse both transitions into chaotic regimes and the energy exchange between modes.

In spite of the application oriented and so far briefly described papers, the existence of global attractors and inertial manifolds exhibited by von Kármán equations for various types of damping laws was rigorously analysed by Chuesov and Lasiecka [66–68].

15.9.2 One Layer Shell

This section is devoted to the investigation of plates/shells subjected to harmonic load actions of their parameters, as it is shown in Fig. 15.42 (see [23]).

We consider flexible one-layer thin shells of length a, width b and height h, made from an isotropic and homogeneous material. The shell is loaded via continuous p_x and p_y loads distributed along its perimeter. The following hypothesis are applied: arbitrary shell's cross-section, being normal to the shell middle surface deformation remains normal after the deformation, and the cross-section height is not changed; although rotational inertia of shell elements is not taken into account, inertial forces associated with displacements along a normal to the middle shell surface are taken

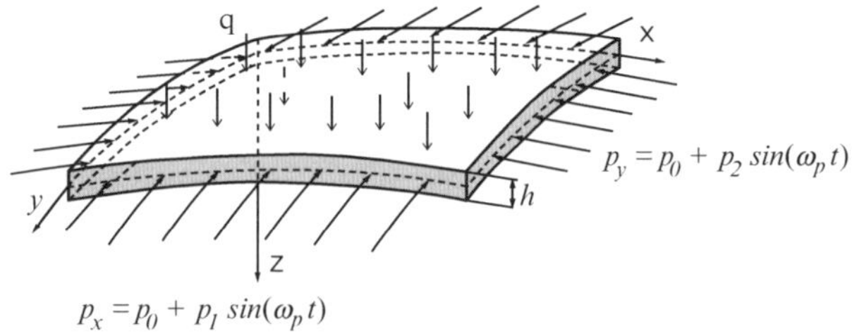

Fig. 15.42 Shell with normal and longitudinal harmonic loads

into consideration; external forces do not change their directions during the shell deformation; geometric nonlinearity is taken in the Kármán form [240].

The so far listed hypotheses are based on the Kirchhoff–Love ideas, and they can be understood as the first approximation approach to build a mathematical model of the shell. The governing non-dimensional PDEs have the following form [239]:

$$\frac{1}{12(1-\mu^2)}{\nabla_\lambda}^4 w - \nabla_k^2 F - L(w,F) - \frac{\partial^2 w}{\partial t^2} - \varepsilon\frac{\partial w}{\partial t} - q(x,y,t) = 0,$$
$${\nabla_\lambda}^4 F + {\nabla_k}^2 w + \frac{1}{2}L(w,w) = 0, \tag{15.161}$$

where

$${\nabla_\lambda}^4 = \frac{1}{\lambda^2}\frac{\partial^4}{\partial x^4} + \lambda^2\frac{\partial^4}{\partial y^4} + 2\frac{\partial^4}{\partial x^2 \partial y^2}, \quad L(w,F) = \frac{\partial^2 w}{\partial x^2}\frac{\partial^2 F}{\partial y^2} + \frac{\partial^2 w}{\partial y^2}\frac{\partial^2 F}{\partial x^2} - 2\frac{\partial^2 w}{\partial x \partial y}\frac{\partial^2 F}{\partial x \partial y},$$
$${\nabla_k}^2 = k_y\frac{\partial^2}{\partial x^2} + k_x\frac{\partial^2}{\partial y^2}.$$

Here w and F are the deflection and stress functions, respectively; $\lambda = a/b$, where a, b are the shell dimensions regarding x and y, respectively; μ is Poisson's coefficient and ε denotes the damping coefficient. The initial conditions follow

$$w(x,y)|_{t=0} = \varphi_1(x,y), \qquad \left.\frac{\partial w}{\partial t}\right|_{t=0} = \varphi_2(x,y), \tag{15.162}$$

and the boundary conditions have the form

$$\begin{aligned} &w = 0, \quad \frac{\partial^2 w}{\partial x^2} = 0, \quad F = 0, \quad \frac{\partial^2 F}{\partial x^2} = p_y \quad \text{for} \quad x = 0, 1, \\ &w = 0, \quad \frac{\partial^2 w}{\partial y^2} = 0, \quad F = 0, \quad \frac{\partial^2 F}{\partial y^2} = p_x \quad \text{for} \quad y = 0, 1. \end{aligned} \tag{15.163}$$

System (15.161)–(15.163) is transformed to its non-dimensional counterpart form using the following parameters: $\lambda = a/b$, $x = a\overline{x}$, $y = b\overline{y}$, $w = h\bar{w}$—deflection; $F = Eh^3\overline{F}$—Airry's function; $t = \bar{t}\frac{ab}{h}\sqrt{\frac{\gamma}{Eg}}$—time; $q = \frac{Eh^4}{a^2b^2}\overline{q}$—transversal load; $\varepsilon = \frac{h}{ab}\sqrt{\frac{Eg}{\gamma}}\bar{\varepsilon}$—damping coefficient; g—Earth acceleration; $\rho = \gamma h$ (γ—unit weight density); $\bar{k}_x = k_x\frac{a^2}{h}$, $\bar{k}_y = k_y\frac{b^2}{h}$, $k_x = \frac{1}{r_x}$, $k_y = \frac{1}{r_y}$(k_x, k_y—shell curvature regarding x and y, respectively); r_x, r_y—curvature radius of the middle shell surface regarding x and y, respectively. We have also

E—elasticity modulus, $p_y = \frac{Eh^3}{b^2}, p_x = \frac{Eh^3}{a^2}$—longitudinal loads regarding x and y respectively. Bars over the non-dimensional quantities are omitted. Harmonic load $p_x = p_0 + p_1 \sin \omega_p t$, $p_y = p_0 + p_2 \sin \omega_p t$, where $p_0 = \text{const}$, ω_p is the frequency of harmonic excitation, and p_1 and p_2 are the amplitudes of the excitation. In addition, $q = q_0 + q_1 \sin \omega_q t$, where $q_0 = \text{const}$, ω_q, q_1 are the frequency and amplitude of the transversal harmonic load, respectively.

The system of PDEs (15.161) is reduced to ODEs via the FDM (Finite Difference Method) with approximation $O(h^2)$ regarding the spatial coordinates x and y. First, equations of nonlinear ODEs in time are solved via the fourth-order Runge–Kutta method with respect to the deflection w. Then, the values w are substituted into the right-hand side of the second system of ODEs. Therefore, the second equation becomes linear, and it is solved using the method of inversed matrix regarding the Airy's function F on each time step. The latter is chosen via the Runge principle. The number of FDM partitions $n = 14$. Discussion of the influence of n on the obtained results can be found in [31], where the rectangular plate is studied. It is shown, among others, that convergence of the results can be obtained in the averaged meaning, i.e. via the estimation of wavelets spectra and Lyapunov exponents. In the case of chaotic vibrations only the integral convergence is achieved, whereas for small amplitudes of the exciting loads also the convergence regarding regular vibrations can be obtained.

(i) *Reliability of the Results*

Reliability of the results is examined via the relaxation method applied for the first time for shells by Feodos'ev [88]. Since stability loss of any deformed system is a process which takes place in time, it should be studied from the point of view of dynamics. However, in many cases stability of the majority of constructions carrying the load can be estimated by a static method (in the case of a conservative system it yields the same results as those obtained using a dynamical approach [215]).

Solving the Cauchy problem for $\varepsilon = \varepsilon_{cr}$, for a series of constant load $\{P_i\}$, we get a sequence of deflection $\{w_i\}$. The value of deflection w should tend to steady-state. Then the dependencies $p_x(w_{\text{st}})$ and $p_y(w_{\text{st}})$ are constructed and the strain-stress system state is investigated. Observe that in order to initiate vibrations the shell had initially introduced imperfection of the magnitude of $q_0 = 0.001$.

We compared the dependencies $p_x(w_{\text{st}})$ and $p_y(w_{\text{st}})$ for the fixed parameters $k_x = 12, k_y = 0$ and $k_x = 0, k_y = 12$ (Fig. 15.43a); $k_x = 24, k_y = 0$ and $k_x = 0$, $k_y = 24$ (Fig. 15.43b); $k_x = 48$, $k_y = 0$ and $k_x = 0$, $k_y = 48$ (Fig. 15.43c). In order to solve the second governing equation via the inversed matrix method, it is necessary to build this equation in the corresponding matrix form and to construct the matrix A $[n^2 \times n^2]$ and the column matrix B containing n^2 rows. Boundary conditions regarding x and y appear in B in a different way. Comparing the obtained results reported in the mentioned figures it is seen that the curves coincide (the difference is less than 1%). The difference appears only for the shells with two geometric parameters $k_x = 12;\ 24$, $k_y = 0$ and $k_x = 0$, $k_y = 12;\ 24$. It happens when the same longitudinal load along the shell perimeters is applied in unstable

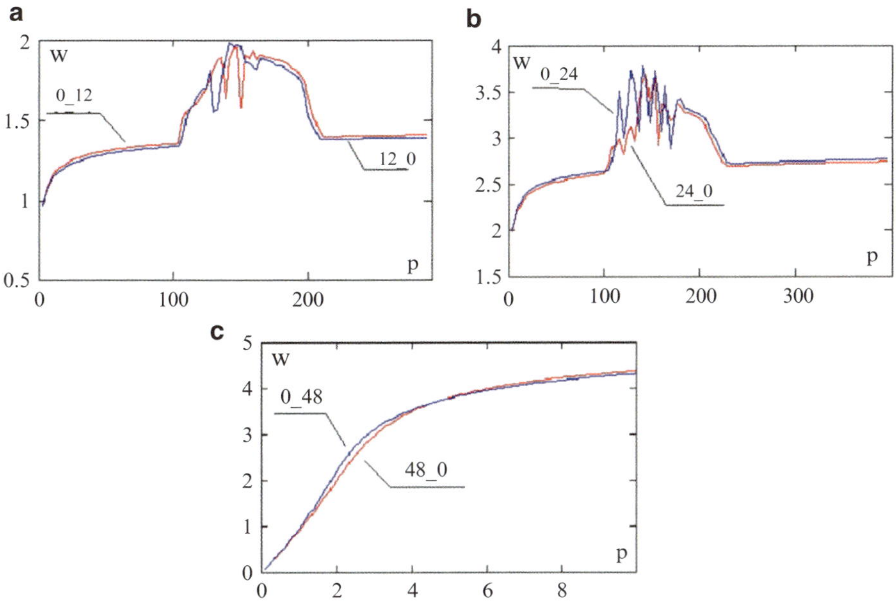

Fig. 15.43 Dependencies $w(p)$ for $k_x = 12$, $k_y = 0/k_x = 0$, $k_y = 12$ (**a**); $k_x = 24$, $k_y = 0/k_x = 0$, $k_y = 24$ (**b**); $k_x = 48$, $k_y = 0/k_x = 0$, $k_y = 48$ (**c**)

zones, i.e. when the load intensity belongs to (100; 200). In the latter case non-unique solutions are observed, and this case is described and studied in [26, 140].

The zone of instability occurs only for high order loads. Dependencies $p_x(w_{st})$ and $p_y(w_{st})$ for curvatures $k_x = 48$, $k_y = 0$ and $k_x = 0$, $k_y = 48$ are computed for the load intensity $p_0 \in (0; 10)$. For the longitudinal loads of $p_0 > 10$, the deflection values are not within the assumed hypotheses regarding the introduced shell model. In the reported results we do not observe zones of stability loss and the difference in results does not reach 1%. This validates the reliability of the results and correctness of the applied algorithms.

Since we studied the squared shell with $a = b$, intensity of the applied loads in both directions is the same for each experiment. In other words, the given pairs of curves describe in fact the same physical models. Therefore, the analysis carried out using the Feodos'ev method with respect to the geometric parameters shows good coincidence with the physical aspects of the investigated process. This validates the reliability of the results and correctness of the applied algorithms.

(ii) *Wavelet Analysis*

Observe that signals obtained as a result of the numerical experiments are presented in time domain. To visualize the signal we need time (independent variable) as one coordinate, and amplitude as a dependent variable, i.e. we should get an amplitude-time signal representation. For the purpose of a qualitative investigation we need to study the frequency spectrum of a signal, i.e. the set of its frequency components. The Fourier transformation has been applied for a long time to study frequencies of a signal. However, from the point of view of exact analysis and detection of the local signal properties, the Fourier series has a lot of limitations and drawbacks. Being well localized in the frequencies domain, it does not yield time representation. It is well known that practically all signals obtained while studying dynamics of nonlinear systems are non-stationary. This fact indicates difficulties while applying the standard Fourier approach. The theory of wavelets, which is an alternative approach to the Fourier analysis, offers deeper techniques of signal analysis. The main advantage of the wavelet analysis relies on a possibility of monitoring of the signal localized properties, whereas the Fourier analysis fails to solve the latter task. The Fourier coefficients express characteristic features of the studied signal within the whole time interval. In other words, if we study a complex signal using the Fourier analysis, i.e. a signal whose characteristics change in time, then in the output we will get the sum of all features exhibited by its local behaviour.

Signals produced by numerical simulations while investigating the continuous mechanical systems often have a complex structure. Their frequency characteristics strongly change in time. Therefore, in this paper in spite of the classical Fourier analysis, the wavelet analysis is applied, which allows us to detect a number of interesting peculiarities of vibrations of the studied systems.

A first key point requiring a serious investigation concerns the choice of a wavelet, which entirely depends on the character of the studied problem. In order to solve the given problem, we consider a non-stationary signal obtained in a numerical experiment. Here we consider the shell with parameters $k_x = 24$, $k_y = 0$, we apply the harmonic longitudinal load in the directions of axis x and y with $\omega_p = 6.7 < \omega_0$, and amplitude $p_1 = 4.9$ (ω_0 is the natural shell frequency). For a given signal various wavelet spectra are constructed [75, 76, 163, 173, 174, 220].

The Haar wavelet is badly localized in the frequency domain, whereas the Shannon wavelet is badly localized in time. Analysis of the wavelets spectra obtained with the help of the Daubechies wavelets, coiflets and symlets shows that an increase of the order of the applied filter implies an increase of the wavelet resolution regarding frequency.

In spite of the differences in the wavelets forms, the wavelet spectra obtained on the basis of the Daubechies wavelets, coiflets and symlets of the same order are practically the same. However, they do not allow us to get a sufficient frequency localization of the investigated vibrating continuous mechanical systems. Considering the results obtained on the basis of the Gauss function derivatives, the accuracy of frequency estimation increases with an increase of the derivative order.

Table 15.1 gives results obtained via Meyer, Morlet (real and complex), Gauss (real and complex) wavelets from 16 up to 32 derivative order, and the Daubechies 16 wavelet.

The data given in Table 15.1 show that the localization with respect to frequency increases with an increase in the number of zero-order moments of an applied wavelet. Complex Morlet and Gauss wavelets exhibit better localization regarding the frequencies than their real counterparts, but the time localization is better in the

Table 15.1 Frequency vs. time (wavelet spectra of different wavelets)

Meyer	Daubechies 16
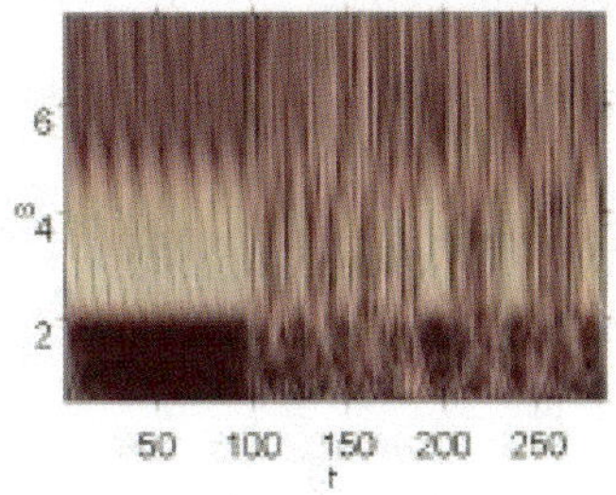	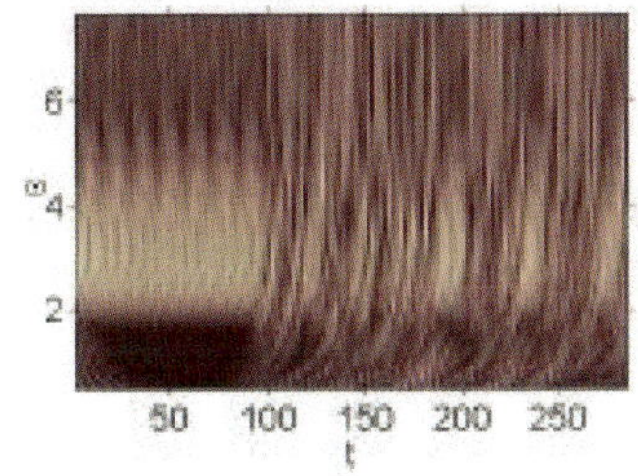
Morlet	**Complex Morlet**
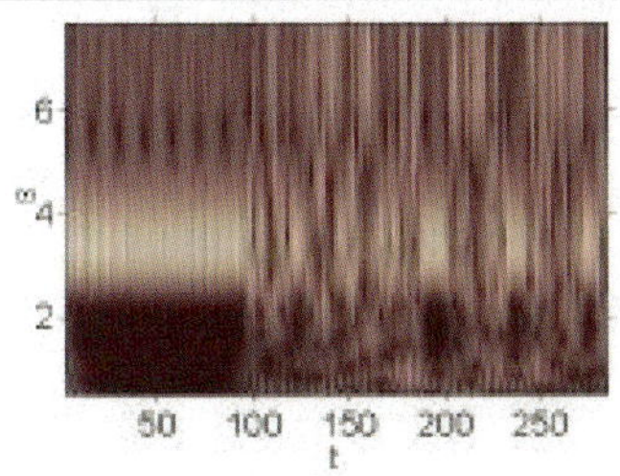	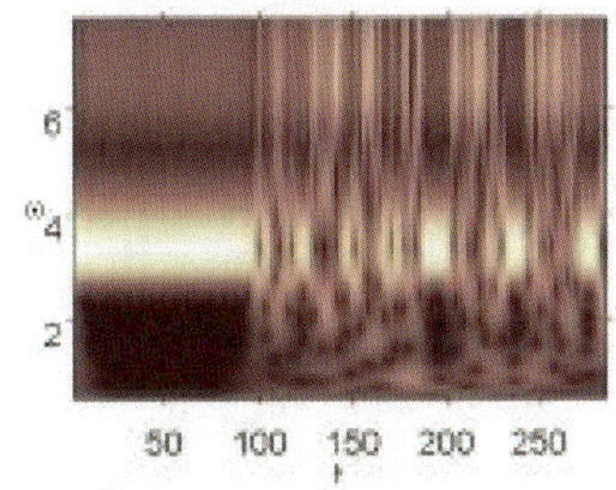
Gauss 16	**Complex Gauss 16**
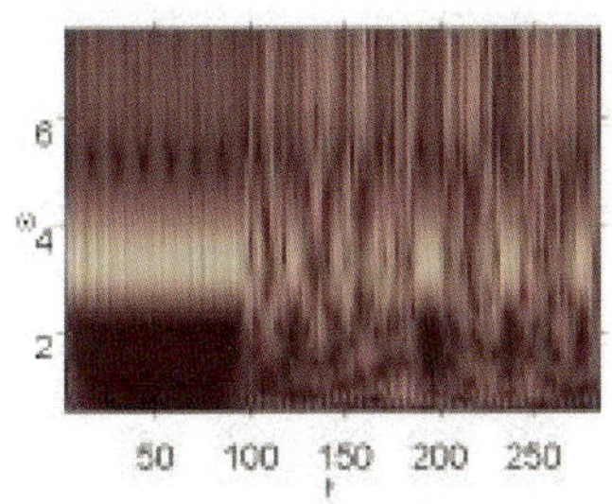	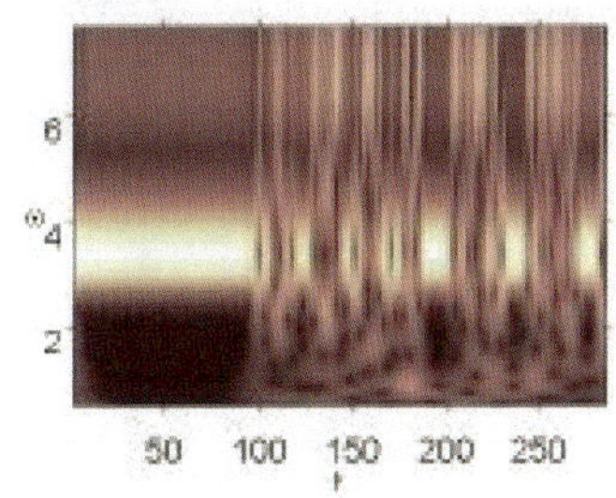
Gauss 32	**Complex Gauss 32**
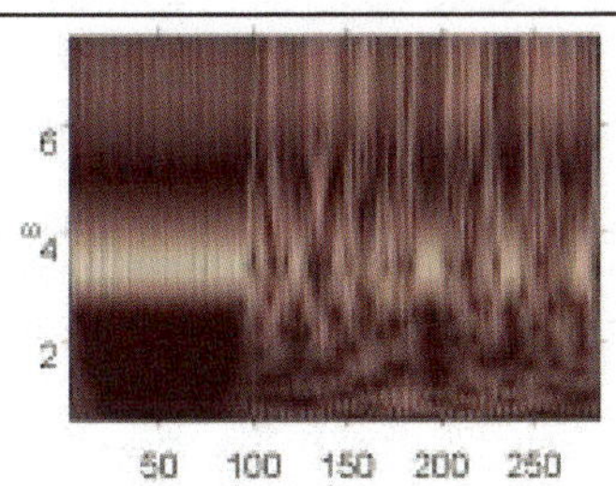	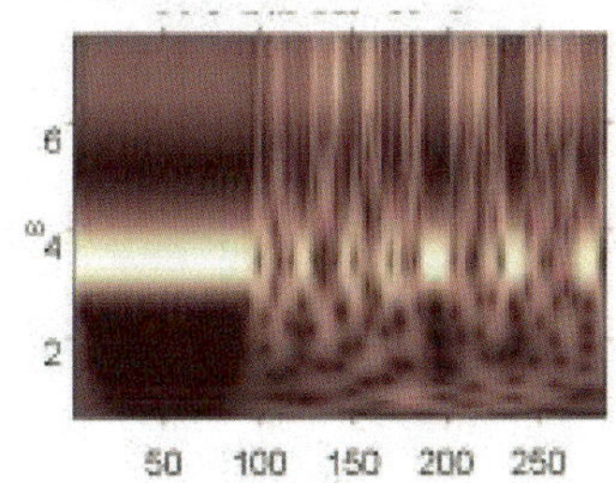

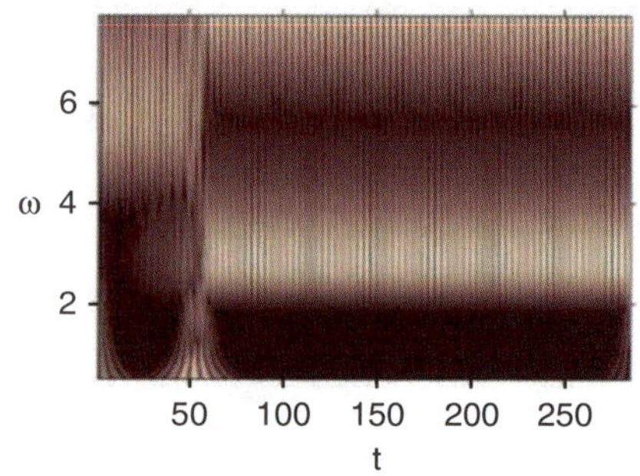

Fig. 15.44 2D wavelet spectrum of the plate ($\omega_p = \omega_0 = 5.8$, $k_x = k_y = 0$)

case of real wavelets. Therefore, in order to study chaotic vibrations of plates and shells one can apply either complex or real Morlet wavelets, as well as the real and complex wavelets obtained via high order differentiation of the Gauss function.

(iii) *Numerical Results*

Numerical experiment carried out for ($\omega_p = \omega_0 = 5.8$, $k_x = k_y = 0$) yielded a bifurcation for a small excitation amplitude with a sudden reconstruction of the plate vibrations character, which is shown by 2D wavelet spectrum in Fig. 15.44. In the initial time interval excitation frequency is exhibited, and then beginning from $t \approx 50$ the first subharmonic $\omega_1 = 2.9 = \omega_p/2$ dominates. Since in this case the change of the vibration character takes place through a narrow chaotic window, therefore the application of Fourier transformation in the whole time interval is not feasible to monitor peculiarities of evolution of the frequency characteristics in time.

However, as expected, the application of Fourier analysis in each of shorter time interval coincides with the results obtained by the wavelets spectrum (second bifurcation for $p_1 = 1.1$ takes place for $t \geq 120$). The monitored scenario is a kind of modification of the Feigenbaum scenario (see Table 15.2).

While investigating a shell with geometric parameters $k_x = k_y = 12$ for $\omega_p = 5.7 < \omega_0$ the following scenario of transition from periodic to chaotic vibrations is observed. For the excitation amplitude 1.7999, the Fourier spectrum exhibits a pair of dependent frequencies $\omega_2 = \omega_p - \omega_1$ ($\omega_1 = 1.644$, $\omega_2 = 4.36$). For $p_1 = 1.8$ the Fourier spectrum consists already of two pairs of non-commensurable frequencies and one more frequency of a third pair with small amplitude (Table 15.3a). The increase of control parameter p_1 makes the Fourier spectrum noisy implying chaotic vibrations of the shell. The monitored wavelet spectra approve that the transition into chaotic vibrations is realized via the Pomeau–Manneville route.

We pay more attention to the information obtained by the wavelet spectra. The wavelet spectra register frequency $\omega_1 = 1.644$ for $p_1 = 1.81$, but only in the initial time interval (Table 15.3b). This low information property of the applied apparatus is caused by the domination of excitation frequency over the remaining frequencies. However, this drawback is removed while constructing a wavelet spectrum with the frequency constrained. Namely, considering the interval of frequencies for $\omega < 4$, the mentioned phenomenon is now well reported (Table 15.3c).

Table 15.2 The Fourier $S(\omega)$ and wavelet 2D spectra for $k_x = k_y = 0$, $\omega_p = \omega_0 = 5.8$, $p_1 = 1.1$

$S(\omega)$ $t \in [0,286]$	$S(\omega)$, $t \in [0,125]$
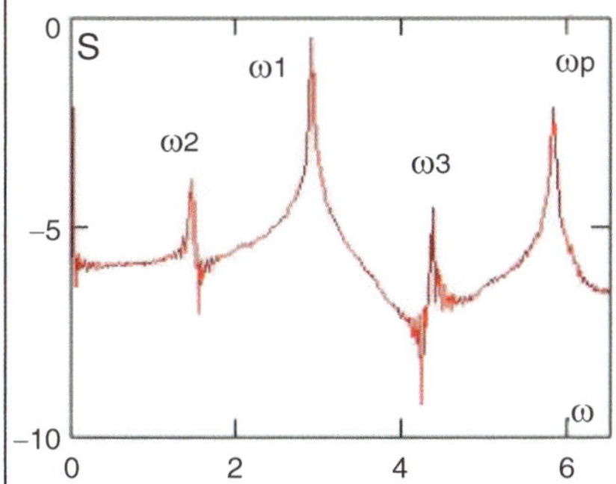	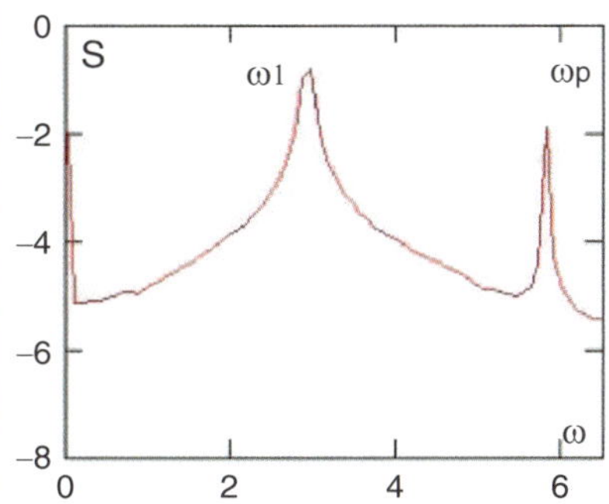
$S(\omega)$, $t \in [126,286]$	2D wavelet spectrum
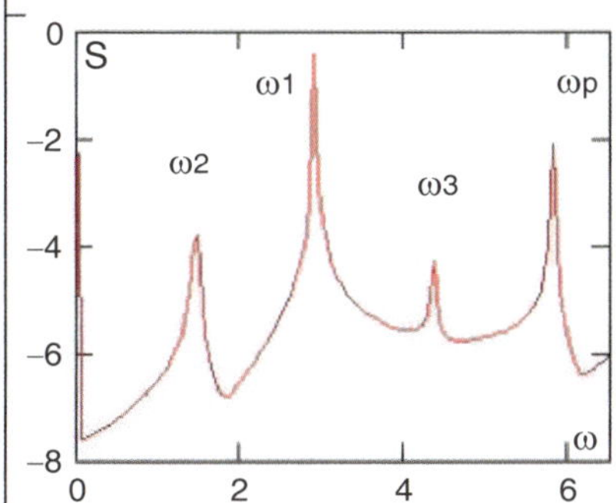	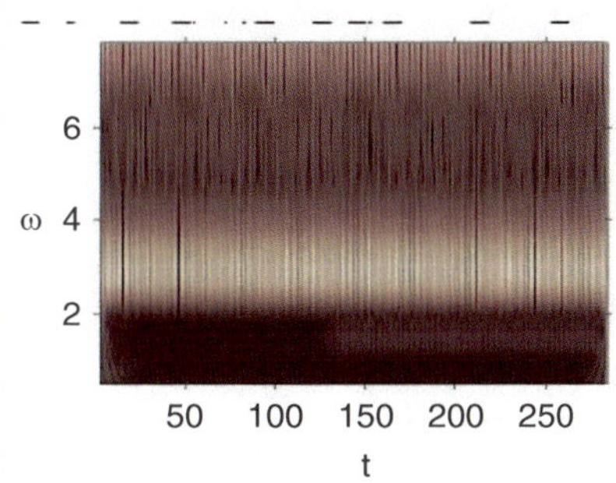

The wavelet spectrum corresponding to the periodic Fourier spectrum ($p_1 = 0.1$) exhibits two pairs of linearly independent frequencies which are detected by the Fourier spectrum for $p_1 = 1.81$. Therefore, the frequency spectrum constructed on the basis of the wavelet transformation allows for the detection and monitoring of frequency characteristics of vibrations.

The numerical simulation for the fixed parameters: $k_x = k_y = 24$, $\omega_p = \omega_0 = 24.8$ shows that a number of linearly dependent frequencies may increase not only due to the increase of a control parameter, but even for its fixed value ($p_1 = 0.1$) owing to the modified Ruelle–Takens scenario. Namely, in the interval $t < 150$ we have two pairs of frequencies $\omega_2 = \omega_p - \omega_1$ and $\omega_3 = \omega_p - \omega_5$, where $\omega_1 = 3.927$, $\omega_2 = 20.873$, $\omega_3 = 7.854$, $\omega_4 = 16.946$. In time interval $t \in [150, 250]$ the frequency $\omega_5 = 10.21$ appears which belongs to a third pair. Then the number of frequencies increases more in the finite time interval. The signal time evolution is well demonstrated by the 2D wavelet-spectrum in Table 15.4.

Table 15.3 Fourier $S(\omega)$ and wavelet $W(\omega)$ spectra for different p_1

<table>
<tr><td>**a** $S(\omega)$ (p_1=1.81)</td><td>**b** $W(\omega)$ (p_1=1.81)</td><td>**c** $W(\omega)$ (p_1=0.1)</td></tr>
<tr><td>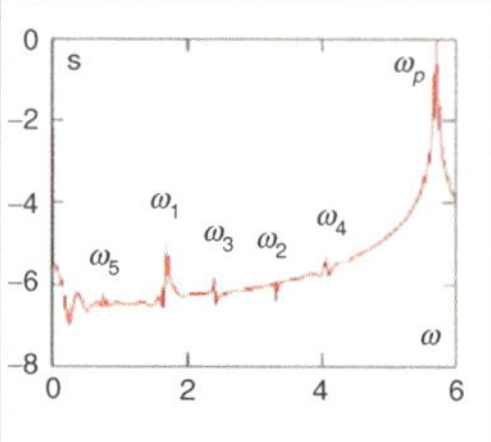
</td><td>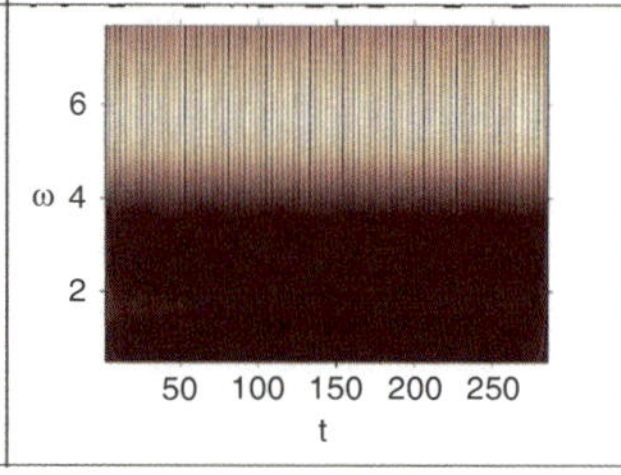
</td><td>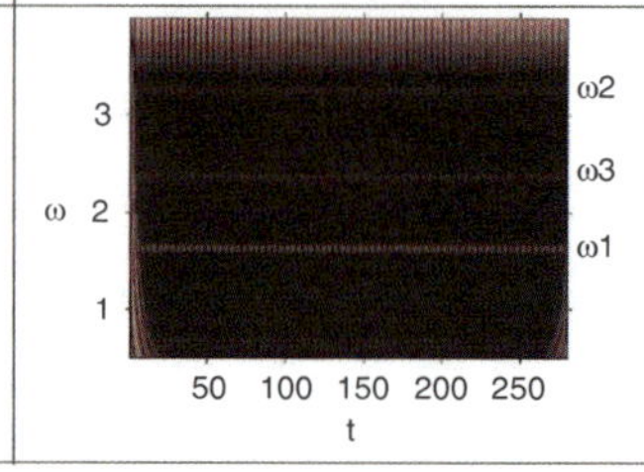
</td></tr>
</table>

Table 15.4 Fourier $S(\omega)$ and wavelet spectrum for $k_x = k_y = 24, \omega_p = \omega_0 = 24.8, p_1 = 0.1$.

<table>
<tr><td>$S(\omega)$ $t < 150$</td><td>$S(\omega)$, $t \in [150;250]$</td></tr>
<tr><td>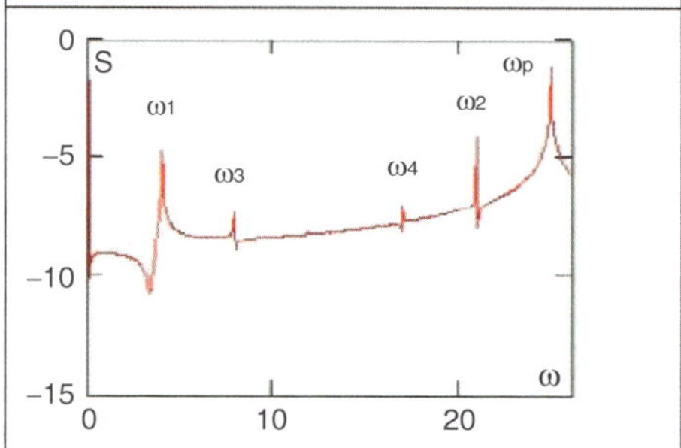
</td><td>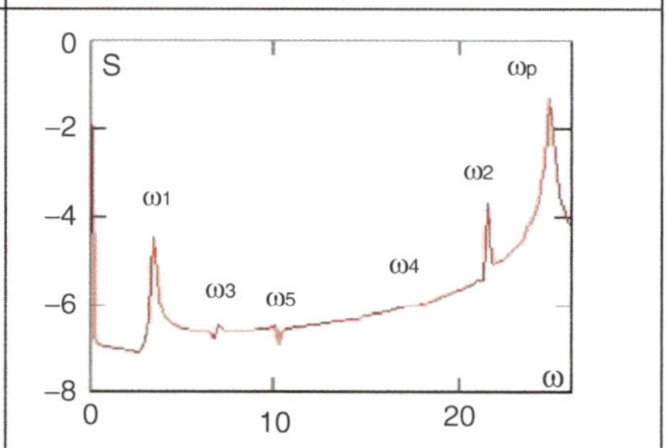
</td></tr>
<tr><td>$S(\omega)$, $t \in [250,286]$</td><td>2D wavelet spectrum</td></tr>
<tr><td>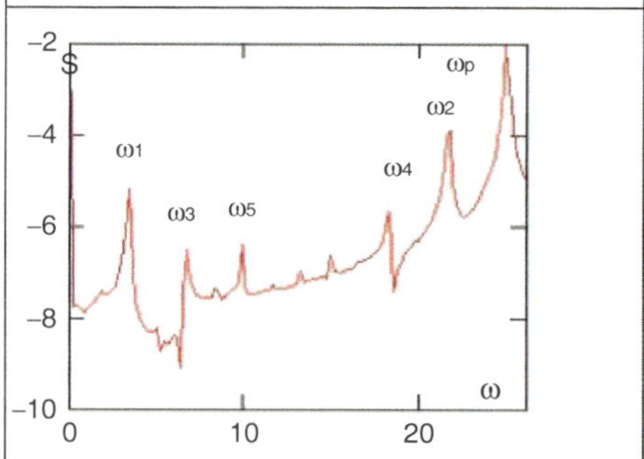
</td><td>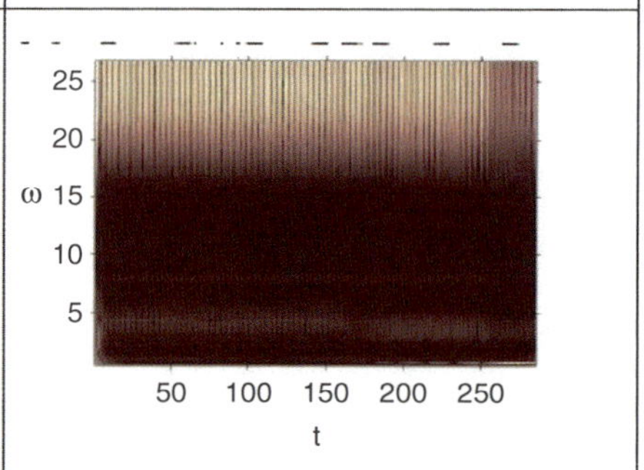
</td></tr>
</table>

15.9.3 *Two-Layer Shell*

(i) *Problem Formulation*

In this section we study a two-layer spherical flexible isotropic elastic shell with constant stiffness and density which is subjected to the action of harmonic longitudinal load (Fig. 15.45) (see [23]). The constant load q is applied only to the

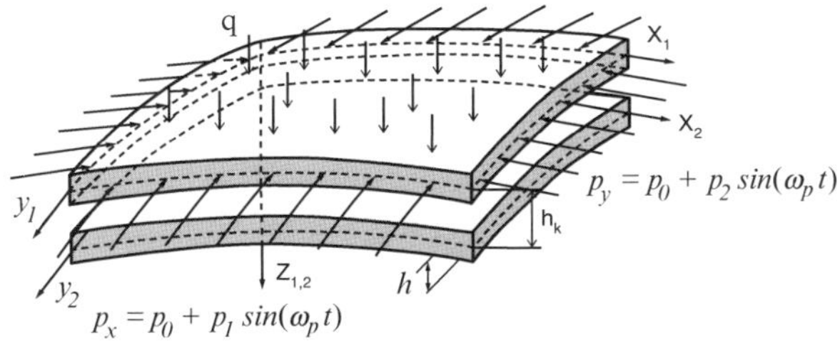

Fig. 15.45 Computational scheme of a two-layer shell

upper shell layer. The layers move freely or slide without friction. Due to small values of the contact pressure, zones of permanent sticks rather do not occur. Contact conditions occurring between the layers can depend on the coordinates, and they include all possible cases of a one-sided contact. The hypotheses are the same as for the one-layer shell. The contact pressure function is excluded from a number of unknowns. The shell occupies the following space: $\Omega_1 = \{x_1, y_1, z_1 | (x_1, y_1) \in [0; a] \times [0; b], z_1 \in [-h_1; h_1]\}$; $\Omega_2 = \{x_2, y_2, z_2 | (x_2, y_2) \in [0; a] \times [0; b], z_2 \in [-h_2; h_2]\}$, $0 \le t < \infty$.

The governing equations of the theory of flexible shallow shells [239] taking into account a contact between the layers [133] have the following non-dimensional form:

$$\frac{1}{12(1-\mu^2)}\nabla_\lambda^4 w_m - L(w_m, F_m) - \nabla_k^2 F_m + q \pm K(w_1 - h_k - w_2)\Psi = \frac{\partial^2 w_m}{\partial t^2} + \varepsilon_1 \frac{\partial w_m}{\partial t},$$

$$\nabla_\lambda^4 F_m = -\frac{1}{2} L(w_m, w_m) - \nabla_k^2 w_m, \tag{15.164}$$

where

$$\nabla_\lambda{}^4 = \frac{1}{\lambda^2}\frac{\partial^4}{\partial x_m{}^4} + \lambda^2 \frac{\partial^4}{\partial y_m{}^4} + 2\frac{\partial^4}{\partial x_m{}^2 \partial y_m{}^2}, \quad \nabla_k{}^2 = k_{x_m}\frac{\partial^2}{\partial x_m{}^2} + k_{x_1}\frac{\partial^2}{\partial y_m{}^2},$$

$$L(w_m, F_m) = \frac{\partial^2 w_m}{\partial x_m^2}\frac{\partial^2 F_m}{\partial y_m^2} + \frac{\partial^2 w_m}{\partial x_m^2}\frac{\partial^2 F_m}{\partial y_m^2} - 2\frac{\partial^2 w_m}{\partial x_m \partial y_m}\frac{\partial^2 F_m}{\partial x_m \partial y_m},$$

$$\psi = \frac{1}{2}\left[1 + \operatorname{sign}(w_1 - h_k - w_2)\right].$$

Here w_m and F_m are the deflection and stress functions, respectively; $m = 1, 2$; K is the stiffness coefficient of the transversal contact zone. We have $\Psi = 1$, if $w_1 > w_2 + h_k$, i.e. contact between panels occurs, or $\Psi = 0$; w_1, w_2 denote deflection of the upper and lower panel, respectively. The following boundary conditions are attached:

$$
\begin{aligned}
&w_1 = 0, \quad \frac{\partial^2 w_1}{\partial x_1^2} = 0, \quad F_1 = 0, \quad \frac{\partial^2 F_1}{\partial {x_1}^2} = p_{y_1}(t) \quad \text{for} \quad x_1 = 0, 1, \\
&w_1 = 0, \quad \frac{\partial^2 w_1}{\partial y_1^2} = 0, \quad F_1 = 0, \quad \frac{\partial^2 F_1}{\partial y_1^2} = p_{x_1}(t) \quad \text{for} \quad y_1 = 0, 1, \\
&w_2 = 0, \quad \frac{\partial^2 w_2}{\partial x_2^2} = 0, \quad F_2 = 0, \quad \frac{\partial^2 F_2}{\partial {x_2}^2} = 0 \quad \text{for} \quad x_2 = 0, 1, \\
&w_2 = 0, \quad \frac{\partial^2 w_2}{\partial y_2^2} = 0, \quad F_2 = 0, \quad \frac{\partial^2 F_2}{\partial y_2^2} = p_{x_2}(t) \quad \text{for} \quad y_2 = 0, 1,
\end{aligned}
\tag{15.165}
$$

where $p_{x_1}(t) = p_0 + p_1 \sin(\omega_p t$, $p_{y_1}(t) = p_0 + p_2 \sin(\omega_p t)$ denote the longitudinal loads. The initial conditions are as follows:

$$
w_m(x_m, y_m)|_{t=0} = \phi_1(x_m, y_m), \frac{\partial w_m}{\partial t} = \phi_2(x_m, y_m). \tag{15.166}
$$

Equation (15.164) is transformed to non-dimensional using the following relations: $x_m = a\bar{x}_m$, $y_m = b\bar{y}_m$; $\bar{k}_{x_m} = k_{x_m}\frac{a^2}{h_m}$, $\bar{k}_{y_m} = k_{y_m}\frac{b^2}{h_m}$, $k_{x_m} = \frac{1}{r_{xm}}$, $k_{y_m} = \frac{1}{r_{ym}}$, $q_m = \bar{q}_m\frac{E_m h_m^4}{a^2 b^2}$, $\tau_m = \frac{ab}{h_m}\sqrt{\frac{\rho_m}{E_m g_m}}$, $\lambda_1 = \frac{a}{b}$, where a, b are the dimensions of the rectangular cylindrical panel regarding x_m and y_m, respectively; h_m denotes the shell thickness; g_m is the Earth acceleration; $\rho_m = \gamma_m h_m$, where γ_m is the volume weight density; r_{x_m}, r_{y_m} is the curvature radius of the shell regarding x_m and y_m, respectively. Furthermore, t is time, ε_m is the damping coefficient, $\mu = 0.3$ is Poisson's coefficient for the isotropic material, E_m is the elasticity modulus, $q_m(x, y, t)$ denotes the transversal load, and $K = \overline{K}\frac{h^4 b}{a^4}$ is the stiffness coefficient of the contact zone. Bars over non-dimensional quantities are omitted.

In order to reduce PDEs (15.164) to ODEs we apply FDM (Finite Difference Method) with approximations $O(c^2)$ regarding spatial coordinates. The obtained Cauchy problem is solved via the fourth-order Runge–Kutta method. Simultaneously, on each time step a linear system of algebraic equations is solved.

(ii) *Phase Chaotic Synchronization*

We introduce phase $\phi(t)$ of a chaotic signal [188,199], with its frequency denoted as an averaged phase velocity $\langle\dot{\phi}(t)\rangle$. There is no universal way to introduce the phase of a chaotic signal which gives correct results for an arbitrary dynamical system. Here, we apply wavelets to detect a regime of chaotic synchronization of mechanical dynamical systems with a badly defined phase. Dynamics of the mentioned systems can be characterized with the help of a continuous set of phases which are defined by a continuous wavelet transformation of the chaotic signal $w(t)$ [147] in the following form

$$V(S,t_0)=\int_{-\infty}^{+\infty} w(t)\psi^*_{s,t_0}(t)dt,\quad \psi_{s,t_0}(t)=\frac{1}{\sqrt{s}}\psi_0\left(\frac{t-t_0}{s}\right),$$

where $\psi_{s,t_0}(t)$ is the wavelet function, obtained from wavelet $\psi_0(t)$, where $(*)$ denotes a complex conjugate. Time scale s defines wavelet width, and t_0 is the time shift of the wavelet function along time axis. We take the Morlet wavelet of the form $\psi_0(\eta)=\pi^{-1/4}\exp(j\omega_0\eta)\exp\left(-\eta^2/2\right)$. Owing to the choice of $\omega_0=2\pi$, we keep the $s\approx 1/\omega$ ratio between time scale s of the wavelet transformation and the frequency ω of the Fourier transformation. Therefore, the time scale s within the wavelet analysis corresponds to the frequency yielded by the Fourier analysis. The wavelet surface $V(s,t_0)=|V(s,t_0)|\exp(j\phi_S(t_0))$ characterizes the system behaviour on each time scale s in the arbitrary time instant t_0. Magnitude $|V(s,t_0)|$ characterizes the time scale s in the given time instant t_0. Analogously is defined the phase $\phi_s(t)=\arg V(s,t)$ for each time scale s. Therefore, behaviour of each time scale s is characterized by the corresponding phase $\phi_s(t)$.

If the structural members are out of the synchronization regime, their behaviour is asynchronous on all time scales s. If synchronization takes place on certain time scales, then the phase synchronization occurs. It is clear that firstly these time scales are synchronized where the largest system energy is transmitted via the wavelet spectrum. Consequently, the phase synchronization implies the phase locking in the synchronized time scales: $|\phi_{s1}(t)-\phi_{s2}(t)|$ <const, where $\phi_{s1}(t)$ and $\phi_{s2}(t)$ are the continuous phases of the first and second shells respectively, corresponding to the synchronization of the time scales s.

(iii) *Numerical Experiment*

We consider nonlinear dynamics of the flexible two-layer shell (plate) with curvatures $k_{x_m}=0$, $k_{y_m}=0$, where the first plate is subjected to harmonic longitudinal load $p_{x_1}(t)=p_{y_1}(t)=p_1\sin(\omega_p t)$, and $\omega_p=5.6$, $K=1.75\times 10^4$. In the initial time interval $0\le t\le 0.001$ we apply the uniform and constant load of $q=0.001$. The amplitude of excitation p_1 changes in the interval $p_1\in(0.5;0.5358)$, and the gap between panels is $h_k=0.5$. The obtained time histories $w_m(t)$ (Fig. 15.46), and phase portraits as well as Fourier spectra are shown in Table 15.5 for periodic vibrations for $p_1=0.5$.

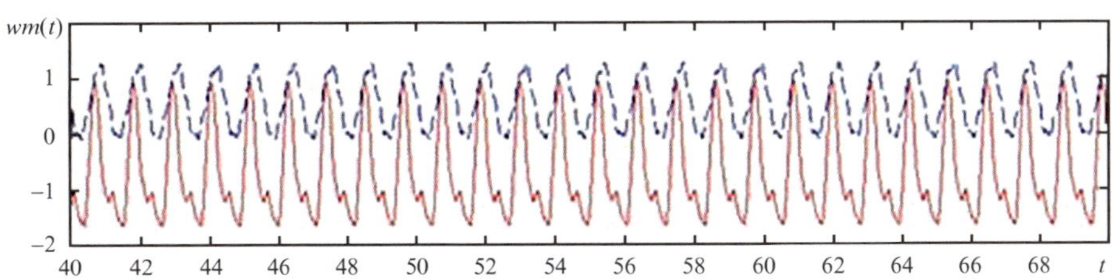

Fig. 15.46 Signals $w_m(t)$ for periodic vibrations ($p_1=0.5$)

Table 15.5 Fourier spectra and phase portraits ($p_1 = 0.5$)

Upper plate		Lower plate	
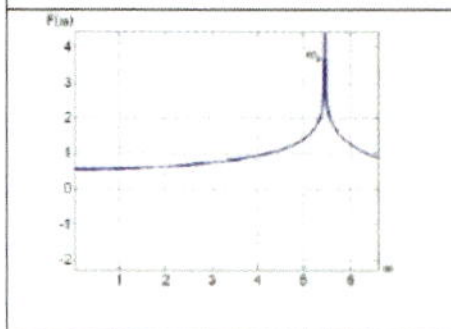	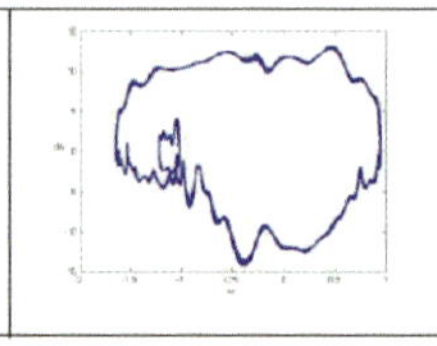	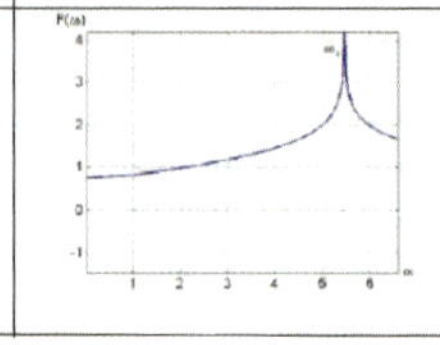	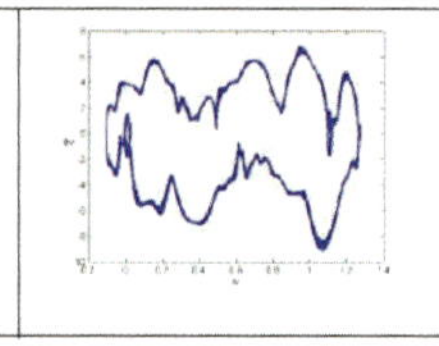

Table 15.6 Fourier spectra and phase portraits ($p_1 = 0.534$)

Upper plate		Lower plate	
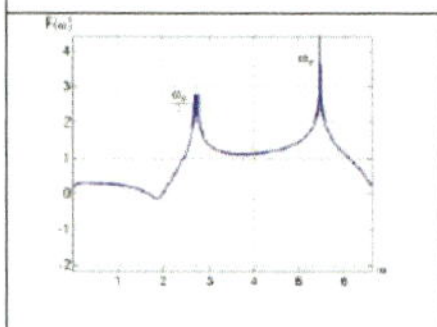	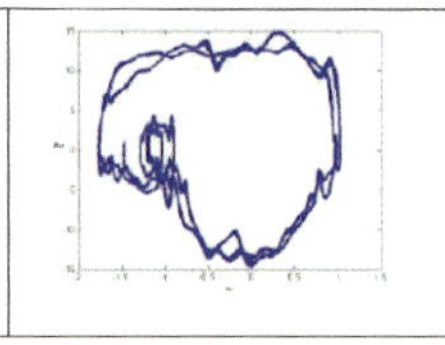	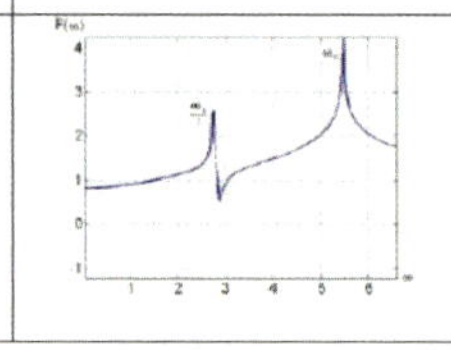	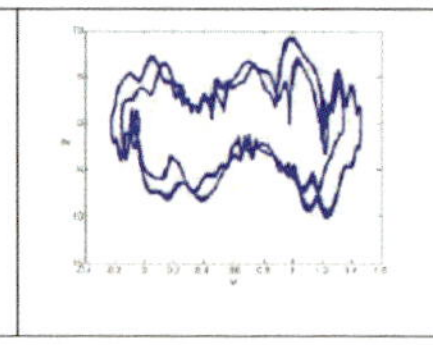

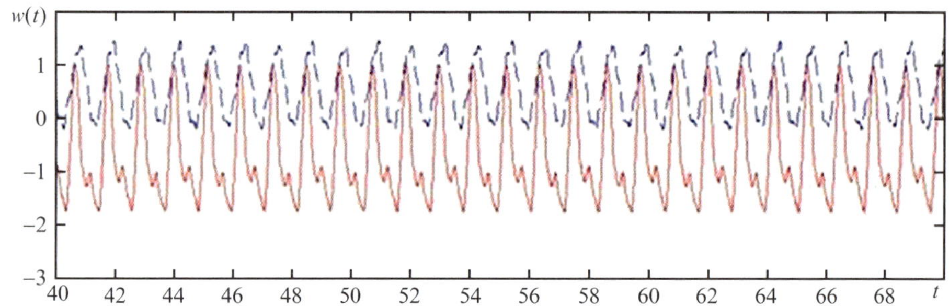

Fig. 15.47 Signals $w_m(t)$ for periodic vibrations ($p_1 = 0.534$)

Increasing the excitation amplitude implies the Hopf bifurcation. Table 15.6 presents the power spectra (time histories are shown in Fig. 15.47) and phase portraits for the upper and lower plate obtained for $p_1 = 0.534$. Hopf bifurcations are easily recognized in the power spectra, whereas phase portraits exhibit two limit cycles.

A further increase of the amplitude of the longitudinal load causes intermittency, i.e. periodic vibrations interacting with chaotic vibrations. Figure 15.48 shows time histories obtained for $p_1 = 0.534$. After an initial chaotic burst for $0 \leq t \leq 100$, time history $w_m(t)$ becomes periodic (see Table 15.7). All characteristics besides the phase difference given in Table 15.7 hold for the upper plate. The Fourier analysis carried out in time interval $t \in (100; 330)$ implies periodic vibrations. Then the system changes its dynamics in time. Power spectrum and phase portrait constructed for $t \in (330; 450)$ again exhibit chaotic vibrations. Further, for $450 < t < 532$, the signal again becomes laminar. The phase difference shows that the frequencies synchronization of both plates takes place only on the excitation frequency. A further increase of the excitation amplitude increases development of

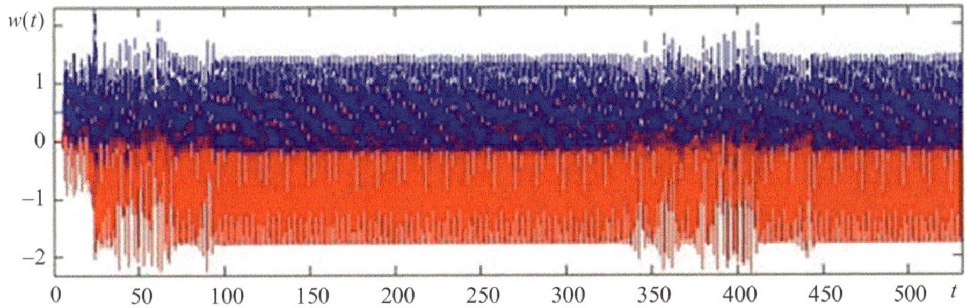

Fig. 15.48 Time histories of plate vibrations for $p_1 = 0.534$

Table 15.7 Time histories, phase portraits, power spectra and phase differences for different time intervals

$0 \le t \le 100$	$100 < t \le 330$	$330 < t \le 450$	$450 < t \le 532$
Time history $w_1(t)$			
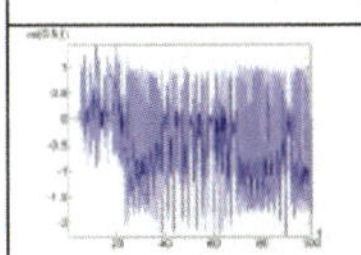	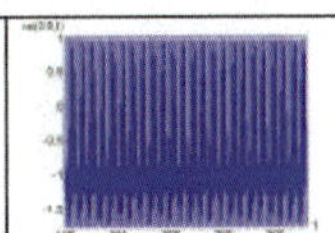	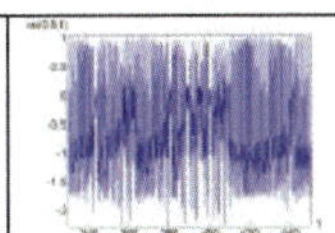	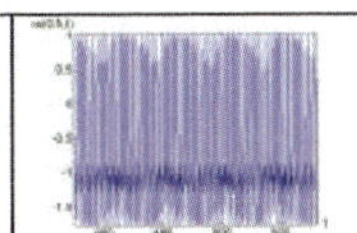
Phase portrait $w_1(\dot{w}_1)$			
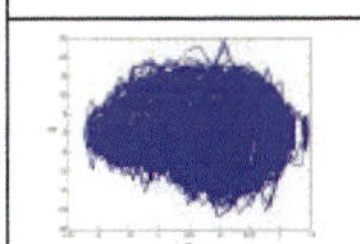	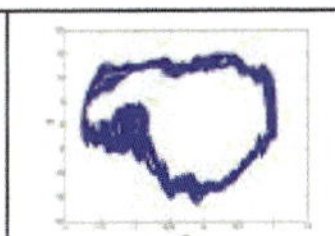	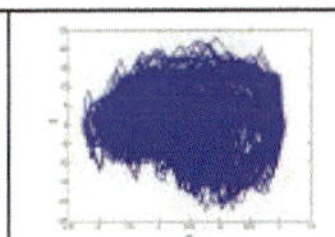	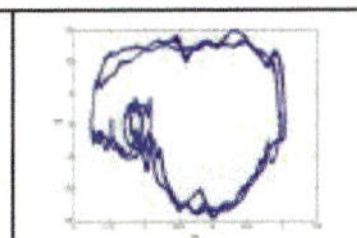
Power spectrum $F(\omega)$			
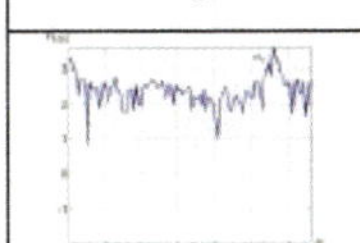	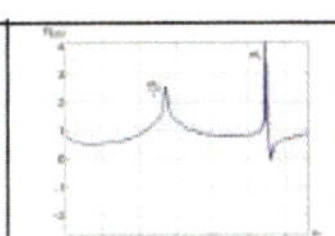	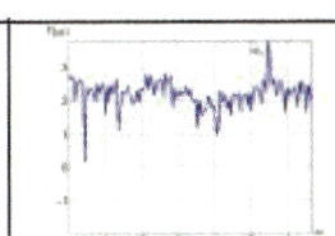	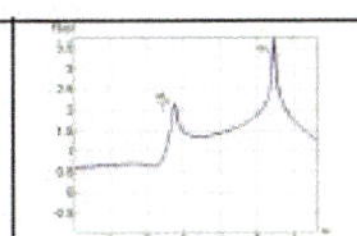
Phase difference $\varphi_1(t) - \varphi_2(t)$			
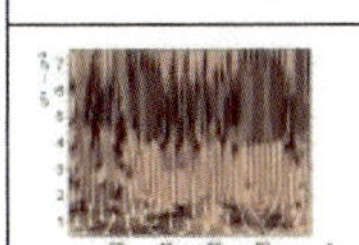	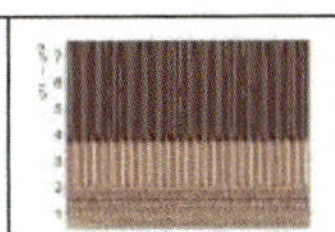	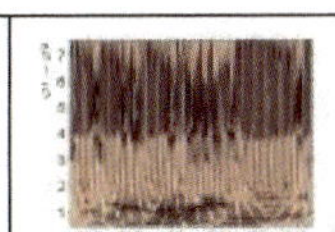	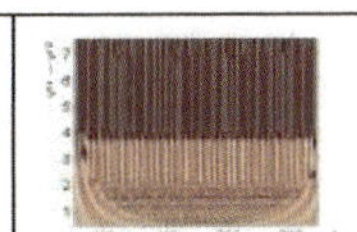

Table 15.8 Time histories, phase portraits, power spectra and phase differences for $q_1 = 0.15$, $\omega_p = 8.4$

Time history	Power spectrum (upper shell)	Power spectrum (lower shell)
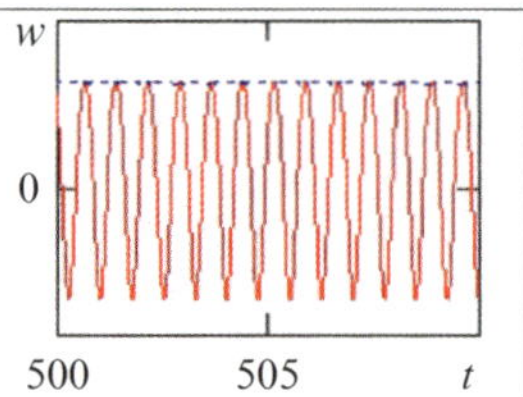	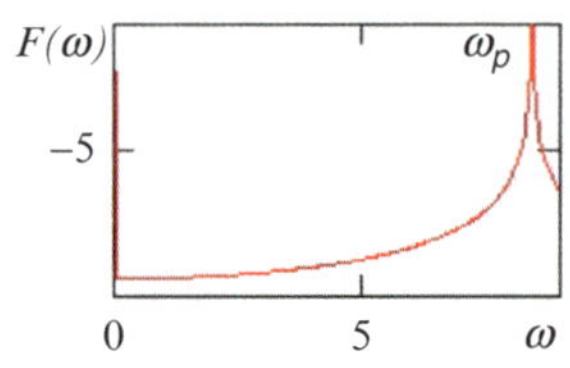	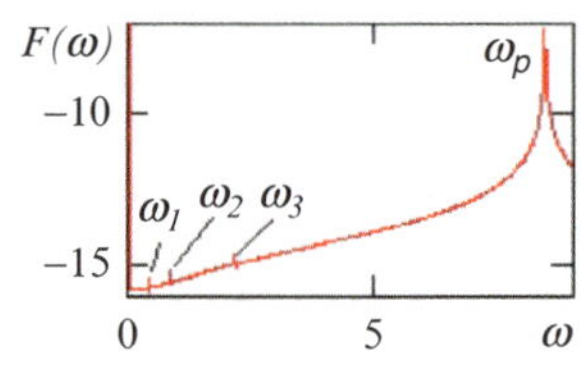

Phase portrait (upper shell)	Phase portrait (upper shell)
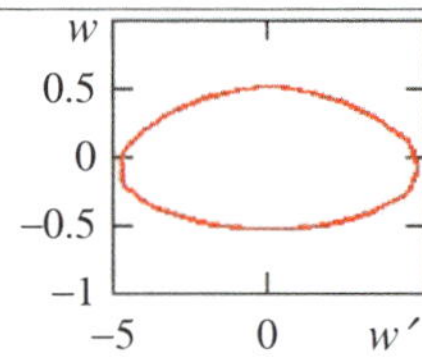	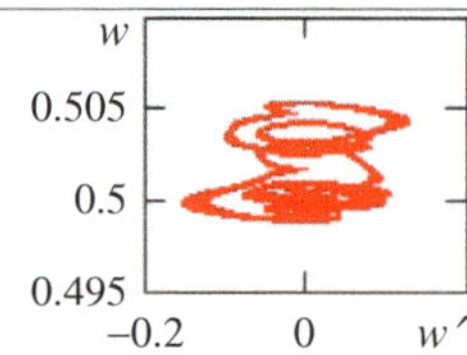

the intermittency effect. Therefore, our mechanical signal exhibits a transition from periodic to chaotic vibrations via the classical Pomeau–Manneville scenario.

Next, we investigated a two-layer shell with $k_{x_m} = 12$, $k_{y_m} = 12$, the first shell being subjected to the periodic load $p_{x_1}(t) = p_{y_1}(t) = p_1 \sin(\omega_p t)$, where $\omega_p = 8.4$, $K = 1.75 \times 10^4$. In the initial time interval $0 \le t \le 0.001$ we applied the transversal constant load with $q = 0.001$. The amplitude of excitation p_1 is changed in the interval $p_1 \in (0.15; 0.178)$, and the gap between shells is $h_k = 0.5$. Similar characteristics as in the previous case are shown in Table 15.8 for $p_1 = 0.15$. In the time instant of a contact between the shells, the first shell continues to vibrate periodically. Power spectrum of the first shell has one frequency ω_p, whereas in the power spectrum of the lower shell two frequencies $\omega_1 = 0.41724$, $\omega_3 = 2.1476$ and the linear combination of ω_1, $\omega_2 = 2\omega_1 = 0.83448$ appear. Phase portraits well coincide with the power spectra. Phase portrait for the upper (lower) shell presents a limit cycle (torus). The phase difference shows that the action of small pressure on the lower shell implies its vibration asynchronously with the upper shell (black (white) color corresponds to synchronous (asynchronous) vibrations).

A further increase of the excitation amplitude up to $p_1 = 0.177$ (Table 15.9) yields shell vibrations of the same frequencies. In the power spectrum of the upper shell one independent frequency $\omega_1 = 0.41724$ appears, and the linear combination of ω_1 is $\omega_2 = 2\omega_1 = 0.83448$, $\omega_4 = 3\omega_1 = 1.25172$. Power spectrum of the upper shell does not have the frequency ω_3. Phase portraits of both plates show tori. Chaotic synchronization of the frequencies takes place only on the frequency of excitation ω_p, and synchronization (black areas) appears in the interval of $6 < \omega < 10$. This is confirmed by the character of their simultaneous vibrations.

A further increase of $p_1 = 0.178$ (Table 15.10) forces the system to reach chaos. Power spectra exhibit broad band regions, and phase portraits exhibit black

Table 15.9 Time histories, phase portraits, power spectra and phase differences for $q_1 = 0.177$, $\omega_p = 8.4$

Time history	Power spectrum (upper shell)	Power spectrum (lower shell)
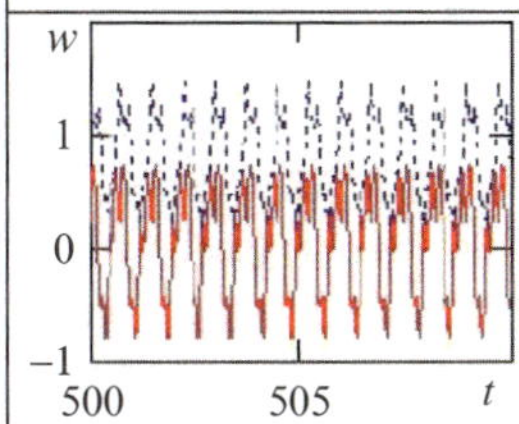	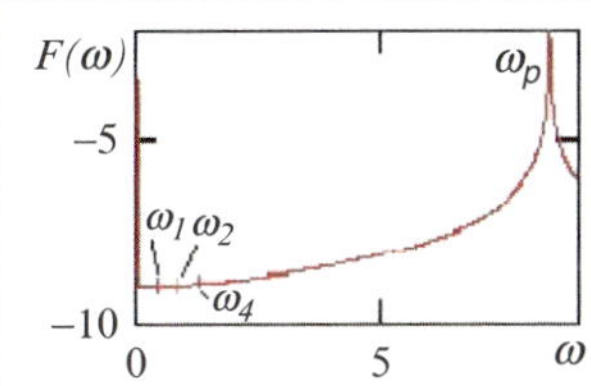	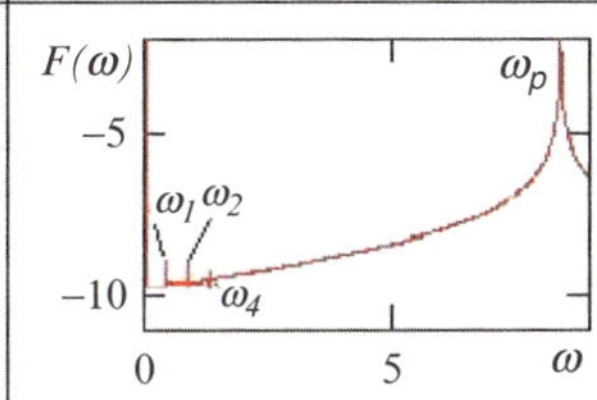
Phase difference	Phase portrait (upper shell)	Phase portrait (lower shell)
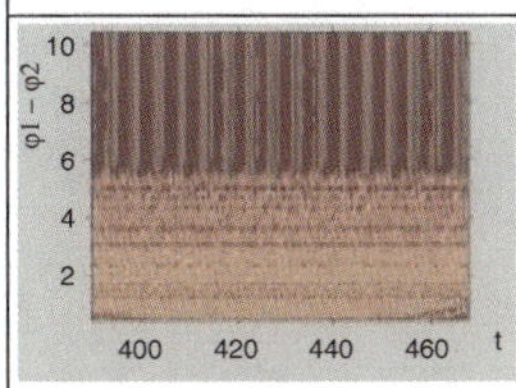	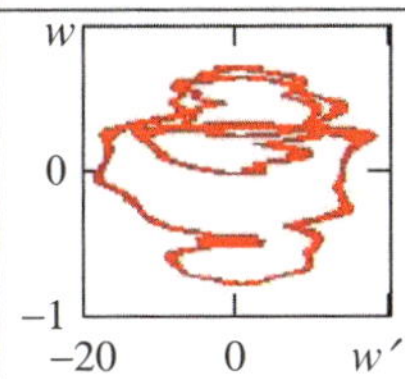	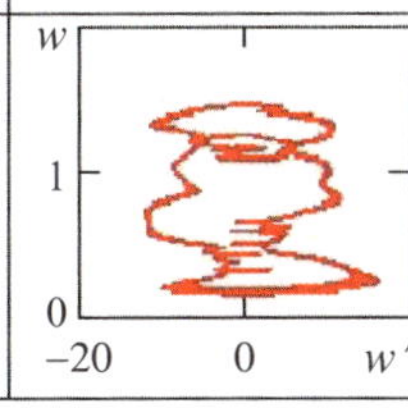

Table 15.10 Time histories, phase portraits, power spectra and phase differences for $q_1 = 0.178$, $\omega_p = 8.4$

Time histories	Power spectrum (upper shell)	Power spectrum (upper shell)
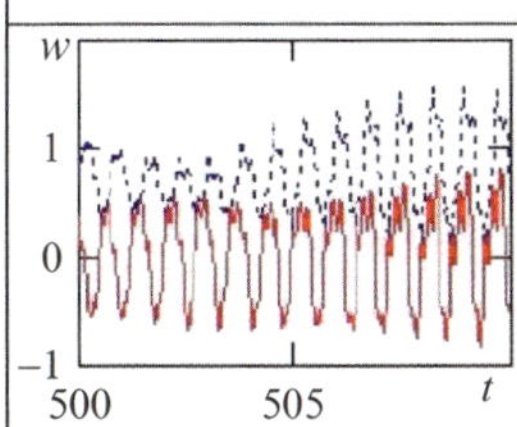	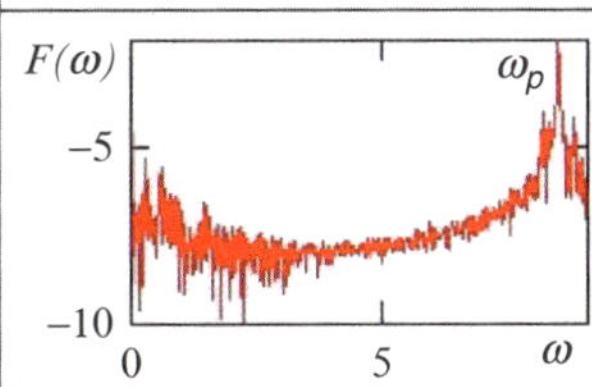	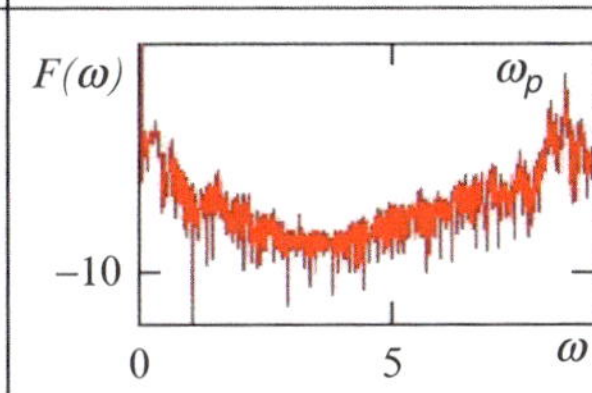
Phase difference	Phase portrait (upper shell)	Phase portrait (lower shell)
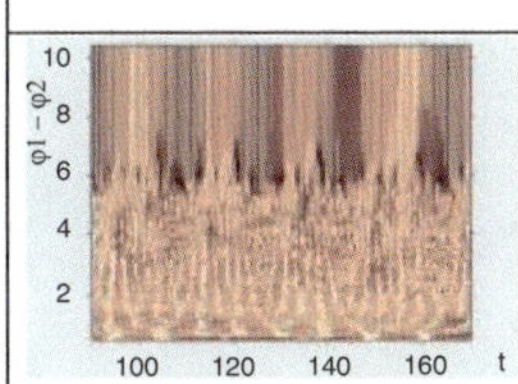	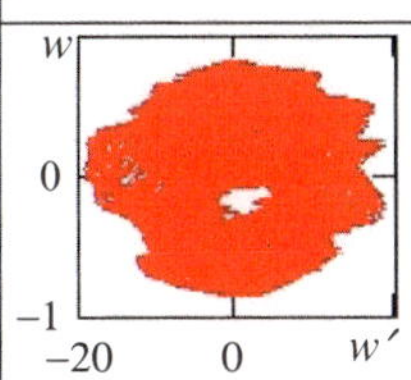	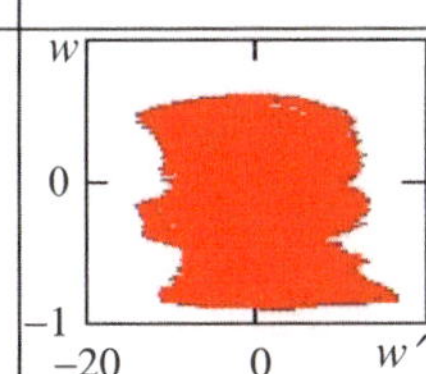

areas. Synchronization has only a local-timing character and is associated with the excitation frequency. Therefore, transition into chaos takes place for the upper and lower shells within the different scenarios: the upper shell exhibits Ruelle–Takens–Newhouse scenario.

In conclusion, the studied simply supported shells being harmonically excited along its perimeter exhibit mainly subharmonic vibrations with the frequency $\omega_p/2$. The analysed vibrations are mainly transitional exhibiting sequences of bifurcations typical for the classical scenario of transition from regular to chaotic vibrations. In addition, we have reported the modified scenarios, where the qualitative change of system vibrations appears in time, i.e. in the illustrated modified Feigenbaum scenario bifurcations appear for the fixed values of the control parameters. The modified Ruelle–Takens scenario stands for another example, where the increase of a number of dependent frequencies takes place in time.

The second part concerns dynamics of a two-layer flexible isotropic elastic shells harmonically excited taking into account the contact interaction between them. In particular, various phase chaotic synchronizations have been detected and studied. We have also reported different scenarios of a transition from periodic to chaotic vibrations of both interacting shells, among others.

References

1. V.S. Afraimovich, L.P. Shilnikov, On some global bifurcations connected with disappearance of fixed points of a saddle-node type. Soviet Math. Dokl. **15**, 1761–1765 (1974)
2. V.S. Afraimovich, L.P. Shilnikov, On a bifurcation on codimension-one leading to the appearance of fixed points of countable set of tori. Soviet Math. Dokl. **25**, 101–105 (1982)
3. M.A. Aizerman, F.R. Gantmakher, On the stability of periodic motions. J. Appl. Math. Mech. **22**, 1065–1078 (1958)
4. J.M. Almira, P.J. Torres, Invariance of the stability of Meissner's equation under a permutation of its intervals. Ann. Math. **180**, 245–253 (2001)
5. M. Amabili, Nonlinear vibrations of doubly-curved shallow shells. Int. J. Nonlin. Mech. **40**(5), 683–710 (2005)
6. M. Amabili, A. Sarkar, M.P. Paidoussis, Chaotic vibrations of circular cylindrical shell: Galerkin versus reduced-order models via the proper orthogonal decomposition method. J. Sound Vib. **290**(3–5), 736–762 (2006)
7. I.V. Andrianov, J. Awrejcewicz, New trends in asymptotic approaches: summation and interpolation methods. Appl. Mech. Rev. **54**(1), 69–92 (2001)
8. A.A. Andronov, E.A. Leontovitch, I.L. Gordon, A.G. Mayer, *Theory of Bifurcation of Dynamical Systems in Plane* (Nauka, Moscow, 1967, in Russian)
9. R. Andrzejewski, *Stability of Motion of Vehicles* (WNT, Warsaw, 1997, in Polish)
10. V.I. Arnold, *Ordinary Differential Equations* (MIT, Cambridge, 1978)
11. V.I. Arnold, *Geometric Methods in the Theory of Differential Equations* (Springer, New York, 1988)
12. J. Awrejcewicz, *Classical Mechanics. Kinematics and Statics* (Springer, New York, 2012)
13. J. Awrejcewicz, *Classical Mechanics. Dynamics* (Springer, New York, 2012)
14. J. Awrejcewicz, *Bifurcation and Chaos in Coupled Oscillators* (World Scientific, Singapore, 1991)
15. J. Awrejcewicz, Bifurcation portrait of the human vocal cord oscillations. J. Sound Vib. **136**(1), 151–156 (1990)
16. J. Awrejcewicz, Bifurcations of the oscillations of the vocal cords. J. Appl. Math. Mech. ZAMM **70**(4), 100–101 (1990)
17. J. Awrejcewicz, Numerical investigations of the constant and periodic motions of the human vocal cords including stability and bifurcation phenomena. Dyn. Stab. Syst. **5**(1), 11–28 (1990)
18. J. Awrejcewicz, Analysis of double Hopf bifurcations. Nonlin. Vib. Prob. **24**, 123–140 (1991)
19. J. Awrejcewicz, Analysis of the biparameter Hopf bifurcation. Nonlin. Vib. Prob. **24**, 63–76 (1991)

J. Awrejcewicz, *Ordinary Differential Equations and Mechanical Systems*,
DOI 10.1007/978-3-319-07659-1

20. J. Awrejcewicz, Hopf bifurcation in Mathieu-Duffing's oscillator. Nonlin. Vib. Prob. **24**, 161–172 (1991)
21. J. Awrejcewicz, On the Hopf bifurcation. Nonlin. Vib. Prob. **24**, 15–31 (1991)
22. J. Awrejcewicz, M. Holicke, *Smooth and Nonsmooth High Dimensional Chaos and the Melnikov-type Methods* (World Scientific, Singapore, 2007)
23. J. Awrejcewicz, V.A. Krysko Jr., I.V. Papkova, E.Yu. Krylova, A.V. Krysko, Spatio-temporal non-linear dynamics and chaos in plates and shells. Int. J. Nonlin. Stud. (to appear)
24. J. Awrejcewicz, V.A. Krysko, *Introduction to Asymptotic Methods* (Chapman and Hall, Boca Raton, 2006)
25. J. Awrejcewicz, V.A. Krysko, *Nonclassical Thermoelastic Problems in Nonlinear Dynamics of Shells* (Springer, Berlin, 2003)
26. J. Awrejcewicz, V.A. Krysko, I.V. Kravtsova, Dynamics and statics of flexible axially-symmetric shallow shells. Math. Prob. Eng. **2006**, 1–25 (2006)
27. J. Awrejcewicz, G. Kudra, Analytical and numerical investigations of time-periodic mechanical systems, in *Book of Abstracts of the 1st EUROMECH Colloguium 532 - Time Periodic Systems (TPS), Current Trends in Theory Application*, TU Darmstadt, 27–30 August 2012, pp. 9–12
28. J. Awrejcewicz, G. Kudra, Stability analysis and Lyapunov exponents of a multi-body mechanical system with rigid unilateral constraints. Nonlin. Anal. **63**(5–7), 909–918 (2005, Special Issue)
29. J. Awrejcewicz, C.-H. Lamarque, *Bifurcation and Chaos in Nonsmooth Mechanical Systems* (World Scientific, Singapore, 2003)
30. J. Awrejcewicz, J. Mrozowski, Bifurcations and chaos of a particular Van der Pol - Duffing oscillator. J. Sound Vib. **132**(1), 89–100 (1989)
31. J. Awrejcewicz, I.V. Papkova, E.U. Krylova, V.A. Krysko, Wavelet-based analysis of the regular and chaotic dynamics of rectangular flexible plates subjected to shear-harmonic loading. Shock Vib. **19**, 979–994 (2012)
32. J. Awrejcewicz, A.G. Petrov, On the normal forms of Hamiltonian systems. Nonlin. Dyn. **48**, 185–197 (2007)
33. J. Awrejcewicz, D. Sendkowski, Geometric analysis of the dynamics of a double pendulum. J. Mech. Mater. Struct. **2**(8), 1421–1430 (2007)
34. J. Awrejcewicz, D. Sendkowski, Lyapunov exponents and geometric instability exponents of a double physical pendulum. Appl. Mech. Mater. **9**, 19–29 (2008)
35. J. Awrejcewicz, D. Sendkowski, M. Kaźmierczak, Geometrical approach to the swinging pendulum dynamics. Comput. Struct. **84**, 1577–1583 (2006)
36. J. Awrejcewicz, B. Supeł, C.-H. Lamarque, G. Kudra, G. Wasilewski, P. Olejnik, Numerical and experimental study of regular and chaotic motion of triple physical pendulum. Int. J. Bifurcations Chaos **18**(10), 2883–2915 (2008)
37. J. Awrejcewicz, W. Wodzicki, *Introduction to Automation - Theory and Examples* (TUL Press, Lodz, 2001, in Polish)
38. A.K. Bajaj, P.R. Sethna, T.S. Lundgren, Hopf bifurcation phenomena in tubes carrying fluid. SIAM J. Appl. Mech. **39**, 213–230 (1980)
39. P. Bak, T. Bohr, M.H. Jensen, D.V. Christiansen, Josephson junctions and circle maps. Solid State Commun. **51**, 231–236 (1984)
40. P. Bak, M.H. Jensen, Mode locking and the transitions to chaos in dissipative systems. Phys. Scripta T **9**, 50–55 (1985)
41. S. Banarjee, C. Grebogi, Border collision bifurcations in two-dimensional piecewise smooth maps. Phys. Rev. E **59**, 4052–4061 (1999)
42. M. Di Bari, P. Cipriani, Geometry and chaos on Riemann and Finsler manifolds. Planet. Space Sci. **46**, 1543–1555 (1998)
43. N.N. Bautin, *Behaviour of Dynamical Systems Close to Stability Loss Boundaries* (Nauka, Moscow, 1984, in Russian)
44. G.L. Benettin, L. Galgani, A. Giorgilli, J.-M. Strelcyn, Lyapunov characteristic exponents for smooth dynamical systems and for Hamiltonian systems: a method for computing all of them. Part 1: Theory. Part 2: Numerical application. Mecanica **15**, 9–29 (1980)

45. E.C. Berkeley, *Symbolic Logic and Intelligent Machines* (Reinhold Publishing Corp., New York, 1960)
46. M. di Bernardo, The complex behavior of switching devices. IEEE Circ. Syst. Soc. News **10**, 1–13 (1999)
47. M. di Bernardo, F. Garofalo, L. Glielmo, F. Vasca, Switching, bifurcations and chaos in DC/DC convert. IEEE Trans. Circ. Syst. I **1**, 133–141 (1998)
48. L. von Bertalanffy, *General System Theory. Foundations, Development, Applications* (George Brazziler, Inc., New York, 1973)
49. E.O. Birgham, *FFT - Schnelle Fourier - Transformation* (Oldenburg-Verlag, München, Wien, 1987)
50. I.I. Blekhman, P.S. Landa, M.G. Rosenblum, Synchronization and chaotization phenomena in oscillatory and rotatory dynamical systems, in *Nonlinear Dynamics: New Theoretical and Applied Results*, ed. by J. Awrejcewicz (Akademie Verlag, Berlin, 1995)
51. N.N. Bogoliubov, Yu.A. Mitropolskiy, *Asymptotic Methods in the Theory of Nonlinear Oscillations* (Gordon and Breach, New York, 1961)
52. W. Bogusz, *Stability of Nonlinear Systems* (IPPT, PAN, 1966, in Polish)
53. W. Bogusz, *Technical Stability* (PWN, Warsaw, 1972, in Polish)
54. I.N. Bronshtein, K.A. Semendayev, *Handbook of Mathematics* (Springer, Berlin, 2004)
55. M.P. do Carmo, *Riemannian Geometry* (Birkhäuser, Boston, 1992)
56. L. Casetti, C. Clementi, M. Pettini, Riemannian theory of Hamiltonian chaos and Lyapunov exponents. Phys. Rev. E **54**(6), 5969–5984 (1996)
57. L. Casetti, M. Pettini, E.G.D. Cohen, Geometric approach to Hamiltonian dynamics and statistical mechanics. Phys. Rep. **337**, 237–341 (2000)
58. M. Casetti, M. Pettini, Analytic computation of the strong stochasticity threshold in Hamiltonian dynamics using Riemannian geometry. Phys. Rev. E **48**(6), 4320–4332 (1993)
59. J. Casey, On the advantages of a geometrical viewpoint in the derivation of Lagrange's equations for a rigid continuum. J. Appl. Mech. Phys. **46**, 805–847 (1995, Special Issue)
60. M. Cerruti-Sola, M. Pettini, Geometric description of chaos in self-gravitating systems. Phys. Rev. E **51**, 53–64 (1995)
61. M. Cerruti-Sola, M. Pettini, Geometric description of chaos in two-degrees-of-freedom Hamiltonian systems. Phys. Rev. E **53**, 179–188 (1996)
62. P.L. Chalfman, *Dynamics* (Nauka, Moscow, 1977)
63. C. Chicone, *Ordinary Differential Equations with Applications* (Springer, New York, 2006)
64. S.N. Chow, J. Mallet-Paret, Integral averaging and Hopf bifurcation. J. Diff. Eq. **26**, 1112–1159 (1977)
65. L.O. Chua, A. Ushida, Algorithms for computing almost periodic steady-state response of nonlinear systems to multiple input frequencies. IEEE Trans. Circ. Syst. **CAS-28**, 953–988 (1981)
66. I. Chueshov, I. Lasiecka, Inertial manifolds for von Kármán plate equations. Appl. Math. Optim. **46**, 179–206 (2002)
67. I. Chueshov, I. Lasiecka, Global attractors von Kármán evolutions with a nonlinear boundary dissipation. J. Diff. Eq. **198**, 196–231 (2004)
68. I. Chueshov, I. Lasiecka, Attractors and long time behavior of Kármán thermoelastic plates. Appl. Math. Optim. **58**, 195–241 (2008)
69. R.M. Clever, F.H. Busse, Instabilities of longitudinal convection rolls in an inclined layer. J. Fluid Mech. **81**, 107 (1977)
70. E.A. Coddington, N. Levinson, *Theory of Ordinary Differential Equations* (McGraw-Hill Book Co., New York, 1982)
71. J.W. Cooley, J.W. Tukey, An algorithm for the machine calculation of complex Fourier. Math. Comput. **19**, 297–301 (1965)
72. T.K. Coughey, F. Ma, Complex modes and solvability of non-classical linear systems. ASME J. Appl. Mech. **60**, 26–28 (1990)
73. M.G. Crandall, P.H. Rabinowitz, The Hopf bifurcation theorem in infinite dimensions. Arch. Rational Mech. Anal. **45**, 79–109 (1973)

74. M.C. Cross, A.C. Newell, Convection patterns in large aspect ratio systems. Physica D **10**, 299 (1984)
75. L. Debnath, *Wavelet Transforms and Their Applications* (Birkhäuser Basel, Boston, 2002)
76. L. Debnath, D. Bhatta, *Integral Transforms and Their Applications* (Chapman & Hall/CRC, 2007)
77. B.P. Demidovich, *Lectures on Mathematical Theory of Stability* (Nauka, Moscow, 1967, in Russian)
78. M. di Bernardo, C. Budd, Corner collision implies border collision. Physica D **154**, 171–194 (2001)
79. M. di Bernardo, C. Budd, A.R. Champneys, Normal form maps for grazing bifurcations in n-dimensional piecewise-smooth dynamical systems. Physica D **160**, 222–254 (2001)
80. M. di Bernardo, M.I. Feigin, S. Hogan, M. Homer, Local analysis of C-bifurcations in n-dimensional piecewise-smooth dynamical systems. Chaos Solit. Fract. **10**, 1881–1908 (1999)
81. M. di Bernardo, K.M. Johansson, F. Vasca, Self-oscillations in rely feedback systems: symmetry and bifurcations. Int. J. Bifurcations Chaos **11**, 1121–1140 (2001)
82. M. di Bernardo, P. Kowalczyk, A.B. Nordmark, Bifurcation of dynamical systems with sliding: derivation of normal form maps. Physica D **170**(3–4), 175–204 (2001)
83. L. Dieci, J. Lorenz, R.D. Russel, Numerical calculation of invariant tori. SIAM J. Sci. Stat. Comput. **12**, 607–622 (1996)
84. C. H. Edwards, D.E. Penney, *Elementary Differential Equations* (Prentice Hall, Las Vegas, 2007)
85. T. Endo, Y. Aruga, K. Yamauchi, Transient dynamics and chaos observed in strongly nonlinear mutually-coupled oscillators, in *Proceedings of the International Symposium on Circuits and Systems ISCAS* (2003)
86. D.A. Evensen, Nonlinear vibrations of circular cylindrical shells, in *Thin-Shell Structures, Theory, Experiment and Design*, ed. Y.C. Fung, E.E. Sechler (Prentice Hall, Englewood, Cliffs, New York, 1974), pp. 133–155
87. I. Fawzy, R.E.D. Bishop, On the dynamics of linear non-conservative systems. Proc. R. Soc. Lond. **1352**, 25–40 (1976)
88. V.I. Fedos'ev, On a certain method of solution to the nonlinear problem of stability of deformable systems. Appl. Math. Mech. **27**, 265–274 (1963)
89. M.I. Feigin, Doubling the oscillation period with C-bifurcations in piecewise continuous systems. Prikl. Matem. Mekhan. (J. Appl. Math. Mech.) **34**, 861–869 (1970)
90. M.I. Feigin, On the generation of sets of subharmonic modes in piecewise continuous system. Prikl. Matem. Mekhan. (J. Appl. Math. Mech.) **38**, 810–818 (1974)
91. M.I. Feigin, On the structure of C-bifurcations boundaries of piecewise continuous system. Prikl. Matem. Mekhan. (J. Appl. Math. Mech.) **42**, 820–829 (1978)
92. M.I. Feigin, The increasingly complex structure of the bifurcation tree of a piecewise-smooth system. J. Appl. Math. Mech. **59**, 853–863 (1995)
93. A. Filippov, *Differential Equations with Discontinuous Righthand Side* (Kluwer Academic, Dortrecht, 1988)
94. W. Findeisen, J. Szymanowski, A. Wierzbicki, *Theory and Computational Methods of Optimization* (PWN, Warsaw, 1977, in Polish)
95. U. Galvanetto, Some discontinuous bifurcations in two-block stick-slip system. J. Sound Vib. **248**, 653–669 (2001)
96. R.F. Ganiyev, L.E. Ukrainskiy, *Dynamics of Particles Subjected to Vibrations* (Naukova Dumka, Kiev, 1975, in Russian)
97. F.R. Gantmacher, *Theory of Matrices*, 2 vols. (Chelsea, New York, 1998)
98. I.M. Gelfand, S.V. Fomin, *Calculus of Variations* (Dover Publications, Dover, 2000)
99. P. Glendinning, *Stability, Instability and Chaos: An Introduction to the Theory of Nonlinear Differential Equations* (Cambridge University Press, Cambridge, 1996)
100. M. Golubitsky, D.G. Schaeffer, *Singularities and Groups in Bifurcation Theory*, vol. 1 (Springer, New York, 1985)

101. M. Golubitsky, I. Stewart, D.G. Schaeffer, *Singularities and Groups in Bifurcation Theory*, vol. 2 (Springer, New York, 1988)
102. H. Górecki, *Optimization of Dynamic Systems* (Engineer Research Library PWN, Warsaw, 1993, in Polish)
103. A Gray, M. Mezzino, M.A. Pinsky, *Introduction to Ordinary Differential Equations with Mathematica* (Springer, New York, 1997)
104. C. Grebogi, E. Ott, S. Pelikan, J.A. Yorke, Strange attractors that are not chaotic. Physica D **11**, 261–268 (1984)
105. R. Grimshaw, *Nonlinear Ordinary Differential Equations* (Blackwell Scientific Pub., Oxford, 1990)
106. J. Guckenheimer, P. Holmes, *Nonlinear Oscillations, Dynamical Systems, and Bifurcations of Vector Fields* (Springer, New York, 1983)
107. J. Gutenbaum, *Mathematical Modeling of Systems* (PWN, Warsaw, 1997, in Polish)
108. R.S. Guter, A.P. Janpolskiy, *Differential Equations* (Vysshaya Shkola, Moscow, 1976, in Russian)
109. G. Hadley, *Nonlinear and Dynamics Programming* (Addison-Wesley, Reading, 1968)
110. E. Hairer, C. Lubich, G. Wanner, *Geometric Numerical Integration: Structure-Preserving Algorithms for Ordinary Differential Equations* (Springer, Berlin, 2006)
111. J.K. Hale, *Ordinary Differential Equations* (Krieger Pub. Co., Malabar, 1980)
112. J.K. Hale, *Topics in Dynamics Bifurcation Theory*. Regional Conference Series in Mathematics, vol. 47 (American Mathematical Society, Rhode Island, 1981)
113. J.K. Hale, J.C. Oliviera, Hopf bifurcation for functional equations. J. Math. Anal. Appl. **74**, 41–59 (1980)
114. P. Hartman, *Ordinary Differential Equations* (Wiley, New York, 1996)
115. B.D. Hassard, N.D. Kazarinoff, Y.H. Wan, *Theory and Applications of Hopf Bifurcation* (Cambridge University Press, Cambridge, 1981)
116. J.-H. He, Homotopy perturbation method: a new nonlinear analytical technique. J. Appl. Math. Comput. **135**(1), 73–79 (2003)
117. J.-H. He, An improved amplitude–frequency formulation for non-linear oscillators. J. Nonlin. Sci. Numer. Simul. **9**(2), 211–212 (2008)
118. J.-H. He, Max-min approach to non-linear oscillators. J. Nonlin. Sci. Numer. Simul. **9**(2), 207–210 (2008)
119. J.-H. He, D.-H. Shou, Application of parameter-expanding method to strongly nonlinear oscillators. Int. J. Nonlin. Sci. Numer. Simul. **8**(1), 121–124 (2007)
120. M. Hénon, A two-dimensional map with a strange attractor. Commun. Math. Phys. **50**, 69–75 (1976)
121. M.S. Heutmaker, P.N. Fraenkel, J.P. Gollub, Convection patterns: time evaluation of the wave-vector field. Phys. Rev. Lett. **54**, 1369 (1985)
122. M. Hirsch, S. Smale, *Differential Equations, Dynamical Systems and Linear Algebra* (Academic, New York, 1974)
123. P.J. Holmes, Bifurcations to divergence and flutter in flow induced oscillations: a finite dimensional analysis. J. Sound Vib. **53**, 471–503 (1977)
124. E. Hopf, Bifurcation of a periodic solution from a stationary solution of system of differential equations. Akad. Wissensch. **94**, 3–22 (1942, in German)
125. K. Huseyin, M.M.E. Leipholtz, Divergence instability of multiple-parameter circulatory systems. Quart. Appl. Math. **31**, 185–197 (1973)
126. D. Inman, Dynamics of asymmetric non-conservative systems. ASME J. Appl. Mech. **50**, 199–200 (1983)
127. G. Iooss, D.D. Joseph, *Elementary Stability and Bifurcation Theory* (Springer, New York, 1980)
128. M.V. Jacobson, Absolutely continuous invariant measures for one-parameter families of one dimensional maps. Commun. Math. Phys. **81**, 39–88 (1981)
129. M.H. Jensen, P. Bak, T. Bohr, Complete devil's staircase, fractal dimension and university of mode locking structures. Phys. Rev. Lett. **50**, 1637–1646 (1983)

130. D.W. Jordan, P. Smith, *Nonlinear Ordinary Differential Equations. An Introduction to Dynamical Systems* (Oxford University Press, Oxford, 1999)
131. C. Kaas-Petersen, Computation, continuation and bifurcation of torus solutions for dissipative maps and ordinary differential equations. Physica D **25**, 288–306 (1987)
132. T. Kaczorek, *Theory of Control and Systems* (PWN, Warsaw, 1993, in Polish)
133. B.Ya. Kantor, *Contact Problems of Nonlinear Theory of Rotational Shells* (Naukova Dumka, Kiev, 1990)
134. J. Kevorkian, J.A. Cole, *Multiple Scale and Singular Perturbation Methods* (Springer, New York, 1996)
135. A.V. Khinchin, *Continued Functions* (University of Chicago Press, Chicago, 1964)
136. V.O. Kononenko, N.P. Plakhtienko, *Methods of Identification of Mechanical Nonlinear Oscillatory Systems* (Naukova Dumka, Kiev, 1976, in Russian)
137. G. Kotowski, Lösungen der inhomogenen Mathieuschen Differentialgleichung mit periodischer Störfunktion beliebiger Frequenz. ZAMM **23**, 213–229 (1943)
138. A.A. Krasowski, G.S. Pospielow, *Basics of Automation and Technical Cybernetics* (WNT, Warsaw, 1965, in Polish)
139. A. Krokiewicz, *Outline of Greek Philosophy* (PAX, Warsaw, 1971, in Polish)
140. V.A. Krysko, J. Awrejcewicz, S.A. Komarov, Nonlinear deformations of spherical panels subjected to transversal load action. Comput. Meth. Appl. Mech. Eng. **194**, 3108–3126 (2005)
141. V.D. Kubenko, P.S. Koval'shuk, N.P. Podshasov, *Nonlinear Oscillations of Cylindrical Shells* (Visha Shkola, Kiev, 1989, in Russian)
142. J. Kudrewicz, *Fractal and Chaos* (WNT, Warsaw, 1993, in Polish)
143. M. Kunze, *Non-smooth Dynamical Systems* (Springer, Berlin, 2000)
144. W. Kurnik, *Divergent an Oscillatory Bifurcations* (WNT, Warsaw, 1997, in Polish)
145. W. Kurnik, Hopf bifurcation in the dynamics of a rotor/bearing system, in *IUTAM Symposium on Nonlinear Dynamics in Engineering Systems* (Springer, Berlin, 1990), pp. 181–188
146. Yu.A. Kuznetsov, *Elements of Applied Bifurcation Theory* (Springer, New York, 1998)
147. J.-P. Lacaux, E. Rodriguez, M. Le Van Quyen, A. Lutz, J. Martinerie, F.J. Varela, Studying single-trials of phase synchronous activity in the brain. Int. J. Bifurcations Chaos **10**(10), 2429–2439 (2000)
148. C. Lanczos, *The Variational Principles of Mechanics* (Dover Publications, New York, 1986)
149. P.S. Landa, *Self-oscillations in Systems with Lumped Parameters* (Nauka, Moscow, 1980, in Russian)
150. E. Laszlo, *The Systems View of the World: A Holistic Vision for Our Time* (Hampton Press, New York, 1996)
151. R.I. Leine, D.H. Van Campen, Discontinuous fold bifurcations in mechanical systems. Arch. Appl. Mech. **72**, 138–146 (2002)
152. R.I. Leine, D.H. van Campen, B.L. van de Vrande, Bifurcations in nonlinear discontinuous systems. Nonlin. Dyn. **23**, 105–164, (2000)
153. M.M.E. Leipholtz, On the sufficiency of the energy criterion for the stability of certain nonconservative systems of the follower-load type. ASME J. Appl. Mech. **39**, 717–722 (1972)
154. G.A. Leonov, *Strange Attractors and Classical Theory of Motion Stability* (Izd. Sankt-Petersburg Universitata, 2004, in Russian)
155. A.J. Lichtenberg, M.A. Lieberman, *Regular and Chaotic Dynamics*. App. Math. Sci., vol. 38 (Springer, New York, 1992)
156. F.H. Ling, Quasi-periodic solutions calculated with the simple shooting technique. J. Sound Vib. **144**(2), 293–304 (1991)
157. E.N. Lorenz, Deterministic nonperiodic flow. J. Atmos. Sci. **20**, 130–141 (1963)
158. M. Lowe, J.P. Gollub, Solitons and the commensurate-incommensurate transition in a convecting nematic fluid. Phys. Rev. A **31**, 3893 (1985)
159. A.M. Lyapunov, The general problem of stability of motion. Izd. Akad. Nauk SSSR **2**, 7–256 (1956)
160. F. Ma, T.K. Caughey, Analysis of linear nonconservative vibrations. J. Appl. Mech. **62**, 685–691 (1995)

161. G.M. Maggio, M. di Bernardo, Nonsmooth bifurcations in a piecewise-linear model of the colpitts oscillator. IEEE Trans. Circ. Syst. I **47**, 1160–1177 (2000)
162. W. Magnus, S. Winkler, *Hill's Equation* (Interscience Publishers, a Division of John Wiley and Sons, New York, 1966)
163. S.G. Mallat, *A Wavelet Tour of Signal Processing* (Academic, San Diego, 1998)
164. J.E. Marsden, M. Mc Cracken, *The Hopf Bifurcation and Its Applications* (Springer, New York, 1976)
165. S.W. McDonald, C. Grebogi, E. Ott, J.A. Yorke, Fractal basin boundaries. Physica D **17**, 125–153 (1985)
166. L. Meirovitch, *Analytical Methods in Vibration* (MacMillan, New York, 1967)
167. E. Meissner, On shaking phenomena in systems with periodically varying elasticity. Schweizer Bauzeitung **72**, 95–98 (1918, in German)
168. M.D. Mesarowic, Y. Takahara, General systems theory: mathematical foundations. Bull. Am. Math. Soc. **81**(6), 1042–1044 (1975)
169. A.V. Metrikine, H. Askes, One-dimensional dynamically consistent gradient elasticity models derived from a discrete microstructure. Part 1: Generic formulation. Eur. J. Mech. A/Solids **21**, 555–572 (2002)
170. A.V. Metrikine, H. Askes, One-dimensional dynamically consistent gradient easticity models derived from a discrete microstructure. Part 2: Static and dynamic response. Eur. J. Mech. A/Solids **21**, 573–588 (2002)
171. K.R. Meyer, G.R. Hall, *Introduction to Hamiltonian Dynamical Systems and the N-Body Problem* (Springer, New York, 1992)
172. W.E. Milne, Numerical determination of characteristic numbers. Phys. Rev. **35**, 863 (1930)
173. J. Morlet, G. Arens, E. Fourgeau, D. Glard, Wave propagation and sampling theory, Part I: Complex signal land scattering in multilayer media. Geophysics **47**, 203–221 (1982)
174. J. Morlet, G. Arens, E. Fourgeau, D. Glard, Wave propagation and sampling theory, Part II: Sampling theory and complex waves. Geophysics **47**, 222–236 (1982)
175. P.C. Müller, Calculation of Lyapunov exponents for dynamic system with discontinuities. Chaos Solit. Fract. **5**, 1671–1691 (1995)
176. P.C. Müller, W.O. Schiehlen, *Linear Vibrations* (Martinus Nijhoff, Dordrecht, 1985)
177. K. Nagai, S. Maruyama, T. Murata, T. Yamaguchi, Experiments and analysis on chaotic vibrations of a shallow cylindrical shell-panel. J. Sound Vib. **305**, 492–520 (2007)
178. K. Nagai, S. Maruyama, M. Oya, T. Yamaguchi, Chaotic oscillations of a shallow cylindrical shell with a concentrated mass under periodic excitation. Comput. Struct. **82**, 2607–2619 (2004)
179. M. Nakahara, *Geometry, Topology and Physics* (Taylor & Francis Group, Bristol, 1990)
180. A.H. Nayfeh, *Perturbation Methods* (Wiley, New York, 1973)
181. J. Neimark, *Dynamical Systems and Controled Processes* (Nauka, Moscow, 1978, in Russian)
182. R.I. Nigmatulin, *Dynamics of Multi-Plate Media* (Nauka, Moscow, 1987, in Russian)
183. H.E. Nusse, E. Ott, J.A. Yorke, Border-collision bifurcations: an explanation for observed bifurcation phenomena. Phys. Rev. E **49**, 1073–1076 (1994)
184. H.E. Nusse, J.A. Yorke, *Dynamics: Numerical Explorations* (Springer, New York, 1994)
185. H.E. Nusse, J.A. Yorke, Border-collision bifurcations including period two to period three for piecewise smooth systems. Physica D **57**(1–2), 39–57 (1992)
186. H.E. Nusse, J.A. Yorke, Border-collision bifurcations for piecewise smooth one-dimensional maps. Int. J. Bifurcations Chaos **5**, 189–207 (1995)
187. V.I. Oseledec, A multiplicative ergodic theorem. Lyapunov charakteristic numbers for dynamical systems. Trans. Moscow Math. Soc. **19**, 197–231 (1968)
188. G.V. Osipov, A.S. Pikovsky, M.G. Rosenblum, J. Kurths, Phase synchronization effects in a lattice of nonidentical Rössler oscillators. Phys. Rev. E **55**, 2253–2361 (1997)
189. E. Ott, *Chaos in Dynamical Systems* (Cambridge University Press, Cambridge, 2002)
190. M.P. Paidoussis, Dynamics of tubular cantilevers conveying fluid. J. Mech. Eng. Sci. **12**, 85–103 (1980)
191. A. Palczewski, *Ordinary Differential Equations* (WNT, Warsaw, 2004, in Polish)

192. P.Y. Papalambros, D.J. Wilde, *Principles of Optimal Design. Modeling and Computation* (Cambridge University Press, Cambridge, 1995)
193. L. Perko, *Differential Equations and Dynamical Systems* (Springer, New York, 1996)
194. W. Pesch, L. Kramer, Nonlinear analysis of spatial structures in two-dimensional anisotropic pattern forming systems. Z. Phys. B **63**, 121 (1986)
195. A.G. Petrov, Modification of the method of invariant normalization of Hamiltonians by parameterizing canonical transformations. Dokl. Phys. **47**, 742 (2002)
196. A.G. Petrov, Asymptotic methods for solving the Hamilton equations with the use of a parametrization of canonical transformations. J. Diff. Eq. **40**(5), 672–685 (2004)
197. A.G. Petrov, Invariant normalization of non-autonomous Hamiltonian systems. J. Appl. Math. Mech. **68**, 357–367 (2004)
198. M. Pettini, Geometrical hints for a nonperturbative approach to Hamiltonian dynamics. Phys. Rev. E **47**(2), 828–850 (1993)
199. A.S. Pikovsky, M.G. Rosenblum, J. Kurths, *Synchronization: A Universal Concept in Nonlinear Sciences* (Cambridge University Press, Cambridge, 2001)
200. E. Pinney, The nonlinear differential equation. Proc. Am. Math. Soc. **1**, 681 (1950)
201. H. Poincaré, Memory of the curves defined by differential equations. J. Math. Pures Appl., 151–217 (1886, in French)
202. M. Poincaré, *New Methods of Celestial Mechanics* (Gauthier-Villars, Paris, 1899, in French)
203. B. Radziszewski, *Mathematical Methods in Dynamics of Machine* (PAN Press, Warsaw, 1974, in Polish)
204. B. Radziszewski, *On the Best Lyapunov Function and Its Application for Investigation of Stability of Motion* (IFTR Reports, Warsaw, 1977, in Polish)
205. B. Radziszewski, Study of stability of motion by the “Best” Lyapunov function, in *Stability and Sensivity in Mechanical Systems* (Ossolineum, Wrocław, 1978, in Polish)
206. B. Radziszewski, A. Sławiński, Comparative analysis of some criteria of stability of motion. Nonlin. Vib. Prob. **23**, 123–135 (1989)
207. A.M. Samoilenko, *Elements of the Mathematical Theory of Multifrequency Oscillations. Invariant Tori* (Nauka, Moscow, 1987, in Russian)
208. A.M. Samoylenko, S.A. Krivosheya, N.A. Perestyuk, *Differential Equations. Examples and Problems* (Vysshaya Shkola, Moscow, 1989, in Russian)
209. A. Sławiński, On the problem of perturbations of nonlinear systems of differential equations. Rev. Math. Appl. Univ. Chile, 45–62 (1990)
210. A. Sławiński, On control of an autonomous system with application of Lyapunov function. Nonlin. Vib. Prob. **24**, 283–295 (1991)
211. A. Sławiński, On the comparative analysis of some criteria of stability of motion. Int. J. Nonlin. Mech. **33**, 783–799 (1998)
212. R.M. Scalan, R. Rasenbaum, *Aircraft Vibration and Flutter* (Macmillan, New York, 1951)
213. H.G. Schuster, *Deterministic Chaos. An Introduction*, 2nd revised version (VCH, Weinheim, 1989)
214. R. Seydel, *From Equilibrium to Chaos. Practical Bifurcation and Stability Analysis* (Elsevier, New York, 1999)
215. A.C. Shian, T.T. Soong, D.S. Roth, Dynamic buckling of conical shells with imperfections. Am. Inst. Aeronautics Astronautics J. **12**(6), 24–30, (1974)
216. L.P. Shilnikov, A.L. Shilnikov, D.T. Turaev, L.O. Chua, *Methods of Qualitative Theory in Nonlinear Dynamics. Part I* (World Scientific Series on Nonlinear Science, Singapore, 1998)
217. L.P. Shilnikov, A.L. Shilnikov, D.T. Turaev, L.O. Chua, *Methods of Qualitative Theory in Nonlinear Dynamics. Part II* (World Scientific Series on Nonlinear Science, Singapore, 2001)
218. X. Shu, Q. Han, G. Yang, The double mode model of chaotic motion for a large defection plate. Appl. Math. Mech. **20**, 360–364 (1992)
219. C. Simo, On the Hénon-Pomeau attractor. J. Stat. Phys. **21**, 465–494 (1979)
220. H.-G. Stark, *Wavelets and Signal Processing* (Springer, Berlin, 2005)
221. W.H. Steeb, *Chaos and Fractals. Algorithms and Computations* (B.I. Wissenschaftsverlag, Mannheim, Leipzig, Wien, Zürich, 1992)

222. H. Steinhaus, Approximation by means of the convex-linear function. Math. Mech. **8**, 414–415 (1928, in German)
223. A.S.J. Suiker, A.V. Metrikine, R. de Borst, Comparison of wave propagation characteristics of the Cosserat continuum model and corresponding discrete lattice models. Int. J. Solids Struct. **38**, 1563–1583 (2001)
224. W. Szemplińska-Stupnicka, *The Behaviour of Nonlinear Vibrating Systems. Vols. I, II* (Kluwer Academic, Dordrecht, 1990)
225. K. Szpunar, Modified concept of technical stability. Nonlin. Vib. Prob. **21**, 141–145 (1983)
226. M. Szydłowski, The Eisenhart geometry as an alternative description of dynamics in terms of geodesics. Gen. Relat. Gravit. **30**, 887–914 (1998)
227. M. Szydłowski, The general relativity dynamics in the Eisenhart geometry. Chaos Solit. Fract. **11**, 685–695 (2000)
228. R.K. Tavakol, A.S. Tworkowski, On the occurrence of quasi-periodic motion on three tori. Phys. Lett. A **100**, 65–69 (1984)
229. K.-E. Thylwe, H.J. Korsch, Harmonic oscillator subject to parametric pulses: an amplitude (Milne) oscillator approach. J. Phys. A Math. Gen. **34**, 3497 (2001)
230. K.-E. Thylwe, H.J. Korsch, On pulse-induced transition amplitudes in two-state quantum system without level crossings. J. Phys. A Math. Gen. **35**, 7507 (2002)
231. S.P. Timoshenko, J.M. Gere, *Theory of Elastic Stability* (McGraw Hill Book Company, New York, 1961)
232. C. Touzé, O. Thomas, M. Amabili, Transition of chaotic vibrations for harmonically forced perfect and imperfect circular plates. Int. J. Nonlin. Mech. **46**(1), 234–246 (2011)
233. Y. Ueda, Explosion of strange attractors exhibited by Duffing's equation. Ann. NY Acad. Sci. **357**, 422–423 (1980)
234. M. Urabe, A. Reiter, Numerical computation of nonlinear forced oscillations by Galerkin's procedure. J. Math. Anal. Appl. **14**, 107–140 (1966)
235. A.J. Uyemov, *System Approach and General Theory of Systems* (Mysl, Moscow, 1978, in Russian)
236. F. Verhulst, *Nonlinear Differential Equations and Dynamical Systems* (Springer, New York, 1990)
237. G. Vilasi, *Hamiltonian Dynamics* (World Scientific, Singapore, 2001)
238. L.N. Virgin, C.J. Begley, Grazing bifurcations and basins od attraction in an impact-friction oscillator. Physica D **130**, 43–57 (1999)
239. S.A. Volmir, *Nonlinear Dynamics of Plates and Shells* (Nauka, Moscow, 1972, in Russian)
240. T. von Kármán, Festigkeitsprobleme in Maschinenbau. Encyklopedie Math. Wissensch. **IV**(4), 311–385 (1910)
241. Y.-G. Wang, H.-F. Songand, D. Li, J. Wang, Bifurcations and chaos in periodic time-varying temperature-excited bimetallic shallow shell of revolution. Arch. Appl. Mech. **80**, 815–828 (2010)
242. E.T. Whittaker, *A Treatise on the Analytical Dynamics of Particles and Rigid Bodies* (Dover Publications, New York, 1944)
243. S. Wiggins, *Introduction to Applied Nonlinear Dynamical Systems and Chaos* (Springer, New York, 1990)
244. A. Wolf, Quantifying chaos with Lyapunov exponents, in *Chaos*, ed. by A.V. Holden (Manchester University Press, Manchester, 1986)
245. A. Wolf, J.B. Swift, H.L. Swinney, J.A. Vastano, Determining Lyapunov exponents from a time series. Physica D **16**, 285–317 (1985)
246. X.L. Yang, P.R. Sethna, Local and global bifurcations in parametrically excited vibrations of nearly square plates. Int. J. Nonlin. Mech. **26**, 199–220 (1991)
247. X.L. Yang, P.R. Sethna, Non-linear phenomena in forced vibrations of a nearly square plate: antisymmetric case. J. Sound Vib. **155**, 413–441 (1992)
248. M. Ye, Y. Sun, W. Zhang, X. Zhan, Q. Ding, Nonlinear oscillations and chaotic dynamics of an antisymmetric cross-ply laminated composite rectangular thin plate under parametric excitation. J. Sound Vib. **287**, 723–758 (2005)

249. T.L. Yeh, C.K. Chen, H.Y. Lai, Chaotic and bifurcation dynamics of a simply supported thermo-elastic circular plate with variable thickness in large deflection. Chaos Solit. Fract. **15**, 811–829 (2003)
250. D. Yu, Analysis of double Hopf bifurcation using computer algebra with the aid of multiple scales. Nonlin. Dyn. **27**, 15–53 (2002)
251. B.P. Zeigler, *Theory of Modelling and Simulation* (Wiley Interscience, New York, 1976)
252. W. Zhang, J. Yang, Y. Hao, Chaotic vibrations of an orthotropic FGM rectangular plate based on third-order shear deformation theory. Nonlin. Dyn. **59**, 619–660 (2010)
253. N.E. Zhukovskiy, On solidity of motion, in *Collection of Works*, vol. 1 (Gostekhizdat, Moscow, Leningrad, 1948), pp. 67–16
254. V.F. Zhuravlev, *Introduction to Theoretical Mechanics* (Nauka Fizmatlit, Moscow, 1997, in Russian)
255. V.P. Zhuravlev, Invariant normalization of non-autonomous Hamiltonians. J. Appl. Math. Mech. **66**(3), 347–356 (2002)
256. Zh.T. Zhusubaliyev, E.A. Soukhoterin, E. Mosekilde, Border-collision bifurcations on a two-dimensional torus. Chaos Solit. Fract. **13**, 1889–1915 (2002)
257. D.G. Zill, M.R. Cullen, *Differential Equations with Boundary-Value Problems* (Brooks/Cole, Pacific Grove, 2001)

Printed by Printforce, the Netherlands